CHEMISTRY OF ADVANCED MATERIALS

CHEMISTRY OF ADVANCED MATERIALS

AN OVERVIEW

Edited by

Leonard V. Interrante
Department of Chemistry
Rensselaer Polytechnic Institute
Troy, New York

and

Mark J. Hampden-Smith
Department of Chemistry
University of New Mexico
Albuquerque, New Mexico

WILEY-VCH

New York • Chichester • Weinheim • Brisbane • Singapore • Toronto

This text is printed on acid-free paper. ♾

Published simultaneously in Canada.

Library of Congress Cataloging in Publication Data:

Chemistry of advanced materials : an overview / edited by Leonard V. Interrante and Mark J. Hampden-Smith

p. cm.

Includes bibliographical references and index.

ISBN 0-471-18590-6 (cloth : alk. paper)

1. Materials. 2. Chemical processes. 3. Polymers. 4. Nanotechnology. I. Interrante, Leonard V., 1939– II. Hampden-Smith, Mark J.

TA403.C43 1998

620.1'1—dc21 97-11898

Printed in the United States of America

10 9 8 7 6 5 4 3 2 1

CONTRIBUTORS

Florence Babonneau, Université Pierre et Marie Curie, Paris, France

Jean-Rémi Butruille, Kansas State University, Manhattan, Kansas

Patrick Cassoux, CNRS, Toulouse, France

Michael E. Hagerman, Northern Arizona University, Flagstaff, Arizona

Mark J. Hampden-Smith, University of New Mexico, Albuquerque, New Mexico

Leonard V. Interrante, Rensselaer Polytechnic Institute, Troy, New York

Kenneth J. Klabunde, Kansas State University, Manhattan, Kansas

Toivo T. Kodas, University of New Mexico, Albuquerque, New Mexico

Jacques Livage, Université Pierre et Marie Curie, Paris, France

Audunn Ludviksson, University of New Mexico, Albuquerque, New Mexico

Seth R. Marder, California Institute of Technology, Pasadena, California

Thomas O. Mason, Northwestern University, Evanston, Illinois

Fabienne Meyers, California Institute of Technology, Pasadena, California

Joel S. Miller, University of Utah, Salt Lake City, Utah

Cathy Mohs, Kansas State University, Manhattan, Kansas

T. X. Neenan, Bell Laboratories, Lucent Technology, Murray Hill, New Jersey

Carole C. Perry, The Nottingham Trent University, Nottingham, England

Joseph W. Perry, California Institute of Technology, Pasadena, California

Thomas J. Pinnavaia, Michigan State University, East Lansing, Michigan

Kenneth Poeppelmeier, Northwestern University, Evanston, Illinois

Elsa Reichmanis, Bell Laboratories, Lucent Technologies, Murray Hill, New Jersey

Paul A. Salvador, Northwestern Univesity, Evanston, Illinois

Clement Sanchez, Université Pierre et Marie Curie, Paris, France

Bashir M. Sheikh-Ali, Rensselaer Polytechnic Institute, Troy, New York

Peter T. Tanev, Kansas State University, Manhattan, Kansas

Gary E. Wnek, Virginia Commonwealth University, Richmond, Virginia

CONTENTS

PREFACE

The purpose of this book is to introduce the concept of "materials chemistry" to the broader chemistry and materials science communities. This book may also be useful as a text or supplemental reading in graduate or advanced undergraduate courses in materials chemistry. In the introductory chapter, we have attempted to define materials chemistry and to outline its long history in the context of the development of what we now call "chemistry" and "materials science and engineering" (MS & E). Taking the broad perspective of materials chemistry as "chemistry related to preparation, processing and analysis of materials,"[1] it is apparent that "materials chemistry" has always been an integral part of chemistry and that a substantial fraction[2] of the chemistry workforce could be described as "materials chemists." Although this label is not always applied to the wide range of activities that constitute the component subjects of materials chemistry, many individuals who refer to their main area of interest by using such labels as, "polymer science," "surface science," "solid state chemistry," etc., could also be considered as "materials chemists." While quite different in orientation, approach and focus, all of these different areas of materials chemistry have a common goal: the generation, study and/or application of a "material." Most definitions of "materials"[3] connect substance with function and utility. As opposed to "chemicals," whose utility lies primarily in their consumption, materials are generally useful because they can be used repeatedly or continuously for an application that does not irreversibly convert them to something else. In this sense materials and their use underlie every aspect of human activity. Whether one considers the practical aspects of living and functioning in a society or the more aesthetic side of life, it is hard to imagine life without materials.

Chemistry and chemical processes have always had a major role to play in the development and use of materials. However, these processes have become so well integrated into technology or other disciplines, such as metallurgy, that many chemists no longer think of them as part of the subject matter of chemistry. Yet, individuals who have received their academic training in chemistry or chemical engineering are becoming increasingly involved in R & D activities whose main purpose is the development, modification or application of materials. The applications of materials, often derived from an understanding of materials chemistry, have a major impact on our lives, including our economic security. For example, current generations of energy storage devices (batteries), displays, electronic components, information exchange devices, sensors and medical technologies are benefiting from a better understanding of the relevant materials chemistry. We hope that this book will serve to remind chemists and chemical engineers about their intimate connections with MS & E and to inform them about recent progress in applying chemistry and chemical concepts to this area. We also hope that it will convey to the broader MS & E community a sense of the excitement in the field and the opportunities for new materials technologies that materials chemistry continues to offer.

Chapters 2–11 highlight some of the areas of materials chemistry research that are currently being pursued by chemists, chemical engineers and materials scientists worldwide. This survey is intended to illustrate the breadth and scope of materials chemistry research. The general progression is from molecular materials toward extended inorganic solids; however, the focus of most of the chapters is primarily the approach employed to obtain the material or its end application. We end with a chapter that emphasizes the connections between materials chemistry and biotechnology. Each chapter has been written by experts in these areas who have been asked to provide a general overview of their subject suitable for non-experts with a basic background in chemistry. While by no means complete in its coverage of materials chemistry, we feel that this book demonstrates the scope and the importance of this subject as an area of research and as a source of new technology for the 21st century and provides a valuable resource for those receiving training in this area.

LEONARD V. INTERRANTE
MARK J. HAMPDEN-SMITH

REFERENCES

1. L. V. Interrante, "Materials Chemistry-A New Subdiscipline?", MRS Bulletin, January 1992, p. 4.
2. "National Materials Policy," Proceedings of a Joint Meeting of the National Academy of Sciences–National Academy of Engineering, Natl. Acad. of Sciences Press, Washington, D.C. (1975) p. 125.
3. "The Random House Dictionary of the English Language, The Unabridged Edition." J. Stein and L. Urdang, eds., Random House, N.Y. (1969) p. 884 ["materials: the articles or apparatus needed to make or do something"]; "Materials and Man's Needs," Supplementary Report of the Committee on the Survey of Materials Science and Engineering, National Academy of Sciences, Volume 1, "The History, Scope and Nature of Materials Science and Engineering," Washington, D.C. (1975).

CHAPTER 1

Introductory Terms and Concepts

MARK J. HAMPDEN-SMITH

University of New Mexico
Albuquerque, New Mexico

LEONARD V. INTERRANTE

Rensselaer Polytechnic Institute
Troy, New York

1.1 MATERIALS CHEMISTRY: INTRODUCTION

Stone, bronze, iron; civilization has been defined by Man's relationship with materials. The utilization of materials for feeding, clothing, and protection not only distinguished human beings from other animal life but ensured the survival and ascendancy of our species. Materials have now become so thoroughly ingrained into society that they are often overlooked or underappreciated. When we turn on a light switch, the fact that we are effectively employing hundreds of different materials is hardly apparent to the casual observer. From the glass, metals, and polymers that make up the light bulb to the wires that connect it, via switches and meters, to the power plant that generates the electricity, we are intimately dependent on a wide range of materials which we ordinarily take for granted.

It is clear that materials have had a profound influence on the cultural, socioeconomic, demographic, and geographic development of society. The definition of *materials* as "substances having properties which make them useful in machinery, structures, devices and products" clearly connects materials with function and, through that function, utility.[1] *Chemistry* has been defined as "the study of the composition, structure, and properties of substances and the transformations by which substances are changed into other substances."[2] It is logical that the effective integration of materials

Chemistry of Advanced Materials: An Overview, Edited by Leonard V. Interrante and Mark J. Hampden-Smith.
ISBN 0-471-18590-6

science with its macroscopic perspective, and chemistry, which focuses on atomic- molecular-level interactions, could provide the opportunity to understand and control the fundamental connections between structure and function from the molecular level to a macroscopic scale. This understanding could lead to improved composition, structure, or synthetic methods and enable the development of new types of advanced-performance materials which have superior properties and performance.

1.2 MATERIALS CHEMISTRY: PAST[3,4]

Historically, the earliest ages of civilization are classified by the key materials that were in use at that time: stone, bronze, and iron. The later industrial age was characterized by the production of materials on a large scale for commercial purposes. In most cases the utility of the material was of primary importance, although there are also many examples of the aesthetic application of a material being the primary origin of its value (e.g., gold and silver as ornamentation, a symbol of wealth, or a medium for trade).

Prior to the discovery that metals could be obtained from rocks by certain (chemical) processes, all metals were relatively rare. Even when the technology for extracting copper and bronze from their ores was developed, the fact that these ores were not widely available presented significant limitations on the widespread use of these metals. It was not until the technology for extracting iron from its much more widely distributed and abundant ores became available that metals became the material of choice for many more practical applications. Clearly, metals have been among the most influential materials in the early development of civilization, and chemistry has been intimately involved in their production.

The role of chemistry in the production of metals is illustrated by the case of iron. Unlike gold or copper, iron is not available in elemental form in nature but must be isolated from iron ore (iron oxide) by chemical means. When iron ore, Fe_2O_3 (hematite), is heated together with a source of carbon, the following reactions can occur:

$$3Fe_2O_3 + 11C \rightarrow 2Fe_3C + 9CO \quad (1)$$

$$Fe_2O_3 + 3C \rightarrow 2Fe + 3CO \quad (2)$$

The Fe_3C (cementite, a hard, brittle material) and iron (relatively soft and malleable) form a composite material known as (carbon) steel. When the concentration of carbon in this steel is too high, the high proportion of cementite in the composite leads to a product that is not very useful, due to its brittle nature. This led to repeated folding, heating, and beating, resulting in removal of some of the carbon as well the silicate "slag," and oxidation of the surface of the iron to black iron oxide, FeO (hence the name *blacksmith*). The iron carbide and iron oxide react to give iron metal,

$$Fe_3C + FeO \longrightarrow 4Fe + CO \tag{3}$$

which, combined with the remaining cementite, give carbon steel its useful properties.

Of course, not only were these specific reactions unknown to early humans, but the basic concepts on which they are based were far from human understanding. Steel, as with all early materials, was developed by an entirely empirical process. As the understanding of the basic chemistry improved, and as the microstructural consequences of the thermal and physical processing became recognized, various refinements of the steelmaking process occurred, leading to the production of carbon and alloy steels with improved properties and therefore higher value.

This example of the successful enhancement of an important materials technology through a better understanding of the chemistry involved is illustrative of the advancement experienced by many material technologies in the late nineteenth and early twentieth centuries as our understanding of the basic science underlying these technologies increased. Other examples can be cited in which advances in chemical understanding, coupled with advancements in analytical methods, led to substantial improvements in materials technologies. These include the dye industry, where the development of synthetic dyes made colored fabrics readily available to most people, and the fibers and plastics industries, which were built on advances in polymer synthesis (see Section 1.3). Thus it is clear that chemistry has always been intimately involved in the development of materials for technology and that a basic understanding of this chemistry was often essential for the optimal (or even successful) production of the material.

Despite its intimate connection with materials, as chemistry became an independent branch of science in the late nineteenth and early twentieth centuries, its study in universities became increasingly disconnected from the technology of materials synthesis and processing. Such "applied" subjects as the chemistry of the earth (geology), of metals (metallurgy), and ceramics were separated off to form new departments in the 1930s to 1950s. In many universities, the study of materials became viewed as an activity more appropriate for engineers, or at least applied scientists, than for "chemists." Few, if any, universities in the United States in the mid-twentieth century undertook to instruct chemistry students in the basic science of materials. In industry, however, chemistry continued to play an important role in the development of new materials and in materials processing, where a growing understanding of the chemical processes involved contributed to the rapid pace of progress. A particularly good example of this key role of chemistry in the production of new materials is provided by the development of synthetic polymers, which started in earnest early in the twentieth century.

Historically, natural organic polymers, in the form of wood and other plant fibers, were among the first materials to be used by humans. The use of wood as a structural material continues to the present, although it is becoming increasingly common to combine wood with synthetic organic polymers so as to provide materials of enhanced utility and/or decreased cost (e.g., particle board and plywood). Perhaps even more important than the structural application of wood was its conversion into paper. This key technology was a contribution of the Chinese that was transmitted westward by the Arabs in the period A.D. 750–800.[4] Even now, the central role of paper as a medium for recording and transmitting information is only gradually giving way to plastics, ceramics, and semiconductors in the form of CDs, magnetic recording media, and integrated circuits.

Among the first natural polymeric materials to attract scientific interest were silk and cobwebs. Robert Hooke (1665) noted the interesting properties of silk and spiders' webs, some of which have properties that are yet to be duplicated in synthetic polymers. The first significant impact of chemistry on the development of useful synthetic polymeric materials came in 1839, when Charles Goodyear found that the elastic properties of natural rubber could be improved, and its tackiness eliminated, by heating it with sulfur. While this discovery was in the process of development into a commercial product, cellulose nitrate, another product resulting from the chemical conversion of a natural material, was discovered and eventually became the basis for both gun cotton and cellophane film. These polymeric materials can be characterized as semisynthetic because they are derived from natural polymers. The first commercial development of fully synthetic polymers came in the early part of the twentieth century, when Leo Baekeland produced Bakelite, a thermoset polymer derived from the reaction of phenol with formaldehyde. However, real progress in synthetic polymer production came only after the development of a scientific understanding of macromolecular structure, bonding, and reactions starting in the 1920s with the work of Hermann Staudinger and continuing into the 1950s with the work of Herman Mark, Kurt Meyer, Wallace Carothers, and Paul Flory.[5] This revolution in synthetic materials production continues to date and is marked by the introduction of such commercial products as nylon, rayon, teflon, and lexan, product names (among many others) that have become part of the vocabulary of modern technology and society. Now organic polymers are part of our everyday life in such a wide variety of forms because they can display the largest spectrum of properties of any class of materials, from nonstick coatings on pots and pans to recording media.

The development of new inorganic materials followed a similar pathway of first using natural materials (such as rocks, minerals, and metals) directly and then later, through experimentation, learning to modify natural substances chemically so as to obtain new materials not known, or not readily available, in nature. An early example of this approach, which undoubtedly preceded the deliberate extraction of metals from their mineral sources, was the development of ceramics, such as pottery, glass, and later, hydraulic cement. The term *ceramics* has now expanded to include virtually all inorganic engineering materials other than metals and semiconductors, where both the material type and its characteristic physical properties provide a basis for its classification.

1.3 MATERIALS CHEMISTRY: PRESENT

1.3.1 Classification of Materials

In most textbooks of material science and engineering, materials are classified into broad categories, based on both their chemical constitution and their typical physical properties.[6, 7] Solid materials are generally grouped into three basic categories: metals, ceramics, and polymers. In addition, there are two other groups of important engineering materials: composites and semiconductors. *Composites* consist of combinations of two or more different materials, whereas *semiconductors* are distinguished by their unusual electrical characteristics. In addition to this classification based on their structure, bonding, and properties, materials have increasingly come to be classified by their function [e.g., electronic, biomedical, structural, and optical (and nonlinear optical) materials].

Metallic materials are usually combinations of one or more metallic elements. Metals are characterized by the existence of large numbers of delocalized electrons; that is, these electrons are not bound to particular atoms. Many properties of metals are directly attributable to these delocalized electrons. Metals are generally good conductors of both electricity and heat and are not transparent to visible light; a polished metal surface typically reflects light and has a lustrous appearance. Furthermore, while metals can be quite strong, they are generally malleable and easily formed into desired shapes, making them particularly useful for structural applications.

Ceramics are compounds formed between metallic and nonmetallic elements; important examples include the oxides, sulfides, nitrides, and carbides. The wide range of materials that fall within this classification includes most of the natural minerals of the earth, such as the silicates, oxides, carbonates, and sulfides, as well as glasses and glass ceramics. These materials are typically insulative to the passage of electricity and heat and are resistant to degration at high temperatures and harsh environments. Ceramics are hard but very brittle. These materials are used in structural, optical, and electronic applications.

Polymers or *molecular materials* include the familiar plastic and rubber materials. They are typically comprised of macromolecules that range from linear polymers having amorphous or quasicrystalline structures to extensively cross-linked networks. In recent years, this class of materials has been effectively expanded to include virtually all types of materials that are comprised of discrete molecules (i.e., molecular materials). In addition to solids comprised of small molecules or molecular ions, this category of molecular materials would logically include certain liquids, such as the silicones (which are used as lubricants and dielectric fluids), along with those that exhibit long-range order, such as liquid crystals, Langmuir–Blodgett (LB) films, and self-assembled monolayers. Moreover, when one considers that wood (or plant fiber), in all forms from paper to structural components, can also be placed in this category, this is clearly one of the largest and most important of the material categories. Polymers and most other molecular materials typically have low electrical and thermal conductivities, are lower in strength than metals and ceramics, and are generally not suitable for use at high temperatures. Thermoplastic polymers in which the long molecular chains are not rigidly connected have good ductility and formability; thermosetting polymers are stronger but more brittle because the molecular chains are tightly linked. Polymers and other molecular materials are used in many applications, including dielectrics in electronic devices, displays, lithography, recording media, fibers for clothing, and in food and other product packaging.

Composites are combinations of different materials. Fiberglass is a familiar example, in which glass fibers are embedded within a polymeric matrix. A composite is designed to display a combination of the best characteristics of each of the component materials. Fiberglass acquires strength from the glass and flexibility from the polymer. Most natural materials, such as stone, concrete, and wood are also logically placed in this category. Many of the recent materials developments in aerospace engineering and sports equipment have involved composite materials where advantage is taken of the combination of high strength and light weight.

Semiconductors have electrical properties that are intermediate between the electrical conductors and insulators. Furthermore, the electrical characteristics of these materials are extremely sensitive to the presence of minute concentrations of impurity atoms, the concentrations of which can be controlled over very small spatial regions. Semiconductors have made possible the advent of integrated circuitry that has revolutionized the electronics and computer industries over the last 50+ years.

1.3.2 The Role of Chemistry

Chemistry is intimately involved in the generation and processing of materials from each of the various categories listed above, as well as in their environmental degradation by both natural and unnatural causes. In the case of metals, most are derived from ores by chemical reduction of their oxides or silicates or by oxidation (and then reduction) of their sulfides. This historical, and largely empirical, application of chemistry in the generation of metals has benefited in more recent times from an understanding of the basic chemistry involved in these processes (e.g., the oxygen furnace in steel production) as well as the corrosion processes that are experienced by many metals during their useful lifetime. The fact that some of these metals act as poisons toward biological organisms when discarded extends even further the need to understand and control the basic chemistry and biochemistry of metallic substances.

The role of chemistry in ceramic processing is somewhat less obvious but equally profound, as most ceramic materials undergo extensive purification and/or processing involving chemical reactions prior to their consolidation as powders. Even their consolidation and their use as powders often requires the addition of organic polymers and surfactants as binders and dispersing agents. In recent years, chemistry has become increasingly involved in the processing of ceramics in special forms; examples include the preparation of powders with well-defined particle sizes, size distributions and shapes; fibers; films; and membranes. Here methods such as spray drying, vapor-phase synthesis, sol-gel processing, chemical vapor deposition and infiltration, and polymer precursor pyrolysis are finding increasing use in various applications.

As with metals, semiconductors are not found in nature in the purity or even the chemical form that is required for most applications in electronics. Silicon, the mainstay material of the electronics industry, is derived from silica by reduction and then further purified by distillation as silicon tetrachloride before being converted back to the element and then zone refined and crystallized by physical processes. The role of chemistry by no means ends here, as processes such as oxidation, doping, patterning, and etching are all either largely or entirely chemical in nature.[8] One additional chemical process that is making an increasing impact on the processing of semiconductors is chemical vapor deposition (CVD), where everything from the actual growth of the semiconductor material (including Si) to the application of insulators and metallic conductors to its surface can be accomplished.

Apart from a few natural materials, the polymers currently in use as materials are all the end products of chemical processes. Even in those cases where natural products such as wood, paper, or cotton are employed, chemical processes are usually employed at some stage in the conversion of the natural material to a more useful final product (e.g., bleaching, dyeing, and painting). Synthetic polymers, derived from natural materials such as hydrocarbons (petroleum), are used increasingly in place of natural fibers in fabrics as well as in a wide range of other products, creating a new set of problems when it comes to the disposal or reuse of the product.

In general, it is becoming widely recognized that the entire enterprise of materials fabrication and processing is intimately interconnected with both the material's origins and its ultimate disposal or recycling. As a result, no new method for extracting or processing a material can be considered without some understanding of the real costs of obtaining it, as well as its eventual fate after its useful lifetime as a material in a product has ended. The concept of a total materials cycle has been used in this context (Fig. 1.1) to indicate the intimate relationship between our continuing need for new

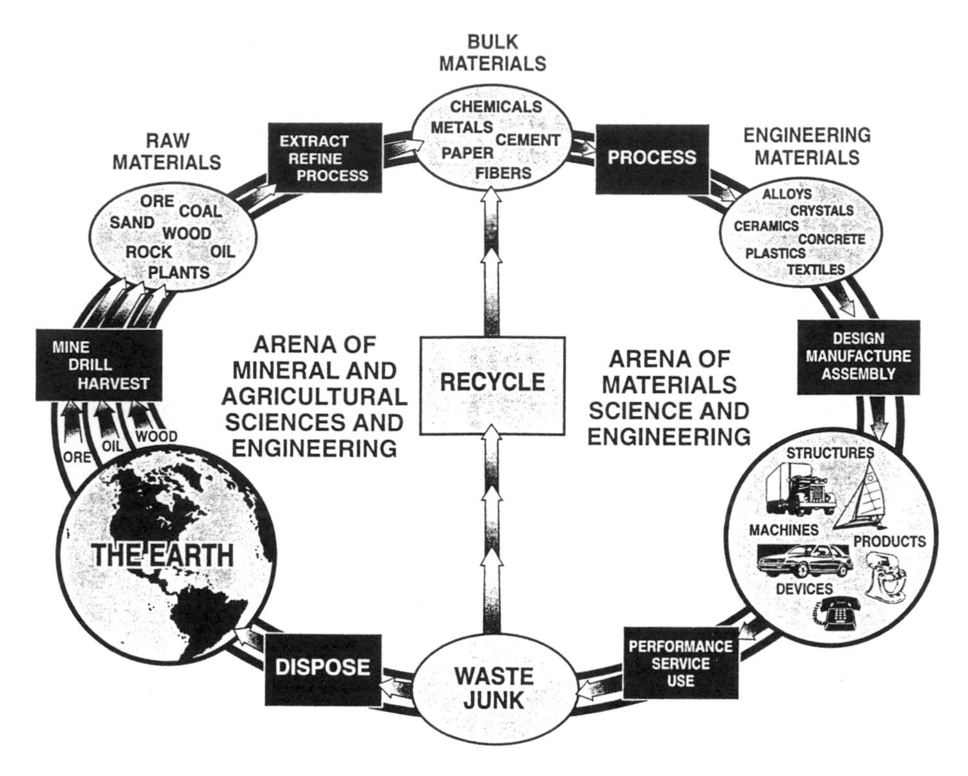

Figure 1.1 The total materials cycle.

materials and the effects of materials development on the environment and on the diminishing supply of basic raw materials.[9a] Indeed, it is irresponsible, if not impossible, these days to bring a new material into the marketplace without consideration of factors such as the supply of the material and the true costs of its development in terms of the potential impact on the environment. As stated in a U.S. National Academy of Sciences report on materials science and engineering published in 1974, "it is vital to learn how to move materials carefully around the materials cycle, from production of raw materials to use and eventual disposal, in ways that minimize strain on natural and environmental resources."[9a] Considering the fact that the processes included in this cycle are typically chemical processes, chemists have an important role to play in the continuing efforts to understand the fundamental nature, and the practical consequences, of these processes.

The role of chemists in the solution of fundamental materials science problems is becoming more important with time. Recent examples of the role of materials chemists in key materials discoveries include (1) the synthesis of the synthetic zeolites in the mid-1960s and their subsequent widespread use as petroleum cracking catalysts and (2) the development of the high-resolution, polymer-based, lithographic imaging (resist) materials that have made possible the amazing reductions in electronic device dimensions and corresponding increases in device density that have characterized the field of electronic processing for the past 25+ years. Both of these efforts are ongoing and in the past few years have resulted in (1) the discovery of new micro- and mesoporous molecular sieves based on both aluminosilicate and other framework structures and (2) increasingly more effective resist materials and mechanisms for resolution enhancements that are pushing the theoretical limits of light-based lithographic processes. These and many other recent developments in materials chemistry are described in more detail in the following chapters of this book as well as in an ever-increasing number of journal publications and books. Between 1988 and 1990 three journals specifically dedicated to research in materials chemistry have appeared: *Advanced Materials* in Germany, *Chemistry of Materials*, published by the

TABLE 1.1 Categories of Papers in *Chemistry of Materials*

Bio- (and biomimetic) materials	Metals
Catalytic materials	Microstructure
Chemical vapor deposition	Molecular crystals (organic, metal–organic, etc.)
Clusters and colloids	Optical (including NLO) materials
Coatings	Nanophase materials
Composites	Phase transitions
Corrosion	Polymeric materials
Defects	Precursor routes to materials
Electronic materials	Self-assembled materials
Fluids and liquid crystals	Semiconductors Sensors
Glasses	Sol-gel chemistry/processing
Inorganic and organic networks	Surface and interfacial phenomena
Inorganic solids/ceramics	Superconductors
Layered materials (including intercalation compounds)	Structural materials
Magnetic materials	Theory
	Thin films and monolayers

American Chemical Society, and the *Journal of Materials Chemistry*, published by the Chemical Society of Great Britain. As an example of the types of articles that appear in these journals, in Table 1.1 are listed the subject headings used by *Chemistry of Materials* to categorize the scientific papers that appear in its monthly issues.

While chemists have made and, undoubtedly, will continue to make important contributions to the development of materials science and engineering, these contributions have rarely involved chemists alone. For real progress, the inherent complexity and interdisciplinary nature of materials science and engineering virtually requires effective cooperation between scientists and engineers from various disciplines. The need for such cooperative, interdisciplinary efforts is well recognized in industry, where teams of scientists and engineers are typically involved in most efforts to develop new or improved materials or to solve other materials-related problems.

1.3.3 Political and Socioeconomic Factors

The field of materials science and engineering (MS & E) in general, and materials chemistry in particular, has progressed to the present Microelectronics Age. Advances in information processing, communications, transportation, and construction are revolutionizing our environment and our daily lives. The computing and communications capability that is currently available on the desktop (often at home!) would have been hard to imagine 20 years ago. The increasing demand by the consumer for advanced-performance products requires an increasing level of sophistication in the workforce. This in turn demands a greater level of recognition on the part of the chemistry profession and on the part of policymakers of the role of materials chemistry in technology. Also needed is the identification and effective organization of the skills that are required for effective training in this area.

The need to nurture and support the development of new materials and the personnel resources that this requires is becoming increasingly recognized at the governmental, industrial, and academic levels. Governments are becoming involved because of the increasingly high costs of materials development, which have severely limited "blue-sky" efforts to prepare and investigate new materials in industry except under rather specific and well-focused circumstances. Government funding agencies have responded to this need in a variety of ways. In the United States the National Research Council (NRC) conducted a broad survey of the needs and opportunities in the area of MS & E. The reports of this survey examined in detail the impact of MS & E on U.S. national competitiveness.[9] Despite their differing needs for particular materials, the industries surveyed expressed a clear need to produce both new and traditional materials more economically and with a higher reproducibility and quality than was currently possible. In particular, the industry survey revealed a serious weakness in United States research efforts involving the synthesis and processing of materials. In response to these reports, new programs were initiated across the United States government to support new efforts and enhance existing efforts to overcome these deficiencies. Within the National Science Foundation (NSF), a cross-divisional materials synthesis and processing (MS & P) initiative was started to enhance research efforts along with other programs that were directed at improvements in education in this area.[10]

From an industrial perspective, the development of advanced-performance materials requires a readily identifiable and economically feasible market need. There are many factors that affect the research, development, and commercialization of materials-based products in an industrial setting, and the deciding factor is often a question of economics and the cost savings or perceived competitive advantage to be gained by using new material rather than a science or engineering problem. In most cases these advanced-performance materials are components of a complex system and are often direct replacements for another material. In some cases, however, a new material can lead to a totally new type of product or even a new technology. The role of tungsten filaments in incandescent lamps, silicon in electronics, optical fibers in communications, and photoconductors in copy machines provide just a few examples of such enabling materials in our current technology.

One difficulty, however, is that the cost of the development and commercialization of new advanced-performance materials has risen considerably, while the probability of recovering this investment and realizing a net profit has decreased. The result is that the economic risks inherent in bringing new materials into the marketplace have increased enormously. An example is provided by the high-strength polyimide-based fibers and plastics, where many years after commercialization, the volume applications associated with system-for-system substitution had still not yet occurred at a level necessary to repay the development and commercialization costs that have been expended.[11]

In some situations, legal risk factors can become of critical importance in decisions to bring a product to the marketplace or maintain it in production. Such legal risk factors have led to the removal of certain biocompatible materials, which are commercially feasible, from the market for fear of future litigation (see Chapter 11). Another consideration here is the fact that most materials commonly serve a range of technologies and tend to be less proprietary than are the products made of them. On the one hand, this could decrease the incentive for a particular company to invest heavily in the development of a new material unless that material is critical to the function of the end product; however, on the other hand, this consideration makes efforts to develop new materials potential fruitful ground for cooperative research and development, particularly involving universities and industries or industry consortia (e.g., in the United States, SEMATECH for development of integrated-circuit technology).

1.3.4 Educational Issues

Despite its long history as a technological activity, the recognition of materials science and engineering as a separate and distinct academic discipline is a relatively recent phenomenon. In United States universities, academic departments of materials science and engineering were virtually unknown before the 1950s, and even today there are far fewer such departments than there are universities. Where such departments exist, they are almost always located in schools of engineering and are often separated from the science departments by differences in administration and philosophy. Within the profession of chemistry, the recognition of materials-related chemistry as a distinct research discipline, let alone as a subject to be taught in either a graduate or undergraduate curriculum, is still not widespread. On the other hand, the subjects of solid-state and polymer chemistry, two major components of materials chemistry under any system of classification, are well recognized as both areas of research and subjects for courses and curricula in a number of universities.

The exceptions to this typical separation of department of MS & E and chemistry in United States academic institutions are mainly among those institutions that have obtained support, usually from the federal government, for the establishment of centers of excellence in materials science and engineering. These centers are usually organized around common research themes and do not often have a major impact on undergraduate teaching at the university, or universities, concerned. They include the long-standing NSF materials research laboratories (MRLs), which usually combine several major "thrust" areas of research, as well as other types of centers which are typically organized around one particular thrust area (e.g., composites and structures, electronic materials and devices, vapor-phase synthesis of materials, etc.). Many of these centers include chemists as well as physicists and other materials scientists/engineers and provide a good opportunity for collaboration across the traditional academic boundaries.

In general, materials chemistry can be viewed as a subdiscipline of chemistry that includes solid-state and polymer chemistry as well as other areas of research and development where the emphasis is placed on materials synthesis, processing or analysis. There is increasing evidence that such a broad perspective on materials chemistry is gradually developing, in the United States as well as in other countries. Examples can be found in the growing popularity of the research journals in this area as well as in the increasing frequency of symposia or entire meetings that are dedicated to materials chemistry.[12]

In summary, although materials chemistry has been important for many years, it has not been clearly defined from an academic viewpoint. There are some compelling reasons to do so, the strongest being educational. Materials chemistry is widely practiced in industrial and governmental laboratories, and a large fraction of those who identify themselves as chemists are, in fact, materials chemists. Current estimates have placed approximately one-third of all U.S. chemists as working in the materials-related area. In the United States much of this work has typically been performed in industrial laboratories by people who received their academic training in chemistry departments and have had little or no formal training in either materials science or such materials chemistry subjects as solid-state and polymer chemistry. Until quite recently, coverage of these topics or even the use of materials-related examples in chemistry courses was a relatively rare occurrence. This situation is apparently beginning to change in some locations, at least partly as a result of the greater availability of resources for education in this area.[13]

One problem has been the lack of a generally accepted definition for materials chemistry and its scope of interest. We will employ here the working definition that one of us has advanced previously,[14] that materials chemistry is "chemistry directed at the preparation, processing and analysis of materials." This includes any and all chemistry involved in the overall materials cycle (see above). As is perhaps clear from this definition, materials chemistry is an extremely broad area. It encompasses all the traditional areas of chemistry—inorganic, organic, physical, analytical, and biochemistry—and has strong connections with most other science and engineering disciplines. The distinguishing feature can be found in the perspective and approach employed. For example, a materials chemist would be more likely to take a molecular (or atomic)-level perspective and approach to the solution of a ceramic processing problem than would a ceramic engineer. Such an understanding of the microscopic nature of materials could enable the effective control over their structure and composition. This has been demonstrated in the ability to achieve lower-temperature synthesis of many inorganic solids, as compared to the traditional "ceramic" methods and to control their particle size, particle size distribution, pore structure, density, and microstructure (see Chapters 5, 7 to 10).

In addition to a thorough grounding in the basic science and engineering of solids, at least a rudimentary familiarity with the tools used to characterize and analyze materials (i.e., materials characterization methods) is necessary for anyone who intends to become proficient in materials chemistry. It probably makes most sense to focus specific efforts to train materials chemists at the graduate level while giving undergraduate chemistry students a broad education in the traditional disciplines of chemistry that incorporates relevant examples which might be considered materials chemistry. For example, *Teaching General Chemistry: A Materials Science Companion* is a volume written for teachers to provide examples from the solid state to complement molecular examples commonly used in general chemistry courses.[13] Elective courses taught in chemistry or other departments on such subjects as polymer science and the inorganic solid state could provide an opportunity for further input to interested chemistry undergraduates. Specialization in the area of materials chemistry could then occur at higher degree levels, but the question remains, which topics should be covered?

1.4 MATERIALS CHEMISTRY: FUTURE

Despite the increasing costs and complexities involved, it seems safe to assume that new materials and processes will continue to be needed and that chemistry will continue to play a large role in their development. What is not so clear is where these breakthroughs in new materials will occur. As illustrated by the discovery of the cuprate superconductors and C_{60} and related fullerences in recent years (see Chapters 2 and 10), breakthroughs can and probably will continue to occur in widely divergent areas of materials science and are likely to involve the full spectrum of material types. What is even less clear is whether or not these breakthroughs in materials science will have a major impact on materials technology. At the time of this writing, the cuprate superconductors and fullerences are still under active investigation in this context.

To illustrate the breadth and the potential for future developments in materials chemistry research, we have arbitrarily chosen two of the many areas of current research for further discussion here. These are the fields of nanotechnology and biomaterials. *Nanotechnology*, which is essentially materials science on the nanometer size scale, involves two rather different but complementary types of materials, both of which are described in subsequent chapters (7 and 8) of this book as well as in a special issue of the journal *Chemistry of Materials*.[15]

One type of nanomaterial might be termed *nanosized materials* or *nanoparticles*, where the relevant or active part of the material is in the nanometer size range in one or more of its dimensions (typically, 1 to 100 nm). Materials in this size regime often exhibit size-dependent properties (see Chapter 7). These properties can be classified into four categories: physical, compositional and phase behavior, mechanical, and chemical properties. Physical characteristics that are size dependent include electronic, optical, and magnetic properties. Size-dependent compositional and phase behavior characteristics include crystal lattice distortions, phase transitions and temperatures, vapor pressure, internal pressure, melting point, and solubility. Mechanical characteristics that are size dependent include hardness, ductility, toughness, creep, and fatigue. Size-dependent chemical characteristics are based on the high surface energy and surface area of nanometer-sized particles, which can result in enhanced sintering rates as well as chemical reactivity. Size-dependent properties are found in all types of materials. Quantum-confinement effects have been studied, in particular, for certain metals and semiconductors when the individual crystallites are in this size range and are effectively separated from one another. The size quantization effect occurs when the size of the particle (or crystallite/grain) becomes comparable to the deBroglie wavelength of the charge carriers in the material. For semiconducting materials this size range is 2 to 50 nm; for metals this effect occurs at subnanometer particle sizes. In the case of ceramics, crystallites in this size range can undergo sintering at faster rates, thereby allowing densification at lower temperatures than would otherwise be required. In some cases, dense ceramic objects that are comprised of such particles undergo superplastic flow under shear stress, allowing superplastic forming of otherwise brittle materials in a manner much like that done for some metals.

The second type of nanomaterial constitutes the negative image of the nanosized material, where the relevant nanoscale portion is a space in a material that can be accessed and/or filled by a molecule of specific size and shape. These materials are often referred to as *nanoporous* (see Chapter 8). The relevant applications for such materials are in the separation of gases, solutes, ions, and so on, as well as in the area of heterogeneous catalysis, where reactions might be carried out in a selective manner on the appropriate molecules. The use of nanoporous materials constitutes the oldest and most widely investigated area of nanoscale technology, where materials such as the zeolites have been used for years on a large scale in industry.

The field of nanotechnology is being approached from two directions by physicists, chemists, and other materials scientists. One is from the top down (i.e., by making things smaller and smaller). The other is from the molecular/atomic (or angstrom) level by building larger clusters of atoms, molecules, or connections between molecules. A goal of both approaches is the fabrication of *quantum dots*, defined as particles that exhibit a size quantization effect in at least one dimension. One potential application for quantum dots is as memory storage devices that store information at a much higher density than can be achieved with electronic transistors. The ultimate in nanoscale memory storage units is probably the individual molecule or, in the case of extended solids, small clusters of atoms, where information might be stored, and accessed, by electrical or magnetic methods. Such molecular storage elements might employ molecular or atomic wires, perhaps fabricated by using self-assembly methods, to address them. As is often the case in materials science, connections with natural systems can be drawn here that might suggest potentially fruitful lines of research. The relevant analog from nature in this case is the human brain, where chemistry on a molecular level is used to store and retrieve information on a scale that has not even been approached by existing computer memory devices. This brings us naturally to the area of biomaterials, where the connections between materials and nature, in the form of biological organisms, are perhaps the most obvious and direct.

The area of biomaterials encompasses an enormous range of topics, which includes everything from materials produced by biological organisms, or as a result of biological processes, to synthetic materials whose design, form, or construction is inspired by analogies to biological systems. The example of wood and plant fibers as a source of materials that was discussed early in this chapter illustrates just one of a whole host of materials of much technological importance that are derived from biology. If we now include cotton, wool, animal skins (leather), bone, and other ceramics or ceramic composites formed as a result of biological processes, we can begin to see the enormous range of materials that we are dealing with in this context. Chemistry at both the biotic and postbiotic level is intimately involved with both the construction and application of all these materials. The understanding, development, and effective use of these chemical processes constitutes one of the major areas of active research in the biomaterials area.

One important aspect of biomaterials is the need to replace some of them on occasion to ensure the continuing functioning of the host organism. An example is provided by the needs of medicine for temporary or "permanent" replacements of parts of the body (i.e., bones, teeth, skin, etc.). The search for such materials constitutes another major area of current research in biomaterials. Chemistry has long been and will continue to be a major source of new materials for use in medicine and chemists will continue to be needed, along with other materials scientists, to design and prepare appropriate replacements for materials needed by surgeons or dentists. The third major area of biomaterials research is that of biomimetic materials, where natural (bio) materials are used as models for the design of new functional or structural materials. For example, the structures and properties of clamshells have stimulated the construction of analogous laminar composite structures comprising ceramics and polymers in the laboratory.

As with most areas of material science, research in biomaterials is broad in scope and multifaceted. To succeed in this area, a broad perspective and base of knowledge relating to materials and their behavior will be needed on the part of the chemists involved as well as a willingness to learn from and cooperate with other scientists, engineers, and medical practitioners having complementary skills in the subject area.

In this context it seems appropriate to close this chapter with a quotation from the 1974 NAS report *Materials and Man's Needs*, which called for "the purposeful mobilization of the technical expertise of materials scientists and engineers" to engage in "an intensive effort on the part of industry to exploit *and nourish* the reservoir of materials knowledge" and on the part of academics to establish "a strengthened academic base to inject new knowledge into that reservoir."[9a] We hope that this book will serve to increase the growing recognition within the chemistry and the materials science professions that materials chemistry is not only an integral part of the field of chemistry but that it is also a vital source of the materials and materials processes on which our society and our civilization so strongly depend.

ACKNOWLEDGMENT

The authors are grateful to Vincent (Don) Interrante for his outstanding rendition of the total materials cycle figure from Ref. 9a (our Fig. 1.1).

REFERENCES

1. M. Cohen, Ed., *Materials Science and Engineering: Its Evolution, Practice and Prospects, Mater. Sci. Eng.* **37**(1) (1974). See also Ref. 6.
2. R. J. Gillespie, D. A. Humphreys, N. C. Baird, E. A. Robinson, *Chemistry*, 2nd ed., Prentice Hall, Upper Saddle River, NJ, 1989.
3. *Materials and Man's Needs*, Supplementary Report of the Committee on the Survey of Materials Science and Engineering, Vol. 1, *The History, Scope and Nature of Materials Science and Engineering*, National Academy of Sciences, Washington, DC, 1975.
4. H. W. Salzberg, *From Caveman to Chemist*, American Chemical Society, Washington, DC, 1995, 269 pp.
5. R. J. Young, P. A. Lovell, *Introduction to Polymers*, 2nd ed., Chapman, Hall, London, 1991, 432 pp.
6. M. B. Bever, Ed., *Encyclopedia of Materials Science and Engineering*, Vol. 1, Pergamon Press, Oxford, 1986.
7. W. D. Callister, Jr., *Materials Science and Engineering: An Introduction*, 3 rd ed., Wiley, New York, 1994; D. R. Askeland, *The Science and Engineering of Materials* 3rd ed., PWS, Boston, 1994.
8. H. B. Pogge, Ed., *Electronic Materials Chemistry*, Marcel Dekker, New York, 1996.
9. (a) *Materials and Man's Needs*, Summary Report of the Committee on the Survey of Materials Science and Engineering, National Academy of Sciences, Washington, DC, 1974; (b) P. Chaudari, M. Flemings, *Materials Science and Engineering for the 1990s: Maintaining Competitiveness in the Age of Materials*, National Academy Press, Washington, DC, 1989.
10. K. G. Hancock, in *Materials Chemistry: An Emerging Discipline*, ACS Advances in Chemistry Series, Vol. 245, L. V. Interrante, L. A. Caspar, A. B. Ellis, Eds., American Chemical Society, Washington, DC, 1995, pp. 18–27.
11. M. L. Good, in *Materials Chemistry: An Emerging Discipline*, ACS Advances in Chemistry Series, Vol. 245, L. V. Interrante, L. A. Caspar, A. B. Ellis, Eds., American Chemical Society, Washington, DC, 1995, pp. 28–36.
12. For example, since 1993, the Chemical Society of Great Britain has been holding a biennial international conference of materials chemistry, and numerous symposia have been held at both regional and national meetings of the American Chemical Society, in the latter case partly through the agency of the Materials Chemistry Secretariat, an organization within the ACS that was developed to help coordinate programs in this area at national meetings.
13. A. B. Ellis, M. J. Geselbracht, B. J. Johnson, G. C. Lisensky, W. R. Robinson, *Teaching General Chemistry: A Materials Science Companion*, American Chemical Society, Washington, DC, 1993, 575 pp.
14. L. V. Interrante, *MRS Bull.* (Jan. 1992).
15. Special Issue on Nanostructured Materials, *Chem. Mater.* **8**, 1569–2193 (Aug. 1996).

GENERAL BIBLIOGRAPHY

- D. W. Bruce and D. O'Hare, Eds., *Inorganic Materials*, Wiley, Chichester, West Sussex, England, 1992, 543 pp.
- L. V. Interrante, L. A. Caspar, A. B. Ellis, Eds., *Materials Chemistry: An Emerging Discipline*, ACS Advances in Chemistry Series, Vol. 245, American Chemical Society, Washington, DC, 1995, pp. 28–36.
- H. B. Pogge, Ed., *Electronic Materials Chemistry*, Marcel Dekker, New York, 1996.
- C. N. R. Rao, Ed., *Chemistry of Advanced Materials*, an IUPAC "Chemistry for the 21st Century" monograph, Blackwell, Oxford, 1993, 388 pp.

CHAPTER 2

Electron-Transfer Salt-Based Conductors, Superconductors, and Magnets

PATRICK CASSOUX

Lab de Chimie Coordination du CNRS, Toulouse UP 8241
Université Paul Sabatier Toulouse Cedex, France

JOEL S. MILLER

University of Utah, Salt Lake City, Utah

2.1 INTRODUCTION

Solids comprised of molecules (or molecular ions) typically exhibit physical properties that are similar to those of the isolated molecule (or ion). For example, the vibrational or electronic absorption bands as well as the nuclear magnetic resonance (NMR) spectra of a molecular solid are typically similar to that of the isolated molecule. Hence molecular orbital theory can be used to predict most properties of molecular solids. Thus the concepts and methods are quite different from those used in atom-based solid-state chemistry or for some polymers. Nonetheless, in some cases, solids comprised of molecules exhibit properties that are distinct from those of the isolated molecules or ions. These include electrical, optical, and magnetic properties. Additionally, in a few cases, molecular solids exhibit cooperative properties such as superconductivity and ferromagnetism.

2.1.1 Electron-Transfer Salt-Based Conductors and Superconductors

In the classical electron conduction model for metals, the valence electrons of the atoms are considered as freely moving between all the atoms in the

Chemistry of Advanced Materials: An Overview, Edited by Leonard V. Interrante and Mark J. Hampden-Smith.
ISBN 0-471-18590-6

solid. In the energy-band model for partially filled bands (metals), very little energy is required to excite the valence electrons so that they can become mobile or are free to attain a directed drift velocity in the presence of an applied electrical potential. For filled bands, much greater energy is required to excite the valence electrons over an energy gap (semiconductors and insulators) and to make them mobile (see Section 2.2). In both cases, electronic conduction (electrical current) involves the movement of electrons (and their associated electrical charge) between atom sites in the solid lattice.

Thus the extension of electrical conductivity models from atom- to molecule-based materials requires that the molecular units interact in the solid. Therefore, for molecule-based materials (1) the interactions between the molecules are essential for electrical conduction and (2) this property can only be observed in the solid state. Molecule-based materials having electrons tightly bound to the constituent units are insulators with low conductivities. In contrast, those with electrons delocalized over the entire solid may exhibit high metal-like conductivities. As early as 1911 molecule-based compounds were anticipated to have metal-like conductivities.[1, 2] These predictions were fulfilled in 1954[3] (see Section 2.3). In the past 30 years there has been an extraordinarily intense research effort in this area, as evidenced by the number of papers, reviews, books, and scientific meetings devoted to highly conducting molecule-based materials.

The primary motivation behind research in this area is the development of novel materials with unusual electrical properties and their technological exploitation. In addition to chemists, these materials have aroused great interest from the solid-state physicists as a wealth of new phenomena has been identified and a new specific condensed-matter physics has evolved. The potential technological applications are a consequence of the low density of these materials (about 1.5 g/cm^3, compared to that of conventional metals, about 9 g/cm^3 for copper), their unusual physical properties, as well as ease of fabrication.[254a, b]

2.1.2 Electron-Transfer Salt-Based Magnets

Classical ferri- or ferromagnets are atom-based (with *d* or *f* orbitals unpaired electron spin sites) and possess extended network bonding in three dimensions and in rare cases two dimensions. In electron-transfer salt molecule–based magnets, which were first postulated in 1963[4] and realized in 1985,[5] the unpaired electron spin sites ($\uparrow$ or $\downarrow$) reside in *p* orbitals, and cooperative magnetic phenomena are required. Hence spins residing on molecular units must couple with all other spins in the molecule-based solid. The requirement for spins necessitates that the molecules or molecular ions are radicals with one or more spins each. Hence the molecular repeat unit must be a free radical, such as a nitroxide or a radical cation or anion. The magnetic susceptibility, χ, of noninteracting spins can be modeled by the

Curie law:

$$\chi = \frac{C}{T} \qquad \text{where} \quad C = \frac{Ng^2\mu_B^2 S(S+1)}{3k_B} \tag{1}$$

where μ_B is the Bohr magneton, N is Avogadro's number k_B the Boltzmann constant, and g the Landé g value.

Interacting spins either reduce or enhance the susceptibility and can be modeled by the Curie–Weiss equation:

$$\chi = \frac{C}{T - \Theta} \tag{2}$$

where $\Theta > 0$ signifies an enhanced susceptibility or ferromagnetic ($\uparrow\uparrow$) coupling and $\Theta < 0$ signifies a reduced susceptibility or antiferromagnetic ($\uparrow\downarrow$) coupling. Ferromagnetic ordering may occur below an ordering or critical temperature, T_c, similar to that for superconductivity. Antiferromagnetically coupling between sites with a differing number of spins cannot cancel to zero and can lead to ferrimagnetic ordering below a T_c. Molecule-based materials possessing $\Theta > 0$ are rare and those that order are exceptionally rare.

2.1.3 Scope of This Chapter

Only electron-transfer salt-based metals, superconductors, and magnets are presented in this chapter (for electroactive polymers, see Chapter 4; for high-temperature superconductors, see Chapter 10). We restrict this overview to the synthesis, structure, and transport or magnetic properties of these materials. Although numerous other techniques may be also used for studying the physics of metals, superconductors, and magnets (NMR, vibrational spectroscopy and polarized reflectance, specific heat measurements, etc.), they will not be discussed here, due to space restrictions.

2.2 DEFINITIONS AND UNITS

2.2.1 Electrical Conductivity, Superconductivity, and Electronic Instabilities

Electrical conduction in metallic solids can be understood in terms of the free-electron model, in which the valence electrons move freely between the nuclei in the lattice under a potential field assumed to be constant.[6] In the absence of an electrical potential, the motion of the electrons is random and restricted. However, in the presence of an applied electrical potential, V, the electrons attain a directed drift velocity proportional to this potential. According to Ohm's law, the resulting current, i, is proportional to V and inversely proportional to the resistance, R, of the material (i.e., $i = V/R$). The resistance R is proportional to the length, l, inversely proportional to the cross section of the sample, s, and proportional to a normalized constant characteristic of the material, the resistivity, ρ (i.e., $R = \rho l/s$). The electrical conductivity, σ, is the reciprocal of ρ, and its SI unit is the S/m (S = Siemens). The S/cm unit (or Ω^{-1}/cm) is more commonly used and is used in this chapter. Electrical conductivities for simple metals are in the range 10^4 to 10^5 S/cm at room temperature.

Most materials (in particular, molecule-based compounds) exhibit much lower conductivities, and an extremely large variation has been observed when spanning the conductivity of insulators ($\sigma < 10^{-9}$ S/cm) to semiconductors and metals ($\sigma > 10^3$ S/cm). The nearly free electron model is an improvement over the classical free-electron model, in which the constant potential field is replaced by a periodic potential generated by the metal nuclei in the lattice.[1] The formation of regions of energy levels in which the electrons can (allowed bands) or cannot (forbidden bands) be accommodated is a consequence of this model. In this energy-band model (Fig. 2.1) metals (and semimetals) have partially filled bands. As a result, very little energy is required to excite electrons from the highest filled level (the Fermi level, E_F) to the lowest empty ones, hence the high conductivity of metals.

The energy-band model for a semiconductor or insulator consists of filled lower valence bands and an empty upper conduction band separated by an energy gap E_g (typically, 0.7 to 1 eV for the semiconductors),[6] which is larger for the insulators (typically, > 6 eV).[6] Thus high energy is required to excite the valence electrons over this gap into the conduction band, and only a few electrons have sufficient energy to do this. This explains the low conductivity

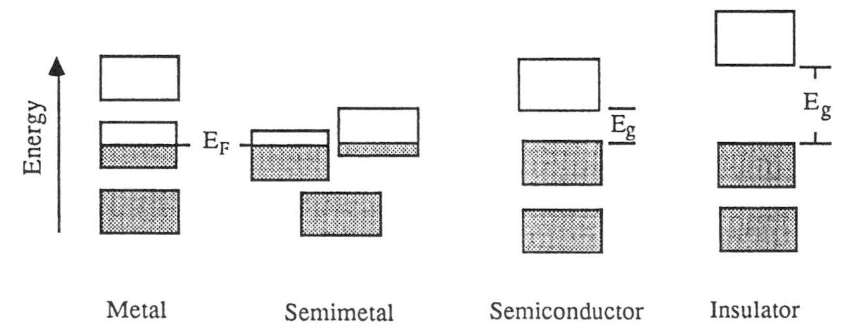

Figure 2.1 Schematic illustration of the occupation of the allowed energy bands for a metal, semimetal, semiconductor, and insulator, which are shaded to the Fermi energy, E_F.

of the semiconductors and the even lower one of the insulators ($< 10^{-9}$ S/cm).[6] Moreover, due to this energy activation mechanism, the conductivity of semiconductors and insulators increases with increasing temperature, whereas the conductivity of metals, which is limited only by lattice vibrations (phonons), decreases when increasing the temperature.

When delocalized electrons interact strongly with the atomic or molecular sites or, as occurs for most molecule-based conductors, the tight-binding band theory is a better representation of the electrons delocalized throughout the crystal, which involves the combination of atomic or molecular orbitals localized on each site. Hence a band is formed from each atomic or molecular orbital of the isolated atom or molecule.[7-10]

Superconductivity was discovered by Kammerlingh-Onnes for mercury.[11] Below a critical temperature, T_c, a superconducting material has zero resistivity. Moreover, if this material is placed in a magnetic field at temperatures below T_c (i.e., in the superconducting state), it becomes perfectly diamagnetic [i.e., the magnetic flux lines are expelled from the bulk of the material (zero induction)]. This is called the Meisser effect.[12] However, if a sufficiently high magnetic field, the critical field H_c, is applied to a superconducting material, the normal metallic state is restored. In addition, a sufficiently high electrical current density, J_c, also destroys superconductivity. The most commonly used theory for explaining these phenomena was developed by Bardeen, Cooper, and Schrieffer (BCS theory),[13] and involves the pairing of two electrons with opposite spins and wave vectors (i.e., a Cooper pair). The BCS coupling is due to electron–phonon interactions.

Other electronic instabilities may be observed, especially in low-dimensional systems such as electron-transfer salt-based conducting materials.[14-18] The spin density wave (SDW) instability is related to magnetic exchange interactions. Below a given temperature, T_s, the material shows an itinerant antiferromagnetic state, as two electrons with opposite spins couple in pairs. This coupling induces a spatial modulation of the spin density, which results in a gap opening at the Fermi level and a transition from a metal to an insulating state (MI).[15] Coulombic interactions also tend to localize electrons and lead to an insulator (Mott insulator) with the magnetic moments localized on each site which are antiferromagnetically coupled.[6] The charge density wave instability (CDW) usually results from an important electron–phonon interaction which induces a periodical distortion of the crystal lattice and simultaneously a spatial modulation of the charge density.[16] In some one-dimensional electron-transfer salt-based conductors (see Section 2.4.1), a spontaneous CDW instability due to the formation of electron–hole pairs may induce a periodical lattice distortion and thus drive an activated metal–insulator Peierls transition.[17, 18] In low-dimensional, poorly conducting electron-transfer salts, interchain spin–phonon coupling drives a magnetoelastic distortion of the lattice [i.e., the spin-Peierls transition (SP)], which induces a magnetic ordering at $T_c > 0$ K.[19]

2.2.2 Ferromagnets and Ferrimagnets

Magnetic behavior is detected by a response of a material, attraction or repulsion, to a magnet. It is due to the quantum mechanical spin of an electron and how nearby spins interact with each other. Each orbital can have two electrons: a *spin-up* $m_s = +\frac{1}{2}$ (↑) and a $m_s = -\frac{1}{2}$ *spin-down* (↓), whereby the spins cancel when both electrons are present in an orbital. When a molecule has an odd number of electrons, at least one orbital has only one electron and a net spin. Such spins are usually sufficiently far apart that their spin–spin coupling energy, J, is small compared to the coupling–breaking thermal energy. These spins do not couple and form a very weak paramagnet (Fig. 2.2*a*). When the spins get closer to each other, the J can become sufficient to enable an effective parallel (ferromagnetic) [or antiparallel (antiferromagnetic)] coupling due to the interactions. Albeit rare, pairwise ferromagnetic coupling can lead to long-range ferromagnetic order (Fig. 2.2*b*) throughout the solid. In contrast, long-range antiferromagnetic order can result from pairwise antiferromagnetic coupling (Fig. 2.2*c*). Ferrimagnets, such as magnetite (Fe_3O_4), result from antiferromagnetic coupling that does not lead to complete cancellation and thus have a net magnetic moment (Fig. 2.2*d*). This ordering to a ferro-, antiferro-, or ferrimagnetic state only occurs below a critical or magnetic ordering temperature, T_c.

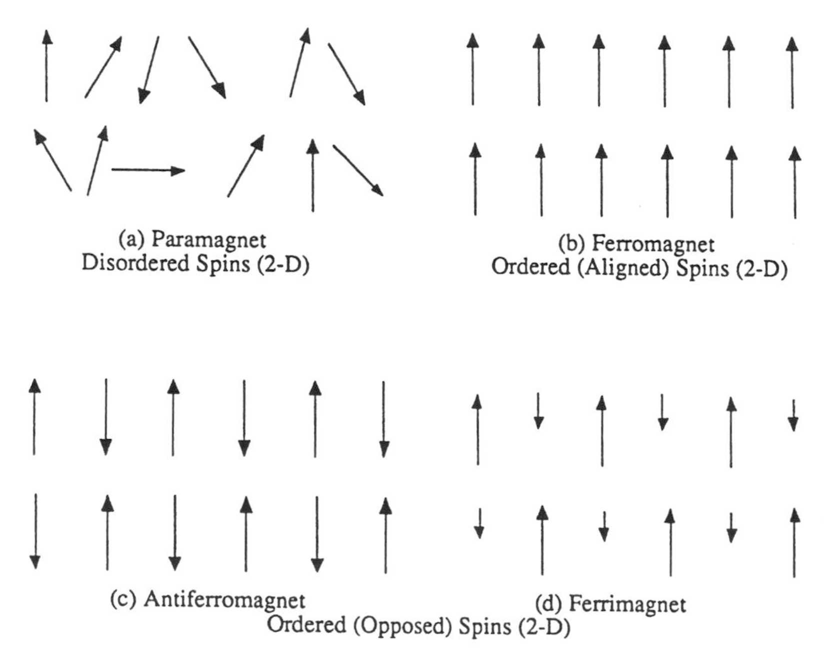

Figure 2.2 Common spin coupling behaviors.

2.3 HISTORICAL BACKGROUND

2.3.1 Electron-Transfer Salt-Based Conductors and Superconductors

The first reports on the observation of a high conductivity for an electron-transfer salt-based compound were made by Akamatu et al.[3] in a perylene bromide salt and by workers at DuPont[20] in several TCNQ (7,7,8,8-tetra-cyano-*p*-quinodimethane)-containing electron-transfer salts. However, the first molecule-based metallic conductor was inadvertently prepared more than 150 years ago by Knop[21] by oxidizing potassium tetracyanoplatinate $K_2[Pt(CN)]_4$ with chlorine or bromine, but the metal-like conductivity of the resulting "copper-shining" material, later called KCP[22] (Fig. 2.3*a*), was measured only 25 years ago[23] (see Section 2.4.1).

Predictions concerning the possible existence of conducting organic solids were made by McCoy[1] in 1911 and Kraus[2] in 1913, and extended by Little[24] in 1964, who suggested that appropriate one-dimensional organic polymers could become superconductive at high (room) temperatures. Together with the electron transfer salts of TCNQ,[20] salts of TTF (tetrathiafulvalene)[25] were also shown to exhibit high conductivities.[26, 27] Metal-like conductivity was first reported for [NMP][TCNQ] (NMP = *N*-methylphenazenium) near room temperature,[28] and [TTF][TCNQ] extended the temperature range for metal-like conductivity down to 53 K[29, 30] (see Section 2.4.2).

Following the discovery of [TTF][TCNQ], a large number of electron-transfer salts were prepared and studied, and the first electron-transfer salt-based superconductor was obtained in 1980 by using a modification of TTF [i.e., TMTSF (tetramethyltetraselenafulvalene)] and closed-shell inorganic monoanions in place of TCNQ. $(TMTSF)_2(PF_6)$ was the first electron-transfer salt to become superconductive under pressure,[31] and $(TMTSF)_2(ClO_4)$ was the first ambient-pressure electron-transfer salt-based superconductor[32] (see Section 2.5.1). Other modifications of TTF (see Fig. 2.3*b*)—BEDT–TTF,[33–38] DMET,[39] MDT–TTF,[40] BEDO–TTF,[41] and BEDT–TSF[42]—have also yielded electron-transfer salt-based superconductors.[43]

The use of electron-transfer salts based on inorganic coordination compounds, especially transition metal complexes, as components for metals and superconductors has been explored extensively[44–54] (see Section 2.4.3). Several electron-transfer salt-based superconductors based on the $M(dmit)_2$ complexes ($dmit^{2-}$ = 2-thioxo-1,3-dithiole-4,5-dithiolato) have been characterized[55] (see Section 2.5.3).

More recently, a new class of electron-transfer salt-based superconductors, the fullerides, A_xC_{60} (A = K, Rb, Cs), discovered by Haddon et al.,[56a] has aroused worldwide interest.[56–59] The highest reported T_c value for a molecule-based superconductor is 33 K, observed for $RbCs_2C_{60}$[59] (see Section 2.6).

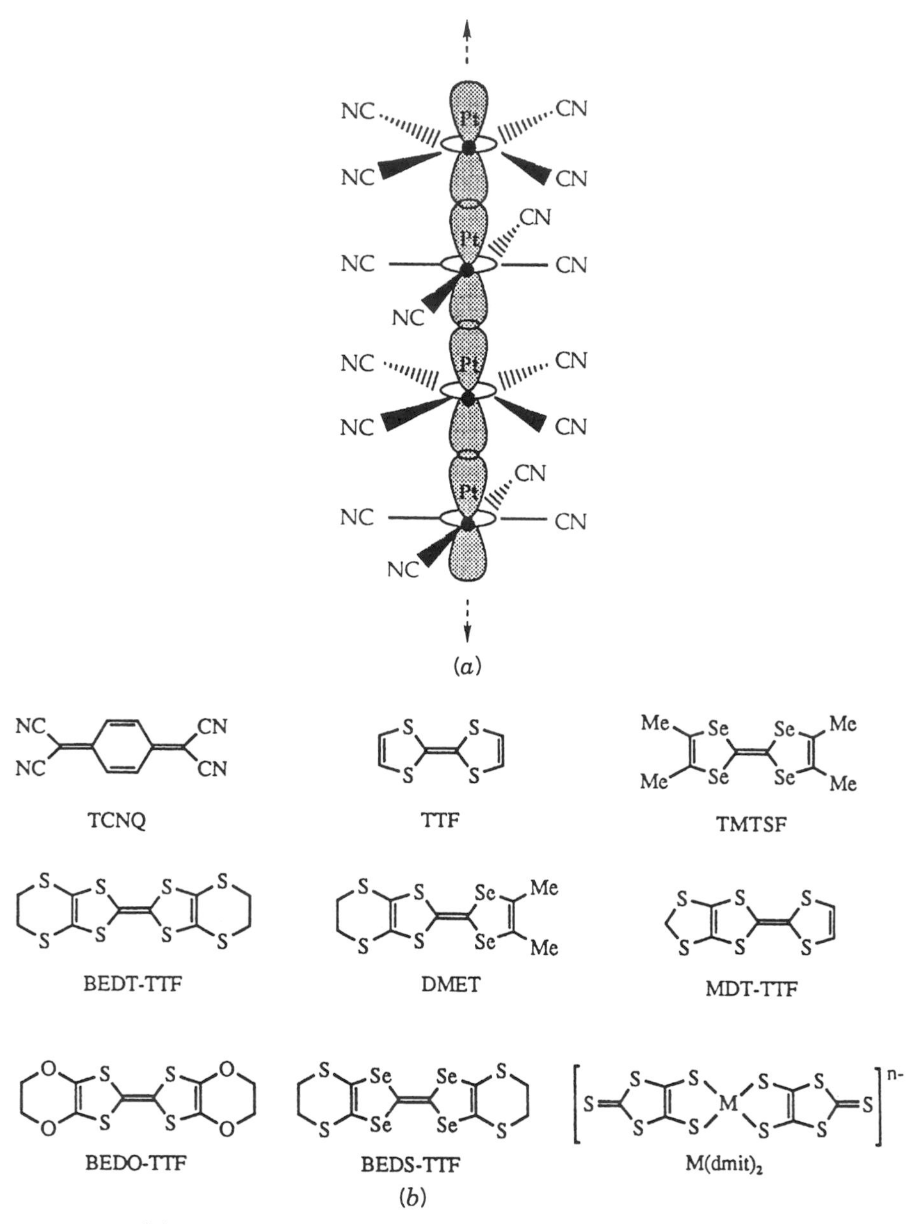

Figure 2.3 (*a*) Stacking of $Pt(CN)_4$ anions in KCP systems; (*b*) precursor molecules to conducting and superconducting electron transfer salts.

2.3.2 Electron-Transfer Salt-Based Magnets

Concepts for molecule-based magnets were first discussed by McConnell in the 1960s;[4, 60] however, the experimental observation of a molecule-based system containing spins on organic species and exhibiting hysteresis was first reported in 1985.[5] In 1991 several additional classes of electron-transfer salt-based magnets—organic nitronyl nitroxides,[61] [TDAE][C_{60}] [TDAE = tetrakis(dimethylamino)ethylene],[62] V[TCNE]$_x \cdot y$(solvent)[63] and others—were introduced. The latter demonstrated that electron-transfer salt-based systems could have ordering temperatures exceeding room temperatures. In 1992, an additional class of electron-transfer salts, comprised of metallo-macrocyclic cations and stable organic radical anions, was reported.[64] Detailed reviews on molecule-based magnets are available.[214, 215]

2.4 ONE-DIMENSIONAL CONDUCTORS

2.4.1 Salts of Partially Oxidized Tetracyanoplatinate Complexes

Two types of partially oxidized tetracyanoplatinate complexes occur: anion-deficient salts with the general formula (cation)$_2$[Pt(CN)$_4$]$X_x \cdot yH_2O$ ($x < 0.4$) or cation-deficient salts of general formula (cation)$_x$[Pt(CN)$_4$]$\cdot yH_2O$ ($1 < x < 2$). For example, the prototypical KCP(X), $K_2[Pt(CN)_4]X_{0.3} \cdot 3H_2O$ with X = Cl or Br, may be obtained (1) by oxidation with the corresponding halogen,[21, 22] (2) by co-crystallization,[22, 65, 66] or diffusion growth[67] of the corresponding Pt(II) and Pt(IV) complexes, and (3) by electrochemical oxidation.[68, 69] Cation-deficient salts can be prepared by aqueous chemical[9, 70, 71] or electrochemical[68] oxidation of the corresponding divalent salt.

The metallic luster of the partially oxidized tetracyanoplatinate complexes is associated with large electrical conductivities, as high as 2300 S/cm for $Rb_2[Pt(CN)_4](FHF)_{0.40}$.[72] The experimental values have been reported to vary from 4 to 1050 S/cm for the most studied compound, KCP(Br),[71] and reflects the sensitivity of the transport measurements to impurities and defects. Many KCP systems behave as metals at temperatures near room temperature [e.g., ≥ 270 K for KCP(Br)],[73] as evidenced by the increase in the electrical conductivity with decreasing temperature. The electrical behavior of the KCP systems is highly anisotropic. For example, for the needle-shaped crystals of KCP(Br), the anisotropic ratio of conductivity parallel to the Pt–Pt chain axis to that perpendicular to the axis is 10^5 at room temperature.[73] Metallic luster, metal-like conductivity, and anisotropy in the conductivities are illustrated simultaneously by the highly anisotropic reflectance spectra of KCP(X), which show a typical plasma edge for light polarized only parallel to the needle axis.[74]

The structure of the KCP systems is characterized by parallel chains of $[Pt(CN)_4]^{z-}$ anions stacked along the direction perpendicular to the $[Pt(CN)_4]^{z-}$ planes (the needle axis is parallel to this stacking axis), with short intrachain Pt–Pt distances ranging from 2.86 to 2.97 Å.[71] The solvent, cations, and where appropriate, the anions are located between the $[Pt(CN)_4]^{z-}$ anion chains, and adjacent $[Pt(CN)_4]^{z-}$anion chains are parallel and about 10 Å apart.

Although the Pt atoms are not crystallographically equivalent,[75, 76] all the Pt atoms may be considered as equivalent and in a nonstoichiometric oxidation state, as confirmed by Mössbauer and ESCA studies.[77, 78] From these and other studies[9, 71] it has been shown that the metallic behavior of the KCP systems near room temperature arises from electron delocalization along a delocalized partially filled electron-energy band comprised of overlapped platinum $5d_{z^2}$ orbitals.

The KCP systems undergo a metal–insulator (MI) transition upon lowering the temperature.[73] This transition, which is due to a Peierls distortion associated with the condensation of a CDW, has been studied extensively.[79–83]

A number of partially oxidized cation-deficient bis(oxalato)platinate salts, $(cation)_x[Pt(C_2O_4)_2] \cdot yH_2O$, have been prepared.[84] Most of these salts exhibit room-temperature conductivity along the direction parallel to the Pt atom chain in the range 1 to 100 S/cm. However, none of the compounds reported have a clearly characterized metallic state, and the various temperature dependencies of the conductivity are related to Peierls as well as non-Peierls distortions.

2.4.2 [TTF][TCNQ]

TCNQ is obtained by bromine–pyridine dehydrogenation of 1,4-bis-(dicyanomethylene)cyclohexane.[20a] In general, the major procedure for preparing TTF and derived donor molecules involves coupling of molecular half-units. These and other methods of preparation of TTF-based systems have been the subject of detailed reviews.[85–87] Both TCNQ and TTF are now commercially available. [TTF][TCNQ] can be prepared[29, 30] either by direct combination of the neutral TTF and TCNQ molecules or by metathesis between appropriate TTF and TCNQ salts.

The structure of [TTF][TCNQ] consists of parallel segregated chains of planar TTFs and TCNQs along the crystallographic *b*-axis of a monoclinic unit cell.[88] The TTF interplanar chain separation, 3.47 Å, is significantly shorter than observed for neutral TTF (3.62 Å).[89] Similarly, the uniform 3.17-Å TCNQ interplanar chain separation is substantially shorter than that observed for neutral TCNQ (3.45 Å).[90]

[TTF][TCNQ] exhibits a high room-temperature conductivity (about 500 S/cm) and a metallic behavior down to about 53 K,[28, 29, 92, 93] and undergoes a metal–insulator transition at lower temperatures (Fig. 2.4). At 38 K a second phase transition is evidenced by an anomaly in the temperature dependency of the conductivity.[92, 93] The conductivity is highest along the chain axis (i.e., the axis of the needle) and the anisotropy in the conductivities is typical of a one-dimensional conductor (10^3 to 10^4).

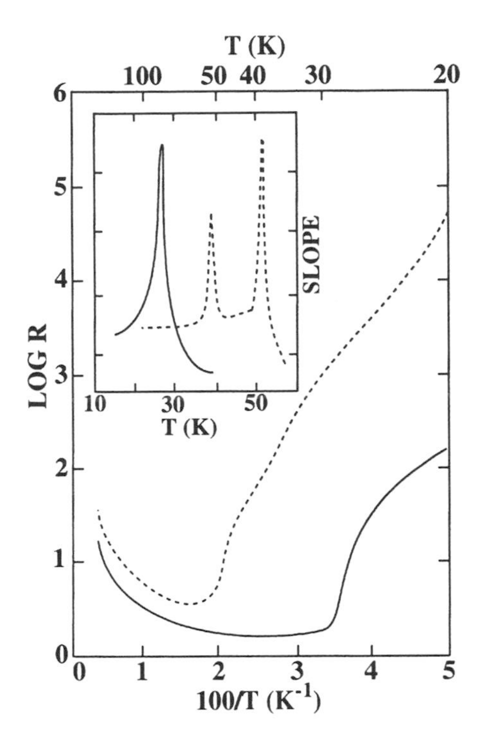

Figure 2.4 Plot of log(resistivity) versus inverse of temperature of [TTF][TCNQ] (dashed curve) and [TSF][TCNQ] (solid curve) (TSF = tetraselenafulvalene). Inset plots slope of log(resistivity) versus temperature. (Redrawn Ref. 93.)

From the structure of [TTF][TCNQ] (i.e., chain arrangement, short intrachain distances) and electronic band structure calculations, narrow bands arising from the one-dimensional overlap of open-shell π molecular orbitals independently on both TTF and TCNQ have been inferred.[94, 95] Note that the 1 : 1 stoichiometry of [TTF][TCNQ] does not reveal the charge distribution in this compound. Neutron scattering showed a partial electron transfer of 0.59e^- between the donor (TTF) and the acceptor (TCNQ) units.[96, 97] The partial occupation of both the TTF and TCNQ conduction bands leads to semimetallic behavior (see Fig. 2.1).

For segregated linear chains of donors and/or acceptors, partial electron transfer may be obtained by using donors and acceptors of intermediate strength.[98, 99] The donor or acceptor strength (i.e., the ionization potential of the donor or the electron affinity of the acceptor) can be estimated from the solution redox potential values of the donor and acceptor. Donors are considered intermediate in strength when their oxidation potential is in the range 0.1 to 0.4 V vs. SCE, whereas acceptors are considered as intermediate in strength when their reduction potential is in the range -0.02 to 0.35 V vs. SCE.[98, 99] More important, the difference in redox potentials of the donor and the acceptor should be less than 0.25 ± 0.15 V.[98, 99]

X-ray diffuse scattering[100, 101] and neutron scattering[102] confirmed that the metal–insulator transition at 53 K is due to a Peierls distortion associated with a crystal lattice distortion. In fact, in this two-component system, where both TTF and TCNQ chains participate in conduction, two instabilities occur, the first at 53 K and a second at 38 K (see Fig. 2.4).[92, 93] The phase transition at 53 K is driven by a CDW (Peierls) instability associated with the TCNQ anion chains, whereas the transition at 38 K (as well as another more subtle transition at 49 K) is driven by an ordering of the TTF chains.[95]

Following the discovery of the unique electrical properties of [TTF][TCNQ], a remarkable number of similar compounds with TTF-derived donor and TCNQ-derived acceptor molecules have been prepared and studied.[93, 103–105] Among these compounds, a majority of those having a segregated chain molecular arrangement similar to [TTF][TCNQ] are highly conductive; those having a mixed chain arrangement in which the donors and acceptors alternate along the chains are poorly conductive, as a partially occupied delocalized electron energy band is not present.

Substitution of selenium for sulfur atoms in TTF led to the formation of TSF (tetraselenafulvalene) and enhanced conductivity. For example, [TSF][TCNQ] has a higher room-temperature conductivity ($\approx$ 800 S/cm) than [TTF][TCNQ], and the metal–insulator transition is shifted to lower temperatures (28 K) (see Fig. 2.4).[93] [HMTSF][TCNQ] (HMTSF = hexamethylenetetraselenafulvalene) at high pressure retains metal-like behavior down to 1K.[106] Nevertheless, none of the [TTF][TCNQ]-like donor acceptor electron-transfer compounds exhibit superconducting behavior.

2.4.3 Other One-Dimensional Conductors

Linear chain transition metal complex systems other than KCP and [TTF][TCNQ] have been studied extensively. Partially oxidized dihalodicarbonyliridate(I) complexes, $A_x[Ir(CO)_2X_2]$ ($A = H^+, K^+, Na^+, [TTF]^+$; $0 < x < 1$, X = Cl, Br),[107, 108] have a gold or copper luster, pressed-pellet conductivity in the range 0.1 to 5 $S\,cm^{-1}$, and behave as semiconductors except for $K_{0.6}[Ir(CO)_2Cl_2] \cdot 0.5H_2O$, for which metal-like behavior near room temperature has been claimed.[108] Poor reproducibility of the stoichiometry, low chemical stability, and poor crystallinity have hampered the characterization of these compounds.[44]

Halotricarbonyliridium complexes, $IrX(CO)_3$ (X = Cl, Br, and I),[109] exhibit high-room-temperature conductivities (up to 0.2 S/cm). Detailed analysis led to the formulation of $IrX(CO)_3$ as $IrX_x(CO)_{3-x}$ ($x \approx 0.07$), with Ir being in the 1.07 formal oxidation state.[44]

Halogen-containing one-dimensional systems with moderate conductivities (up to 10^{-2} S/cm) have been obtained from nickel or palladium complexes of glyoximate ligands (glyoximate, diphenylglyoximate, and *o*-benzoquinonedioximate).[45, 48, 110, 111] Halogen-containing metallophtalocyanines, $(MPc)I_x$, and metal complexes of porphyrins and other tetraazaannulenes also yield conductive species.[45, 48, 112]

Several metal complexes of *cis*-1,2-disubstituted ethylene-1,2-dithiol, $[M(S_2C_2R_2)_2]^{n-}$ (M = Ni, Pd, Pt, and Cu; R = H, C, and CF_3) form electron-transfer salts with several donors.[98, 113-116] Bis(dithiolene) complexes combine two key features: an extended π-electron system similar to TTF[117] and a transition metal similar to KCP, thus introducing possible metal–metal interactions as in KCP although they have not been encountered to date. Unfortunately, very few metal bis(dithiolene)-based compounds, with the exception of the $M(dmit)_2$-based compounds (see Section 2.5.3), exhibit high conductivity and metallic behavior.[118-120] Except for $[H_3O]_{0.33}[Li]_{0.82}[Pt(S_2C_2(CN)_2)_2]$[110, 121] and $M(dmit)_2$-based compounds, only the organic donor component contributes to conductivity.

2.4.4 Concluding Remarks

A linear chain conductor requires a uniform partially occupied delocalized electron energy band. To achieve this, the following guidelines have evolved:[9, 99, 104, 105, 119] (1) use of planar molecules for one-dimensional chain formation; (2) good overlap of metal and/or ligand orbitals, which requires close interplanar separations, minimized steric hindrance, and appropriate symmetry and large extension of these orbitals; and (3) partial filling of the conduction band through either partial oxidation or partial electron transfer. Electron-transfer salt-based materials meeting these criteria and exhibiting high conductivity and metallic behavior have been prepared. However, almost all of the one-dimensional molecular systems studied until the late 1980s undergo a metal–insulator transition when lowering the temperature. This transition is often associated with a structural distortion transition and was predicted by Peierls[17] as being inherent to any one-dimensional system. Ironically, the first guideline listed above seemed to prevent the possibility of metal-like conductivity at low temperature, let alone superconductivity.

A decrease in the metal–insulator transition temperature, even a suppression of this transition, was achieved by applying pressure on selenium-substituted TTF derivatives such as [HMTSF][TCNQ] (HMTSF = hexamethylene-TSF) and [TMTSF][DMTCNQ] (TMTSF = tetramethyl-TSF; DMTCNQ = 2,5-dimethyl-TCNQ).[122-125] It should be noted that in these compounds stronger interactions between the chains were observed as compared to the purely one-dimensional systems studied previously. These results prompted the synthesis and studies of several TMTSF derivatives, in which, as described in the next section, the ubiquitous metal–insulator transition encountered upon cooling can be avoided.

2.5 QUASI ONE- AND TWO-DIMENSIONAL SUPERCONDUCTORS

2.5.1 TMTSF Salts

The general method of obtaining TMTSF (tetramethyltetraselenafulvalene) involves self-coupling of the 4,5-dimethyl-1,3-diselenole-2-selenone or of the dimethyl(4,5-dimethyl-1,3-diselenolium) salt.[126–131] TMTSF may be used in the preparation of electron-transfer salts with TCNQ, derivatives, but it is also used to prepare 2:1 $(TMTSF)_2X$ salts, with a closed-shell anion (X^- = PF_6-, AsF_6-, ClO_4-, NO_3-, etc.).[132] The $(TMTSF)_2X$ salts are typically prepared by electrocrystallization from solutions of TMTSF and a salt (mostly tetraalkylammonium) of the appropriate anion X^-.

All $(TMTSF)_2X$ compounds are isomorphous with a triclinic (space group $P1^-$) unit cell at room temperature.[132] The structure consists of zigzag columnar chains of nearly planar TMTSF moieties parallel to the *a*-axis of the unit cell (Fig. 2.5). The TMTSF chains form infinite two-dimensional molecular sheets that extend in the *ab*-plane. Within these sheets, the TMTSF moieties are connected through intra- and interchain Se–Se contacts shorter than the sum of the van der Waals radii (4.0 Å).[133] Structural studies at different temperatures reveal that the interchain distances decrease twice as much as the intrachain distances as the temperature is lowered.[134] The TMTSF sheets are separated from each other by the anions X^- along axis *c*. There are also short Se ··· F or Se ··· O contacts in salts containing F or O atoms.

The quasi-one-dimensional $(TMTSF)_2X$ salts have a quarter-filled hole band arising from the $\frac{1}{2}$ + charge on each TMTSF moiety which is achieved by charge delocalization along the $(TMTSF)^{1/2+}$ chains. Most of the $(TMTSF)_2X$ salts are highly conductive along the *a*-chain axis (typically, about 600 S/cm), and metallic, for example, down to 12 K in the case of $(TMTSF)_2PF_6$. Application of 6.5 kbar pressure to $(TMTSF)_2PF_6$ leads to a transition from a metal to a superconducting state below 0.9 K.[31] Therefore, $(TMTSF)_2PF_6$ is the first electron-transfer salt-based superconductor. The superconducting transition was confirmed by the observation of the Meissner effect.[135] This major breakthrough was soon followed by the observation of superconductivity at ambient pressure at 1.4 K in $(TMTSF)_2ClO_4$.[32]

Several TMTSF-based systems have been studied, and seven different superconductors salts with octahedral (e.g., PF_6^-, AsF_6^-, SbF_6^-, TaF_6^-), tetrahedral (e.g., ClO_4^-, ReO_4^-), and low-symmetry (e.g., FSO_3^-) counter-anions have been characterized.[105] All but $(TMTSF)_2ClO_4$ require pressure for exhibiting superconductive behavior. In addition to superconductivity, the $(TMTSF)_2X$ systems exhibit various interesting and unusual properties.[105, 136] In the salts with octahedral anions (e.g., PF_6^-, AsF_6^-), a metal–insulator (MI) transition observed at about 15 K at ambient pressure is due to a spin density wave (SDW), but not a charge density wave (CDW), instability.[137] The salts with tetrahedral anions have a structural phase transition at relatively high temperatures [e.g., 180 K for $(TMTSF)_2ReO_4$], which is associated with anion ordering.[138]

Shubnikov de Haas oscillations have been observed in the magnetoresistance of $(TMTSF)_2PF_6$,[139] and explained in terms of a field-induced spin density wave.[14, 140, 141] Current interest in this field, recently raised again by the observation of new series of oscillations in $(TMTSF)_2NO_3$,[142] is devoted

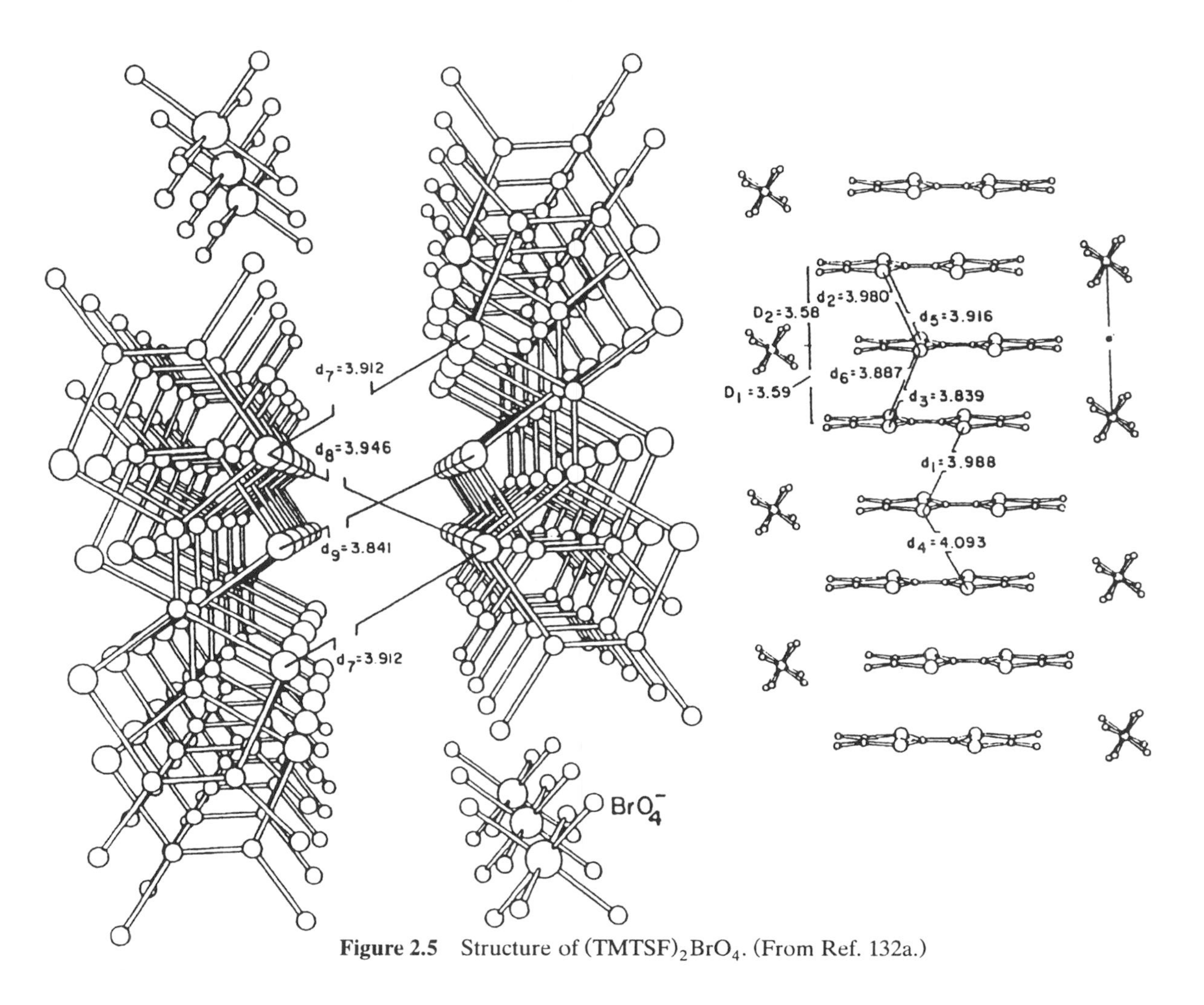

Figure 2.5 Structure of $(TMTSF)_2BrO_4$. (From Ref. 132a.)

to these peculiar field-induced phenomena.[143] Structure–property relationships, based on, for example, anion size and symmetry,[105, 132, 144] have been studied. The most significant empirical relationship correlates the occurrence of superconductivity and the interchain interactions observed. For this class of superconductors the anisotropy in the conductivities is relatively large (about 200);[145] thus these salts are quasi-one-dimensional rather than two-dimensional, as confirmed by band structure calculation.[146] Nevertheless, this observation led to the strategy of enhancing interchain interaction, which eventually resulted in study of the BEDT-TTF-based two-dimensional systems.

2.5.2 BEDT-TTF (ET) Salts

The BEDT-TTF molecule (BEDT-TTF = bis(ethylenedithio)tetrathiafulvalene; Fig. 2.3), also referred to as ET, was first prepared by Mizuno et al.[147] by the self-coupling of the corresponding thione half-unit. The most convenient route for obtaining this thione involves reduction of CS_2 by sodium, affording the Na_2(dmit) intermediate ($dmit^{2-}$ = 2-thioxo-1,3-dithiole-4,5-dithiolato).[148] Electrocrystallization procedures for preparing the $(BEDT\text{-}TTF)_2X$ salts are similar to those used for the $(TMTSF)_2X$ salts.[43] As for the $(TMTSF)_2X$ salts, $(BEDT\text{-}TTF)_2X$ salts with octahedral (e.g., PF_6^-) and tetrahedral (e.g., ClO_4^-) anions have been prepared. In addition, linear (e.g., I_3^-), planar (e.g., $AuCl_4^-$), and polymeric [e.g., $Cu(NCS)_2^-$] anions have also been used extensively. This class of salts frequently have several stoichiometries and/or crystal phases for the same anion.

More than 130 crystal structures of $(BEDT\text{-}TTF)_2X$ phases have been reported.[43] As a function of some characteristic structural features, they have been classified into several families:

1. The α and β families, whose most representative members are the α- and β-$(BEDT\text{-}TTF)I_3$ phases. Both α- and β-type structures are triclinic with one BEDT-TTF layer per unit cell and alternating sheets of parallel BEDT-TTF ions in a honeycomb fashion and X^- anions[43] (Fig. 2.6).
2. The κ family, which has the highest superconductive critical temperatures. In κ-type structures [e.g., κ-$(BEDT\text{-}TTF)_2Cu(NCS)_2$] (Fig. 2.7), the unit cell contains one BEDT-TTF layer, but no chains can be identified within the layer. Instead, the BEDT-TTF molecules form face-to-face dimers rotated by about 90° with respect to each other.
3. The θ family, which is orthorhombic, belonging to the *Pnma* space group, contains BEDT-TTF layers in which the BEDT-TTF ions form uniform chains which are tilted from the BEDT-TTF plane by about 20°.
4. Several less well characterized other families, designated γ, δ, ε, η, ζ, and so on, have also been reported.[43]

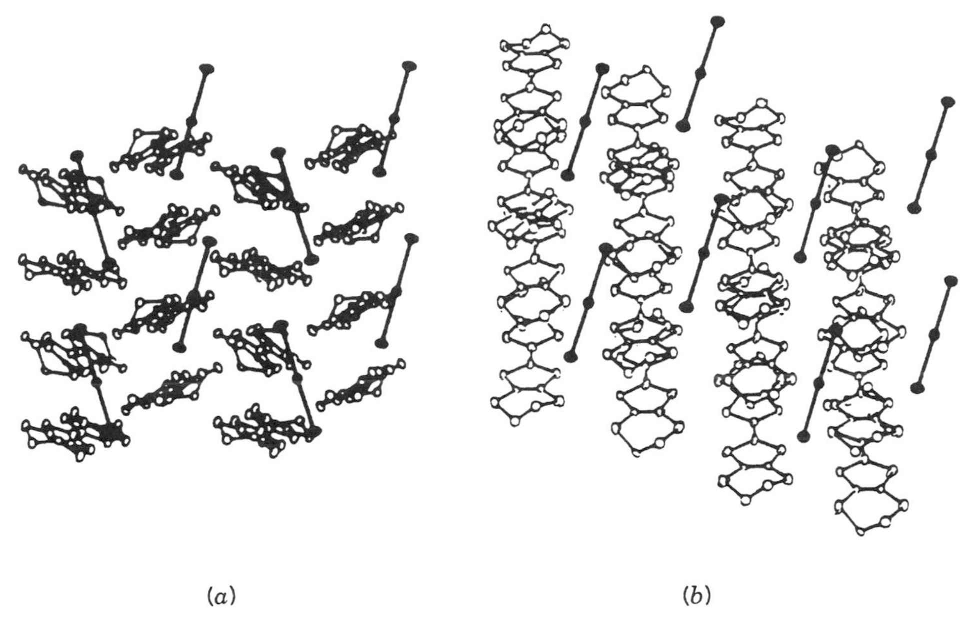

Figure 2.6 Views of the two-dimensional network of BEDT-TTF molecules and the adjacent anion sheet of (*a*) α-(BEDT-TTF)$_2$I$_3$ and (*b*) β-(BEDT-TTF)$_2$I$_3$. (Redrawn from Ref. 43.)

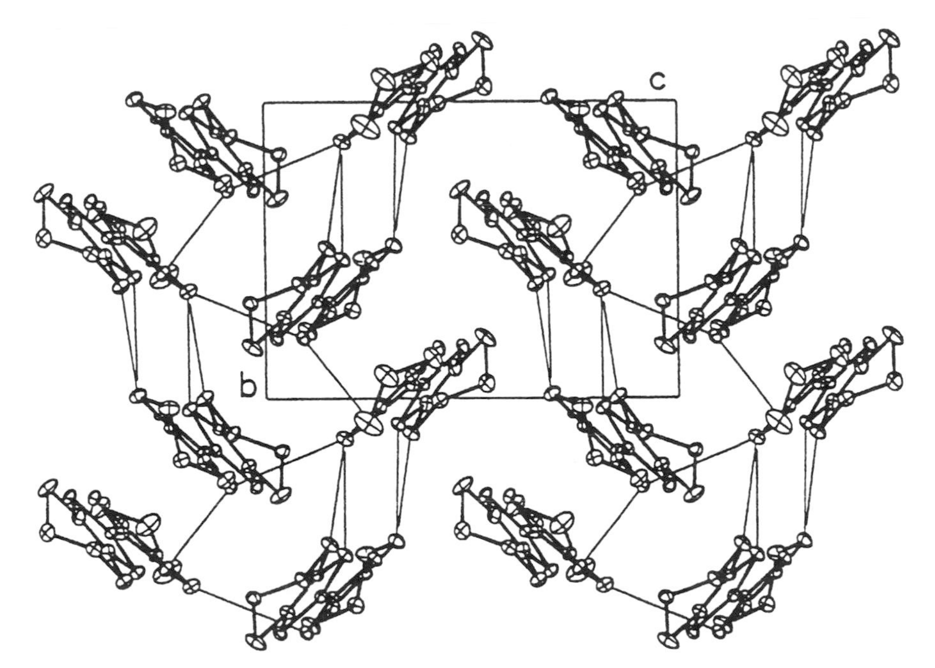

Figure 2.7 Structure of κ-(BEDT-TTF)$_2$[Cu(NCS)$_2$] at 118 K. S ··· S contacts shorter than 3.6 Å are indicated by thin lines. (From Ref. 43.)

The structures of the (BEDT-TTF)$_2$X salts are so diverse that it is difficult to establish structure–property relationships within this series. Attempts have been made for the β- and κ-salts.[43, 149] The most important structural feature common to all BEDT-TTF superconductors is their layered structures, in which the BEDT-TTF layers alternate with the anion layers. Additionally, the presence of an extended network of S···S interactions within the BEDT-TTF layers is observed. The steric effect of the ethylene groups tends to reduce the face-to-face interactions compared to those observed in the quasi-one-dimensional TMTSF salts. Therefore, side-by-side interactions may be as important as face-to-face interactions, and this competition may be the reason for the variety of structural types and packing modes observed in BEDT-TTF compounds. The layered molecular arrangement and the extent of these interactions, leading from quasi-one-dimensional to more two-dimensional systems, has a strong influence on the conductive and superconductive properties of these compounds. This tendency toward a two-dimensional electronic configuration is consistent with band structure calculations which show partially filled conduction bands leading to open (one-dimensional) and closed (two-dimensional) Fermi surfaces within the BEDT-TTF layer plane.[150, 151]

The differing structures of the (BEDT-TTF)$_2$X salts have led to a variety of conductivity behaviors, ranging from semiconductive to metallic and superconductive.[43] β-(BEDT-TTF)$_2$I$_3$ was the first ambient pressure superconductor in this family, with a T_c value of 1.4 K, identical to that of (TMTSF)$_2$ClO$_4$.[33b] The BEDT-TTF-based superconductors form the largest family of electron-transfer salt-based superconductors (i.e., > 50 salts) studied to date. The recent examples are BEDT-TTF salts possessing the $[Cu^{III}(CF_3)_4]^-$[152] and tris(oxalato)iron(III) anions.[153] The T_c value has been improved by one order of magnitude, and the highest T_c phases are κ-(BEDT-TTF)$_2$Cu(NCS)$_2$ ($T_c \approx 10.4$ K),[34] and κ-(BEDT-TTF)$_2$Cu[N(CN)$_2$]Y, where Y = Br ($T_c \approx 11.6$ K)[37] and Cl ($T_c \approx 12.8$ K at 0.3 kbar).[38]

As expected from the crystal structures and band structure calculations,[150, 151] the anisotropy in the conductivities within the BEDT-TTF layers is often quite low, with the ratio of conductivities along two perpendicular directions approaching unity, as occurs for (BEDT-TTF)$_2$ClO$_4$ · (1,1,2-trichloroethane)$_{0.5}$,[154] indicating a quasi-two-dimensional electronic structure. Four phases of (BEDT-TTF)$_2$I$_3$ are known. The β-phase is the best characterized, together with similar β-phases with X = IBr_2^- and AuI_2^-. As is the case with most electron-transfer salt-based superconductors, the superconducting transition temperature decreases with pressure. β*-(BEDT-TTF)$_2$I$_3$ is metastable with an enhanced T_c value of 8 K that can be obtained by low-temperature annealing under pressure of the β-phase, which initially had a T_c value of 1.4 K.[155]

The T_c value increases with the length of the anion for the I_3^-, IBr_2^-, and AuI_2^- series. This empirical correlation was explained by decreasing the density, resulting in a narrower bandwidth and an increase in the density of states at the Fermi level.[34] Following this strategy, the $[Cu(NCS)_2]^-$ salt was prepared and an enhanced T_c value of 10.4 K was reported for the κ-phase.[34] In addition to the typical arrangement of a κ-phase consisting of orthogonal face-to-face $[BEDT\text{-}TTF]_2^+$ dimers, the $[Cu(NCS)_2]^-$ anions form a zigzag polymer network involving both N and S bonding tricoordinated Cu(I) ions (Fig. 2.8). The κ-type phases having the highest T_c values are the most extensively studied. In addition to $[Cu(NCS)_2]^-$, other polymeric anions have been used, such as $Cu[N(CN)_2]Cl$, $Ag(CN)_2$, $Hg_{2.89}Br_8$, $Hg(SCN)_{3-n}Cl_n$, and so on.[43]

Although BEDT-TTF remains the best donor for generating electron-transfer salt-based superconductors, other TTF derivatives (see Fig. 2.3) also led to superconductors. Two unsymmetrical modifications of TTF and/or BEDT-TTF have led to electron-transfer salt-based superconductors. The first, DMET, is a hybrid of TMTSF and BEDT-TTF.[39] The behavior of the derived $(DMET)_2X$ salts is intermediate between those of $(TMTSF)_2X$ and $(BEDT\text{-}TTF)_2X$.[156] Only salts of linear anions are superconductive at low temperatures [e.g., κ-$(DMET)_2I_3$ (T_c = 0.47 K)]. Another unsymmetrical molecule, MDT-TTF, has also been used.[40] The κ-$(MDT\text{-}TTF)_2AuI_2$ phase is an ambient-pressure superconductor with a T_c value of 4.5 K.[40, 157] More recently, oxygen- and selenium-containing modifications of BEDT-TTF, the BEDO-TTF[41] and BEDT-TSF[42] molecules, respectively, have also led to the characterization of superconductive compounds.

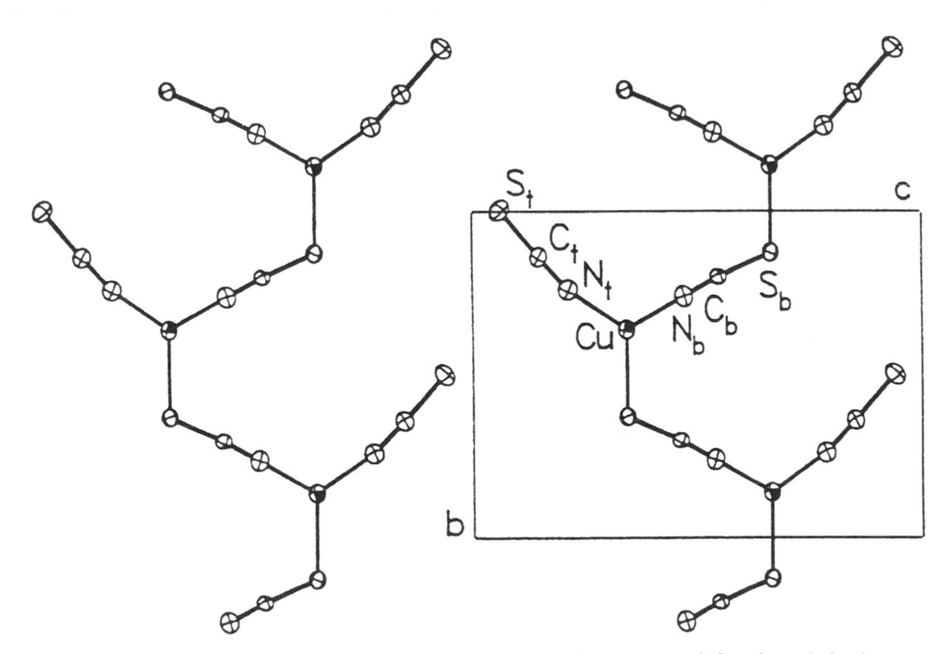

Figure 2.8 Polymeric $[Cu(NCS)_2]^-$ anion in κ-$(BEDT\text{-}TTF)_2[Cu(NCS)_2]$. (From Ref. 43.)

The BEDT-TTF and related compounds form the largest and most important family of electron-transfer salt-based conductors and superconductors. The variety of their structures and properties is still beyond any rationalizing classification scheme. More detailed and complete information on this family of compounds may be found in a book by Williams et al.[43]

Due to the high probability of finding salts with interesting conductive and superconductive properties, this family of materials is still a very active research area.[158] The main area of current interest is directed toward *fermiology*, the use of magneto-oscillatory effects in the resistance (Shubnikov–de Haas) or magnetization (de Haas–van Alphen) to study the Fermi surface topology of these materials.[159]

2.5.3 ($M(dmit)_2$ Compounds

The $dmit^{2-}$ ligand (2-thioxo-1,3-dithiole-4,5-dithiolate) is prepared by reducing CS_2 with sodium in dimethylformamide.[160] The divalent and monovalent $[A_n][M(dmit)_2]$ complexes ($n = 2, 1$) can readily be obtained by reaction of the appropriate M^{II} metal salt and a salt of the appropriate cation A, typically a tetraalkylammonium salt.[161–163] Two kinds of electron transfer salts may be obtained by either standard electrocrystalization or by methathesis reaction: (1) the cation-deficient $[A]_x[M(dmit)_2]$ nonintegral oxidation state complexes, and (2) donor–acceptor $[D][M(dmit)_2]_y$ compounds in which D is an appropriate donor molecule.[54, 55]

In both the $(TMTSF)_2X$ and $(BEDT\text{-}TTF)_2X$ salts described previously, the superconductive properties were related to the extended interactions between selenium or sulfur atoms on the periphery of the molecules. The $M(dmit)_2$ systems possess 10 sulfur atoms on their periphery, and indeed an extended network of S ··· S interactions is observed, as for example in $TTF[Ni(dmit)_2]_2$ (Fig. 2.9). Short S ··· S contacts are observed between not only $Ni(dmit)_2$ moieties belonging to adjacent stacks, but also between the TTFs, extending the electronic interactions into the third direction. Nonetheless, band structure calculations[164] indicate that the $M(dmit)_2$ systems should be considered best as quasi-one-dimensional systems.

Several $M(dmit)_2$ salts exhibit high conductivities and semiconductive or metal-like behavior.[54, 55] To date, seven superconductive phases have been characterized, including α-$[EDT\text{-}TTF][Ni(dmit)_2]$, which is the only ambient-pressure superconductor of this series (Table 2.1). This class of superconductors is based on metal complexes as electron acceptors instead of organic donors, and their physical properties show some marked differences:[55] (1) the $M(dmit)_2$ moiety is responsible for superconductivity; (2) T_c increases with increasing pressure (contrary to most organic superconductors[43]); (3) CDW states (instead of SDW states as occurs for most organic superconductors) are in competition or coexistence with the superconducting state; (4) in $TTF[Ni(dmit)_2]_2$, a CDW instability does not induce a metal–insulator transition; and (5) many of the properties may be explained by a

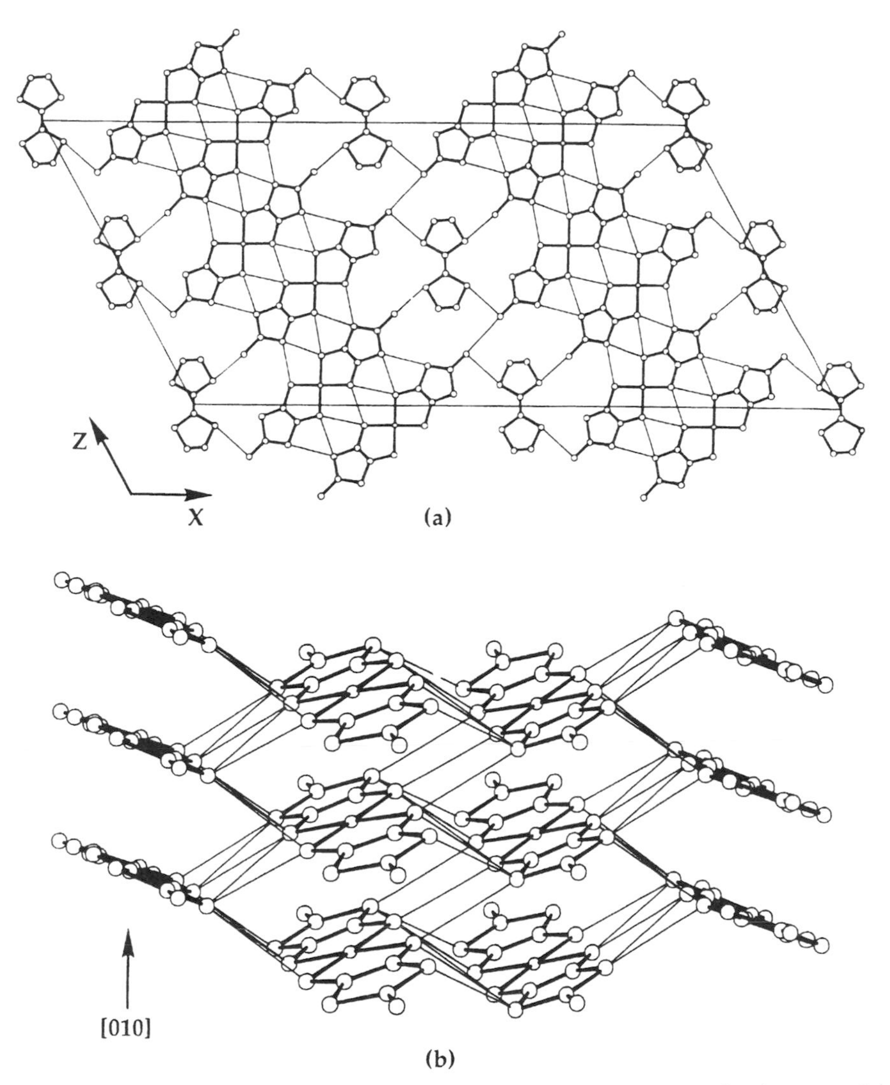

Figure 2.9 Structure of TTF[Ni(dmit)$_2$]$_2$: (*a*) projection onto the (010) plane; (*b*) parallel view along the [101] direction. Thin lines indicate S ··· S short contacts.

unique multisheet Fermi surface based on both the HOMO and the LUMO bands[54, 164, 172] (not only the HOMO band, as in organic superconductors).

The interesting properties of M(dmit)$_2$-based salts have prompted extensive studies of related systems obtained by changing not only the nature of the metal, the cation or donor, but also by using other ligands resembling dmit^{2-}.[173-177] However, to date, no superconductor has been characterized in these series and superconductivity seems to be restricted to salts derived from the dmit^{2-} ligand.

TABLE 2.1 List of $M(dmit)_2$-Based Superconductors

Compound	Year Reported	T_c (K)	P (kbar)	Ref.
$[TTF][Ni(dmit)_2]_2$	1986	1.62	7	165
$[Me_4N]_{0.5}[Ni(dmit)_2]$	1987	5	7	166
α'-$[TTF][Pd(dmit)_2]_2$	1989	5.93	24	167
α-$[TTF][Pd(dmit)_2]_2$	1990	1.7	22	168
β-$[Me_4N]_{0.5}[Pd(dmit)_2]$	1991	6.2	6.5	169
$[Me_2Et_2N]_{0.5}[Pd(dmit)_2]$	1992	4	2.4	170
α-$[EDT$-$TTF][Ni(dmit)_2]$	1993	1.3	Ambient	171

2.6 FULLERIDES

Kroto et al. experimentally detected the first fullerene, C_{60}, by mass spectral analysis of a laser-initiated plasma.[178] Preparation and separation of C_{60} in sizable quantities[179] involves soot generation from an electrical arc between graphite rods, or their plasma ignition, followed by column chromatography in an organic solvent such as toluene. These techniques have been improved continuously, and C_{60} has now become commercially available.

Intercalation of C_{60} with alkali (A) or alkaline-earth metals (A′) can be achieved by reaction of the metal vapor and solid C_{60} or by the heterogeneous reaction of the metal and C_{60} dissolved in hot toluene. Both thin films and single crystals of reduced A_xC_{60} or A'_xC_{60} may be obtained, but their high chemical reactivity toward oxygen and moisture makes it difficult to study their properties. The degree of intercalation can be varied progressively from $C_{60}(x = 0)$ up to $x = 6$.[180]

The structure of C_{60} is face-centered cubic (fcc; $a = 14.1$ Å).[181] As is the case for all close-packed fcc unit cells, two vacant tetrahedral sites (< 4.2 Å) and one vacant octahedral site (< 2.2 Å) per molecule are available for occupancy of small electron donors. Molecular orbital calculations[182] show that threefold-degenerate t_{1u} and t_{1g} LUMO levels are low-lying, such that C_{60} behaves as acceptor, as also determined by electron affinity determinations.[183] Therefore, C_{60} is easily reducible by electron transfer up to six electrons, as confirmed by electrochemical studies.[184] Upon reduction, all the vacant interstitial sites are progressively occupied by to three alkali or alkaline-earth ions, maintaining the fcc structure.[185] Further reduction, however, induces structure transformations to body-centered-tetragonal A_4C_{60} and body-centered-cubic A_6C_{60} phases.[186]

The electrical behavior of reduced fullerenes can be studied by standard transport measurements on thin films or single crystals. However, due to small sample sizes, the characterization of superconductivity is preferably carried out by diamagnetic susceptibility measurements (Meissner effect). With increasing reduction with alkali metal atoms, the conductivity of A_xC_{60}

increases to a maximum for $x \approx 3$ and then decreases again. Superconductivity is observed when all the vacant interstitial sites of the fcc structure are occupied. Further reduction yields nonsuperconducting phases. This corresponds to progressive filling of the threefold degenerate t_{1u} orbital, leading to a half-filled conduction band for $x = 3$ and a fully filled band for $x = 6$. The situation is somewhat different for reaction with alkaline-earth metals. In this case the t_{1g} orbital-derived band may also be partially filled, and, for example, the bcc phases Ca_5C_{60} and Ba_6C_{60} are superconducting.

The superconductive T_c values for the A_3C_{60} phases are remarkably high, 18 K in K_3C_{60}[56] up to a maximum of 33 K in $RbCs_2C_{60}$.[59] For alkali-based superconductors, T_c is nearly proportional to the lattice constant a_0 (Fig. 2.10). Na_3C_{60} is an exception in the A_3C_{60} series, as it is not a superconductor. This is due to the small size of the Na^+ ion compared to the available volume of the octahedral vacant site, which leads to an insufficient stabilization energy and a disproportionation at low temperatures into two cubic phases with lattice parameters close to those of NaC_{60} and Na_6C_{60}.[56b, 59]

The mechanism of conduction in the A_xC_{60} phases is based on the partial filling of the t_{1u} and t_{1g} orbital-derived bands. The mechanism of superconductivity, however, is not well understood and may be attributed to either electron–phonon coupling or electron–electron correlations. Clearly, a better understanding of the properties of the known A_xC_{60} phases is a contemporary research challenge.

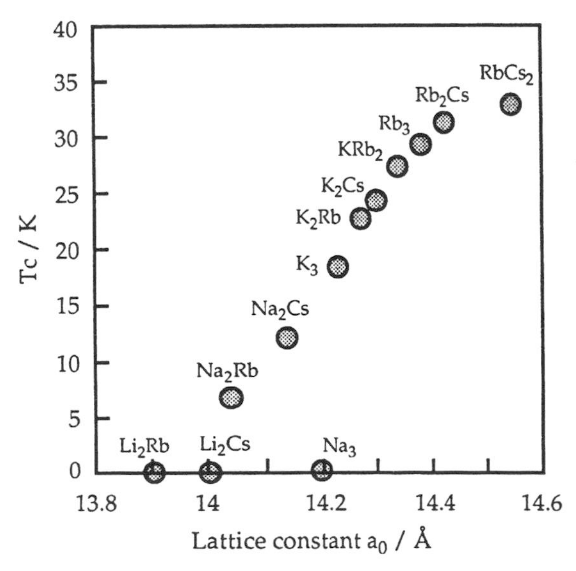

Figure 2.10 Relationship between lattice constant and T_c in A_3C_{60} phases. (Redrawn from Ref. 187.)

The reduction of C_{60} seems to have been restricted to alkali and alkaline earth metals. The chemical instability and poor crystallinity hinder characterization of the C_{60} compounds. To date, no electron-transfer salts with organic donors exhibit metal-like conducting properties. The chemistry and physical properties of the other fullerenes, C_n ($n = 70, 76, 78, 82$, etc.), remain to be explored.

In conclusion, there are three principal structural families of electron-transfer salt-based superconductors: (1) the one-dimensional $(TMTSF)_2X$ family, with T_c values $\approx$ 1 to 2 K (although T_c values up to 6 K occur for the quasi-one-dimensional $M(dmit)_2$-based systems); (2) the two-dimensional $(BEDT\text{-}TTF)_2X$ family, with T_c values $\leq$ 13 K; and (3) the three-dimensional A_xC_{60} family, with T_c values up to 33 K.

2.7 RELATED RESEARCH AREAS

Except for the fullerides, all electron-transfer salt-based conductors and superconductors may be considered as being derived from TTF (see Fig. 2.3) or TCNQ. Small chemical modifications may result in drastic changes in the physical properties, as is noted when going from TTF to TMTSF to BEDT-TTF or when changing the counteranion X in the $(BEDT)_2X$ series.[43] Moreover, the preparation of new compounds has led to new properties, concepts, and theories (e.g., SDW, CDW, FISDW, oscillations in the magnetoresistance, and fermiology)[14, 43, 159] and is the impetus for the continuation of work in this area.[158]

The rather restricted range of compounds exhibiting superconductivity and the leveling off of T_c presently observed suggests that this research is reaching some limits (this, of course, was thought to be the case for the inorganic superconductors prior to the discovery of the high-T_c cuprate superconductors). Among other interesting properties that compounds designed and studied primarily for their conducting behavior may exhibit, in the next section we discuss the intramolecular mixed valency evidenced in bis-TTF systems and the coupling of conducting and magnetic properties in several molecule-based conductors and superconductors.

2.7.1 Bis-TTF Intramolecular Mixed-Valence Compounds

A challenge for molecular electronics is to design molecules in which long-distance electron transfer is realized between a donor site and an acceptor site through a conjugated bridging ligand. These systems, often designated as "molecular wires," are considered as key elements in the making of molecule-based electronic devices such as switches.[188]

Inorganic mixed-valence systems in which two transition-metal ions in different oxidation states are connected by a bridging ligand are the simplest model examples for studying such intramolecular electron transfer. The prototypical Creutz–Taube complex, $[(NH_3)_5Ru^{II}(pyrazine)Ru^{III}(NH_3)_5]^{5+}$, has been studied extensively.[189] These are *intramolecular* mixed-valence systems in which the electronic interactions between the two redox sites are propagated within the molecule through the bridging group. Such systems are different from the *intermolecular* mixed-valence systems, such as KCP, [TTF][TCNQ], or $(TMTSF)_2X$ compounds, in which electronic interactions occur in the solid state between different molecules. A classification of mixed-valence systems into three classes according to the extent of the electron delocalization has been proposed by Robin and Day.[190]

Intramolecular mixed-valence systems in which the two redox sites are purely organic are not very common.[191] Several bis-TTF systems (i.e., compounds containing two TTF moieties connected by aromatic bridging groups) have been prepared.[192] Electrochemical studies of these compounds show four successive, reversible one-electron oxidations, consistent with the sequential formation of the mono-, di-, tri-, and tetracationic species.[192–194] Intense intervalence transition bands characteristic of intramolecular mixed-valence species[189] have recently been observed in the near-infrared spectra of $(bis\text{-}TTF)^+$ monocations (Fig. 2.11), indicating that they are organic equivalents of the well-known inorganic transition-metal mixed-valence complexes.[193, 194]

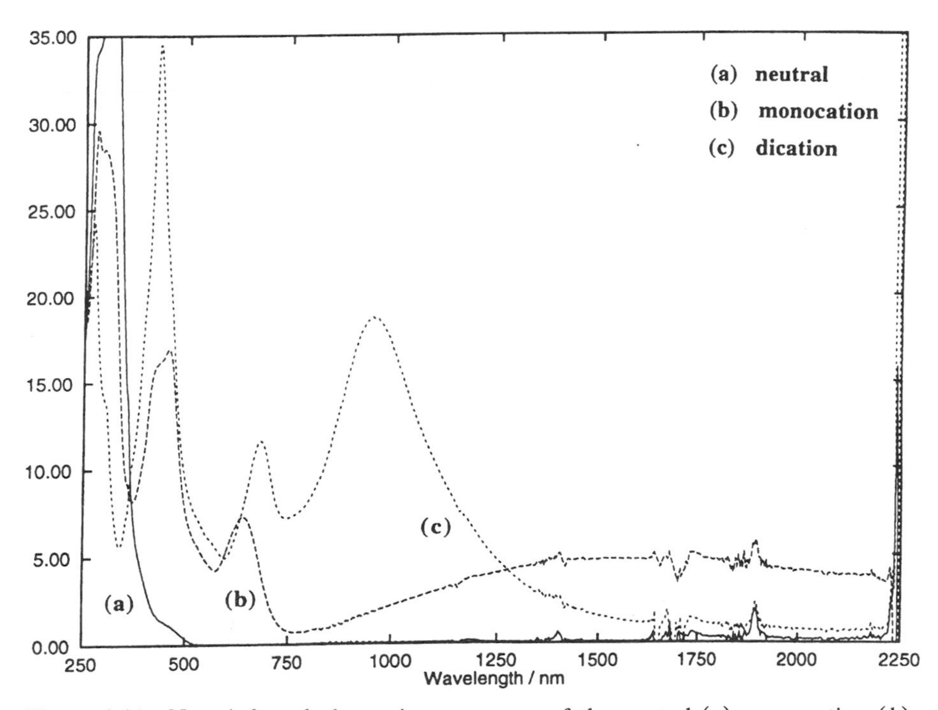

Figure 2.11 Near-infrared absorption spectrum of the neutral (*a*), monocation (*b*), and dication (*c*) of 2,6-bis(4,5-dimethyl-1,3-dithiole-2-ylidene)-4,8-difluorobenzo[1,2-d; 4,5-d′] bis-1,3-dithiole. (From Ref. 193.)

The shape and the intensity of the intervalence band indicates that the (bis-TTF)$^+$ radical cations belong to class II of the Robin and Day classification,[190] corresponding to a partial electron delocalization over the two redox sites. Moreover, theoretical calculations at the semiempirical as well as the ab initio levels show that an excess charge is localized on only one of the two TTFs of the (bis-TTF)$^+$ radical cations, as evidenced by significant elongation of the central C=C double bound of only one of the TTFs, 1.427 Å, as compared to 1.351 Å as in neutral TTF.[194] Hence, due to the large number of available, or readily prepared, bis-TTF compounds, this work opens an entirely new area for the study of purely organic intramolecular mixed-valence systems based on such compounds, whose properties can be tuned by changing the nature and the length of the bridge between the two TTF or TTF-like redox sites.

2.7.2 Paramagnetic Conductors and Superconductors

Most electron-transfer salt-based organic metals and superconductors are essentially derived from delocalized π-electron systems. These salts have open-shell structures and there is little electronic interaction with the counterion (see Section 2.5). It is of interest to prepare electron-transfer salt-based superconductors exhibiting interactions between the conduction electrons and localized magnetic moments, because the superconductivity mechanism in such systems might differ from that observed in compounds studied previously. Hence π-delocalized electrons (as in TMTSF- and BEDT-TTF-derived compounds) should produce metallic conducting properties, while localized unpaired electrons (as in magnetic transition metal complex anions), will act as a magnetic center.

A prototypical example of electrically and magnetically active molecule-based compounds are the DCNQI-Cu^{II} systems (DCNQI = N,N'-dicyanoquinone-p-diimine).[49, 195] Several [2,5-XYDCNQI]$_2$Cu compounds (X, Y = Me, Cl, Br) exhibit high conductivities and metal-like properties (e.g., 1000 S/cm for [2,5-Me_2DCNQI]$_2$Cu[196]) down to low temperatures (e.g., < 1K for [2,5-Me_2DCNQI[$_2$Cu).[196] Antiferromagnetic ordering of the $S = \frac{1}{2}$ Cu(II) spins [2,5-Me_2DCNQI]$_2$Cu occurs at 10 K.[197, 198] Moreover, this compound undergoes a pressure-induced reentrant metal–insulator–metal transition.[52, 199] This behavior has been explained by admixing of the copper $3d$ orbitals with the π^* conduction electrons of [2,5-Me_2DCNQI]$\cdot^-$, resulting in a formal $Cu^{1.3+}$ species.[50, 51, 200] In this system, however, the coexistence of π conduction electrons and localized magnetic moments could not be achieved.[52]

The (perylene)$_2$[M(mnt)$_2$] electron-transfer salts (M = transition metal; mnt^{2-} = maleononitriledithiolato, $[C_4N_2S_2]^{2-}$) form a unique family of conductors exhibiting conducting and magnetic properties.[201] In these compounds, chains of perylene cations provide highly anisotropic metallic properties, whereas segregated $[M(mnt)_2]^{z-}$ chains with appropriate metals (M =

Ni, Pd, Pt, Fe) supply localized magnetic moments. Although these two chains are prone to exhibit the instabilities typical of one-dimensional conducting (Peierls, for the perylene chains) and magnetic (spin-Peierls, for the $[M(mnt)_2]^{z-}$ chains) systems, π conduction electrons in perylene chains and magnetic moments of the $[M(mnt)_2]^{z-}$ chains coexist to low temperatures ($>$ 8 K for α = (perylene)$_2$[Pt(mnt)$_2$]). Moreover, both instabilities occur at the same temperature, but the nature of their coupling is still unresolved.[201, 202]

Several metal bis(dithiolene)-complexes have been used for the formation of semiconducting electron-transfer salts with TTF or TTT (see Section 2.4.3)[119, 120] In this series, several members of the type [TTF][$MS_4C_4(CF_3)_4$] (M = Cu, K = S, Se) exhibit quasi-one-dimensional Heisenberg antiferromagnetic spin behavior above 12 K but undergo a unique second-order transition at lower temperatures ($<$ 12 K for M = Cu, X = S; $<$ 2K for M = Cu, XS; $<$ 6 K for M = Cu, XSE) to a diamagnetic state that has been characterized as a spin-Peierls transition.[19, 53]

The first example of a BEDT-TTF-based metal-incorporating magnetic anions has been reported by Day et al.[203] They found that (BEDT-TTF)$_3$CuCl$_4$ · H$_2$O retains a metallic state down to 0.4 K. ESR measurements indicate weak interactions between the itinerant spins on the BEDT-TTF chains and the localized spins on the $CuCl_4{}^{2-}$ anions and weak ferromagnetic exchange between the Cu(II) ions.[203, 204] Other analogous, but semiconducting BEDT-TTF salts with simple magnetic anions such as $[FeCl_4]^-$, $[FeBr_4]^-$, $[FeCl_4]^-$, $[CuBr_4]^{2-}$, and $[CuCl_2Br_2]^{2-}$ have also been reported.[205] More important, the first electron-transfer salt-based superconductor, β''-(BEDT-TTF)$_4$[(H$_2$O)Fe-(C$_2$O$_4$)$_3$] · C$_6$H$_5$CN (T_c = 8.5 K), with a magnetic counterion has been prepared.[153] The magnetic susceptibility of this compound is dominated from 9 to 300 K by the high-spin Fe(III) and obeys the Curie–Weiss law with a weak antiferromagnetic Θ value of -0.2 K.

Several BEDT-TTF salts with paramagnetic polyoxometallate polyanions, such as $[Co^{II}W^{VII}{}_{12}O_{40}]^{6-}$, have been studied.[206, 207] These salts are semiconducting and BEDT-TTF and polyanion spin systems coexist but do not interact significantly.[207] Substitution of Se for S atoms in the TTF fragment of BEDT-TTF results in several BEDT-TSF [BEDT-TS = bis(ethylenedithio)tetraselenafulvalene, also referred to as BETS; see Fig. 2.3], conducting salts with stable metallic states.[208] BEDT-TSF salts with magnetic $[MX_4]^{n-}$ anions (M = Fe, Co, Mn; X = Cl, Br, etc.) have been specifically prepared for studying the interaction between the BEDT-TSF π conduction electrons and the localized anion spins.[209, 210]

The most striking result was reported for the λ-(BETS)$_2$FeCl$_4$ phase, which undergoes a metal–insulator transition at 8 K; the isostructural analog λ-(BETS)$_2$GaCl$_4$, with the nonmagnetic $[GaCl_4]^-$ anion, undergoes a superconducting transition at the same temperature.[42, 209] The behavior of λ-(BETS)$_2$FeCl$_4$ is related to the presence of the magnetic Fe(III) ion ($S = \frac{5}{2}$), as shown by electron spin resonance (ESR) studies which indicate that magnetic transition of the $[FeCl_4]^-$ anions into a nonmagnetic state and condensation of the conduction electrons into an insulating state take place simultaneously.[210] Moreover, in contrast to the SDW condensation induced by the magnetic field (FISDW) in TMTSF salts,[14, 140] the low-temperature insulating state of λ-$(BETS)_2FeCl_4$ is suppressed with application of a magnetic field higher than 9 T, and a field-restored highly conducting state is stabilized at high fields (Fig. 2.12).[211]

The isostructural alloy λ-$(BETS)_2(GaCl_4)_{0.5}(FeCl_4)_{0.5}$ has also been studied. This compound, containing disordered magnetic $[FeCl_4]^-$ anions, undergoes a superconducting transition at 4.6 K.[210] It should be noted that the presence of tris(oxalato)iron(III) magnetic ions does not prevent either of the derived BEDT-TTF salts from being superconductive.[153]

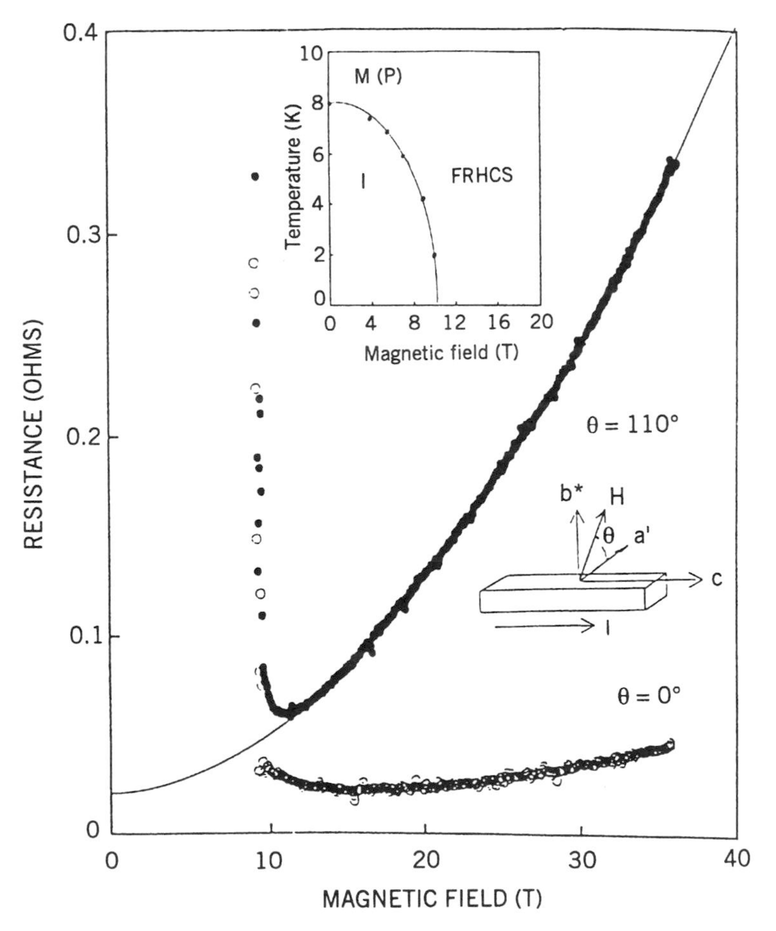

Figure 2.12 Field dependence of the resistance of λ-$(BETS)_2FeCl_4$. Inset: Temperature–magnetic field phase diagram. (Redrawn from Ref. 211.)

2.7.3 Concluding Remarks

Two new research directions that may lead to new classes of materials with novel properties have been identified. Much work remains to obtain a better understanding of the novel properties. For example, the length of the bridging unit in bis-TTF systems may be increased without losing the intramolecular electron transfer. Several mechanisms for magnetic-conducting systems have been proposed for describing the nature of the coupling between conduction electrons and localized moments, for explaining the stabilization of the metallic state under a high magnetic field in λ-$(BETS)_2FeCl_4$, and for understanding the existence of superconductivity in the presence of magnetic ions, but they must be verified.

Other novel properties may be observed in electron-transfer salts derived from those studied initially for their electrical properties: compounds with nonlinear optical properties (see Chapter 6), (conducting) liquid crystals, (conducting) Langmuir–Blodgett films, and ferromagnets (see below).

2.8 ELECTRON-TRANSFER SALT-BASED FERROMAGNETS

The idea that a magnet can be comprised of an electron-transfer salt-based solid in addition to classical solids comprised of metal atoms or ions was first presented by McConnell. In 1963 he suggested a mechanism for ferromagnetic exchange that involved a specific spatial arrangement of neighboring radicals possessing both positive and negative spin densities.[4] In 1967 he suggested another mechanism for ferromagnetic exchange for $\cdots D\cdot^{+}A\cdot^{-} D\cdot^{+}A\cdot^{-}D\cdot^{+}A\cdot^{-} \cdots$ (A = donor; D = acceptor) chains.[60] A third hypothesis for another mechanism for ferromagnetic exchange was introduced in 1968 when Itoh and Mataga suggested[212] that large, planar, alternate hydrocarbons comprised of *meta*-substituted triplet diphenylcarbene moieties will have a high-spin (ferromagnetically coupled) ground state, in accord with Hund's rule and spin polarization.

The experimental realization of ferromagnetic ordering in an electron-transfer salt-based material occurred in 1979 with the discovery that $[FeCp^*_2]\cdot^{+}[TCNQ]\cdot^{-}$ (Cp* = cyclopentadienide) was a metamagnet (i.e., it had an antiferromagnetic ground state) but that above a 1.6-kOe critical applied magnetic field it had a high moment state.[213] In 1985, via deliberate replacement of TCNQ with the smaller TCNE (TCNE = tetracyanoethylene), $[FeCp^*_2]\cdot^{+}[TCNE]\cdot^{-}$ was reported to be a bulk ferromagnet.[5] With worldwide interest in this class of materials, several electron-transfer salt-based magnets have subsequently been prepared.[214, 215] These magnets have a variety of bonding/structural motifs, including isolated molecules (zero-dimensional), chains (one-dimensional), layers (two-dimensional), and networks (three-dimensional). Table 2.2 presents examples of the classes of

TABLE 2.2 Representative Spin Sites of Electron-Transfer Salt-Based Magnets as a Function of Orbital in Which Spin Resides and Type of Bonding or Structure

		Bonding Motif		
Spin-Residing Orbital	Isolated Molecule/Ion (0-D)	Chains (1 − D)	Layers (2-D)	Network (3-D)
p	Nitroxides			
p and *d*	$[FeCp_2^*]^{\cdot +}[TCNE]^{\cdot -}$	$[MnTPP]^{+}[TCNE]^{\cdot -}$	Mn^{II}/nitroxide	V/$[TCNE]^{\cdot -}$
d	Cr^{III}; Cr^{III}/Fe^{III}	Mn^{II}/Cu^{II}, Cr^{III}/Ni^{II}	Cr^{IV}	Cr^{III}/Ni^{II}

materials based on the orbitals in which the spins reside and the associated dimensionality of the structure. Two features, not present for classical, metallurgically processed magnets, surface as being unique to some electron-transfer salt-based magnets: magnets comprised of spins (1) residing in *p*-orbitals on organic species and (2) not coupled through direct (covalent) bonding.

This overview focuses on zero- and one-dimensional electron-transfer salt-based magnets where three types of magnetic behaviors have been reported: ferromagnets, ferrimagnets, and weak ferromagnets. Reviews are available for a comprehensive discussion of all the electron-transfer salt-based magnet systems as well as the molecule-based magnets.[214, 215] The ferromagnets are either organic nitroxides or electron-transfer salts of decamethylmetallocenes and TCNE or TCNQ. The ferrimagnets are comprised of chains of alternating $S > \frac{1}{2}$ donors and $S = \frac{1}{2}$ acceptors. The weak ferromagnets are electron-transfer salts with structures that result in the partial cancellation of spins.

2.8.1 Nitroxide-Based Ferromagnets

The earliest studied organic magnet is layered-structure tanol suberate (Fig. 2.13), **1**, which exhibits Curie–Weiss behavior at higher temperature ($\Theta = +0.7$ K). It is metamagnetic below a T_c value of 0.38 K; that is, below the critical field of 100 Oe **1** is an antiferromagnet, while above 100 Oe it has a high moment state.[216] The best characterized organic ferromagnet is β-*p*-$NO_2C_6H_4$NIT (NIT = nitronyl nitroxide), **2**, which has a θ value of 1 K and is a ferromagnet below a T_c value of 0.6 K. (In contrast, γ-*p*-$NO_2C_6H_4$NIT is an antiferromagnet and the α and δ phases do not exhibit magnetic ordering.)[61, 217] Several additional examples or neutral organic magnets have been reported, with 1,3,5,7-tetramethyl-2,6-diazaadamantane-*N*,*N*′-doxyl, **3**, having the highest Θ (10 K) and T_c (1.48 K) values.[218]

2.8.2 Metallocene-Based Ferromagnets

$[Fe^{III}(C_5Me_5)_2]^{\cdot +}[TCNE]^{\cdot -}$[5d, 219] was the first electron-transfer salt-based material reported to have hysteresis and a ferromagnetic ground state.[220] The 4.8-K Curie temperature, T_c, was determined from magnetization[219] and heat

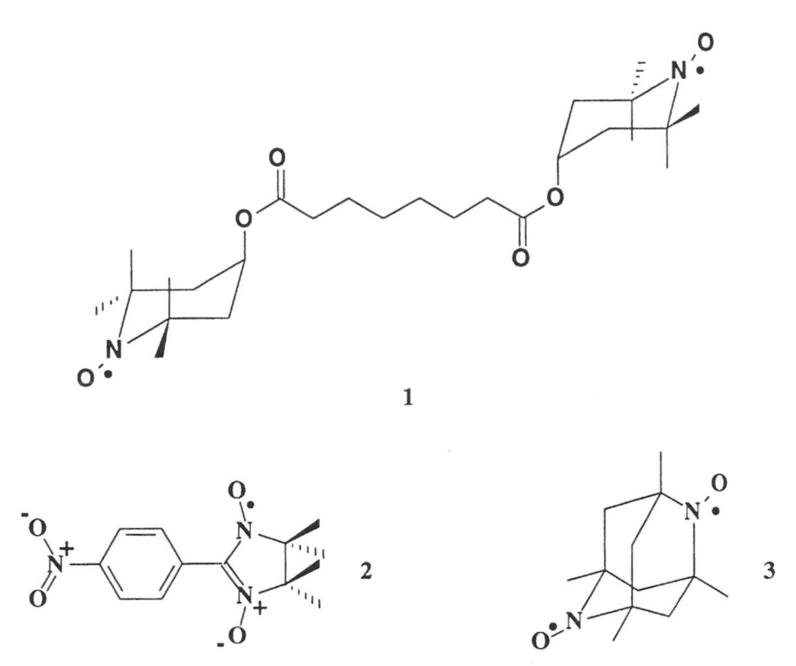

Figure 2.13 Tanol suberate, **1**; β-*p*-$NO_2C_6H_4$NIT (NIT = nitronyl nitroxide), **2**; and 1,3,5,7-tetramethyl-2,6-diazaadamantane-*N*,*N*'-doxyl, **3**.

capacity data.[221] This observation led to systematic study of the structure–function relationship of metallocenium electron-transfer salts of polycyano anions in our and other laboratories.[215, 219]

$[Fe^{III}(C_5Me_5)_2]^{\cdot +}[CNE]^{\cdot -}$ has the alternating $\cdots D^{\cdot +}A^{\cdot -}D^{\cdot +}A^{\cdot -}D^{\cdot +}A^{\cdot +}\cdots$ crystal [$D = Fe^{II}(C_5Me_5)_2$; A = TCNE] (Fig. 2.14) and electronic structures necessary for the stabilization of ferromagnetic coupling by the configuration mixing mechanism.[5, 60, 222] The high-temperature reciprocal

Figure 2.14 Alternating donor/acceptor $\cdots D^{\cdot +}A^{\cdot -}D^{\cdot +}A^{\cdot -}D^{\cdot +}A^{\cdot -}\cdots$ linear chain structure of $[Fe^{III}(C_5Me_5)_2]^{\cdot +}[A]^{\cdot -}$[$A$ = TCNQ, TCNE, DDQ, $C_4(CN)_6$, etc.] and $[Fe^{II}(C_5H_5)_2]$[TCNE], and $[Fe^{III}(C_5Me_5)_2]^{\cdot +}[C_3(CN)_5]^-$ as well as several $[M^{III}(C_5Me_5)_2]^{\cdot +}[A]^{\cdot -}$(M = Mn and Cr) structures. The structure above shows a chain for A = TCNE. (From Refs. 5c, d and 215a, c, d.)

molar susceptibility data for single crystals of $[Fe^{III}(C_5Me_5)_2]^{\cdot +}[TCNE]^{\cdot -}$ aligned parallel to the C_5 molecular axis fits the Curie–Weiss expression with $\theta = +30$ K, consistent with dominant ferromagnetic interactions.[219] The saturation magnetization calculated is 16,700 emu · Oe/mol, which agrees well with the observed value of 16,300 emu · G/mol for single crystals aligned parallel to the chain axis and is 36% greater than that of iron metal on either a per mole or a per iron basis.[219] A spontaneous magnetization is observed for polycrystalline samples below 4.8 K T_c in the earth's magnetic field.[219] Hysteresis loops reveal a coercive field of 1 kOe at 2K.[219]

To identify the key structural features required for the stabilization of bulk ferromagnetic behavior for $[Fe^{III}(C_5Me_5)_2]^{\cdot +}[TCNE]^{\cdot -}$ the differing magnetic properties of metallocenium-based electron-transfer salts with dif-

TABLE 2.3 Summary of Values of Curie–Weiss Θ and Critical Temperature T_c for Representative $\cdots D^{\cdot +}A^{\cdot -}D^{\cdot +}A^{\cdot -}D^{\cdot +}A^{\cdot -}\cdots$ Structures with A = TCNE and TCNQ

Salt with $\cdots D^{\cdot +}A^{\cdot -}D^{\cdot +}A^{\cdot -}D^{\cdot +}A^{\cdot -}\cdots$ Structure	Θ (K)	T_c^{a} (K)
$[MnCp^*_2]^{\cdot +}[TCNE]^{\cdot -}$	+22.6	8.8
$[CrCp^*_2]^{\cdot +}[TCNE]^{\cdot -}$	+22.2	3.65
$[FeCp^*_2]^{\cdot +}[TCNE]^{\cdot -}$	+16.8[b]	4.8[c]
$[FeCp^*_2]_{0.955}{}^{\cdot +}[CoCp^*_2]_{0.045}{}^{\cdot +}[TCNE]^{\cdot -}$		4.4
$[FeCp^*_2]_{0.923}{}^{\cdot +}[CoCp^*_2]_{0.077}{}^{\cdot +}[TCNE]^{\cdot -}$		3.8
$[FeCp^*_2]_{0.915}{}^{\cdot +}[CoCp^*_2]_{0.085}{}^{\cdot +}[TCNE]^{\cdot -}$		2.75
$[FeCp^*_2]_{0.855}{}^{\cdot +}[CoCp^*_2]_{0.145}{}^{\cdot +}[TCNE]^{\cdot -}$		0.75[d]
$[CrCp^*_2]^{\cdot -}[TCNQ]^{\cdot -}$	+12.8	3.5[e]
$[FeCp^*_2]^{\cdot +}[TCNQ]^{\cdot -}$	+12.3	2.55[f]
$[FeCp^*_2]^{\cdot +}[TCNQ]^{\cdot -}$	+3.8	3.1[g]
$[MnCp^*_2]^{\cdot +}[TCNQ]^{\cdot -}$	+10.5	6.5[g]
$[Fe(C_5Et_5)_2]^{\cdot +}[TCNE]^{\cdot -}$	+7.5	
$[Fe(C_5Et_5)_2]^{\cdot +}[TCNQ]^{\cdot -}$	+6.1	
$[FeCpCp^*]^{\cdot +}[TCNE]^{\cdot -}$	+3.3	
$[Fe(C_5Me_4H)_2]^{\cdot +}[TCNQ]^{\cdot -}$	+0.8	
$[Fe(C_5Me_4H)_2]^{\cdot +}[TCNE]^{\cdot -}$	−0.3	
$[CoCp^*_2]^{\cdot +}[TCNE]^{\cdot -}$	−1	
$([Fe(C_5Me_5)_2]^{\cdot +})_2([TCNQ]^{\cdot -})_2$	−3	

[a] Determined from a linear extrapolation of the steepest slope of the $M(T)$ data to the temperature at which $M = 0$.
[b] For crystals aligned parallel and perpendicular to the applied magnetic field: $\Theta_{\parallel} = +30$ K and $\Theta_{\perp} = +10$ K.
[c] 7.8 K at 14 kbar.
[d] Ac (100 Hz) measurement.
[e] 15 Oe; T_c from the maximum slope of dM/dT is reported as 3.1 K.
[f] Metamagnetic.
[g] 50 Oe; T_c from the maximum slope of dM/dT is reported as 6.2 K.

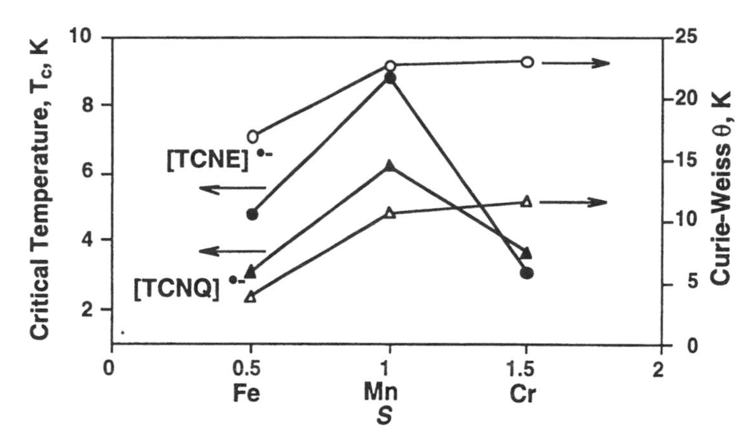

Figure 2.15 Values of Θ and T_c as a function of S for $[M^{III}(C_5Me_5)_2]^{\cdot +}[TCNE]^{\cdot -}$ and $[M^{III}(C_5Me_5)_2]^{\cdot +}[TCNQ]^{\cdot -}$(M = Fe, Cr, Mn).

fering radical anions were determined. Metallocenes studied included those with H or Et substituted for Me groups on the cyclopentadienide ring, as well as those with Mn, Cr, Ni, Ru, and Os substituted for Fe. Numerous nonisomorphous 1:1 salts with the alternating $\cdots D^{\cdot +}A^{\cdot -}D^{\cdot +}A^{\cdot -}D^{\cdot +}A^{\cdot +} \cdots$ crystal structure (see Fig. 2.14), as observed for $[Fe^{III}(C_5Me_5)_2]^{\cdot +}[TCNE]^{\cdot -}$ and diamagnetic $[Fe^{II}Cp_2][TCNE]$, were prepared and their Θ and (where observed) T_c values are listed in Table 2.3.[215a, 223, 224]

The only salts observed to order magnetically are the $[TCNE]^{\cdot -}$ and $[TCNQ]^{\cdot -}$ salts, where the cations are also radicals. The T_c values increase as expected with S for M = Fe ($S = \frac{1}{2}$) and Mn ($S = 1$), but are unexpectedly reduced for $S = \frac{3}{2}$ Cr (Table 2.3 and Fig. 2.15). Removal of the D spins leads to a dramatic reduction in T_c with increasing $[Co(C_5Me_5)_2]^+$ content; a 14.5% replacement of spinless Co(III) for $S = \frac{1}{2}$ Fe(III) reduces T_c by 84%, to 0.75 K from 4.8 K.[225, 226] For M = Fe(III) and A = TCNQ three phases have been isolated: para-,[227, 228] meta-,[213, 217] and ferromagnetic phases.[224] The origin of the failure of T_c to increase with S may be due in part to the fact that $[Cr^{III}(C_5Me_5)_2]^{\cdot +}[TCNE]^{\cdot -}$ is not isostructural to the Mn and Fe analogs and that the magnetic couplings are very sensitive to small structure changes; hence a drop in T_c may be not unexpected. These results support the need for a one-dimensional $\cdots D^{\cdot +}A^{\cdot -}D^{\cdot +}A^{\cdot -}D^{\cdot +}A^{\cdot -} \cdots$ chain structure for achieving ferromagnetic coupling and ultimately bulk ferromagnetic behavior as observed for $[M(C_5Me_5)_2{}^{\cdot +}[TCNE]^{\cdot -}$ and $[M(C_5Me_5)_2]^{\cdot +}$ $[TCNQ]^{\cdot -}$ (M = Fe, Mn, Cr).

2.8.3 Ferrimagnets Based on Alternating $S > \frac{1}{2}$ Donors and $S = \frac{1}{2}$ Acceptors

Several ferrimagnets having the structural feature of being comprised of covalently bound alternating D's and A's with a differing number of antifer-

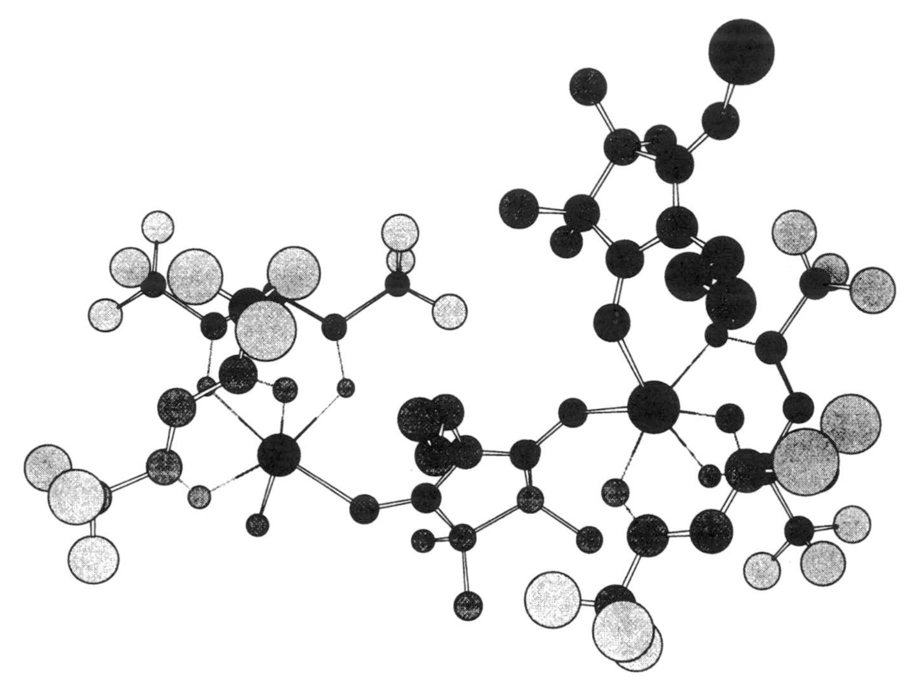

Figure 2.16 Segment of the chain structure of $Mn^{II}(hfac)_2NITPr^n$ showing the *cis* arrangement (the H's have been omitted for clarity).

romagnetically coupled spins per site have been characterized. The donor/acceptor complex with $S = \frac{5}{2}D = Mn^{II}(hfac)_2$ (hfac = hexafluoroacetylacetonate) and $S = \frac{1}{2}A = NITPr^n$ ($NITPr^n$ = *n*-propyl nitronyl nitroxide) has the alternating $\cdots DADADA \cdots$ structure with a *cis* geometry about the Mn(II) (Fig. 2.16). The material is a ferrimagnet with a T_c value of 8.6 K.[229] Replacement of the *n*-Pr group with Et leads to a ferrimagnet with a T_c value of 8.1 K and a *trans* geometry about the Mn(II).[229] In contrast, the rare earth–containing analog, $Dy^{III}(hfac)_3NITEt$, exhibits antiferromagnetic ordering at 4.4 K.[230]

Electron-transfer salts with D = mangano(II)macrocycles and A = TCNE also form an alternating $\cdots D^{\cdot+} A^{\cdot-} D^{\cdot+} A^{\cdot-} D^{\cdot+} D^{\cdot-} \cdots$ structure with a *trans* geometry about both the Mn(II) and the $[TCNE]^{\cdot-}$, as the $[TCNE]^{\cdot-}$ binds to two $S = 2$ mangano(II)macrocycles in a uniform *trans*-μ_2-*N*-σ-bound manner. These salts exhibit magnetic ordering at higher T_c values than those of magnets described previously. $[MnTPP]^+[TCNE]^{\cdot-}$ (TP = *meso*-tetraphenylporphinato), **4**, is the prototype salt (Figs. 2.17 and 2.18).[64] Between 115 and 250 K the magnetic susceptibility of $[MnTPP]^+[TCNE]^{\cdot-}\cdot 2PhMe$ can fit the Curie–Weiss equation with $\Theta = +61$ K. A minimum in the $\chi T(T)$ plot characteristic of ferrimagnetic behavior occurs at about 310 K.[231] The magnetization approaches 30,000 emu · G/mol at low temperature when cooled in a 2-T applied magnetic field, and hysteresis with a coercive

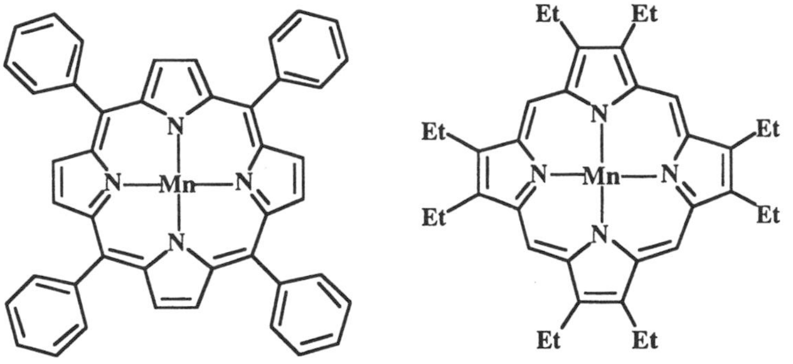

Manganese tetraphenylporphine
MnTPP, **4**

Manganese octaethylporphine
MnOEP, **5**

Figure 2.17 MnTTP, **4**, and MnOEP, **5**, molecules.

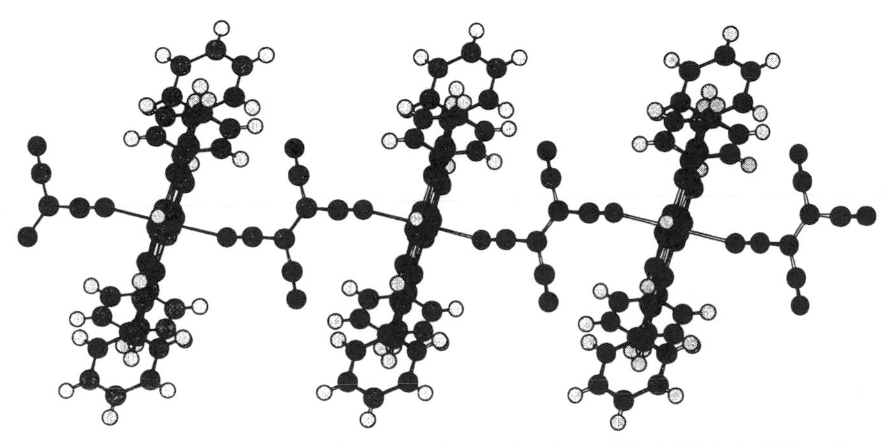

Figure 2.18 Segment of a uniform one-dimensional $\cdots D^{+}A\cdot^{-}D^{+}A\cdot^{-}D^{+}A\cdot^{-}\cdots$ chain of [MnTPP][TNCE]. (From Ref. 64.)

field of 375 G was observed at 5 K.[64] The magnetic ordering temperature is 14 K.[232] Refinement of the structural data reveals that disorder occurs with one-sixth of the $[TCNE]\cdot^{-}$ having the central C—C bond nominally perpendicular to the C—C bond of the $[TCNE]\cdot^{-}$ in its dominant orientation.[233]

The *meso*-tetraphenylporphinatomanganese(III)-based magnetic system was extended to TCNE electron-transfer salts with other metallomacrocycles [e.g., the easier-to-oxidize Mn^{II}OEP (OEP = octaethylporphine), **5**]. $[MnOEP]^{+}[TCNE]\cdot^{-}$ exhibits weak magnetic coupling and does not exhibit evidence for magnetic ordering down to 2 K.[234] Both $[MnTPP]^{+}$ and $[MnOEP]^{+}$ electron-transfer salts of $[TCNE]\cdot^{-}$ form parallel one-dimensional $\cdots D^{+}A\cdot^{-}D^{+}A\cdot^{-}D^{+}A\cdot^{-}\cdots$ chains (Figs. 2.18 and 2.19); however,

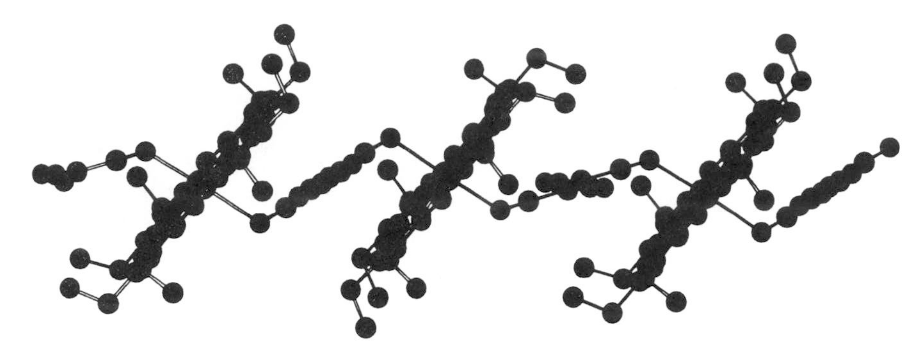

Figure 2.19 Segment of a nonuniform one-dimensional $\cdots D^{+}A\cdot^{-}D^{+}A\cdot^{-}D^{+}A\cdot^{-}$ $\cdots$ chain of [MnOEP][TCNE]. (From Ref. 234.)

in the latter $[\text{TCNE}]\cdot^{-}$ has nondisordered differing orientations. Thus a uniform chain is not present and appears necessary for achieving strong magnetic coupling. For example, $[\text{MnOEP}]^{+}[C_4(CN)_6]\cdot^{-}$ possesses a uniform chain and a Θ value of 6 K.[234]

A third example of an electron-transfer salt-based ferrimagnet is formed by the reaction of $V(C_6H_6)_2$ and TCNE. Since $S = 1$ $[\text{MnCp}_2^*]\cdot^{+}$ is isoelectronic to $S = 1$ $[V(C_6H_6)_2]\cdot^{+}$ and $[\text{MnCp}_2^*]\cdot^{+}[\text{TCNE}]\cdot^{-}$ is ferromagnetic with a T_c value of 8.8 K,[235] $[V(C_6H_6)_2][\text{TCNE}]$ was expected to be a ferromagnet with a comparable T_c value. The reaction of $V(C_6H_6)_2$ and TCNE in dichloromethane results in an immediate black solid of nominal composition of $V(\text{TCNE})_x \cdot y\text{CH}_2\text{Cl}_2$ ($x \approx 2$, $y \approx \frac{1}{2}$) stoichiometry.[236] $V(\text{TCNE})_x \cdot y(\text{CH}_2\text{Cl}_2)$ exhibits hysteresis at room temperature and is the first and only example of an organic molecule-based material that orders magnetically. Its critical temperature exceeds the thermal decomposition temperature of 350 K and is estimated to be about 400 K.

The $V(\text{TCNE})_x \cdot y$(solvent) magnet has been prepared with a variety of $S = 0$ solvents, including THF, acetonitrile (MeCN), diethyl ether, benzene, 1,2-dichloroethane, and hexane. The T_c value decreases systematically with the ability of the solvent to coordinate with V. The detailed structure of $V(\text{TCNE})_x \cdot y$(solvent) has yet to be elucidated. TCNE may bond to metals in many ways. The bonding of $[\text{TCNE}]\cdot^{-}$ with up to four V's linearly (or bent) is the structure proposed (Fig. 2.20), **6**, which can lead to significant overlap and exchange and hence strong magnetic coupling. The $V(\text{TCNE})_x \cdot y$(solvent) magnets are semiconductors with moderate conductivities of order 10^{-5} to 10^{-3} S/cm at room temperature, depending on the solvent from which they are prepared. The dc conductivity varies as $\exp[-(T_0/T)^{1/4}]$, consistent with Mott variable range hopping.[237]

2.8.4 Weak Ferromagnets

Several electron-transfer salt-based weak (or canted) magnets have been reported. A weak ferromagnet results from the relative canting of ferromag-

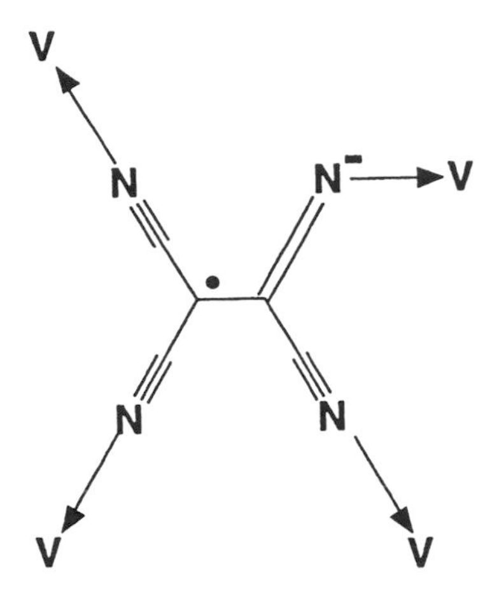

Figure 2.20 Proposed structure, **6**, of $V(TCNE)_x \cdot y$(solvent) compounds.

netically coupled spins, which reduces the moment (Fig. 2.21). Manganese phthalocyanine (Fig. 2.22), **7**, is an example of a canted ferromagnet, with a canting angle of 57°.[238] Other examples of weak ferromagnets are listed in Table 2.4.

κ-(BEDT-TTF)$_2$Cu[N(CN)$_2$]Cl is a superconductor at 12.8 K and 300 bar and undergoes a transition to a antiferromagnetic state at 45 K. At 22 K there is an additional transition to a weak ferromagnetic state with hysteresis and a saturation magnetization corresponding to 0.08% of one $S = \frac{1}{2}$ spin per formula unit and a canting angle of 0.046°.[240] The electron-transfer salt-based conductor perdeutero-[2,5-Me$_2$DCNQI]$_2$Cu (2,5-Me$_2$DCNQI = 2,5-dimethyl-N,N'-dicyano-p-quinonediimine) also exhibits weak ferromagnetic behavior, with a spontaneous magnetization perpendicular to the c-axis below 8 K. The saturation magnetization corresponds to 0.05% of one $S = \frac{1}{2}$ spin per formula unit.[198] Weak ferromagnetic behavior is also observed for [2,5-Br$_2$DCNQI]$_2$Cu.[243] [TDAE][C$_{60}$] (Fig. 2.23) is a semiconductor with a room-temperature conductivity of 10^{-4} S/cm and temperature dependence

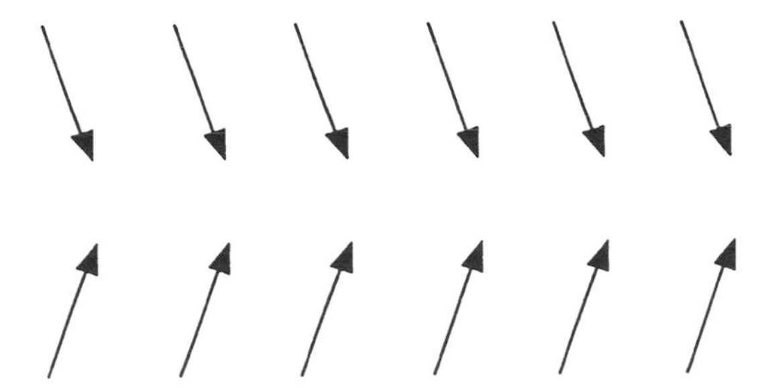

Figure 2.21 Canted arrangement of ferromagnetically coupled spins leading to a weak ferromagnet.

7

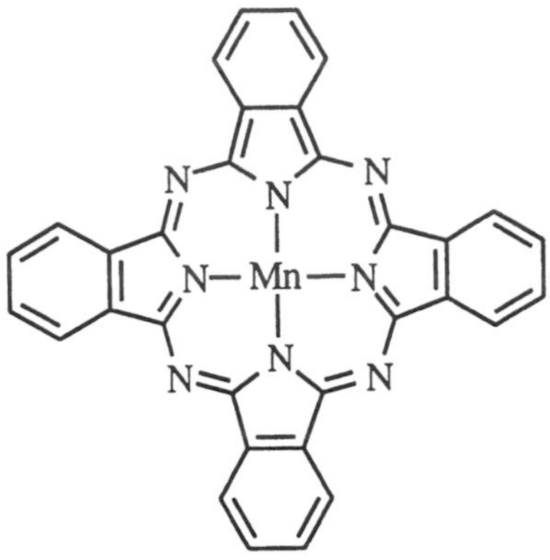

Figure 2.22 Manganese phthalocyanine.

TABLE 2.4 Representative Examples of Weak (Canted) Ferromagnets

Weak (Canted) Ferromagnet	T_c (K)	Canting Angle (deg)	Ref.
β-Mn(II)Pc	8.3	37	238
Fe(octaethyltetraazaporphyrin)	5.6	57	239
[κ-ET]$_2$Cu[N(CN)$_2$]Cl	22	0.046	240
Li[TCNQF$_4$]	12	0.7	241
Cu(Me$_2$DCNI)$_2$-d_{16}	8		198
TDAE(C_{60})	16.1		242

TDAE

C_{60}

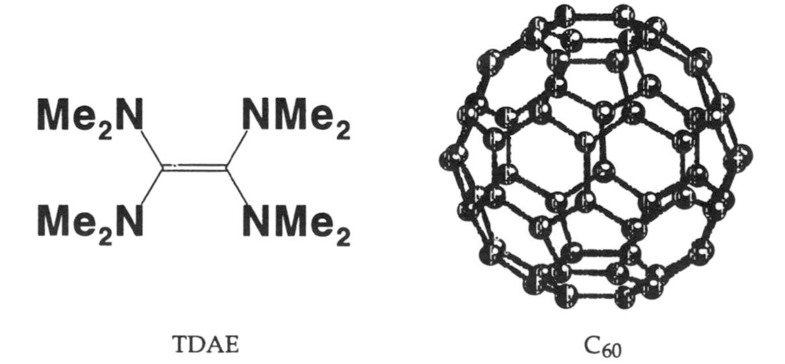

Figure 2.23 TDAE and C_{60} molecules.

proportional to $T^{2.3}$[244] and exhibits magnetic ordering transition at 16.1 K[62, 245, 246] which decreases with increasing pressure.[246a] A Rietvelt analysis of the powder diffraction data has been reported. However, due to the large R-factor (15.7%), details of the molecular structure that would provide insight into the charge distribution were not reported.[247] The saturation magnetization corresponds to only 4.2% of one $S = \frac{1}{2}$ spin per formula unit.[246a]

^{1}H NMR studies suggest that this system possesses isolated, ferromagnetically correlated clusters.[248] EPR studies have indicated that the system comprises ferromagnetically correlated spins.[246] Pulsed EPR studies reveal polarized spin clusters or nonclassical ferromagnetic order,[249] whereas EPR time decay of isothermal magnetization is consistent with spin glass behavior.[250] The 16.1-K ordering temperature has been confirmed by muon spin resonance.[251] This material has been described as a weak ferromagnet.[242] In contrast to C_{60}, TDAE adducts with C_{70}, C_{84}, C_{90}, and C_{96} do not exhibit ferromagnetic behavior above 4.5 K.[246] Solid solutions of $[TDAE][C_{60}]_{1-x}[C_{70}]_x$, however, have a suppressed T_c value that decreases linearly with x.[246d] This suppression is weak compared to the depression of T_c with x for superconducting $K_3[C_{60}]_{1-x}[C_{70}]_x$.[252]

2.9 CONCLUSIONS

This chapter provides a short introduction to the field of electron-transfer salt-based conductors, superconductors, and magnets and is not intended to be a detailed account of these subjects. For additional and more detailed information, the reader is directed to the more specialized books and reviews referred throughout. Nonetheless, several conclusions may be drawn from the present overview.

1. The concept of dimensionality plays a key role in the development of an understanding of the properties of electron-transfer salt-based conductors and superconductors. One-dimensional systems (e.g., [TTF][TCNQ], two-dimensional systems [e.g., $(BEDT\text{-}TTF)_2X$], and three-dimensional systems (e.g., the A_xC_{60} fullerides) were studied successively. The electronic band structure, and subsequently the physical properties, are strongly dependent on the dimensionality. For example, one-dimensional systems are prone to Peierls distortion, two-dimensional systems have closed Fermi surfaces, and the mechanism of superconductivity of three-dimensional A_xC_{60} fullerides may be due to electron–electron correlations.

2. Breakthroughs in the field of electron-transfer salt-based conductors, superconductors, and magnets have more often resulted from serendipitous procedures than from deliberate strategies. The shape, packing arrangement,

and redox properties have usefully directed the investigation toward potential donors and acceptors, but unfailing rules based on a clear-cut theory are still to be found. Consequently, progress in this field and future breakthroughs may still be based on trial-and-error procedures, involving the design, preparation, and evaluation of new systems using empirical selection criteria.

3. There are few examples where such narrowly focused research has mobilized such a large and diverse scientific community. This field continues to be very active and has evolved toward conducting polymers and their applications. This lasting interest is probably due to the unique multidisciplinary character of this work, which must necessarily involve close interactions among scientists in various fields, from synthetic chemists to experimental physicists and theoreticians.

4. As a consequence, in addition the design and preparation of new molecules, a key asset in research on electron-transfer salt-based conductors and superconductors has been better experimental and theoretical understanding of the physics of low-dimensional systems. Although a well-defined

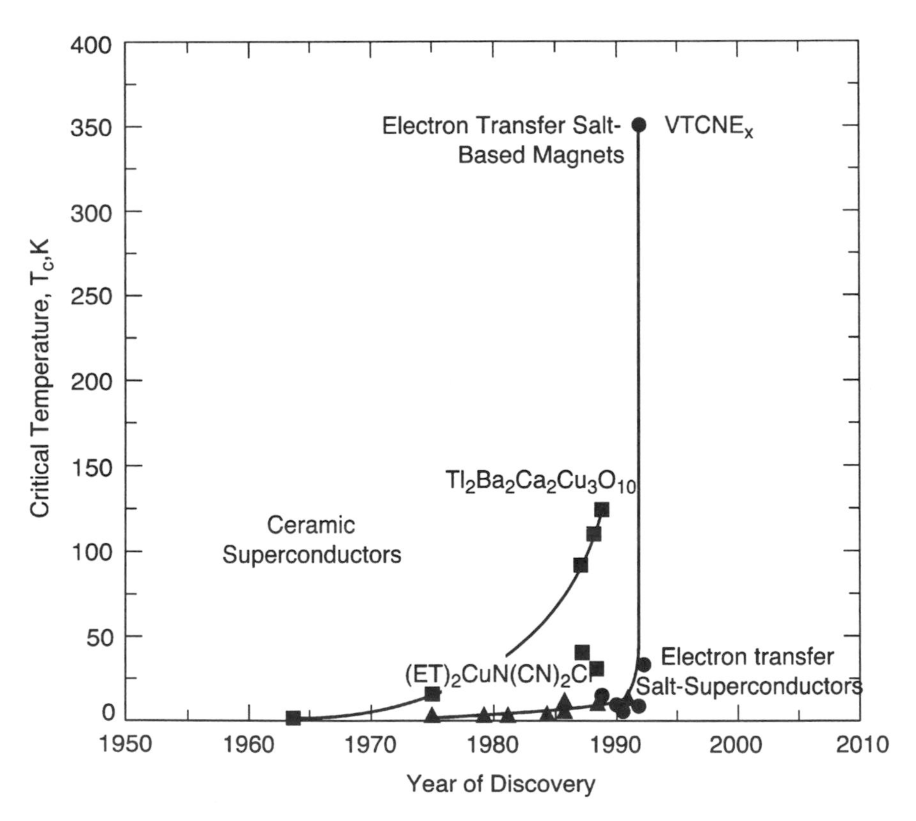

Figure 2.24 Time evolution of the discovery of increasing critical temperatures, T_c, for electron-transfer salt-based superconductors and magnets and oxide superconductors.

mechanism for superconductivity in electron-transfer salt-based and high-T_c-value ceramic superconductors is still not available, confirmation or development of new theories and concepts in the field of condensed-matter physics remains a contemporary research challenge.

5. Electron-transfer salt-based materials prepared by low-temperature synthesis rather than the high-temperature metallurgical methods currently utilized lend themselves to simplified fabrication techniques which would promote their use in electronic devices. This prospect attracted the early attention of several manufacturers, and tentative applications have been proposed.[253, 254] Few technological and commercially useful applications for electron-transfer salt-based conductors, superconductors, and magnets have been realized but, hopefully, are on the horizon. The T_c value needs to be raised and advances in processing (e.g., deposition of thin films) are also needed. Nevertheless, the field is embryonic (less than 15 years old for electron-transfer salt-based superconductors and less than 10 years old for electron-transfer salt-based magnets), and the future should witness important progress. This is evident in the chronology of increases in T_c value for both ceramic superconductors and electron-transfer salt-based superconductors and magnets (Fig. 2.24). From these examples of materials exhibiting cooperative ordering, the field of electron-transfer salt-based superconductors and magnets has evolved to include a wide range of phenomena for materials at T_c values significantly in excess of room temperature.

ACKNOWLEDGMENTS

J. S. M. gratefully acknowledges the continued support of the U.S. Department of Energy, Division of Materials Sciences (Grant DE-FG03-93ER45504) as well as the U.S. National Science Foundation (Grant CHE9320478). P.C. acknowledges the support of the Commission of the European Communities, Directorate General XII (Grants ERBCHBICT920042 and ERB4050PL930085) as well as the Conseil Régional Midi-Pyrénées, CCR-RDT (Grants 8800630 and 9300690). P.C. also thank all the members, past and present, of the Equipe Précurseurs Moleculaires et Matériaux, LCC/CNRS, for their contributions, which made this chapter possible.

REFERENCES

1. H. N. McCoy, W. C. Moore, *J. Am. Chem. Soc.* **33**, 273 (1911).
2. H. J. Kraus, *J. Am. Chem. Soc.* **34**, 1732 (1913).
3. H. Akamatu, H. Inokuchi, Y. Matsunaga, *Nature* **173**, 168 (1954).
4. H. M. McConnell, *J. Chem. Phys.* **1963**, 39 (1910).
5. (a) J. S. Miller, A. J. Epstein, W. M. Reiff, *Mol. Cryst. Liq. Cryst.* **120**, 27 (1985). (b) J. S. Miller, A. J. Epstein, W. M. Reiff, *Mol. Cryst. Liq. Cryst.* **120**, 234

(1986). (c) J. S. Miller, J. C. Calabrese, A. J. Epstein, R. W. Bigelow, J. H. Zhang, W. M. Reiff, *J. Chem. Soc. Chem. Commun.*, 1026 (1986). (d) J. S. Miller, J. C. Calabrese, H. Rommelmann, S. Chittapeddi, J. H. Zhang, W. M. Reiff, A. J. Epstein, *J. Am. Chem. Soc.* **109**, 769 (1987).

6. C. Kittel, *Introduction to Solid State Physics*, Wiley, New York, 1976.
7. B. C. Gerstein, *J. Chem. Educ.* **50**, 316 (1973).
8. D. Allender, J. Bray, J. Bardeen, *Phys. Rev. B* **9**, 119 (1974).
9. J. S. Miller, A. J. Epstein, in *Progress in Inorganic Chemistry*, Vol. 20, S. J. Lippard, Ed., Wiley, Chichester, West Sussex, England, 1979, pp. 1–151.
10. M. H. Whangbo, in *Electron Transfer in Biology and the Solid State: Inorganic Compounds with Unusual Properties*, M. K. Johnson, R. B. King, D. M. Kurtz, Jr., C. Kutal, M. L. Norton, R. A. Scott, Eds., American Chemical Society, Washington, DC, 1990, pp. 269–298.
11. H. Kammerlingh-Onnes, *Akad. Wetenschappen* **14**(113), 818 (1911).
12. W. Meissner, P. Ochsenfeld, *Naturwissenshaften* **21**, 787 (1933).
13. J. Bardeen, L. N. Cooper, J. R. Schrieffer, *Phys. Rev.* **106**, 162 (1957); *Phys. Rev.* **108**, 1175 (1957).
14. T. Ishiguro, K. Yamaji, in *Organic Superconductors*, Springer-Verlag, Berlin, 1990.
15. (a) A. W. Overhauser, *Phys. Rev. Lett.* **4**, 462 (1960). (b) W. M. Lomer, *Proc. Phys. Soc. (London)* **80**, 489 (1962).
16. P. Monceau, in *Electronic Properties of Quasi 1-D Compounds*, D. Reidel, Dordrecht, The Netherlands, 1985.
17. R. E. Peierls, in *Quantum Theory of Solids*, Oxford University Press, London, 1955.
18. (a) J.-P. Pouget, S. K. Khanna, F. Denoyer, R. Comes, A. F. Garrito, A. J. Heeger, *Phys. Rev. Lett.* **37**, 437 (1976). (b) G. A. Toombs, *Phys. Rev. C* **40**, 181 (1978).
19. J. W. Bray, L. V. Interrante, I. S. Jacobs, J. C. Bonner, in *Extended Linear Chain Compounds*, Vol. 3, J. S. Miller, Ed., Plenum Press, New York, 1983, pp. 353–415.
20. (a) D. S. Acker, R. J. Harder, W. R. Hertler, W. Mahler, L. R. Melby, R. E. Benson, W. E. Mochel, *J. Am. Chem. Soc.* **82**, 6408 (1960). (b) L. R. Melby, R. J. Harder, W. R. Hertler, W. Mahler, R. E. Benson, W. E. Mochel, *J. Am. Chem. Soc.* **84**, 3374 (1962). (c) W. R. Hertler, W. Mahler, L. R. Melby, J. S. Miller, R. E. Putscher, O. W. Webster, *Mol. Cryst. Liq. Cryst.* **171**, 205 (1989).
21. (a) W. Knop, *Justus Liebigs Ann. Chem.* **43**, 111 (1842). (b) W. Knop, G. Schnedermann, *J. Prakt. Chem.* **37**, 461 (1846).
22. K. Krogmann, H. D. Hausen, *Z. Anorg. Allge Chem.* **358**, 67 (1968).
23. H. Zeller, *Phys. Rev. Lett.* **28**, 1452 (1972).
24. W. A. Little, *Phys. Rev. A* **134**, 1416 (1964).
25. B. A. Scott, S. J. LaPlaca, J. B. Torrance, B. B. Silverman, B. Welber, *Ann. N.Y. Acad. Sci.* **313**, 369 (1978).
26. F. Wudl, D. Wobschall, E. J. Hufnagel, *J. Am. Chem. Soc.* **94**, 671 (1972).

27. (a) J. S. Miller, *Ann. N.Y. Acad. Sci.* **313**, 25 (1978). (b) L. C. Isett, E. A. Perez-Albuerne, *Ann. N.Y. Acad. Sci.* **313**, 395 (1978).
28. A. J. Epstein, S. Etemad, A. F. Garito, A. J. Heeger, *Phys. Rev. B* **5**, 952 (1972).
29. L. B. Coleman, M. J. Cohen, D. J. Sandman, F. G. Yamagishi, A. F. Garito, A. J. Ferraris, *Solid State Commun.* **12**, 1125 (1973).
30. J. P. Ferraris, D. O. Cowan, V. Valatka, J. H. Perlstein, *J. Am. Chem. Soc.* **95**, 948 (1973).
31. (a) D. Jérome, A. Mazaud, M. Ribault, K. Bechgaard, *J. Phys. Lett. (Orsay, France)* **41**, 95 (1980). (b) S. S. P. Parkin, M. Ribault, D. Jérome, K. Bechgaard, *J. Phys. C.* **14**, 5305 (1981).
32. K. Bechgaard, K. Carneiro, F. B. Rasmussen, H. Olsen, G. Rindorf, C. S. Jacobsen, H. Pedersen, J. E. Scott, *J. Am. Chem. Soc.* **103**, 2440 (1981).
33. (a) S. S. P. Parkin, E. M. Engler, R. R. Schumaker, R. Lagier, V. Y. Lee, J. C. Scott, R. L. Greene, *Phys. Rev. Lett.* **50**, 270 (1983). (b) E. B. Yagubskii, I. F. Schegolev, V. N. Laukhin, P. A. Konovich, M. V. Kartsvnik, A. V. Zvarykina, L. I. Buravov, *JETP Lett.* **39**, 12 (1984). (c) G. W. Crabtree, K. D. Carlson, L. N. Hall, P. T. Copps, H. H. Wang, T. J. Emge, M. A. Beno, J. M. Williams, *Phys. Rev. B* **30**, 2958 (1984). (d) J. M. Williams, T. J. Emge, H. H. Wang, M. A. Beno, P. T. Copps, L. N. Hall, K. D. Carlson, G. W. Crabtree, *Inorg. Chem.* **23**, 2558 (1984).
34. (a) H. Urayama, H. Yamochi, G. Saito, K. Nozawa, T. Sugano, M. Kinoshita, S. Sato, K. Oshima, A. Kawamoto, J. Tanaka, *Chem. Lett.* 55 (1988). (b) H. Urayama, H. Yamochi, G. Saito, S. Sato, A. Kawamoto, J. Tanaka, T. Mori, Y. Maruyama, H. Inokuchi, *Chem. Lett.*, 463 (1988). (c) K. Nozawa, K. Sugano, H. Urayama, H. Yamochi, G. Saito, M. Kinoshita, *Chem. Lett.*, 617 (1988).
35. S. Gärtner, E. Gogu, I. Heinen, H. J. Keller, T. Klutz, D. Schweitzer, *Solid State Commun.* **65**, 1531 (1988).
36. K. D. Carlson, U. Geiser, A. M. Kini, H. H. Wang, L. K. Montgomery, W. K. Kwok, M. A. Beno, J. M. Williams, C. S. Cariss, G. W. Crabtree, M.-H. Whangbo, M. Evain, *Inorg. Chem.* **27**, 965 (1988).
37. A. M. Kini, U. Geiser, H. H. Wang, K. D. Carlson, J. M. Williams, W. K. Kwok, K. G. Vandervoort, J. E. Thomson, D. L. Stupka, D. Jung, M.-H. Whangbo, *Inorg. Chem.* **29**, 2555 (1990).
38. (a) J. M. Williams, A. M. Kini, H. H. Wang, K. D. Carlson, U. Geiser, L. K. Montgomery, G. J. Pyrka, D. M. Watkins, J. M. Kommers, S. J. Boryschuk, A. V. Strieby Crouch, W. K. Kwok, J. E. Schirber, D. L. Overmyer, D. Jung, M.-H. Whangbo, *Inorg. Chem.* **29**, 3262 (1990). (b) H. H. Wang, K. D. Carlson, U. Geiser, A. M. Kini, A. J. Schultz, J. M. Williams, L. K. Montgomery, W. K. Kwok, U. Welp, K. G. Vandervoort, S. J. Boryschuk, A. V. Strieby Crouch, J. M. Kommers, D. M. Watkins, J. E. Schirber, D. L. Overmyer, D. Jung, J. J. Novoa, M.-H. Whangbo, *Synth. Met.* **41–43**, 1983 (1991).
39. (a) K. Kikuchi, M. Kikuchi, T. Namiki, K. Saito, I. Ikemoto, K. Murata, T. Ishiguro, K. Kobayashi, *Chem. Lett.* 931 (1987). (b) K. Kikuchi, K. Murata, Y. Honda, T. Namiki, K. Saito, H. Anzai, K. Kobayashi, I. Ikemoto, *J. Phys. Soc. Jpn.* **56**, 4241 (1987).
40. G. C. Papavassiliou, G. A. Mousdis, J. S. Zambounis, A. Terzis, A. Hountas, B. Hilti, C. W. Mayer, J. Pfeiffer, *Synth. Met.* **27**, B379 (1988).

41. (a) M. A. Beno, H. H. Wang, K. D. Carlson, A. M. Kini, G. M. Frankenbach, J. R. Ferraro, N. F. Larson, G. D. McCabe, J. Thompson, C. Purnama, M. Vashon, J. M. Williams, D. Jung, M.-H. Whangbo, *Mol. Cryst. Liq. Cryst.* **181**, 145 (1990). (b) M. A. Beno, H. H. Wang, A. M. Kini, K. D. Carlson, U. Geiser, W. K. Kwok, J. E. Thomson, J. M. Williams, J. Ren, M.-H. Whangbo, *Inorg. Chem.* **29**, 1599 (1990).
42. H. Kobayashi, T. Udagawa, H. Tomita, K. Bun, T. Naito, A. Kobayashi, *Chem. Lett.*, 1559 (1993).
43. J. M. Williams, J. R. Ferraro, R. J. Thorn, K. D. Carlson, U. Geiser, H. H. Wang, A. M. Kini, M.-H. Whangbo, *Organic Superconductors (Including Fullrenes)*, Prentice Hall, Upper Saddle River, NJ, 1992.
44. A. H. Reis, Jr., in *Extended Linear Chain Compounds*, Vol. 1, J. S. Miller, Ed., Plenum Press, New York, 1982, pp. 157–196.
45. T. J. Marks, D. W. Kalina, in *Extended Linear Chain Compounds*, Vol. 1, J. S. Miller, Ed., Plenum Press, New York, 1982, pp. 197–331.
46. (a) P. Cassoux, L. V. Interrante, J. Kasper, *C.R. Acad. Sci. (Paris), Ser. C* **291**, 25 (1980). (b) P. Cassoux, L. V. Interrante, J. Kasper, *Mol. Cryst. Liq. Cryst.* **81**, 293 (1982).
47. M. S. McClure, L. S. Lin, T. C. Whang, M. T. Ratajack, C. R. Kannewurf, T. J. Marks, *Bull. Am. Phys. Soc.* **25**, 314 (1980).
48. J. A. Ibers, L. J. Pace, J. Martinsen, B. M. Hoffman, *Struct. Bonding (Berlin)* **50**, 1 (1982).
49. S. Hünig, P. Erk, *Adv. Mater.* **3**, 225 (1991).
50. A. Kobayashi, R. Kato, H. Kobayashi, T. Mori, H. Inokuchi, *Solid State Commun.* **64**, 45 (1987).
51. H. Kobayashi, A. Miyamoto, R. Kato, F. Sakai, A. Kobayashi, Y. Yamakita, Y. Furukawa, M. Tasumi, T. Watanabe, *Phys. Rev. B* **45**, 3500 (1993).
52. H. Kobayashi, H. Sawa, S. Aonuma, R. Kato, *J. Am. Chem. Soc.* **115**, 7870 (1993).
53. P. Cassoux, L. V. Interrante, *Comments Inorg. Chem.* **12**, 47 (1991).
54. P. Cassoux, L. Valade, H. Kobayashi, A. Kobayashi, R. A. Clark, A. Underhill, *Coord. Chem. Rev.* **110**, 115 (1991).
55. P. Cassoux, L. Valade, in *Inorganic Materials*, D. W. Bruce, D. O'Hare, Eds., Wiley, Chichester, West Sussex, England, 1992, pp. 1–58.
56. (a) A. F. Hebard, M. J. Rosseinsky, R. C. Haddon, D. W. Murphy, S. H. Glarum, T. T. M. Palstra, A. P. Ramirez, A. R. Kortan, *Nature* **350**, 600 (1991). (b) M. J. Rosseinsky, D. W. Murphy, R. M. Fleming, R. Tycko, A. P. Ramirez, T. Siegrist, G. Dabbagh, S. E. Barrett, *Nature* **356**, 416 (1992).
57. A. F. Hebard, *Phys. Today* **11**, 26 (1992).
58. H. Ehrenreich, F. Spaepen, Eds., *Solid State Physics: Advances in Research and Applications*, Vol. 48, Academic Press, San Diego, CA, 1990, pp. 1–434.
59. A. F. Hebard, *Physica B* **197**, 544 (1994).
60. H. M. McConnell, *Proc. R.A. Welch Found. Chem. Res.* **11**, 144 (1967).
61. M. Tamura, Y. Nakazawa, D. Shiomi, K. Nozawa, Y. Hosokoshi, M. Ishikawa, M. Takahashi, M. Kinoshita, *Chem. Phys. Lett.* **186**, 401 (1991).

62. P. Allemand, K. Khemani, A. Koch, F. Wudl, K. Holczer, S. Donovan, G. Grüner, J. D. Thompson, *Science* **253**, 301 (1991).
63. J. M. Manriquez, G. T. Yee, R. S. McLean, A. J. Epstein, J. S. Miller, *Science* **252**, 1415 (1991).
64. J. S. Miller, J. C. Calabrese, R. S. McLean, A. J. Epstein, *Adv. Mater.* **4**, 498 (1992).
65. R. B. Saillant, R. C. Jaklevic, D. Bedford, *Mater. Res. Bull.* **9**, 289 (1974).
66. J. A. Abys, N. P. Enright, H. M. Gerdes, T. L. Hall, J. M. Williams, J. Ackerman, A. Wold, in *Inorganic Syntheses*, Vol. XIX, D. F. Shriver, Ed., Wiley, Chichester, West Sussex, England, 1979, pp. 1–5.
67. H. J. Guggenheim, D. Bahnck, *J. Cryst. Growth* **26**, 29 (1974).
68. (a) J. S. Miller, *Science* **194**, 189 (1976). (b) J. S. Miller, A. L. Balch, C. Hartman, in *Inorganic Syntheses*, Vol. XIX, D. F. Shriver, Ed., Wiley, Chichester, West Sussex, England, 1979, pp. 13–18.
69. J. M. Williams et al., in *Inorganic Syntheses*, Vol. XXI, J. P. Fackler, Ed., Wiley, Chichester, West Sussex, England, 1982, pp. 141–156.
70. L. A. Levy, *J. Chem. Soc.* 1081 (1912).
71. (a) J. M. Williams, A. J. Schultz, in *Molecular Metals*, W. E. Hatfield, Ed., Plenum Press, New York, 1979, pp. 337–368. (b) J. M. Williams, A. J. Schultz, A. E. Underhill, K. Carneiro, in *Extended Linear Chain Compounds*, Vol. 1, J. S. Miller, Ed., Plenum Press, New York, 1982, pp. 73–118. (c) J. M. Williams, in *Advances in Inorganic Chemistry and Radiochemistry*, Vol. 26, H. J. Emeléus, A. G. Sharpe, Eds., Academic Press, San Diego, CA, 1983, pp. 235–268.
72. D. J. Wood, A. E. Underhill, A. J. Schultz, J. M. Williams, *Solid State Commun.* **30**, 501 (1979).
73. H. R. Zeller, A. Beck, *J. Phys. Chem. Solids* **35**, 77 (1974).
74. H. P. Geserich, H. D. Hausen, K. Krogmann, P. Stampel, *Phys. Status Solidi A* **9** (1972).
75. J. M. Williams, P. L. Johnson, A. J. Schulz, C. C. Coffey, *Inorg. Chem.* **17**, 834 (1978).
76. J. M. Williams, J. L. Petersen, H. M. Gerdes, S. W. Peterson, *Phys. Rev. Lett.* **33** 1079 (1974).
77. W. Rügge, D. Kuse, H. R. Zeller, *Phys. Rev. B* **8**, 952 (1973).
78. M. A. Butler, D. L. Rousseau, D. N. E. Buchanan, *Phys. Rev. B* **7**, 61 (1973).
79. B. Renker, H. Rietschel, L. Pintschovius, W. Gläser, P. Brüesch, D. Kuse, M. J. Rice, *Phys. Rev. Lett.* **30**, 114 (1973).
80. R. Comès, M. Lambert, H. R. Zeller, *Phys. Status Solidi B* **58**, 587 (1973).
81. B. Renker, L. Pintschovius, W. Gläser, H. Rietschel, R. Comès, L. Liebert, W. Drexel, *Phys. Rev. Lett.* **32**, 836 (1974).
82. (a) J. W. Lynn, M. Iizumi, G. Shirane, S. A. Werner, R. B. Saillant, *Phys. Rev. B* **12**, 1154 (1975). (b) J. W. Lynn, M. Iizumi, G. Shirane, S. A. Werner, R. B. Saillant, *Bull. Am. Phys. Soc.* **20**, 439 (1975).
83. R. K. Brown, J. M. Williams, *Inorg. Chem.* **18**, 1922 (1979).
84. A. E. Underhill, D. M. Watkins, K. Carneiro, in *Extended Linear Chain Compounds*, Vol. 1, J. S. Miller, Ed., Plenum Press, New York, 1982, pp. 119–156.

85. M. Narita, C. U. Pittman, Jr., *Synthesis*, 489 (1976).
86. A. Krief, *Tetrahedron* **42**, 1209 (1986).
87. G. Schukat, A. M. Richter, E. Fanghänel, *Sulfur Rep.* **7**, 155 (1987).
88. T. J. Kistenmacher, T. E. Phillips, D. O. Cowan, *Acta Crystallogr. B* **30** 763 (1974).
89. W. F. Cooper, N. C. Kenney, J. W. Edmonds, A. Nagel, F. Wudl, P. Coppens, *J. Chem. Soc. Chem. Commun.*, 889 (1971).
90. R. E. Long, R. A. Sparks, K. N. Trueblood, *Acta Crystallogr.* **18**, 932 (1965).
91. A. J. Epstein, J. S. Miller, *Sci. Am.* **241**(4), 48 (1979).
92. (a) D. Jérome, W. Müller, M. Weger, *J. Phys. Lett. Paris* **35**, L77 (1974). (b) J. R. Cooper, D. Jérome, M. Weger, S. Etemad, *J. Phys. Lett. Paris* **36**, L219 (1975).
93. E. M. Engler, V. V. Patel, J. R. Andersen, Y. Tomkiewicz, R. A. Craven, B. A. Scott, S. Etemad, *Ann. N.Y. Acad. Sci.* **313**, 343 (1978).
94. F. Herrman, D. R. Salahub, R. P. Messmer, *Phys. Rev. B* **16**, 2453 (1977).
95. D. Jérome, H. J. Schulz, *Adv. Phys.* **31**, 299 (1982).
96. R. Comès, S. M. Shapiro, G. Shirane, A. F. Garito, A. J. Heeger, *Phys. Rev. Lett.* **35**, 1518 (1975).
97. S. Megtert, J. P. Pouget, R. Comès, *Ann. N.Y. Acad. Sci.*, **313**, 235 (1978), and references therein.
98. R. C. Wheland, *J. Am. Chem. Soc.* **98**, 3926 (1976).
99. (a) J. B. Torrance, *Acc. Chem. Res.* **12**, 79 (1979). (b) J. B. Torrance, *Mol. Cryst. Liq. Cryst.* **126**, 55 (1985).
100. F. Denoyer, R. Comès, A. F. Garito, A. J. Heeger, *Phys. Rev. Lett.* **35**, 445 (1975).
101. S. Kagoshima, H. Anzai, K. Kajimura, T. Ishiguro, *J. Phys. Soc. Jpn.* **39**, 1143 (1975).
102. R. Comès, S. M. Shapiro, G. Shirane, A. F. Garito, A. J. Heeger, *Phys. Rev. Lett.* **35**, 1512 (1975).
103. R. C. Wheland, J. L. Gillson, *J. Am. Chem. Soc.* **98**, 3916 (1976).
104. J. B. Torrance, in *Molecular Metals*, W. E. Hatfield, Ed., Plenum Press, New York, 1979, pp. 7–14.
105. J. R. Ferraro, J. M. Williams, *Introduction to Synthetic Electrical Conductors*, Academic Press, San Diego, CA, 1987.
106. A. Andrieux, C. Duroure, D. Jérome, K. Bechgaard, *J. Phys. Lett.* **40**, 381 (1979).
107. L. Malatesta, M. Angoletta, *J. Inorg. Nucl. Chem.* **8**, 273 (1958).
108. A. P. Ginsberg, J. W. Koepke, J. J. Hauser, K. W. West, F. J. Di Salvo, C. R. Sprinkle, R. L. Cohen, *Inorg. Chem.* **15**, 514 (1976).
109. E. O. Fischer, K. S. Brenner, *Z. Naturforsch.* **17**, 774 (1962).
110. A. E. Underhill, in *Comprehensive Coordination Chemistry*, Vol. 6, G. Wilkinson, Ed., Pergamon Press, Oxford, 1987, pp. 133–154.
111. M. Cowie, A. Gleizes, G. W. Grynkevitch, D. W. Kalina, M. S. McClure, R. P. Scaringe, R. C. Teitelbaum, S. L. Ruby, J. A. Ibers, C. R. Kannewurf, T. J. Marks, *J. Am. Chem. Soc.* **101**, 2921 (1979).

112. C. J. Schramm, R. P. Scaringe, D. R. Stojakovic, B. M. Hoffman, J. A. Ibers, T. J. Marks, *J. Am. Chem. Soc.* **102**, 6702 (1980).
113. L. Alcacer, H. Novais, in *Extended Linear Chain Compounds*, Vol. 3, J. S. Miller, Ed., Plenum Press, New York, 1983, pp. 319–351.
114. J. A. McCleverty, *Prog. Inorg. Chem.* **10**, 49 (1968).
115. (a) G. N. Schrauzer, *Trans. Met. Chem.* **4**, 299 (1968). (b) G. N. Schrauzer, *Acc. Chem. Res.* **2**, 72 (1969).
116. E. Hoyer, W. Dietzsch, W. Schroth, *Z. Chem.* **11**, 41 (1971).
117. S. Alvarez, R. Vicente, R. Hoffmann, *J. Am. Chem. Soc.* **107**, 6253 (1985).
118. (a) L. J. Alcacer, A. H. Maki, *J. Phys. Chem.* **78**, 215 (1974). (b) L. J. Alcacer, A. H. Maki, *J. Phys. Chem.* **80**, 1912 (1976). (c) L. J. Alcacer, H. M. Novais, F. P. Pedroso, in *Molecular Metals*, W. E. Hatfield, Ed., Plenum Press, New York, 1979, pp. 415–422.
119. L. V. Interrante, J. W. Bray, H. R. Hart, J. S. Kasper, P. A. Piacente, G. D. Watkins, *Ann. N.Y. Acad. Sci.* **313**, 407 (1978).
120. J. W. Bray, H. R. Hart, L. V. Interrante, I. S. Jacobs, J. S. Kasper, P. A. Piacente, G. D. Watkins, *Phys. Rev. B* **16**, 1359 (1977).
121. M. M. Ahmad, D. J. Turner, A. E. Underhill, C. S. Jacobsen, K. Mortensen, K. Carneiro, *Phys. Rev. B* **29**, 4796 (1984).
122. A. N. Bloch, D. O. Cowan, K. Bechgaard, R. E. Pyle, R. H. Banks, T. O. Poehler, *Phys. Rev. Lett.* **34**, 1561 (1975).
123. J. R. Cooper, M. Weger, D. Jérome, D. Lefur, K. Bechgaard, A. N. Bloch, D. O. Cowan, *Solid State Commun.* **19**, 749 (1976).
124. C. S. Jacobsen, K. Mortensen, J. R. Andersen, K. Bechgaard, *Phys. Rev. B* **18**, 905 (1978).
125. A. Andrieux, P. M. Chaikin, C. Duroure, D. Jerome, C. Weyl, K. Bechgaard, J. R. Andersen, *J. Phys. (Paris)* **40**, 1199 (1979).
126. (a) K. Bechgaard, D. O. Cowan, A. N. Bloch, *J. Chem. Soc. Chem. Commun.*, 937 (1974). (b) J. R. Anderson, K. Bechgaard, *J. Org. Chem.* **40**, 2016 (1975). (c) K. Bechgaard, D. O. Cowan, L. Henriksen, *J. Org. Chem.* **40**, 746 (1975).
127. P. Shu, A. N. Bloch, T. F. Carruthers, D. O. Cowan, *J. Chem. Soc. Chem. Commun.*, 505 (1977).
128. (a) F. Wudl, D. Nalewajek, *J. Chem. Soc. Chem. Commun.*, 866 (1980). (b) F. Wudl, E. Aharon-Shalom, S. H. Bertz, *J. Org. Chem.* **46**, 4612 (1981).
129. J. M. Braam, C. D. Carlson, D. A. Stephens, A. E. Rehan, S. J. Compton, J. M. Williams, in *Inorganic Syntheses*, Vol. 24, J. M. Shreeve, Ed., Wiley, New York, 1986, pp. 130–134.
130. A. Moradpour, V. Peyrussan, I. Johansen, K. Bechgaard, *J. Org. Chem.* **48**, 388 (1983).
131. A. Moradpour, K. Bechgaard, M. Barrie, C. Lenoir, K. Murata, R. C. Lacoe, M. Ribault, D. Jérome, *Mol. Cryst. Liq. Cryst.* **119**, 69 (1985).
132. (a) J. M. Williams, K. Carneiro, in *Advanced Inorganic Chemistry and Radiochemistry*, Vol. 29, H. J. Emeléus, A. G. Sharpe, Eds., Academic Press, San Diego, CA, 1985, pp. 248–296. (b) J. M. Williams, in *Progress in Inorganic Chemistry*, Vol. 33, S. J. Lippard, Ed., Wiley, New York, 1985, pp. 183–220.

133. L. Pauling, *The Nature of the Chemical Bond*, Cornell University Press, Ithaca, NY, 1960.
134. M.-H. Whangbo, J.-M. Williams, M. A. Beno, J. R. Dorfman, *J. Am. Chem. Soc.* **105**, 645 (1983).
135. K. Andres, F. Wudl, D. B. McWhan, G. A. Thomas, D. Nalewajek, A. L. Stevens, *Phys. Rev. Lett.* **45**, 1449 (1980).
136. K. Bechgaard, *Conjugated Polymers and Related Materials*, 81st Nobel Symposium Series, No. 431, Oxford Univ. Press, Oxford, UK, 1993.
137. S. S. P. Parkin, F. Creuzet, M. Ribault, D. Jérome, K. Bechgaard, J.-M. Fabre, *Mol. Cryst. Liq. Cryst.* **79**, 249 (1982).
138. R. Moret, J.-P. Pouget, R. Carnes, K. Bechgaard, *J. Phys. (Paris) Colloq.* **44**, C3-957 (1983).
139. J. F. Kwak, J. E. Schirber, R. L. Greene, E. M. Engler, *Phys. Rev. Lett.* **46**, 1296 (1981).
140. L. P. Gor'kov, A. G. Lebed', *J. Phys. (Paris) Lett.* **45**, 433 (1984).
141. (a) H. Bando, K. Oshima, M. Suzuki, H. Kobayashi, G. Saito, *J. Phys. Soc. Jpn.* **51**, 2711 (1982). (b) J.-P. Ulmet, P. Auban, A. Khmou, S. Askenazy, *J. Phys. Lett.* **46**, 1535 (1985).
142. A. Audouard, F. Goze, S. Dubois, J.-P. Ulmet, L. Brossard, S. Askenazy, S. Tomic, J.-M. Fabre, *Europhys. Lett.* **25**, 363 (1994).
143. (a) M. J. Naughton, O. H. Chung, M. Chaparala, X. Bu, P. Coppens, *Phys. Rev. Lett.* **67**, 3712 (1991). (b) W. Kang, S. T. Hannahs, P. M. Chaikin, *Phys. Rev. Lett.* **69**, 2827 (1992).
144. F. Wudl, *Acc. Chem. Res.* **17**, 227 (1984).
145. C. S. Jacobsen, K. Mortensen, N. Thorup, D. B. Tanner, M. Weger, K. Bechgaard, *Chem. Scr.* **17**, 103 (1981).
146. (a) P. M. Grant, *J. Phys. (Paris) C3* **44**, 847 (1983). (b) L. Ducasse, M. Abderrabba, B. Gallois, *J. Phys. C* **18**, L947 (1985).
147. M. Mizuno, A. F. Garito, M. P. Cava, *J. Chem. Soc. Chem. Commun.*, 18 (1978).
148. P. F. Reed, J. M. Braam, L. M. Sowa, R. A. Barkhau, G. S. Blackman, D. D. Cox, G. A. Ball, H. H. Wang, J. M. Williams, in *Inorganic Syntheses*, Vol. 26, H. D. Kaesz, Ed., Wiley, New York, 1989, pp. 388–390.
149. H. Yamochi, T. Komatsu, G. Saito, M. Kusunoki, K.-I. Sakaguchi, *Mol. Cryst. Liq. Cryst.* **234**, 137 (1993).
150. (a) T. Mori, A. Kobayashi, Y. Sasaki, H. Kobayashi, G. Saito, H. Inokuchi, *Chem. Lett.*, 1963 (1982). (b) H. Mori, S. Tanaka, M. Oshima, G. Saito, T. Mori, Y. Maruyama, H. Inokuchi, *Bull. Soc. Chim. Jpn.* **63**, 2183 (1990).
151. (a) T. J. Emge, H. H. Wang, U. Geiser, M. A. Beno, K. S. Webb, J. M. Williams, *J. Am. Chem. Soc.* **108**, 3849 (1986). (b) D. Jung, M. Evain, J. J. Novoa, M.-H. Whangbo, M. A. Beno, A. M. Kini, A. J. Schulz, J. M. Williams, P. J. Nigrey, *Inorg. Chem.* **28**, 4516 (1989).
152. J. A. Schlueter, U. Geiser, J. M. Williams, H. H. Wang, W.-K. Kwok, J. A. Fendrich, K. D. Carlson, C. A. Achenbach, J. D. Dudek, D. Naumann, T. Roy, J. E. Schirber, W. R. Bayless, *J. Chem. Soc. Chem. Commun.*, 1599 (1994).

153. P. Day, personal communication; M. Kurmoo, A. W. Graham, P. Day, S. J. Coles, M. B. Hursthouse, J. L. Caulfield, J. Singleton, F. L. Pratt, W. Hayes, L. Ducasse, P. Guionneau, *J. Am. Chem. Soc.* **117**, 12209 (1995).

154. G. Saito, T. Enoki, K. Toriumi, H. Inokuchi, *Solid State Commun.* **42**, 557 (1982).

155. (a) F. Creuzet, G. Creuzet, D. Jérome, D. Schweitzer, H. J. Keller, *J. Phys. Lett. (Paris)*, L1079 (1985). (b) F. Creuzet, D. Jérome, D. Schweitzer, H. J. Keller, *Europhys. Lett.* **1**, 461 (1986).

156. (a) K. Saito, Y. Ishikawa, K. Kikuchi, K. Kobayashi, I. Ikemoto, *Bull. Chem. Soc. Jpn.* **63**, 1865 (1990). (b) Y. Ishikawa, K. Saito, K. Kikuchi, K. Kobayashi, I. Ikemoto, *Bull. Chem. Soc. Jpn.* **64**, 212 (1991).

157. A. M. Kini, M. A. Beno, D. Son, H. H. Wang, K. D. Carlson, L. C. Porter, U. Welp, B. A. Vogt, J. M. Williams, D. Jung, M. Evain, M.-H. Whangbo, D. L. Overmyer, J. E. Schirber, *Solid State Commun.* **69**, 503 (1989).

158. *Proceedings of the International Conference on Science and Technology of Synthetic Metals*, Seoul, Korea, *Synth. Met.* **70** (1995).

159. J. S. Brooks, *MRS Bull.* **8**, 29 (1993).

160. G. Steimecke, R. Kirmse, E. Hoyer, *Z. Chem.* **15**, 28 (1975).

161. G. Steimecke, H. J. Sieler, R. Kirmse, E. Hoyer, *Phosphorus Sulphur* **7**, 49 (1979).

162. K. Hartke, T. Kissel, J. Quante, R. Matusch, *Chem. Ber.* **113**, 1898 (1980).

163. L. Valade, J.-P. Legros, M. Bousseau, P. Cassoux, M. Garbauskas, L. V. Interrante, *J. Chem. Soc. Dalton Trans.*, 783 (1985).

164. E. Canadell, E. I. Rachidi, S. Ravy, J.-P. Pouget, L. Brossard, J.-P. Legros, *J. Phys. (Paris)* **50**, 2967 (1989).

165. (a) L. Brossard, M. Ribault, M. Bousseau, L. Valade, P. Cassoux, *C.R. Acad. Sci. (Paris) Ser. 2* **302**, 205 (1986). (b) L. Brossard, M. Ribault, L. Valade, P. Cassoux, *Physica B & C (Amsterdam)* **143**, 378 (1986).

166. (a) A. Kobayashi, H. Kim, Y. Sasaki, R. Kato, H. Kobayashi, S. Moriyama, Y. Nishio, K. Kajita, W. Sasaki, *Chem. Lett.* 1819 (1987). (b) K. Kajita, Y. Nishio, S. Moriyama, R. Kato, H. Kobayashi, W. Sasaki, *Solid State Commun.* **65**, 361 (1988).

167. L. Brossard, H. Hurdequint, M. Ribault, L. Valade, J.-P. Legros, P. Cassoux, *Synth. Met.* **27**, B157 (1988).

168. L. Brossard, M. Ribault, L. Valade, P. Cassoux, *J. Phys. (Paris)* **50**, 1521 (1989).

169. A. Kobayashi, H. Kobayashi, A. Miyamoto, R. Kato, R. A. Clark, A. E. Underhill, *Chem. Lett.*, 2163 (1991).

170. H. Kobayashi, K. Bun, T. Naito, R. Kato, A. Kobayashi, *Chem. Lett.*, 1909 (1992).

171. H. Tajima, M. Inokuchi, A. Kobayashi, T. Ohta, R. Kato, H. Kobayashi, H. Kuroda, *Chem. Lett.*, 1235 (1993).

172. E. Canadell, S. Ravy, J.-P. Pouget, L. Brossard, *Solid State Commun.* **75**, 633 (1990).

173. (a) J. P. Cornelissen, J. G. Haasnoot, J. Reedijk, C. Faulmann, J.-P. Legros, P. Cassoux, P. Nigrey, *Inorg. Chim. Acta* **202**, 131 (1992). (b) R.-M. Olk, R.

Kirmse, E. Hoyer, C. Faulmann, P. Cassoux, *Z. Anorg. Allg. Chem.* **620**, 90 (1994). (c) T. Naito, A. Sato, K. Kawano, A. Taneto, H. Kobayashi, A. Kobayashi, *J. Chem. Soc. Chem. Commun.*, 351 (1995).

174. (a) K. Awaga, T. Okuno, Y. Maruyama, A. Kobayashi, H. Kobayashi, S. Schenk, A. E. Underhill, *Inorg. Chem.* **33**, 5598 (1994). (b) O. A. Dyachenko, S. V. Konovalikhin, A. I. Kotov, G. V. Shilov, E. B. Yagubskii, C. Faulmann, P. Cassoux, *J. Chem. Soc. Chem. Commun.*, 508 (1993).
175. D.-Y. Noh, M. Mizuno, J.-H. Choy, *Inorg. Chim. Acta* **216**, 147 (1994).
176. (a) E. B. Yagubskii, *Mol. Cryst. Liq. Cryst.* **230**, 139 (1993). (b) C. Faulmann, P. Cassoux, E. B. Yagubskii, L. V. Vetoshkina, *New J. Chem.* **17**, 385 (1993).
177. A. Errami, Thèse de Doctorat de l'Université Paul Sabatier, Toulouse, France, 1995.
178. H. W. Kroto, J. R. Heath, S. C. O'Brien, R. F. Curl, R. E. Smalley, *Nature* **318**, 162 (1985).
179. W. Krätschmer, L. D. Lamb, K. Fostiropoulos, D. R. Huffman, *Nature* **347**, 354 (1990).
180. R. C. Haddon, A. F. Hebard, M. J. Rosseinsky, D. W. Murphy, S. J. Duclos, K. B. Lyons, B. Miller, J. M. Rosamilia, R. M. Fleming, A. R. Kortan, S. H. Glarum, A. V. Makhija, A. J. Muller, R. H. Eick, R. H. Zahurak, R. Tycko, G. Dabbagh, F. A. Thiel, *Nature* **350**, 320 (1991).
181. J. E. Fischer, P. A. Heiney, A. R. McGhie, W. J. Romanow, A. M. Denenstein, J. P. McAuley, A. B. Smith, *Science* **252**, 1288 (1991).
182. (a) R. C. Haddon, L. E. Brus, K. Raghavachari, *Chem. Phys. Lett.* **125**, 459 (1986). (b) S. Saito, A. Oshiyama, *Phys. Rev. Lett.* **66**, 2637 (1991).
183. S. H. Yang, C. L. Pettiete, J. Conceicao, O. Chesnovsky, R. E. Smalley, *Chem. Phys. Lett.* **139**, 233 (1987).
184. (a) D. Dubois, G. Moninot, W. Kutner, M. T. Jones, K. M. Kadish, *J. Phys. Chem.* **96**, 7137 (1992). (b) Q. Xie, E. Pérez-Cordero, L. Echegoyen, *J. Am. Chem. Soc.* **114**, 3978 (1992).
185. D. W. Murphy, M. J. Rosseinsky, R. M. Fleming, R. Tycko, A. P. Ramirez, R. C. Haddon, T. Siegrist, G. Dabbagh, J. C. Tully, R. E. Walstedt, *J. Phys. Chem. Solids* **53**, 1321 (1992).
186. A. F. Hebard, *Annu. Rev. Mater. Sci.* **23**, 159 (1993).
187. K. Tanigaki, in *Polymeric Materials for Microelectronic Applications*, H. Ito, S. Tagawa, K. Horie, Eds., ACS Symposium Series, Vol. 579, American Chemical Society, Washington, DC, 1994, p. 343.
188. (a) F. L. Carter, Ed., *Molecular Electronic Devices*, Marcel Dekker, New York, 1982. (b) F. L. Carter, Ed., *Molecular Electronic Devices II*, Marcel Dekker, New York, 1987. (c) S. Woitellier, J.-P. Launay, C. W. Spangler, *Inorg. Chem.* **28**, 758 (1989).
189. (a) C. Creutz, *J. Am. Chem. Soc.* **91**, 3988 (1969); *Prog. Inorg. Chem.* **30**, 1 (1983). (b) N. S. Hush, *Coord. Chem. Rev.* **64**, 135 (1985). (c) K. Prassides, Ed., *Mixed Valency Systems: Applications in Chemistry, Physics and Biology*, NATO ASI Series, Kluwer Academic Publishers, Dordrecht, The Netherlands, 1991.
190. M. B. Robin, P. Day, *Adv. Inorg. Radiochem.* **10**, 247 (1967).

191. (a) A. H. Schroeder, S. Mazur, *J. Am. Chem. Soc.* **100**, 7339 (1978). (b) S. Utamapanya, A. J. Rajca, *J. Am. Chem. Soc.* **113**, 9242 (1978). (c) S. F. Nelsen, J. Adamus, J. J. Wolff, *J. Am. Chem. Soc.* **116**, 7339 (1994). (d) L. L. Miller, C. A. Liberko, *Chem. Mater.* **2**, 339 (1994). (e) J. Bonvoisin, J.-P. Launay, C. Rovira, J. Veciana, *Angew. Chem. Int. Ed. Engl.* **33**, 2106 (1994).

192. U. Scherer, Y.-J. Shen, A. Adam, W. Bietsch, J. U. von Schütz, K. Müllen, *Adv. Mater.* **5**, 109 (1993).

193. P. Cassoux, C. J. Bowlas, K. Lahlil, A. Moradpour, J. Bonvoisin, J.-P. Launay, *Acta Phys. Polonica A* **87**, 743 (1995).

194. K. Lahlil, A. Moradpour, C. J. Bowlas, F. Menou, P. Cassoux, J. Bonvoisin, J.-P. Launay, G. Dive, D. Dehareng, *J. Am. Chem. Soc.* **117**, 9995 (1995).

195. A. Aumüller, P. Erk, G. Klebe, S. Hünig, J. U. von Schütz, H.-P. Werner, *Angew. Chem. Int. Ed. Engl.* **25**, 740 (1986).

196. P. Erk, S. Hünig, J. U. von Schütz, H.-P. Werner, H. C. Wolf, *Angew. Chem. Int. Ed. Engl.* **27**, 267 (1988).

197. H. P. Werner, J. U. von Schütz, H. C. Wolf, R. K. Kremer, M. Gehrke, A. Aumüller, *Solid State Commun.* **65**, 809 (1988).

198. M. Tamura, H. Sawa, S. Aonuma, R. Kato, M. Kinoshita, H. Kobayashi, *J. Phys. Soc. Jpn.* **62**, 1470 (1993).

199. S. Tomic, D. Jérome, A. Aumüller, P. Erk, S. Hünig, J. U. von Schütz, *J. Phys. C* **21**, L203 (1988).

200. R. Kato, H. Kobayashi, A. Kobayashi, *J. Am. Chem. Soc.* **111**, 5224 (1989).

201. (a) C. Bourbonnais, R. T. Henriques, P. Wzietek, D. Köngeter, J. Voiron, D. Jérome, *Phys. Rev. B* **44**, 641 (1991). (b) M. Almeida, V. Gama, R. T. Henriques, L. Alcacer, in *Inorganic and Organometallic Polymers with Special Properties*, R. M. Laine, Ed., Kluwer, Dordrecht, The Netherlands, 1992, pp. 163–177. (c) V. Gama, R. T. Henriques, G. Bonfait, M. Almeida, S. Ravy, J.-P. Pouget, L. Alcacer, *Mol. Cryst. Liq. Cryst.* **234**, 171 (1993).

202. (a) V. Gama, R. T. Henriques, M. Almeida, L. Alcacer, *J. Phys. Chem.* **98**, 997 (1994). (b) E. B. Lopes, M. J. Matos, R. T. Henriques, M. Almeida, J. Dumas, *Europhys. Lett.* **27**, 241 (1994). (c) E. B. Lopes, M. J. Matos, R. T. Henriques, M. Almeida, J. Dumas, *Phys. Rev. B* **52**, 2237 (1995).

203. P. Day, M. Kurmoo, T. Mallah, I. R. Marsden, R. H. Friend, F. L. Pratt, W. Hayes, D. Chasseau, J. Gaultier, G. Bravic, L. Ducasse, *J. Am. Chem. Soc.* **114**, 10722 (1992).

204. P. Day, *Phys. Scri.* **T49**, 726 (1993).

205. (a) T. Mallah, C. Hollis, S. Bott, M. Kurmoo, P. Day, M. Allan, R. H. Friend, *J. Chem. Soc. Dalton Trans.*, 859 (1990). (b) M. Kurmoo, P. Day, M. Allan, R. H. Friend, *Mol. Cryst. Liq. Cryst.* **234**, 199 (1993).

206. (a) C. J. Gomez-Garcia, L. Ouahab, C. Gimenez, S. Triki, E. Coronado, P. Delhaes, *Angew. Chem. Int. Ed. Engl.* **33**, 223 (1994). (b) C. J. Gomez-Garcia C. Gimenez, S. Triki, E. Coronado, P. Le Magueres, L. Ouahab, L. Ducasse, C. Sourisseau, P. Delhaes, *Inorg. Chem.* **34**, 4139 (1995).

207. E. Coronado, C. J. Gomez-Garcia, *Comments Inorg. Chem.* **17**, 255 (1995).

208. (a) R. Kato, H. Kobayashi, A. Kobayashi, *Synth. Met.* **42**, 2093 (1991). (b) T. Naito, A. Miyamoto, H. Kobayashi, R. Kato, A. Kobayashi, *Chem. Lett.* 1945

(1991). (c) A. Kobayashi, R. Kato, T. Naito, H. Kobayashi, *Synth. Met.* **56**, 2078 (1993).

209. A. Kobayashi, T. Udagawa, H. Tomita, T. Naito, H. Kobayashi, *Chem. Lett.* 2179 (1993).
210. H. Kobayashi, H. Tomita, T. Naito, A. Kobayashi, F. Sakai, T. Watanabe, P. Cassoux, *J. Am. Chem. Soc.* **118**, 368 (1996).
211. F. Goze, V. N. Laukhin, L. Brossard, A. Audouard, J.-P. Ulmet, S. Askenazy, T. Naito, H. Kobayashi, A. Kobayashi, M. Takumoto, P. Cassoux, *Europhys. Lett.* **28**, 427 (1994).
212. (a) S. Morimoto, K. Itoh, F. Tanaka, N. Mataga, *Preprints of the Symposium on Molecular Structure*, Tokyo, 1968, p. 6. (b) N. Mataga, *Theor. Chim. Acta* **10**, 372 (1968).
213. G. A. Candela, L. Swartzendruber, J. S. Miller, M. J. Rice, *J. Am. Chem. Soc.* **101**, 2755 (1979).
214. (a) J. S. Miller, D. A. Dougherty, Eds., *Proceedings of the Conference on Ferromagnetic and High Spin Molecular Based Materials, Mol. Cryst. Liq. Cryst.* **176** (1989). (b) O. Kahn, D. Gatteschi, J. S. Miller, F. Palacio, Eds., *Proceedings of the Conference on Molecular Magnetic Materials: NATO ARW Molecular Magnetic Materials*, Kluwer, London, 1991. (c) H. Iwamura, J. S. Miller, Eds., *Proceedings of the Conference on Chemistry and Physics of Molecular Based Magnetic Materials, Mol. Cryst. Liq. Cryst.* **232 / 233** (1993). (d) J. S. Miller, A. J. Epstein, Eds., *Proceedings of the Conference on Molecule-Based Magnets, Mol. Cryst. Liq. Cryst.* **271** (1995).
215. (a) J. S. Miller, A. J. Epstein, *Angew. Chem.* **106**, 399 (1994); *Angew. Chem. Int. Ed. Engl.* **33**, 385 (1994). (b) D. Gatteschi, *Adv. Mater.* **6**, 635 (1994). (c) O. Kahn, *Molecular Magnetism*, VCH, New York, 1993. (d) J. S. Miller, A. J. Epstein, *Chem. Eng. News* **73**(40), 30 (1995). (e) J. S. Miller, A. L. Buchachenko, *Russ. Chem. Rev.* **59**, 307 (1990).
216. G. Chouteau, Cl. Veyret-Jeandey, *J. Phys.* **42**, 1441 (1981).
217. (a) M. Kinoshita, *Jpn. J. Appl. Phys.* **33**, 5718 (1994). (b) K. Awaga, Y. Maruyama, *Chem. Mater.* **2**, 535 (1990); *J. Chem. Phys.* **173**, 33 (1990).
218. (a) R. Chirelli, A. Rassat, P. Rey, *J. Chem. Soc. Chem. Commun.* 1081 (1992). (b) R. Chirelli, M. A. Novak, A. Rassat, J. L. Tholence, *Nature* **363**, 147 (1993).
219. S. Chittipeddi, K. R. Cromack, J. S. Miller, A. J. Epstein, *Phys. Rev. Lett.* **58**, 2695 (1987).
220. S. Chittippeddi, M. A. Selover, A. J. Epstein, D. M. O'Hare, J. Manriquez, J. S. Miller, *Synth. Met.* **27**, B417 (1989).
221. A. Chackravorty, A. J. Epstein, W. N. Lawless, J. S. Miller, *Phys. Rev. B* **40**, 11422 (1989).
222. J. S. Miller, A. J. Epstein, *J. Am. Chem. Soc.* **109**, 3850 (1987).
223. W. E. Broderick, D. M. Eichorn, X. Lu, P. J. Toscano, S. M. Owens, B. M. Hoffman, *J. Am. Chem. Soc.* **117**, 3641 (1995).
224. (a) W. E. Broderick, J. A. Thompson, E. P. Day, B. M. Hoffman, *Science* **249**, 410 (1990). (b) W. E. Broderick, B. M. Hoffman, *J. Am. Chem. Soc.* **113**, 6334 (1991).

225. K. S. Narayan, K. M. Kai, A. J. Epstein, J. S. Miller, *J. Appl. Phys.* **69**, 5953 (1991).

226. K. S. Narayan, B. G. Morin, J. S. Miller, A. J. Epstein, *Phys. Rev. B* **46**, 6195 (1992).

227. J. S. Miller, A. H. Reis, Jr., E. Gerbert, J. J. Ritsko, W. R. Saleneck, L. Kovnat, T. W. Cape, R. P. Van Duyne, *J. Am. Chem. Soc.* **101**, 7111 (1979).

228. J. S. Miller, J. H. Zhang, W. M. Reiff, L. D. Preston, A. H. Reis, Jr., E. Gerbert, M. Extine, J. Troup, M. D. Ward, *J. Phys. Chem.* **91**, 4344 (1987).

229. A. Caneschi, D. Gatteschi, J. P. Renard, P. Rey, R. Sessoli, *Inorg. Chem.* **28**, 3314 (1989).

230. (a) C. Bellini, A. Caneschi, D. Gatteschi, R. Sessoli, *Adv. Mater.* **4**, 504 (1992). (b) J. Seiden, *J. Phys. Lett.* **44**, L947 (1983).

231. (a) M. Verdaguer, M. Julve, A. Michalowicz, O. Kahn, *Inorg. Chem.* **22**, 2624 (1983). (b) M. Drillon, J. C. Gianduzzo, R. Georges, *Phys. Lett. A* **96**, 413 (1983).

232. A. Böhm, C. Vazquez, R. S. McLean, J. C. Calabrese, S. E. Kalm, J. L. Manson, A. J. Epstein, J. S. Miller, *Inorg. Chem.* **35**, 3083 (1996).

233. B. M. Burkhart, B. G. Morin, A. J. Epstein, J. C. Calabrese, J. S. Miller, M. Sundaralingham, submitted for publication.

234. J. S. Miller, C. Vazquez, N. L. Jones, R. S. McLean, A. J. Epstein, *J. Mater. Chem.* **5**, 707 (1995).

235. G. T. Yee, J. M. Manriquez, D. A. Dixon, R. S. McLean, D. M. Grroski, R. B. Flippen, K. S. Narayan, A. J. Epstein, J. S. Miller, *Adv. Mater.* **3**, 309 (1991).

236. (a) A. J. Epstein, J. S. Miller *Conjugated Polymers and Related Materials: The Interconnection of Chemical and Electronic Structure*, Proceedings of Nobel Symposium NS-81, W. R. Salaneck, I. Lündstrom, B. Rånby, Eds., Oxford University Press, Oxford, 1993, p. 475. (b) *Chim. Ind.* **75**, 185 (1993). (c) J. S. Miller, G. T. Yee, J. M. Manriquez, A. J. Epstein, in *Conjugated Polymers and Related Materials: The Interconnection of Chemical and Electronic Structure*, Proceedings of Nobel Symposium NS-81, W. R. Salaneck, I. Lündstrom, B. Rånby, Eds., Oxford University Press, Oxford, 1993, p. 461. (d) J. S. Miller, A. J. Epstein, *Mol. Cryst. Liq. Cryst.* **233**, 133 (1993).

237. (a) G. Du. J. Joo, A. J. Epstein, J. S. Miller, *J. Appl. Phys.* **73**, 6566 (1993). (b) B. G. Morin, C. Hahm, A. J. Epstein, J. S. Miller, *J. Appl. Phys.* **75**, 5782 (1994).

238. S. Mitea, A. K. Gregson, W. E. Hatfield, R. E. Weller, *Inorg. Chem.* **22**, 1729 (1983).

239. B. C. Conklin, S. P. Sellers, S. Fitzgerald, G. T. Yee, *Adv. Mater.* **6**, 836 (1994).

240. U. Welp, S. Fleshler, W. K. Kwok, G. W. Crabtree, K. D. Carlson, H. H. Wang, U. Geiser, J. M. Williams, V. M. Hitsman, *Phys. Rev. Lett.* **69**, 840 (1992).

241. T. Sugimoto, M. Tsujii, H. Matsuura, N. Hosoito, *Chem. Phys. Lett.* **235**, 183 (1995).

242. D. Milhailovic, private communication.

243. M. Tamura, H. Sawa, Y. Kashimura, S. Aonuma, R. Kato, M. Kinoshita, *J. Phys. Soc. Jpn.* **64**, 425 (1994).

244. A. Schilder, H. Klos, I. Rystau, W. Schulz, B. Gotschy, *Phys. Rev. Lett.* **73**, 1299 (1994).

245. J. D. Thompson, G. Sparn, F. Dierdrich, G. Grüner, K. Holczer, R. B. Kaner, R. L. Whetten, P.-M. Allemand, F. Wudl, *Mater. Res. Soc. Symp.* **247**, 315 (1992).

246. (a) K. Tanaka, A. A. Zakhidov, K. Yoshizawa, K. Okahara, T. Tamabe, K. Yakushi, K. Kikuchi, S. Suzuki, I. Ikemoto, T. Achiba, *Phys. Lett. A* **164**, 221 (1992). (b) K. Tanaka, A. A. Zakhidov, K. Yoshizawa, K. Okahara, T. Tamabe, K. Yakushi, K. Kikuchi, S. Suzuki, I. Ikemoto, T. Achiba, *Phys. Rev. B* **47**, 7554 (1993). (c) K. Tanaka, A. A. Zakhidov, K. Yoshizawa, K. Okahara, T. Tamabe, K. Yakushi, K. Kikuchi, S.Suzuki, I. Ikemoto, T. Achiba, *Solid State Commun.* **85**, 69 (1993). (d) K. Tanaka, T. Sato, T. Tamabe, K. Yoshizawa, K. Okahara, A. A. Zakhidov, *Phys. Rev. B* **51**, 990 (1995).

247. P. W. Stephens, D. Cox, J. W. Lauher, L. Mihaly, J. B. Wiley, P.-M. Allemand, A. Hirsch, K. Holczer, Q. Li, F. Wudl, J. Thompson, *Nature* **355**, 331 (1992).

248. R. Blinc, J. Dolinsek, D. Arcon, D. Mihailovic, P. Venturini, *Solid State Commun.* **89**, 487 (1994).

249. P. Cevc, R. Blinc, D. Arcon, D. Mihailovic, P. Venturini, S. K. Hoffman, W. Hilczer, *Europhys. Lett.* **26**, 707 (1994).

250. R. Blinc, P. Cevc, D. Arcon, D. Mihailovic, P. Venturini, *Phys. Rev. B* **50**, 13051 (1994).

251. A. Lappas, K. Prassides, K. Vavekis, D. Arcon, R. Blinc, P. Cevc, A. Amato, R. Feyerherm, F. N. Gygax, A. Schenk, *Science* **267**, 1799 (1995).

252. (a) A. A. Zakhidov, K. Yakushi, K. Imaeda, H. Inokuchi, K. Kikuchi, S. Suzuki, I. Ikemoto, Y. Achiba, *Mol. Cryst. Liq. Cryst.* **218**, 299 (1992). (b) A. A. Zakhidov, K. Imaeda, A. Ugawa, K. Yakushi, H. Inokuchi, Z. Iqbal, R. Baughman, B. L. Ramakrishna, Y. Achiba, *Physica C* **185–189**, 411 (1991).

253. (a) S. Yoshimura, in *Molecular Metals*, W. E. Hatfield, Ed., Plenum Press, New York, 1979, pp. 471–489. (b) R. S. Potember, T. O. Poehler, R. C. Benson, *Appl. Phys. Lett.* **41**, 548 (1982). (c) Y. W. Park, H. Lee, Eds., *Proceedings of the Conference on Science and Technology of Synthetic Metals*, *Synth. Met.* **71**, 2117–2272 (1995).

254. (a) J. S. Miller, *Adv. Mater.* **5**, 587 (1993). (b) J. S. Miller, *Adv. Mater.* **5**, 671 (1993). (c) J. S. Miller, *Adv. Mater.* **6**, 322 (1993).

CHAPTER 3

Advanced Polymeric Materials: Functional Electroactive Polymers

BASHIR M. SHEIKH-ALI[†] and GARY E. WNEK*
Department of Chemistry, Rensselaer Polytechnic Institute, Troy, New York

3.1 INTRODUCTION

Polymeric materials are serious candidates for a number of applications that require good electronic and/or ionic conductivity. As in the case with many materials systems, outstanding properties in one area (e.g., conductivity) is not necessarily a guarantee of practical utility, as several factors, such as cost, processability, mechanical integrity, and the quality of polymer–electrode contacts need to be broadly considered. Many of these issues are addressable with organic polymers as the result of the ability to fine-tune properties through backbone and/or chain functionalization. Indeed, the current resurgence in interest in electroactive (or electrically responsive) polymers stems from the clever application of synthetic chemistry to the design of useful materials. We limit our review to conjugated (electronically conducting) and ionically conducting polymers. A brief mention of other types of electroactive polymers, including field-responsive gels and electrostrictive materials, is given at the end of the chapter.

3.2 BRIEF STRUCTURAL OVERVIEW OF CONJUGATED POLYMERS

The structures of electronically conductive polymers are characterized by a backbone in which each atom is involved in a π-bond (an exception to this is

[†]Present address: Dais Corp., Palm Harbor, FL 34685
*Present address: Dept. of Chemical Engineering, Virginia Commonwealth University, Richmond, VA 23284-3028

Chemistry of Advanced Materials: An Overview, Edited by Leonard V. Interrante and Mark J. Hampden-Smith.
ISBN 0-471-18590-6 © 1998 Wiley-VCH, Inc.

polysilanes which are σ-conjugated systems). Defects and rotations around single bonds in the backbone impart a nonplanarity to the polymer and reduce delocalization of the π-electrons. This effect is manifested as a blue shift in the absorption spectrum of the conducting polymer when heated.[1]

Defect-free conjugated polymers can be considered as a one-dimensional system with one electron per carbon atom. The typical energy gap between the conduction and valence bands is 1 to 3 eV. Since it is highly unlikely that excitation of carriers would occur across a gap of 2 eV, these systems would be insulators or, at best, semiconductors. It has been shown that the conductivity of semiconducting polymers becomes metallic if they are doped with strong oxidizing or reducing agents. Introduction of dopants gives rise to strong electron–phonon interactions, leading to new quasiparticles: solitons, polarons, and bipolarons. The electrical conductivity of conducting polymers results from these mobile charge carriers.

From the point of view of metallic conduction, conducting polymers are not comparable to inorganics. The superiority of conducting polymers over their inorganic counterparts resides in their structural versatility, flexibility, and processibility. Ironically, one of the most challenging problems of conducting polymer technology is processibility. The presence of strong interchain electron transfer interactions, a desirable property for electronic conduction, makes them, nevertheless, frequently insoluble and infusible.

3.3 SYNTHESIS, PROCESSING, AND DOPING OF CONJUGATED POLYMERS

Since optical properties and oxidation/reduction potentials of mobile charge carriers vary for short chains, and eventually become invariant for longer ones, it is desirable to have control over the molecular weight of the conducting polymer. This control can be achieved by choosing an appropriate polymerization method. Polymerizations can be classified into step-growth and chain-growth reactions. Polyesters, polyamides, polyurethanes, polycarbonates, and many other commonly used polymers are synthesized by the step-growth polymerization method. The molecular weights of polymers prepared in this manner depend on the conversion, and high polymers are obtained only in conversions on the order of 99% or higher. Most organic reactions do not have yields higher than 99%, and this deficiency limits the applicability of step-growth methods to the preparation of conjugated polymers.

Many conjugated polymers are synthesized by chain growth. This method involves three steps: (1) initiation, (2) propagation, and (3) termination. The initiator reacts with a monomer, generating a reactive species that can react with another monomer and result in a new reactive species. This process (propagation) continues until the monomer is depleted, the polymer precipitates, or the growing end of the polymer is deactivated (termination). The rates of the three steps determine the character of the polymerization reaction. In some chain-growth reactions, the propagation step is much faster than the initiation ($R_p \gg R_i$), such that high polymers form very quickly, with little control over the molecular weight. In such cases (free-radical polymerization is a typical example), the chain is said to be "short lived." However, conditions can be established such that $R_i \gg R_p$ and all the chains are initiated at about the same time. If the termination step can be made much slower than propagation ($R_t \ll R_p$), the propagating species will "live" until the reaction is quenched, yielding strong control over molecular weight, with narrow molecular weight distribution.

Polyacetylene Polyacetylene (PA), $\text{-(}CH\text{)-}_n$, is the most extensively studied conjugated polymer. Although a large number of publications was devoted to the synthesis of PA,[2] detailed characterization of the polymer was inhibited by intractability until Shirakawa and co-workers were able to obtain it in a film form using soluble Ziegler–Natta catalysts.[3] In Shirakawa's method, acetylene gas is blown onto a solution of the catalyst [$Ti(OC_4H_9)_4$–$Al(C_2H_5)_3$, $Ti(acac)_3$–$Al(C_2H_5)_3$, and $Cr(acac)_3$–$Al(C_2H_5)_3$, typically 0.3 to 0.5 mol/L] and the thickness of the resulting film is controlled by the reaction time. A variant method, developed by Naarmann and Theophilou,[4] uses $Ti(OC_4H_9)_3$–$Al(C_2H_5)_3$ (with butyl lithium added to the aged catalyst in silicone oil) as catalyst, and the polymerization reaction is done at room temperature. Both methods and their modifications can be used to obtain a uniform film with good physical properties.

Since Shirakawa et al.[5] reported that the semiconducting PA becomes metallic (with several orders of magnitude increase in conductivity) upon exposure to vapors of halogens, a great deal of effort has been directed toward understanding the mechanism of the doping process.[6] Both electron donors and acceptors react with PA, forming highly conducting derivatives.[7] Structural data on doped PA indicate that the conducting derivatives are indeed ionic, hence doping appears to be a redox reaction.[5] The amount of the counterion that can build up in the crystal structure of PA limits the extent of doping. Dopants are generally singly charged: If PA is in its oxidized state, it is said to be *p*-doped, and the counterion is negatively charged. Halogen-doped PA falls into the category of *p*-type, and doping is carried out either with the halogen gas[5,8] or with their solutions in nonpolar solvents.[6a,9] The conductivity of the PA increases as the extent of doping is increased (see Fig. 3.1).

Strong reducing agents, such as alkali metals, generate *n*-doped PA. However, the low vapor pressure of alkali metals at room temperature limits their use as direct dopants. Direct doping of PA by alkali metals can, however, be achieved by immersing PA in a liquid sodium–potassium alloy.

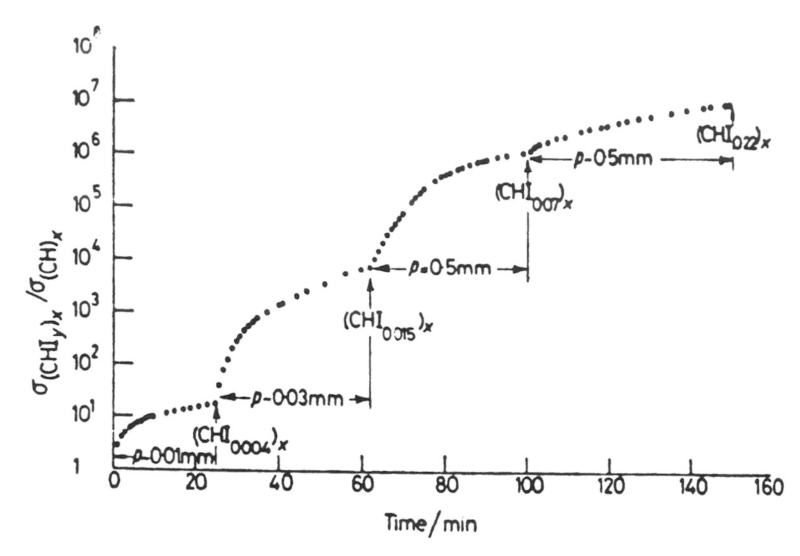

Figure 3.1 Conductivity of PA as a function of time at fixed iodine pressure. σ PA, undoped = 3.2×10^{-6} S/cm. (From Ref. 5.)

An elegant method that uses alkali salts (such as naphthalides) has been introduced by Heeger and co-workers.[7] In addition, alkali-doped PA contains no traces of metallic alkali metal, and the PA consists of polyenic anions.[10]

Polyaniline (PANI) Although polyaniline can be synthesized electrochemically, it is often obtained from aqueous acid solution of the monomer and an oxidant [e.g., $(NH_4)_2S_2O_8$] via a step-growth polymerization reaction.[11] Recently, Wei and co-workers introduced the concept of nonclassical chain polymerization[12] for the polymerization of aniline and related compounds (toluidine and other alkyl anilines), in which the process is neither classical step nor classical chain polymerization. The mechanism proposed[13] (Scheme 3.1) starts with the oxidation of the monomer to a radical cation, the rate-limiting step.[14] The radical cations dimerize, forming *p*-aminodiphenylamine, *N*,*N*′-diphenylhydrazine, and benzidine. Oxidation of these dimeric species (all have oxidation potential lower than that of aniline[14]) yields the corresponding quinoidal diiminium ions. Reactions between these quinoidal ions and aniline provide the growth step of the polymerization. Since the subsequent oligomers have lower oxidation potentials than aniline, growth proceeds, leading to the final polymer. Substantial enhancement of the polymerization rate has been observed when the rate-limiting step was eliminated by deliberately introducing the dimers. An observed lack of correlation between the oxidation potentials of the additives and the resulting reaction rates is explained in terms of steric accessibility of the aromatic amino groups. When steric factors are included, rate-enhancement effects of the additives are easily justified.

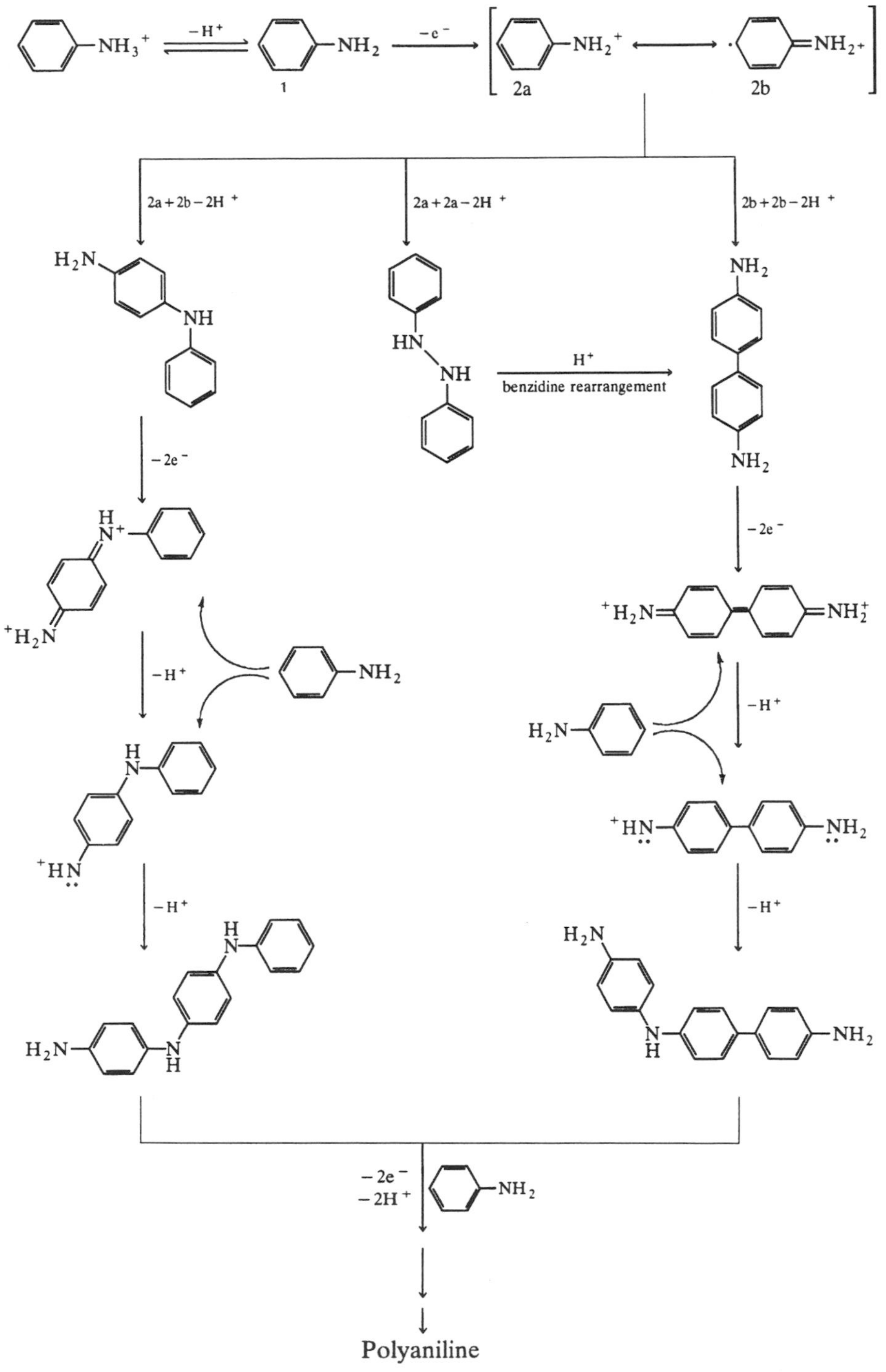

Polyaniline

Scheme 3.1 Wei's mechanism for polymerization of aniline.

PANI can exist in a variety of forms. It is soluble in *N*-methylpyrrolidinone (NMP) and concentrated strong acids (such as H_2SO_4, CF_3SO_3OH, and CH_3SO_3OH). Insulating films and fibers of the emeraldine base form of PANI has been made from NMP solutions of the polymer. A conducting emeraldine salt form of PANI can be processed from solutions in strong acids to form oriented films and fibers. The acid protonates the imine nitrogen, converting the quinoidal ring to a benzonoid structure. This modification leads to a structure with one unpaired electron per repeat unit. A "self-doped" PANI has recently been obtained by partial sulfonation of PANI.[15]

In addition, counterion-induced processibility of PANI has recently been introduced.[16] In this method, PANI is rendered soluble in common solvents by doping with an anionic surfactant such as dodecylbenzene sulfonate (DBS). The DBS/PANI complex was codissolved with polyolefins, nylon, polycarbonate, PS, PVA, PVC, PMMA, ABS, and so on, in a common solvent. Solution processed polyblends (e.g., fiber spinning, spin coating, dip coating) with 2% conducting polymer have electrical conductivities on the order of 1 S/cm, with no indication of a percolation threshold. More recently, a DBS/PANI complex was melt blended with conventional thermoplastic polymers.[17]

From electro-oxidation of aniline in aqueous H_2SO_4 solution using a constant anodic potential, aniline black is obtained as a powder that adheres poorly to the electrode. However, electroactive PANI films were obtained from electro-oxidation of aniline with a continuously cycling potential between -0.2 and $+0.8$ V (vs. SCE).[18]

Polythiophene (PT) Thiophene has been polymerized with a variety of catalysts: protic acids, Lewis acids, and Ziegler catalysts.[19] Catalytic coupling of Grignard reagents of 2,5-dihalothiophenes by nickel salts has also been reported.[20] An intractable black PT has also been obtained from thiophene in the presence of AsF_5.[19]

Poly(3-alkylthiophene) (P3AT) has been obtained from the Grignard reagent of the corresponding 2,5-dihalothiophenes with Ni salts as catalysts.[21] P3AT with alkyl side chains longer than butyl are both solution and melt processible. Polyblends with a percolation threshold of 5 to 10 wt % P3AT have been reported.[21, 22] Conductivities measured for P3AT (with hexyl or longer as a side chain) are about an order of magnitude smaller than that of PT (10^2 to 10^3 S/cm).

PT has also been obtained in a single-cell, three-electrode electrochemical system[23] with Pt-, Au-, or In_2O_3-coated glass as a working electrode, Pt, Ni, or C as a counterelectrode, and SCE as a reference electrode. The electrolyte medium typically consists of an organic solvent, a supporting electrolyte, and the monomer, degassed thoroughly before electrolysis. Since PT has a lower oxidation potential than thiophene, an electrochemically generated PT is obtained in its oxidized conducting state. The film thickness is controlled by electrolysis time.

Polypyrrole (PPy) Polypyrrole (PPy) is typically synthesized electrochemically from an aqueous, aqueous/organic, or organic solution of pyrrole.[24] Both single-compartment and two- or three-electrode electrochemical cells have been used, with a PT- or indium–tin oxide-coated glass slide as the anode. A thin bronze-to-blue PPy film, deposited on the working electrode, is removed with a razor blade. The working electrode can be of any material that is not electroactive at the operating potential of 0.5 V. Significant amounts of oxygen (from aqueous electrolyte solutions) and hydrogen (from organic electrolyte solutions) incorporation have been reported for electrochemically synthesized PPy films. Almost irrespective of the electrolyte medium, electrochemically synthesized PPy has an electrical conductivity of 40 to 100 S/cm.

Poly(p-phenylenevinylene) (PPV) Scheme 3.2 shows a synthetic procedure for PPV from a soluble prepolymer as reported by Burroughes and co-workers.[25] The PPV film synthesized following this method (originally by Wessling and Zimerman[26]), however, suffers from several disadvantages as far as electroluminescence (EL) device performance is concerned (PPV is the most used polymer in EL devices; see below). The tetrahydrothiophene group is readily displaced by a hydroxy group,[27] which is then oxidized to a carbonyl group.[28] This results in potentially uncontrollable polymer properties and performances. Also, the method produces a predominantly *trans*, and therefore crystalline, form of PPV.

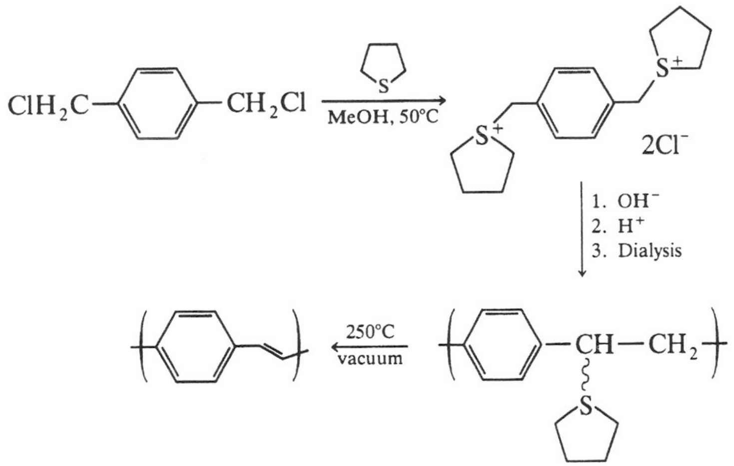

Scheme 3.2 Synthesis of PPV.

Son et al.[29] have introduced a variant method of synthesis of PPV that uses the same principle of a soluble prepolymer but eliminates the problematic step that introduces the hydroxyl groups. This method produces a mixture of *cis* and *trans*-PPV, and thus amorphous material (Scheme 3.3). The *cis*-linkages serve as the controlling factor in determining the conjugation length.

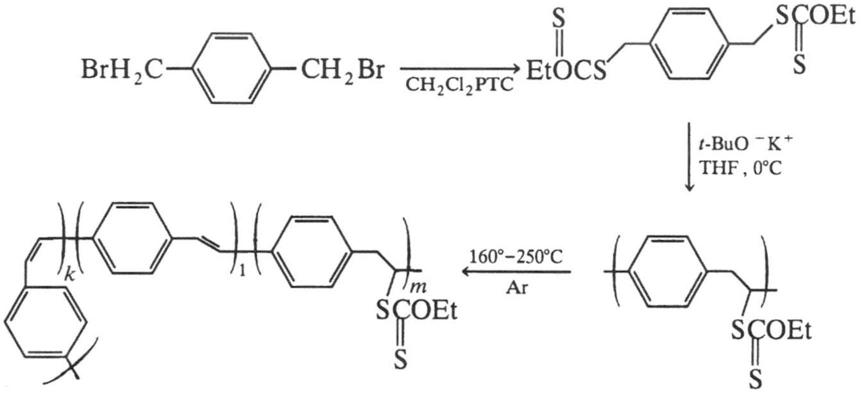

Scheme 3.3 Synthesis of amorphous PPV.

3.4 IONICALLY CONDUCTING POLYMERS

Polymer-based ionic conductors (i.e., polymer–salt complexes) have been a subject of great interest since the pioneering studies of materials based on alkali metal salts and polyethylene oxide (PEO) by Wright and co-workers.[30] The majority of the work on polymer electrolytes has concentrated on PEO. At room temperature, high-molecular-weight PEO exists as a mixture of crystalline and amorphous regions. Since ionic conduction is predominantly through the amorphous phase,[31] PEO has limited applicability for the development of electrochemical devices. At room temperature, PEO-based electrolytes (with Li^+) have typical conductivities of 10^{-8} S/cm. Reasonably high conductivity (10^{-5} S/cm) is obtained only at temperatures near 100°C. Several salts (including RbSCN, RbI, CsSCN, and CsI) form amorphous complexes.[32] However, the interest in polymer electrolytes lies in lithium-polymer-based electrochemical devices (see below). A comprehensive review on solid polymer electrolytes has been published.[33]

Since the dielectric constant of PEO ($\varepsilon \sim 5$) is comparable to that of diethyl ether ($\varepsilon = 4.2$), the observed low conductivity of Li-PEO electrolytes (about 10^{-8} vs. $> 10^{-3}$ S/cm for liquid-ether electrolyte) is attributed to the low mobility of ions in the polymer matrix rather than charge carrier concentration.[34] In an attempt to improve the Li^+ conductivity of polymer-based electrolytes, several approaches have been explored, including (1) blending (i.e., the addition of a plasticizer in the form of a small molecule or liquid oligomer that gels the host polymer), (2) enhancement of chain flexibility, and (3) reducing the crystallinity by incorporating secondary units in the PEO chain.[35]

Using propylene carbonate (PC) as a plasticizer, Cameron et al.[36] have shown that the conductivity of PEO increases continuously as the concentration of PC is increased. Because of the higher permittivity, PC was thought to decrease ion pairing and decouple the ionic motion from that of the polymer chains.

In addition, polymer electrolytes with low T_g and improved ionic conductivity can be obtained by increasing the main-chain flexibility. Polyphosphazene-based polymers of the type shown below have been synthesized in several laboratories.[37] The most successful devices using such electrolytes are those obtained from poly(methoxyethoxy ethoxyphosphazines) (MEEP).[37] MEEP is obtained by reacting the sodium salt of 2-(2-methoxyethoxy)ethanol with poly(dichlorophosphazine). The modest enhancement in conductivity ($\sigma \sim 10^{-4}$ S/cm) of MEEP is undermined by the poor physical properties of the polymer. In fact, MEEP is a highly viscous liquid that flows under pressure.[38] Tonge and Shriver[39] have described a chemically cross-linked MEEP by addition of 1 mol% poly(ethylene glycol) (PEG) during the derivatization of the parent polymer; the conductivities of cross-linked MEEP are comparable to that of the corresponding uncross-linked polymer. Radiation cross-linking of MEEP[40] does not affect the conductivity and improves the physical properties of the polymer. The dimensional stability of MEEP has been observed to increase by forming composites with more rigid polymers (such as PEO, PVP, etc.) that also complex Li salts.[38] A MEEP/PEO–$LiCF_3SO_3$ electrolyte system[41] has the same conductivity as that of the MEEP/PEO–$LiN(SO_2CF_3)_2$ system[34] ($\sigma = 6.7 \times 10^{-5}$ S/cm at 20°C). This value is slightly higher than that of a radiation cross-linked MEEP/PEG–$LiCF_3SO_3$ electrolyte system ($\sigma = 4.1 \times 10^{-5}$ S/cm at 30°C).[39]

Polysiloxane-based polymer electrolytes have also been explored. Zinc octanoate[42]- or triethylamine[43]-catalyzed hydrosilation of poly(methyl siloxane) with ω-methoxyoligo(oxyethylene)ethanol yields PMMS-m, which has a conductivity of 10^{-4} S/cm. However, the hydrolytic susceptibility of Si—O—C bond limits the practical application of polysiloxane-containing electrolytes. To circumvent this problem, Fish and co-workers[44] synthesized poly[(ω-methoxyoligo(oxyethylene)propyl methylsiloxane]s (PAGS-m), which have conductivities similar to those of PMMS-m but lack Si—O—C bonds. PAGS-m is prepared by hydrosilating poly(methylsiloxane) with an allyl ether, $(CH_2{=}CHCH_2O(CH_2CH_2O)_nCH_3$, in the presence of platinum complex catalyst.

Poly {[3-[2,3-(carbonyldioxy)propoxyl]propyl]methylsiloxane} (PCPS), a polysiloxane with a cyclic carbonate on the side chain, has recently been reported by Zhu et al.[45] The high dielectric constant associated with these cyclic carbonates was anticipated to increase ion-pair dissociation and thus ionic conductivity. However, PCPS containing $LiSO_3CF_3$ has a rather low conductivity at room temperature but exhibits a conductivity of 1.7×10^{-6} S/cm at 40°C. The conductivity increases by two orders of magnitude when the polymer is plasticized with acetonitrile.

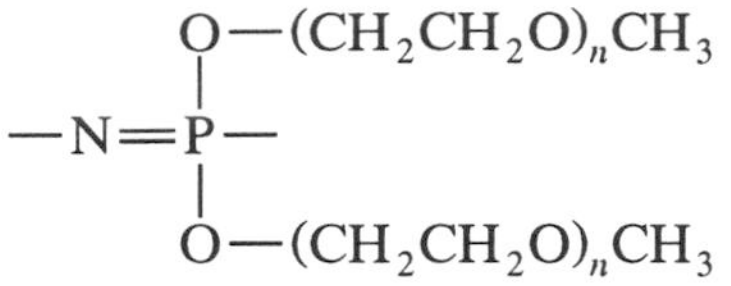

$$-N{=}P(O{-}(CH_2CH_2O)_nCH_3)_2-$$

$n = 3$; MEEP

$$-Si(CH_3)(O{-}(CH_2CH_2O)_nCH_3){-}O-$$

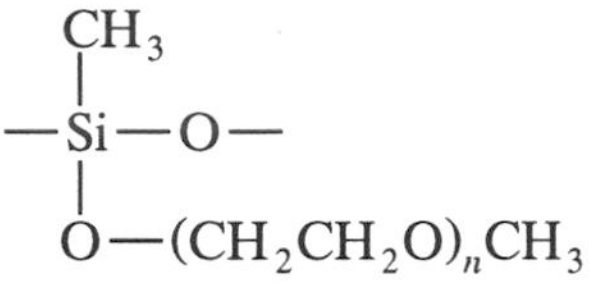

PMMS-m; m = $n + 1$

$$-Si(CH_3)((CH_2)_3{-}O{-}(CH_2CH_2O)_nCH_3){-}O-$$

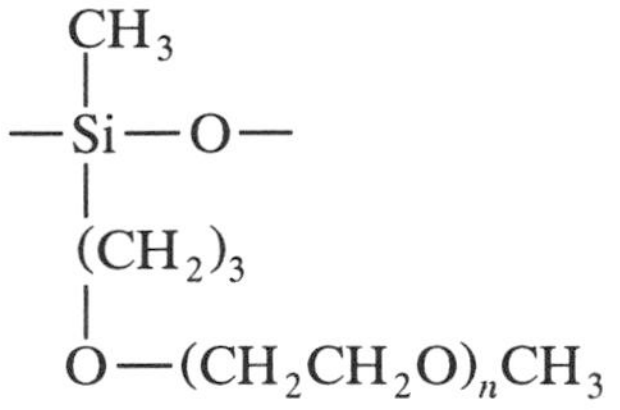

PAGS-m

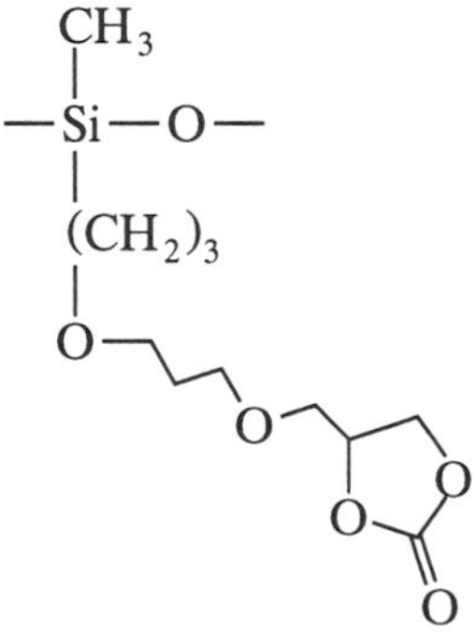

PCPS

Conductivities of about 10^{-3} S/cm and higher, similar to values observed for liquid-ether electrolytes, have been reported for gelled and plasticized polymer electrolytes. For instance, the gel electrolyte containing poly(acrylonitrile) (PAN), ethylene carbonate (EC), PC, and $LiClO_4$ or $LiN(SO_2CF_3)_2$ have conductivities of 1.7 and 1.8×10^{-3} S/cm at 20°C, respectively.[46] PVC-PC-EC-$LiClO_4$ gel electrolyte (15 : 40 : 40 : 5 weight ratio) has a conductivity of 1.2×10^{-3} S/cm at 20°C.[47] 3 M Li^+ Nafion 1100 in a 50 : 50 PC/dimethoxyethane mixture exhibits a conductivity of 7×10^{-4} S/cm at 25°C.[48] Polymer electrolytes based on PAN or PVC with EC and PC plasticizing agents that contain $LiAsF_6$ or $LiN(SO_2CF_3)_2$ exhibit conductivities in the range 1 to 2×10^{-3} S/cm at 20°C.[34] However, gelled or plasticized blends tend to give electrolytes with conductivities in the range 10^{-5} to 10^{-7} S/cm near room temperature.[49]

A Li-conducting ionic rubber with a low T_g value and conductivity as high as 10^{-2} S/cm at room temperature has been described by Angell and co-workers.[50] Although this ionomer has a wide electrochemical window, it contains $LiClO_3$ or $LiClO_4$, neither of which is suitable for lithium-ion battery applications. It would be interesting if one could find a chemically and electrochemically stable replacement for these salts and preserve the rubbery property of the electrolyte. A truly solid-state battery requires the replacement of the presently used pseudosolid electrolytes (polymer electrolytes either plasticized or gelled with low-molecular-weight liquids) by a polymer electrolyte with high ionic conductivity and a wide electrochemical window.

3.5 APPLICATIONS OF CONJUGATED POLYMERS

3.5.1 Light-Emitting Diodes

Although not used for device fabrication, electroluminescence (EL) from organic semiconductors has been known for a considerable time.[51] Since Tang et al.[52] have reported efficient (0.025 photon/electron) electroluminescence from an organic heterojunction using tris(8-hydroxy)quinoline aluminum (Alq), organic light-emitting diodes have attracted considerable interest for display applications. A typical EL device is made of a semiconducting film (inorganic, organic, or polymeric) sandwiched between a metal electrode (typically of low work function, such as calcium) and a semitransparent electrode (typically, an ITO-coated glass plate). When bias voltage is applied, electrons are injected from the metal electrode and holes from the high-work-function ITO electrode. Recombination of the electrons and holes injected from opposite electrodes excites the luminescent material to an excited singlet state. Radiative decay of the excited singlet state results in emission that corresponds to the energy gap of the semiconductor.

Burroughes et al.[25] first demonstrated that PPV can be used in EL device fabrication. Since then, a large number of publications have been devoted to the search for semiconducting polymers that will exhibit EL.[53] Prior to the use of macromolecules in EL devices, inorganic and low-molecular-weight organics had been investigated for EL applications. Inorganic semiconductors are the most efficient of all EL materials studied (about 3%) but suffer from poor stability and require relatively high ac drive voltage. Low-molecular-weight organic semiconductors, on the other hand, have efficiencies of 1 to 2% under low-dc-voltage conditions, as well as better tunability than that of inorganic systems but are prone to crystallization.

In the EL device of Burroughes et al.,[25] PPV was synthesized from a soluble prepolymer as shown in Scheme 3.2. A film of the prepolymer deposited on an electrode was converted thermally to PPV, resulting in a homogeneous, dense, and uniform film about 100 nm thick. The second electrode was deposited on the thermally stable PPV. For electron injection, a low-work-function material (such as aluminum, magnesium–silver alloy, or amorphous silicon–hydrogen alloy) was used, and high-work-function materials (such as oxidized aluminum, gold, or indium oxide) were used as hole-injecting electrodes. Figure 3.2 shows the current–voltage characteristics of an EL device with indium oxide and aluminum as electrodes. The EL spectra of the device at various temperatures is presented in Fig. 3.3. In view of the

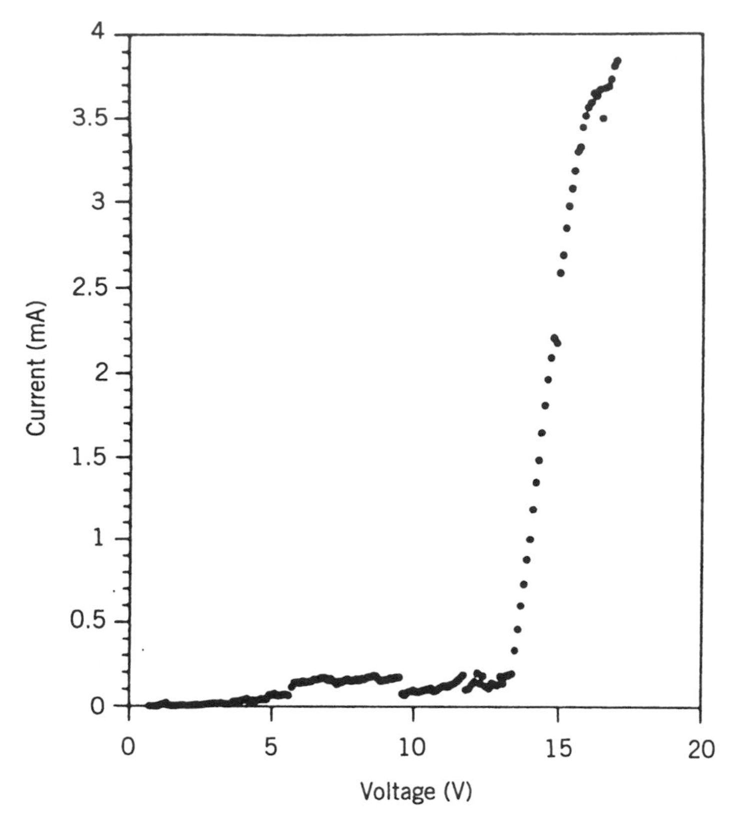

Figure 3.2 Current–voltage characteristic (forward-bias regime; indium oxide positive with respect to the aluminum electrode) for an EL device fabricated from a 70-nm-thick PPV, and indium oxide and aluminum electrodes. (From Ref. 25.)

similarity between the photoluminescence and EL of PPV, the two bands presumably originate from the same excited state. The typical quantum efficiency for this device was reported to be about 0.01%, about two orders of magnitude smaller than the photoluminescence of PPV. In contrast, EL devices (ITO-coated glass and aluminum as electrodes) fabricated from Son's[29] PPV (converted at 170°C) has an efficiency of 0.22%, about 20 times higher than that reported by Burroughes et al.[25] At higher conversion temperatures for PPV, the device efficiency drops, presumably due to higher *trans* content.

Conducting polymers provide the required good charge transport properties as well as structural stability, flexibility, and processibility. The EL quantum efficiencies of conducting polymers are, however, an order of magnitude smaller than those of low-molecular-weight organics. This ineffi-

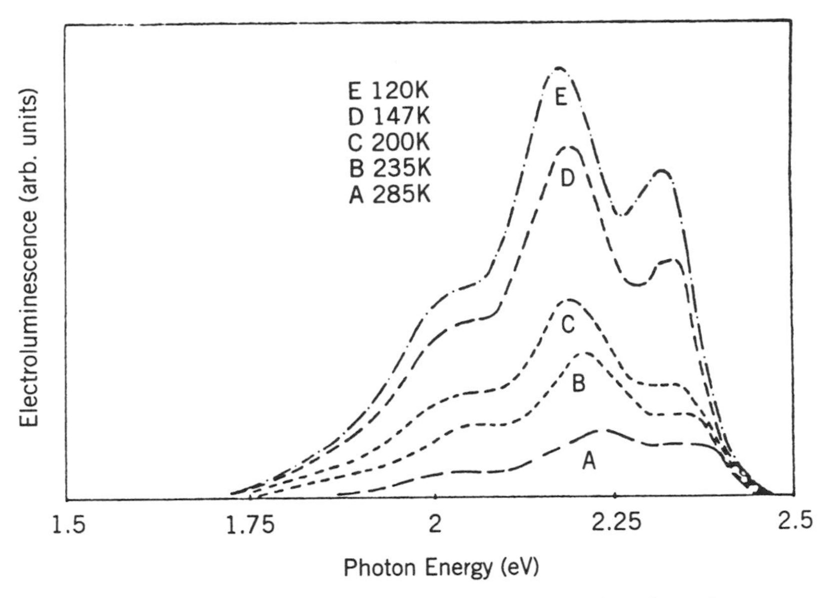

Figure 3.3 Luminescence spectra of the EL device as a function of temperature. (From Ref. 25.)

ciency is presumably due to the high barrier to electron–hole injection at the polymer–electrode interfaces. Low-work-function electrodes such as calcium do yield higher efficiency (0.2%),[53b] but susceptibility to atmospheric degradation makes their use impractical.

Nevertheless, use of hole and electron transport layers in conjunction with the light-emitting layer allows for a reasonable efficiency.[54] For instance, insertion of an electron transport layer such as 2-(4-biphenyl)-5-(4-*t*-butylphenyl)-1,3,4-oxadiazole (PBD)[55] between the light emitter and the electron-injecting electrode increases the device efficiency by a factor of 5 to 10. A bilayer EL device with a poly(cyanophthalylidene) (PCP) layer inserted between PPV and the electron-injecting electrode exhibits a red emission from PCP at a significantly reduced field and with a quantum efficiency of 4%.[56] The comparable efficiencies obtained with aluminum and calcium as the electron-injecting electrodes indicate that charge injection is improved in the bilayer device. Thus PPV acts as a hole-transporting layer and PCP as an electron-transporting layer as well as the emitting layer. This result indicates that exciton transfer is probably occurring and emission is from the lowest-energy traps present (PCP). PPV is, however, generally used as the light-emitting (active) layer.[25, 57]

Yang and Heeger[58] have reported an improved EL device efficiency (by about 30 to 40%) and with reduced operating voltage (by about 30 to 50%) when the ITO-glass hole-injecting electrode is replaced with PANI. In this device, poly[2-methoxy-5-(2′-ethylhexyloxy)-1,4-phenylenevinylene] (MEH-PPV) is the light-emitting layer. The barrier height at the interface between MEH-PPV and the hole-injecting electrode is estimated to be twice as large in an ITO (about 0.2 eV) as in a PANI (about 0.1 eV) electrode.

A very important aspect of fabricating LEDs is the ability to tune the color of emitted light. Since the emitted light is due to an electron jump from a π^* state to a π state, color tuning is essentially synthesizing polymers with the desired π–π^* energy gap. Polymers that emit in the red [poly(2-methoxy-5-(2′-ethylhexoxy-*p*-phenylenevinylene[53b] and poly(cyanophthalylidene)[56]], yellow-green (PPV[25]), blue [poly(*p*-pyridine)[59]], violet-blue [poly(*N*-vinylcarbazole),[60] poly(phenylene) with broken conjugation either by meta linkages or steric reasons,[61] and stepladder poly(*p*-phenylenes)[62]], yellow-orange [poly(*p*-pyridylvinylene)[63]], orange-red (poly(3-alkylthiophenes)[64]], and green [poly(3-cyclohexythiophene)[65]] have been reported. A different approach for developing an EL device with a specific color has been suggested by Osaheni and Jenekhe.[66] This approach involves use of conjugated polymer exciplexes to enhance and tune the fluorescence behavior of the polymer. By using a bilayer system comprising a layer of poly[*p*-phenylene-2,6-benzobis(oxazole)] (PBO) and a second one of tris(*p*-tolyl)amine (TTA) dispersed in polycarbonate, they observed a blue emission (λ_{max} = 474 nm) with an enhanced fluorescence quantum yield. By comparison, the PBO and TTA layers emit at 500 nm (green) and 380 nm (UV), respectively.

Burn et al.[67] have reported copolymers that allow the wavelength of the emitted light to be varied from the blue–green to the orange–red region of the spectrum. These copolymers were synthesized from a PPV-precursor polymer with two different leaving groups (methoxy and sulfonium); control of the conversion conditions afforded the preferential removal of one to produce conjugated segments isolated by saturated units.

By controlling (with steric hindrance of side groups) the torsion of the main chain and thus the conjugation length of 3-substituted poly(thiophenes), Berggren and co-workers[68] were able to obtain electroluminescence polymers with colors ranging from blue to near-IR, with green, orange, and red as intermediate steps. LEDs incorporating blends of different 3-substituted PTs were shown to exhibit emission of each different polymer. The relative intensities of the peaks are determined by the voltage applied to the diode. The presence of exciton transfer (from the higher band polymer to one of the lower band gap) in the blends, however, limits the color control.

A white-light-emitting organic electroluminescence device has recently been reported by Berggren and co-workers.[69] This device was fabricated from an electron transport layer (PBD) and a luminescent polymer [poly(3-(4-octylphenyl)-2,2′-bithiophene)] sandwiched between ITO-coated glass and an aluminum electrode. The luminescence spectrum consists of peaks at 410 (blue), 530 (green), and 610 nm (yellow–orange) which to the eye appear as bluish white.

The ability of random-coil polymers to become anisotropic by stretching has been exploited in an attempt to make polarized LE devices. The first such device was reported by Dyreklev and co-workers.[70] When a film of poly[3-(4-octylphenyl)-2,2′-bithiophene] is elongated to twice its original length, the ratio of the intensities of the light emitted parallel and perpendicular to the elongation (orientation) axis is 2.4 (Fig. 3.4). Unidirectional rubbing of the emitting layer of an EL device was shown to induce molecular orientation. Hamaguchi and Yoshino[71] used this technique to fabricate a polarized EL device. In this method, poly(2,5-dinonyoxy-1,4-phenylene-vinylene) was spin coated on ITO-coated glass plate. After drying, the film was rubbed with rayon cloth. The dichroic ratio of the EL device was evaluated to be 1.6 from electroluminescence measurements.

Heeger and co-workers[72] recently introduced a polymer light-emitting electrochemical device that uses MEH-PPV, which is luminescent in its neutral state and is capable of both *p*- and *n*-type doping. An ionically conducting polymer (PEO) provides the required ion mobility for the device. A 1 : 1 (by weight) blend (about 2500 Å) of MEH-PPV/PEO that contains $LiCF_3SO_3$ is sandwiched between an ITO-coated glass and aluminum electrodes. Propagating electrons (generated at the anode) meet the holes in a region that defines the electrochemically induced *p-n* junction. Annihilation of the electron–hole pair results in excitation followed by radiative decay.

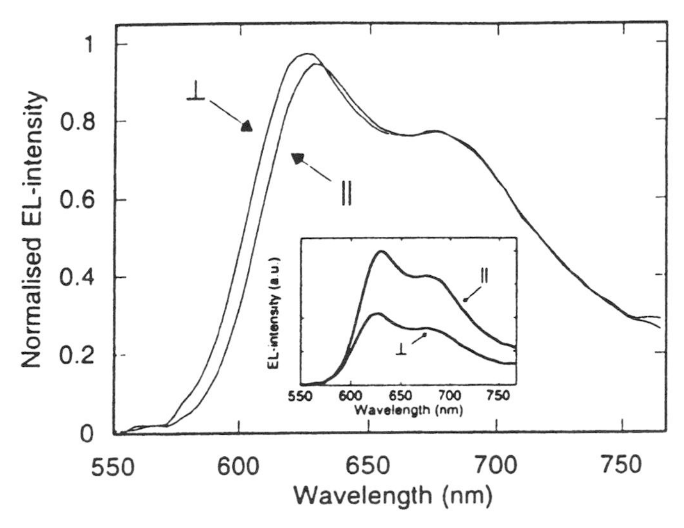

Figure 3.4 Normalized luminescence spectra from an oriented PTOPT EL device. Absolute intensity spectra are shown in the inset. (From Ref. 70.)

3.5.2 Electronically Conducting Polymers in Lithography

The precise registration capability of electron-beam writing in lithography is limited by the charging of the resist substrate. The delay in bleeding off the accumulated charge through the underlying silicon wafer causes the image quality to deteriorate.[73] Recently, Angelopoulos and co-workers[74] and Huang[75] demonstrated that an electrically conducting organic polymer [PA[74] or poly(3-alkylthiophenes), P3AT[75]] underlayer between the silicon wafer and the resist eliminates this charging problem. When an interlayer conductor is not present, a pattern displacement (with respect to a reference pattern) of as much as 5 μm is observed. Incorporation of a nonconducting emeraldine PA or undoped P3AT between the resist and the wafer does not eliminate the charging problem. However, zero-pattern displacements are obtained when a 0.1 to 0.2-μm conducting PA or P3AT interlayer is present. A conductivity of 10^{-4} S/cm for the interlayer appears to be the minimum requirement for the complete elimination of pattern displacements.

3.5.3 Electronically Conducting Polymer-Based Sensors

The electronic conductivity of conjugated polymers decreases as much as 10^5 or more upon introducing defects and rotations around single bonds that impart nonplanarity of the polymer backbone.[76] Swager and co-workers have recently shown that this large dynamic range can be exploited for sensor

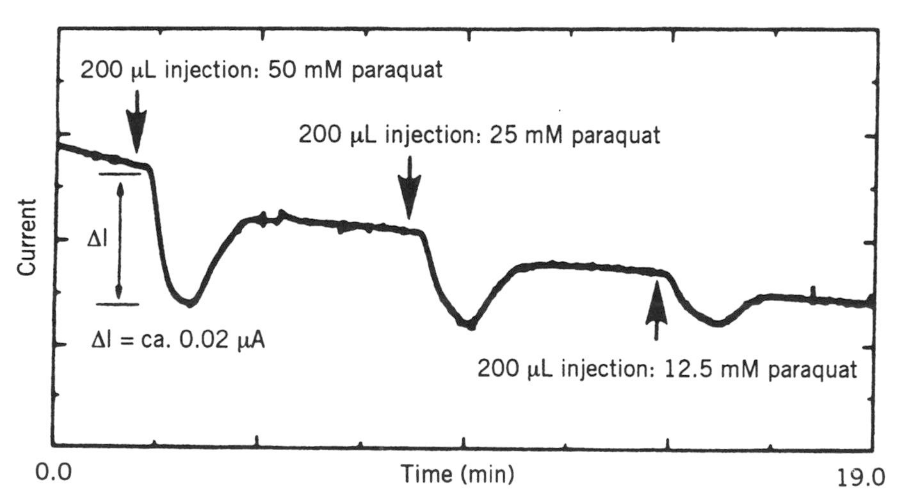

Figure 3.5 Chemoresistive response (current versus time; 50-mV offset potential) of a flow cell device fabricated from CEPT electrochemically oxidized to -0.1 V (vs. Ag wire) when 200 μL of paraquat solutions of variable concentrations is injected into the mobile phase [$0.1 M \, LiCO_4$/MeCN : H_2O (1 : 1), 0.2 mL/min flow rate]. (From Ref. 77c.)

applications.[77] Central to the proposed devices is the attachment of molecular recognition sites (crown ether) to the conducting polymer backbone (PT). The choice of crown ether–containing polythiophenes (an example, CEPT, is shown below) as chemoresistive sensory materials are based on the high conductivity, environmental stability, and relative ease of structural modification of PT.

The arrival of the chemical to be sensed at a molecular recognition site would induce a conformational change. Because of the cooperative nature of carrier transport in conducting polymers (each receptor-containing repeat unit is "sampled" by the carrier in the conduction process), the resistivity of the molecular wire is increased by the occurrence of a single molecular recognition event. Figure 3.5 shows the chemoresistive response of a flow cell device fabricated from CEPT electrochemically oxidized to -0.1 V (vs. Ag wire). Injection of a solution of paraquat clearly produces chemoresistive responses that qualitatively correspond to its concentration. A slight loss of electrochemical activity is indicated by the sloping of the baseline. Despite these shortcomings, the potential for these materials as real-time sensory devices is evident.

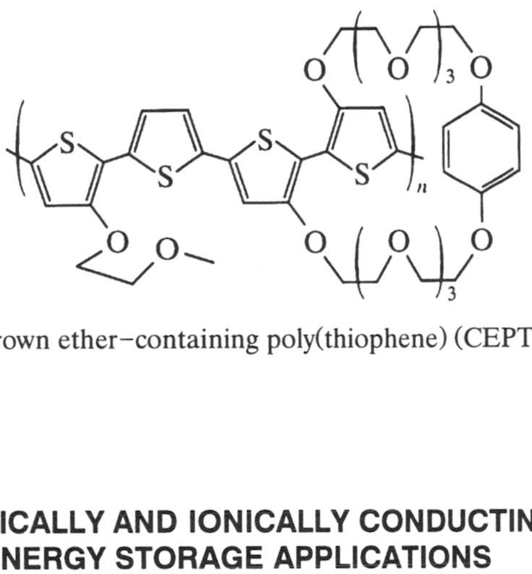

Crown ether–containing poly(thiophene) (CEPT).

3.6 ELECTRONICALLY AND IONICALLY CONDUCTING POLYMERS IN ENERGY STORAGE APPLICATIONS

Batteries are devices in which electrical energy is generated as a result of an electrochemical reaction. A secondary battery is one in which the chemical–electrochemical reactions are reversible, and as a result, the battery is rechargeable. Rechargeable batteries generally consist of two electrodes and an electrolyte/separator.

Nigrey et al.[78] first reported that electrochemical doping of PA is a reversible process and that PA can, in principle, be used as a cathode as well as anode in a rechargeable battery. Chiang [79] described a cell that uses PA as the electroactive material and PEO–NaI as the solid electrolyte: $(CH)_x/P(EO)_5–NaI/(CH)_x$. These discoveries initiated the search for conducting polymers for secondary batteries.[80] In a somewhat parallel development, it was recognized that ionically conducting polymers could be used as electrolytes in batteries if their conductivities could be raised to the level commonly obtained with liquid electrolytes (typically, 10^{-3} S/cm). In this section we discuss several current examples of batteries that use either of the two types of conducting polymers (electronic or ionic) or a combination of them.

Organic conducting polymers have several advantages over inorganic electroactive materials such as higher cell voltages (2 to 4 V vs. 1 V), higher energy density, longer lifetime, potentially less toxicity, and ease of molding to the desired shape. Other factors that should be considered are cost, rechargeability, cycleability, and energy density.

The need to store more energy per unit weight (or volume) demands use of light materials with high potentials, such as lithium.[81] Lithium has a high standard electrode potential but is reactive to solution/atmospheric moisture, and batteries using them must be fabricated from nonaqueous electrolytes in a water-free environment. Typically, a lithium polymer electrolyte battery is fabricated from a lithium-ion conducting membrane (separator and electrolyte) sandwiched between a Li metal strip and a composite cathode.

The prospects of developing new rechargeable batteries from conducting polymers have been explored at several laboratories. While the advantages of conducting polymers over inorganic counterparts have not materialized fully, commercial development of a polymeric secondary battery by BASF/VARTA[82] is encouraging. This battery can be written schematically as $Li/LiClO_4$–PC/PPy. The cathodic reaction is the reduction of oxidized PPy to neutral PPy. A 10% cathode loss after 400 cycles (which increases to only 20% loss after 1000 cycles) and an energy density of 30 Wh/kg at discharge times of 20 h has been reported for this cell.

Kakuda et al.[83] recently reported a lithium/PPy battery that uses a PAN-based gel electrolyte (Li/PAN-PC-EC/PPy). What is remarkable about the combination of this gel electrolyte with electropolymerized PPy as a cathode is that the electropolymerization potential of PPy does not affect the battery's performance. Depending on the electropolymerization potential, PPy with a rough and porous surface (4.2 V vs. Li/Li^+ polymerization potential) or a dense surface (3.9 V vs. Li/Li^+) are obtained. EC/PC components of the gel electrolyte presumably soak into the PPy film and enhance the transfer of ClO_4^- anions in and out of the polymer, irrespective of the morphology of the surface. Passivation of the lithium–electrolyte interface has been cited as a possible source of the observed decline of battery efficiency.

Recently, Oyama et al.[84] described a battery that uses a gel electrolyte [poly(acrylonitrile-co-methylacrylate) (PAcoMA) and EC] and a composite cathode that contains PANI, written schematically as

$$\text{Li/EC–PAcoMA–LiBF}_4\text{/DMcT–PANI–C}$$

where DMcT,2,5-dimercapto-1,3,4-thiadiazole, is also a cathode-active material. DMcT polymerizes upon oxidation and depolymerizes upon reduction.

$$n\left(\text{LiS–}\underset{\text{S}}{\overset{\text{N–N}}{\diamond}}\text{–SLi}\right) \underset{+e^-,\ +Li^+}{\overset{-e^-,\ -Li^+}{\rightleftharpoons}} \left[\text{S–}\underset{\text{S}}{\overset{\text{N–N}}{\diamond}}\text{–S}\right]_n$$

DMcT_{red} $\qquad$ DMcT_{ox}

The cathode (DMcT-PANI-C) is prepared by casting a thin film of a DMcT/PANI mixture in NMP on a carbon film. The film was then dried in a vacuum oven at 80°C for 1 h (about 30% NMP is retained). A film of the polymer electrolyte (a solution of EC, PC, PAcoMA, and $LiBF_4$ in acetonitrile cast on a glass plate and dried under vacuum at 60°C) is sandwiched between a lithium foil and the composite cathode. The observed capacity of the composite cathode exceeds the theoretical value of PANI in the modified electrode by a factor of more than 2, confirming that not only PANI but also DMcT functions as a cathode-active material. After 100 cycles, 80 to 90% of the initial capacity is retained. The low physical density of the cathode ($1.5\ g/cm^3$), however, will offset the high gravimetric energy density, making its application limited to devices where weight rather than volume is critical. An undesirably low maximum current density of about $0.1\ mA/cm^2$ was attributed to the decreased conductivity of PANI at potentials above 3.8 V vs. Li/Li^+.

Lithium batteries, however, suffer from uncontrollable side reactions at the negative electrode interface, the most troublesome of which is the nonuniform lithium plating during the charging process, which can lead to total failure and, in some cases, hazards. The future of Li-polymer batteries in the commercialization of electric vehicles has been discussed.[85] The dendritic nature of a plated lithium electrode has been cited as a possible source of a chain of events that eventually could lead to an explosion. Another disadvantage cited is the higher manufacturing cost (due to large electrode area per ampere-hour and extremely tight tolerance levels to produce thin pinhole-free, metal plates) compared to that of competing batteries (such as Ni–Cd).

The most effective way of overcoming these problems is by replacing the metal with an insertion compound that is capable of accepting lithium ions. In this case the anode accepts Li^+ during the charging process and releases them during discharge. This type of batteries is called a lithium-ion or rocking-chair battery (lithium ions rock from one electrode to the other).[86] Carbon-type insertion compounds are the most commonly used lithium-sink negative electrodes in lithium-ion batteries. The overall process of charge–discharge is

$$C_6 + LiMO_2 \underset{\text{discharge}}{\overset{\text{charge}}{\rightleftharpoons}} Li_xC_6 + Li_{1-x}MO_2$$

Croce et al.[87] reported a lithium-ion battery that uses a PAN-based gel electrolyte (PAN–EC/PC–$LiClO_4$) with TiS_2 as an anode (unlike the more commonly used Li_xC_6 lithiated carbon) and $LiCoO_2$ as a cathode. The cell has an open-circuit voltage of 0.3 V. After a charging current of 0.02 mA cm^{-2}, a charged state voltage of 2.1 V is obtained. The electrolyte has an electrochemical stability that exceeds 5 V (vs. Li). The charge–discharge behavior of the electrodes was studied using liquid[88] ($LiAsF_6$–acetonitrile) and inorganic solid electrolytes.[89]

$$TiS_2 + LiCoO_2 \underset{\text{discharge}}{\overset{\text{charge}}{\rightleftharpoons}} Li_xTiS_2 + Li_{1-x}CoO_2$$

A major advantage of lithiated TiS_2 electrodes over their carbon counterparts is less voltage fluctuation over cycling in the former than in the latter. Also, TiS_2 can accept more than one equivalent of Li^+, ensuring that slight overcharge does not cause short-circuiting and consequent hazards. However, a decline in capacity upon cycling (due to cathode loss) is a major drawback for rocking-chair batteries.

3.7 OTHER ELECTROACTIVE POLYMER SYSTEMS

Polymeric electrolyte gels immersed in an electrolyte solution deform under an electric field (i.e., they are capable of transducing a stimulus into mechanical energy). Since the electrochemical phenomenon in synthetic polymer gels was first observed in 1982 by Tanaka et al.,[90] several workers have studied the volume changes and deformation behavior of polymer gels in an electric field. Applications such as artificial muscles,[91] actuators,[92] and controlled drug delivery[93] for these gels have been explored. The polymers used are mostly random copolymer or polymer blends based on acrylic acid or sulfonic acid. The major problems with these polymers are their weak mechanical strength and long copolymer response time. More recently, the electrochemical behavior of a novel block copolymer gel, sulfonated styrene–ethylene–butylene–styrene, has been investigated.[94]

Polyvinylidene fluoride (PVDF or PVF2) continues to be a focus of attention as a prototypical polymer exhibiting usefully large piezoelectric, pyroelectric, and ferroelectric responses.[95] This polymer also exhibits electrostriction,[96] which is a field-induced elastic strain in which the signs of the strains are unchanged upon reversal of direction of the field. Large electrostrictive responses have been observed in polyurethane thermoplastic elastomer,[97] indicating that microphase-separated elastomers may be useful as actuators and artificial muscles.

REFERENCES

1. (a) S. D. D. V. Rughooputh, S. Hotta, A. J. Heeger, F. Wudl, *J. Polym. Sci. Polym. Phys.* **25**, 1071 (1987). (b) O. Inganas, W. R. Salaneck, J.-E. Osterholm, J. Laasko, *Synth. Met.* **22**, 395 (1988). (c) K. Yoshino, D. H. Park, P. K. Park, M. Onoda, R. Sugimoto, *Jpn. J. Appl. Phys.* **27**, L1612 (1988).
2. For instance, see: (a) M. Hatano, S. Kambara, S. Okamoto, *J. Polym. Sci.* **55**, 137 (1961). (b) L. B. Luttinger, *J. Org. Chem.* **27**, 1591 (1962). (c) W. E. Daniels, *J. Org. Chem.* **29**, 2936 (1964). (d) I. V. Nicolescu, E. Anglescu, *J. Polym. Sci. A* **3**, 1227 (1965); *J. Polym. Sci. A-1* **4**, 2963 (1966). (d) D. J. Berets, D. S. Smith, *Trans. Faraday Soc.* **64**, 823 (1968).
3. T. Ito, H. Shirakawa, S. Ikeda, *J. Polym. Sci. Polym. Chem. Ed.* **12**, 11 (1974).
4. H. Naarmann, N. Theophilou, *Synth. Met.* **22**, 1 (1987).
5. H. Shirakawa, E. J. Louis, A. G. MacDiarmid, C. K. Chiang, A. J. Heeger, *J. Chem. Soc. Chem. Commun.* 578 (1977).
6. (a) S. L. Hsu, A. J. Signorelli, G. P. Pez, R. H. Baughman, *J. Chem. Phys.* **69**, 1098 (1978). (b) T. C. Clarke, R. H. Geiss, W. D. Gill, P. M. Grant, H. Morawitz, G. B. Street, D. E. Sayers, *Synth. Met.* **1**, 21 (1979). (c) T. C. Chung, A. Feldblum, A. J. Heeger, A. G. MacDiarmid, *J. Chem. Phys.* **74**, 5504 (1981).
7. C. K. Chiang, E. J. Louis, A. G. MacDiarmid, C. K. Chiang, A. J. Heeger, *J. Am. Chem. Soc.* **100**, 1013 (1978).
8. (a) C. K. Chiang, C. R. Fincher, Y. W. Park, A. J. Heeger, H. Shirakawa, E. J. Louis, S. C. Gau, A. G. MacDiarmid, *Phys. Rev. Lett.* **39**, 1098 (1977). (b) Y. W. Park, A. J. Heeger, M. A. Druy, A. G. MacDiarmid, *J. Chem. Phys.* **73**, 946 (1980).
9. F. Beniere, S. Haridos, J. P. Louboutin, M. Aldissi, J. M. Fabre, *J. Phys. Chem. Solids* **42**, 649 (1981).
10. T. C. Chung, A. Feldblum, A. J. Heeger, A. G. MacDiarmid, *J. Chem. Phys.* **74**, 5504 (1981).
11. J.-C. Chiang, A. G. MacDiarmid, *Synth. Met.* **13**, 193 (1986).
12. Y. Wei, K. F. Hsueh, G.-W. Jang, *Polymer* **35**, 3572 (1994).
13. Y. Wei, G.-W. Jang, C.-C. Chan, K. F. Hseuh, R. Hariharan, S. A. Patel, C. K. Whitecar, *J. Phys. Chem.* **94**, 7716 (1990).
14. Y. Wei, Y. Sun, X. Tang, *J. Phys. Chem.* **93**, 4878 (1989).
15. J. Yue, A. J. Epstein, *J. Am. Chem. Soc.* **112**, 2800 (1990).
16. A. J. Heeger, *Synth. Met.* **55–57**, 3471 (1993).
17. S. J. Davies, T. G. Ryan, C. J. Wilde, G. Beyer, *Synth. Met.* **69**, 209 (1995).

18. A. Diaz, J. A. Logan, *J. Electroanal. Chem.* **111**, 111 (1980).
19. (a) G. Kossmehl, G. Chatzitheodorou, *Makromol. Chem. Rapid Commun.* **2**, 551 (1981). (b) G. Kossmehl, G. Chatzitheodorou, *Mol. Cryst. Liq. Cryst.* **83**, 291 (1982).
20. T. Yamamoto, K. Sanechika, A. Yamamoto, *J. Polym. Sci. Lett. Ed.* **18**, 9 (1980).
21. J.-E. Osterholm, J. Laasko, P. Nyholm, H. Isotalo, H. Stubb, O. Inganas, W. R. Salaneck, *Synth. Met.* **28**, C435 (1989).
22. J. Laasko, J.-O. Osterholm, P. Nyholm, *Synth. Met.* **28**, C467 (1989).
23. For instance, see: (a) G. Tourillon, F. Garnier, *J. Electroanal. Chem.* **135**, 173 (1982). (b) K. Kaneto, K. Yoshino, Y. Inuishi, *Jpn. J. Appl. Phys.* **21**, L567 (1982). (c) S. Hotta, T. Hosaka, W. Shimotsuna, *Synth. Met.* **6**, 317 (1983).
24. K. K. Kanazawa, A. F. Diaz, W. D. Gill, P. M. Grant, G. B. Street, G. P. Gardini, J. F. Kwak, *Synth. Met.* **1**, 329 (1980).
25. J. H. Burroughes, D. D. C. Bradley, A. R. Brown, R. N. Marks, K. Mackay, R. H. Friend, P. L. Burns, A. B. Holmes, *Nature* **347**, 539 (1990).
26. R. A. Wessling, R. G. Zimerman, U.S. patent 3,401,152 (1968).
27. B. R. Hsieh, H. Antoniadis, M. A. Abkwowitz, M. Stolka, *Polym. Prep. (2)* **33**, 414 (1992).
28. F. Papadimitrakopoulos, M. Yan, L. J. Rothberg, H. E. Katz, E. A. Chandross, M. E. Galvin, *Mol. Cryst. Liq. Cryst.* **256**, 663 (1994).
29. S. Son, A. Dodabalapur, A. J. Lovinger, M. E. Galvin, *Science* **269**, 376 (1995).
30. (a) P. V. Wright, *Polym. J.* **7**, 319 (1975). (b) D. E. Fenton, J. M. Parker, P. V. Wright, *Polymer* **14**, 589 (1973).
31. C. Berthier, W. Gorecki, M. Minier, M. B. Armand, J. M. Chabagno, P. Rigaud, *Solid State Ion.* **11**, 91 (1983).
32. (a) C. R. A. Catlow, A. V. Chadwick, G. N. Greaves, L. M. Moroney, M. R. Worboys, *Solid State Ion.* **9–10**, 1107 (1983). (b) P. G. Bruce, F. Krock, J. Evans, C. A. Vincent, *Polym. J.* **20**, 193 (1988). (c) R. D. Armstrong, M. D. Clarke, *Electrochem. Acta* **29**, 443 (1984).
33. F. M. Gray, *Solid Polymer Electrolytes: Fundamentals and Technological Applications* VCH, New York, 1991.
34. K. M. Abraham, M. Alagmer, *Solid State Ion.* **70–71**, 20 (1994).
35. For instance, see: (a) Y. Yamaguchi, S. Aoki, M. Watanabe, K. Sanui, N. Ogata, *Solid State Ion.* **40–41**, 628 (1990). (b) C. V. Nicholas, D. J. Wilson, C. Booth, J. R. M. Giles, *Br. Polym. J.* **20**, 289 (1988). (c) E. Linden, J. R. Owen, *Solid State Ion.* **28–30**, 994 (1988). (d) J. S. Foos, S. M. Erker, *J. Electrochem. Soc.* **134**, 1724 (1987). (e) J. R. Graven, R. H. Mobbs, C. Booth, J. R. M. Giles, *Makromol. Chem. Rapid Commun.* **7**, 81 (1986).
36. G. G. Cameron, M. D. Ingram, K. Sarmouk, *Eur. Polym. J.* **26**, 197 (1990).
37. For instance, see: (a) P. M. Blonsky, D. F. Shriver, H. A. Alcock, *Solid State Ion.* **18–19**, 258 (1989). (b) H. R. Alcock, P. E. Austin, T. X. Neenan, J. T. Sisko, D. F. Shriver, *Macromolecules* **19**, 1508 (1986). (c) P. M. Blonsky, D. F. Shriver, P. Austin, H. A. Alcock, *J. Am. Chem. Soc.* **106**, 6854 (1984).
38. K. M. Abraham, M. Alamgir, *Chem. Mater.* **3**, 339 (1991).
39. J. S. Tonge, D. F. Shriver, *J. Electrochem. Soc.* **134**, 269 (1987).

40. J. L. Bennet, A. A. Dembek, H. A. Alcock, B. J. Heyen, D. F. Shriver, *Polym. Prepr.* **30**(1), 437 (1989).
41. K. M. Abraham, M. Alamgir, S. J. Perotti, *J. Electrochem. Soc.* **135**, 535 (1988).
42. D. Fish, I. M. Khan, J. Smid, *Makromol. Chem. Rapid Commun.* **7**, 115 (1986).
43. P. G. Hall, G. R. Davies, J. E. McIntyre, I. M. Ward, D. J. Bannister, K. M. F. Le Brocq, *Polym. Commun.* **27**, 98 (1986).
44. D. Fish, I. M. Khan, E. Wu, J. Smid, *Br. Polym. J.* **20**, 281 (1988).
45. Z. Zhu, A. G. Einset, C.-Y. Yang, W.-X. Chen, G. E. Wnek, *Macromolecules* **27**, 4076 (1994).
46. (a) K. M. Abraham, *Electrochem. Acta* **38**, 1233 (1993). (b) K. M. Abraham, M. Alagmir, *J. Electrochem. Soc.* **137**, 1657 (1990).
47. K. M. Abraham, M. Alagmir, *J. Electrochem. Soc.* **140**, L96 (1993).
48. P. Aldebert, M. Guglielme, M. Pineri, *Polym. J.* **23**, 399 (1991).
49. (a) K. M. Abraham, M. Alagmir, S. J. Perotti, *J. Electrochem. Soc.* **135**, 535 (1988). (b) K. M. Abraham, M. Alagmir, R. K. Reynolds, *J. Electrochem. Soc.* **136**, 3576 (1989).
50. C. A. Angell, J. Fan, C. Liu, Q. Lu, E. Sanchez, K. Xu, *Solid State Ion.* **69**, 343 (1994).
51. (a) W. Helfrich, W. G. Schneider, *Phys. Rev. Lett.* **14**, 229 (1965). (b) M. Pope, H. P. Kallmann, P. Magnante, *J. Chem. Phys.* **38**, 2042 (1963).
52. (a) C. W. Tang, S. A. VanSlyke, *Appl. Phys. Lett.* **51**, 913 (1987). (b) C. W. Tang, S. A. VanSlyke, C. H. Chen, *J. Appl. Phys.* **65** 3610 (1989).
53. (a) N. Leventis, L.-Y. Huang, *Polym. News* **20**, 307 (1995). (b) D. Braun, A. J. Heeger, *Appl. Phys. Lett.* **58** 1982 (1991). (c) Y. Ohmari, M. Uchida, K. Muro, K. Yoshino, *Jpn. J. Appl. Phys.* **30**, L1938 L1941 (1991). (d) G. Gren, G. Leditzky, B. Ullrich, G. Leising, *Adv. Mater.* **4**, 36 (1992). (e) A. B. Holmes et al., *Synth. Met.* **55–57**, 4031 (1993).
54. (a) S. Aratani, C. Zhang, K. Pakbaz, S. Hoger, F. Wudl, A. J. Heeger, *J. Electron. Mater.* **22**, 745 (1993). (b) A. R. Brown, D. D. C. Bradley, J. H. Burroughes, R. H. Friend, N. C. Greenham, P. L. Burn, A. B. Holmes, *Appl. Phys. Lett.* **61**, 2793 (1992).
55. (a) C. Zhang, S. Hoger, K. Pakbaz, F. Wudl, A. J. Heeger, *J. Electron. Mater.* **23**, 453 (1994). (b) A. R. Brown, D. D. C. Bradley, J. H. Burroughes, R. H. Friend, N. C. Greenham, P. L. Burn, A. B. Holmes, A. Kraft, *Appl. Phys. Lett.* **61**, 2793 (1992).
56. N. C. Greenham, S. C. Moratti, D. D. C. Bradley, R. H. Friend, A. B. Holmes, *Nature* **365**, 628 (1993).
57. P. L. Burn, A. B. Holmes, A. Kraft, D. D. C. Bradley, A. R. Brown, R. H. Friend, R. W. Gymer, *Nature* **356**, 47 (1992).
58. Y. Yang, A. J. Heeger, *Appl. Phys. Lett.* **64**, 1245 (1994).
59. D. D. Gebler, Y. Z. Wang, J. W. Blatchford, S. W. Jessen, L.-B. Lin, T. L. Gustafason, H. L. Wang, T. M. Swager, A. G. MacDiarmid, A. J. Epstein, *J. Appl. Phys.* **78**, 4264 (1995).
60. (a) B. Hu, Z. Yang, F. E. Karatz, *J. Appl. Phys.* **76**, 2419 (1994). (b) J. Kido, K. Hongawa, K. Okuyama, K. Nagai, *Appl. Phys. Lett.* **63**, 2627 (1993).

61. J. L. Musfeldt, J. R. Reynolds, T. B. Tanner, J. B. Ruiz, J. Wang, M. Pomerantz, *J. Polym. Sci. B. Polym. Phys.* **32**, 2395 (1994).
62. G. Crem, C. Paar, J. Stampfl, G. Leising, J. Huber, U. Scherf, *Chem. Mater.* **7**, 2 (1995).
63. M. Onada, *J. Appl. Phys.* **78**, 1327 (1995).
64. D. Braun, G. Gustafasson, D. McBranch, A. J. Heeger, *J. Appl. Phys.* **72**, 564 (1992).
65. M. Berggren, G. Gustafason, O. Inganas, M. R. Andersson, O. Wennerstrom, T. Hjertberg, *Adv. Mater.* **6**, 488 (1995).
66. (a) J. A. Osaheni, S. A. Jenekhe, *Macromolecules* **27**, 739 (1994). (b) S. A. Jenekhe, J. A. Osaheni, *Science* **265**, 765 (1994).
67. P. L. Burn, A. B. Holmes, A. Kraft, A. R. Brown, D. D. C. Bradley, R. H. Friend, *Mater. Res. Symp. Proc.* **247**, 647 (1992).
68. (a) M. Berggren, O. Inganas, G. Gustafasson, J. Rasmusson, M. R. Andersson, T. Hjertberg, O. Wennerstrom, *Nature* **372**, 444 (1994). (b) M. R. Andersson, M. Berggren, G. Guatafasson, T. Hjertberg, O. Inganas, O. Wennerstrom, *Synth. Met.* **71**, 2183 (1995). (c) O. Inganas, M. Berggren, M. R. Andersson, G. Gustafason, T. Hjertberg, O. Wennerstrom, P. Dyreklev, M. Granstrom, *Synth. Met.* **71**, 2121 (1995).
69. M. Berggren, G. Gustafsson, O. Inganas, M. R. Andersson, T. Hjertberg, O. Wennerstrom, *J. Appl. Phys.* **76**, 7530 (1994).
70. P. Dyreklev, M. Berggren, O. Inganas, M. R. Andersson, O. Wennerstrom, T. Hjertberg, *Adv. Mater.* **7**, 43 (1995).
71. M. Hamaguchi, K. Yoshino, *Jpn. J. Appl. Phys.* **34**, L712 (1995).
72. Q. Pei, G. Yu, C. Zhang, Y. Yang, A. J. Heeger, *Science* **269**, 1086 (1995).
73. K. D. Cummings, M. Kiersh, *J. Vac. Sci. Technol.* **B7** 1536 (1989).
74. (a) M. Angelopoulos, J. M. Shaw, R. D. Kaplan, S. Perreault, *J. Vac. Sci. Technol.* **B7**, 1519 (1989). (b) M. Angelopoulos, J. M. Shaw, K.-L. Lee, W.-S. Huang, M.-A. Lecorre, M. Tissier, *J. Vac. Sci. Technol.* **B9**, 3428 (1991).
75. W.-S. Huang, *Polymer* **35**, 4057 (1994).
76. T. J. Skotheim, Ed., *Handbook of Conducting Polymers*, Marcel Dekker, New York, 1986.
77. (a) M. J. Marsella, T. M. Swager, *J. Am. Chem. Soc.* **115**, 12214 (1993). (b) M. J. Marsella, P. J. Carroll, T. M. Swager, *J. Am. Chem. Soc.* **116**, 9347 (1994). (c) M. J. Marsella, P. J. Carroll, T. M. Swager, *J. Am. Chem. Soc.* **117**, 9832 (1995). (d) M. J. Marsella, R. J. Newland, P. J. Carroll, T. M. Swager, *J. Am. Chem. Soc.* **117**, 9842 (1995).
78. (a) P. J. Nigrey, A. G. MacDiarmid, A. J. Heeger, *J. Chem. Soc. Chem. Commun.* 594 (1979). (b) P. J. Nigrey, D. MacInnes, Jr., D. P. Nairns, A. G. MacDiarmid, A. J. Heeger, *J. Electrochem. Soc.* **128**, 1651 (1981).
79. C. K. Chiang, *Polym. Commun.* **22**, 1454 (1981).
80. For instance, see: (a) L. W. Shacklette, R. R. Chance, D. M. Ivory, G. G. Miller, R. H. Baugman, *Synth. Met.* **1**, 307 (1979). (b) A. F. Diaz, K. K. Kanazawa, G. P. Gardini, *J. Chem. Soc. Commun.*, 535 (1979). (c) T. Yamamoto, K. Saneckika, A. Yamamoto, *J. Polym. Sci. Lett. Ed.* **18**, 9 (1980). (d) A. G. MacDiarmid,

J. C. Chiang, M. Halpern, W. S. Huang, S. L. Mu, N. L. D. Somasiri, W. Wu, S. Yaniger, *Mol. Cryst. Liq. Cryst.* **121**, 173 (1985), and references therein.

81. (a) G. L. Henricksen, W. L. De Luca, D. R. Vissers, *Chemtech*, 32 (1994). (b) E. J. Cains, *Interface* **1**, 24 (1992). (c) S. Megahed, B. Scrosati, *Interface* **4**, 34 (1995).
82. D. Naegele, R. Bittihn, *Solid State Ion.* **28–30**, 983 (1988).
83. S. Kakuda, T. Momma, T. Osaka, G. B. Appetecchi, B. Scrosati, *J. Electrochem. Soc.* **142**, L1 (1995).
84. N. Oyama, T. Tatsuma, T. Sato, T. Sotomura, *Nature* **373**, 598 (1995).
85. M. Anderman, *Solid State Ion.* **69**, 336 (1994).
86. B. Scorsati, *J. Electrochem. Soc.* **139**, 2276 (1992).
87. F. Croce, S. Passerini, B. Scrosati, *J. Electrochem. Soc.* **141**, 1405 (1994).
88. E. J. Plitcha, W. K. Behl, *J. Electrochem. Soc.* **139**, 1509 (1992).
89. E. J. Plitcha, W. K. Behl, *J. Electrochem. Soc.* **140**, 46 (1993).
90. T. Tanaka, I. Nishio, S.-T. Sun, S. Ueno-Nishio, *Science* **218**, 467 (1982).
91. (a) T. Kurauchi, T. Shiga, Y. Hirose, A. Okada, in *Polymer Gels: Fundamentals and Biomedical Applications*, D. DeRossi, K. Kajiwara, Y. Osada, A. Yamauchi, Eds., Plenum Press, New York, 1991, p. 237. (b) T. Shiga, T. Kurauchi, *J. Appl. Polym. Sci.* **39**, 2305 (1990). (c) K. Osada, H. Okuzaki, H. Hori, *Nature* **355**, 242 (1992). (d) I. C. Kwon, Y. H. Bae, S. W. Kim, *J. Polym. Sci. B. Polym. Phys.* **32**, 1085 (1994).
92. D. E. DeRossi, P. Chiarell, G. Buzzigoli, C. Domenci, *Trans. Am. Soc. Artif. Internal Organs* **32**, 157 (1986).
93. K. Sawahata, M. Hata, H. Yasunaga, Y. Osada, *J. Control Rel.* **14**, 253 (1990).
94. Y. Ye, J. N. Rider, A. Sekhar, G. Wong, K. Trout, K. Graczyk, W. Brown, J. Gross, M. Stewart, M. Kamler, G. E. Wnek, *Polym. Prepr.*, in press.
95. M. G. Broadhurst, G. T. Davis, *Ferroelectrics* **60**, 3 (1984).
96. T. Furukawa, N. Seo, *Jpn. J. Appl. Phys.* **29**, 675 (1990).
97. M. Zhenyi, J. I. Scheinbeim, J. W. Lee, B. A. Newman, *J. Polym. Sci. B Polym. Phys.* **32**, 2721 (1994).

CHAPTER 4

Polymers in Electronics

E. REICHMANIS and T. X. NEENAN*

Bell Laboratories, Lucent Technologies
Murray Hill, New Jersey

4.1 INTRODUCTION

Polymeric materials have found widespread use in the electronics industry in both the manufacturing processes used to generate today's integrated circuits and as component structures in the completed devices. The broad applicability of polymers arises from the ability to design and synthesize such materials with the precise functionalities and properties required for a given application. Notably, polymeric materials have been used as lithographic imaging materials (resists) and as dielectric, passivation, and insulating materials.

Radiation sensitivity is the key property required of materials used for imaging the individual elements of an integrated circuit. Known as the microlithographic process, it is the linchpin technology used to fabricate electronic devices and is critically dependent on the polymer–organic materials chemistry used to generate the radiation-sensitive imaging material known as the resist. As the lithographic technologies evolve to allow fabrication of smaller and more compact circuit elements, new resist chemistries and processes will be needed.

Low-dielectric polymeric materials are used as adhesives, encapsulants, substrates, and thin-film dielectric layers. The ease of processing polymeric materials is stimulating the design and development of materials with improved performance. Polymers with lower dielectric constants, higher thermal stability, photo-imageability, low dielectric loss, and low thermal expansion coefficient are especially sought after. In this chapter we discuss the diversity

* Present address: GelTex Pharmaceuticals Inc., Waltham, Massachusetts

Chemistry of Advanced Materials: An Overview, Edited by Leonard V. Interrante and Mark J. Hampden-Smith.
ISBN 0-471-18590-6

of the chemical challenges facing the microelectronics industry as well as the rapid progress and evolution of research in advanced polymeric materials for microelectronics.

4.2 LITHOGRAPHIC MATERIALS

Significant advances are continually being made in microelectronic device fabrication, especially in lithography, the technique that is used to generate the high-resolution circuit elements characteristic of today's integrated circuits.[1] It was not that long ago that state-of-the-art devices contained only a few thousand transistor elements and had 5- to 6-μm minimum features. Today, devices with several million transistor cells are commercially available and are fabricated with minimum features as small as 0.25 μm. These accomplishments have been achieved using conventional photolithography (photolithography employing 350 to 450-nm light) as the technology of choice. Figure 4.1 provides a schematic representation of the lithographic process. Incremental improvements in tool design and performance have allowed continued use of 350- 450-nm light to produce even smaller features.[2]

This progress has been evolutionary, and the same basic unit processes and resist materials have been employed in manufacture throughout this period. The wavelength of the light (or radiation) used to form the image will ultimately determine the resolution of a given printing technique, with

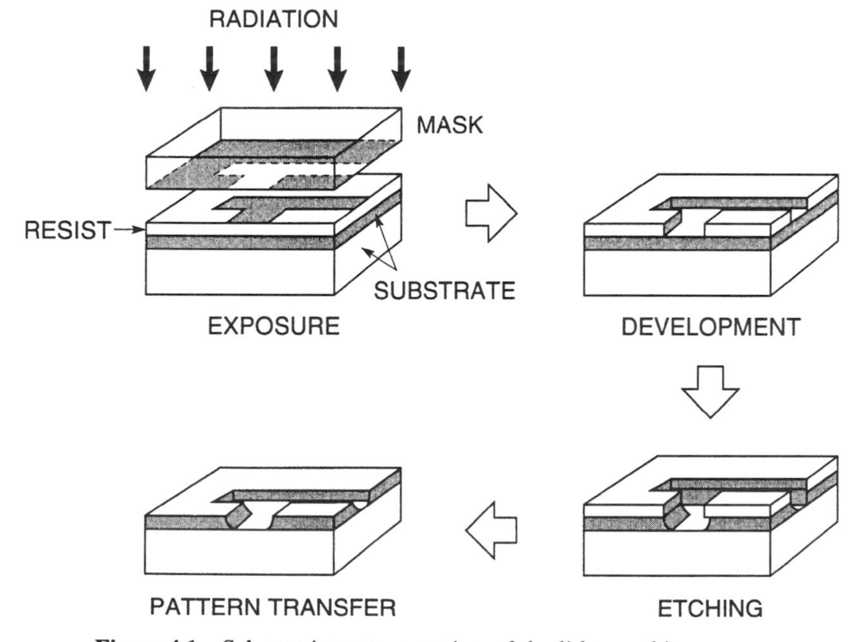

Figure 4.1 Schematic representation of the lithographic process.

shorter wavelengths yielding higher resolution.[3] The resist, a positive-acting material consisting of a photoactive diazonaphthoquinone compound and a novolac resin (Fig. 4.2) that has been in pervasive use since the mid-1970s, will probably be the resist of choice for some more years.[4,5] The cost of introducing a new technology, which includes the cost associated with the development and implementation of new hardware and resist materials, is a

OH
CH_2
CH_3
NOVOLAC STRUCTURE

O
N_2
AQUEOUS
ALKALI
INSOLUBLE
R
hν
COOH
AQUEOUS
ALKALI
SOLUBLE
R
H_2O
O
C
O
R
R
DIAZONAPHTHOQUINONE PHOTOCHEMISTRY

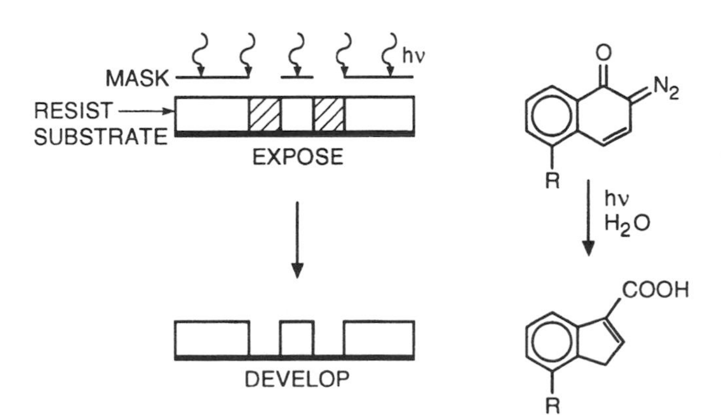

Figure 4.2 Schematic representation of conventional, diazonaphthoquinone–novolac, positive-resist chemistry.

strong driving force pushing conventional photolithography to its absolute resolution limit and extending its commercial value.

The technological alternatives to conventional photolithography are largely the same as they were a decade ago: short-wavelength (220 to 280 nm) photolithography, scanning or projection electron-beam, 1 : 1 proximity x-ray, or ion-beam lithography.[3,6] Unfortunately, conventional photoresists are not appropriate for use with these new, alternative lithographic technologies. The most notable deficiencies of the conventional novolac–diazonaphthoquinone resists are their inherent low sensitivity and absorption characteristics that are too high for shorter-wavelength exposure to allow uniform imaging through practical resist film thicknesses (about 0.7 to 1 μm). Thus no matter which technology becomes dominant after photolithography has reached its resolution limit (0.3 to 0.5 μm), new resists and processes will be required.[7] The introduction of new resist materials and processes will also require considerable lead time to bring them to the performance level currently realized by conventional positive photoresists.

There are significant trade-offs between optimum process performance and the chemical design of new resists. The ultimate goal of any lithographic technology is to be able to produce the smallest possible features with the lowest cost per level and with wide process latitude. The best solution will invariably require compromises, and an understanding of materials and process issues is essential to select the correct compromise.

4.3 RESIST DESIGN REQUIREMENTS

Resist chemistry must be designed carefully to meet the specific requirements of a given lithographic technology. Although these requirements vary according to the radiation source, device process requirements, and exposure tool design, the following are ubiquitous: sensitivity, contrast, resolution, etching resistance, purity, and manufacturability.[8] These properties can be achieved by careful manipulation of polymer structure, molecular properties, and synthetic methods.[4]

The materials issues that must be considered in designing resists with the appropriate properties are given below. The polymer resins must (1) exhibit solubility in solvents that allow the coating of uniform, defect free, thin films; (2) be sufficiently thermally stable to withstand the temperatures and conditions used with standard device processes; (3) exhibit no flow during pattern transfer of the resist image into the device substrate; (4) possess a reactive functionality that will facilitate pattern differentiation after irradiation; and (5) for photoexposure, have absorption characteristics that will permit uniform imaging through the thickness of a resist film. In general, thermally stable (> 150°C), high-glass-transition-temperature (T_g > 90°C) materials with low absorption at the wavelength of interest are desired. If other additives are to be employed to effect the desired reaction, similar criteria apply. Specifically, they must be nonvolatile, be stable up to at least 175°C, possess a reactive functionality that will allow a change in solubility after irradiation, and have low absorbance. The sections that follow outline many of the chemistries that have been applied to the design of resist materials for microlithography. The reader is referred to Refs. 1, 3, 5, and 9–11 for additional major overviews of the field.

4.4 SOLUTION-DEVELOPED RESIST CHEMISTRY

4.4.1 Negative-Resist Chemistry

The traditional radiation-induced chemical processes that effect reduced solubility of the exposed areas of a resist leading to a negative image are cross-linking between polymer chains, image reversal, and use of a radiation-generated species as a catalyst for cross-linking reactions in the resist matrix.

Two-Component Cross-linking Resists During the early stages of the semiconductor industry (1957–1970), the minimum size of circuit features exceeded 5 μm, and the primary resist used during this time consisted of cyclized poly(*cis*-1,4 isoprene) and an aromatic azide cross-linking compound.[11] The bisaryldiazide, 2,6-bis(4-azidobenzal)-4-methylcyclohexanone, effectively initiated cross-linking of the matrix resin upon exposure to near-ultraviolet (UV) light. The resolution of this highly sensitive two-component resist was limited due to solvent-induced swelling followed by stress relaxation of the developed resist images. Enhanced resolution was achieved by Iwayanagi et al.[12, 13] When an arylazide was employed in conjunction with an aqueous-base-soluble poly(hydroxystryrene) matrix, submicron patterns were defined upon 250-nm exposure. However, the high optical density of the material at 250 nm afforded undercut resist profiles. Appropriate choice of the cross-linking agent allows extension of the chemistry of this system into the mid-UV range.[14]

Single-Component Cross-linking Resists Concurrent with rapid development of the electron-beam lithographic tools for both optical mask making and direct-write application was the commercialization of single-component negative resists. The electron-beam exposure requirements of these materials were compatible with the dose outputs (about 1 $\mu C/cm^2$ at 10 kV) of raster scan machines developed by Bell Laboratories.[15] These resists typically contained epoxy, vinyl, and halogen functionalities (Fig. 4.3).[16, 17]

Exposure of epoxy-,[18–20] vinyl-, and allyl[21]-containing polymers affords a radical or cationic species that can react with the same (intramolecular cross-link) or neighboring (intermolecular cross-link) polymer chain. This process continues via a chain reaction leading to the formation of an insoluble polymer network. Although this reaction sequence affords high

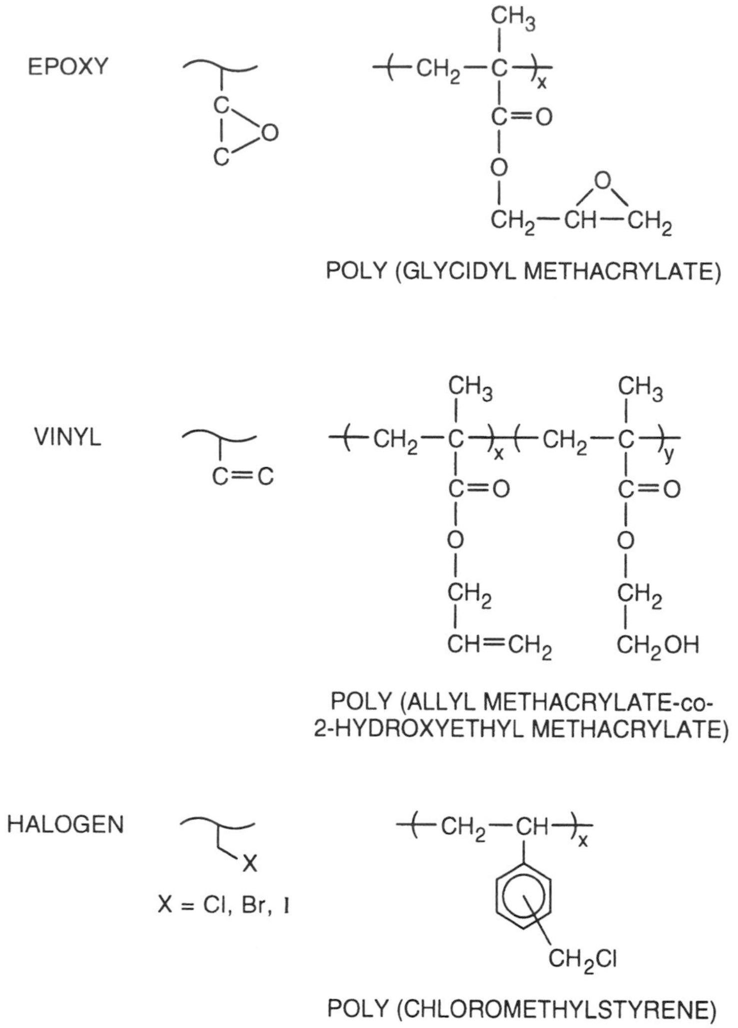

EPOXY

POLY (GLYCIDYL METHACRYLATE)

VINYL

POLY (ALLYL METHACRYLATE-co-2-HYDROXYETHYL METHACRYLATE)

HALOGEN

X = Cl, Br, I

POLY (CHLOROMETHYLSTYRENE)

Figure 4.3 Examples of negative, single-component cross-linking resists containing epoxy, vinyl, or halogen functionalities.

resist sensitivity, the propagation of radiation-generated reactive species continues in the vacuum environment of an electron-beam exposure tool. The consequence is that the features that were exposed first will have different dimensions from those exposed last.[22] In certain cases, the feature-size difference can exceed the maximum allowable variation specified for a particular device level, and thus these chemistries have limited use today.

On the other hand, the halogenated styrene negative resists cross-link by a radiation-induced reaction which involves radicals that recombine and do not propagate. The mechanism is based on experimental evidence obtained by Tabata et al.[23] Since the reaction sequence does not involve a postexposure curing reaction, and additionally, the incorporation of styrene into the resist improves the dry etching characteristics of the polymers, these materials are often favored over aliphatic-based resists. Specific examples of halogenated styrene-based resists are chlorinated[24] or chloromethylated[25] polystyrene, and poly(chloromethylstyrene).[26, 27] In addition to being sensitive electron-beam resists, the latter also have sensitivity to deep-UV[28] and x-ray exposure.[29] In one case, the halogenated material, chlorostyrene, was copolymerized with glycidyl methacrylate to afford a very sensitive electron-beam resist that exhibits little of the curing phenomenon typically observed with epoxy cross-linking reactions.[30-32]

While the aromatic styrene ring affords improved dry-etching resistance, further improvement in this parameter can be achieved through incorporation of silicon. For instance, Hatzakis et al.[33] showed that polysiloxane polymers such poly(vinylmethyl siloxane) readily provide electron-beam sensitivity in the range 1 to 2 $\mu C/cm^2$. These materials have a high silicon content (> 30 wt %), and as such have found use in bilevel (see Ref. 11 for a definition of bilevel) lithographic processes. Other resists that exhibit acceptable thermal properties and etching resistance for bilevel applications are copolymers of trimethylsilyl-[34,35] and trimethylstanylstyrene[34] with chlorostyrene. Additionally, copolymerization of trimethylsilylmethyl methacrylate with chloromethylstyrene yields a workable electron-beam and deep-UV negative resist even though the homopolymer of the silicon-containing methacrylate is a positive-acting material.[36] In this case, at the exposure dose employed, cross-linking of the chloromethylstyrene units predominate.

For the resists discussed in this section, lithographic performance is hindered by the extent to which the materials swell in organic developing solvents. This phenomenon is shown in Fig. 4.4, which depicts a bridged 0.75-μm line and pad resist features in a styrene-based negative resist. The type of pattern distortion shown in this figure ultimately limits the resolution capability of these resists. Novembre and co-workers[32] found that the extent of swelling could be minimized by proper choice of developer using a method based on the Hansen three-dimensional solubility parameter model. They determined that developers found to be thermodynamically poor, and kinetically good solvents, afford resist materials such as poly(glycidyl methacrylate-co-3-chlorostyrene) (GMC) enhanced resolution capability (Fig. 4.4). This methodology facilitates selection of an optimal developer without the tedious trial-and-error approach commonly used.

Image-Reversal Chemistry Through creative chemistry and resist processing, schemes have been developed that produce negative tone images in positive photoresist. One embodiment of these image-reversal processes requires the addition of small amounts of base additives such as monazoline, imidazole, or triethanolamine to diazoquinone–novolac resists.[37-39] The

SINGLE COMPONENT DEVELOPMENT

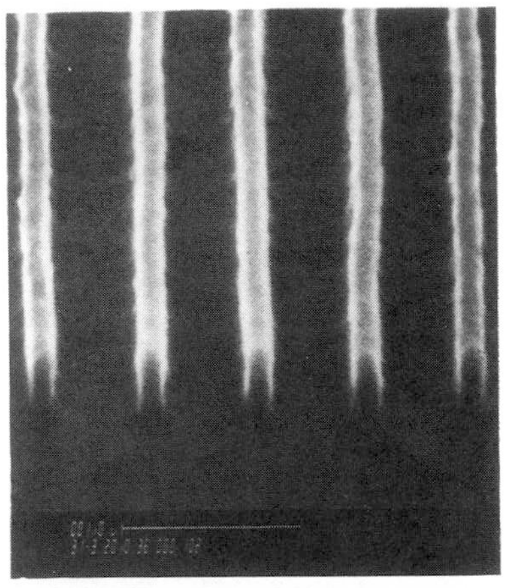

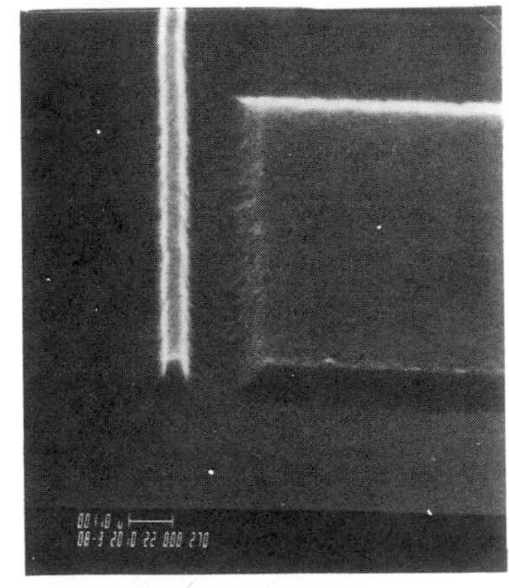

CONVENTIONAL TWO-COMPONENT DEVELOPMENT

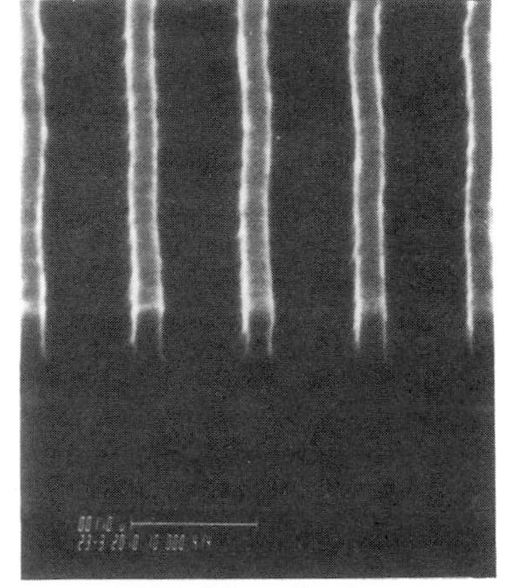

coded 0.5μm line/space images prior to postbake

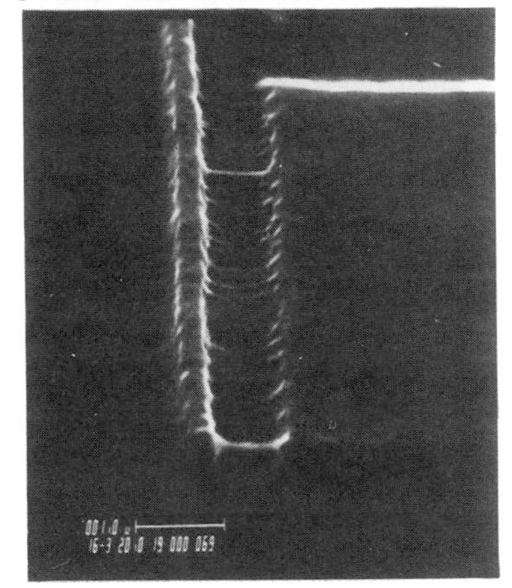

coded 0.75μm gap prior to postbake

Figure 4.4 Comparison of the effect of developer selection on the lithographic performance of GMC.

doped resist is exposed through a mask, baked after exposure, flood exposed, and finally developed in aqueous base to generate high-quality negative-tone images. The chemistry and processes associated with this system are shown in Fig. 4.5. Thermally induced base-catalyzed decarboxylation of the indene carboxylic acid destroys the aqueous-base solubility of the exposed resist. Subsequent flood exposure renders the previously masked regions soluble in aqueous base, allowing generation of negative tone patterns. It is not always necessary to add the base to the resist prior to exposure. Alternative image-reversal processes have been developed involving treatment of exposed photoresist with a gaseous amine in a vacuum environment.[40]

Chemically Amplified Resist Chemistry Chemical amplification through acid-catalyzed cross-linking for negative working resist applications has been achieved through various mechanisms. These include cationic polymerization, condensation polymerization, electrophilic aromatic substitution, and acid-catalyzed rearrangement. The acid species may be generated from a variety of materials.[41]

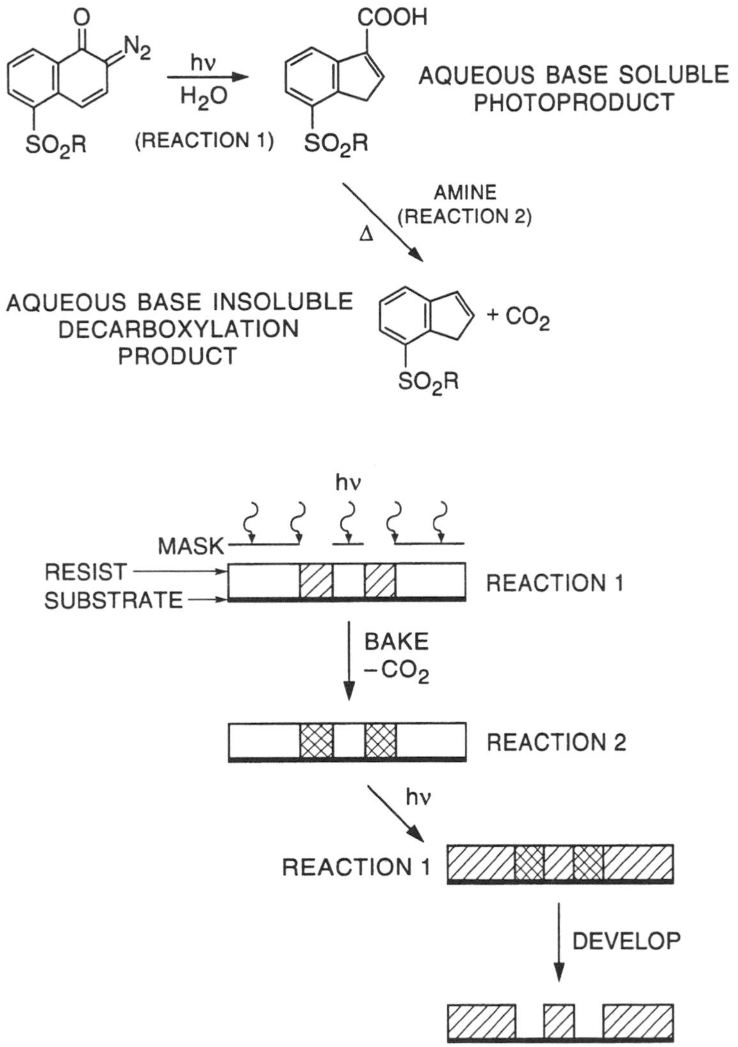

Figure 4.5 Chemistry and process sequence for a negative-tone resist operating via an image-reversal process.

Cationic Polymerization Mechanisms The first chemically amplified resist systems to be developed were those based on the cationic polymerization of epoxy materials.[42, 43] In general, resolution of sub-0.5-μm features in resists based on this mechanism is difficult due to distortion resulting from solvent-induced swelling of the irradiated regions. Recently, Allen described a highly sensitive, aqueous-base soluble 365 nm and electron beam resist based on acid-catalyzed cross-linking of the epoxy group of poly(hydroxystyrene-dicyclopentyloxy methacrylate).[44]

Condensation Polymerization Mechanisms Condensation polymerization mechanisms are probably the most prevalent in the design of chemically amplified negative resists. Such resist systems consist of three key components: (1) a polymer resin with reactive site(s) (also called a binder) for cross-linking reactions (e.g., a polymer containing a hydroxy functionality); (2) a radiation-sensitive acid generator; and (3) an acid-activated cross-linking agent.[45-49] Figure 4.6 depicts some of the alternative structures for the components above. The photogenerated acid catalyzes the reaction between the resin and cross-linking agent to afford a highly cross-linked polymer network that is significantly less soluble than the unreacted polymer resin. A postexposure bake step prior to development is required to complete the condensation reaction as well as to amplify the cross-linking yield to enhance sensitivity and improve image contrast. Sub-0.5-μm features could be resolved with deep-UV[50] and electron-beam[51] radiation with wide process latitude and high sensitivity using this chemistry. Very sensitive x-ray and electron-beam resist formulations based on similar chemistry using melamine and benzyl alcohol derivatives as cross-linking agents, formulated with onium salt photoacid generators in novolac or poly(hydroxystyrene) binders, have shown 0.2-μm line and space resolution.[45, 46] Other additives that undergo acid-induced condensation include compounds such as diphenylsilanediol,[52, 53] polysiloxanes,[54] and diphenylcarbinol.[55]

Electrophilic Aromatic Substitution Mechanisms Photo-induced cross-linking can be achieved in styrene polymers that are susceptible to electrophilic aromatic substitution by addition of a latent electrophile (i.e., a carbocation precursor) and a photoacid generator.[56, 57] The photogenerated acid reacts with the latent electrophile during a postexposure bake step to generate a reactive carbocation that then reacts with an aromatic moiety in the matrix affording a cross-linked network. The latent electrophile may be either an additive or a monomer that is copolymerized into the polymer binder. Examples of latent electrophiles include dibenzylacetate and copolymers of acetyloxymethylstyrene[56, 57] and 1,3-dioxlane-blocked benzaldehyde.[58]

4.4.2 Positive-Resist Chemistry

Positive resists exhibit enhanced solubility after exposure to radiation. The principles leading to this increased solubility are chain scission and/or

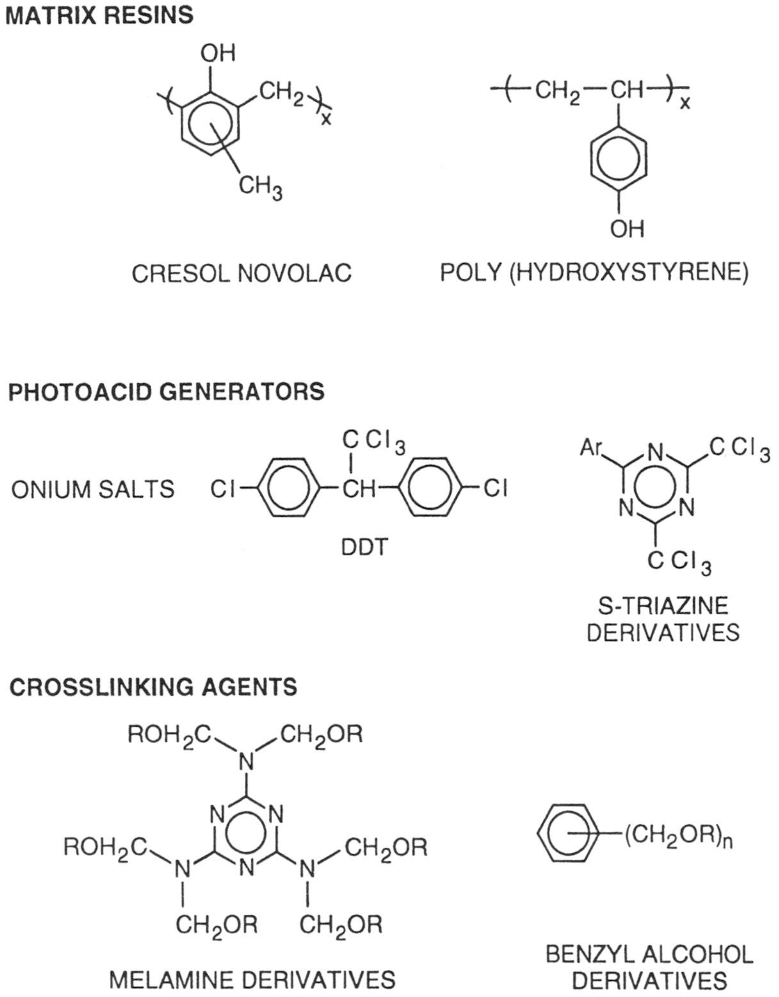

MATRIX RESINS

OH
CH_2
x
CH_3
CRESOL NOVOLAC

$-(CH_2-CH)_x-$
OH
POLY (HYDROXYSTYRENE)

PHOTOACID GENERATORS

ONIUM SALTS

CCl_3
Cl—CH—Cl
DDT

Ar
N
CCl_3
N
N
CCl_3
S-TRIAZINE DERIVATIVES

CROSSLINKING AGENTS

ROH_2C CH_2OR
N
N N
ROH_2C N N N CH_2OR
CH_2OR CH_2OR
MELAMINE DERIVATIVES

$(CH_2OR)_n$
BENZYL ALCOHOL DERIVATIVES

Figure 4.6 Selected chemical structures that may be employed as components of a chemically amplified negative resist operating via a condensation polymerization mechanism.

polarity change. Positive photoresists that operate on the polarity change principle have been widely used for the fabrication of VLSI devices because of their high resolution and excellent dry etching resistance. The chain scission mechanism typically operates at photon energies below 300 nm, where the energy is sufficient to break main-chain bonds.

Main-Chain Scission Resists The substituted methacrylates are probably the most extensively investigated polymers which undergo radiation-induced main-chain scission. Poly(methyl methacrylate) (PMMA) undergoes chain scission when exposed to electron-beam,[59] x-ray,[60] and deep-UV radiation,[61, 62] swells only minimally during the wet development process, and exhibits extremely high resolution. The ability to delineate submicron images in PMMA is, however, offset by its low sensitivity and relatively poor dry etching resistance. Electron-beam and deep-UV exposure doses in excess of 50 $\mu C/cm^2$ at 10 kV[59] and 1 J/cm^2,[61, 62] respectively, have been required for patterning purposes. The mechanism of PMMA main-chain cleavage has been studied extensively and is described elsewhere.[63]

Efforts to improve the sensitivity of PMMA have spawned many research efforts. These include introduction of electronegative substituents,[64-67] incorporation of bulky substituents,[68] incorporation of fluorine into the ester group,[69-71] generation of inter- and intramolecular anhydride linkages,[72-74] and copolymerizing methyl methacrylate (MMA) with other methacrylate-based monomers to effect improved absorption characteristics[75-78] or etching resistance.[79-81] Materials based on poly(methyl isopropenyl ketone) (PMIPK) are also known to undergo radiation-induced chain scission.[82, 83]

Copolymers of an alkene or vinylaryl compound and sulfur dioxide are another major class of chain scission resists. Interest in these materials as resists arose from results of Brown and O'Donnell, who reported a $G(s)$ value of about 10 for poly(1-butene sulfone).[84, 85] The proposed radiation-induced degradation of the poly(olefin sulfones) involves cleavage of the relatively weak main-chain carbon–sulfur bond[86] followed by depolymerization to yield the starting monomers as the major products. The copolymer containing 1-butene (PBS) is a highly sensitive (< 1 $\mu C/cm^2$ at 10 kV) electron-beam resist[87] and is the predominant positive resist used in the fabrication chromium photomasks.[88] An O_2 reactive ion etching-resistant material can be obtained using silyated monomers such as *p*-trimethylsilystyrene and *p*-pentamethyldisilylstyrene.[89, 90]

Polymers such as polysilanes which contain silicon atoms in the backbone have been reported to function as positive-acting deep-UV bilevel resists.[91-93] More recently, such materials have found application in 193-nm eximer laser lithography.[94] Exposure results in cleavage of the main-chain Si—Si bond, resulting in a decrease in molecular weight.[91-93] Additionally, these materials exhibit high quantum yields for scission, nonlinear photobleaching, and submicron resolution.

Dissolution-Inhibition Resists Photolithography represents the workhorse technology for device manufacture and has traditionally used a Hg or Hg–Xe discharge lamp as the radiation source. Resist systems that have been developed to respond favorably to this energy spectrum (250–450 nm) are often called conventional photoresists (Fig. 4.2) and typically are comprised of two components: an aqueous alkali-soluble resin and a photosensitive dissolution inhibitor. The alkaline soluble resin, a novolac, is prepared via condensation polymerization of a substituted phenol and formaldehyde.[5] These resins, and modifications thereof, are formulated to exhibit low absorbance in the near- and mid-UV region, are glassy amphorous materials at room temperature, and can be dissolved in a variety of organic solvents useful for spin coating applications. Additionally, for oxygen-reactive ion etching applications, silicon may be readily incorporated into the polymer.[95, 96] The second component of conventional photoresists is a hydrophobic, substituted diazonaphthoquinone (DNQ) dissolution inhibitor. Addition of this component to the novolac renders the polymer matrix insoluble in aqueous-base developers. Upon irradiation, the diazonaphthoquinone undergoes a Wolff rearrangement followed by hydrolysis to generate the base-soluble indene carboxylic acid. The photogenerated acid allows the exposed regions of the film to be dissolved in aqueous alkaline solution. The remaining, nonexposed regions are unaffected and do not swell in the developer. The dissolution mechanism has long attracted the attention of many workers. Most recently, Reiser has postulated a mechanism based on percolation theory. In this model, the diffusion of base is regulated by the density of hydrophilic percolation sites in the solvent penetration zone. This density can be changed by the introduction of additives (i.e., the inhibitor) which block some of the hydrophilic sites, and it can be changed by irradiation (i.e., conversion of inhibitor to indene carboxylic acid.[97, 98] The novolac–diazonaphthoquinone chemistry affords high resolution, and as a consequence of the aromatic nature of the resin, good dry-etching resistance for pattern transfer processes.

The performance of conventional resists depends on the precise structure of the photosensitive dissolution inhibitor and the novolac resin.[99-105] Most photoresists designed for 365 to 436-nm exposure utilize a 1,2-diazonaphthoquinone-5-sulfonate ester that exhibits absorbance maxima at about 340, 400, and 430 nm. Changes in the position of the aryl substitutent can lead to variations in sensitivity and light absorption. For example, the 4-aryl sulfonate analogs exhibit absorption characteristics that are more appropriate for shorter-wavelength exposure (i.e., 313 and 365 nm.[99] This substitution pattern leads to the appearance of a bleachable absorbance at about 315 and 385 nm, extending the sensitivity of conventional resists to shorter wavelengths. Optimization of resist sensitivity for a particular exposure tool requires an understanding of the effect of substituents on the absorption characteristics of the materials. This approach was used effectively by Miller and co-workers,[99] who coupled such studies with semiempirical calculations to facilitate the design of diazonaphthoquinone dissolution inhibitors for mid-UV applications.

The high absorbance of conventional photoresists prevents their application to shorter-wavelength (< 280 nm) lithographies. As a result, alternative resins and dissolution inhibitors have been proposed. Examples include dissolution inhibitors based on 5-diazo-Meldrums's acid chemistry for novolac resins[106, 107] and 2-nitrobenzyl carboxylates.[108] In the case of the nitrobenzyl carboxylates, optimum results were obtained for ester derivatives of large-molecule organic acids such as cholic acid. These esters are nonvolatile and allow conversion of a relatively large volume fraction of resist from an alkali-insoluble to an alkali-soluble state. While these resists exhibit high sensitivity and contrast, and are capable of submicron resolution, their aliphatic nature provides only marginal dry etching resistance.[109, 110] Improvements in the transparency at sub-250 nm of the matrix resin used in dissolution inhibition resists while maintaining the good dry-etching characteristics can be obtained by replacing the novolac with poly(4-hydroxystyrene) (PHS) and its substituted analogs.[111, 112] Alternatively, alicylic methacrylates have been used to develop etching resistant 193-nm resist materials.[113-116]

Conventional positive photoresists based on novolac–DNQ chemistry have application not only to photoresist technology, but also to electron-beam[117] and x-ray lithography.[118, 119] Their use is limited, however, by poor sensitivity; sensitivities of about 1800 and 300 mJ/cm^2 have been reported for synchrotron x-radiation ($\lambda = 0.8$ nm)[118] and x-radiation ($\lambda = 1.4$ m)[119] obtained from a pulsed laser source, respectively. The use of polymeric dissolution inhibitors in novolac resins has been shown to overcome this issue and produce electron-beam-sensitive positive-resist materials. Bowden et al.[120] have shown that a two-component system consisting of a novolac and a poly(olefin sulfone) as the dissolution inhibitor can generate a resist having electron-beam sensitivities in the range 3 to 5 $\mu C/cm^2$ at 20 kV. The specific sulfone is poly(2-methyl-1-pentene sulfone) (PMPS), which when formulated with a novolac resin renders the novolac insoluble in alkaline media. Exposure to electrons results in spontaneous depolymerization of PMPS to its volatile monomers.[121-124] The dissolution inhibitor is thus effectively vaporized, thereby allowing dissolution of the remaining aqueous-base-soluble novolac resin (Fig. 4.7). Similar materials have been reported by other workers.[125, 126]

Chemically Amplified Resist Chemistry

Deprotection Chemistry The pioneering work relating to the development of chemically amplified resists based on deprotection mechanisms was carried out by Ito et al.[127] These initial studies dealt with the catalytic deprotection of poly(4-*t*-butoxycarbonyloxystyrene) (PTBS), in which the thermally stable, acid-labile *t*-butoxycarbonyl group is used to mask the hydroxyl functionality of poly(vinylphenol).[128-130] As shown in Fig. 4.8, irradiation of PTBS films containing small amounts of an onium salt such as diphenyliodonium hexafluoroantimonate with UV light liberates an acid species that, upon subsequent baking, catalyzes cleavage of the protecting group to generate poly(*p*-hydroxystyrene). Loss of the *t*-butoxycarbonyl group results in a large polarity change in the exposed areas of the film. While the substituted phenol polymer is a nonpolar material soluble in nonpolar lipophilic solvents, poly(vinylphenol) is soluble in polar organic solvents and aqueous base. These resists have been used successfully in the manufacture of integrated-circuit devices.[131]

BASE INSOLUBLE

e^-

EXPOSE

NOVOLAC + $-\!\!(CH_2-C(CH_3)((CH_2)_2CH_3)-SO_2)\!\!-_x$

$\downarrow e^-$

DEVELOP

NOVOLAC + $SO_2\uparrow$ + $CH_2=C(CH_3)((CH_2)_2CH_3)\uparrow$

BASE SOLUBLE

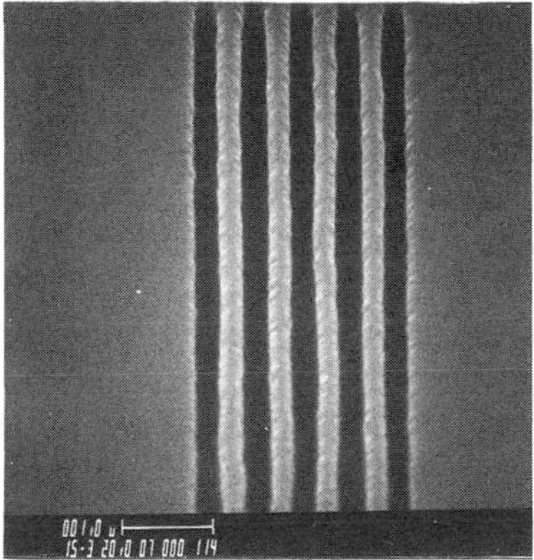

Figure 4.7 Chemistry and process sequence for the novolac–PMPS-positive electron-beam resist.

Alternative resins have been investigated for chemically amplified resist applications. The parent polymer is typically an aqueous-base-soluble high-T_g resin. Examples include poly(hydroxystyrene),[131] poly(vinyl benzoate),[132] poly(methacrylic acid),[133] N-blocked maleimide/styrene resins,[134, 135] and poly(hydroxyphenyl methacrylate).[136]

A matrix polymer that combines the advantages of chemically amplified deprotection with conventional chain-scission processes is poly(4-*t*-butoxycarbonyloxystyrene-sulfone) (PTBSS).[137, 138] As in the case of PTBS, the

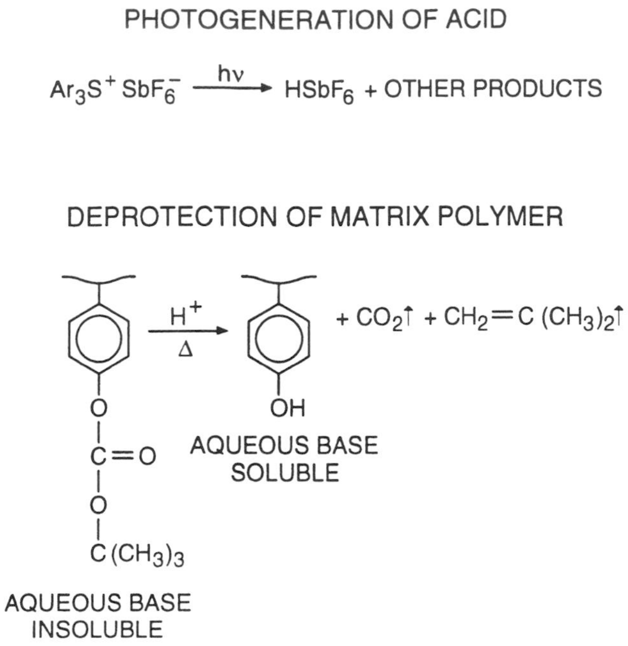

PHOTOGENERATION OF ACID

$$Ar_3S^+ SbF_6^- \xrightarrow{h\nu} HSbF_6 + \text{OTHER PRODUCTS}$$

DEPROTECTION OF MATRIX POLYMER

$$\xrightarrow[\Delta]{H^+} + CO_2\uparrow + CH_2{=}C(CH_3)_2\uparrow$$

O–C(=O)–O–C(CH$_3$)$_3$ AQUEOUS BASE INSOLUBLE

OH AQUEOUS BASE SOLUBLE

Figure 4.8 Chemistry associated with the PTBS-positive, chemically amplified resist.

t-butoxycarbonyl moiety is used as the acid-labile protective group. The inclusion of sulfur dioxide into the backbone of the polymer affords a high T_g that gives greater flexibility for processing. Additionally, introduction of sulfur dioxide into similar polymers (see above) has effectively improved their sensitivity to electron-beam radiation due to C — S bond scission.[139] When exposed to x-ray irradiation, PTBSS is an effective single-component chemically amplified resist.[140, 141] Radiation-induced C — S bond scission leads to generation of either sulfinic or sulfonic acid end groups that subsequently induce the deprotection reaction (Fig. 4.9).

The choice of protective group is equally wide. Clearly, this group should be thermally stable and acid labile. Examples of substituents that have been employed include *t*-butyl,[142] tetrahydropyranyl,[143-148] dihydropyranyl, and α,α-dimethylbenzyl.[133, 144, 149] Hydrolyzable groups such as trimethylsilyl[150-153] and various acetals and ketals[143, 144, 154, 155] have also been employed. Issues related to acetal chemistry include a decrease in the linewidth of unirradiated patterns with increasing delay intervals.[156] This phenomenon arises from acid migration at room temperature and may be alleviated through the use of bulky acids and/or organic base additives.

The concept of acid-catalyzed deprotection may also be applied to resist formulations utilizing a small molecule acting as a dissolution inhibitor for an aqueous alkali-soluble resin. This approach possesses a key advantage, reduced shrinkage. By using a small-molecule dissolution inhibitor, the content

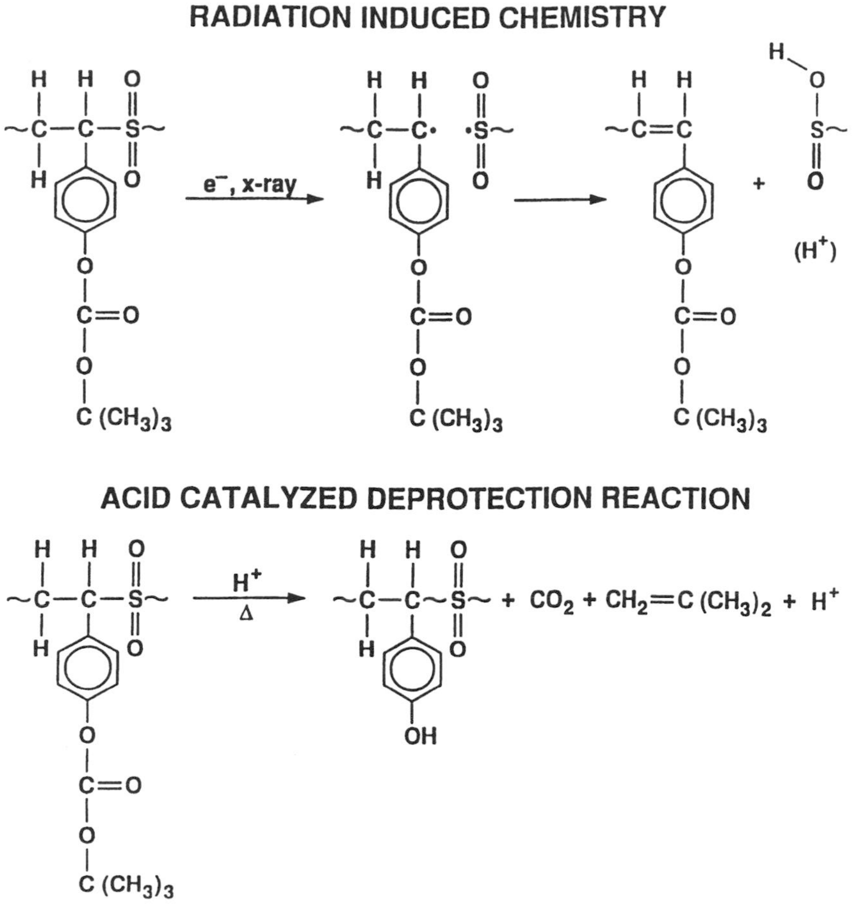

RADIATION INDUCED CHEMISTRY

ACID CATALYZED DEPROTECTION REACTION

Figure 4.9 Radiation-induced chemistry of PTBSS, a single-component, chemically amplified positive resist.

of the volatile, acid-cleavable group can be minimized, thus increasing the thermal stability of developed images. Materials that may be used effectively in dissolution inhibitor processes include carbonates or ethers of phenols,[143, 157–159] esters of carboxylic acids,[160–161] acetals,[162] or orthocarboxylic acid esters.[162] In one example, the *t*-butyl ester of cholic acid is used as a dissolution inhibitor for a phenol–formaldehyde matrix resin[160] (Fig. 4.10). Alternatively, a 193-nm resist has been developed using a substituted methacrylate resin in connection with a cholate-based inhibitor.[116] When formulated with an acid generator, irradiation generates a strong acid, which upon mild heating, liberates cholic acid. The irradiated regions may then be removed by dissolution in aqueous base. Workers at Fuji Film have applied their knowledge of traditional novolac diazonaphthoquinone dissolution inhibitor chemistry to the design of improved inhibitors for deep-UV

NOVOLAC RESIN + [cholate ester, C—O—C(CH₃)₃] + $Ar_3S^+X^-$

AQUEOUS BASE INSOLUBLE

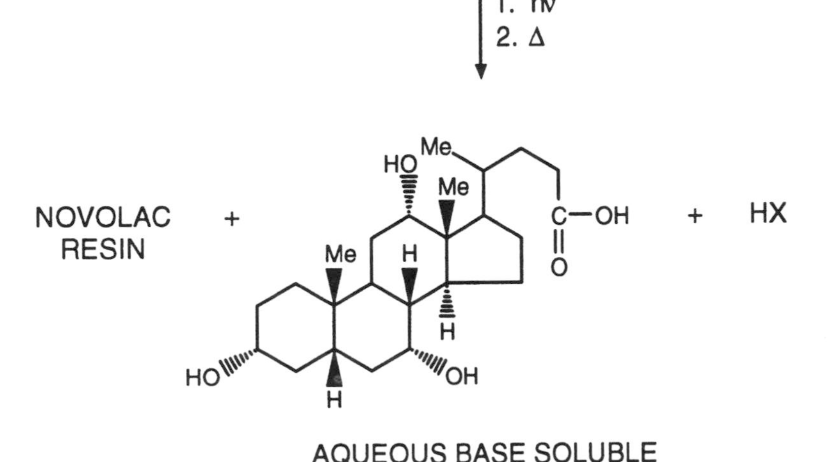

AQUEOUS BASE SOLUBLE

Figure 4.10 Schematic representation of cholate ester chemistry employed in a dissolution inhibition chemically amplified photoresist.

applications.[163] Notably, hydrophobicity, molecular size, and dispersivity of the acid-cleavable groups were influential in defining performance. The dissolution inhibitor may also be combined with the acid generator functions into a single chemically amplified resist additive.[164, 165]

Depolymerization Chemistry Chemically amplified resists that act through a polymer depolymerization mechanism can be broadly divided into two classes: those that act through a thermodynamically induced depolymerization mechanism and those requiring catalytic cleavage of a polymer backbone. The former process depends on the use of low-ceiling-temperature polymers that have been stabilized by suitable end capping. Introduction of a photocleavable moiety either at the end cap or along the polymer backbone may then allow depolymerization to take place after irradiation and mild heating. A

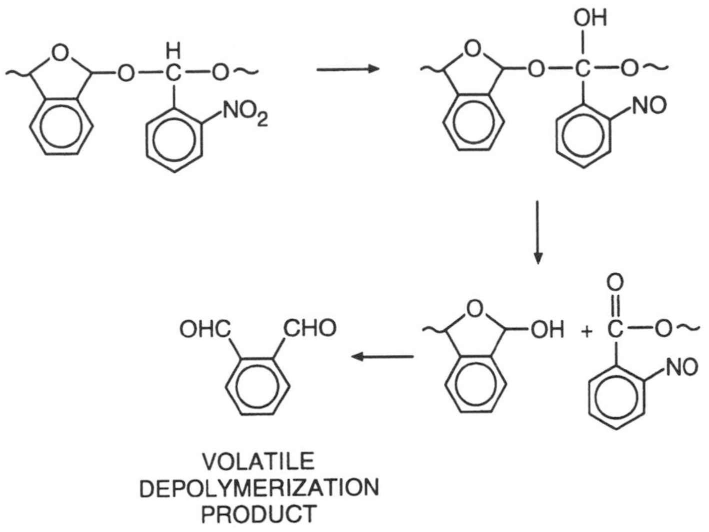

VOLATILE
DEPOLYMERIZATION
PRODUCT

Figure 4.11 Schematic representation of the radiation-induced depolymerization of polyphthalddehyde–nitrobenzyl ester copolymers.

variant of this approach utilizes an end cap or polymer chain that may be cleaved by photogenerated acid.[130, 132, 166–171] An example of this approach is presented in Fig. 4.11.

4.5 DRY-DEVELOPED RESIST CHEMISTRY

The increasing complexity and miniaturization of integrated-circuit technology is pushing conventional single-layer resist processes to their limit. The demand for improved resolution requires imaging features with increasingly higher aspect ratios and smaller linewidth variations over steep substrate topography. A number of schemes have been proposed to address this issue: the use of polymeric planarizing layers, antireflection coatings, and contrast-enhancement materials. Since this topic has been the subject of several recent articles, in this section we concentrate on the development of newer dry-developed resist chemistries, many of which are associated with top-surface-imaging mechanisms rather than delving into the more traditional, multilevel approaches.[79, 172, 173]

4.5.1 Dry-Developed Negative-Acting Resists

Simplification of both tri- and bilevel resist processes can be achieved by incorporating all of the desirable features of these multilevel technologies into a single layer of resist. In one case, the advantages of multilayer processing can be extended to a single resist layer by selective incorporation of a vinyl organometallic monomer into the exposed regions of a resist film.[174, 175] Specifically, the x-ray resist poly(2,3-dichloropropyl acrylate) was used as the host polymer, and diphenyldivinylsilane was used as the organometallic monomer.[174] X-ray exposure of the two-component resist film results in cross-linking of the polymer host and incorporation of the polymerizable metal (silicon)-containing monomer into the cross-linked network. In the subsequent process step, the organometallic monomer is thermally removed from the nonexposed regions of the film. Development is performed in an O_2 plasma where the covalently bound organometallic reagent present in the exposed region forms a refractory oxide (SiO_x) protecting the underlying regions. The unexposed regions that contain little or no metal are easily eroded in the oxygen environment.

Incorporation of inorganic species into the resist can also be accomplished through gas-phase functionalization processes.[176-179] For this process to work effectively, it is desirable to have an organic polymer that both contains a reactive functionality and is sufficiently absorbant to limit deposition of the irradiation dose to the topmost part of the resist film, allowing higher-resolution imaging. One example involves the use of poly(*t*-butyl-*p*-vinylbenzoate).[180] When formulated with an onium salt, this material has an optical density $\gg 1$ at 248 nm. Exposure and postexposure bake generates poly(*p*-vinylbenzoic acid) in the near-surface regions of the resist, which may then be functionalized with a variety of reagents. After irradiation, treatment with a reactive inorganic or organometallic reagent such as $SiCl_4$, $TiCl_4$, or hexyamethyldisilazane (HMDS) results in a reaction between the exposed regions of the matrix and incorporation of the metal atom into these regions. When such films are subjected to a reactive ion etching environment, an etch barrier is created in the exposed areas, and a negative image is generated. Similar results were obtained with PTBS as the matrix polymer,[178] a schematic representation of which is shown in Fig. 4.12.

Conventional positive photoresist can be used in similar processes.[181-184] The high absorption of a DNQ modified/novolac resist formulation at 248 nm limits exposure to the near surface. Heating the film after exposure effectively cross-links only the nonirradiated regions of the film, allowing selective silylation in the exposed regions. Patterning is subsequently accomplished via O_2 RIE processes. A modification of this process to generate positive tone patterns has been reported by Pierrat et al.[185]

A recent modification of the gas-phase silylation process involves liquid-phase silylation, after either exposure or development. For example, the bifunctional silylating agent hexamethylcyclotrisilazane (HMCTS), which acts as a cross-linking agent when incorporated into a standard diazonaphthoquinone–novolac resist, has been reported as a means to incorporate silicon into a patterned resist.[186, 187] Silylation of a styrene maleic anhydride copolymer containing a diazoquinone photoactive compound in an aqueous solution containing a bisaminosilazane has also been reported.[188]

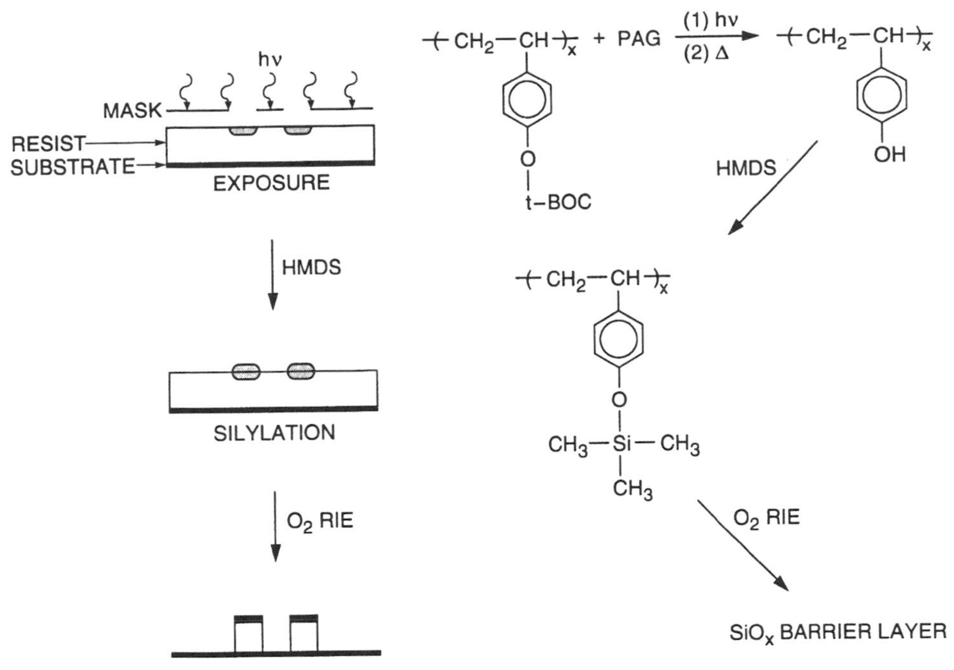

Figure 4.12 Schematic representation of the gas-phase silylation process utilizing PTBS as the matrix resist polymer.

A recent comparative study between gas- and liquid-phase silylation using the DESIRE process has been performed.[189] The study indicated that use of the liquid-phase silylation agent bis(dimethylaminodimethylsilane) dissolved in xylene containing NMP as a diffusion promoter resulted in higher silicon incorporation into the exposed area than could be achieved using a gas-phase silylation process. The improved silylation selectivity afforded improved resolution (0.25 μm · l/s), wide focus latitude, and enhanced dry-etching resistance and thermal stability of the patterned resist.[190]

An all-dry lithographic process has been proposed by Weidman and co-workers.[191, 192] In this scheme, a conformal coating of polymethylsilane is generated on a substrate surface via plasma deposition. These plasma-deposited films are highly sensitive to UV light and can be patterned in an entirely dry process. On exposure to light, photo-oxidation of the plasma-polymerized methylsilane (PPMS) films converts it to a form of silicon dioxide. The unexposed portions of the film are easily removed by subsequent treatment with a halogen plasma, which does not attack the exposed silicon oxide–like regions (Fig. 4.13). The resolution achievable is as good as the mask that defines the exposure. The PPMS film can be deposited onto any substrate layer. Because the films bleach during exposure, no antireflection layer is required. Further, because a thin (0.2-μm) conformal film is

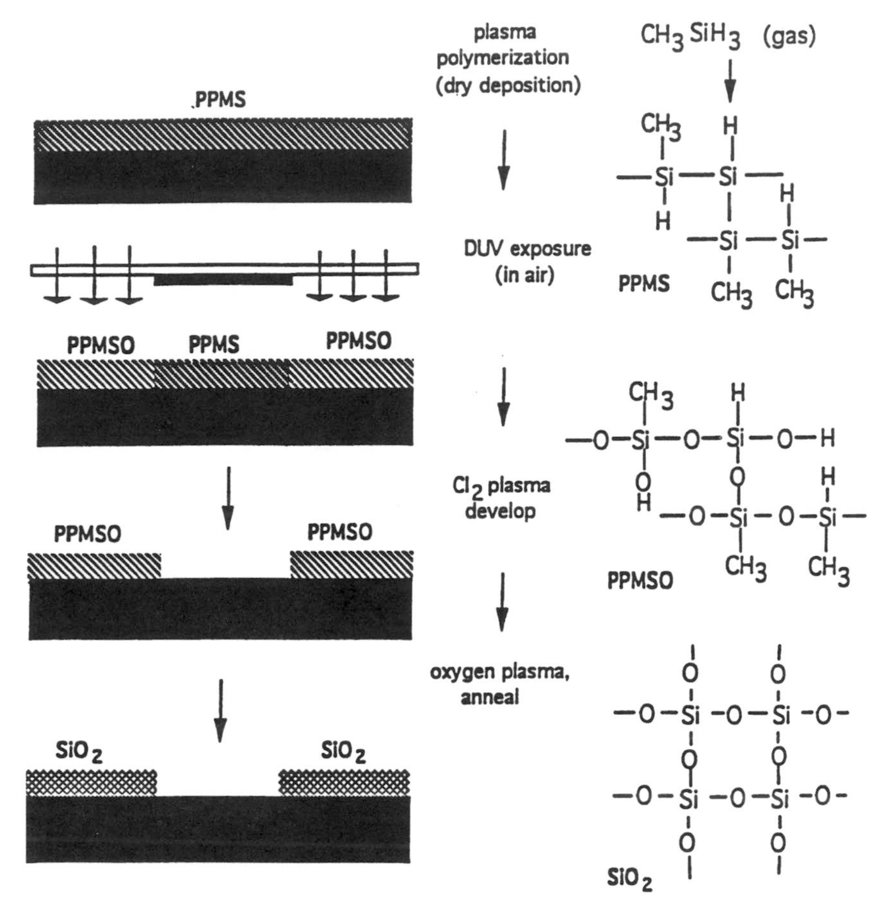

Figure 4.13 Process sequence for the plasma polymerized methyl silane resist system.

employed, there is no depth-of-focus problem in creating a pattern over topographic features on the substrate. As no organic solvent is required for the deposition step and there are no liquid wastes generated in the pattern development process, this is a "green" technology.

4.5.2 Dry-Developed Positive-Acting Resists

The first positive-resist process in which the development and pattern transfer step can be carried out via O_2 reactive ion etching has been reported by Meyer et al.[193] Deep-UV exposure of an acrylate-based copolymer containing silicon substituents effects cleavage of the side chains, which are subsequently removed from the film by heating. The exposed areas are then deficient in silicon and may be removed in an O_2 RIE environment.

In an alternative approach, Taylor et al. incorporated an etching-resistant moiety into the resist after the exposure step.[194] In one embodiment, a polyisoprene–bisazide resist was cross-linked upon UV irradiation. The exposed resist was then subjected to a $SiCl_4$ atmosphere, resulting in selective diffusion of the silicon reagent into the unexposed regions of the film. The exposed, cross-linked, and silicon-deficient areas were then developed via O_2 RIE. This concept has been extended to other resin systems, such as novolacs.[195] Upon 193-nm irradiation, the novolac undergoes cross-linking in the uppermost regions of the film. Subsequent treatment of the film with gaseous dimethylsilyl–dimethyl amine at 100°C resulted in selective diffusion of silicon into the nominally unexposed areas. This near-surface imaging silylation process readily produced sub-0.5-μm features in a wide variety of novolac-based conventional resist systems. Positive tone patterns were obtained in three-component chemically amplified–based systems in processes involving acid-catalyzed cross-linking followed by gas-phase silylation.[196]

4.6 POLYMERS AS PACKAGING MATERIALS AND ENCAPSULANTS

Polymers are central to microlithography, the process by which patterns are drawn onto substrates such as silicon, and the ever-increasing complexity and density of microelectronic devices remains a driving force for new materials research. Organic resists are materials that play an active role in device fabrication, but there are additional passive roles for polymers in microelectronics. These applications include the use of polymers as thin-film dielectric layers, adhesives, encapsulants, and substrates. An important distinction exists between resist polymers and those used in passive roles. Resists are removed from the device after pattern formation has been achieved, but the latter class of materials remain throughout the lifetime of that device. Thus considerations such as long-term thermal, hydrolytic, mechanical, and photostability become an issue. As performance demands increase, coefficient of thermal expansion (CTE) and Young's modulus are also of concern. In the following sections we discuss the uses of polymeric materials in the context of these passive components.

4.6.1 Organic Dielectric Materials

Apart from resists, dielectric materials are by far the most important class of organic materials used in microelectronics. Dielectric materials fulfill two critical roles: as interlayer materials where the polymer serves to separate small metal lines and permits high chip-to-carrier connection and in systems in which the dielectric material is used as a printed circuit board (PCB), encapsulant, or substrate. The most important class of polymeric materials to fulfill these roles are the polyimides, although inorganic dielectric materials are also important.

Polyimides Polyimides are a class of polymers that possess a unique combination of chemical, electrical, and physical properties, including high thermal decomposition temperatures (up to 500°C), high glass transition temperatures, and high mechanical strength.[197, 198] Additionally, polyimides have suitable coefficients of thermal expansion that can be matched to a variety of substrates and a unique combination of electrical properties. It is these properties, high resistivity, high breakdown voltages, and low dielectric constant, that have ensured wide acceptance for polyimides in the microelectronics industry. The importance of polyimides as dielectric materials for microelectronic devices can be understood by considering some elementary relationships. In a generic device, the propagation velocity (V) of a pulsed signal in a line is inversely proportional to the square root of the dielectric constant of the propagation medium,[199] or $V = c/\sqrt{\epsilon}$. It is clear from this equation that the smaller the dielectric constant (ϵ), the faster the signal can travel in the medium, leading to shorter device cycle time. An added advantage of using a low-dielectric-constant insulator is that crosstalk, electronic communication between levels within a multilevel device, is reduced. With the development of ever-faster and ever-denser electronic devices, this second advantage becomes of increasing importance.[200]

Polyimide Synthesis Prototypical of a simple polyimide dielectric is the polymer derived from pyromellitic dianhydride (PMDA) and 4,4′-oxidianiline (ODA); the synthesis of this material is shown schematically in Fig. 4.14. Polyimides are normally prepared in two steps.[201] In the first step, an aromatic dianhydride is reacted with an aromatic diamine in a solvent such as *N*-methylpyrolidinone (NMP) to form an intermediate polyamic acid. Next, the polyamic acid is converted to a polyimide via either chemical treatment or thermal dehydration at high temperature. NMP loss starts at 120°C. Imidization begins at about 100°C and is not complete until temperatures much higher than 200°C are reached. Reproducible preparation of the polyamic acid is the most critical factor to be considered, since polyamic acids with controlled molecular weights are necessary for reliable and uniform film coverage. The molecular weight of the polyamic acid depends on environmental factors, the amount of H_2O present, the solvent, the monomer purity, and the monomer addition sequence. Most preparations call for slow addition of a dianhydride solution or a powdered dianhydride to the diamine in solution. Polar solvents such as NMP, dimethylsulfoxide (DMSO), and dimethylformamide (DMF) are necessary to form homogeneous solutions. Solution viscosity changes markedly with time; a highly viscous solution is formed initially, but the viscosity drops until the molecular weight of the polyamic acid stabilizes. Initially, this behavior was attributed to hydrolysis

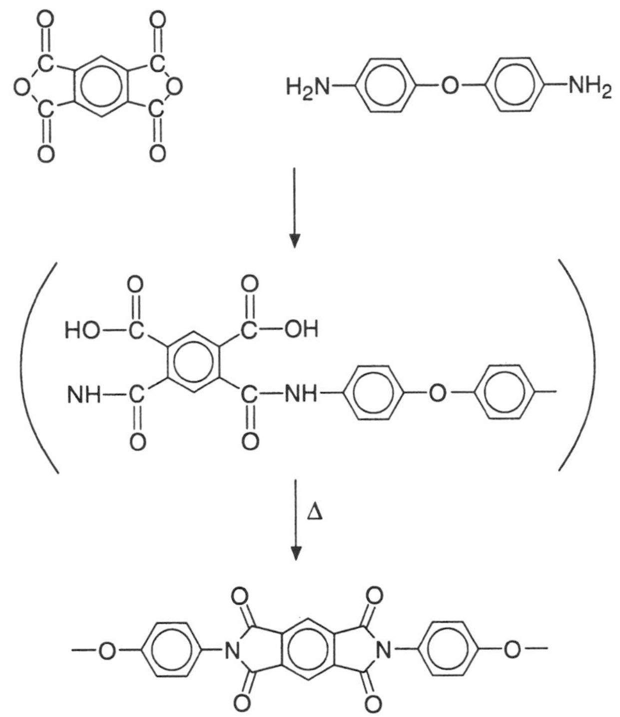

Figure 4.14 Schematic representation of the polyamic acid obtained form PMDA and ODA.

from trace amounts of H_2O present in the solvent.[202, 203] Later work has shown that this phenomenon is due to the reversibility of the propagation reaction.[204]

The polymer derived from PMDA and ODA (Fig. 4.14) was developed by DuPont under the tradename Kapton. It has an excellent combination of physical, electrical, and chemical properties.[205] Electronic applications for this material include its use as an interlayer dielectric ($\epsilon = 3.1$), for α-particle protection, as a passivation overcoat, as a planarizing layer, and as an encapsulant of IC circuits.

More complex polyimide structures are clearly possible and have been developed. For example, polyimide–isoindoloquinazolinedione (PIQ), developed by Sato et al.,[206] was the first polyimide designed exclusively for multilevel applications. PIQ was synthesized from an aromatic diaminocarbonamide, an aromatic diamine, and two aromatic dianhydrides. As a result, in addition to the usual polyimide functionality, it contains an isoindoloquinazolinedione group (Fig. 4.15) in the polymer backbone that significantly enhances the heat resistance of the polymer. PIQ is stable to temperatures as

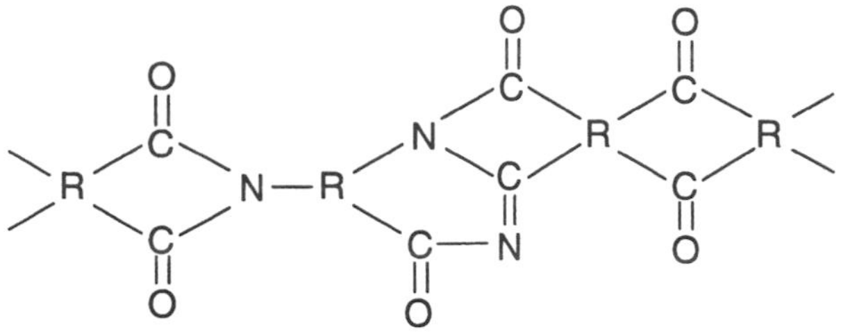

Figure 4.15 Schematic representation showing the isoindoloquinazolendione group of PIQ.

TABLE 4.1 Comparison of Properties of Polyimide and Inorganic Dielectrics

	Polyimide	SiO_2	Si_3N_4
Process temperature (°C)	300–350	350–450[a]	700–900[a]
Decomposition temperature (°C)	450	1710[b]	1900[b]
Dielectric strength (MV/cm)	3–7	5–8	5–10
Volume resistivity (Ω · cm)	10^{15}–10^{16}	$> 10^{16}$	10^{14}–10^{16}
Dielectric constant	3.2–3.8	3.5–4.0	7–10
Dissipation factor	0.01–0.02	0.001	—
Expansion coefficient (10^{-6} °C^{-1})	20–70	0.3–0.5	4
Thermal conductivity (W/cm · °C)	0.0017	0.021	0.12
Density (g/cm^3)	1.42	2.2	2.8–3.1
Planarization (%)	60–100	0	0
Refractive index	1.6–1.8	1.45	2.0

[a]Decomposition temperature.
[b]Melting point.

high as 450°C before it suffers any appreciable weight loss and has excellent planarization behavior.[207]

When fully cured, polyimides are resistant to chemical and physical attack by most acids, common organic solvents, and weak bases. Adhesion to silicon dioxide and silicon nitride is good. The physical properties of a conventional polyimide are compared to silicon dioxide and silicon nitride in Table 4.1. Polyimides have dielectric constants ranging from 3.8 for conjugated examples to as low as 2.2 for highly fluorinated materials. Efforts to reduce the dielectric constant remain an area of intense interest, since this goal is related to the ability to reduce circuit dimensions. Reduced circuit dimensions lead in turn to smaller chip design and faster circuit speed.

Advanced Polyimides Much of the driving force in polyimide research in recent years has been aimed at (1) decreasing the dielectric constant and (2) developing mechanisms to introduce photodefinibility into polyimide systems. We address each of these in turn.

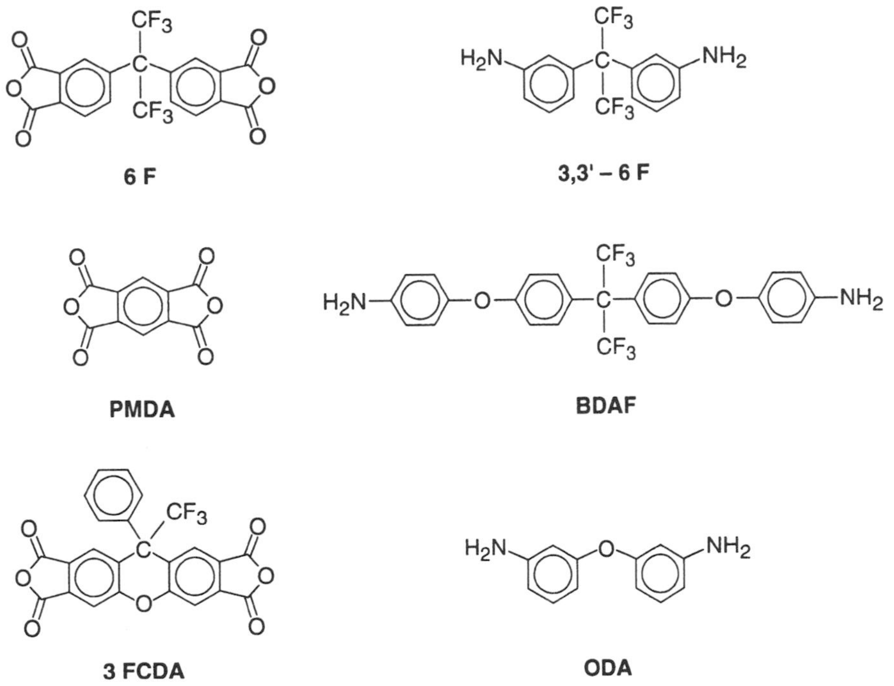

6 F

3,3' – 6 F

PMDA

BDAF

3 FCDA

ODA

Figure 4.16 Polyimide monomer structures.

Reduction of the Dielectric Constant Much effort has been directed to lowering the dielectric constant (ϵ) of polyimides. The introduction of fluorine as a substituent into polyimide has received much attention in this regard. Typical fluorinated polyimide precursors are shown in Fig. 4.16; use of the hexafluoroisoprpopyl group is a popular strategy to introduce fluorine into the polymer backbone. Polymers have been prepared in which the hexafluoroisopropyl group is in either the anhydride or the amine component of the reactive components. The use of fluorinated monomers has a dramatic effect on the value of the dielectric constant of the resulting polymers, as evidenced by the data in Table 4.2[208, 209] Note that the polymer containing the hexafluoroisopropyl group in both the anhydride and the diamine monomers has the lowest dielectric constant, 2.39. In certain cases, fluorinated monomers impart other desirable properties to polyimides. For example, when 9-trifluoromethyl-9-phenyl-2,3,6,7-xanthenetetracarboxylic dianhydride (3FCDA) is paired to certain rodlike diamines, fully soluble quasi rodlike polyimides were obtained which yielded low-thermal-expansion polyimide films.[210, 211] The major disadvantages of these fluorinated polyimides are their cost, but in certain applications where a low dielectric constant coupled with low moisture uptake and high T_g is an issue, the performance of complex fluorinated polyimides justifies their cost.

TABLE 4.2 Dielectric Constants for Fluorinated Polyimides

Polymer	Dielectric Constant (at 10 MHz)
PMDA + ODA (Kapton)	3.22
6F + ODA	2.79
6F + BDAF	2.50
6F + 3,3′-6F	2.39
3F LDA + ODA	2.80[a]

[a]At 1 MHz.

A second recent innovation to lowering the dielectric constant of polyimide films is to introduce microscope bubbles into the polymer matrix.[212-214] This approach is based on the realization that air, with a defined dielectric constant of 1, can significantly lower the dielectric constant of a polyimide nanofoam. Typical synthetic approaches to this class of materials involves preparing microphase-separated block copolymers where a thermally labile component is dispersed in a thermally stable matrix. Upon thermolysis, the thermally labile component undergoes decomposition, leaving pores in the film commensurate with the initial copolymer morphology. For thin-film applications, the pore sizes must be significantly smaller (i.e., $\ll$ 1 mm) to maintain the integrity of the dielectric layer. This approach is likely to see wider application in the future.

Photosensitive Polyimides The ability to photodefine polymers with greater and greater resolution is the key to advanced resists and is becoming increasingly important in low-dielectric polymers such as polyimides. For multilevel applications, in which vias for metal connections between levels are needed, subtractive photolithography techniques in which a photoresist is applied on top of a polyimide are usually employed. These techniques[215-217] can be rather cumbersome, since they require etching of a cured polyimide film with corrosive agents such as hydrazine hydrate, using a photoresist as an etching mask. Photosensitive polyimides that can be patterned directly are attractive alternatives in that they minimize the number of processing steps and give better control over via sidewalls.

Kerwin and Goldrich[218] were first to design a photosensitive polyimide. They reported that polyamic acids sensitized with potassium dichromate could be used as negative-working high-temperature photoresists. They were able to obtain good resolution, but the resist solutions suffered from a short shelf life and had to be mixed immediately before use. This work, because of the potential of reducing the number of processing steps associated with the use of polyimides as protective and insulating layers, generated a great deal of interest in developing photosensitive polyimides. Rubner and coworkers[219, 220] developed a series of photosensitive polyimides that contained

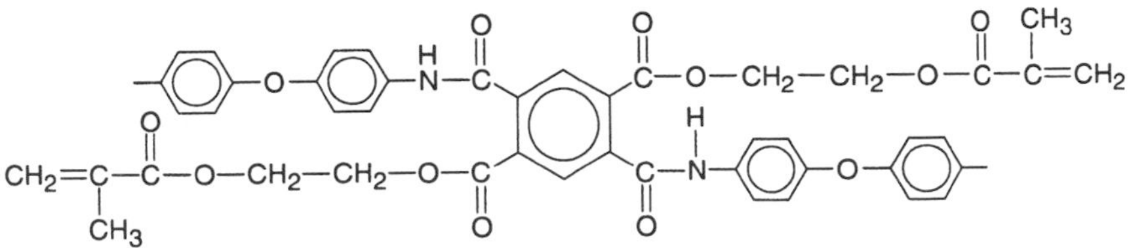

Figure 4.17 Photosensitive polyamides containing methacrylate esters.

carboxylic acid methacrylate esters (Fig. 4.17). Exposure times depend on the initial film thickness with typical exposures requiring 40 to 180 s. Resolution of 20-μm lines and spaces were obtained in 2-μm-thick films. After exposure and development, the esters are converted into polyimides by thermal treatment at 275°C in N_2. This general approach of esterifying a polyamic acid with a polymerizable group such as acrylate has been adopted widely. Incorporation of a photoinitiator into a film of the resulting polymer allows it to be gelled in the exposed areas. Unexposed regions remain soluble and may be washed away by a suitable developer. The cross-linkable moities are then volatilized upon thermal treatment, leaving behind patterned polyimide. The problem with this approach is that a significant fraction of the mass of the original film must be volatilized, resulting in significant shrinkage.

Davis[221] has described a photosensitive siloxane polyimide that has excellent sensitivity and adhesion. The material contains polyaminoesters from diamines that contain an aminodisiloxane. Thermal curing of the polymer, after exposure and development, forms a polyimide siloxane that has good insulation resistance and better adhesion than that of most commercial polyimides. Films as thick as 12 μm were patterned, and 40-μm features were resolved.

A second approach to photoimaged polyimides relies upon the incorporation of photosensitive moities directly into the polyimide backbone. Photoexcitation results in the abstraction of hydrogen atoms from benzylic groups incorporated into the polymer backbone. The resulting radicals combine to yield cross-links. These materials are typically soluble in their imidized state, have relatively low T_g values, and are less solvent resistant than typical polyimides. They also suffer from the disadvantage of being considerably less sensitive to UV light than a conventional deep-UV resist (often by an order of magnitude) and can be partially opaque because of the high concentration of chromophore required. Despite these limitations, the development of polyimides that are intrinsically photoimagable continues to be an area of intense interest.

Other High-Temperature Polymers Several high-temperature polymers besides polyimides have been evaluated for use as dielectrics. These include cyclized rubber photoresists, poly-*p*-xylylene (PPX), and novolac resins. Easy application, good adherence, high thermal stability, excellent insulation resistance, and high mechanical strength are attractive attributes of these materials. Several of these materials are discussed in detail below.

Poly-p-xylylene and Related Aromatic Polymers Poly-*p*-xylylene (PPX) is a crystalline polymer that has several advantages as a dielectric material.[222-224] PPX films are formed via a three-step vacuum process (Fig. 4.18), which includes (1) vaporization of di-*p*-xylylene, (2) cleavage of the dimer to form monomeric diradicals, and (3) vapor deposition of the diradicals onto substrates at room temperature. PPX has lower water absorption, a lower dielectric constant (about 2.65), and higher dielectric strength than those of most polyimides.[225] Unfortunately, the material is insoluble and can be deposited only by the vacuum deposition process. The latter consideration may not itself be a problem since chemical vapor deposition has become a preferred method for producing inorganic thin-film devices. The generation of organic thin films by similar processes may be an attractive methodology if the materials can be generated without the production of unwanted side products. Fluorinated PPX, in which all the methylene protons have been replaced by fluorine, has higher oxidative thermal stability than PPX (up to 500°C), is resistant to UV irradiation even in the presence of oxygen, and has a dielectric constant of 2.36. Fluorinated PPX has as yet only been reported in the literature,[226] but efforts continue to develop commercially feasible routes to this and analogous polymers.

Other promising materials for microelectronic applications are based on the ring opening of benzocyclobutenes to form *o*-quinodimethane (*o*-xylylene) intermediates, which then undergo cycloaddition or dimerization reactions[227] (Fig. 4.19). When the monomers contain one or more cyclobutene groups per molecule, this reaction can be applied to generate polymers. Polymers prepared from these starting materials have a low dielectric constant, little moisture sensitivity, form conformal films on a variety of substrates, and have good thermal stability.[228] Workers at Dow Chemical have

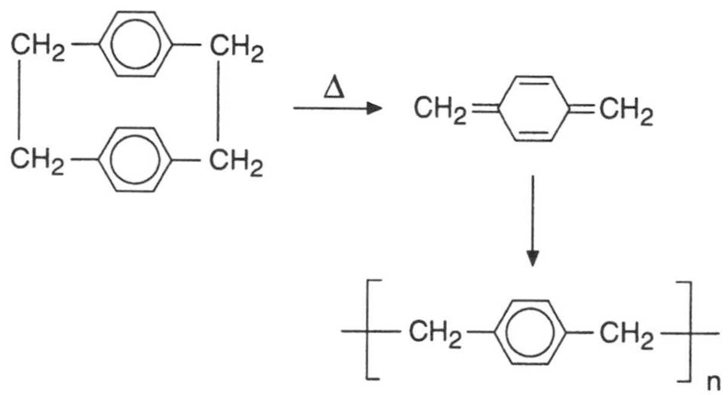

Figure 4.18 Preparation of poly(*p*-xylylene) (PPX).

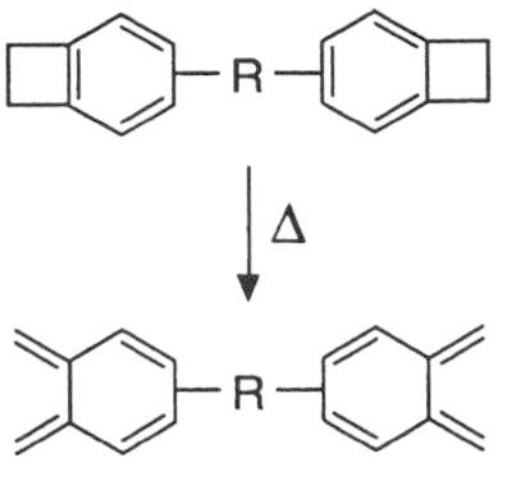

Figure 4.19 Schematic representation of the generation of *o*-quinodimethane.

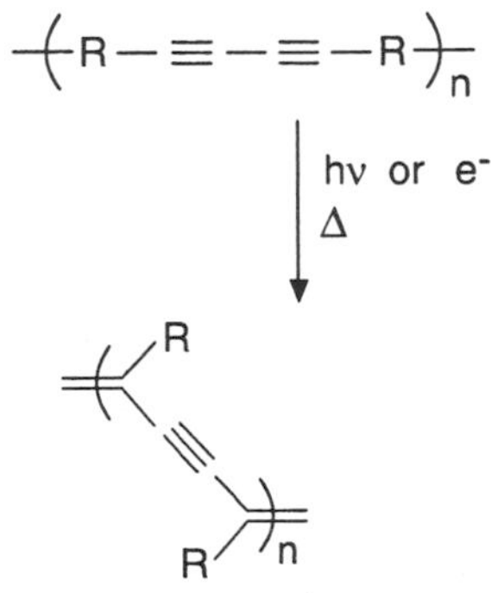

Figure 4.20 Schematic representation of the cross-linking reaction of polydiacetylenes.

used the reactivity of cyclobutene monomers toward Diels–Alder reactions to develop maleimide benzobutene monomers, which undergo self-condensation to prepare a wide variety of polymers possessing a broad range of physical properties.[229]

Miller and co-workers have prepared intrinsically photochemically cross-linkable polymers based on diaryl diacetylenes.[230, 231] It has long been known that poly(diaryl diaceylenes) can be thermally or photochemically cross-linked (Fig. 4.20), although the thermal reactivity and solubility of such systems was too high to be of practical importance.[226-234] Miller has shown that the preparation of polymers having greater chain flexibility leads to photodefinable polymers with dielectric constants as low as 2.82.

Cyclized Polybutadiene Dielectrics Photosensitive cyclized polybutadiene (CBR) is a negative-working photoresist that has been evaluated for use as an insulating dielectric in multilevel metallization schemes. The material is thermally and chemically stable enough to withstand wet- and dry-etching processes. Ito[235] reported that smooth uniform films with thicknesses from submicrometer to several micrometers are easily obtained. CBR consists of cyclized polybutadiene and a bisazide cross-linking agent dissolved in xylene (Fig. 4.21). This material is extremely photosensitive and is chemically and thermally stable. In addition, it can easily be deposited in films up to 10 mm thick and is simple to process.

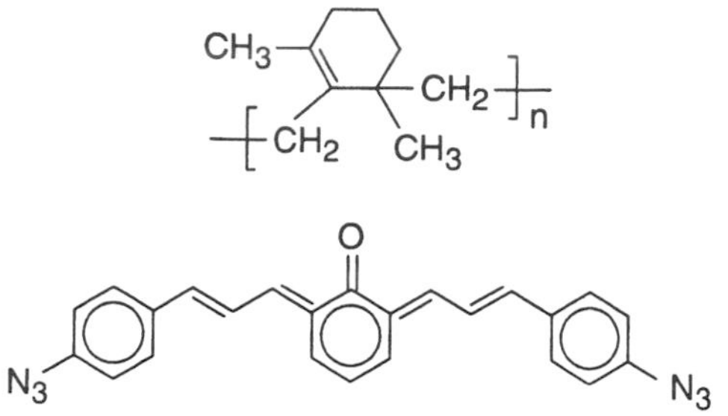

Figure 4.21 Schematic representation of the structure of a cyclized polybutadiene dielectric.

Novalac Resins The use of commercial novolac photoresists (described earlier in this chapter) as photodefinable dielectrics was demonstrated by Lyons and co-workers.[236, 237] These materials are easy to process, thermally stable to about 270°C, good insulators ($p \sim 10^{15}$ $\Omega \cdot$ cm), and have a low dielectric constant.

4.6.2 Planarization

An important advantage of organic dielectrics is their ability to planarize a surface. Planarization is the process by which a flat topography is generated on a device surface after a number of processing steps. The degree of planarization is determined by the ratio of the step height with the organic coating (typically, a polyimide) to the initial step height of the underlying metal coating. The small depth of focus of step-and-repeat exposure tools make the generation of planar surfaces a vital step in the fabrication of a multilevel device. Because of their solubility in organic solvents, polyimides may be spin-coated onto an underlying topography and are widely used. Inorganic dielectrics, deposited from the vapor phase, although of high purity, suffer from the fact that they form conformal coatings and are difficult to apply in thick films without producing mechanical defects. New generations of inorganic spin-on glasses are being produced which promise to address these issues.

4.6.3 Encapsulants

The rapid development of integrated-circuit (IC) technology has imposed stringent requirements not only on IC physical design and fabrication but also on IC encapsulants. The purpose of an encapsulant is to protect the electronic device from moisture, mobile ions, UV and visible irradiation, α-particles, and hostile environmental conditions. Furthermore, the encapsulant should improve the mechanical properties of the fragile IC device and increase device reliability. Encapsulants must have high purity, good electrical and mechanical properties, be resistant to solvents, fluxes, and detergents, and like interlayer dielectrics, form low-stress films. The commonly used organic encapsulants include thermosets such as silicones, polyimides, polyesters, epoxies, and alkyl resins; thermoplastics such as polystyrene, polyethylene, poly(vinyl chloride); and fluorocarbon polymers and elastomers such as silicone rubbers, silicone gel, and polyurethanes.

Silicones Room-temperature-vulcanized (RTV) silicone elastomers are among the best encapsulants for moisture- and temperature-cycling protection of IC devices.[238, 239] They consist of linear, hydroxy-terminated siloxane fluids which are cross-linked or vulcanized by condensation cure reactions. Typically, inorganic fillers are added to reduce the coefficient of thermal expansion and to control the color of the encapsulant. Intense purification techniques are needed to keep the Na^+, K^+, and Cl^- mobile ion concentrations to less than a few ppm. RTV elastomers suffer from poor solvent resistance and have weak mechanical properties. Highly fluorinated, alkyl-substituted siloxanes have been developed that have better solvent-resistant properties. Heat-curable silicone gels[239] have better thermal properties than RTV silicones. Because of their very low modulus, the gels are attractive candidates for encapsulating wire-bonded IC devices. They are solvent-free systems and cure without any by-product formation. Silicone gels are expected to see increasing use in electronic applications.

Epoxies The unique thermal and physical properties of epoxies have made them among the most widely utilized polymeric materials in microelectronics.[240, 241] They have excellent adhesion, insulation resistance, low shrinkage, and low cost. Typical commercial formulations are based on bisphenol A, which reacts with epichlorohydrin to produce bisglycidyl ethers. Novolac epoxies produce materials with higher cross-link density and better chemical and thermal resistance than those of bisphenol A derivatives. Anhydrides and amines are the most frequently used curing agents. The reactivity of anhydrides with epoxies is slow; therefore, tertiary amine elastomer is added to a rigid epoxy to reduce its cross-link modulus.

REFERENCES

1. M. J. Bowden, S. R. Turner, Eds., *Electronic and Photonic Applications of Polymers*, ACS Advances in Chemistry Series, Vol. 218, American Chemical Society, Washington, DC, 1988.
2. J. H. McCoy, W. Lee, G. L. Varnell, *Solid State Technol.* **32**(3), 87 (1989).
3. L. F. Thompson, C. G. Willson, M. J. Bowden, Eds., *Introduction to Microlithography*, ACS Symposium Series, Vol. 219, American Chemical Society, Washington, DC, 1983.

4. E. Reichmanis, L. F. Thompson, *Chem. Rev.* **89**, 1273 (1989).
5. C. G. Willson, in *Introduction to Microlithography*, ACS Symposium Series, Vol. 219, L. F. Thompson, C. G. Willson, M. J. Bowden, Eds., American Chemical Society, Washington, DC, 1983, pp. 88–159.
6. T. Takigawa, *J. Photopolym. Sci. Technol.* **5**(1), 1 (1992).
7. E. Reichmanis, L. F. Thompson, in *Polymers in Microlithography: Materials and Processes*, ACS Symposium Series, Vol. 412, E. Reichmanis, S. A. MacDonald, T. Iwayanagi, Eds., Washington, DC, 1989, pp. 1–24.
8. L. F. Thompson, M. J. Bowden, in *Introduction to Microlithography*, ACS Symposium Series, Vol. 219, L. F. Thompson, C. G. Willson, M. J. Bowden, Eds., American Chemical Society, Washington, DC, 1983, pp. 162–214.
9. W. M. Moreau, *Semiconductor Lithography: Principles, Practices and Materials*, Plenum Press, New York, 1988.
10. E. Reichmanis, S. A. MacDonald, T. Iwayanagi, Eds., *Polymers in Microlithography*, ACS Symposium Series, Vol. 412, American Chemical Society, Washington, DC, 1989.
11. A. Reiser, *Photoreactive Polymers: The Science and Technology of Resists*, Wiley, New York, 1989, pp. 22–65.
12. T. Iwayanagi, T. Kohashi, S. Nonogaki, T. Matsuzawa, K. Douta, H. Yanazawa, *IEEE Trans. Electron. Dev.* **28**(11), 1306 (1981).
13. S. Nonagaki, H. Hashimoto, T. Iwayanagi, H. Shiraishi, *Proc. SPIE* **539**, 189 (1985).
14. H. Hashimoto, T. Iwayanagi, H. Shiraishi, S. Nonogaki, *Proceedings of the Regional Conference on Photopolymers*, Mid-Hudson Section, SPE, Ellenville, NY, 1985, p. 11.
15. D. R. Heriott, R. J. Collier, D. S. Alles, J. W. Stafford, *IEEE Trans. Electron. Dev.* **22**, 385 (1975).
16. L. F. Thompson, R. E. Kerwin, in *Annual Review of Materials Science*, Vol. 6, R. A. Huggins, R. H. Bube, R. W. Roberts, Eds., 1976, pp. 267–301.
17. S. Tagawa, in *Polymers for High Technology: Electronics and Photonics*, ACS Symposium Series, Vol. 346, L. F. Thompson, C. G. Willson, J. M. J. Frechet, Eds., American Chemical Society, Washington, DC, 1987, pp. 37–45.
18. T. Hirai, Y. Hatano, S. Nonogaki, *J. Electrochem. Soc.* **118**(4), 669 (1971).
19. E. D. Feit, L. F. Thompson, R. D. Heidenreich, *ACS Div. Org. Coat. Plast. Chem. Prepr.* 383 (1973).
20. Y. Taniguchi, Y. Hatano, H. Shiraishi, S. Horigome, S. Nonogaki, K. Naraoka, *Jpn. J. Appl. Phys.* **28**, 1143 (1979).
21. Z. C. Tan, C. C. Petropoulos, F. J. Rauner, *J. Vac. Sci. Technol.* **19**(4), 1348 (1981).
22. A. E. Novembre, M. J. Bowden, *Polym. Eng. Sci.* **23**, 977 (1983).
23. Y. Tabata, S. Tagawa, M. Washio, in *Materials for Microlithography*, ACS Symposium Series, Vol. 266, L. F. Thompson, C. G. Willson, J. M. J. Frechet, Eds., American Chemical Society, Washington, DC, 1984, pp. 161–163.
24. M. A. Hartney, R. G. Tarascon, A. E. Novembre, *J. Vac. Sci. Technol.* **B3**, 360 (1985).
25. S. Imamura, *J. Electrochem. Soc.* **126**(9), 1268 (1979).

26. H. S. Choong, F. J. Kahn, *J. Vac. Sci. Technol.* **19**(4), 1121 (1981).
27. E. D. Feit, L. F. Thompson, C. W. Wilkins, Jr., M. E. Wurtz, E. M. Doerries, L. E. Stillwagon, *J. Vac. Sci. Technol.* **16**(6), 1987 (1979).
28. S. Imamura, S. Sugawara, *J. Appl. Phys.* **21**, 776 (1982).
29. N. Yoshioka, Y. Suzuki, T. Yamazaki, *Proc. SPIE* **537**, 51 (1985).
30. L. F. Thompson, E. M. Doerries, *J. Electrochem. Soc.* **126**(10), 1699 (1978).
31. L. F. Thompson, L. Yau, E. M. Doerries, *J. Electrochem. Soc.* **126**(10), 1703 (1979).
32. A. E. Novembre, L. M. Masakowski, M. A. Hartney, *Polym. Eng. Sci.* **26**(6), 1158 (1986).
33. M. Hatzakis, J. Paraszczak, J. M. Shaw, in *Microcircuit Engineering* 81, A. Oosenburg, Ed., Swiss Federal Institute of Technology, Lausanne, Switzerland, 1981, pp. 386–396.
34. S. A. MacDonald, F. Steinman, H. Ito, W.-Y. Lee, C. G. Willson, *Prepr. ACS Div.*, Polym. Mater. Sci. Eng. **50**, 104 (1983).
35. M. Suzuki, K. Saigo, H. Gokan, Y. Ohnishi, *J. Electrochem. Soc.*, **30**, 1962 (1983).
36. A. E. Novembre, M. J. Jurek, A. Kornblit, E. Reichmanis, *Polym. Eng. Sci.* **23**, 920 (1989).
37. H. Moritz, *IEEE Trans. Electron. Dev.* **32**, 672 (1985).
38. Y. Takahashi, F. Shinozaki, T. Ikeda, *Jpn. Kokai Tokyo Koho* **88**, 8032 (1980).
39. S. A. MacDonald, H. Ito, C. G. Willson, *Microelectron. Eng.* **1**, 269 (1983).
40. E. Alling, C. Stauffer, *Proc. SPIE* **539**, 194 (1985).
41. E. Reichmanis, F. M. Houlihan, O. Nalamasu, T. X. Neenan, *Chem. Mater.* **3**, 394 (1991).
42. J. V. Crivello, in *Polymers in Electronics*, ACS Symposium Series, Vol. 242, T. Davidson, Ed., American Chemical Society, Washington, DC, 1984, pp. 3–10.
43. K. J. Stewart, M. Hatzakis, J. M. Shaw, D. E. Seeger, E. Neumann, *J. Vac. Sci. Technol.* **B7**, 1734 (1989).
44. R. D. Allen, W. E. Conley, J. D. Gelorme, *Proc. SPIE* **1672**, 513 (1992).
45. J. Lingnau, R. Dammel, J. Theiss, *Solid State Technol.* **32**(9), 105–112 (1989).
46. J. Lingnau, R. Dammel, J. Theiss, *Solid State Technol.* **32**(10), 107 (1989).
47. W. E. Feely, European patent application 232, 972 (1980).
48. G. Buhr, U.S. patent 4, 189, 323 (1980).
49. A. K. Berry, K. A. Graziano, L. E. Bogen, Jr., J. W. Thackeray, in *Polymers in Microlithography*, ACS Symposium Series, Vol. 412, E. Reichmanis, S. A. MacDonald, T. Iwayanagi, Eds., American Chemical Society, Washington, DC, 1989, pp. 87–99.
50. J. W. Thackeray, G. W. Orsula, E. K. Pavelchek, D. Canistro, *Proc. SPIE Adv. Resist Technol. Process. VI*, **1086**, 34 (1989).
51. H.-Y. Liu, M. P. deGrandpre, W. E. Freely, *J. Vac. Sci. Technol.* **B6**, 379 (1988).
52. M. Toriumi, H. Shiraishi, T. Ueno, N. Hayashi, S. Nonogaki, *J. Electrochem. Soc.* **134**, 334 (1987).

53. H. Shiraishi, E. Fukuma, N. Hayashi, K. Tadano, T. Ueno, *Chem. Mater.* **3**, 621 (1991).
54. M. Sakata, T. Ito, Y. Yamashita, *Jpn. J. Appl. Phys.* **30**(11B), 3116 (1991).
55. S. Uchino, M. Katoh, T. Sakamizu, M. Hashimoto, *Microelectron. Eng.* **17**, 261 (1992).
56. B. Reck, R. E. Allen, R. J. Twieg, C. G. Willson, S. Matsusczak, H. D. H. Stover, N. H. Li, J. M. J. Frechet, *Polym. Eng. Sci.* **29**, 960 (1989).
57. J. M. J. Frechet, S. Matsusczak, H. D. H. Stover, C. G. Willson, B. Reck, in *Polymers in Microlithography, ACS Symposium Series*, Vol. 412, E. Reichmanis, S. A. MacDonald, T. Iwanagi, Eds., American Chemical Society, Washington, DC, 1989, pp. 74–85.
58. U. Schaedeli, H. Holzwarth, N. Muenzel, R. Schulz, *Proceedings of the Regional Technical Conference on Photopolymers*, Mid-Hudson Section, SPE, Ellenville, NY, Oct. 28–30, 1991, p. 145.
59. M. Hatzakis, *J. Electrochem. Soc.* **116** 1033 (1969).
60. M. W. Moreau, R. R. Schmidt, *138th Electrochemical Society Meeting, Extended Abstracts*, 1970, p. 459.
61. Y. Mimura, T. Ohkubo, T. Takanichi, K. Sekikawa, *Jpn. Appl. Phys.* **17**, 541 (1978).
62. B. Lin, *J. Vac. Sci. Technol.* **12**, 1317 (1975).
63. B. Ranby, J. F. Rabek, *Photodegradation, Photooxidation and Photostabilization of Polymers*, Wiley, New York, 1975, pp. 156–159.
64. J. N. Helbert, C. Y. Chen, C. U. Pittman, Jr., G. L. Hagnauer, *Macromolecules* **11**, 1104 (1978).
65. J. H. Lai, J. N. Helbert, C. F. Cook, Jr., C. U. Pittman, Jr., *J. Vac Sci. Technol.* **16**(6), 1992 (1979).
66. J. N. Helbert, B. E. Wagner, J. P. Caplan, E. H. Poindexter, *J. Appl. Polym. Sci.* **19**, 1201 (1975).
67. C.-Y. Chen, C. U. Pittman, Jr., J. N. Helbert, *J. Polym. Sci. Polym. Chem. Ed.* **18**, 169 (1980).
68. W. M. Moreau, *Proc. SPIE* **333**, 2 (1982).
69. M. Kakuchi, S. Sugawara, K. Murase, K. Matsuyama, *J. Electrochem. Soc.* **224**, 1648 (1977).
70. T. Tada, *J. Electrochem. Soc.* **126**, 1829 (1979).
71. T. Tada, *J. Electrochem. Soc.* **130**, 912 (1983).
72. E. D. Roberts, *ACS Div. Org. Coat. Plastics Chem. Prepr.* **37**(2), 36 (1977).
73. W. Moreau, D. Merritt, W. Moyer, M. Hatzakis, D. Johnson, L. Pederson, *J. Vac. Sci. Technol.* **16**(6), 1989 (1979).
74. Y. M. N. Namastse, S. K. Obendorf, C. C. Anderson, P. D. Krasicky, F. Rodriquez, R. Tiberio, *J. Vac. Sci. Technol. B* **1**(4), 1160 (1983).
75. C. W. Wilkins, Jr., E. Reichmanis, E. A. Chandross, *J. Electrochem. Soc.* **127**(11), 2510 (1980).
76. E. Reichmanis, C. W. Wilkins, Jr., E. A. Chandross, *J. Electrochem. Soc.* **127**(11), 2514 (1980).

77. E. Reichmanis, C. W. Wilkins, Jr., *Polymer Materials for Electronics Applications*, ACS Symposium Series, Vol. 184, E. D. Feit, C. W. Wilkins, Jr., Eds., American Chemical Society, Washington, DC, 1982, pp. 29–43.
78. R. L. Hartless, E. A. Chandross, *J. Vac Sci. Technol.* **19**, 1333 (1981).
79. E. Reichmanis, G. Smolinsky, C. W. Wilkins, Jr., *Solid State Technol.* **28**(8), 130 (1984).
80. E. Reichmanis, G. Smolinsky, *Proc. SPIE* **469**, 38 (1984).
81. E. Reichmanis, G. Smolinsky, *J. Electrochem. Soc.* **132**, 1178 (1985).
82. M. Tsuda, S. Oikawa, Y. Nakamura, H. Nagata, A. Yokota, H. Nakane, T. Tsumori, Y. Nakane, *Photogr. Sci. Eng.* **23**, 1290 (1979).
83. S. A. MacDonald, H. Ito, C. G. Willson, J. W. Moore, H. M. Charapetian, J. E. Guillet, in *Materials for Microlithography*, L. F. Thompson, C. G. Willson, J. M. J. Frechet, Eds., ACS Symposium Series, Vol. 266, American Chemical Society, Washington, DC, 1984, p. 179.
84. J. R. Brown, J. H. O'Donnell, *Macromolecules* **3**, 265 (1970).
85. J. R. Brown, J. H. O'Donnell, *Macromolecules* **5**, 109 (1972).
86. T. N. Bowmer, J. H. O'Donnell, *Radiat. Phys. Chem.* **17**, 177 (1981).
87. M. J. Bowden, L. F. Thompson, J. P. Ballantyne, *J. Vac. Sci. Technol.* **126**(6), 1294 (1975).
88. M. J. Bowden, L. F. Thompson, *Solid State Technol.* **22**, 72 (1979).
89. A. S. Gozdz, *Solid State Technol.* **30**(6), 75 (1987).
90. A. S. Gozdz, H. Ono, S. Ito, J. A. Shelbourne III, *Proc. SPIE* **1446**, 200 (1991).
91. D. C. Hofer, R. D. Miller, C. G. Willson, *Proc. SPIE* **469**, 16 (1984).
92. R. D. Miller, D. Hofer, D. R. McKean, C. G. Willson, R. West, R. T. Trefonas III, in *Materials for Microlithography*, ACS Symposium Series, Vol. 266, L. F. Thompson, C. G. Willson, J. M. J. Frechet, Eds., American Chemical Society, Washington, DC, 1984, p. 283.
93. J. M. Zeigler, L. A. Harrah, A. W. Johnson, *Proc. SPIE* **539**, 166 (1985).
94. R. R. Kunz, M. W. Horn, *Proceedings of the Regional Technical Conference on Photopolymers*, Mid-Hudson Section, SPE, Ellenville, NY, Oct. 28–30, 1991, p. 291.
95. R. G. Tarascon, A. Shugard, E. Reichmanis, *Proc. SPIE* **631**, 40 (1986).
96. Y. Saotome, H. Gokan, K. Saigo, M. Suzuki, Y. Ohnishi, *J. Electrochem. Soc.* **132**, 909 (1985).
97. T. F. Yeh, A. Reiser, R. R. Dammel, G. Pawlowski, H. Roeschert, *Macromolecules* **26**, 3862 (1993).
98. H.-Y. Shih, T. F. Yeh, A. Reiser, R. Dammel, H. J. Merrem, G. Pawlowski, *Proc. SPIE* **2195**, 514 (1994).
99. R. D. Miller, C. G. Willson, D. R. McKean, T. Tompkins, N. Clecak, J. Michl, J. Downing, *Proceedings of the Regional Technical Conference on Photopolymers*, Mid-Hudson Section, SPE, Ellenville, NY, Nov. 8–10, 1982, p. 111.
100. M. Hanabata, A. Furuta, Y. Uemura, *Proc. SPIE* **631**, 76 (1986).
101. M. Hanabata, A. Furuta, Y. Uemura, *Proc. SPIE* **771**, 85 (1987).
102. M. Hanabata, F. Oi, A. Furuta, *Proc. SPIE* **1446**, 132 (1991).

103. M. K. Templeton, C. R. Szmanda, A. Zampini, *Proc. SPIE* **777**, 136 (1987).

104. P. Trefonas, III, B. K. Daniels, R. L. Fischer, *Solid State Technol.* **30**, 131 (1987).

105. M. Hanabata, F. Oi, A. Furuta, *Proceedings of the Regional Technical Conference on Photopolymers*, Mid-Hudson Section, SPE, Ellenville, NY, Oct. 28–30, 1991, p. 77.

106. B. D. Grant, N. J. Clecak, R. J. Twieg, C. G. Willson, *IEEE Trans. Electron. Dev.* **25**(11), 1300 (1981).

107. C. G. Willson, R. D. Miller, D. R. McKean, *Proc. SPIE* **771**, 2 (1987).

108. E. Reichmanis, C. W. Wilkins, Jr., E. A. Chandross, *J. Vac. Sci. Technol.* **19**(4), 1338 (1981).

109. L. A. Pederson, *J. Electrochem. Soc.* **129**, 205 (1982).

110. H. Gokan, S. Esho, Y. Ohnishi, *J. Electrochem. Soc.* **130**, 143 (1983).

111. G. Pawlowski, T. Sauer, R. Dammel, D. J. Gordon, W. Hinsberg, D. McKean, C. R. Lindley, H.-J. Merrem, R. Vicari, C. G. Willson, *Proc. SPIE* **1262**, 391 (1990).

112. K. Przybilla, H. Roeschert, W. Spiess, C. Eckes, S. Chatterjee, D. Khanna, G. Pawlowski, R. Dammel, *Proc. SPIE* **1466**, 174 (1991).

113. Y. Kaimoto, K. Nozaki, S. Takechi, N. Abe, *Proc. SPIE* **1672**, 66 (1992).

114. R. R. Kunz, R. D. Allen, W. D. Hinsberg, G. M. Wallraff, *Proc. SPIE* **1925**, 167 (1993).

115. N. Abe, S. Takechi, Y. Kaimoto, M. Takahashi, K. Nozaki, *J. Photopolym. Sci. Technol.* **8**(4), 637 (1995).

116. R. D. Allen, I. Y. Wan, G. M. Wallraff, R. A. DiPietro, D. C. Hofer, *J. Photopolym. Sci. Technol.* **8**(4), 623 (1995).

117. J. M. Shaw, M. Hatzakis, *J. Electrochem. Soc.* **126**, 2026 (1979).

118. H. L. Huber, M. Betz, A. Heuberger, S. Pongrate, *Microelectron. Eng.* **84**, 325 (1984).

119. D. W. Peters et al., *Proc. SPIE* **923**, 36 (1988).

120. M. J. Bowden, L. F. Thompson, S. R. Fahrenholtz, E. M. Doerries, *J. Electrochem. Soc.* **128**, 1304 (1981).

121. M. J. Bowden, D. L. Allara, W. I. Vroom, S. Frackoviak, L. C. Kelly, D. R. Falcone, in *Polymers in Electronics*, ACS Symposium Series, Vol. 242, T. Davidson, Ed., American Chemical Society, Washington, DC, 1984, pp. 135–152.

122. T. N. Bowner, M. J. Bowden, in *Polymers in Electronics*, ACS Symposium Series, Vol. 242, T. Davidson, Ed., American Chemical Society, Washington, DC, 1984, pp. 153–166.

123. R. G. Tarascon, J. Frackoviak, E. Reichmanis, L. F. Thompson, *Proc. SPIE* **771**, 54 (1987).

124. J. Frackoviak, R. G. Tarascon, S. Vaidya, E. Reichmanis, *Proc. SPIE* **771**, 120 (1987).

125. H. Shiraishi, A. Isobe, F. Murai, S. Nongaki, in *Polymers in Electronics*, ACS Symposium Series, Vol. 242, T. Davidson, Ed., American Chemical Society, Washington, DC, 1984, pp. 167–176.

126. Y. Y. Chang, B. D. Grant, L. A. Pederson, C. G. Willson, U.S. patent 4,398,001 (1983).
127. H. Ito, C. G. Willson, *Polymers in Electronics*, ACS Symposium Series, Vol. 242, T. Davidson, Ed., American Chemical Society, Washington, DC, 1984, pp. 11–23.
128. J. M. J. Frechet, E. Eichler, H. Ito, C. G. Willson, *Polymer* **24**, 995 (1980).
129. H. Ito, C. G. Willson, J. M. J. Frechet, M. J. Farrall, E. Eichler, *Macromolecules* **16**, 1510 (1983).
130. H. Ito, C. G. Willson, *Polym. Eng. Sci.* **23**, 1012 (1983).
131. J. G. Maltabes et al., *Proc. SPIE* **1262**, 2 (1990).
132. H. Ito, C. G. Wilson, *Proc. SPIE* **771**, 24 (1987).
133. H. Ito, M. Ueda, M. Ebina, in *Polymers in Microlithography*, ACS Symposium Series, Vol. 412, E. Reichmanis, S. A. MacDonald, T. Iwayanagi, Eds., American Chemical Society, Washington, DC, 1989, p. 57.
134. C. E. Osuch, K. Brahim, F. R. Hopf, M. J. McFarland, A. Mooring, C. J. Wu, *Proc. SPIE* 631 68 (1986).
135. S. R. Turner, K. D. Ahn, C. G. Willson, in *Polymers for High Technology*, ACS Symposium Series, Vol. 346, M. J. Bowden, S. R. Turner, Eds., American Chemical Society, Washington, DC, 1987, pp. 200–210.
136. K. J. Przybilla, R. Dammel, G. Pawlowski, J. Roschert, W. Spiess, *Proceedings of the Regional Technical Conference on Photopolymers* Mid-Hudson Section, SPE, Ellenville, NY, Oct. 28–30, 1991, p. 131.
137. R. S. Kanga, J. M. Kometani, E. Reichmanis, J. E. Hanson, O. Nalamasu, L. F. Thompson, S. A. Heffner, W. W. Tai, P. Trevor, *Chem. Mater.* **3**, 660 (1991).
138. F. M. Houlihan, E. Reichmanis, L. F. Thompson, R. G. Tarascon, in *Polymers in Microlithography*, ACS Symposium Series, Vol. 412, E. Reichmanis, S. A. MacDonald, T. Iwayanagi, Eds., American Chemical Society, Washington, DC, 1989, pp. 39–56.
139. M. J. Bowden, E. A. Chandross, *J. Electrochem. Soc.* **122**, 1370 (1975).
140. A. E. Novembre, W. W. Tai, J. M. Kometani, J. E. Hanson, O. Nalamasu, G. N. Taylor, E. Reichmanis, L. F. Thompson, *Chem. Mater.* **4**, 278 (1992).
141. A. E. Novembre, W. W. Tai, J. M. Kometani, J. E. Hanson, O. Nalamasu, G. N. Taylor, E. Reichmanis, L. F. Thompson, D. N. Tomes, *J. Vac. Sci. Technol.* **B9**, 3338 (1991).
142. J. V. Crivello, *J. Electrochem. Soc.* **189**, 1453 (1989).
143. N. Hayashi, S. M. A. Hesp, T. Ueno, M. Toriumi, T. Iwayanagi, S. Nonogaki, *Proc. Polym. Mater. Sci. Eng.* **61**, 417 (1989).
144. J. M. J. Frechet, N. Kallman, B. Kryczka, E. Eichler, F. M. Houlihan, C. G. Willson, *Polym. Bull.* **20**, 427 (1988).
145. J. M. J. Frechet, W. Eichler, S. Gauthier, B. Krychzka, C. G. Willson, in *The Effects of Radiation on High-Technology Polymers*, ACS Symposium Series, Vol. 381, E. Reichmanis, J. H. O'Donnell, Eds., American Chemical Society, Washington, DC, 1989, pp. 155–171.
146. S. A. M. Hesp, N. Hayashi, T. Ueno, *J. Appl. Polym. Sci.* **42**, 877 (1991).

147. L. Schlegel, T. Ueno, H. Shiraishi, N. Hayashi, T. Iwanyanagi, *Microelectron. Eng.* **14**, 227 (1991).

148. G. N. Taylor, L. E. Stillwagon, F. M. Houlihan, T. M. Wolf, D. Y. Sogah, W. R. Hartler, *J. Vac. Sci. Technol.* **B9**, 3348 (1991).

149. H. Ito, M. Ueda, *Macromolecules* 21, 1475 (1988).

150. T. Yamaoka, M. Nishiki, K. Koseki, M. Koshiba, *Proceedings of the Regional Technical Conference on Photopolymers*, Mid-Hudson Section, SPE, Ellenville, NY, Oct. 30–Nov. 2, 1988, p. 27.

151. M. Murata, E. Kobayashi, M. Yamachika, Y. Kobayashi, Y. Yamota, T. Miura, *J. Photopolym. Sci. Technol.* **5**(1), 79 (1992).

152. K. E. Uhrich, E. Reichmanis, S. A. Heffner, J. M. Kometani, O. Nalamasu, *Chem. Mater.* **6**, 287 (1994).

153. F. Schue, L. Giral, *Makromol. Chem. Macromol. Symp.* **24**, 21 (1989).

154. M. Padmanabhan, Y. Kinoshita, T. Kudo, T. Lynch, S. Masuda, Y. Nozaki, H. Okazaki, G. Pawloski, K. T. Przybilla, H. Roeschert, W. Spiess, N. Suehiro, H. Wengenroth, *Proc. SPIE* **2195**, 61 (1994).

155. R. Schwalm, H. Binder, T. Fischer, D. Funhoff, M. Goethals, A. Grassmann, H. Moritz, M. Reuhman-Huisken, F. Vinet, H. Dijkstra, A. Krause, *Proc. SPIE* **2195**, 2 (1994).

156. T. Hattori, L. Schlegel, A. Imai, N. Hayashi, T. Ueno, *J. Photopolym. Sci. Technol.* **6**, 497 (1993).

157. L. Schlegel, T. Ueno, H. Shiraishi, N. Hayashi, T. Iwayanagi, *Chem. Mater.* **2**, 299 (1990).

158. D. R. McKean, S. A. MacDonald, N. J. Clecak, C. G. Willson, *Proc. SPIE* **920**, 60 (1988).

159. L. Schlegel, T. Ueno, H. Shiraishi, N. Hayashi, T. Iwayanagi, *J. Photopolym. Sci. Technol.* **3**, 281 (1990).

160. M. J. O'Brien, *Polym. Eng. Sci.* **29**, 846 (1989).

161. M. J. O'Brien, J. V. Crivello, *Proc. SPIE* **920**, 42 (1988).

162. J. Lingnau, R. Dammel, J. Theiss, *Proceedings of the Regional Technical Conference on Photopolymers*, Mid-Hudson Section, SPE, Ellenville, NY, Oct. 20–Nov. 2, 1988, pp. 87–97.

163. T. Aoai, T. Yamanaka, T. Kokubo, *Proc. SPIE* **2195**, 111 (1994).

164. R. Schwalm, *Proc. Polym. Mater. Sci. Eng.* **61**, 278 (1989).

165. F. M. Houlihan, E. Chin, O. Nalamasu, J. M. Kometani, *Proc. ACS Div. Polym. Mater. Sci. Eng.*, **66** 38 (1992).

166. H. Ito, C. G. Willson, *Proceedings of the Regional Technical Conference on Photopolymers*, Mid-Hudson Section, SPE, Ellenville, NY, 1982, p. 331.

167. H. Ito, M. Ueda, R. Schwalm, *J. Vac. Sci. Technol.* **B6**, 2259 (1988).

168. H. Ito, R. Schwalm, *J. Electrochem. Soc.* **136**, 241 (1989).

169. H. Ito, N. Ueda, A. F. Renaldo, *J. Electrochem. Soc.* **136**, 245 (1988).

170. H. Ito, E. Flores, A. F. Renaldo, *J. Electrochem. Soc.* **135**, 2328 (1988).

171. J. M. J. Frechet, C. G. Willson, T. Iizawa, T. Nishikubo, K. Igarashi, J. Fahey, in *Polymers in Microlithography*, ACS Symposium Series, Vol. 412, E. Reichmanis, S. A. MacDonald, T. Iwayanagi, Eds., American Chemical Society, Washington, DC, 1989, pp. 100–112.
172. B. J. Lin, in *Introduction to Microlithography*, ACS Symposium Series, Vol. 219, L. F. Thompson, M. J. Bowden, C. G. Willson, Eds., American Chemical Society, Washington, DC, 1983, pp. 287–350.
173. C. G. Willson, M. J. Bowden, in *Electron and Photonic Applications of Polymers*, M. J. Bowden, S. R. Turner, Eds., ACS Advances in Chemistry Series, Vol. 218, American Chemical Society, Washington, DC, 1988, pp. 75–108.
174. G. N. Taylor, T. M. Wolf, *J. Electrochem. Soc.* **127**, 2665 (1980).
175. G. N. Taylor, *Solid State Technol.* **23**(5), 73 (1980).
176. L. E. Stillwagon, P. J. Silverman, G. N. Taylor, *Proceedings of the Regional Technical Conference on Photopolymers*, Mid-Hudson Section, SPE, Ellenville, NY, Oct. 28–30, 1985, pp. 87–103.
177. O. Nalamasu, F. A. Baiocchi, G. N. Taylor, in *Polymers in Microlithography*, ACS Symposium Series, Vol. 412, E. Reichmanis, S. A. MacDonald, T. Iwayanagi, Eds., American Chemical Society, Washington, DC, 1989, pp. 189–209.
178. S. A. MacDonald, H. Schlosser, H. Ito, N. J. Clecak, C. G. Willson, *Chem. Mater.* **3**, 435 (1991).
179. M. A. Hartney, D. C. Shaver, M. I. Shephard, J. Melngailis, V. Medvedev, W. P. Robinson, *J. Vac. Sci. Technol.* **B9**, 3432 (1991).
180. H. Ito, *J. Photopolym. Sci. Technol.* **5**(1), 123 (1992).
181. F. Coopmans, B. Roland, *Proc. SPIE* **631**, 34 (1986).
182. B. Roland, R. Lombaerts, C. Jakus, F. Coopmans, *Proc. SPIE* **777**, 69 (1987).
183. C. M. Garza, *Proc. SPIE* **920**, 233 (1988).
184. B. Roland, J. Vandendriessche, R. Lombaerts, B. Denturck, C. Jakus, *Proc. SPIE* **920**, 120 (1988).
185. C. Pierrat, S. Tedesco, F. Vinet, M. Lerme, B. Dal'Zotto, *J. Vac. Sci. Technol.* **B7**, 1782 (1989).
186. J. M. Shaw, M. Hatzakis, E. D. Babich, J. R. Parasczak, D. F. Witman, K. J. Stewart, *J. Vac. Sci. Technol.* **B7**, 1209 (1989).
187. D. C. LaTulipe, A. T. S. Pomerene, J. P. Simons, D. S. Seeger, *Microelectron. Eng.* **17**, 265 (1992).
188. R. Sezi, M. Sebald, R. Leuscher, H. Ahne, S. Birkle, H. Burndörfer, *Proc. SPIE*, **1262**, 84 (1990).
189. K. H. Baik, L. Van den hove, B. Roland, *J. Vac. Sci. Technol. B* **B9**, 3399 (1991).
190. K. H. Baik, K. Ronge, L. Van den hove, B. Roland, *Proc. SPIE* **1672**, 362 (1992).
191. T. W. Weidman, O. Joubert, A. M. Joshi, R. L. Kostelak, *Proc. SPIE* **2438** (1995).
192. T. W. Weidman, O. Joubert, A. M. Joshi, J. T.-C. Lee, D. Boulin, E. A. Chandross, R. Cirelli, F. P. Klemens, H. L. Maynard, V. M. Donnelly, *J. Photopolym. Sci. Technol.* **8**(4), 679 (1995).
193. W. H. Meyer, B. J. Curtis, H. R. Brunner, *Microelectron. Eng.* **1**, 29 (1983).

194. G. N. Taylor, L. E. Stillwagon, T. Venkatesan, *J. Electrochem. Soc.* **131**, 1658 (1984).

195. M. A. Hartney, R. R. Kunz, D. J. Ehlrich, D. Shaver, *Proc. SPIE* **1292**, 119 (1990).

196. J. P. W. Schellekens, R.-J. Visser, *Proc. SPIE* **1086**, 220 (1989).

197. H. Satou, H. Suzuki, D. Makino, in *Polyimides*, D. Wilson, H. D. Stenzenberger, P. M. Hergenrother, Eds., Chapman & Hall, New York, 1990, Chapter 8.

198. D. S. Stone, Z. Martynenko, *Polymers in Electronics: Fundamentals and Applications*, Elsevier, New York, 1989.

199. J. H. Lai, Ed., *Polymers for Electronic Applications*, CRC Press, Boca Raton, FL, 1989.

200. T. Rucker, V. Murali, R. Shukis, H. Neuhaus, in *Electronic Packaging Materials Science VI, MRS Symposium Proceedings*, Vol. 264, P. S. Ho, K. A. Jackson, C.-Y. Li, G. F. Lipscomb, Eds., Materials Research Society, Pittsburgh, PA, 1992, p. 71.

201. S. R. Sandler, W. Karo, in *Polymer Syntheses*, Vol. I, 2nd ed., Academic Press, San Diego, CA, 1992, Chapter 9.

202. G. M. Bower, L. Frost, *J. Polym. Sci.* A13135 (1963).

203. R. A. Dine-Hart, W. W. Wright, *J. Appl. Polym. Sci.* **11**, 609 (1967).

204. W. Volksen, P. M. Cotts, in *Polyimides: Synthesis, Characterization and Properties*, Vol 1, (K. L. Mitsl), Ed., Plenum Press, New York, 1984, pp. 163–170.

205. Y. K. Lee, J. D. Craig, *Polymer Materials for Electronic Applications*, ACS Symposium Series, Vol. 184, E. D. Feit, C. W. Wilkins, Jr., Eds., American Chemical Society, Washington, DC, 1983, p. 106.

206. K. Sato, S. Harada, A. Saiki, T. Kimura, T. Okubo, K. Mukai, *IEEE Trans. Parts Hybrids Pack.* **9**(3), 176 (1973).

207. L. B. Lothman, *J. Electrochem. Soc. Solid State Sci. Technol.* **127**, 2216 (1980).

208. T. S. St. Clair, in *Polyimides*, D. Wilson, H. D. Stenzenberger, P. M. Hergenrother, Eds., Chapman & Hall, New York, 1990, Chapter 3.

209. A. K. St. Clair, T. S. St. Clair, W. P. Winfree, *Polym. Mater. Sci. Eng.* **59**, 28 (1988).

210. B. C. Auman, S. Trofimenko, *Polym. Prepr.* **33**, 244 (1992).

211. B. C. Auman, *Advances in Polyimide Science and Technology, Proceedings of the 4th International Conference on Polyimides*, C. Feger, M. M. Khojasteh, M. S. Htoo, Eds., Technomic, Lancaster, PA, 1993, p. 15.

212. K. R. Carter, J. W. Labadie, R. A. DiPietro, M. I. Sanchez, T. P. Russell, S. A. Swanson, B. C. Auman, P. Lakshmanan, J. E. McGrath, *Polym. Mater. Sci. Eng.* **72**, 383 (1995).

213. J. L. Hedrick, J. W. Labadie, T. P. Russell, D. C. Hofer, V. Wakharkar, *Polymer* **34**, 4717 (1993).

214. J. L. Hedrick, J. W. Labadie, T. P. Russell, V. Wakharkar, *Polymer* **34**, 22 (1993).

215. S. J. Rhodes, in *Polyimides: Synthesis, Characterization, and Applications*, Vol. 2, K. L. Mittal, Ed., Plenum Press, New York, 1984, p. 795.

216. A. Saiki, T. Nishida, H. Suzuki, and D. Makino, in *Polyimides: Synthesis, Characterization, and Applications*, Vol. 2, K. L. Mittal, Ed., Plenum Press, New York, p. 827.
217. T. O. Herndon, R. L. Burke, *Proceedings of the Kodak Interface Conference*, Oct. 1979.
218. R. E. Kerwin, M. R. Goldrich, *Polym. Eng. Sci.* **11**, 426 (1971).
219. H. Ahne, H. Kruger, E. Pammer, R. Rubner, in *Polyimides: Synthesis, Characterization, and Applications*, Vol. 2, K. L. Mittel, Ed., Plenum Press, New York, 1984, p. 905.
220. R. Rubner, *Photogr. Sci. Eng.* **23**, 305 (1979).
221. G. C. Davis, in *Polymers in Electronics*, T. Davidson, Ed., ACS Symposium Series, Vol. 242, American Chemical Society, Washington, DC, 1984, p. 259.
222. M. Szwarc, *Polym. Eng. Sci.* **16**(7), 473 (1976).
223. W. F. Gorham, *Adv. Chem. Ser.* **91**, 643 (1966).
224. W. F. Gorham, *J. Polym. Sci.* A-1 **4**, 3027 (1966).
225. J. A. Moore, C.-I. Lang, T.-M. Lu, G.-R. Yang, *Polym. Mater. Sci. Eng.* **72**, 437 (1995).
226. B. L. Joesten, *J. Appl. Polym. Sci.* **18**, 439 (1974).
227. W. Oppolzer, *Synthesis* **973** (1978).
228. D. Burdeaux, P. Townsend, J. J. Carr, P. E. Garrou, *J. Electron. Mater.* **19**, 1357 (1990).
229. R. A. Kirch, K. J. Bruza, *Chemtech* **23**, 22 (1993).
230. T. M. Miller, E. W. Kwock, T. Baird, Jr., A. Hale, *Chem. Mater.* **6**, 1569 (1994).
231. E. W. Kwock, T. Baird, Jr., T. M. Miller, *Macromolecules* **26**, 2935 (1993).
232. A. S. Hay, *J. Org. Chem.* **25**, 1275 (1960)
233. R. A. Nallicheri, M. F. Rubner, *Macromolecules* **24**, 517 (1991).
234. J. L. Stanford, R. J. Young, R. J. Day, *Polymer* **32**, 1713 (1991).
235. H. Ito, Y. Mizushima, A. Takepa, *Jpn. J. Appl. Phys.* **19**(8), 2460 (1980).
236. A. M. Lyons, D. M. Barnes, C. W. Wilkins, Jr., unpublished results.
237. M. L. While, *Proc. IEEE* **57**, 1610 (1969).
238. C. P. Wong, *Polym. Mater. Sci. Eng.* **55**, 803 (1986).
239. C. P. Wong, U.S. patent 4,592,959 (June 3, 1986).
240. H. Lee, K. Neville, *Handbook of Epoxy Resins*, McGraw-Hill, New York, 1967.
241. C. A. May, Y. Tanaka, *Epoxy Resins*, Marcel Dekker, New York, 1973.

CHAPTER 5

Chemical Vapor Deposition

MARK J. HAMPDEN-SMITH

Department of Chemistry and Center for Micro-Engineered Materials, University of New Mexico, Albuquerque, New Mexico

TOIVO T. KODAS and AUDUNN LUDVIKSSON

Department of Chemical Engineering and Center for Micro-Engineered Materials, University of New Mexico, Albuquerque, New Mexico

5.1 INTRODUCTION

The goal of this chapter is to describe the fundamental aspects and technological significance of chemical vapor deposition (CVD). The chapter is not meant to provide a comprehensive review of the field but rather to focus on the underlying principles. This should provide the reader with an appreciation for the key issues in the CVD of materials, how the field has developed to its current status, and which issues need to be addressed to achieve significant advances.

Chemical vapor deposition is a method in which a volatile molecular species (often metal containing) is transported into an apparatus (reactor) that contains a substrate where the molecular species adsorbs and reacts to deposit a film of a particular material (often metal containing) as shown in Fig. 5.1. With this definition, chemical vapor deposition is a broad term that incorporates a variety of more specific processes.[1] For example, MOCVD refers to CVD using a metal–organic (MO) compound, while OMCVD refers to CVD using organometallic (OM) compounds as precursors. This broad definition encompasses other processes, such as vapor-phase epitaxy (VPE), chemical vapor transport, and atomic-layer epitaxy (ALE), which are usually not considered as classical CVD processes. Other variations are described in Section 5.3. The relationship between film deposition processes in general

Chemistry of Advanced Materials: An Overview, Edited by Leonard V. Interrante and Mark J. Hampden-Smith.
ISBN 0-471-18590-6 © 1998 Wiley-VCH, Inc.

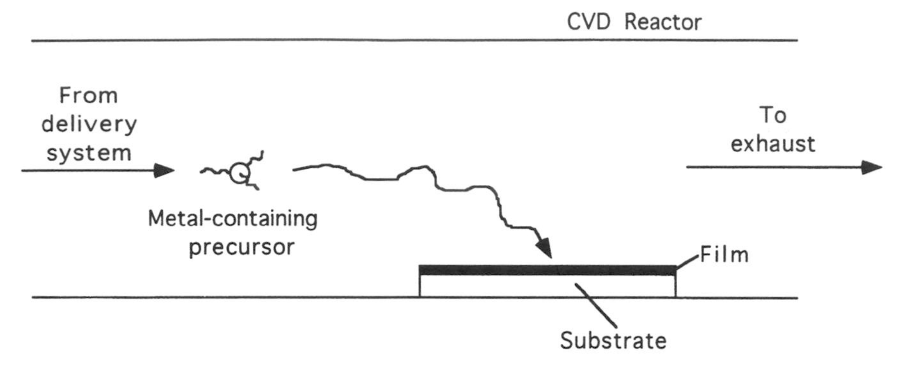

Figure 5.1 Schematic representation of an idealized CVD process for an elemental material.

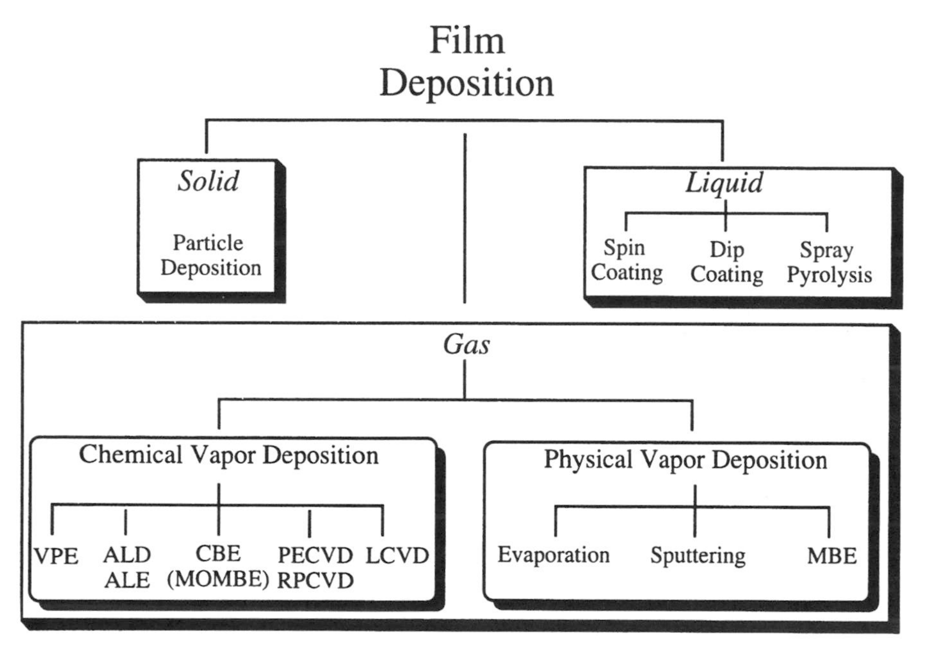

Scheme 5.1 Relationship of different film deposition methods.

and chemical vapor deposition processes in particular are represented in Scheme 5.1.

A broad range of science and engineering concepts are crucial to gain a complete understanding of the steps involved in CVD, which makes this field highly interdisciplinary. Chemical concepts of molecular precursor design are important to control precursor properties such as volatility and reactivity. Chemical engineering concepts such as reaction engineering, transport phenomena, and reactor design, and materials science aspects that involve the appropriate selection of materials, choice of deposition conditions to obtain films with controlled microstructure (density, porosity, crystallinity, etc.), and properties are important. Characterization of the films also requires analytical expertise and often sophisticated instrumentation. As a result, CVD is an excellent educational vehicle for students to develop an appreciation for the importance of key concepts in disciplines outside their major field. In addition, it can be argued that due to the highly interdisciplinary nature of this field, it is likely that significant research advances are most likely to come from interdisciplinary approaches to solve problems.

The discussion that follows focuses on the deposition of inorganic films from metal-containing precursors. Similar principles apply to the deposition of organic thin films, but this is outside the topic of this discussion.[2]

5.1.1 Overview of Techniques to Deposit Films

Films can be deposited by a wide variety of techniques that can be classified according to the phase from which they are deposited (the solid phase, liquid phase, and gas phase). The relationship between different film deposition methods is represented in Scheme 5.1. Solid-state routes involve the deposition of powders on a surface either directly[3] or entrained in a paste through a processes such as screen printing or doctor blading,[4] which upon subsequent sintering can result in formation of a continuous film (Fig. 5.2). Films made by this method are typically thick ($> 10\ \mu m$), require relatively high substrate temperatures to achieve dense films, and do not allow for selective deposition or conformal coverage of the substrate except in special cases. A conformal coating is one that follows the contours of the substrate with a uniform thickness. Solid-state routes generally employ powders of the final desired material rather than molecular species as precursors to the final material.

Liquid-phase routes generally involve the dissolution of a soluble precursor for the desired film in a suitable solvent and deposition on a substrate surface by spin coating or dip coating to obtain a uniform film followed by a thermal treatment. When metal-containing compounds are used as precursors to inorganic materials, this process is often termed "metal–organic decomposition (MOD). Liquid-phase routes have a number of advantages over other deposition routes. A wider variety of precursors are available compared to CVD since solubility in a particular solvent is the principal requirement. Complex structures or particles can be created in solution and delivered to a surface to control film microstructure (as in sol-gel routes). The apparatus required to deposit films is simpler than that often required for CVD processes (Figs. 5.3 and 5.4). However, liquid-phase routes have the disadvantages that conformal coverage and selective deposition are hard to achieve and the liquid waste associated with this route can be an environmental problem. Liquid-phase routes are notorious for depositing contaminated films as a result of the difficulty in purifying solvents. The crystallization of solutes during solvent evaporation can also be a problem.

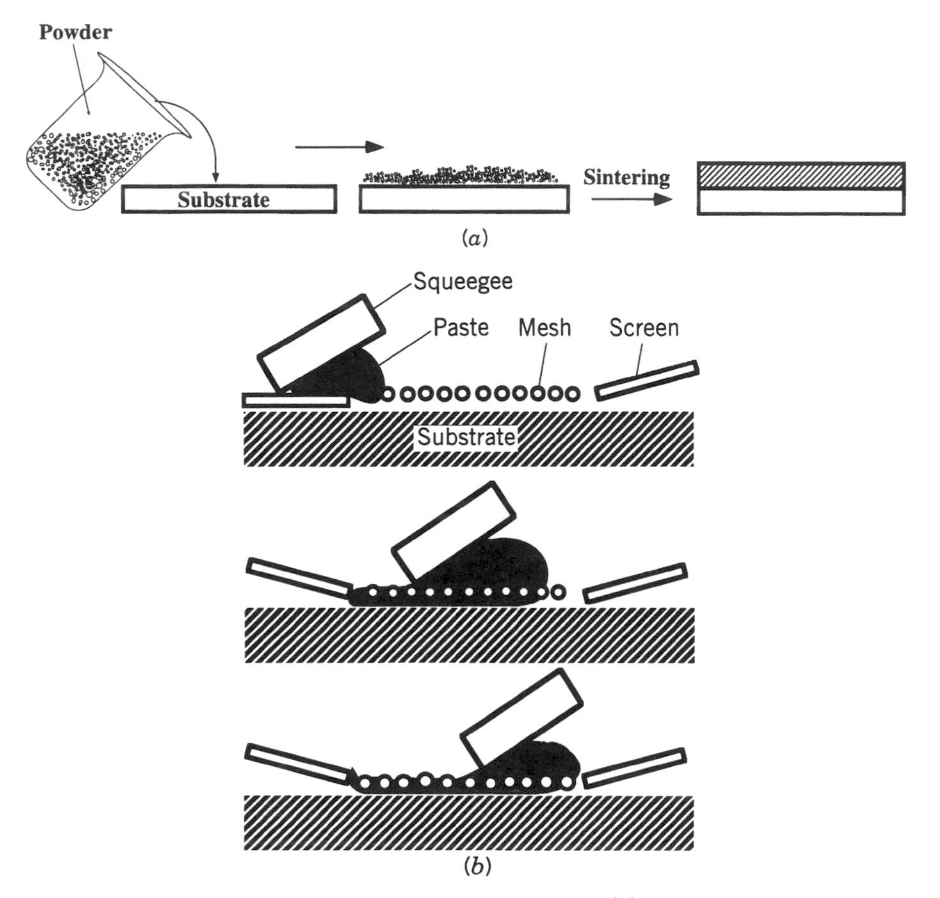

Figure 5.2 Schematic representation of an example of (*a*) solid-state deposition and (*b*) screen printing.

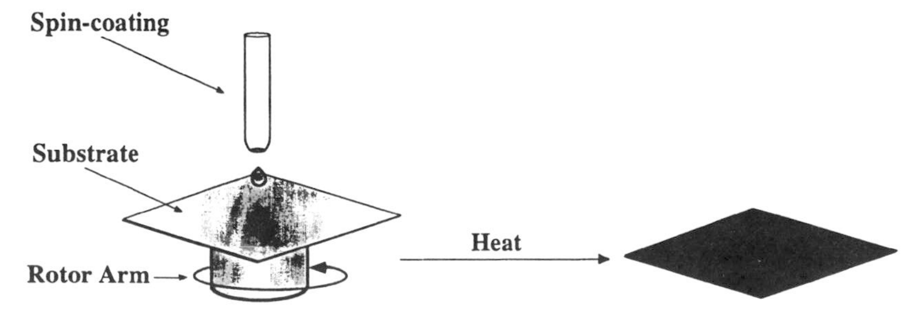

Figure 5.3 Schematic representation of liquid-phase deposition by spin coating.

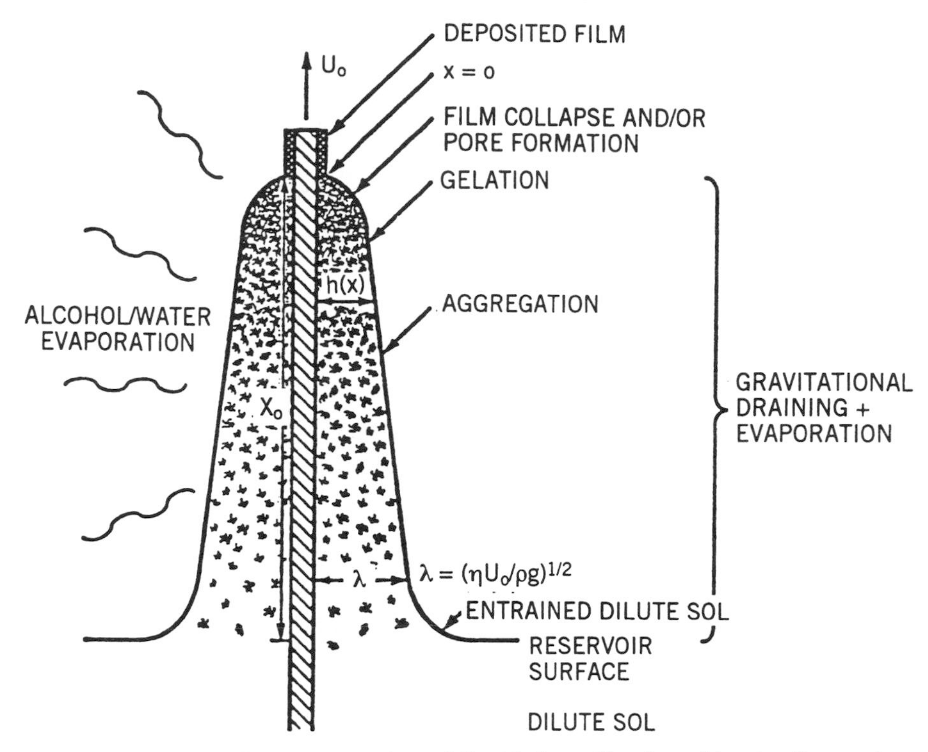

Figure 5.4 Schematic representation of liquid-phase film deposition by dip coating. (From Ref. 5)

There is a class of routes that are often mistaken as being related to chemical vapor deposition but are in fact better described as solid- or liquid-phase routes. These are methods that rely on spraying onto a surface liquids comprised of solutions of involatile compounds. The metal-containing compounds are either delivered to the heated surface still entrained in the solvent or as solid particles as a result of solvent evaporation. This class of routes does not possess the attributes of CVD routes because the precursor does not evaporate. It is often difficult to distinguish this literature from spray MOCVD and aerosol-assisted (AA) CVD, in which volatile precursors are delivered to a heated substrate and do evaporate prior to reaching the surface (see Section 5.2.6).

Gas-phase routes include a variety of physical and chemical methods to deposit films that are reviewed later in this chapter. Chemical vapor deposition has the disadvantages over other deposition routes that the number of molecular species that are suitable as precursors is relatively low due to the requirements of high volatility and suitable reactivity, and that the apparatus used to deposit these films can be complicated and/or expensive. However, CVD has the advantages that deposition can be achieved at relatively low substrate temperatures and that the coatings can be dense, conformal, and

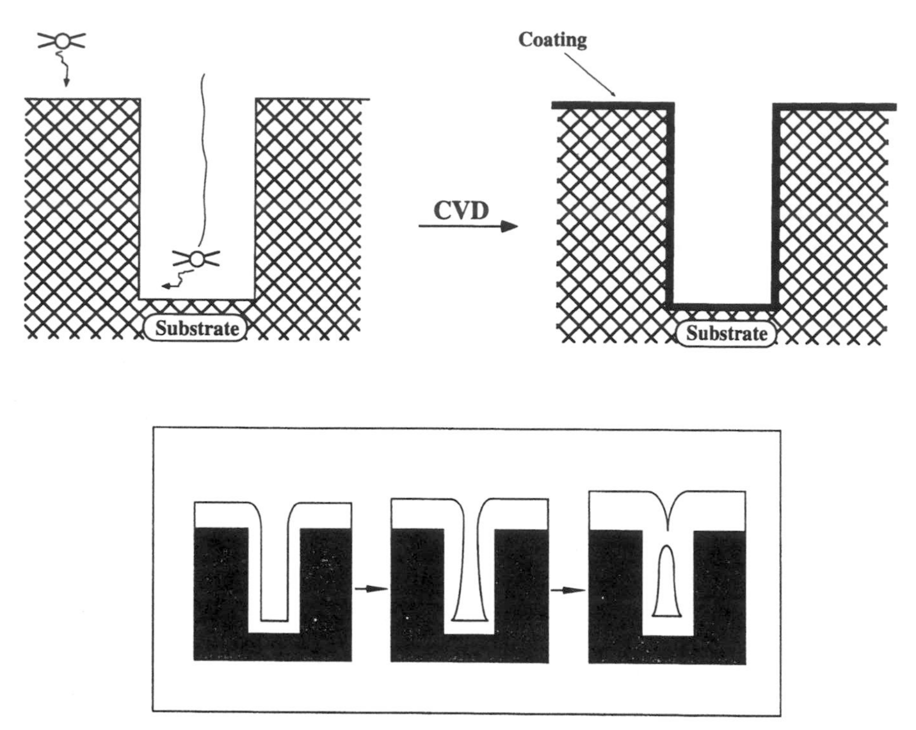

Figure 5.5 Schematic representation of conformal coverage and the formation of nonconformal films, leading to keyhole void formation.

deposited in an area-selective fashion. This is possible by CVD because the molecular precursors diffuse to the substrate surface where they react, and as a result, substrates with complex topography can be conformally coated. Highly reactive CVD precursors can result in a problem called keyhole formation, which describes a void formed during via (hole) filling. This is the result of the high reaction efficiency of the precursors which causes them to react near the top of the via forming a restriction that limits the access of precursors to the base of the via under certain deposition conditions (Fig. 5.5). High-purity materials can be deposited because gases can be purified to extremely high purity levels, especially when compared to liquids.

Area-selective deposition is possible with, in principle, molecular-level resolution as a result of the chemical origin of the deposition process.[1,6] It is reasonable to expect that if a molecular precursor is delivered to a substrate that already contains coatings of two different materials (a growth surface and a nongrowth surface; Fig. 5.6), at a given temperature, the precursor may react at a different rate on each surface. To achieve complete selectivity (i.e., no deposition on the nongrowth surface), it is often necessary to modify the reactivity of this surface to inhibit adsorption of the precursor on that surface. This aspect is described in more detail later. As a result of these aspects, and the development of photolithographic processes for film pattern-

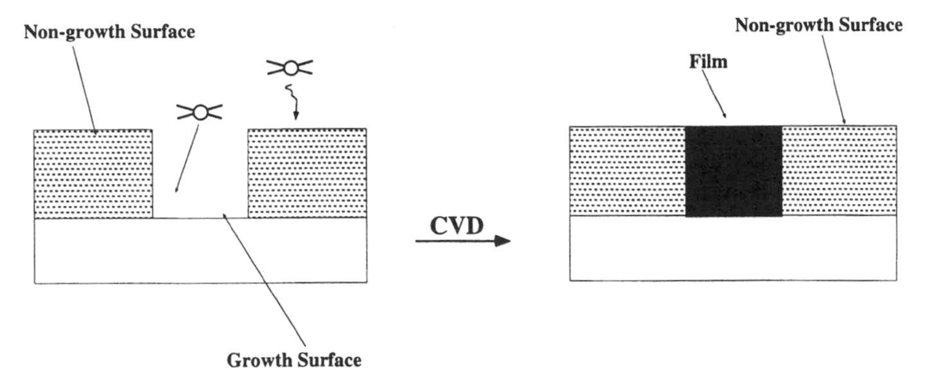

Figure 5.6 Schematic representation of selective deposition.

ing, CVD is used extensively in the microelectronics industry to manufacture microelectronic circuits.[7]

5.1.2 Technological Importance of CVD in Industrial Processes

Films of various materials deposited on a substrate are of increasing technological importance because a thin coating can impart different chemical and physical properties to the substrate. As a result, unique combinations of properties can be obtained through particular combinations of substrate and coating, and intricate structures can be built on substrates to perform increasingly complex tasks as witnessed by the miniaturization of electronic components in the computer industry.[7]

Chemical vapor deposition is used heavily in the microelectronics industry but is also used for coating glass used in houses and automobiles, forming protective coatings, fabricating optical fibers, forming some of the layers in solar cells, and forming coatings for catalysis, membranes, and a variety of other applications.

In the microelectronics industry, CVD is used primarily for applications where epitaxial or conformal films are required. Materials including Si, SiO_2, TiN, W, and GaAs are routinely deposited by CVD.[1,7,8] Processes have been developed for other materials, including Cu[9,10] and Al,[11] and are actively being considered. Overall, deposition temperatures can be 500 to 1000°C before any metal or dielectric layers are deposited and must less than 500°C after metal and dielectric layers are deposited. Both hot- and cold-wall systems are utilized. There is a premium on the performance of precursors (deposit purity, volatility, deposition rate) relegating cost issues to a lower status than for most other CVD processes.

In the glass industry, CVD is used to coat large panels of glass with metal oxide materials including SnO_2, TiN, and SiO_2. Typically, the glass is formed on a float line in which the hot glass is transported on top of molten tin. As the glass is passed down the float line, it passes through several CVD reactors in which reactants are directed toward the hot substrate, where they form coatings. Deposition temperatures are often near 600°C. The glass is eventually removed at the end of the system completely coated. Because of the relatively low selling price of glass and the large area to be covered, only inexpensive precursors are economical. This requires the use of simple precursors such as metal halides and alkyl halides. A variety of protective coatings are formed by CVD. These include hard coatings for tools in which metal carbide materials are often of greatest interest.

Optical fiber generation is carried out by a modified chemical vapor deposition process, often in a hot-wall reactor.[12] In this approach, $SiCl_4$, $GeCl_4$, and other metal halides are reacted with O_2 in the gas phase at high temperatures near 1000°C. The gas-phase reaction results in formation of metal oxide particles which deposit on the walls of a tube to form a porous coating. Eventually, the tube is consolidated and drawn into an optical fiber.

5.1.3 Manufacturing Issues

A major consideration in utilizing CVD processes for a particular application is the cost of the process compared to film deposition alternatives such as physical vapor deposition.[13] A manufacturable CVD process is one that provides a cost-effective method to produce a product that has additional value. Cost of ownership (COO) models, which are general for most CVD processes, have been developed to determine cost-effectiveness for a given process. Parameters that are used to determine the COO of a CVD process include precursor costs, tool costs, process yield, failure rates, deposition rate, throughput, and process stability. These issues are often discussed in the context of the cost of constructing fabrication facilities for smaller critical dimension microelectronic device technologies and have been described in detail elsewhere.[1]

5.2 FUNDAMENTAL ASPECTS OF CVD PROCESSES

5.2.1 Steps Involved in General CVD Processes

To identify and understand the fundamental aspects of CVD processes, CVD can be considered as a number of steps in series, any of which may be rate-limiting under a particular set of circumstances. In this section, these steps are identified and described without consideration of other factors that may influence these steps, such as the nature of the reactor used, the pressure regime, and the method of precursor delivery. These aspects are considered later. The key steps involved in the CVD of any film from a molecular precursor are identified in Fig. 5.7.

Chemical vapor deposition of an inorganic film involves the transport of a metal-containing compound or precursor into a reactor. The precursor first physisorbs and then chemisorbs and reacts on the heated substrate to liberate the supporting ligands. The supporting ligands liberated may desorb or react further before ultimately desorbing. Desorbed species are then

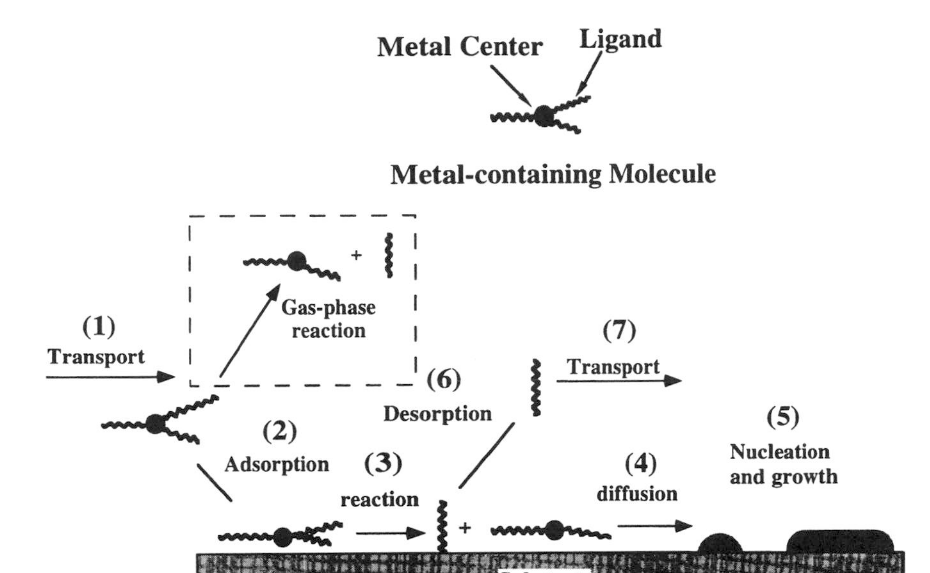

Figure 5.7 Schematic representation of key steps involved in CVD.

transported out of the reactor. Alternatively, the precursor may react with another reagent (such as H_2 or another metal-containing compound) to form a volatile product. The chemisorbed precursor may react further to eliminate the remaining supporting ligands to form atoms of a metal or other element which can diffuse over the surface to a nucleation site, leading to growth of the film. Any of these steps may be rate limiting, including the feed rate of the precursors to the reactor, the evaporation rate of the precursor where it is supplied by a liquid or solid delivery method, the diffusion of the precursor through the boundary layer to the substrate, the adsorption of the precursor, the chemical reaction of the precursor on the substrate (which generally is not an elementary process), or the rate of desorption of either the ligands or other products. The position of the rate-limiting step influences not only the growth rate but also film properties such as morphology, composition, purity, selective-deposition, uniformity, and conformality. As a result it is valuable to understand the consequence of the position of the rate-limiting step and how to influence it during a deposition process. It is worth noting at this stage that it is common in the literature (although rarely stated) that the film deposition rate reported is the *apparent* growth rate because the time for nucleation (induction period) is not normally measured.

Some potential problems associated with CVD are also illustrated in Fig. 5.6, which shows that premature reaction of the precursor in the gas-phase above the substrate surface can occur. This is particularly likely in a hot-wall deposition system with high total pressure and high precursor concentrations. Gas-phase reactions are generally detrimental because they can result in particle formation and deposition, incomplete reaction of the precursor, and depletion of the precursor concentration at the surface, all causing a decrease in the rate of deposition and formation of by-products which may react on the surface to introduce surface impurities. However, there are examples of deposition processes (e.g., CVD of Si and GaAs) where gas-phase reactions are considered beneficial.

5.2.2 Precursor Requirements

Metal-containing compounds that are to be used for the CVD of metal-containing films should exhibit a number of chemical and physical properties. They should have a reasonable vapor pressure at a temperature [see equation (1)] that does not lead to thermal decomposition, and they should undergo thermal decomposition in a controlled manner (by eliminating organic by-products, thus avoiding impurity incorporation) to give a film with the desired composition, morphology, and purity. These aspects of precursor requirements are described in this section.

Precursor Volatility As a rule of thumb, a vapor pressure of at least 100 mTorr is desirable at the delivery temperature to achieve reasonable deposition rates (e.g., hundreds of nm/min). From a purely practical point of view, a higher vapor pressure is required to use standard gas mass flow controllers (about 10 torr) to ensure controlled delivery of the precursor to the deposition chamber. This aspect is particularly important when employing conventional precursor delivery using bubblers (see later), where it is often necessary to heat the precursor for long periods at elevated temperatures. Even partial thermal decomposition of the precursor can result in changes in precursor delivery rates, changes in films properties, and irreproducible results between depositions. This requirement of long-term thermal stability at ordinary temperatures (shelf life), combined with high thermal reactivity at elevated temperatures (to achieve high deposition rates), creates a dichotomy because it is often desirable to deposit a film at the minimum possible substrate temperature but at high deposition rates. As a result, there is often a narrow temperature window where at the low-temperature end, the precursor should be thermally inert for long periods, while at the high-temperature end, the precursor should react at high rates to deposit pure films. Most industrially important CVD processes utilize simple precursors such as metal hydrides (SiH_4, AsH_3), metal alkyls [$Al(\textit{i}\text{-Bu})_3$, $GaEt_3$], metal halides (WF_6, $TiCl_4$), or mixed-ligand systems such as Me_2SnCl_2, because they have high vapor pressures and deposit films in a highly reproducible and controlled fashion, avoiding impurity incorporation [often with the sacrifice of relatively high substrate temperatures (about 500°)]. However, a wide variety of materials exhibit interesting properties in thin-film form, and this is creating a challenge to develop molecular metal-containing compounds, specifically as precursors for CVD. Some examples of precursors, conditions, and deposition techniques used to deposit a variety of materials are presented in Table 5.1.

TABLE 5.1 Deposition Characteristics

#	Precursor(s)	Substrate	T_s/°C	Carrier Gas	Deposition Rate	Deposit	Purity	Comments	Reference
METALS									
1	Al(*i*-Bu)$_3$	Si(111)	400	Ar	900 nm/min	Al(111)	Epitaxial films	RHEED Gas temperature controlled CVD	14
2	AlH_3•NMe_3	Cu	85-200	140 - 1000 nm/min	up to 1μm/min.	Al	High purity	highly oriented (111) films	15
3	Cu(hfac)$_2$	SiO_2, TiN, Al_2O_3	250-350	H_2	120-180 nm/min	Cu	XRD, RBS, AES. XPS	Near bulk resistivity	16
4	(hfac)Cu(vtms)	SiO_2, W	150-200	None	up to 1 μm/min	Cu	2 μΩcm	Low pressure differential reactor	17
5	Cu(acac)$_2$, H_2	Si	30 - 400	He	up to 20 nm/min.	Cu	AES, XRD, 2 μΩcm	RPECVD	18
6	WF_6	Si(100)	250-500	SiH_4	90 μm/hr	W	Low Si content	7.5 - 15 μΩcm	19
CARBON									
7	CF_n	SiC, WC	900	H_2/O_2	~1.5 μm/min	C diamond	Raman	PECVD non-diamond C etched by F	20
CARBIDES									
8	BCl_3	Si	450	H_2/CH_4	4.5 μm/hr.	B_4C	XPS	HFCVD	21
9	B_2H_6/CH_4, BMe_3	Si	25	CH_4	0.5 Å/sec	B_4C	B_2O_3 contamination	PECVD	22
10	$CrNp_4$	Si(100)	520-570	None	~0.15 μm/hr	Cr_7C_3	XPS, AES, SIMS ERD	Hot-wall low press	23
11	Cr(*t*-Bu)$_4$	Si(100) steel	100-150	He or N_2	up to 850 nm/hr	CrC_x	x=1.2-1.4	XPS	24
12	$HfCl_4$, CH_4	Carbon, Quartz	<1300	H_2	Not reported	HfC	SNMS		25
13	1,3-disilacyclobutane	TiC	810-1285	Atm. press.	4.2 μm/hr.	SiC	SiC primary phase	Cold-wall	26
13	$TiCl_4$, C_2H_4, NH_3	Steel bearing	25	H_2	up to 20 μm/min	TiN/TiC	Stoichiometric	LACVD	27
14	$MeSiCl_3$	Si(100)	1050	H_2	~10^{-6} g/cm^2/min.	SiC	Oriented SiC(111)	Hot-wall LPCVD	28

TABLE 5.1 (*Continued*)

15	Si_2Me_6	Si(111)	900 - 1100	None or H_2	0.9 - 55 nm/min.	SiC_x	x = 0.8 - 2.3 (AES, XPS)	Hot-wall RPECVD	29
16	H_2SiEt_2	Si(100)	700	N_2	~15 mg/hr	SiC	β-SiC	Ea=41Kcal/mol .	30
17	$H_2Si(\text{-}t\text{-}Bu)_2$	Si(100)	<850	None	1 - 7 mg/hr	SiC	Si:C::1:1 (AES)	Ea = 24 Kcal/mol	30
19	$TiCl_4$, C_3H_8	W	up to 1500	None or H_2	3 μm/min.	TiC	AES XPS	Kinetic study	31
20	$(\eta^5\text{-}C_5H_5)_2V$	Glass/alumina	Not reported	H_2	2.2 nm/min.	VC_{1-x} VO_2	XPS	PECVD	32
21	$(\eta^5\text{-}C_5H_5)_2V$	Steel	700	H_2	Not reported	VC	WDS O and amorphous C contaminated		33
22	WF_6, OMe_2	Metals	500	H_2	0.008 g/min	W_3C, W_2C	XRD TEM	Composition not investigated	34
					NITRIDES				
23	$AlMe_3$, NH_3	β-SiC(111)	1100	None	60 μm/hr	AlN	XRD SEM	Morphology Study	35
24	$AlCl_3$, NH_3	β-SiC	1000	$N_2O/H_2/O_2$	Not reported	AlN		Insert O into AlN lattice	36
25	$AlMe_3$, Me_3SiN_3	Si(111)	300-450	H_2	3 μm/hr.	AlN	Rel. high levels of C & O impurities	Atom press CVD	37
26	$Al(t\text{-}Bu)_3$, $^tBuNH_2$		500-600		0.5 μm/hr	AlN			38
27	BH_3NEt_2H	Si or Ni	450	none	Not reported	BN	FTIR, XRD	Claim first single-soruce precursor to BN	39
28	Ga, N_2	SiC(0001)	annealed at 700	some H_2	up to 5000 Å/hr.	GaN	RHEED SEM	MBE	40
29	$GaEt_3$, NH_3	Sapphire (0001)	1030	H_2	~0.5 μm/hr	GaN	XRD	MBE	41
30	$GaMe_3$, NH_3	Sapphire (0001)	1030	NH_3	~0.5 μm/hr	GaN	XPS RED	ArF eximer laser gas phaase reactions	42
31	$InMe_3$,HN_3	Si(100)	308 nm 25-430	None	Not reported	InN	PS, UPS, SEM	LACVD	43
32	$MoCl_2(N\text{-}t\text{-}Bu)_2(NH_2t\text{-}Bu)]_2$	Si and glass	450 - 650	H_2	10 - 70 Å/min.	MoN Mo-C-N	Inc. C at higher temp.	LPCVD	44
33	SiH_4, N_2	Not reported	150-350	None	up to 17 nm/min.	Si_3N_4	Si-H present FTIR	RPECVD	45,46
34	$\eta^5\text{-}C_5H_5)Ti(\eta^7\text{-}C_7H_7)$	Si(100)	300 - 600	$H_2/N_2/NH_3$	2.5 - 40 nm/min	Ti/TiN	Stoichiometric TiN, $TiC_{2.7}$	PECVD	47

35	$TiCl_4$, H_2, N_2	Mo	800 - 1000	$Ar/N_2/H_2$	2 mg/m^2/sec.	TiN	XRD	APCVD Ea = 100 K/mol. mechanistic study	48
36	$TiCl_4$, NH_3	Si(100)	400-700	Ar	400 Å/min	TiN	Cl-contaminated	LPCVD Ea = 35 kj/mol	49
37	$(\eta^5\text{-}C_5H_5)_2Ti(N_3)_2$	Si	600	N_2/Ar	Not reported	TiN	AES	Photo - and thermal CVD	50
38	$Ti(NMe_2)_4/NH_3$	Si(100)	250-500	H_2	5 nm/min	TiN	AES	RPECVD	51
39	$TiCl_4$, $AlCl_3$, NH_3	Steel	450	H_2/Ar	Not reported	$Ti_{(1-x)}Al_xN$	No C, O or Cl contamination	PECVD	52
40	$M(NR_2)x$ M=V, Nb, Ta	Si, SiO_2, C, B	200 - 400	NH_3	up to 2000 Å/min.	VN Ta_3N_5	XPS, RBS	APCVD	53
41	$(t\text{-}BuN)_2W(NH\text{-}t\text{-}Bu)_2$	glass/Si	500 - 700	None	2 - 10 nm/min.	WN_x	x= 0.7 - 1.8 (SIMS, PS)	Cold-wall	54
					OXIDES				
42	$SiCl_4$	Si(100)	50-200	H_2/O_2/Ar	up to 16 Å/min	SiO_2	AES	RPECVD	55
43	$SnCl_4$, $Cu(OAc)_2$	Glass	550	O_2/air	Not reported	SnO_2	Not reported	Spray Pyrolysis/ PECVD	56
44	$TiCl_4$	Quartz	650 - 1,000	none	4 - 22 Å/sec	Ti, Si_x/TiO_2	XPS	LACVD	57
					SULFIDES				
45	Cd(morpholinodithioato)	Quartz	200-400	N_2	Not reported	CdS	RBS, FTIR Cd:S::1:1		58
46	$R_2In(SR')Cu(S_2CNR'')$	Si(111)	400	N_2	Not reported	$CuInS_2$	XPS, XRF	Cold-wall low pressure CVD	59
47	$[t\text{-}BuGaS]_4$	KBr, GaAs SiO_2	380-420	Ar	Not reported	GaS	Oriented cubic GaS by RBS	Phase control by precursor design	60
48	$[R_2In(SR')]_2$	Si(100), KBr	300-400	Ar	Not reported	In, $InS/\beta In_2S_3$	XPS		61
49	MoF_6, $MoCl_5$	Si(100), steel, graphite	400	H_2S/Ar	Not reported	MoS_2	XRD, SEM	Homogenous gas phase reactions	62
50	$ZnMe_2$, H_2S	Glass coated with Indium tin oxide	150-275	H_2	Not reported	ZnS	Electron microprobe $\sim ZnS_{0\text{-}9}$		63

TABLE 5.1 (*Continued*)

51	$M(SAc)_2TMEDA$, M = Zn, Cd	Si	80 - 175	N_2	10 nm/min.	MS and $Zn_{1-x}Cd_xS$	Electron microprobe and AES	claimed lowest temperature deposition	64
52	$M(SAc)_2$(polyether), M = Ca, Sr, Ba	Si	300 - 400	N_2	10 nm/min.	MS	AES	AACVD, claimed first single-source precursor to MS	65
53	$M(S_2CNR_2)_2 \cdot L$	Glass/GaAs	350-450	None	Not reported	ZnS/CdS	1:1M:E (EDS)		66
PHOSPHIDES									
54	$Co(CO)_3NO$, $(t\text{-}Bu)_2EH$ E=P,As, $Co(CO)_2[E(t\text{-}Bu)_2H](NO)$, $Co_2(CO)_6(\mu E_2)$		300-500		4-330 Å/min	CoE	AES XRD	Comparison of single-source & individual sources	67
55	$InMe_3$, PH_3	40 - 150 pore Vycor glass	300	None	Not reported	InP	IR, XRD	Sequential deposition of $InMe_3$ followed by PH_3	68
56	$InMe_3$, PH_3, $P(t\text{-}Bu)_3$	InP(100)	650	H_2/PH_3	2.1 μm/hr	InP	Good quality InP	OMVPE PH_3 better than $P(t\text{-}Bu)_3$	69
57	$ZnMe_2$, PH_3	GaAs ZnSe	320-410	PH_3/H_2	up to 100 nm/min	Zn_3P_2	Epitaxial growth	Photo-CVD	70,71
SILICIDES									
58	$TiCl_4/SiH_4$	Si(100)	500-520	None	Not reported	Ti_5Si_3	XRD TEM	Selective deposition	72
59	WCl_4/SiH_4	Si	500-800	H_2	up to 200 nm/min.	WSi_2	AES	LPCVD	73
TELLURIDES									
60	$Sb(NMe_3)_3$ $Te(SiMe_3)_2$	Si(111)	25	None	1 μm/hr.	Sb_2Te_3	3 at %C 2 at %O (AES)		74

Source: Data from Refs. 14 to 74.

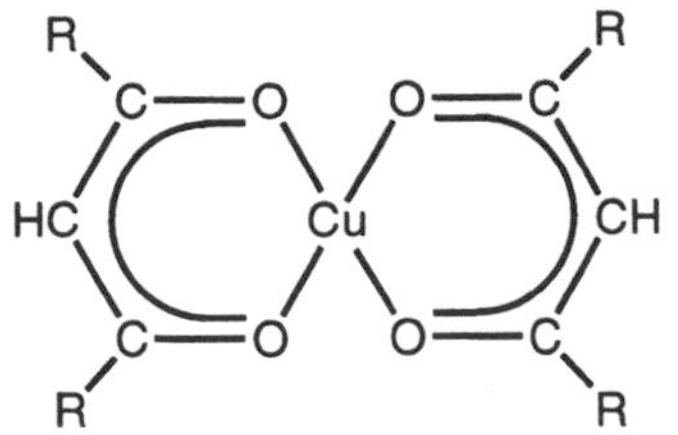

Figure 5.8 Structure of Cu(β-diketonate)$_2$.

The most volatile metal-containing compounds are typically monomeric and often contain fluorinated ligands. An example is shown in Fig. 8. Fluorinated metal β-diketonates such as Cu(hfac)$_2$ have been studied extensively as precursors for a wide variety of metal-containing films, including metals, metal oxides, and metal sulfides. Because β-diketonate ligands are bidentate and generally (but not always!) prefer to chelate to a metal center, the metal β-diketonate complexes of low-coordination-number metals are often monomeric and volatile.[75] However, problems arise (such as insufficient volatility) with metals that prefer higher coordination numbers, such as the group II elements. The group II elements, which are used in a large number of technologically relevant materials such as ferroelectric films (e.g., $BaTiO_3$) and ceramic superconductors, exemplify this problem due to their coordination chemistry. In this case, elements such as Ba prefer coordination numbers in the range 8 to 12, so a chelating bidentate ligand for a low-valent (e.g. 2+) metal coordinatively saturate the metal center if it prefers a coordination number greater than 4. This results in oligomerization of the compound to a point where the metal's coordination number is satisfied. The solid-state structure of "Ba(tmhd)$_2$," better described as Ba_4(tmhd)$_3$, is shown as as example in Fig. 5.9.

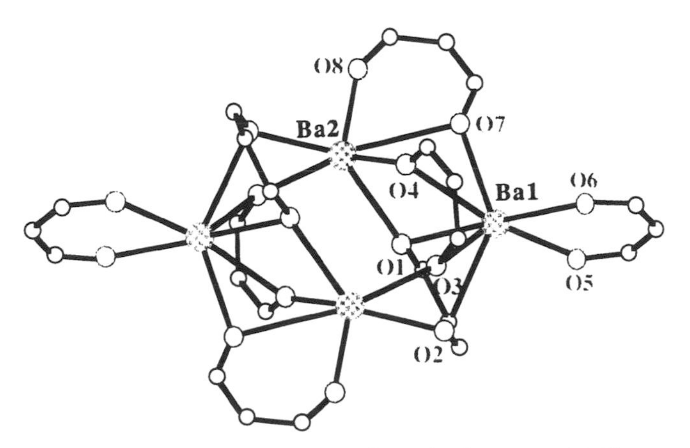

Figure 5.9 Structure of Ba_4(tmhd)$_8$. (From Ref. 76.)

There are two ways to solve this problem and prevent oligomerization. One method is to employ sterically demanding ligands which prevent oligomerization by limiting the steric accessibility of ligands of an adjacent molecule to the metal center. This has been shown to prevent oligomerization in a number of cases, but has some problems. One is that the metal center in sterically encumbered compounds is generally very reactive, due to its unusually low coordination number. As a result, these compounds are typically very reactive toward small molecules, such as O_2 and H_2O, which can penetrate the ligand shell (i.e., protective toward larger ligands). This can make them difficult to handle from a practical viewpoint. The other problem with sterically encumbered metal-containing compounds is that the large ligands generally impart a relatively high molecular weight to the compound, which tends to decrease the volatility. The other method to solve this problem is to introduce chelating multidentate ligands into the coordination sphere of the metal, which can satisfy the preferred higher coordination number of the metal while preventing oligomerization. This method has been used successfully in a number of cases, particularly for the heavier elements such as the lower group II and lanthanide elements, as shown in Figs. 5.10 and 5.11.[77-82] There are also some problems associated with this approach. The multidentate ligands may prefer to bridge two or more metal centers rather than chelate or may dissociate prior to or during transport, which can lead to oligomerization and decreased volatility. However, where these problems are avoided, this approach has led to major developments in the deposition of heavier element-containing films.

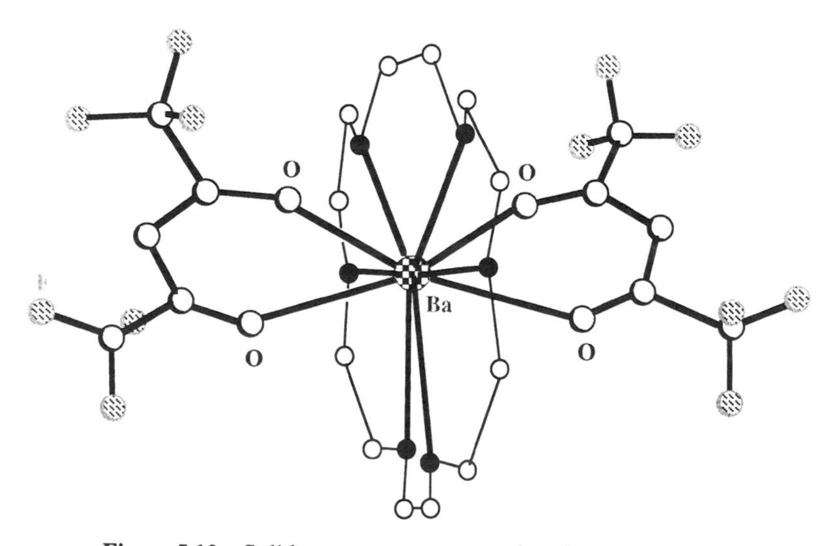

Figure 5.10 Solid-state structure of $Ba(hfac)_2 \cdot$ 18-crown-6.

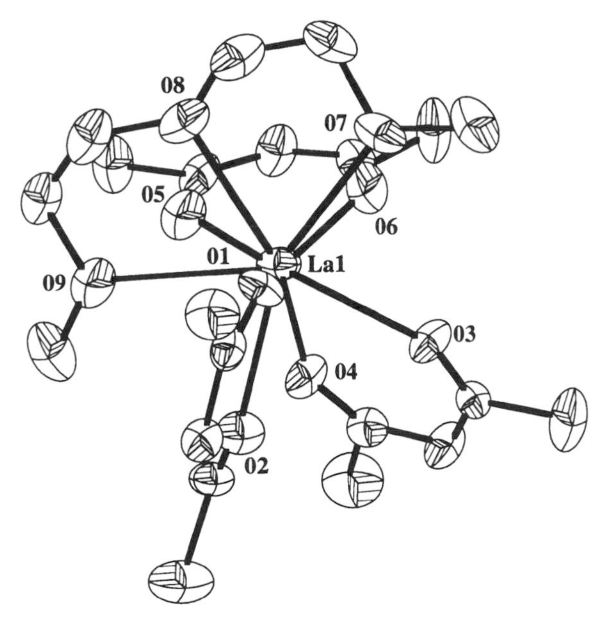

Figure 5.11 Solid-state structure of $La(hfac)_3 \cdot$ triglyme. (From Ref. 83.)

In cases where the precursor is delivered to the CVD reactor from an evaporator such as a bubbler, the phase in which the precursor exists at the bubbler temperature can be important. Gaseous and liquid precursors are likely to be delivered to the deposition chamber more reproducibly and in a more controlled manner than solid precursors. The surface area of a solid precursor is likely to change with evaporation time and result in changes of evaporation rate that may affect reproducibility.

The paucity of metal-containing compounds with suitable thermal stability and sufficient volatility together with the difficulties associated with evaporation of solids has led to significant interest in the development of alternative precursor delivery systems in which precursors with lower volatility and lower thermal stability can be tolerated. These alternative delivery methods generally employ a liquid delivery method, so precursor solubility and compatibility with the solvent and other precursors, if present, become important issues, as described in Section 5.2.6.[84]

Precursor Classification and Reactivity Metal-containing compounds can be classified into three different types, depending on the nature of the ligands attached to the metal center: (1) inorganic compounds, which possess ligands that do not contain carbon; (2) metal–organic compounds, which possess ligands that contain carbon but do not possess metal–carbon bonds; and (3) organometallic compounds, which possess carbon-containing ligands with metal–carbon bonds. Inorganic compounds that have been used as precursors are primarily the metal halides. They typically have low volatility (with a few exceptions, such as $TiCl_4$ and WF_6), require high temperatures to initiate reactions, and generally require a coreagent (such as H_2) to produce the final desired film. Metal–organic and organometallic compounds generally have an advantage over inorganic compounds in that they have higher volatility. However, they often thermally decompose to liberate organic fragments which lead to carbon or other contamination of the substrate. This can be alleviated to some extent by addition of a coreagent that eliminates the organic fragments. Examples of metal-containing compounds used in CVD processes are provided in equations (1) to (13), where their overall reaction pathway is also classified.

Reductive elimination

$$2(CH_3)Au^{I}(PMe_3) \rightarrow 2Au^0 + C_2H_6 + 2PMe_3 \quad (1)$$

$$2Al^{III}H_3 \cdot NR_3 \rightarrow 2Al^0 + 3H_2 + 2NR_3 \quad (2)$$

Bond homolysis

$$2Al^{III}(\text{-}i\text{-Bu})_3 \rightarrow 2Al^0 + CH_2{=}C(CH_3)_2 + 3H_2 \quad (3)$$

Ligand dissociation

$$Fe^0(CO)_5 \rightarrow Fe^0 + 5CO \quad (4)$$

Disproportionation

$$2(\text{hfac})Cu^{I}L \rightarrow Cu^0 + Cu^{II}(\text{hfac})_2 + 2L \quad (5)$$

Reduction

$$WF_6 + 3H_2 \rightarrow W + 6HF \quad (6)$$

$$Cu(\text{hfac})_2 + H_2 \rightarrow Cu + 2H\text{hfac} \quad (7)$$

Redox

$$2WF_6 + 3Si \rightarrow 2W + 3SiF_4 \quad (8)$$

$$Pd(\text{hfac})_2 + Cu \rightarrow Pd + Cu(\text{hfac})_2 \quad (9)$$

Metathesis

$$ZnEt_2 + H_2S \rightarrow ZnS + 2EtH \quad (10)$$

$$BCl_3 + NH_3 \rightarrow BN + 3HCl \quad (11)$$

Metathesis and reduction

$$6TiCl_4 + 8NH_3 \rightarrow 6TiN + 24HCl + N_2 \quad (12)$$

Ligand elimination

$$Zn(SEt)_2 \rightarrow ZnS + SEt_2 \quad (13)$$

One other potential method of deposition, not commonly used for CVD processes, but used more often for preparation of solid-state materials, is that of chemical vapor transport. This process generally involves the vapor-phase transport of a gaseous inorganic material, such as ZnS, through reaction with another reagent to form a more volatile species which is then transported and reacts on a substrate with a lower temperature. The resulting product can be the same as the starting materials or different as a result of a chemical reaction. For example, ZnS can be deposited by chemical vapor transport in the presence of I_2 as a result of the reversible equilibrium[85]

$$ZnS_{(s)} + I_2 \leftrightarrow ZnI_2 + \text{“S”} \quad (14)$$

Once the precursor has been delivered to the substrate it should undergo surface-initiated chemical reactions to deposit the desired material, preferably at high rates and with high purity. The simplest material that can be deposited by CVD is an elemental film such as a metal. As a result, there has been a great deal of fundamental interest in the mechanistic aspects of CVD of metals (in addition to the technological relevance of metal films).[1] However, even in these relatively simple systems, the detailed mechanisms of the surface reactions are not well understood generally because it is difficult to characterize the intermediates during CVD processes or under typical CVD conditions. A full mechanistic understanding of the deposition of more complex materials such as binary or ternary materials requires detailed understanding of the heterogeneous thermal decomposition mechanism for each individual molecular species on a given initial substrate and/or growing surface and an understanding of the reactivity between these species and intermediates. However, despite these difficulties there are a number of systems in which some mechanistic details have been derived and the overall reaction pathway is established through a simple by-product analysis. In general, it is desirable to use metal-containing precursors that undergo surface reactions that lead to complete elimination of the supporting organic ligands to avoid impurity incorporation in the film. In many cases, additional reagents are used as oxidants, to oxidize (and so remove) the ligands or the metal; reductants, to reduce the metal and protonate (and so remove) the ligands; or as a source of one of the elements desired in the film, such as “S” from H_2S, or “O” from H_2O.

The expected reactions of particular metal-containing species (metal halides, metal alkyls, metal alkoxides, metal β-diketonates, etc.) under different conditions will not be described here because this topic is the subject of other texts. Some examples of well-characterized CVD processes are described in more detail in Section 5.4. These examples provide a cross section that illustrate the different reaction mechanisms possible.

Finally, organic polymeric materials have found widespread use in the manufacture of electronic devices and have been employed in increasingly diverse areas. Polymers have been used as lithographic resists and dielectrics in the fabrication of semiconductor devices and as passivation and insulating materials in electronics packaging. As a result, there is considerable interest in developing dry processing techniques, which include CVD, to deposit films of organic polymers. Analogous precursor requirements pertain to organic monomers as metal-containing precursors. Deposition of organic films is described in more detail elsewhere.[86, 87]

Single-Source Precursors One problem associated with the deposition of complex films (i.e., a film comprised of two or more elements) is in the reproducible, controlled delivery of the metal-containing compounds to the surface and control of film stoichiometry. This problem arises because the metal-containing precursors have different volatilities and reactivities. The different volatilities of the precursors result in different evaporation rates and partial pressures at a given temperature such that each precursor must be delivered separately to the CVD reactor through (heated) mass flow controllers. This complexity is compounded by the problem described above, where solid precursors are used which can result in different evaporation rates due to decreasing surface areas as a function of time. Once the precursors are in the vapor phase, they are transported to the substrate, where they can react at different rates which depend on the substrate temperature. Under these conditions, if any of the latter steps is rate limiting, the film stoichiometry is strongly affected. To avoid these limitations, and to ensure control over film composition, the deposition must be carried out under conditions where the feed rate of the least volatile precursor is the rate-limiting step. However, this can impart severe restrictions on the deposition rate and film quality.

Another problem with using a number of individual precursors is that if one of the intermediates formed during film deposition is volatile at the deposition temperature required for the other precursors, this may lead to selective depletion of this component of the film. A typical example is the deposition of lead-containing films such as $PbTiO_3$, which can result in formation and desorption of volatile PbO as an intermediate, which results in depletion of lead in the final material. Using this delivery approach, the only choice is to add additional lead-containing precursor to make up for the loss of this reagent.

To circumvent some of these problems, the potential utility of single-source precursors rather than multiple-source precursors (as described in the preceding paragraph) to deposit complex films has been studied. A single-source precursor is one that contains the (metallic) elements desired in the final film, preferably in the ratio also required in the film. If this precursor can be delivered to the substrate and a suitable mechanism exists to remove the supporting ligands, there is a reasonable chance that the film may exhibit the desired composition and stoichiometry. This strategy can be quite successful when the films to be deposited are binary and composed of a metal and a nonmetal. In these cases, a single metal-containing compound is required in which the ligand is generally the source of the nonmetallic element in the film. Examples of this strategy include the deposition of metal sulfides (ZnS, CdS), metal nitrides (GaN, BN), GaAs, metal phosphides, metal carbides, metal silicides, and metal oxides (see Table 5.2.)

There are, however, some potential problems with this strategy, especially where two or more metals are necessary in the film. Single-source precursors are likely to have higher molecular weights than multiple-source precursors, so they generally exhibit lower volatility, which can lead to either low deposition rates or the inability to deposit films without resorting to nontraditional delivery methods. It is also generally difficult to deposit films with compositions other than those that require integral stoichiometries. This can be a problem where nonintegral stoichiometry or doped films may be required, which is often the case to obtain the desired properties. The deposition of nonintegral stoichiometry films, which are often in the form of solid solutions, from single-source precursors can be alleviated by the development of precursors, which can exist as solid solutions. For example, the species $M(S_2CNEt_2)_2(PMe_3)$, (M = Zn and Cd) forms solid solutions, as do their decomposition products, $Zn_{1-x}Cd_xS$.[88] This put some stringent requirements on the precursor design because it generally requires the same ligand set for the different metals to avoid stoichiometric reactions.

One unique advantage of the use of single-source precursors is the use of molecular clusters as precursors with a particular structural architecture such that the internal structure of the precursor can be used to control the crystallization of the final film. This could potentially be used to deposit new or metastable phases of certain materials; however, there are few examples of this strategy in the literature. Cubic GaS[89] and hexagonal ZnS[90, 91] films have been deposited from molecular clusters, where a relationship between the cluster structure and film crystalline phase has been observed.

5.2.3 Comparison of Hot- and Cold-Wall CVD Reactors

Chemical vapor deposition has been carried out in many types of reactors, primarily hot- and cold-wall reactors, and plasma reactors (see Section 5.3).[8, 92, 93] An account of the characteristics of hot- and cold-wall reactors is

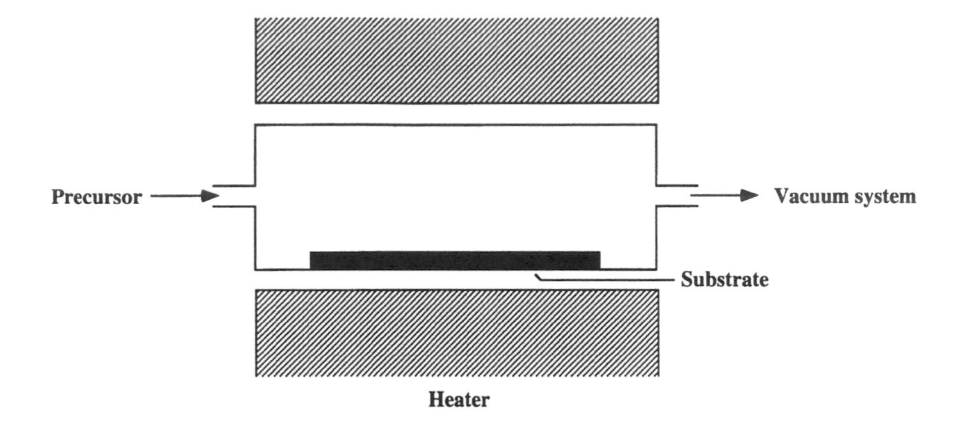

Figure 5.12 Schematic representation of a hot-wall CVD reactor.

relevant to this discussion for two reasons. First, the type of reactor and its operating conditions can have a profound effect on the deposition process, and second, the mode of reactor operation can be used to advantage to derive quantitative kinetic and mechanistic information if care is taken in the interpretation of rate data, as described below.

Hot-wall reactors (Fig. 5.12) have been used extensively for laboratory studies of metal CVD, even though most metal CVD in industry is carried out in single-wafer, cold-wall systems. The advantages of hot-wall systems are that they are simple to operate, can accommodate several substrates, can be operated under a range of pressures and temperatures, and allow different orientations of the substrate relative to the gas flow. The major problem with hot-wall systems is that deposition can occur not only on the substrate but also on the reactor walls. These deposits can eventually fall off the surface and contaminate the substrate. Also, the correspondingly large consumption of the precursor can result in feed-rate-limited deposition (discussed below) at relatively low temperatures compared to cold-wall reactors. Deposition only on the substrate can occur in special cases when deposition is selective and the reactant does not decompose on the reactor walls.[94, 95] However, reactions usually occur on the walls, and in most cases, the fraction of the surface area of the reactor covered with deposit can change with time during an experiment and from one experiment to another, leading to problems in reproducing deposition conditions. In addition, homogeneous gas-phase reactions in the heated gas can occur, which can lead to reduced deposition rates. For this reason, hot-wall reactors are used primarily at the laboratory scale to study the overall feasibility of using a given precursor for CVD. They are also often used to determine reaction-product distributions because the large

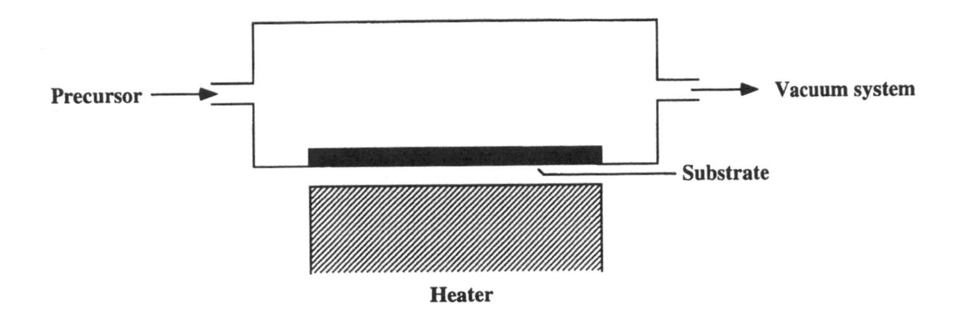

Figure 5.13 Schematic representation of a cold-wall CVD reactor.

heated surface area can consume the precursor completely and provide high yields of the reaction products.[94, 95] Although exceptions exist, hot-wall reactors are generally not used in industry or for quantitative measurements of reaction kinetics. They are, however, used extensively in industry for CVD of semiconductors and oxides, with precursors having high vapor pressures.

On the other hand, cold-wall reactors (Fig. 5.13), are used extensively at the laboratory scale and are also the industry workhorse for CVD. Although cold-wall reactors usually accommodate only a single wafer in various orientations to the flow, pressure and temperature can be controlled, plasmas can be used, deposition does not occur on the reactor walls, homogeneous reactions are less favored, and higher deposition rates can be obtained than with hot-walls reactors. Cold-wall reactors are also more suitable for studies of selective deposition because deposition does not occur on the reactor walls. They are also used for measurements of kinetic parameters because surface-reaction-limited kinetics can be achieved easily.[17, 96, 97] To obtain kinetic data, it is desirable to operate in the differential mode, where conversion of the precursor is low enough to ensure surface-reaction-limited kinetics and to minimize conversion of the reaction products.[17] High deposition rates can be obtained because deposition occurs only on a single substrate, so that the precursor consumption rate is low. For manufacturing applications, single-wafer cold-wall reactors are preferred because they allow better control of deposition behavior.

5.2.4 Overall Reactor Behavior

It is often important to understand the overall behavior of a cold- or hot-wall CVD reactor in order to make improvements and provide better control of the process. This understanding is usually obtained by a variety of measurements. The simplest and most common types of measurements are the dependence of the rate of deposition on the type of substrate, temperature, reactant feed rate, and reactant and total pressures. These measurements are often motivated by attempts to optimize film characteristics while maximizing the deposition rate, which must often be greater than 0.1 $\mu m/min$ for CVD processes used in microelectronics processing and higher for large-area coating processes such as those found in the glass industry. Another motivation for these measurements is that these dependencies can reflect the kinetics of the surface reaction under surface-reaction-limited conditions. This information can provide insight into the surface reaction mechanism, which can allow synthesis of new precursors or modification of operating conditions to provide new reaction pathways for improved purity or selectivity and increased deposition rate. In combination with these measurements, it is also important to know the dependence of film properties, such as morphology, purity, and resistivity on deposition conditions. This task is usually accomplished by analyzing films once they are deposited, although some in situ techniques exist.[8] Another motivation for these measurements is that they can provide insight into the mode of reactor operation.

A problem with the rate measurements described above is that they can give large amounts of complex information which can be difficult to interpret. For example, complex variations of the rate with changes in the temperature and partial pressures of the various species are observed. A classic example is tungsten deposition with SiH_4 as a coreactant. In this case, positive-, zero-, and negative-order WF_6 pressure dependencies are all observed, depending on the operating regime. Similar situations occur for CVD of Cu from Cu^{I} and Cu^{II} precursors. A further complication is that these variations are not necessarily directly related to the reactions occurring on the substrate surface; this is the case for feed-rate-limited or transport-limited deposition. Another complication is that various types of reactors have been used for CVD in industry and in the laboratory. These reactors have been used mainly for measurements of the deposition rate as a function of temperature and pressure and for measurement of film properties. Because of differences in geometry, reactant introduction, and other factors, different results are often obtained for the same precursor, making comparison of results from different systems difficult.

Overall reactor behavior can be understood crudely in terms of the boundary layers for velocity, temperature, and reactant concentration. The first step in understanding the qualitative behavior of CVD reactors is to know whether the reactor is operating in the molecular-flow or continuum-flow regime. In the molecular-flow regime [usually encountered during low-pressure CVD (LPCVD)], the gas mean free path is long compared to reactor dimensions, and collisions of molecules and atoms occur primarily with the walls of the reactor and the substrate. An average concentration can be defined based on the pressure, but the transport of the gas through the reactor is not governed by continuum-fluid-mechanics equations. For continuum flow [usually encountered with atmospheric-pressure CVD (APCVD)], the gas mean free path is much less than the reactor size and collisions occur primarily between gaseous species. In the latter case, the gas behaves like a viscous fluid and the variations of gas velocity, temperature and species concentrations can be obtained from equations based on continuum theory.[8, 93, 98, 99] An overview of boundary conditions has been described elsewhere and the reader is referred to more specialized texts on this subject for detailed equations, more in-depth discussions of these and related concepts, and references to the literature.[8, 92, 93, 98–101]

Rate-Limiting Steps Deposition-rate measurements have been reported under a wide variety of conditions, and it is usually not clear when the deposition is surface-reaction, feed-rate, or transport-rate limited. This distinction is critical if the data are to be used to assist in precursor design. A major concern for the design and operation of CVD reactors is the dependence of the deposition rate on operating variables. For conventional delivery systems, the deposition process involves three steps in series: (1) precursor delivery, (2) precursor transport to the surface of the substrate, and (3) surface reaction of the precursor. Because these steps are in series, the slowest step will limit the overall rate, and three cases are possible. In this section, this problem is quantified. For liquid delivery, precursor vaporization occurs after step 1 and can also be rate limiting.

The maximum deposition rate under any circumstances is given by

$$\text{rate} = \frac{\alpha P_r v_m}{\sqrt{2\pi m_r k_B T}} = \frac{\alpha_p P_r v_m \exp(-E_A/RT)}{\sqrt{2\pi m_r k_B T}} \tag{15}$$

where rate = deposition rate, cm/s
v_m = volume of depositing metal atom, cm^3
P_r = precursor partial pressure immediately above surface, dyn/cm^2
α = fraction of reactant molecules that react after striking surface, dimensionless
m_r = molecular mass of precursor, g
k_B = Boltzmann's constant, 1.38×10^{-16} erg/molecule · K
T = temperature, K
R = gas constant, ergs/(mol · K)
E_A = activation energy, ergs/mol
α_0 = constant

The quantity $P_r/\sqrt{2\pi m_r k_B T}$ is the flux of reactant to the substrate surface (collisions/$cm^2 \cdot s$). The quantity v_m converts the flux to units of deposition rate (cm/s). The parameter α takes into account the details of the adsorption/desorption/reaction phenomena. As a result, α depends exponentially on temperature. The maximum value of α is 1 and is achieved when every collision results in reaction. This expression predicts that approximately 1 monolayer (roughly 1 Å of metal) can be deposited every second at 10^{-6} torr if $\alpha = 1$. At 10^{-2} torr, 10,000 monolayers per second, and at 1 torr, 10^6 monolayers per second can be deposited. In most cases, $\alpha \ll 1$, and the actual deposition rate is much lower than the values mentioned above. The parameter α is also important because values of α near unity result in poor conformality.[99] Conversely, values of α much less than unity provide conformal deposition. The value of α can be controlled through precursor design. At a given temperature, precursors with lower decomposition temperatures will have higher values of α. Therefore, conformality can be improved by using precursors that are more thermally inert (see Fig. 5.5).

For diffusion-limited deposition, the precursor pressure at the surface, P_r, is much lower than the pressure of the precursor in the main gas outside the concentration boundary layer. For this case the deposition rate is again reduced below the maximum possible value. Calculation of the pressure at the substrate surface is complicated and in general requires complex descriptions of the fluid flow, mass transfer, and heat transfer, as well as surface and gas-phase reaction kinetics.

Surface-reaction limited deposition occurs under conditions where one step in the surface reaction is slow or deposition rate limiting. This is typically likely to occur when the substrate temperature is low and could be chemical (a thermally induced chemical reaction) or physical (a diffusion of nucleation step) in origin (see Fig. 5.2). More details on feed-rate, diffusion-rate, and surface-reaction-rate-limited deposition and the pressure and temperature dependence of deposition rates are given elsewhere.[1, 8, 92, 93, 98–101]

5.2.5 Blanket and Selective CVD

Area-selective CVD has been studied most often for a metal or semiconductor (such as Si) as the growth surface and SiO_2 as the nongrowth surface. The origin of selectivity can have kinetic or thermodynamic limitations for the formation of the new phase on the nongrowth surface. Thermodynamic considerations have been discussed by Carlsson.[102] In this review, both limitations are discussed. The results of some representative studies of selective CVD of metals are presented in Table 5.2.

In many CVD studies, however, the exact nature of the initial surface species was either not specified or was now known. For example, Si forms a native oxide surface layer unless care is taken first to remove it and to passivate the surface. Similarly, TiN can have OH groups on the surface.[122, 123] Many metals readily form oxides when exposed to air, which can result in the presence of M—OH groups on the surface. Because nucleation and selectivity are heavily influenced by the nature of the surface layer, great care must be taken in controlling surface properties and characterizing the surfaces before deposition if the origin of the selective deposition behavior is to be understood. In general, the surfaces described above have various functional groups, most commonly including —OH and M—O—M.[124–127] Depending on their environment and the chemical composition of the substrate, the

TABLE 5.2 Summary of Selective Deposition

Mechanism	Precursor Coreactant	Growth Surface	Non-Growth Surface	Ref.
Intrinsic differences in rates	(hfac)Cu(PMe_3)	metals	SiO_2	94, 95
	Cu$(hfac)_2$	Pt	SiO_2	103
	Cu$(nona\text{-}F)_2$	SiO_2	Si/metals	104
	$Co(\eta^5\text{-}C_5H_5)_2/H_2$	SiO_2	Si	105, 106
	$Fe(\eta^5\text{-}C_5H_5)_2/H_2$	SiO_2	Si	105, 106
Sacrificial reaction of growth surface	WF_6/Si	Si	SiO_2	107
	MoF_6/Si	Si	SiO_2	108
	Pd$(hfac)_2$/Cu	Cu	Si, SiO_2, Al, W, Ni, Co	109
Activation of growth surface	WF_6/H_2	W	SiO_2	107, 110
	Cu$(hfac)_2/H_2$	Cu	SiO_2	103
	Cu$(hfac)_2/H_2/H_2O$	Pt, W, Pd, Ag, Au	SiO_2	111, 112
	Cu$(hfac)_2$/He/H_2O	Pt	SiO_2	103
	Al alkyls	$TiCl_4$-treated SiO_2	untreated SiO_2	113, 114
Selective passivation of non-growth surface	(hfac)Cu (2-butyne)	metals	SiO_2	97
	(hfac)Cu(VTMS)	metals	SiO_2	115, 116
	(hfac)Cu(1,5-COD)	Au	alkanethiol SiO_2	117, 118
	Al(-*i*-Bu$)_3$	metals	SiO_2	119
	$(\eta^5\text{-}C_5H_5)PtMe_3$	PTFE	modified PTFE	120
	WF_6	W	SiO_2	107
	$RAuPMe_3$	Cr	BF_3/Cr	121
Gettering	WF_6/H_2	Si	P-doped SiO_2	110

—OH groups will exhibit different pK_a values, which along with variations in their surface concentration and local environment, will lead to different reactivities.[126, 127] Even on the same surface, a variety of different functional groups (including isolated —OH, hydrogen-bonded —OH and Si—O—Si rings in the case of SiO_2) may be present simultaneously,[126, 127] and each will react in a different manner with the same precursor.[128-131] In addition, the types of reactive surface sites available (such as —OH groups) influence strategies to optimize selective deposition (such as passivation via removal of the surface OH groups on the nongrowth surface). These differences are only now beginning to be exploited.[105, 114, 115, 117, 118, 132-134]

A variety of organic polymers, including fluorinated polymers and polyimides, also have applications as substrate materials;[135] however, few studies have examined selective CVD onto these and similar surfaces. Adhesion can be problematic during deposition onto these materials, which is critical for materials such as PTFE. Chemical vapor deposition has been carried out onto modified PTFE,[136, 137] but insufficient characterization data are available to define the reactive sites present on the initial growth surface.[138-140]

Selectivity for modified PTFE in the presence of unmodified PTFE has been observed for (hfac)CuL species[136, 137, 141] and for $(C_5H_5)PtMe_3$.[120]

Reactor operating conditions such as pressure, temperature, reactant concentrations, coreactant concentrations, and reactor type (hot-wall or cold wall) also influence selective deposition. Often, these can be varied to give selective deposition, as in the case of W deposition from WF_6 or Cu deposition from some (β-diketonate)CuL compounds. In addition, care must be taken when comparing reports of selective deposition different laboratories because of these variables. Conditions that promote gas-phase reactions must be avoided because the species formed may be transported to the nongrowth surface, where reaction can lead to loss of selectivity.

Mechanisms of Selective CVD There are a variety of mechanisms for selective deposition that rely on either inhibiting the adsorption and reaction of the precursor on the nongrowth surface or promoting its reaction on the growth surface. These mechanisms can be divided into six groups, with the first five resulting from chemical nucleation and the sixth, physical nucleation:

1. The intrinsic reaction rate of the precursor on the nongrowth surface is slower than its reaction rate on the initial growth surface and on the growing film.
2. The growth surface (e.g., Si) acts as a reducing agent and is selectively, sacrificially consumed by a precursor such as WF_6 or MoF_6, while the nongrowth surface provides, a slower reaction rate because no reducing agent is present.
3. A chemical reaction such as dissociation of a coreactant (a reducing agent, H_2) occurs on the growth surface (e.g., a metal) but not on the adjacent nongrowth surface (SiO_2 or metal oxide).
4. The rate is increased on the growth surface by radiation (often photochemically driven reactions), while the nonirradiated nongrowth surface provides a slow, thermally driven reaction.
5. Selective passivation of the nongrowth surface prevents adsorption and reaction of the precursor on the nongrowth surface, while adsorption and reaction occur readily on the growth surface.
6. Reaction proceeds on both the growth and nongrowth surfaces but a free-energy barrier exists that inhibits nucleation on the nongrowth surface, while a smaller barrier exists on the initial growth surface that allows physical nucleation to occur.

Note that in mechanisms 1 to 5, it is assumed for the sake of discussion that there is no thermodynamic barrier to the formation of metal clusters on the surface, so that even a single product atom remains on the surface after being formed. This indicates that the rate of the surface reaction limits the nucleation rate. In mechanism 6, however, it is assumed that the surface reaction is rapid and does not limit the nucleation rate. Instead, the nucleation rate is limited by the rate of formation of thermodynamically stable clusters on the growth surface. In practice, the surface reactions and the formation of nuclei are coupled phenomena, which makes identifying the limiting step difficult. However, the mechanisms discussed in these sections serve as useful limiting cases for the more complex systems encountered in practice.

Table 5.2 shows a summary of the precursors, mechanisms, growth surfaces, and nongrowth surfaces that have been described in the literature on selective chemical vapor deposition of metals, where most work has been conducted. The mechanisms of selective CVD have been described in more detail in another text.[1]

5.2.6 Precursor Delivery Methods

Precursor delivery methods vary depending on the physical properties of the precursor. These delivery methods play a critical role in the CVD of metals because the overall deposition rate can be limited by the rate of precursor delivery into the reactor if the precursor has a low vapor pressure and is delivered using a carrier gas or by direct evaporation into a vacuum. In the most desirable case, the precursor is a gas at room temperature, as with WF_6. In this case the precursor can be delivered using conventional gas-mass-flow controllers, and feed-rate limitations do not occur in most cases. However, precursors to most metal-containing films are more often liquids or solids, with relatively low vapor pressures, and are generally introduced through bubblers or by direct vaporization. In some cases the precursors are also thermally unstable at temperatures where useful vapor pressures are obtained. For these precursors, conventional delivery methods are either unsuitable or provide delivery rates that are slow and result in feed-rate-limited deposition. Therefore, alternative methods of delivery have been developed which include liquid delivery, aerosol delivery, and supercritical fluid delivery. These aspects have been described in some detail elsewhere.[1]

5.3 OVERVIEW OF VAPOR-PHASE DEPOSITION METHODS, RELATED AND UN-RELATED TO CVD

The objective of this section is to give a perspective of various vapor-phase deposition methods. This includes the limitations and relative advantages and disadvantages of these methods. Most film deposition processes can be divided into two general categories: chemical and physical vapor deposition. In addition to chemical vapor deposition, covered in the preceding sections, there are currently numerous CVD-related deposition methods that have found widespread applications. These include vapor-phase epitaxy (VPE), atomic-layer epitaxy (ALE), metal–organic molecular-beam epitaxy (MOMBE), chemical-beam epitaxy (CBE), plasma-assisted chemical vapor deposition (PACVD), remote-plasma chemical vapor deposition (RPCVD), and laser-assisted chemical vapor deposition (LCVD). These methods are discussed in Section 5.3.1. Non-CVD processes also play important roles in thin-film deposition. These processes include physical vapor deposition (PVD) by evaporation or sputtering of a target material onto a substrate and molecular-beam epitaxy (MBE). These methods are discussed in Section 5.3.2.

5.3.1 CVD-Related Processes

Vapor-Phase Epitaxy Vapor-phase epitaxy (VPE) is a term generally used in conjunction with compound semiconductors and usually refers to any vapor-phase epitaxial process that utilizes metal halides as transport agents.[142, 143] The major distinction between CVD and VPE is that the reactive precursors are generated in situ. An example of chloride VPE of GaAs involves flowing a $AsCl_3/H_2$ mixture over liquid Ga in a source zone (T = 750 to 850°C), forming GaCl, which is then mixed with another flow of $AsCl_3/H_2$ downstream. This is followed by reaction and subsequent GaAs growth in the deposition zone. In hydride VPE a HCl/H_2 mixture reacts with liquid Ga (In), forming GaCl (InCl), which further reacts with AsH_3 downstream and deposits GaAs (InAs) films on a substrate.[144] Because of the ease by which VPE can be scaled, it has been utilized in large-scale commercial operations for production of materials such as GaAsP for light-emitting diodes. VPE has typically been limited to growth of thick, high-purity epitaxial layers. Epitaxy refers to growth of a crystallographically oriented film on a single-crystal substrate with similar lattice constants. When a material is growth epitaxially on a substrate of the same material, such as silicon on silicon, the process is termed *homoepitaxy*. If the layer and substrate are of different materials, such as $Al_xGa_{1-x}As$ on GaAs, the process is termed *heteroepitaxy*. However, in heteroepitaxy the crystal structures of the layer and the substrate should have a good lattice match if oriented crystalline growth is to be obtained.[145] VPE closely resembles CVD and therefore has many of the advantages and disadvantages of CVD. Growth of superlattice structures (multiple layers of films with nanometer dimensions) is difficult with VPE due to transient changes in gas-phase composition when a new gas is introduced, and superlattice structures are therefore normally grown with MBE or MOCVD techniques. Another problem is film contamination originating from the species adsorbed on reactor walls.

Atomic-Layer Epitaxy The versatility and flexibility of CVD originate in part from the simplicity of the reactor and the wide range of metal-containing compounds that are available as precursor molecules. CVD has been used to produce every major type of compound semiconductor device with high film purity and abrupt interfaces. Among the drawbacks of CVD are the need for expensive and often hazardous reactants and the large number of parameters that must be precisely controlled during deposition to obtain the desired film uniformity and reproducibility. Limitations in uniformity of doping, composition, and film thickness are restrictions that are commonly encountered.

These limitations can be reduced in the growth of group 13 to 15 compounds by using the atomic-layer epitaxy (ALE) process, where the film is grown one atomic layer at a time. This is accomplished by *sequentially* introducing the source gases containing group 13 and 15 precursors into the reactor. At each step, self-limiting adsorption takes place because the group 13 molecule (or its dissociation product) chemisorbs only to active sites on the substrate until all sites are occupied. The group 15 molecule then reacts in a site-selective manner with the group 13 centers, which again leads to self-limiting adsorption. In the appropriate temperature regime and under ideal conditions, this sequence results in film deposition one atomic layer at a time. The major advantages of ALE are in the growth of very uniform thin layers over large areas with abrupt interfaces. Growth rates by ALE are inherently low, but the nature of the process makes it particularly effective for creating abrupt interfaces and heterostructures. The ALE growth rate is typically limited by the time required for switching from one precursor to the other, which usually requires several seconds. Consequently, a growth rate of a few atomic layers per minute is obtained by ALE compared to growth rates of several atomic layers per second for a conventional MOCVD process.

A variation on this process is to use the same method as described above for ALE but without employing an oriented or crystalline substrate, which results in polycrystalline films. This method has been termed atomic-layer deposition (ALD). This has been used for the deposition of 2 to 16 films for applications in flat panel displays.

Chemical-Beam Epitaxy and Metal–Organic Molecular-Beam Epitaxy
Chemical-beam epitaxy (CBE) [also called metal–organic molecular-beam epitaxy (MOMBE)] is a major deposition technique in semiconductor technology. CBE combines the beam nature of molecular-beam epitaxy (MBE) (Section 5.3.2) and the advantages of an all-vapor source such as in MOCVD. In contrast to MOCVD, CBE film growth is performed in a UHV chamber under low-pressure conditions (10^{-10} torr base pressure) and the source material is supplied to the deposition zone by molecular beams of reactive gases (typical beam pressure about 10^{-6} torr).[146] The molecular beams are effusive jets of the gaseous source materials. At these low pressures, mass transport proceeds via molecular flow with no collisions in the gas phase, and surface kinetics are solely responsible for the chemical reactions that lead to the film growth. When moving from CBE to MOCVD conditions by increasing the system pressure, additional growth-limiting steps may appear. The

process is then controlled by the gas-phase flow dynamics with associated lateral and vertical gas-phase diffusion. In CBE, a higher precursor growth efficiency is generally obtained than in MOCVD. Another important advantage of CBE over MOCVD is the UHV environment, which allows the application of in situ surface characterization techniques. These techniques [e.g., reflective high-energy electron diffraction (RHEED)] can be used for in situ study of the growth process and also to study structural, morphological, and electronic properties of freshly grown surfaces, interfaces, and thin layers, within the same vacuum conditions. The low pressure used in CBE allows selective area epitaxy (SAE), a technique that has been applied to group 13 to 15 materials. In SAE the film deposition is restricted to surface areas that are not masked by an oxide or a nitride layer. The major drawbacks of CBE and MBE are the very high cost of UHV equipment and the low deposition rates and throughput.

Plasma-Enhanced Chemical Vapor Deposition Plasma-enhanced chemical vapor deposition (PECVD) is also referred to as plasma-assisted CVD (PACVD). PECVD is one of the modifications of conventional CVD and involves generation of reactive species in a plasma.[1, 147, 148] The working pressure is usually in the range 0.1 to 1.0 torr. Inelastic collisions between high-energy electrons and the gaseous precursors generate excited neutrals and free radicals as well as ions and additional electrons. In the plasma, the degree of ionization is typically only 10^{-4}, so the gas in the reactor consists mostly of neutrals.[149] The reactive species have lower energy barriers to chemical reactions than the parent species and consequently, can react at lower temperatures. This is the major advantage of PECVD over a thermally driven CVD along with potential high growth rates. PECVD is the established commercial technique for low-temperature deposition of a number of important materials, especially insulating films such as silicon nitride and silicon oxide. Lack of substrate selectivity, poor conformality (step coverage) of the deposited films, plasma-induced substrate damage, and complexity are potential disadvantages of PECVD.

Remote Plasma-Enhanced Chemical Vapor Deposition In remote plasma-enhanced chemical vapor deposition (RPECVD), the plasma excitation region is separated from the growth region. This is in contrast to PECVD (above) and plasma sputtering (Section 5.3.2), where the substrate is exposed directly to the plasma. RPECVD reduces the potential of plasma damage to the substrate and film, but it is a complex process and is still being developed. Energy transfer occurs through multiple gas-phase collisions between long-lived excited species extracted from the plasma and precursor gas molecules injected downstream from the plasma region. The energy transfer produces electronically excited precursors which remain molecularly intact prior to interacting with the growth surface. The excited molecules in RPCVD have lower barriers to film deposition, and this allows the use of more complex precursors that do not form films under normal PECVD processing conditions.[150]

Laser-Assisted Chemical Vapor Deposition Laser-assisted chemical vapor deposition (LCVD) can be further divided into pyrolysis (thermal) and photochemical LCVD.[151-153] In thermal LCVD a laser is used to heat the substrate analogous to local heating in a cold-wall reactor and induces pyrolysis of the precursor on the surface. Photochemical LCVD involves the use of a laser (or another light source) to activate the precursor through specific electronic excitation in the molecule. A serious drawback is the incompatibility with parallel processes used in microelectronics, since LCVD is a serial process for deposition which requires scanning of the laser over the surface. This results in a slow process that is expensive and complicated. However, the usefulness of LCVD has been demonstrated in localized deposition to repair expensive electronic devices and offers the possibility of depositing thin films on fragile substrates such as polyimide or GaAs.[154] Laser-assisted chemical reactions involved in the deposition of materials used in the microelectronics field has been reviewed by a number of authors.[155-157]

5.3.2 Non-CVD Processes

Most metallization for microelectronics today is performed by physical vapor deposition (PVD), which includes evaporation and sputtering processes. This topic is relevant to discuss here because it is the major competitor to CVD. With both evaporation and sputtering methods, the formation of a layer on a substrate involves three steps: (1) converting a condensed phase material (generally a solid) into the gaseous or vapor phase, (2) transporting the gaseous phase from the source to the substrate, and (3) condensing the gaseous source on the substrate surface, followed by nucleation and growth of a new layer. Step coverage is a serious problem for films deposited by PVD. Step coverage usually refers to the ratio of the film thickness on the step sidewall to the film thickness on the top surface. Inadequate step coverage results from (1) the directionality of the deposition from the evaporating or sputtering sources, (2) the low mobility of the deposited atoms, molecules, or ions, and (3) severe topography (tall surface structures).

Physical Vapor Deposition by Evaporation Deposition of films by thermal evaporation is inherently a simple method.[158] The evaporated material deposits on a substrate that is maintained at a lower temperature than that of the vapor. In a conventional evaporation process, the material to be deposited is heated under high vacuum conditions ($10^{-5} \leq P \leq 10^{-8}$ torr). At such low pressures, the mean free path of the evaporated species is very long (5×10^2 to 5×10^5 cm) compared to the source-to-substrate distance, and this allows an essentially collisionless line-of-sight transport of the evaporated material to the substrate surface. The substrates are located at an appropriate distance facing the evaporation source. Evaporation is widely used for depositing aluminum, gold, and other metallic films for integrated circuit fabrication. The material to be deposited can be heated directly by an electron beam (or laser beam) by directing a stream of high-energy electrons (photons) at the target material to create a molten region at the surface and vaporize the material. Alternatively, a crucible containing the material is heated by induction or by resistive heating. To avoid contamination and to ensure the purity of the deposited film, it is necessary to use inert container material (evaporation cells) such as tungsten, molybdenum, tantalum, or platinum that have high melting points with negligible vapor pressures at the operating temperatures. Furthermore, the possibility of alloying by chemical reactions between the container material and the evaporant must be considered. However, this is less of a problem for target materials that sublime because they have a high vapor pressure below their melting point. Such materials can therefore be readily evaporated with less concern for source/evaporant reaction and contamination. Elements such as Cr, Mo, Pd, V, Fe, and Si reach vapor pressures of 10^{-2} torr well before they melt and hence can sublime to deposit films at sufficiently high rates. For a single-element material, the rate of evaporation is controlled by the temperature of the source. Deposition of alloy films that consist of two or more components can be complicated due to differences in vapor pressures of the various elements and compounds. This makes maintaining stoichiometry of both the target and the deposited films difficult since the target becomes richer in the less volatile species. Multiple-source methods where each element is independently heated is a more versatile process but requires good calibration of evaporation rates. Homogeneity in composition across the wafer surface is an additional potential problem. Simplicity and reliability are advantages using evaporation processes for thin-film deposition, along with high deposition rates and high film purity. Since the evaporation process is inherently directional, films are deposited with poor step coverage. This is one of the major reasons for the increasing interest in sputter deposition processes and CVD.

Physical Vapor Deposition by Sputtering Physical sputter deposition of thin films is the workhorse of microelectronics.[159] Sputter deposited films are widely used for metallization, diffusion barriers, and adhesion promoters in integrated circuits. Physical vapor deposition by sputtering is a process where the surface atoms of a target material are dislodged by bombardment with energetic ions that are generated in a glow discharge or plasma. The sputtered atoms are then ballistically transported to the substrate surface, where they condense to deposit a film. There are basically two different configurations of glow discharges: (1) a direct-current (dc) glow discharge, and (2) a radio-frequency (RF) glow discharge. Unlike evaporation, sputtering processes are very well controlled and are generally applicable to all materials: metals, alloys, semiconductors, and insulators.[145] The deposition of alloys by sputtering of an alloy target is possible because the composition of the film is locked to the composition of the target. This is true even when there is considerable difference between the sputtering rates of the individual alloy elements. At the early stages of sputtering, the component with high sputtering rate is preferentially sputtered from the target, leaving the target deficient in this component. The deficient region soon becomes deficient enough to compensate for the higher sputtering rate, which leads to film deposition that is of similar composition to that of the target. Alloys can also be deposited with very good control of composition by the use of individual component targets in a reactive environment.[149] This technique, known as reactive sputtering, uses reactive gas species such as oxygen or nitrogen. Examples are the use of gases such as methane, ammonia or nitrogen, and diborane to deposit carbide, nitride, and boride materials, respectively. In some applications, sputtering is used to clean the substrate surface by removing surface impurities such as oxide layers prior to film deposition. During the sputtering process, the deposited film is also exposed to ion bombardment, which can cause sputtering damage in the film. However, such damage can often be removed by annealing the film following the deposition process.

The dc plasma configuration (Fig. 5.14) is the simplest type of discharge, having the target material as the cathode and the substrate as the anode. The glow discharge is maintained under the application of dc voltage between the electrodes. Ions are generated by the inelastic collisions of accelerating electrons with inert gas atoms. The ions are accelerated toward the cathode,

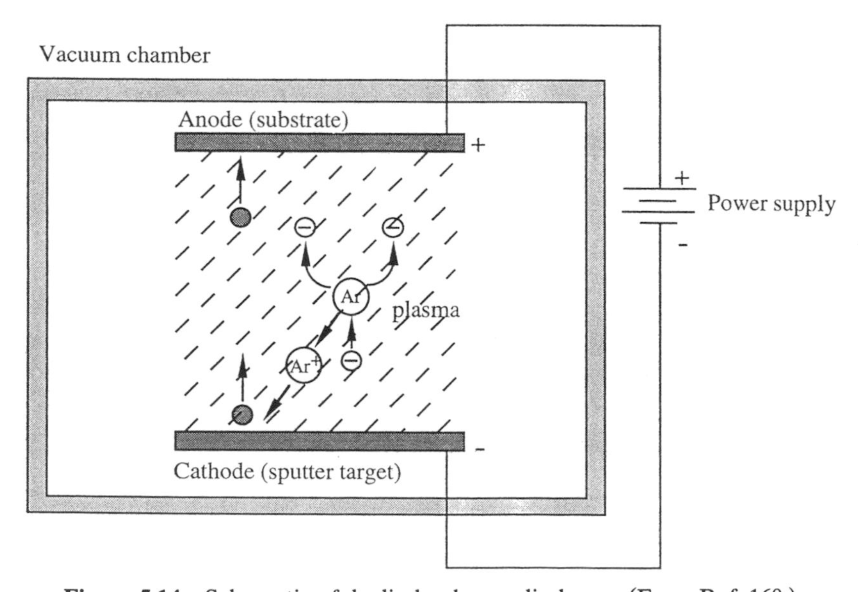

Figure 5.14 Schematic of dc diode plasma discharge. (From Ref. 160.)

where they knock off atoms from the target material by momentum deposition and the target material is subsequently deposited on the substrate. A dc sputtering process, however, possesses certain limitations for depositing films. For instance, it cannot be used to sputter deposit insulators because the glow discharge cannot be maintained if the electrodes are covered with an insulating material because of immediate buildup of a surface charge of positive ions on the front side of the insulator. Application of ac voltage to the electrodes allows regeneration of the lost electrons, and hence the discharge can be sustained without the need for secondary electron emission from the target material. The RF discharge can also be used for dry etching; however, RF glow discharges suffer from the same inefficient use of secondary electrons as dc discharges since most electrons emitted from the target are collected by the anode. As a consequence, the ionization events produced by the electrons are not significantly greater in number in an RF discharge than in a dc glow discharge. Thus high deposition rates cannot be achieved for RF sputtering. This is the major motivation for the use of magnetron-based sputter deposition.

Magnetron-Based Sputter Deposition Magnetrons are diode plasma devices that use a combination of electric and magnetic fields to provide a drift for secondary electrons from the cathode, thereby forming a closed loop. By using these devices in a sputtering environment, it is possible to confine any electrons that might have strayed away from the cathode region of the plasma. The voltage in a magnetron is typically much lower than an RF diode at the same applied power. This causes current densities at the target to be at least an order of magnitude higher than that for nonmagnetron configurations. As a result, more ions are generated in the plasma, which increases the sputtering rate of the target material and thereby increases the deposition rate on the substrate.

Inherently, the step coverage of sputter deposited films is very low. Methods to improve the step coverage of PVD films are (1) to deposit the films at high substrate temperatures, (2) to collimate the incoming flux of the depositing species, or (3) bias sputtering. High-temperature deposition improves step coverage if there is significant surface diffusion. Conventional magnetron sputter deposition, due to the broad angular distribution of the depositing atoms, has generally been considered to be obsolete for high-aspect-ratio semiconductor geometries. Collimation is used to improve the film thickness on the base of features for systems in which high temperature does not provide significant improvement in step coverage or in situations which the thermal duty required for surface diffusion is undesirable.[161] The improved step coverage is attained because components of depositing species that are emitted from the target at wide angles are filtered (removed) by the collimator. Thus a larger fraction of the unfiltered species deposit on the bottoms of surface features, but sidewall coverage is decreased in the process. However, collimated sputtering is an expensive technology due to low deposition rate, poor target utilization efficiency, and concerns such as particle contamination. Similarly, in long-throw sputtering, the large angle flux is reduced dramatically by an extended target to substrate spacing. Another technique is bias sputtering, which enhances step coverage of sputter-deposited films by resputtering material that is already deposited and redepositing it on the sidewall of the contact or a via (vertical connection) hole. As the bias voltage is increased, ion bombardment of the substrate is increased and relatively more material is resputtered onto the sidewalls. In addition, the increased ion bombardment results in an increase in surface mobility of adsorbed target molecules on the substrate. As a result, step coverage is improved significantly. These approaches are competitors to CVD processes.

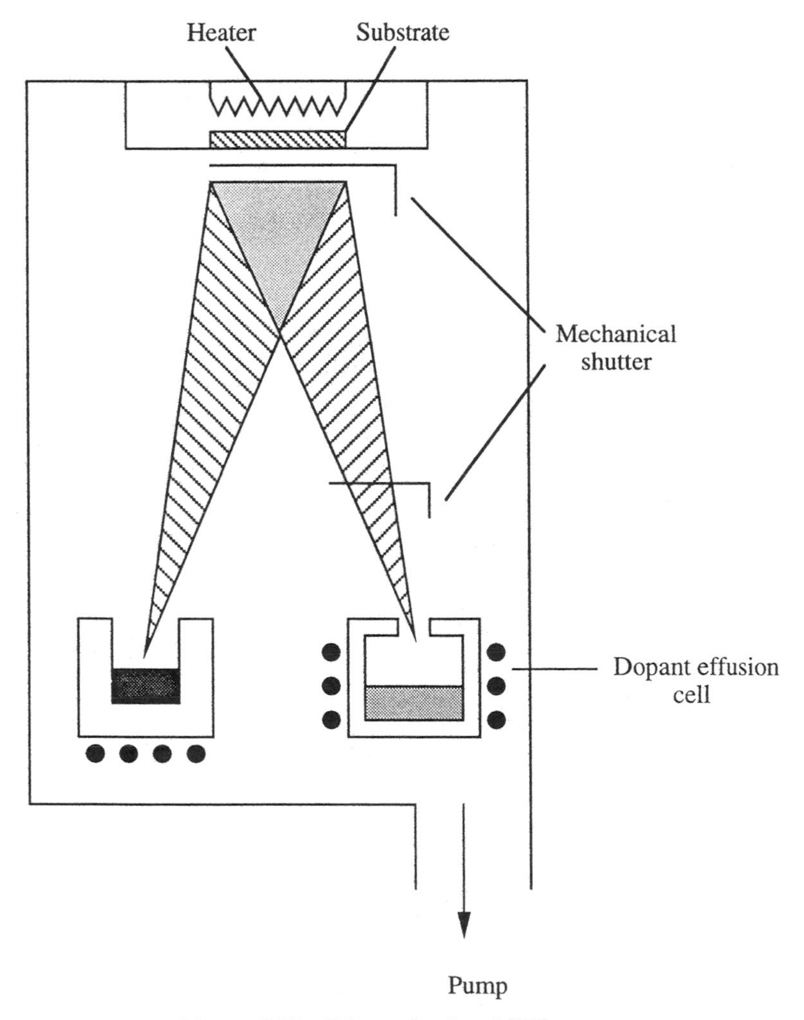

Figure 5.15 Schematic of an MBE system.

Molecular-Beam Epitaxy Molecular-beam epitaxy (MBE) is a non-CVD epitaxial process that uses an evaporation method.[162] MBE is a relatively simple process in which elemental sources are independently evaporated at a controlled rate, forming molecular beams that intercept at the heated substrate (Fig. 5.15). MBE is usually carried out under ultrahigh-vacuum (UHV) conditions (10^{-10} torr) and at low growth rates, where the growth process can be monitored in situ as the material is built up one atomic layer at a time.[142, 163] In contrast to CVD processes, MBE is not complicated by transport effects, nor are there any chemical reactions to consider. MBE has numerous advantages over CVD processes. Films with high purity and very complex layered structures can be deposited (nanostructures) with a precise control of doping of the deposited layers. The low processing temperature with MBE is also important for microelectronic production. A conventional temperature range for Si MBE is from 400 to 800°C. Growth rates in the range 0.01 to 0.3 μm/min have been obtained.[145] The higher value is comparable to those obtained in CVD epitaxy. A slow deposition rate (monolayer at a time) is very useful for creating materials with extremely uniform and abrupt interfaces, but these slow rates are not practical for depositing various simple materials where high throughout is needed. Further drawbacks for MBE are the requirement for very expensive (capital) and complex UHV chambers and surface analytical instruments.

5.4 CASE STUDIES

In this section we discuss selected applications of chemical vapor deposition (and physical vapor deposition) as a means of depositing pure metals, semiconductors, insulators, and mixtures of elements for applications in the microelectronics industry. The reader is referred to review articles in the following sections on the application of metal CVD for thin-film deposition in metallization of integrated circuits. Tungsten metallization (Section 5.4.1) is used in microelectronics for conducting layers and diffusion barriers. CVD of W also illustrates a deposition process that utilizes simple molecular precursors. It also serves as an example to illustrate many of the more practical aspects of CVD. Titanium nitride (TiN) (Section 5.4.2) thin ceramic films are used as diffusion barriers in submicron structures, commonly in contact with tungsten metal layers. Sputtering techniques are usually used for TiN deposition, but poor conformal film coverage is creating a need for new CVD approaches. Diamond (Section 5.4.3) is a material with unique properties. Methods of depositing polycrystalline diamond by CVD techniques use very simple precursors but require extreme process conditions. Lower deposition temperatures are needed to allow for a wide variety of new applications. Silicon (Section 5.4.4) is discussed because it is the most important material in the electronics industry.

5.4.1 Tungsten

Tungsten (W) films are used in applications that range from wear and corrosion protection to conducting layers and diffusion barriers in electronic devices. Examples of the latter are lines, vias, and plugs in integrated circuits (ICs), which are used as the metallization for the source, drain, and gate in transistors, ohmic contacts, and interconnections. Contacts make connections between the device at the silicon level and the first layer of metal. Vias make connections between subsequent metal layers. Tungsten has a number of attractive electrical, mechanical, and chemical properties. Tungsten has excellent resistance toward electromigration (the transport of metal atoms along grain boundaries driven by the force exerted by flowing electrons under high current densities). Furthermore, W has a low thermal expansion coefficient that is a good match to that of silicon and forms a stable interface with silicon. The low resistivity of tungsten films (about 5.6 $\mu\Omega \cdot cm$) has been exploited successfully at the first interconnect level to replace high-resistivity doped polycrystalline silicon (about 500 $\mu\Omega \cdot cm$). At the first interconnect level, contact is made between a metal and silicon, polysilicon, or materials such as GaAs. CVD of tungsten is commonly carried out by silane or hydrogen reduction of WF_6 after a selective, self-limiting reduction by Si. Tungsten deposition has recently been reviewed by Creighton and Parmeter[164] and Gladfelter.[6] Tungsten exhibits excellent barrier properties for Si–Al interactions since aluminum does not react with tungsten below 500°C, and tungsten silicide formation, and the W-Si interface occurs only at much higher temperatures. Tungsten has replaced aluminum in certain key areas of interconnect schemes. However, because tungsten has higher resistance, it is not a useful replacement for aluminum where long distances must be covered.

Physical vapor deposition (PVD) has been used extensively to deposit tungsten films but is becoming less useful as minimum feature sizes become increasingly smaller. Unless it is deposited at elevated substrate temperatures, the resistivity of the tungsten films is significantly higher than bulk and the commonly observed poor conformality (step coverage) is a severe drawback. Since PVD is a line-of-sight deposition technique, it is not capable of adequately depositing thin films on high-aspect-ratio submicron structures without leaving voids.

Chemical vapor deposition of W, on the other hand, has extremely good conformality that allows it to easily fill small contact holes in integrated circuits. Tungsten is typically deposited as a blanket film (controlled nonselective deposition), and etched back to create a W plug. The W CVD/etchback process is well developed and understood (Fig. 5.16). In the future, a chemical mechanical polishing (CMP) step will probably replace the etchback step, due to stricter planarization requirements. Selectively filling contact or via holes is an alternative where the need for etchback can be avoided by filling the shallower vias completely and filling the deeper vias partially.

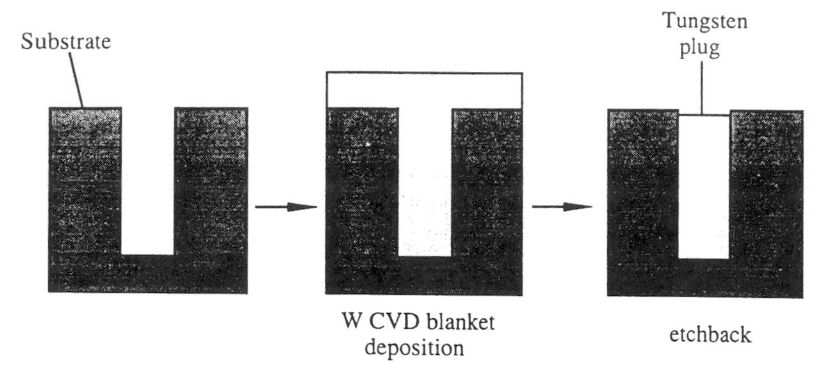

Figure 5.16 Schematic of the W CVD/etchback process to form a W plug in integrated circuits.

Tungsten Halides as CVD Precursors Tungsten halides (WF_6, WCl_6, and WBr_6) are easy to synthesize, are readily available and have sufficiently high vapor pressures to be used in W CVD, (Fig. 5.17). Table 5.3 lists some of the physical properties of the tungsten halides. The most commonly used methods of CVD of tungsten involve reduction of WF_6 using agents such as H_2, Si, and SiH_4 to obtain low deposition temperatures. WF_6 is a gas at room temperature (1000 torr vapor pressure at 25°C), and this is an important advantage over other tungsten CVD precursors since it greatly facilitates the precursor transport into the processing chamber and provides the potential of high deposition rates.

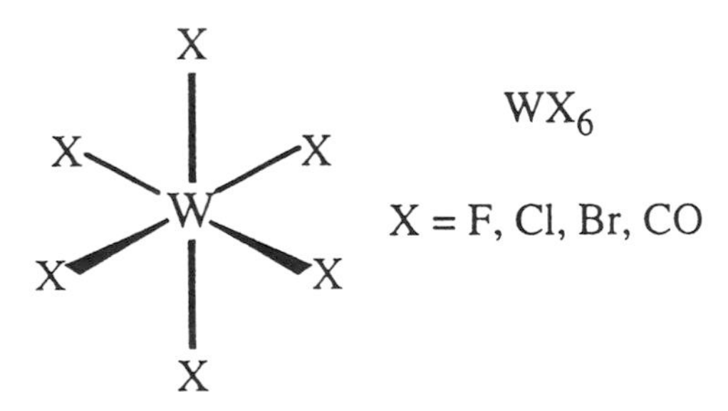

Figure 5.17 Tungsten halides (WX_6) and $W(CO)_6$ have an octahedral structure.

TABLE 5.3

Compound	Melting point (°C)	Boiling point (°C)	Vapor pressure (Torr/°C)	Decomp. temp. (°C)
WF_6	2.0	17	889/21	> 750
WCl_6	275	346	7/200	≈ 600
WBr_6	232	decomp.	—	≈ 600

$WF_6 + H_2$ Blanket chemical vapor deposition using hydrogen reduction of WF_6 has been used to deposit tungsten at temperatures between 300 and 800°C and is presently the major tungsten CVD approach used for thin-film deposition.[165–170] The overall reaction is described by

$$WF_{6(g)} + 3H_{2(g)} \rightarrow W_{(s)} + 6HF_{(g)} \tag{16}$$

The hydrogen reduction of WF_6 is slightly endothermic ($\Delta H = 21$ kcal/mol), but tungsten deposition occurs due to the large net increase in entropy when going from reactants to products and the nonequilibrium nature of the CVD process.[164] Selective tungsten CVD for applications in integrated-circuit technology usually involves process conditions of low pressures (0.1 to 10 torr) and low temperatures (300 to 500°C). The reaction that proceeds according to equation (16) generally obeys the rate law in equation (17), with an apparent activation energy of about 16 kcal/mol under low-pressure (LPCVD) conditions with excess H_2:

$$\text{growth rate} \propto P(H_2)^{1/2} P(WF_6)^0 \tag{17}$$

The observed pressure dependence for H_2 agrees with the fact that the coverage of chemisorbed atomic hydrogen (θ_H) in equilibrium with gas phase H_2 is expected to vary as $P(H_2)^{1/2}$ for low adsorbate coverage.[164]

$$H_{2(g)} + 2* \leftrightarrow 2H_{(a)} \qquad * = \text{vacant surface site} \tag{18}$$

$$WF_{6(g)} + 6* \rightarrow 6F_{(g)} + W_{(s)} \tag{19}$$

$$F_{(a)} + H_{(a)} \rightarrow HF_{(g)} + 2* \tag{20}$$

Similarly, if WF_6 is assumed to dissociate completely on the surface and maintain equilibrium with the gas phase, the adsorbed surface fluorine coverage (θ_F) will vary as $P(WF_6)^{1/6}$, again for low total adsorbate coverage. If the reaction proceeds via an irreversible Langmuir–Hinschelwood step between $F_{(a)}$ and $H_{(a)}$ [where step (20) is rate limiting], the rate law will obey the equation

$$\text{growth rate} \propto \theta_H \theta_F \propto P(H_2)^{1/2} P(WF_6)^{1/6} \tag{21}$$

This equation correctly predicts the observed $P(H_2)$ dependence and gives a very low order ($\frac{1}{6}$) for the $P(WF_6)$ dependence. Typically, $P(WF_6)$ has been varied by one to two orders of magnitude, but this may be insufficient data to differentiate between true zeroth- and $\frac{1}{6}$-order kinetics. As the H_2 pressure is lowered and becomes comparable with the WF_6 pressure, the deposition rate has been observed to drop to zero. This is because hydrogen chemisorption is completely quenched and the surface is saturated with adsorbed fluorine. However, just above the H_2 pressure threshold, the deposition rate becomes first order rather than $\frac{1}{2}$-order with respect to H_2. In this regime the WF_6 pressure dependence becomes strongly negative order rather than zeroth order. Tungsten deposition using WF_6/H_2 gas mixtures in plasma-assisted CVD[171–174] and laser-assisted CVD[175–182] are potential alternatives to the "thermal" or conventional metal CVD.

WF_6 + Si The reduction of WF_6 by silicon is of interest partly because of the potential of area-selective CVD of W.[168, 183–188] Tungsten thin films can be deposited selectively on silicon and metal surfaces, without nucleation and growth on adjacent insulators (e.g., SiO_2 and Si_3N_4). This selective nature can be understood by considering the following reduction (displacement) process:

$$2WF_{6(g)} + 3Si_{(s)} \rightarrow 2W_{(s)} + 3SiF_{4(g)} \tag{22}$$

$$WF_{6(g)} + 3Si_{(s)} \rightarrow W_{(s)} + 3SiF_{(g)} \tag{23}$$

With increasing temperature the transition from (22) to (23) takes place at about 450°C. The Si-reduction step provides the basis for selective deposition on surface areas that provide electrical contacts. Silicon is consumed in these areas and removed while tungsten is deposited. After rapid formation of the thin tungsten layer, further reaction is inhibited because it acts like a diffusion barrier against further reaction between WF_6 and silicon, since the WF_6 source gas can no longer diffuse sufficiently fast through the deposited film to react with the underlying Si. Hence the Si reduction is self-limiting. At typical low-pressure CVD conditions, SiO_2 is inert toward WF_6, and no significant interaction occurs with the insulating surfaces. If hydrogen is present in the reactant mixture, tungsten deposition may be sustained by H_2 reduction of WF_6. This process proceeds on metal surfaces that are capable of activating H_2 and WF_6 through dissociative adsorption, processes that do not occur on SiO_2.

Tungsten deposition by hydrogen reduction of WF_6 is well suited for filling of vias since it is highly selective. This is illustrated in Fig. 5.18, which shows a selectively deposited tungsten plug, connecting through a SiO_2 layer to a molybdenum layer. The WF_6 reaction proceeds initially by Si reduction and then by H_2 reduction. The reduction by Si is very fast and only stops due to solid-state mass transport limitations of W to the interface. This reduction is responsible for the selectivity since it nucleates the tungsten plug only on the exposed silicon surface. Once a tungsten film is deposited, the growth on tungsten inside the via is favored over the growth on SiO_2 and growth on W continues selectively by the second mechanism. The reduction of WF_6 by hydrogen has shown promise for the deposition of diffusion barriers directly on silicon, and for selective contact hole and via fill, in part because the necessity of additional patterning, including lithography and etching, is avoided.

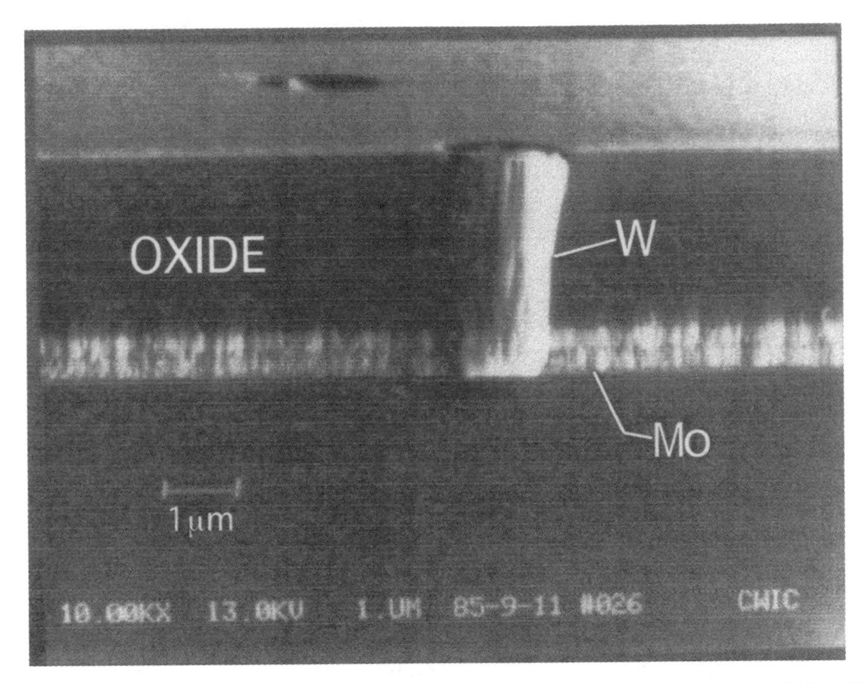

Figure 5.18 Tungsten plug formed by hydrogen reduction in a via in a SiO_2 film making a connection to a Mo layer on a silicon substrate. (From Ref. 163.)

The difficulty in selective CVD is maintaining selectivity. Preserving selectivity can be troublesome, as nucleation on SiO_2 generally occurs after short deposition times, and lateral tungsten growth by Si reduction occurs underneath SiO_2 regions in the Si–SiO_2 interface. Impurities on the SiO_2 surface are potential sites of tungsten nucleation and can contribute to loss of selectivity. A process step for in situ cleaning of the surface is desirable to limit contamination of the silicon dioxide with minimum exposure to the ambient atmosphere before selective tungsten CVD.

An important intrinsic mechanism of selectivity loss that occurs during tungsten CVD involves tungsten transport by the formation and disproportionation of volatile tungsten subfluorides (i.e., WF_5).[164, 189] A protective layer is necessary to prevent corrosive or damaging attack on the substrate. TiN thin films are an example of a suitable diffusion barrier for this purpose.

WF_6 + SiH_4 The electronics industry is in search of alternatives to hydrogen reduction of WF_6 because of the problems associated with this approach. Silane (SiH_4) is a colorless gas that has received much interest as a possible alternative to hydrogen as a reducing agent for WF_6 in tungsten thermal,[190-202] plasma-assisted,[203] and laser-assisted CVD.[154, 176, 204, 205] The addition of SiH_4 into the WF_6/H_2 gas stream effectively suppresses the reduction reaction of

WF_6 by the Si substrate since the following reactions with SiH_4 are thermodynamically and kinetically favored. The overall reaction can be expressed as[197]

$$WF_{6(g)} + 2SiH_{4(g)} \rightarrow W_{(s)} + 2SiHF_{3(g)} + 3H_{2(g)} \qquad 250°C \qquad (24)$$

$$2WF_{6(g)} + 3SiH_{4(g)} \rightarrow 2W_{(s)} + 3SiF_{4(g)} + 6H_{2(g)} \qquad > 600°C \qquad (25)$$

Reduction of WF_6 using silane has some advantages over using H_2. High deposition rates have been achieved at low temperatures (1 μm/min at 210°C[206] in a cold-wall reactor) and high selectivity is maintained continuously. This is in contrast to hot-wall reactors, where it is difficult to maintain selectivity for a long period of time due to gas-phase reactions. Silane reduction of WF_6 results in smoother films compared to that of hydrogen reduction. Depending on the SiH_4/WF_6 ratio and deposition temperature, either tungsten or tungsten silicide is deposited by CVD. Tungsten films can also be deposited selectively on many materials other than Si (e.g., Al, PtSi, TiN). In most commercial reactors it is found that silane is required for the nucleation of W (from WF_6) on TiN. Therefore, W deposition is commonly carried out with initial use of WF_6 and SiH_4 followed by bulk growth of W from WF_6 and H_2. The step coverage of the W "steed" layer is critical in affecting the overall conformality and reliability of the W. However, the SiH_4 reduction process has many disadvantages, such as silicon incorporation in the film, reactor contamination through particle formation, and poor step coverage, in addition to being expensive and difficult to handle.

The reactor design and choice of reactor material are also important for selective tungsten CVD. Although WF_6 and H_2 do not react in the gas phase at room temperature, WF_6 and SiH_4 undergo extensive gas-phase reactions, even at room temperature, that lead to formation of highly reactive reaction intermediates and particle formation. However, it is difficult to maintain selectivity with silane reduction compared to hydrogen reduction. This problem can be minimized by mixing the two gases just above the substrate surface. The use of cold-wall reactors is also effective because it reduces the concentration of reaction intermediates and products that can deposit on the nongrowth surface and degrade selectivity.

Organometallic Precursors The microelectronics industry has a need for the deposition of tungsten at lower temperatures and under less corrosive conditions. This has created interest in other tungsten precursors. Organometallic precursors have several potentially favorable properties as precursors for tungsten CVD. The relatively low thermal stability of these precursors enable thin-film deposition at substrate temperatures as low as 100 to 200°C compared to H_2 or SiH_4 reduction of WF_6, which occurs at higher temperatures. For this reason, these precursors have not been used successfully as alternatives for tungsten halides in microelectronic applications, due to the prevalence of impurity incorporation, especially carbon.

Various tungsten metal–organic precursors have been studied to date for use in W CVD applications.[1] These precursors are volatile, thermally stable, and relatively easy to decompose at low temperatures. $W(CO)_6$[164, 207, 208] is a white volatile solid (1 torr at 65°C) which starts to decompose around 230°C in the solid state according to the idealized equation

$$W(CO)_{6(g)} \rightarrow W_{(s)} + 6CO_{(g)} \tag{26}$$

In principle, no reducing agent is needed for $W(CO)_6$ (because it is already zero-valent), but hydrogen usually necessary to reduce carbon contamination. Methods of tungsten CVD using $W(CO)_6$ also include laser-assisted CVD.[209, 210]

In summary, WF_6 is still the best source for CVD of pure and well-adherent W films with low resistivities. Several exceptional properties of WF_6 put it ahead of all other precursors investigated so far: (1) high vapor pressure (WF_6 is a gas at room temperature), (2) metal films can be 100% pure, and (3) selective deposition can be obtained on silicon surfaces. No other precursor investigated to date fulfills all three of these requirements. Furthermore, WF_6 very high purity is available at low cost.

5.4.2 Titanium Nitride

Titanium nitride (TiN) is the most important metal nitride for microelectronic applications. It has a unique combination of properties, including high hardness, good electrical conductivity, a high melting point (3300°C), and chemical inertness. Thin films of TiN have traditionally found practical uses as tool coatings and bearings due to their superior mechanical properties and wear resistance and also as decorative layers because of their goldlike color (Fig. 5.19). Deposition processes and metal-cutting applications of TiN coatings have been reviewed.[211-213] More recently, TiN has found applications as a diffusion barrier material in microelectronics applications. Diffusion barrier layers are often introduced between the metal and semiconductor to suppress their interaction. The choice of the barrier material depends on the materials that must be separated. Passive diffusion barriers have a large negative free energy of formation and are, therefore, chemically inert, strongly bonded compounds. They do not interact with the overlying or underlying materials and generally have a low solid solubility for silicon and metals such as aluminum. As a result, these are preferred for barrier layer materials.

Titanium nitride acts as a passive barrier because it is chemically and thermodynamically stable. It acts as an impermeable barrier which physically prevents Al metal from diffusing into the Si. Titanium nitride also has a high activation energy of diffusion for other metals. These thin films are also used as an adhesion layer for CVD blanket tungsten deposition, as well as a means of helping preserve the junction integrity during tungsten CVD. TiN films are very stable over a broad compositional range, yet the film performance and

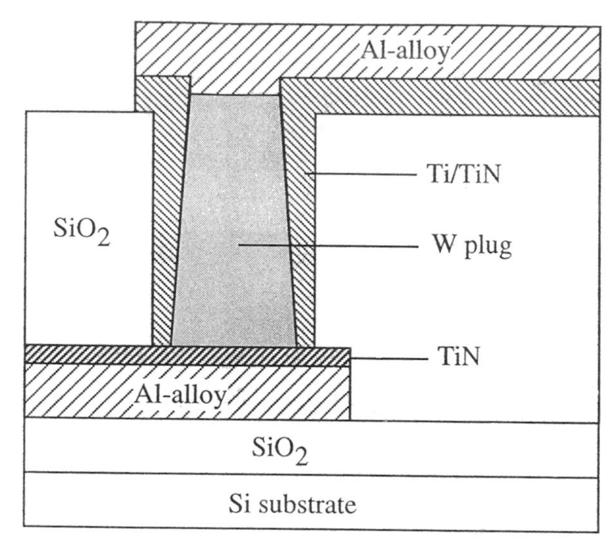

Figure 5.19 Schematic of a W via structure with interfaces of Al-alloy/W and W/TiN/Ti/Al-alloy at the upper and lower sides if the W plug. (From Ref. 214.)

properties depend critically on the film composition (N/Ti ratio) and film impurity levels. The structural consequence of compositional variation in the TiN sodium chloride structure is an excess of either sublattice Ti or N. Precise control of the PVD parameters is required to maintain a N/Ti ratio near unity. A N/Ti ratio of 1.0 is very desirable for TiN films used in microelectronics technology, due to minimum values of film resistivity and stress, and maximum values of the lattice parameter, hardness, and density at that composition.[215] In a CVD TiN film, the N/Ti ratios usually fall between 0.95 and 1.05, irrespective of the process chemistry used, making this technique very attractive for producing high-quality stoichiometric TiN.[49] For state-of-the-art device applications, TiN step coverage of nearly 100% is needed for deep, narrow, submicron structures. For TiN CVD to be successful in manufacturing, the process must be very reproducible and the key parameters affecting the film properties characterized.

PVD of TiN Physical vapor deposition has allowed TiN film deposition at temperatures below 500°C. At these lower deposition temperatures, the use of TiN thin films in advanced metallization schemes for ultralarge-scale integration (ULSI) applications in silicon microelectronic technology has become very attractive. Different deposition methods are used for collimated TiN sputtering: (1) the nitrided mode, in which the target surface is nitrogen-rich; (2) reactive sputter deposition of titanium in a nitrogen environment,[216] and (3) the most efficient, but least pure, nonnitrided metal mode, where Ti from the target is converted to TiN after sputtering by thermal nitridation in nitrogen or ammonia ambients. TiN films formed by thermal nitridation have limited barrier properties and are therefore not used widely in sub-0.5-μm-based technologies. TiN films formed using reactive sputter deposition techniques are stoichiometric and possess superior barrier properties. However, their use in sub-0.5-μm-based technologies is expected to be severely limited by the step coverage of the sputtered films. Collimating techniques during sputter deposition of materials have helped improve step coverage of reactively sputtered titanium nitride films or aspect ratios of more than 2 : 1 (height/width) and contact sizes of less that 0.5 μm. However, PVD techniques are inherently limited and may eventually yield to CVD of titanium nitride. TiN films derived from PVD exhibit poor step coverage, especially for submicron structures, which can cause shoulders or overhangs to form at the contact opening and poor sidewall (vertical) coverage.

The recent developments in collimated sputtering have reduced this problem and extended the applicability of TiN PVD to increasingly smaller features, but inadequate sidewall coverage in deep structures can lead to reliability problems due to poor functionality as a barrier or nucleation layer. In view of this, it is generally accepted that collimated sputtering and CVD TiN will compete based on cost and achievable conformality. TiN deposition into sub-0.5-μm-wide contact holes is currently achieved by collimated reactive sputter deposition of titanium in a nitrogen ambient.

CVD of TiN The CVD of TiN is classified here into methods based on inorganic precursors such as $TiCl_4$ and N_2 or NH_3 and metal–organic precursors such as $Ti(NR_2)_4$.

$TiCl_4 + H_2 + N_2$ Titanium nitride films have traditionally been deposited by thermal CVD from $TiCl_4/N_2/H_2$ gas mixtures at high temperatures ($T > 700°C$):

$$2TiCl_4 + 4H_2 + N_2 \rightarrow 2TiN + 8HCl \quad (27)$$

This high deposition temperature is incompatible with processes used past the first level of metallization because of the lack of thermal stability of the underlying layers.[217-219]

$TiCl_4 + NH_3$ To date, the best understood TiN CVD process uses $TiCl_4$ + NH_3 as reactants:[49, 220-228]

$$6TiCl_4 + 8NH_3 \rightarrow 6TiN + 24HCl + N_2 \quad (28)$$

This reaction is thermodynamically favored above 320°C. When deposition occurs in the surface-reaction-rate-limited regime, high step coverage is achieved. However, if deposition occurs in the gas-phase diffusion-limited regime, poorer step coverage results. The crystallographic orientation of TiN thin films depends on the deposition temperature and the partial pressures of the reactants. A low resistivity of TiN films is desirable in microelectronic applications, because it improves the electrical conductance of an interconnect. Difficulties in achieving a low resistivity at low deposition temperature using $TiCl_4$ is attributed to high levels of chlorine impurities that are incorporated into the TiN film. Chlorine not only affects the film resistivity but, more important, may cause reliability problems due to corrosion of the aluminum interconnects.

Laser-assisted CVD of TiN films can be carried out using reactive gas mixtures of $TiCl_4/NH_3/H_2$[27] and $TiCl_4/N_2/H_2$.[229-232] A CO_2 laser is used to induce pyrolysis of the reactants at the substrate surface and deposit dots and lines of TiN thin films. Other TiN LCVD coatings methods include an excimer laser and TDMAT (below)/N2/H_2 and an argon ion laser and $TiCl_4/N_2/H_2$ and $TiCl_4/NH_3$ gas mixtures.[233] However, it should be noted that there are few commercial applications of laser-based CVD processes.

The reaction of $TiCl_4$ with nitrogen precursors other than NH_3 to achieve lower deposition temperatures has shown promise. The TiN deposition rate at 450°C using $TiCl_4/(CH_3)(HNNH)_2$ is an order of magnitude higher than that from $TiCl_4/NH_3$, but degradation of step coverage to 70% was observed.[215] The film resistivity at 500°C was comparable to that of a film grown using $TiCl_4/NH_3$ at 700°C. The precursors $TiCl_4 + (CH_3)CH_2N$ produced a 200-$\mu\Omega \cdot$ cm film with comparable chlorine impurity levels as $TiCl_4 + NH_3$ chemistry but at a 100°C lower deposition temperature.

TiI_4 TiI_4 is a possible alternative to $TiCl_4$ for TiN deposition. The lower dissociation energy of Ti–I bonds compared to Ti–Cl can potentially result in lower deposition temperatures. TiN films can be grown from $TiI_4/NH_3/H_2$ using thermal CVD at $T > 365$°C (compared to $T > 600$°C using $TiCl_4$,[49, 224] and from $TiI_4/N_2/H_2$ using PECVD at $T > 250$°C.[234]

Metal–Organic Precursors (TDMAT and TDEAT) Deposition of TiN films from metal–organic precursors (MOCVD) offers an attractive route for low-temperature deposition with adequate feature coverage. Although metal–organic TiN CVD has been studied widely, precursors for TiN have been confined primarily to tetrakis(dimethylamino)titanium(IV) (TDMAT) and tetrakis(diethylamino)titanium(IV) (TDEAT) because of commercial availability. TDMAT ($Ti(NMe_2)_4$) is a liquid at room temperature with a vapor pressure of 1 torr at 60°C. TDEAT ($Ti(NEt_2)_4$) has a vapor pressure of about 100 mtorr at 100°C. A higher growth rate can be obtained using TDMAT due to its higher vapor pressure. Several TiN deposition processes are available using metal–organic precursors, including unimolecular pyrolysis, a reaction with another molecule, such as ammonia, and plasma-enhanced reactions. TiN films have been deposited from TDMAT[215, 235] and

TDEAT[235] with NH_3 at temperatures as low as 450°C with low impurity levels. The overall stoichiometry for the TDMAT reaction is

$$(Ti(NMe_2)_4) + NH_3 \rightarrow TiN + 3HNMe_2 + \cdot NMe_2$$

where the dimethylamine radical indicated undergoes further reaction with unspecified species (e.g., it could extract a hydrogen atom from TDMAT in the gas phase).[236] Gas-phase reaction (transamination) between $Ti(NR_2)_4$ and NH_3 occurs to produce HNR_2. This reaction is critical to the process and is required to produce low-resistance, low-carbon films.[237]

In summary, the use of $TiCl_4$ has several advantages over the use of TDMAT and TDEAT for TiN deposition. The cost of $TiCl_4$ is approximately 20% of the cost of the MOCVD precursor. The high vapor pressure of $TiCl_4$ (m.p. −23°C, b.p. 136°C) allows the use of conventional mass flow controllers, while low-vapor-pressure MOCVD precursors require expensive liquid delivery systems. CVD-TiN films grown from $TiCl_4$ provide good conformality and high-quality film properties, while TiN films deposited by MOCVD can exhibit either high-quality film properties (such as low resistivity, high density, and low impurity content) or good conformality—but not both.[238]

5.4.3 Diamond

Diamond is best known as a precious gem, but it has many potential applications as an engineering material, due to its unique properties. In this section we illustrate some of the many aspects of recently developed diamond CVD methods. Currently, diamond finds use as an abrasive, in wear-resistant coatings, and as a heat sink. Table 5.4 lists some of the outstanding properties of diamond. The two most useful forms of carbon as thin films, diamond

TABLE 5.4 Attributes of Diamond Coatings

- Extreme mechanical hardness (~ 90 GPa)
- Strongest known material, highest bulk modulus (1.2×10^{12} Nm^{-2}), lowest compressibility (8.3×10^{-13} m^2N^{-1})
- Highest known value of thermal conductivity at room temperature (2×10^3 Wm^{-1} K^{-1})
- Low thermal expansion coefficient at room temperature (0.8×10^{-6} K^{-1})
- Broad optical transparency for the deep UV to the far IR region
- Good electrical insulator (room temperature resistivity is ~ 10^{16} Ωcm)
- Diamond can be doped to change its resistivity of the range $10 - 10^6$ Ωcm, so becoming a semiconductor with a wide band gap of 5.4 eV
- Very resistant to chemical corrosion
- High radiation hardness

Source: Ref. 239.

and graphite, differ in their physical and chemical properties because of differences in the arrangement and bonding of the atoms. Diamond is denser than graphite (diamond, 3.51 g/cm^3; graphite, 2.22 g/cm^3), but graphite is more stable, by 2.9 kJ/mol at 300 K and 1 atm pressure. At high pressure, diamond is the stable form of solid carbon. This has been the basis for commercial diamond syntheses for many decades (diamond is crystallized from metal solvated carbon at $P \sim 50$ to 100 kbar and $T \sim 1800$ to 2300 K). The increased interest in diamond is due to the appearance of diamond CVD processes as economically feasible alternatives to the traditional high-pressure, high-temperature (HPHT) methods.

The scarcity and high cost of natural diamond has strongly affected its use and prompted new routes for economical synthesis of diamond in the laboratory. Recently, methods have been developed to produce polycrystalline diamond thin films by a wide variety of CVD techniques using hydrocarbon gases (typically methane) in an excess of hydrogen. These CVD techniques for depositing diamond films require methods of activating the gas-phase carbon-containing precursors. This generally involves thermal (e.g., hot filament) or plasma (dc, RF, or microwave) activation, or the use of a combustion flame (Fig. 5.20). Activation of the reactant gas mixture, either thermal or by electron bombardment, involves H-atom abstraction from the methane, resulting in formation of methyl radicals. Plasma-enhanced CVD processes involve different excitation/dissociation mechanisms; these involve a higher concentration of electronically excited species (e.g., H atoms and simple radical species such as CH, OH, and C_2). Normally, the substrate must be at high temperature (1000 to 1400 K) and the precursor gas diluted in an excess of hydrogen (typical CH_4 mixing ratio about 1 to 2%). The resulting films are polycrystalline, with a surface morphology that is sensitive to the precise CVD conditions.

One of the great challenges facing researchers in CVD diamond technology is to increase the growth rates to economically feasible rates at lower temperatures without compromising film quality. There are considerable data which suggest that methyl radicals are the dominant growth species in many variants of diamond CVD. The methyl radicals, along with other gas-phase species, are transported toward the substrate, while the relative species concentrations continue to evolve as a result of further gas-phase reactions. The fact that pure diamond films can be formed by CVD techniques is inextricably linked to the presence of hydrogen atoms. The hydrogen atoms help remove nondiamond phases from the growing surface, since the rate of dissolution (etching) of graphite and amorphous carbon is about 50 times that of diamond.[240] This is a very important characteristic since when these different forms of carbon are deposited together, graphite and amorphous carbon are removed preferentially, while diamond remains. Furthermore, the atomic hydrogen can remain adsorbed on the diamond surface, thus preventing the reconstruction of the carbon dangling bonds. This stabilization of the surface by the hydrogen atoms is also crucial in the promotion of the

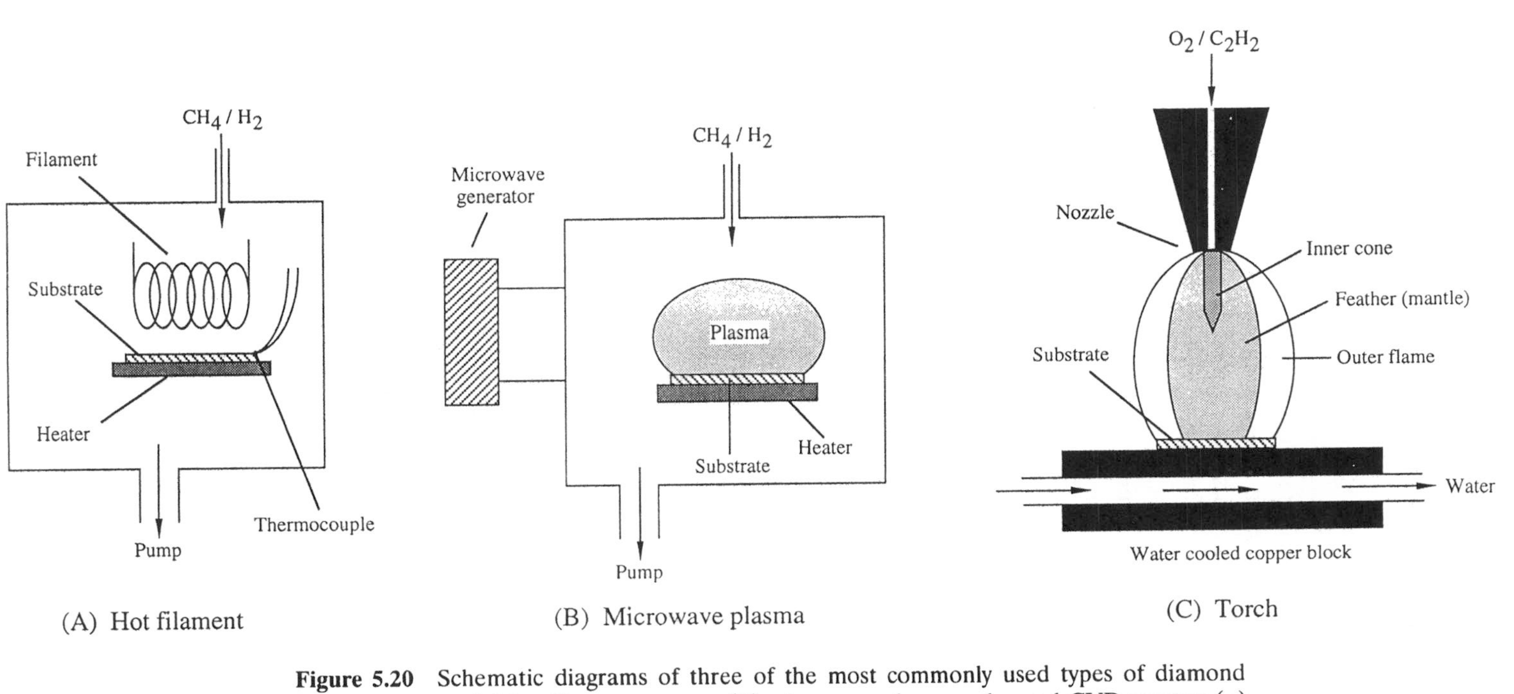

Figure 5.20 Schematic diagrams of three of the most commonly used types of diamond CVD apparatus: (*a*) hot filament reactor; (*b*) microwave-plasma-enhanced CVD reactor; (*c*) oxyacetylene torch.

diamond phase, since the occurrence of the reconstructed surface rules out the possibility of further diamond growth and favors the presence of the graphite phases.

The addition of oxygen-containing gas species to the gas mixture increases the reaction rate and improves the crystal quality of the deposits. The effects of oxygen have been attributed partially to the formation of —OH radicals, detected by optical emission spectroscopy, which are supposed to stabilize the dangling bonds at the diamond surface.

The most widely used technique, both for establishing that a film is indeed diamond and for providing some measure of the film quality, is laser Raman spectroscopy. The Raman spectra of natural diamond shows a sharp, single peak centered at a wavenumber of about 1332 cm^{-1}. As Fig. 5.21 shows, this feature also dominates the Raman spectra of high-quality polycrystalline diamond films grown by CVD methods, although the peak linewidth is usually somewhat greater than for natural diamond and the line center is often slightly shifted in wavenumber—both effects generally attributed to compressive stress in the films. By way of contrast, the Raman spectrum of highly crystalline graphite shows a somewhat broader feature centered around 1580 cm^{-1}.

Natural diamond has a thermal conductivity roughly four times that of Cu, and it is an electrical insulator; therefore, CVD diamond is marketed as a heat sink for laser diodes and for small microwave integrated circuits. CVD diamond is also finding applications as an abrasive and as cutting tool inserts. In both this and the preceding application, CVD diamond is performing a task that could have been fulfilled equally well by natural diamond if economics were not a consideration. However, there are many other applications at, or very close to, the marketplace, where CVD diamond offers new opportunities.

The possibility of doping diamond and so changing it from being an insulator into a semiconductor opens up a whole range of possible electronic applications. Among problems that have to be overcome is the fact that CVD diamond films are polycrystalline and hence contain grain boundaries, stacking faults, and other defects, which all reduce the lifetime and mobilities of carriers. One further difficulty that must be overcome if diamond devices are to be realized is the ability to pattern the diamond films into the required micron or even submicron geometries. Dry-etching processes using O_2-based plasmas can be applied, but etch rates are slow and the masking procedure complex.

Despite the difficulties mentioned above, CVD diamond-based devices are gradually beginning to appear. Another interesting new development in diamond technology is the ability to deposit CVD diamond onto the outer surfaces of metal wires or nonmetallic fibers. Such diamond-coated fibers show increased stiffness and strength over the noncoated fibers. Furthermore, etching out the metal core of the diamond-coated wire using a suitable chemical reagent yields free-standing diamond tubes or hollow diamond fibers.

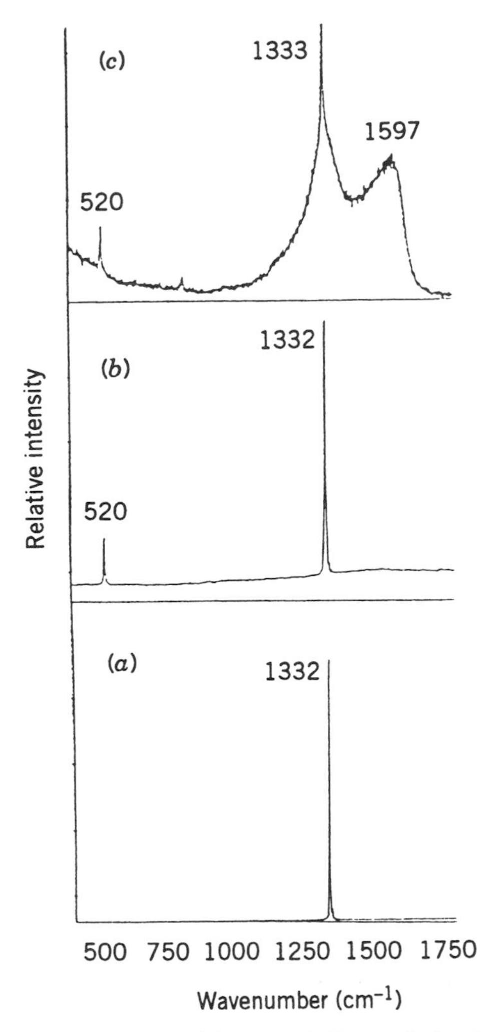

Figure 5.21 Laser Raman spectra. (*a*) Natural diamond showing the characteristic sharp peak at 1332 cm^{-1}. (*b*) Good-quality CVD diamond film grown on Si using 0.5% CH_4 in H_2 gas mixture. The peak at 520 cm^{-1} is due to Si. (*c*) Poor-quality CVD diamond film grown using 1.5% CH_4 in H_2. The broad hump at about 1597 cm^{-1} is evidence that the film contains nondiamond phases, such as graphite or amorphous carbon.

In summary, several issues need to be addressed before CVD diamond technology can make a significant impact in various applications. Growth rates need to be increased (by one or more orders of magnitude) without loss of film quality. Deposition temperatures need to be reduced by several hundred degrees, allowing low-melting-point materials to be coated and to increase the number of substrates onto which adherent diamond films can be deposited. A better understanding of the nucleation process is required, hopefully leading to an elimination of the poorly controlled preabrasion step. Deposition onto larger-area substrates is required without loss of uniformity or film quality. For electronic applications, low-temperature processing of single-crystal diamond films is desperately needed, along with reliable techniques for patterning and controlled *n*- and *p*-type doping.

5.4.4 Silicon

Silicon, a semiconductor with the diamond crystal structure, is the material of choice for fabrication of integrated circuits and is often deposited by CVD. Silicon grown by CVD can be single crystal, polycrystalline, or amorphous. Four silicon sources have been used for growing epitaxial silicon by CVD:[145] silicon tetrachloride ($SiCl_4$), dichlorosilane ($SiCl_2H$), trichlorosilane ($SiCl_3H$), and silane (SiH_4). $SiCl_4$ has been studied most extensively and has seen the widest industrial use. The overall reaction of hydrogen reduction of $SiCl_4$ is

$$SiCl_4 + 2H_2 \rightarrow Si + 4HCl \tag{29}$$

Table 5.5 shows each silicon source used along with characteristic temperature ranges and growth rates. The choice of a silicon source for film deposition is based on several considerations. SiH_4 is often chosen when a rather low deposition temperature is desired. Drawbacks are gas-phase nucleation (the formation of silicon particles in the gas phase above the substrate), which leads to poor film quality, and the high reactivity of silane, which results in rapid coating of the reactor walls, resulting in the need for frequent reactor cleaning. $SiCl_2H_2$ is a liquid at room temperature, with a high vapor pressure, and offers high growth rates at relatively low deposition temperatures and higher purity, and lower defect densities than those of $SiCl_4$. The silicon process is usually carried out at atmospheric pressure (APCVD) and allows production of high-quality films. The high-deposition-temperature restriction on the reactions above has motivated the development of new epitaxial deposition procedures that are performed at lower temperature. An important advantage of LPCVD is the reduction of the minimum temperature at which epitaxial growth is achieved. The price for the advantage of LPCVD is a substantially lower deposition rate than is accomplished with APCVD. Low-temperature deposition minimizes the ther-

TABLE 5.5

Precursor	Temperature range (°C)	Nominal growth rate ($\mu m/min$)
$SiCl_4$	1150–1250°C	0.4–1.5
$SiCl_3H$	1100–1200°C	0.4–2.0
$SiCl_2H_2$	1050–1120°C	0.2–1
SiH_4	950–1250°C	0.2–0.3

Source: Ref. 145.

mal effects on underlying device regions, and therefore LPCVD is the preferred method for deposition of polycrystalline-Si gate contacts and related polycrystalline-Si composite metallization patterns. The lower limit for the growth of polycrystalline Si by LPCVD is 525°C. Below 535°C there is a transition from microcrystalline to amorphous silicon films.

Further advances have been made toward even-lower processing temperature for epitaxial growth by the use of plasma-enhanced CVD. PECVD epitaxial growth of silicon is achieved at temperatures between 475 and 525°C. Remote plasma-enhanced CVD (RPECVD) has produced homoepitaxial growth of silicon epilayers at even lower temperatures, $150 \leq T_s \leq 350$°C.

5.5 SUMMARY, CONCLUSIONS, AND FUTURE DIRECTIONS

In this chapter we have discussed various aspects of CVD of thin films. The key features of the process of chemical vapor deposition and the chemical and physical requirements of the molecular precursors have been defined. Chemical vapor deposition has been described as a number of steps in series, any or all of which need to be controlled and preferably understood to make a significant advance in process development. A number of case studies were chosen as examples to illustrate more detailed aspects of the types of problems and issues encountered during deposition of specific materials. The case studies of diverse materials—metal (W), titanium nitride (TiN), diamond, and Si—were chosen to illustrate the wide variety of compounds that have received attention. In addition, variations on CVD were described as well as competitive physical processes such as PVD.

The unique advantages of CVD over PVD, especially the potential of accomplishing conformal coverage of thin films, ensure that CVD will find a wide range of uses in the future. It is clear that CVD is a valuable method of depositing thin films, as evidenced by its current use in microelectronics and glass coating industries, and it is likely to become important in other coating technologies, such as flat panel displays and high-density magnetic recording media, provided that developments can be made in certain key areas. It is clear that a wider variety of suitable precursors is required or the ability to use a larger variety of precursors through the development of alternative precursor delivery techniques. A better understanding leading to a better level of control over precursor reaction chemistry is required, especially in the case of deposition of complex film stoichiometries and structures. Finally, a better understanding of industrial needs is required to develop and make improvements to technologically important processes.

ACKNOWLEDGMENTS

We thank NSF, ONR, AFOSR, the University of New Mexico, and industrial sponsors for continued support of our funding in this area. M.H.S. thanks Nena Davis for technical support.

REFERENCES

1. T. T. Kodas, M. J. Hampden-Smith, *The Chemistry of Metal CVD*, VCH, Weinheim, Germany, 1994.
2. D. G. Whitten, T. Kajiyama, T. Kunitake, *Organic Thin Films: An Overview*, MRS Bulletin Materials Research Society, Vol. XX, 1995.
3. A. R. West, *Solid State Chemistry and Its Applications*, Wiley, New York, 1989.
4. W. Borland, *Electronic Materials Handbook*, Vol. 1, Packaging, ASM, Materials Park, OH, 1989.
5. C. J. Brinker, G. W. Scherer, *Sol-Gel Science: The Physics and Chemistry of Sol-Gel Processing*, Academic Press, San Diego, CA, 1990.
6. W. L. Gladfelter, *Chem. Mater.* **5**, 1372 (1993).
7. S. Wolf, R. N. Tauber, *Silicon Processing for the VLSI Era*, Lattice Press, Sunset Beach, CA, 1987.
8. M. L. Hitchman, K. F. Jensen, *Chemical Vapor Deposition: Principles and Applications*, Academic Press, San Diego, CA, 1993.
9. M. J. Hampden-Smith, T. T. Kodas, in *The Chemistry of Metal CVD*, T. T. Kodas, M. J. Hampden-Smith, Eds., VCH, Weinheim, Germany, 1994, Chapter 5.
10. A. Maverick, G. L. Griffin, in *The Chemistry of Metal CVD*, T. T. Kodas, M. J. Hampden-Smith, Eds., VCH, Weinheim, Germany, 1994, Chapter 4.
11. Simmonds, W. Gladfelter, in *The Chemistry of Metal CVD*, T. T. Kodas, M. J. Hampden-Smith, Eds., VCH, Weinheim, Germany, 1994, Chapter 2.
12. J. M. Blocher, *Chemically Vapour Deposited Coatings*, American Chemical Society, Washington, DC, 1981.
13. R. Jaraith, A. Jain, R. D. Tolles, M. J. Hampden-Smith, T. T. Kodas, in *The Chemistry of Metal CVD*, T. T. Kodas, M. J. Hampden-Smith, Eds., VCH, Weinheim, Germany, 1994, Chapter 1.
14. A. Sekiguchi, A. Kobayashi, N. Hosokawa, T. Asamaki, *Jpn. J. Appl. Phys.* **27**, 364 (1988).
15. M. E. Gross, K. P. Cheung, C. G. Fleming, J. Kovalchick, L. A. Heimbrook, *J. Vac. Sci. Technol.* **A9**, 57 (1991).
16. A. E. Kaloyeros, A. Feng, J. Garhart, K. C. Brooks, S. K. Gosh, A. N. Saxena, F. Luethrs, *J. Electronic Mater.* **19**, 271 (1990).
17. A. Jain, K.-M. Chi, M. J. Hampden-Smith, T. T. Kodas, M. F. Paffett, J. D. Farr, *J. Electrochem. Soc.* **140**, 1434 (1993).
18. T. Aoki, S. Wickramanayaka, A. M. Wrobel, Y. Nakanishi, Y. Hatanaka, *J. Electrochem. Soc.* **142**, 166 (1995).
19. R. S. Rosler, M. J. Rice, J. Mendonca, *J. Vac. Sci. Technol.* **B6**, 1721 (1988).
20. K. J. Grannen, R. P. H. Chang, *J. Mater. Res.* **9**, 2154 (1994).
21. S. V. Deshpande, E. Gulari, S. J. Harris, A. M. Weiner, *Appl. Phys. Lett.* **65**, 1757 (1994).
22. H. Kunzli, P. Gantenbein, R. Steiner, P. Oelhafen, *J. Nucl. Mater.* **196**, 622 (1992).
23. M. D. Healy, D. C. Smith, R. R. Rubiano, N. E. Elliott, R. W. Springer, *Chem. Mater.* **6**, 448 (1994).

24. F. Maury, F. Ossola, *Thin Solid Films* **219**, 24 (1992).
25. H. F. Ache, J. Goschnick, M. Sommer, G. Emig, G. Schoch, O. Wormer, *Thin Solid Films* **241**, 356 (1994).
26. A. K. Chaddha, J. D. Parsons, J. Wu, H. S. Chen, D. A. Roberts, H. Hockenhull, *Appl. Phys. Lett.* **62**, 3097 (1993).
27. L. X. Cao, Z. C. Feng, Y. Liang, W. L. Hou, B. C. Zhang, Y. Q. Wang, L. Li, *Thin Solid Films* **257**, 7 (1995).
28. C. C. Chiu, S. B. Desu, C. Y. Tsai, *J. Mater. Res.* **8**, 2617 (1993).
29. H. T. Chiu, J. S. Hsu, *Thin Solid Films* **252**, 13, (1994).
30. J. M. Grow, R. A. Levy, Y. T. Shi, R. L. Pfeffer, *J. Electrochem. Soc.* **140**, 851 (1993).
31. E. A. Haupfear, L. D. Schmidt, *J. Electrochem. Soc.* **140**, 1793 (1993).
32. L. Deutschmann, J. Messelhauser, H. Suhr, W. A. Herrmann, P. Harter, *Adv. Mater.* **6**, 392 (1994).
33. L. Poirier, O. Richard, M. Ducarroir, M. Nadal, F. Teyssandier, F. Laurent, O. Cyrathis, R. Choukroun, L. Valade, P. Cassoux, *Thin Solid Films* **249**, 62 (1994).
34. C. M. Kelly, D. Garg, P. N. Dyer, *Thin Solid Films* **219**, 103 (1992).
35. B. Aspar, R. Rodriguezclemente, A. Figueras, B. Armas, C. Combescure, *J. Cryst. Growth* **129**, 56 (1993).
36. B. Aspar, B. Armas, C. Combescure, A. Figueras, R. Rodriguezclemente, A. Mazel, Y. Kihn, J. Sevely, *Mater. Res. Bull.* **28**, 531 (1993).
37. J. Auld, D. J. Houlton, A. C. Jones, S. A. Rushworth, G. W. Critchlow, *J. Mater. Chem.* **4**, 1245 (1994).
38. A. C. Jones, J. Auld, S. A. Rushworth, D. J. Houlton, G. W. Critchlow, *J. Mater. Chem.* **4**, 1591 (1994).
39. A. R. Phani, G. S. Devi, S. Roy, V. J. Rao, *J. Chem. Soc. Chem. Commun.*, 684 (1993).
40. W. Cheng, R. F. Davis, *Appl. Phys. Lett.* **63**, 990 (1993).
41. K. G. Fertitta, A. L. Holmes, J. G. Neff, F. J. Ciuba, R. D. Dupuis, *Appl. Phys. Lett.* **65**, 1823 (1994).
42. S. S. Lee, S. M. Park, P. J. Chong, *J. Mater. Chem.* **3**, 347 (1993).
43. Y. Bu, L. Ma, M. C. Lin, *J. Vac. Sci. Technol. A* **11**, 2931 (1993).
44. H. T. Chiu, G. B. Chang, W. Y. Ho, S. H. Chuang, G. H. Lee, S. M. Peng, *J. Chin. Chem. Soc.* **41**, 755 (1994).
45. S. E. Alexandrov, M. L. Hitchman, S. Shamlian, *J. Phys. IV* **3**, 233 (1993).
46. S. E. Alexandrov, M. L. Hitchman, S. H. Shamlian, *J. Mater. Chem.* **4**, 1843 (1994).
47. R. M. Charatan, M. E. Gross, D. J. Eaglesham, *J. Appl. Phys.* **76**, 4377 (1994).
48. J. P. Dekker, P. J. Vanderput, H. J. Veringa, J. Schoonman, *J. Electrochem. Soc.* **141**, 787 (1994).
49. R. I. Hegde, R. W. Fiordalice, E. O. Travis, P. J. Tobin, *J. Vac. Sci. Technol. B* **11**, 1287 (1993).
50. K. Ikeda, M. Maeda, Y. Arita, *Jpn. J. Appl. Phys. Pt. 1* **32**, 3085 (1993).
51. A. Intemann, H. Koerner, F. Koch, *J. Electrochem. Soc.* **140**, 3215 (1993).
52. S. H. Lee, H. J. Ryoo, J. J. Lee, *J. Vac. Sci. Technol. A* **12**, 1602 (1994).

53. R. Fix, R. G. Gordon, D. M. Hoffman, *Chem. Mater.* **5**, 614 (1993).
54. H. T. Chiu, S. H. Chuang, *J. Mater. Res* **8**, 1353 (1993).
55. A. Ortiz, J. C. Alonso, C. Falcony, M. H. Farias, L. Cotaaraiza, G. Soto, *J. Electrochem. Soc.* **140**, 3014 (1993).
56. S. Manorama, G. S. Devi, V. J. Rao, *J. Appl. Phys. Lett.* **64**, 3163 (1994).
57. R. Alexandrescu, R. Cireasa, B. Dragnea, I. Morjan, I. Voicu, A. Andrei, *J. Phys. IV* **3**, 265 (1993).
58. O. B. Ajayi, O. K. Osuntola, I. A. Ojo, C. Jeynes, *Thin Solid Films* **248**, 57 (1994).
59. R. Nomura, Y. Seki, K. Konishi, H. Matsuda, *Appl. Organometal. Chem.* **6**, 685 (1992).
60. A. N. Macinnes, M. B. Power, A. R. Barron, *Chem. Mater.* **5**, 1344 (1993).
61. A. N. Macinnes, M. B. Power, A. F. Hepp, A. R. Barron, *J. Organometal. Chem.* **449**, 95 (1993).
62. W. Y. Lee, T. M. Besmann, M. W. Stott, *J. Mater. Res.* **9**, 1474 (1994).
63. J. W. Li, J. D. Chiang, Y. K. Su, M. Yokoyama, *J. Cryst. Growth* **137**, 421 (1994).
64. M. Nyman, M. J. Hampden-Smith, E. N. Duesler, *Adv. Mater.* Vol. 2, p. 171, 1995.
65. K. Kunze, L. Bihry, P. Atanasova, M. J. Hampden-Smith, E. N. Duesler, *Adv. Mater.* Vol. 2, p. 105, 1995.
66. M. A. Malik, M. Motevalli, T. Saeed, P. Obrien, *Adv. Mater.* **5**, 653 (1993).
67. Y. Senzaki, W. L. Gladfelter, *Polyhedron* **13**, 1159 (1994).
68. D. G. Hendershot, D. K. Gaskill, B. L. Justus, M. Fatemi, A. D. Berry, *Appl. Phys. Lett.* **63**, 3324 (1993).
69. J. D. Parsons, K. Oatis, J. Wu, A. K. Chaddha, S. R. Hahn, H. S. Chen, S. Wild, C. Deng, T. Plant, J. Marlia, *J. Electrochem. Soc.* **140**, 3280 (1993).
70. K. Kakishita, K. Aihara, T. Suda, *Appl. Surf. Sci.* **80**, 281 (1994).
71. K. Kakishita, K. Aihara, T. Suda, *Solar Energy Mater. Solar Cells* **35**, 333 (1994).
72. K. Saito, Y. Higashi, T. Amazawa, Y. Arita, *J. Electrochem. Soc.* **141**, 1879 (1994).
73. N. Thomas, P. Suryanarayana, E. Blanquet, C. Vahlas, R. Madar, C. Bernard, *J. Electrochem. Soc.* **140**, 475 (1993).
74. T. J. Groshens, R. W. Gedridge, C. K. Lowema, *Chem. Mater.* **6**, 727 (1994).
75. J. C. Kenvin, M. G. White, M. B. Mitchel, *Langmuir* **7**, 1198 (1991).
76. A. Gleizes, S. Sans-Lenain, D. Medus, *C.R. Acad. Sci. Paris* **313**, 761 (1991).
77. P. Van der Sluis, A. L. Spek, K. Timmer, H. A. Meinema, *Acta Crystallogr. C* **46**, 1741 (1990).
78. K. Timmer, C. I. M. A. Spee, A. Mackor, H. A. Meinema, European patent 0,405,634, A2 (1991).
79. D. L. Schulz, B. J. Hinds, D. A. Neumayer, C. L. Stern, T. J. Marks, *Chem. Mater.* **5**, 1605 (1993).
80. D. L. Schulz, B. J. Hinds, C. L. Stern, T. J. Marks, *Inorg. Chem.* **32**, 249 (1993).
81. G. Malandrino, I. L. Fragala, D. A. Neumayer, C. L. Stern, B. J. Hinds, T. J. Marks, *J. Mater. Chem.* **4**, 1061 (1994).

82. J. A. T. Norman, G. P. Pez, *J. Chem. Soc. Chem. Commun.*, 971 (1991).
83. G. Malandrino, R. Licata, F. Castelli, I. L. Fragala, C. Benelli, *Inorg. Chem.* **34**, 6234 (1995).
84. T. T. Kodas, M. J. Hampden-Smith, in *The Chemistry of Metal CVD*, T. T. Kodas and M. J. Hampden-Smith, Eds., VCH, Weinheim, Germany, 1994, Chapter 9.
85. A. K. Cheetham, P. Day, Eds., *Solid State Chemistry: Techniques*, Clarendon Press, Oxford, 1990.
86. M. S. Htoo, Ed., *Microelectronic Polymers*, Marcel Dekker, New York, 1989.
87. G. J. Ashwell, Ed., *Molecular Electronics*, Research Studies Press, Taunton, Somerset, England, 1992.
88. D. Zeng, M. J. Hampden-Smith, T. M. Alam, *Polyhedron* **13**, 2715 (1994).
89. A. N. MacInnes, M. B. Power, A. R. Barron, *Chem. Mater.* **5**, 1344 (1993).
90. D. Zeng, M. J. Hampden-Smith, E. N. Duesler, *Inorg. Chem.* **33**, 5376 (1994).
91. M. Nyman, S. Pennino, M. J. Hampden-Smith, D. Zeng, T. T. Kodas, in *MRS Fall Meeting Abstract Book*, Boston, 1994, p. 179.
92. H. O. Pierson, *Handbook of Chemical Vapor Deposition*, Noyes Publications, Park Ridge, NJ, 1992.
93. A. Sherman, *Chemical Vapor Deposition for Microelectronics*, Noyes Publications, Park Ridge, NJ, 1987.
94. H. K. Shin, K.-M. Chi, M. J. Hampden-Smith, T. T. Kodas, M. F. Paffett, J. D. Farr, *Chem. Mater.* **4**, 788 (1992).
95. H. K. Shin, K.-M. Chi, M. J. Hampden-Smith, T. T. Kodas, M. F. Paffett, J. D. Farr, *Angew, Chem. Adv. Mater.* **3**, 246 (1991).
96. A. Jain, K.-M. Chi, M. J. Hampden-Smith, T. T. Kodas, M. F. Paffett, J. D. Farr, *J. Mater. Res.* **7**, 261 (1992).
97. A. Jain, K.-M. Chi, M. J. Hampden-Smith, T. T. Kodas, M. F. Paffett, J. D. Farr, *Chem. Mater.* **3**, 995 (1991).
98. G. B. Stringfellow, *Organometallic Vapor-Phase Epitaxy: Theory and Practice*, Academic Press, San Diego, CA, 1989.
99. H. H. Lee, *Fundamentals of Microelectronics Processing*, McGraw-Hill, New York, 1990.
100. H.-Y. Lee, H.-G. Kim, *Thin Solid Films* **229**, 187 (1993).
101. J. L. Vossen, W. Kern, Eds., *Thin Film Processes*, Academic Press, San Diego, CA, 1978.
102. J.-O. Carlsson, *Crit. Rev. Solid. State Mater. Sci.* **16**, 161 (1990).
103. B. Lecohier, B. Calpini, J. M. Philippoz, H. van den Bergh, *J. Appl. Phys.* **72**, 2022 (1992).
104. S. J. Fine, P. N. Dyer, J. A. T. Norman, B. A. Muratore, R. L. Iampietro, in *MRS Symposium Proceedings*, Vol. 204, Materials Research Society, Pittsburgh, PA, 1990, p. 415.
105. G. J. M. Dormans, G. J. B. M. Meekes, E. G. J. Staring, *J. Cryst. Growth* **114**, 364 (1991).
106. G. J. M. Dormans, *J. Cryst. Growth* **108**, 806 (1991).
107. C. M. McConica, K. Krishnamani, *J. Electrochem. Soc.* **133**, 2542 (1986).

108. N. Lifshitz, D. S. Williams, C. D. Capio, J. M. Brown, *J. Electrochem. Soc.* **134** 2061 (1987).
109. W. Lin, T. H. Warren, R. G. Nuzzo, G. S. Girolami, *J. Am. Chem. Soc.* **115**, 11644 (1993).
110. J. R. Creighton, J. E. Parmeter, *Crit. Rev. Solid State Mater. Sci.* **18**, 175 (1993).
111. B. Lecohier, J.-M. Philippoz, H. van den Bergh, *J. Vac. Sci. Technol. B* **10**, 262 (1992).
112. B. Lecohier, J.-M. Philippoz, B. Calpini, T. Stumm, H. van den Bergh, *J. Phys.* **NC2**, 279 (1991).
113. R. A. Levy, M. L. Green, P. K. Gallagher, *J. Electrochem. Soc.* **131**, 2175 (1984).
114. M. J. Cooke, R. A. Heinecke, R. C. Stern, J. W. Maes, *Solid State Technol.* **25**, 62 (1982).
115. A. Jain, J. Farkas, M. J. Hampden-Smith, T. T. Kodas, *Appl. Phys. Lett.* **62**, 2662 (1992).
116. J. A. T. Norman, B. A. Muratore, P. N. Dyer, D. A. Roberts, A. K. Hochberg, *J. Phys. IV*, **1**, C2-271 (1992).
117. T. S. Corbitt, R. M. Crooks, C. B. Ross, M. J. Hampden-Smith, J. K. Schoer, *Angew. Chem. Adv. Mater.* **5**, 935 (1994).
118. T. S. Corbitt, R. M. Crooks, M. J. Hampden-Smith, C. B. Ross, J. K. Schoer, *Langmuir* **10**, 615 (1994).
119. D. A. Mantel, *Appl. Phys. Lett.* **53**, 1387 (1988).
120. H. D. Kaesz, unpublished results.
121. M. M. Banaszak Holl, P. F. Seidler, S. P. Kowalczyk, F. R. McFreely, *Appl. Phys. Lett.* **62**, 1475 (1993).
122. M. E. Gross, V. M. Donnelly, in *Advanced Metallization for ULSI Applications*, V. V. S. Rana, R. V. Joshi, I. Ohdomari, Eds., Materials Research Society, Pittsburgh, PA, 1992, p. 355.
123. K. V. Guinn, V. M. Donnelly, M. E. Gross, F. A. Baiocchi, I. Petrov, J. E. Greene, *Mater. Res. Soc. Symp. Proc.* **282**, 379 (1993).
124. H. Knoezinger, in *Surface Organometallic Chemistry: Molecular Approaches to Surface Catalysis*, J.-M. Basset, Ed., Kluwer, Dordrecht, The Netherlands, 1988; Vol. 231, p. 35.
125. M. L. Hair, *Infrared Spectroscopy in Surface Chemistry*, Marcel Dekker, New York, 1967.
126. B. A. Morrow, in *Spectroscopic Characterization of Heterogeneous Catalysis*, Part 57A: *Methods of Surface Analysis*, J. L. G. Fierro, Ed., Elsevier, Amsterdam, 1990, Vol. 57A.
127. R. K. Iler, *The Chemistry of Silica*, Wiley, New York, 1979.
128. J. Farkas, M. J. Hampden-Smith, T. T. Kodas, *J. Electrochem. Soc.* **141**, 3547 (1994).
129. J. Farkas, M. J. Hampden-Smith, T. T. Kodas, *J. Electrochem. Soc.* **141**, 3539 (1994).
130. F. D. Hardcastle, J. Farkas, C. H. F. Peden, T. R. Omstead, R. S. Blewer, M. J. Hampden-Smith, T. T. Kodas, in *Advanced Metallization for ULSI Applications*, V. S. Rana, R. V. Joshi, I. Ohdomari, Eds., Materials Research Society, Pittsburgh, PA, 1992, p. 413.

131. L. H. Dubois, *J. Electrochem. Soc.* **139**, 3295 (1992).
132. R. W. Cheek, J. Prasad, J. A. Kelber, R. S. Blewer, J. Fleming, R. D. Lujan, in *Advanced Metallization for ULSI Applications* V. V. S. Rana, R. V. Joshi, I. Ohdomari, Eds., Materials Research Society, Pittsburgh, PA, 1992, p. 227.
133. J. A. T. Norman, D. A. Roberts, A. K. Hochberg, *Mater. Res. Soc. Symp. Proc.* **282**, 347 (1993).
134. A. Jain, R. Jaraith, T. T. Kodas, M. J. Hampden-Smith, *J. Vac. Sci. Technol. B.* **11**, 2107 (1993).
135. K. L. Mittal, Ed., *Metallized Plastics 3: Fundamental and Applied Aspects*; Plenum Press, New York, 1992.
136. R. R. Rye, K. M. Chi, M. J. Hampden-Smith, T. T. Kodas, in *Advanced Metallization for ULSI Applications* V. V. S. Rana, V. Joshi, I. Ohdomari, Eds., Materials Research Society, Pittsburgh, PA, 1992, p. 421.
137. R. Rye, J. A. Knapp, K. M. Chi, M. J. Hampden-Smith, T. T. Kodas, *J. Appl. Phys.* **72**, 5941 (1992).
138. R. R Rye, G. W. Arnold, *Langmuir* **5**, 1331 (1989).
139. R. R. Rye, *Langmuir* **6**, 338 (1990).
140. R. R. Rye, N. D. Shinn, *Langmuir* **9**, 142 (1990).
141. R. R. Rye, *J. Polymer Sci. Polymer Phsy.*, in press.
142. G. B. Stringfellow, *Organometallic Vapor-Phase Epitaxy: Theory and Practice*, Academic Press, San Diego, CA, 1989.
143. T. F. Kuech, K. F. Jensen, in *Thin Film Processes II*, J. L. Vossen, W. Kern, Eds., Academic Press, San Diego, CA, 1991, p. 369.
144. W. T. Tsang, in *Semiconductors and Semimetals*, W. T. Tsang, Ed., Academic Press, San Diego, CA. 1985, Vol. 22, p. 96.
145. S. M. Sze, *VLSI Technology*, 2nd ed., McGraw-Hill, New York, 1988.
146. H. Luth, *Surf. Sci.* **299–300**, 867 (1994).
147. R. Reif, W. Kern, in *Thin Film Processes II*, J. L. Vossen, W. Kern, Eds., Academic Press, San Diego, CA, 1991, p. 525.
148. A. Sherman, *Chemical Vapor Deposition for Microelectronics: Principles, Technology, and Applications*, Noyes Publications, Park Ridge, NJ, 1987.
149. K. Wasa, S. Hayakawa, *Handbook of Sputter Deposition Technology: Principles, Technology and Applications*, Noyes Publications, Park Ridge, NJ, 1991.
150. G. Lucovsky, G. W. Tsu, R. A. Rudder, R. J. Markunas, in *Thin Film Processes II*,
J. L. Vossen, W. Kern, Eds., Academic Press, San Diego, 1991, p. 565.
151. I. P. Herman, *Chem. Rev.* **89**, 1323 (1989).
152. T. M. Mayer, S. D. Allen, in *Thin Film Processes II*, J. L. Vossen, W. Kern, Eds., San Diego, CA, 1991, p. 621.
153. J. G. Eden, in *Thin Film Processes II*, J. L. Vossen, W. Kern, Eds., Academic Press, San Diego, CA, 1991, p. 443.
154. M. Meunier, P. Desjardins, M. Tabbal, N. Elyaagoubi, R. Izquierdo, A. Yelon, *Appl. Surf. Sci.* **86**, 475 (1995).
155. G. Auvert, *Appl. Surf. Sci.* **86**, 466 (1995).
156. D. Bauerle, *Chemical Processing with Lasers*, Springer-Verlag, New York, 1986.

157. J. Mazumder, A. Kar, *Theory and Application of Laser Chemical Vapor Deposition*, Plenum Press, New York, 1995.

158. C. V. Deshpandey, R. F. Bunshah, in *Thin Film Processes II*, J. L. Vossen, W. Kern, Eds., Academic Press, San Diego, CA, 1991, p. 79.

159. R. Parsons, in *Thin Film Processes II*, J. L. Vossen, W. Kern, Eds., Academic Press, San Diego, CA, 1991, p. 177.

160. S. M. Rossnagel, in *Thin Film Processes II*, J. L. Vossen, W. Kern, Eds., Academic Press, San Diego, CA, 1991, p. 11.

161. D. S. Taylor, M. K. Jain, T. S. Cale, M. G. Fissel, I. J. Raaijmakers, in *Proceedings of the 9th International VLSI Multilevel Interconnection Conference*, (1993).

162. P. P. Chow, in *Thin Film Processes II*, J. L. Vossen, W. Kern, Eds., Academic Press, San Diego, CA, 1991, p. 133.

163. K. J. Bachman, *The Materials of Microelectronics*, VCH, New York, 1995.

164. J. R. Creighton, J. E. Parmeter, *Crit. Rev. Solid State Mater. Sci.* **18**, 175 (1993).

165. N. Desatnik, B. E. Thompson, *Electrochem. Soc.* **141**, 3532 (1994).

166. Y. W. Park, C. O. Park, J. S. Chun, *Thin Solid Films* **201**, 167 (1991).

167. Y. W. Park, I. L. Kim, C. O. Park, J. S. Chun, *J. Mater. Sci.* **26**, 5318 (1991).

168. A. S. Turtsevich, V. Y. Krasnitsky, V. I. Granko, V. P. Lesnikova, I. V. Smal, O. E. Sarychev, O. Y. Nalivaiko, S. V. Kravtsov, *Thin Solid Films* **221**, 191 (1992).

169. A. Katz, A. Feingold, A. Elroy, S. J. Pearton, E. Lane, S. Nakahara, M. Geva, *Appl. Phys. Lett.* **61**, 1522 (1992).

170. J. T. Wang, C. B. Cao, H. Wang, S. L. Zhang, *J. Electrochem. Soc.* **141**, 2192 (1994).

171. S. Mcclatchie, H. Thomas, D. V. Morgan, *Appl. Surf. Sci.* **73**, 58 (1993).

172. Y. Saito, T. Takagi, *Jpn. J. Appl. Phys. Pt. 1* **33**, 4413 (1994).

173. W. W. Lee, R. R. Reeves, J. Halstead, *J. Vac. Sci. Technol. A Vac. Surf. Films* **9**, 653 (1991).

174. A. Balkacem, Y. Arnal, J. Pelletier, E. Andre, J. C. Oberlin, *Thin Solid Films* **241**, 301 (1994).

175. P. Mogyorosi, J. O. Carlsson, *J. Vac. Sci. Technol. A* **10**, 3131 (1992).

176. M. Meunier, R. Izquierdo, P. Desjardins, M. Tabbal, A. Lecours, A. Yelon, *Thin Solid Films* **218**, 137 (1992).

177. A. J. P. Vanmaaren, W. C. Sinke, *J. Appl. Phys.* **73**, 1989 (1993).

178. P. Heszler, J. O. Carlsson, P. Mogyorosi, *J. Vac. Sci. Technol. A* **11**, 2924 (1993).

179. R. L. Krans, C. Brands, W. C. Sinke, *Appl. Surf. Sci.* **54**, 117 (1992).

180. G. Auvert, Y. Pauleau, D. Tonneau, *Jpn. J. Appl. Phys. Pt. 1 Reg. Pap. Short Notes* **31**, 100 (1992).

181. G. Auvert, Y. Pauleau, D. Tonneau, *Appl. Phys.* **71**, 4533 (1992).

182. K. Piglmayer, Z. Toth, Z. Kantor, *Appl. Surf. Sci.* **86**, 484 (1995).

183. J. C. Dupuy, A. Essaadani, A. Sibai, C. Dubois, F. C. Dassapa, Y. Pauleau, *Thin Solid Films* **227**, 167 (1993).

184. P. A. C. Groenen, J. G. A. Holscher, H. H. Brongersma, *Appl. Surf. Sci.* **78**, 123 (1994).

185. U. Jansson, J. O. Carlsson, *Mater. Sci. Eng. B Solid State M* **17**, 131 (1993).
186. R. V. Joshi, V. Prasad, M. L. Yu, G. Scilla, *J. Appl. Phys.* **71**, 1428 (1992).
187. U. Jansson, *Appl. Surf. Sci.* **73**, 51 (1993).
188. S. L. Lantz, W. K. Ford, A. E. Bell, D. Danielson, *J. Vac. Sci. Technol. A* **11**, 911 (1993).
189. J. R. Creighton, *Thin Solid Films* **241**, 310 (1994).
190. P. Gouypailler, P. Lami, R. Morales, *Thin Solid Films* **241**, 374 (1994).
191. S. C. Chen, A. Sakamoto, H. Tamura, M. Yoshimaru, M. Ino, *Jpn. J. Appl. Phys. Pt. 1 Reg. Pap. Short Notes Rev. Pap.* **32**, 1929 (1993).
192. C. Lee, Y. J. Im, J. G. Lee, *J. Phys. IV* **3**, 433 (1993).
193. J. J. Hsieh, *J. Vac. Sci. Technol. A* **11**, 3040 (1993).
194. H. S. Choi, S. W. Rhee, *J. Electrochem. Soc.* **141**, 475 (1994).
195. E. G. Colgan, J. D. Chapplesokol, *J. Vac. Sci. Technol. B Microelectron. Process. Phenomena* **10**, 1156 (1992).
196. W. Grunewald, S. E. Schulz, *Fresenius J. Anal. Chem.* **349**, 239 (1994).
197. N. Kobayashi, H. Goto, M. Suzuki, *J. Appl. Phys.* **69**, 1013 (1991).
198. S. L. Lantz, A. E. Bell, W. K. Ford, D. Danielson, *J. Vac. Sci. Technol. A* **12**, 1032 (1994).
199. Y. Maeda, H. Suzuki, T. Sakoh, K. Morita, M. Morita, T. Ohmi, *J. Electrochem. Soc.* **141**, 566 (1994).
200. E. J. Mcinerney, T. W. Mountsier, B. L. Chin, E. K. Broadbent, *J. Vac. Sci. Technol. B* **11**, 734 (1993).
201. E. J. Mcinerney, T. W. Mountsier, B. L. Chin, E. K. Broadbent, in *Advanced Metallization for ULSI Applications*, Materials Research Society, Pittsburgh, PA, 1992, pp. 69–74.
202. E. J. Mcinerney, T. W. Mountsier, B. L. Chin, E. K. Broadbent, in *Advanced Metallization for ULSI Applications*, Materials Research Society, Pittsburgh, PA, 1992, pp. 61–68.
203. E. G. Colgan, *J. Electrochem. Soc.* **140**, 485 (1993).
204. P. Desjardins, R. Izquierdo, M. Meunier, *Can. J. Phys.* **70**, 898 (1992).
205. P. Desjardins, R. Izquierdo, M. Meunier, *J. Appl. Phys.* **73**, 5216 (1993).
206. H. Suzuki, Y. Maeda, K. Morita, M. Morita, T. Ohmi, *Jpn. J. Appl. Phys. Pt. 1* **33**, 451 (1994).
207. K. A. Gesheva, V. Abrosimova, G. D. Beshkov, *J. Phys. II* **1**, 865 (1991).
208. T. Gessner, B. Hintze, T. Raschke, S. E. Schulz, *Appl. Surf. Sci.* **53**, 41 (1991).
209. F. A. Houle, K. A. Singmaster, *J. Phys. Chem.* **96**, 10425 (1992).
210. A. M. Dhote, S. B. Ogale, *Appl. Phys. Lett.* **64**, 2809 (1994).
211. S. Chatterjee, S. Chandrashekhar, T. S. Sudarshan, *J. Mater. Sci.* **27**, 3409 (1992).
212. H. E. Rebenne, D. G. Bhat, *Surf. Coat. Technol.* **63**, 1 (1994).
213. S. Y. Zhang, W. G. Zhu, *J. Mater. Process. Technol.* **39**, 165 (1993).
214. N. Takeyasu, Y. Kawano, E. Kondoh, T. Katagiri, H. Yamamoto, H. Shinriki, T. Ohta, *Jpn. J. Appl. Phys. Pt. 1* **33**, 424 (1994).

215. J. B. Price, J. O. Borland, S. Selbrede, in *Thin Solid Films*, Elsevier, Lausanne, Switzerland, 1993, pp. 311–318.
216. A. Valentini, F. Quaranta, M. Peza, L. Vasenelly, G. Bhattablin, *J. Appl. Phys.* **69**, 7360 (1991).
217. J. P. Dekker, P. J. Vanderput, H. J. Veringa, J. Schoonman, *J. Electrochem. Soc.* **141**, 787 (1994).
218. K. Glejbol, N. H. Pryds, A. R. Tholen, *J. Mater. Res.* **8**, 2239 (1993).
219. N. Yoshikawa, K. Higashino, A. Kikuchi, *J. Jpn. Inst. Met.* **58**, 442 (1994).
220. F. Pintchovski, T. White, E. Travis, P. J. Tobin, J. B. Price, in *Tungsten and Other Refractory Metals for VLSI Applications*, Materials Research Society, Pittsburgh, PA, 1989.
221. A. Sherman, *J. Electrochem. Soc.* **137**, 1892 (1990).
222. J. T. Hillman, M. J. Rice, Jr., D. W. Studiner, R. F. Foster, R. W. Fiordalice, in *Proc. 9th International VLSI Multilevel Interconnection Conference*, 1992, p. 246.
223. R. I. Hegde, R. W. Fiordalice, P. J. Tobin, *Appl. Phys. Lett.* **62**, 2326 (1993).
224. R. I. Hegde, P. J. Tobin, R. W. Fiordalice, E. O. Travis, *J. Vac. Sci. Technol. A* **11**, 1692 (1993).
225. C. C. Jiang, T. Goto, T. Hirai, *J. Alloys Compounds* **190**, 197 (1993).
226. C. C. Jiang, T. Goto, T. Hirai, *J. Mater. Sci.*, **28**, 6446 (1993).
227. C. Jiang, T. Goto, T. Hirai, *J. Mater. Sci.* **29**, 669 (1994).
228. T. Kaizuka, H. Shinriki, N. Takeyasu, T. Ohta, in *Jpn. J. Appl. Phys. Pt. 1 Reg. Pap. Short Notes Rev. Pap. Jpn. J. Appl. Phys.*, 470 (1994).
229. X. L. Chen, J. Mazumder, *J. Appl. Phys.* **76**, 3914 (1994).
230. O. Conde, M. L. G. Ferreira, P. Hochholdinger, A. J. Silvestre, R. Vilar, *Appl. Surf. Sci.* **54**, 130 (1992).
231. A. J. Silvestre, O. Conde, R. Vilar, M. Jeandin, *J. Mater. Sci.* **29**, 404, (1994).
232. A. J. Silvestri, M. L. G. F. Parames, O. Conde, in *Thin Solid Films*, Elsevier, Lausanne, Switzerland, 1994, pp. 57–60.
233. U. Illmann, R. Ebert, G. Reisse, H. Freller, P. Lorenz, in *Thin Solid Films*, Elsevier, Lausanne, Switzerland, 1994, pp. 71–75.
234. A. Kaloyeros, C. Faltermeier, C. Goldberg, M. Jones, G. Petersen, B. Arkles, in *Advanced Metallization for ULSI Applications*, 1995.
235. S. C. Sun, M. H. Tsai, *Thin Solid Films* **253**, 440–444 (1994).
236. T. S. Cale, M. B. Chaara, G. B. Raupp, I. J. Raaijmakers, in *Thin Solid Films*, Elsevier, Lausanne, Switzerland, 1993, pp. 294–300.
237. B. H. Weiller, B. V. Partido, *Chemistry of Materials* **6**, 260–261 (1994).
238. Hillman, in *Semiconductor International*, 1995, p. 147.
239. M. N. R. Ashfold, P. W. May, C. A. Rego, N. M. Everitt, *Chemical Society Reviews* **23**, 21–30 (1994).
240. C. Gomez-Aleixandre, O. Sanchez, J. M. Albella, J. Santiso, A. Figueras, *Adv. Mater.* **7**, 111 (1995).

CHAPTER 6

Introduction to the Nonlinear Optical Properties of Organic Materials*

FABIENNE MEYERS, SETH R. MARDER, and JOSEPH W. PERRY

Molecular Materials Resource Center, Beckman Institute, California Institute of Technology, Pasadena, California, and the Jet Propulsion Laboratory, California Institute of Technology, Pasadena, California

6.1 INTRODUCTION

Since the advent of lasers, the use of light to carry, store, and process information has increased tremendously. The current push to more fully exploit photonic technologies has created a demand for high-performance nonlinear optical (NLO) materials whose optical properties are very sensitive to an applied electric field or to the electric field of the laser light itself. Nonlinear optical properties enable modification of the amplitude, phase, or frequency of an optical signal: as, for example, in optical second harmonic generation, where laser light is converted to light at twice the frequency of the input beam. Over the past decade there has been growing interest in organic materials for nonlinear optics. Conjugated organic molecules and molecular materials have been shown to exhibit large optical nonlinearities and are amenable to optimization of their properties by rational modification of their structures at the molecular or supramolecular levels.[1–10]

*The work in this paper was performed in part at the Center for Space Microelectronics Technology, Jet Propulsion Laboratory (JPL), California Institute of Technology, under contract with the National Aeronautics and Space Administration (NASA). The work was sponsored by the Ballistic Missile Defense Organization, Innovative Science and Technology Office (BMDO). In addition, the work was supported by the US Office of Naval Research (ONR) through the MURI Center for Advanced Multifunctional Nonlinear Optical Polymers and Molecular Assemblies (CAMP), the US National Science Foundation (NSF), and by the US Air Force Office of Scientific Research (AFOSR).

Chemistry of Advanced Materials: An Overview, Edited by Leonard V. Interrante and Mark J. Hampden-Smith.
ISBN 0-471-18590-6 © 1998 Wiley-VCH, Inc.

The purpose of this chapter is to provide the reader with an introduction to the essential basic concepts of nonlinear optical phenomena and materials to facilitate an understanding of the current literature on nonlinear optics of organic materials. Nonlinear optical effects are a direct manifestation of the nonlinear polarization of materials. Accordingly, our starting point in this chapter is a brief review of linear and nonlinear polarization concepts. In the second section we briefly present quantum-chemical methods that are often used to investigate the electronic structure and NLO properties of molecules. In the third section we describe, at the microscopic level, the structural requirements needed for potent NLO chromophores. In the fourth section we consider the requirements that nonlinear optical materials have to satisfy at the macroscopic level, (e.g., optical absorption, scattering loss, processibility, and mechanical, thermal, and environmental stability) in order to be useful in device applications.

6.2 BASIC CONCEPTS

6.2.1 Microscopic Description of the Polarization[11]

Why does a molecule (or a material) polarize in the presence of an electric field such as the oscillating electric field of a light wave? To understand this process, we need to consider what is happening, at the molecular level, when the electric field, **E**, interacts with the charge distribution of the system. This interaction produces a force ($\mathbf{F} = q\mathbf{E}$, where q is the charge), that causes displacement of the electron density. This displacement of the center of electron density away from the nuclear framework results in a separation of positive and negative charge and consequently, in an induced polarization (Fig. 6.1*b*), or in other words, an *induced dipole* $\boldsymbol{\mu}^{\text{ind}}$. For small fields, the displacement of charge from the equilibrium position (and consequently, the induced dipole) can be treated as being linearly proportional to the strength of the applied field (Fig. 6.1*c*). Since $\boldsymbol{\mu}^{\text{ind}}$ and **E** are vector quantities with both direction and magnitude, $\boldsymbol{\mu}^{\text{ind}}$ is given by

$$\boldsymbol{\mu}_i^{\text{ind}}(\omega) = \alpha_{ij}(\omega)\mathbf{E}_j(\omega) \tag{1}$$

where $\alpha_{ij}(\omega)$ is the *linear polarizability tensor* of the molecule or atom, which defines the linear variation of the induced dipole moment with the electric field. For frequencies far from a resonance of the molecule, the frequency dependence is essentially instantaneous; thus the induced polarization will have the same frequency and phase as the applied field. For fields oscillating with optical frequencies (about 10^{15} s^{-1}), the dominant contribution is due essentially to electronic polarization (i.e., associated with only the displacement of electron density). Otherwise, at lower frequencies, vibrational and orientational mechanisms also contribute to the induced polarization.

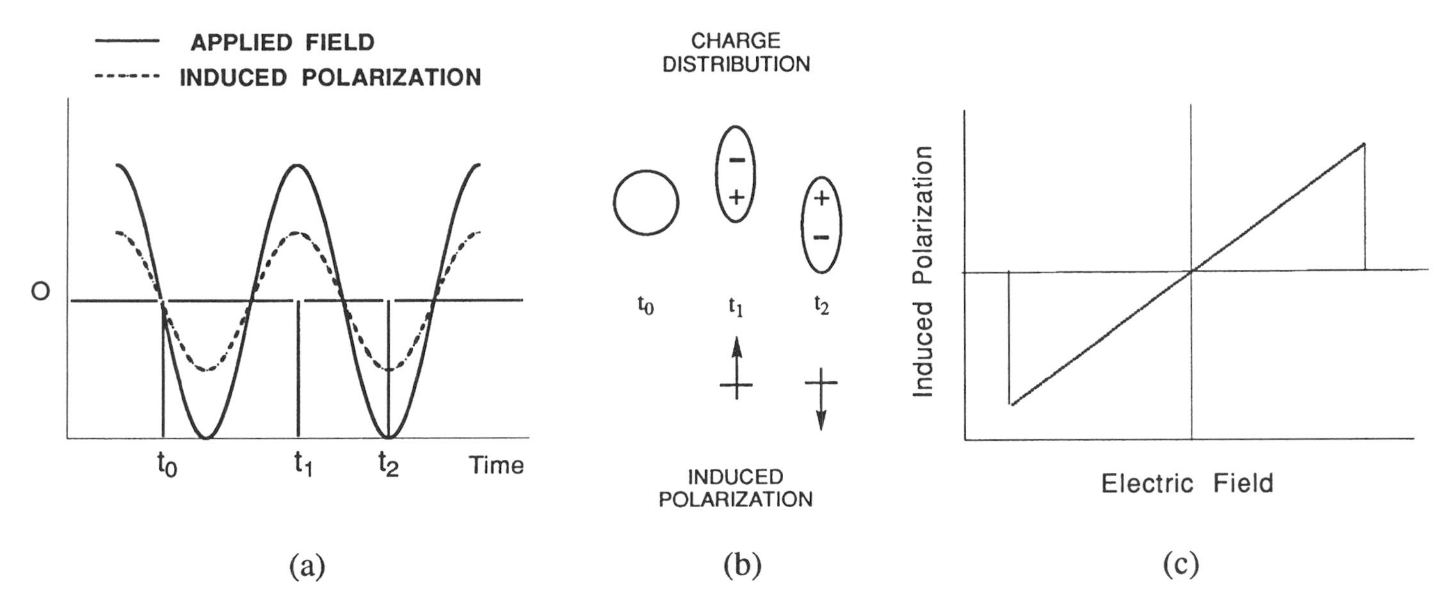

Figure 6.1 (*a*) Plots of the electric field of an applied light wave and the induced polarization wave as a function of time, for a linear material; (*b*) diagram depicting the charge distribution and the polarization in the material as a function of time; (*c*) plot of the induced polarization as a function of the applied field. (From Ref. 11.)

Until now we have assumed that the polarization is linearly proportional to the applied electric field. However, when a molecule or a material is subjected to very high intensity electric fields, it can become sufficiently polarized that its polarizability can change. Thus the induced polarization becomes a nonlinear function of the field strength. For a second-order nonlinear optical molecule (possessing a dipole, as depicted in the Fig. 6.2*b*), application of a sinusoidally oscillating field will lead to an asymmetric oscillating induced polarization (Fig. 6.2*a*) which is stronger in one direction (as in Fig. 6.2*a* at time t_1) and weaker in the opposite direction (at time t_2). This asymmetric induced polarization can be decomposed into a direct-current (dc) component and components at the fundamental and second harmonic frequencies, as discussed below and illustrated in Fig. 6.3.

The common approximation to formulate the nonlinear polarization is to expand the total dipole as a Taylor series:[12]

$$\mu_i = \mu_i^0 + \left(\frac{\partial \mu_i}{\partial \mathbf{E}_j}\right)_0 \mathbf{E}_j + \frac{1}{2}\left(\frac{\partial^2 \mu_i}{\partial \mathbf{E}_j\,\partial \mathbf{E}_k}\right)_0 \mathbf{E}_j\mathbf{E}_k + \frac{1}{3!}\left(\frac{\partial^3 \mu_i}{\partial \mathbf{E}_j\,\partial \mathbf{E}_k\,\partial \mathbf{E}_1}\right)_0 \mathbf{E}_j\mathbf{E}_k\mathbf{E}_1 + \cdots \tag{2}$$

$$\mu_i = \mu_i^0 + \alpha_{ij}\mathbf{E}_j + \tfrac{1}{2}\beta_{ijk}\mathbf{E}_j\mathbf{E}_k + \tfrac{1}{6}\gamma_{ijkl}\mathbf{E}_j\mathbf{E}_k\mathbf{E}_l + \cdots \tag{3}$$

where μ_i^0 is the permanent dipole, α_{ij} the linear (or first-order) polarizability, β_{ijk} the second-order polarizability (also called the first hyperpolarizability), and γ_{ijkl} the third-order polarizability (or second hyperpolarizability). To be exact, this equation should also include the frequency dependence, where, for example, the term $\frac{1}{2}\beta_{ijk}\mathbf{E}_j\mathbf{E}_k$ should be written as $\frac{1}{2}\beta_{ijk}(-\omega_3;\,\omega_1,\,\omega_2)\mathbf{E}_j(\omega_1)\mathbf{E}_k(\omega_2)$ (see below).

The terms beyond $\alpha\mathbf{E}$ are not linear in $\mathbf{E}$ and are referred to as the *nonlinear polarization*; they give rise to *nonlinear optical effects*. Note that nonlinear polarization becomes more important with increasing field strength, since it scales with higher powers of the field. Under normal conditions, $\alpha E > \beta E^2 > \gamma E^3$, which explains why there were few observations of NLO effects prior to the invention of the laser.

6.2.2 Macroscopic Description of the Polarization

At the macroscopic level, the description of the bulk polarization is analogous to the previous description of the microscopic polarizabilities. In bulk materials, the polarization is given by

$$\mathbf{P} = \mathbf{P}^0 + \chi^{(1)}\mathbf{E} + \tfrac{1}{2}\chi^{(2)}\mathbf{EE} + \tfrac{1}{6}\chi^{(3)}\mathbf{EEE} + \cdots \tag{4}$$

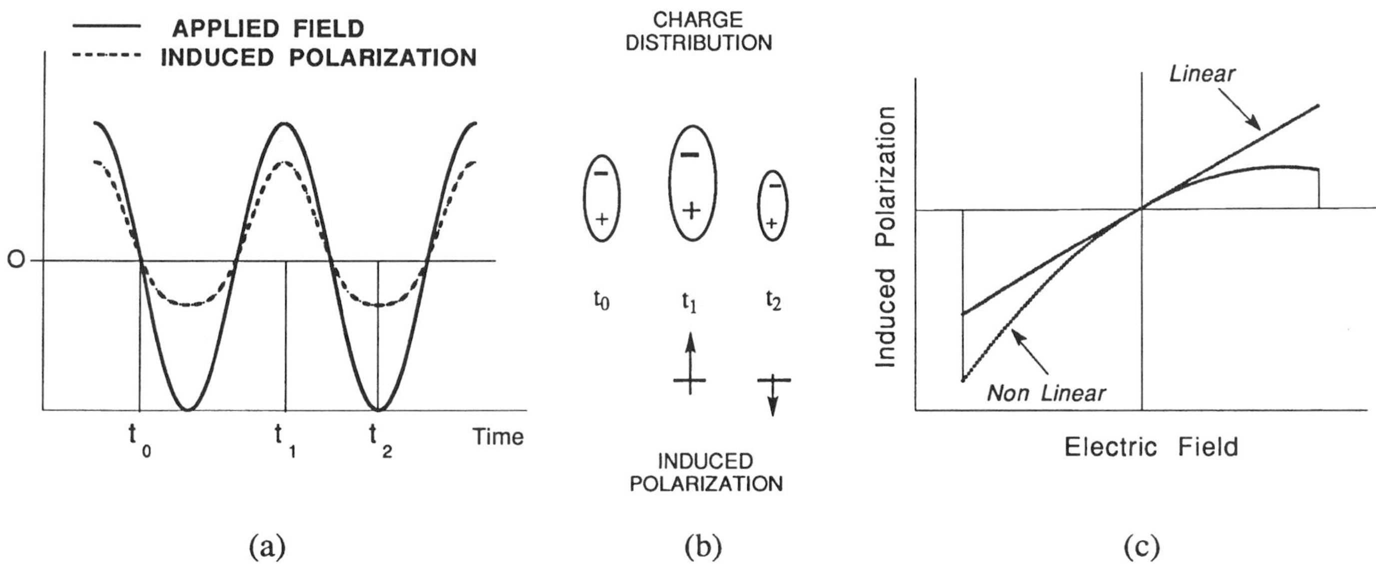

Figure 6.2 (*a*) Plots of the electric field of an applied light wave and the induced polarization wave as a function of time for a second-order nonlinear material; (*b*) diagram depicting the charge distribution and the polarization in the material as a function of time; (*c*) plots of the induced polarization as a function of the applied field, for both linear and nonlinear material. (From Ref. 11.)

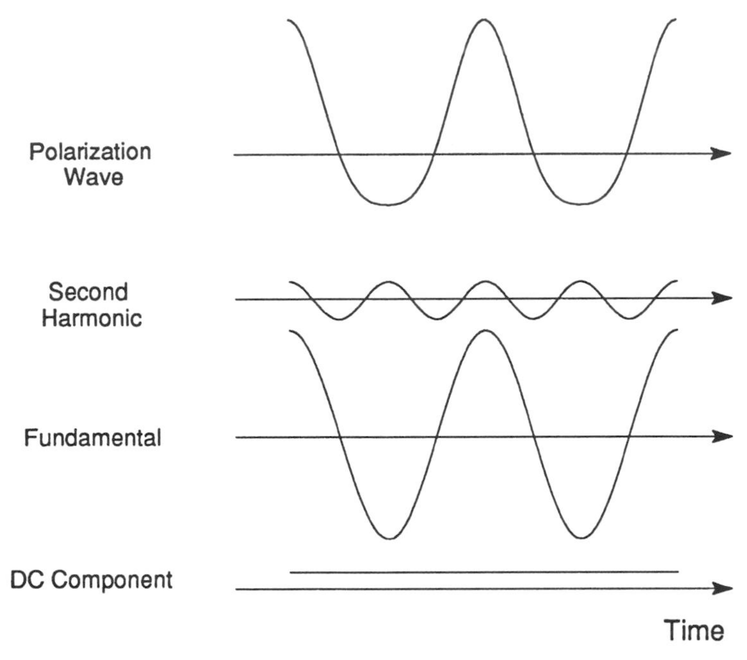

Figure 6.3 Fourier analysis of a second-order polarization wave (see the text); the induced polarization is similar to the one depicted in Fig. 6.2*a*. (From Ref. 11.)

where **P** is the total polarization (the total dipole moment per unit volume), $\mathbf{P}^0$ the permanent polarization, $\chi^{(1)}$ the *linear susceptibility* of an ensemble of molecules, and $\chi^{(2)}$ and $\chi^{(3)}$ are the second- and third-order susceptibility, respectively. $\chi^{(i)}$ are tensor quantities of rank $i + 1$. Similar to equation (3), equation (4) should include the spatial indexes $i, j, k, \ldots$ and the frequency dependence.

To relate the susceptibilities $\chi^{(i)}$ to the microscopic analogs described above, the assumption is frequently made that the atoms or molecules that make up the material are independently polarized by the light with no interatomic or intermolecular coupling. Within this approximation, the relations between the macroscopic and microscopic values depend qualitatively on the average polarizabilities, the number density, and local field factors that account for the electric field strength in the materials compared to that applied to the material.

Linear Term When the electronic charges in a material are displaced by the electric field (**E**) of the light and polarization takes place, the *total electric field* (the "displaced" field, **D**) *within the material* becomes

$$\mathbf{D}(\omega) = \mathbf{E}(\omega) + 4\pi\mathbf{P}(\omega) = \left[1 + 4\pi\chi^{(1)}(\omega)\right]\mathbf{E}(\omega) \tag{5}$$

where the polarization is restricted here to its linear term, and where $4\pi\chi\mathbf{E}$ is the internal electric field created by the induced displacement (polarization) of charges.

Two common bulk parameters that characterize the susceptibility of a material are the *dielectric constant*, $\varepsilon(\omega)$, and the *refractive index*, $n(\omega)$. The dielectric constant is defined as the ratio of the displaced field, **D**, to the applied field, **E**:

$$\varepsilon(\omega) = 1 + 4\pi\chi(\omega) \tag{6}$$

The susceptibility, χ, and the dielectric constant, ε, describe the effect of light on the polarization of the medium. Since the study of NLO addresses how the optical material can change the propagation characteristics of light, we must now ask what is happening to the light as it goes through the medium. The refractive index, n, is in this context the useful parameter to consider. It is defined as the ratio of the speed of light in vacuum, c, to the speed of light in a material, v. At the optical frequencies, in the absence of dispersion (absorption), the square of the refractive index equals the dielectric constant. Consequently, a simple expression links the refractive index to the linear susceptibility:

$$n^2(\omega) = 1 + 4\pi\chi(\omega) \tag{7}$$

Within the approximation that the atoms or molecules are polarized independently (the interatomic or intermolecular coupling is neglected), this equation relates how the optical properties of a material (via n) depend on the electronic density and molecular polarizability (via χ), and therefore on the chemical structure itself.

Second-Order Term The second-order susceptibility, as presented above, should be written more precisely as a tensor of rank 3, including the frequency of the incident light:

$$\chi^{(2)} \equiv \chi^{(2)}_{ijk}(-\omega_3; \omega_1, \omega_2)$$

The second-order phenomena are referred to as three-wave mixing processes. The frequency ω_1, ω_2, and ω_3 of these waves are related by the energy conservation principle, $\omega_3 = \omega_1 \pm \omega_2$, where ω_1 and ω_2 are associated with the incident photons (written on the right side of the semicolon) and ω_3 is the frequency of the radiated photons.

A classification of the NLO phenomena is based on the frequency values ω_1 and ω_2. The first example presented below refers to second-harmonic generation (SHG), where ω_1 and ω_2 are equal to ω and the frequency generated is 2ω. Sum-frequency generation (SFG) or difference-frequency generation (DFG) processes are the more general cases where ω_1 and ω_2

are different and the sum ($\omega_1 + \omega_2$) and the difference ($\omega_1 - \omega_2$) frequencies are generated. Optical parametric generation (OPG) is the case where one input frequency (ω_3) is split into two, such that ω_1 and ω_2 are generated and $\omega_3 = \omega_1 + \omega_2$. Finally, we will see that a static dc electric field that polarizes a material also encompasses a second-order process, called the Pockels effect. Since only the time-averaged asymmetrically induced polarization leads to second-order NLO effects, only molecules and materials lacking a center of symmetry can exhibit them.

Second-Harmonic Generation The first SHG experiment was realized in 1961.[13] When a crystal of quartz was exposed to a ruby laser ($\lambda = 694$ nm), two transmitted beams were detected: first, a red beam, the transmitted incident beam, and second, a weak ultraviolet beam at a frequency exactly twice ($\lambda = 347$ nm) of the incident beam. To understand the origin of the frequency doubling, we can simply rewrite the electric field as a plane light wave, expressed as $\mathbf{E} = \mathbf{E}_0 \cos \omega t$, and replacing $\mathbf{E}$ in equation (4), we find

$$\mathbf{P} = \mathbf{P}_0 + \chi^{(1)}\mathbf{E}_0 \cos \omega t + \tfrac{1}{2}\chi^{(2)}\mathbf{E}_0^2 \cos^2 \omega t + \cdots \tag{8}$$

Since $\cos^2 a = \frac{1}{2} + \frac{1}{2}\cos 2a$, equation (8) becomes

$$\mathbf{P} = \left(\mathbf{P}_0 + \tfrac{1}{4}\chi^{(2)}\mathbf{E}_0^2\right) + \chi^{(1)}\mathbf{E}_0 \cos \omega t + \tfrac{1}{4}\chi^{(2)}\mathbf{E}_0^2 \cos 2\omega t + \cdots \tag{9}$$

This equation states that the polarization consists of a second-order dc field contribution to the static polarization (first term), a frequency component ω corresponding to the incident light frequency (second term), and a new *frequency-doubled* component, 2ω (third term). Therefore, an intense light beam passing through a second-order NLO material produces light at twice the input frequency, as well as a static electric field. The first process is the second-harmonic-generation process described above, and the second is called optical rectification. In Fig. 6.3, the Fourier analysis of the resultant second-order polarization is presented and the induced polarization wave is identical to the one depicted in Fig. 6.2*a*.

To become useful, the SHG signal should be relatively intense; in other words, the frequency conversion ($\omega + \omega \rightarrow 2\omega$) should be efficient. At this point, the quality of the material is critical, and the largest susceptibility $\chi^{(2)}$ is desirable. From a microscopic point of view, this concern underscores the importance of designing molecules with a large first hyperpolarizability β. Beside the material's susceptibility, an important requirement is that the fundamental and second-harmonic light waves propagate throughout the material in a synchronous manner. The propagation will, in fact, be optimum when the two light waves have equal speed, or in other words, when the refractive indexes at the fundamental and harmonic frequencies satisfy the condition $n(\omega) = n(2\omega)$; this is the phase-matching condition. Efficient SHG has been achieved primarily in birefringent crystals, where the proper crystal orientation can satisfy the phase-matching condition.

Parametric Processes The general picture of second-order NLO effects involves the interaction of two laser beams. These waves are associated with the frequencies ω_1 and ω_2 and have amplitudes $\mathbf{E}_1$ and $\mathbf{E}_2$. With these two input waves, the second-order term of the induced polarization becomes

$$\mathbf{P}^{(2)} = \chi^{(2)}\mathbf{E}_1 \cos \omega_1 t \mathbf{E}_2 \cos \omega_2 t \tag{10}$$

Since $\cos a \times \cos b = \frac{1}{2}\cos(a + b) + \frac{1}{2}\cos(a - b)$, equation (10) becomes

$$\mathbf{P}^{(2)} = \tfrac{1}{2}\chi^{(2)}\mathbf{E}_1\mathbf{E}_2 \cos[(\omega_1 + \omega_2)t] + \tfrac{1}{2}\chi^{(2)}\mathbf{E}_1\mathbf{E}_2 \cos[(\omega_1 - \omega_2)t] \tag{11}$$

This equation shows that in sum- or difference-frequency generation (SFG or DFG) processes, two light beams of frequency ω_1 and ω_2 interact in an NLO material, and the resulting polarization occurs at sum $(\omega_1 + \omega_2)$ and difference $(\omega_1 - \omega_2)$ frequencies. SHG is a special case of SFG, where the two frequencies are equal and where the sum is the second-harmonic and the difference is the dc component.

Another important second-order process is the optical parametric generation (OPG), which consists of the decay of input (pump) photons (ω_3) into signal and idler photons such that $\omega_3 = \omega_1 + \omega_2$. The intensity of the generated fields at frequencies ω_1 and ω_2 are determined by the phase-matching condition, which itself is controlled by the orientation of the material, the temperature, or by applying an external potential; variation of these parameters allows for alteration of the refractive indexes.

Pockels Effect The process associated with the Pockels effect, $\omega + 0 \rightarrow \omega$, can be viewed as a particular case of SFG where one light beam is substituted by a static dc electric field of magnitude $\mathbf{E}_2$ (and $\omega_2 = 0$). The second-order term of the induced polarization simply becomes

$$\chi^{(2)}\mathbf{E}_1\mathbf{E}_2 \cos \omega_1 t$$

$\mathbf{E}_2$ is associated with a voltage applied to the nonlinear material. This contribution is added to the linear polarization term [second term of equation (8)]:

$$\mathbf{P} = [\chi^{(1)} + \chi^{(2)}\mathbf{E}_2]\mathbf{E}_1 \cos \omega_1 t \tag{12}$$

This expression demonstrates that an applied field can change the effective linear susceptibility by an amount linearly proportional to the second-order susceptibility $\chi^{(2)}$ multiplied by the applied field $\mathbf{E}_2$. As we have seen earlier, the linear susceptibility and refractive index are related [equation (7)]. Accordingly, an applied voltage modulates the refractive index of the material and therefore the speed of the light beam traveling through the material. In other words, the application of a voltage to the material causes the optical beam to "see" a different refractive index than in the absence of the voltage. This, the linear electro-optic (LEO) or Pockels effect, is used to modulate the phase or the polarization state of light by changing the applied voltage.

The LEO modulation is one of the main techniques used to determine second-order susceptibility.[14] In this measurement a light beam is passed through a material that is subjected to an electric field via a voltage applied with attached electrodes. The electric field causes a change in the refractive index, n, or birefringence (the difference between n in the two directions perpendicular to the direction of propagation). This change in n leads to a change in phase or polarization state of the light beam that can be converted into a change in intensity of the beam using interference with a reference beam or an analyzing polarizer, respectively. This measurement gives a value of the electro-optic coefficient, r, that is related to $\chi^{(2)}(-\omega; 0, \omega)$. Like $\chi^{(2)}$, r is a third-rank tensor. In practice, a contracted notation as r_{ij} is used, where $i = 1$ to 3 (with $1 = x$, $2 = y$, and $3 = z$), and where the second index $j = 1$ to 6 (with $1 = xx$, $2 = yy$, $3 = zz$, $4 = yz$, $5 = xz$, and $6 = xy$).

Third-Order Term Third-order phenomena are referred to as four-wave mixing processes. At the microscopic level, they correspond to the following term:

$$\gamma_{ijkl}(-\omega; \omega_1, \omega_2, \omega_3)\mathbf{E}_j(\omega_1)\mathbf{E}_k(\omega_2)\mathbf{E}_l(\omega_3)$$

which is the microscopic equivalent of the fourth term of equation (4), and where three incident beams have the frequency ω_1, ω_2, and ω_3, and the output beam has a resultant frequency $\omega_4 = \omega_1 \pm \omega_2 \pm \omega_3$.

In contrast to second-order processes, there is no symmetry restriction at third order: Both centrosymmetric and noncentrosymmetric materials will have third-order optical nonlinearities. As for the second order, classification of the third-order processes is based on the combination of input frequencies.

Third-Harmonic Generation The third-harmonic-generation process (THG) is a particular case where the three incident beams have the same frequency ω. If we reconsider equation (3) for the induced polarization of a centrosymmetric molecule (i.e., the even-order terms are zero), we see that

$$\mu = \alpha \mathbf{E}_0 \cos \omega t + \tfrac{1}{6}\gamma \mathbf{E}_0^3 \cos^3 \omega t + \cdots \tag{13}$$

Since $\cos^3 a = \frac{3}{4}\cos a + \frac{1}{4}\cos 3a$, equation (13) becomes

$$\mu = \left(\alpha + \tfrac{1}{8}\gamma \mathbf{E}_0^2\right)\mathbf{E}_0 \cos \omega t + \tfrac{1}{24}\gamma \mathbf{E}_0^3 \cos 3\omega t \tag{14}$$

The last term indicates that the interaction of light with third-order NLO molecules (or material, at a macroscopic scale) will create a polarization component at its third harmonic and hence radiate at light at 3ω.

Kerr Effects The first term of equation (14) describes another important manifestation of the interaction of an intense beam with a third-order NLO: modification of the refractive index. In fact, on the macroscopic level, this term, which depends on the frequency ω, correlates to the first-order susceptibility, which itself is related to the refractive index [see equation (7)]. Thus if we refer only to optical fields $\mathbf{E}(\omega)$, equation (14) shows that the refractive index variation depends on the optical field intensity; this refers to the optical Kerr effect and can be illustrated by the relation

$$n(\omega) = n_0(\omega) + n_2(\omega)I(\omega) \tag{15}$$

where n_0 is the refractive index when the intensity is weak, n_2 is the nonlinear refractive index, which depends on the third-order susceptibility, and where $I = cn_0|\mathbf{E}(\omega)|^2/8\pi$; where c is the speed of light. Consequently, we can see that the sign of $\chi^{(3)}$ (or γ) will determine if the third-order contribution to the refractive index is positive or negative. Materials with positive $\chi^{(3)}$ are self-focusing, which means that the light-induced index change focuses the beam and it becomes more intense as it propagates through the material. On the contrary, materials with a negative $\chi^{(3)}$ are self-defocusing.

Similar to the Pockels effect at second order, one can imagine that if the material is submitted to an electric field and a laser beam of frequency ω simultaneously, the linear susceptibility is changed by an amount proportional to the third-order susceptibility $\chi^{(3)}$ and the square of the amplitude of the applied electric field (i.e., the intensity). This effect is known as the dc Kerr effect, or quadratic electro-optic effect.

Degenerate Four-Wave Mixing Another interesting manifestation of the third-order NLO polarization is degenerate four-wave mixing (DFWM). Briefly, the associated process is $\omega + \omega - \omega \rightarrow \omega$. Two beams of light interacting within a third-order NLO material will create a refractive index grating or interference pattern (Fig. 6.4) that leads to a spatially periodic variation in light intensity across the material. When a third beam is incident on this grating, a fourth beam, the phase conjugate, is diffracted from the grating.[15] This process is called four-wave mixing: Two writing beams and a probe beam result in a fourth, phase-conjugate beam.

A potential use of this phenomenon is in phase-conjugate optics.[16] Phase-conjugate optics takes advantage of a special feature of the diffracted beam: Its path exactly retraces the path of one of the writing beams. As a result, a pair of diverging beams impinging on a phase-conjugate mirror will converge after reflection. In contrast, a pair of diverging beams reflected from an ordinary mirror will continue to diverge. Thus, using a phase-conjugate mirror, distorted optical wavefronts can be reconstructed using phase-conjugate optical systems (Fig. 6.5).

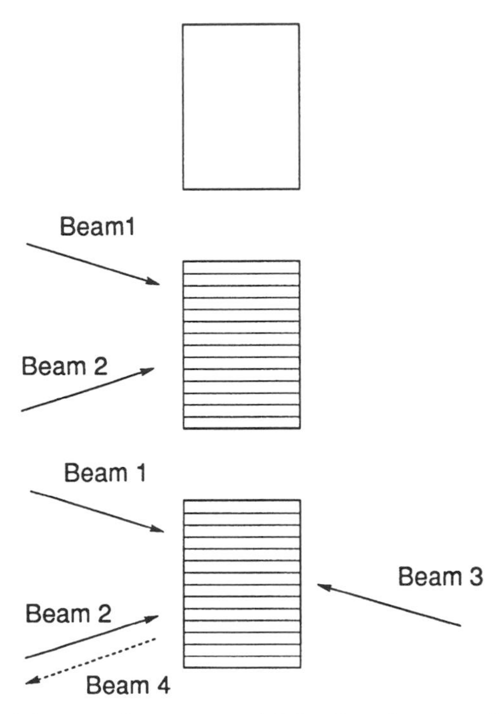

Figure 6.4 (Top) Phase-conjugate material in the absence of an applied field; (center) beams 1 and 2 create a refractive index grating; (bottom) beam 3 interacts with the grating, creating beam 4, the *phase conjugate* of beam 2. (From Ref. 11.)

Electric-Field-Induced Second-Harmonic Generation Electric-field-induced second-harmonic generation (EFISH) can be denoted simply as $\omega + \omega + 0 \rightarrow 2\omega$. This process is the basis for the most common technique used to evaluate the hyperpolarizabilities β and γ.[17-20] The third-order term of the induced polarization $\mathbf{P}^{(3)}$ is

$$\mathbf{P}^{(3)} = \chi^{(3)}\mathbf{E}_1^2\mathbf{E}_2 \cos^2 \omega_1 t = \chi^{(3)}\mathbf{E}_1^2\mathbf{E}_2\left(\tfrac{1}{2} + \tfrac{1}{2}\cos 2\omega_1 t\right) \tag{16}$$

where we can recognize a term that oscillates at the second harmonic frequency 2ω.

When a measurement is performed in solution, the field $\mathbf{E}_1$ induces a macroscopic reorganization such that the molecular dipoles align with the direction of the applied dc field, leading to a net dipole moment. A fundamental wave proceeding through this resulting asymmetric medium generates second harmonic light. The resulting third-order signal is

$$\gamma^{\mathrm{EFISH}} = \gamma_{\mathrm{int}} + \gamma_{\mathrm{orient}}$$

where the intrinsic hyperpolarizability (γ_{int}) is mostly electronic in origin and

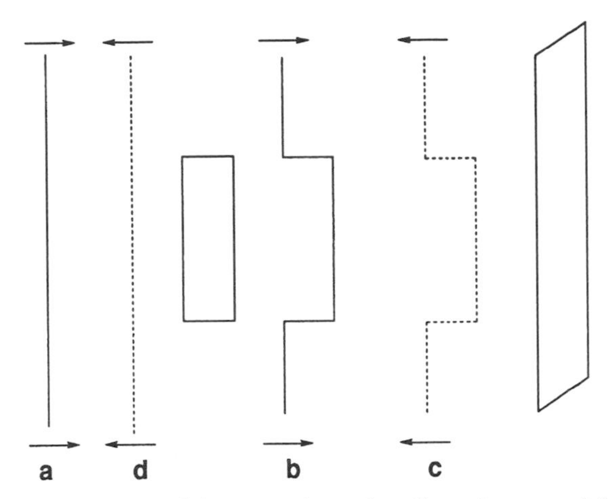

Figure 6.5 (a) A planar wave (b) passes through a distorting material that introduces an aberration and (c) the material interacts with a phase-conjugate mirror, creating the phase-conjugate image. (d) When the phase-conjugate wave passes through the distorting material on the reverse path, the original aberration is canceled, producing an undistorted image. (From Ref. 11.)

can be evaluated separately via a THG measurement,[21] and the orientational contribution (γ_{orient}) is equal to $\boldsymbol{\mu\beta}/5kT$, where $\boldsymbol{\beta}$ is the vector component of the β tensor along the dipole moment k is Boltzmann's constant, and T is temperature.

Many NLO materials absorb in the visible and near infrared. These absorptions introduce difficulties for assessing the electronic hyperpolarizability of the material. The THG technique is powerful but suffers from the problem that the third harmonic is often in or near an absorbing region of the spectrum. (Even when using 1.907-μm light from a H_2 Raman-shifted Nd:YAG laser, the third harmonic occurs at 636 nm.) This does not preclude obtaining a value for γ; however, interpretation of this value is often not straightforward, since the γ measured is often a complex quantity and/or dispersively enhanced. Thus extreme caution must be taken when comparing hyperpolarizabilities for materials measured at different frequencies. Using DFWM[22] to determine γ circumvents some of the problems associated with absorption since one usually does not need to be concerned with three photon resonances, although two photon resonance may still be important and complicate interpretation of the DFWM experiment.

6.3 QUANTUM-CHEMICAL DESCRIPTION

In this section we present a qualitative and brief discussion of how quantum-chemical calculations can help in the design of NLO molecules and/or help to explain experimental observations. Such a theoretical approach provides a description of the geometric and electronic structures as well as information on the evolution of the microscopic polarizability and hyperpolarizabilities. The actual approaches used in the field have been reviewed several times.[8, 23-25]

One important point worth emphasizing is that most of the calculations are usually performed in the gas phase (i.e., on an isolated molecule). This may appear contradictory with respect to the ideas developed in the next section, which emphasize the influence of the medium in determining the extent of ground-state polarization, and consequently, the NLO properties. It is only recently that this problem has been considered in the context of NLO.[26-31]

The quantum chemistry techniques include a wide range of approaches, including various degrees of approximation, that allow resolution of the Schrödinger equation. The choice of technique has direct consequences on the quality and validity of the results, but such a discussion is beyond the scope of this chapter. There are several approaches to calculating the molecular polarizabilities.

The molecular total energy W, under the perturbation of a field $\mathbf{E}$, is

$$W(\mathbf{E}) = W^0 - \mu^0\mathbf{E} - \frac{1}{2!}\alpha\mathbf{E}^2 - \frac{1}{3!}\beta\mathbf{E}^3 - \frac{1}{4!}\gamma\mathbf{E}^4 + \cdots \qquad (17)$$

where W^0 is the energy of the unperturbed system, μ^0 the permanent dipole, and the higher terms translate the induced dipole, as defined earlier. The validity of this expression is restricted to the case where $\mathbf{E}$ is approximately uniform on the molecular scale. The wavelength of the electromagnetic wave to which the electric field $\mathbf{E}$ is associated is in the optical domain (i.e., from two to four orders of magnitude larger than the molecular size); thus the expression above is appropriate. This restriction of a uniform field also means that the variation in induced energy is related to the variation of the dipole, which is evaluated using the Hellman–Feynman theorem,

$$\mu(\mathbf{E}) = -\frac{\delta W(\mathbf{E})}{\delta \mathbf{E}} \qquad (18a)$$

This is consistent with the concepts developed in the first section, where the total dipole was expressed as

$$\mu(\mathbf{E}) = \mu^0 + \alpha\mathbf{E} + \frac{\beta}{2}\mathbf{EE} + \frac{\gamma}{6}\mathbf{EEE} + \cdots \qquad (18b)$$

Also, at the limit where the field approaches zero, the energy W can be written as a Taylor series:

$$W(\mathbf{E}) = W^0 + \left(\frac{dW}{d\mathbf{E}}\right)_0 \mathbf{E} + \frac{1}{2!}\left(\frac{d^2W}{d\mathbf{E}^2}\right)_0 \mathbf{E}^2 + \frac{1}{3!}\left(\frac{d^3W}{d\mathbf{E}^3}\right)_0 \mathbf{E}^3 + \cdots \qquad (19a)$$

We can therefore introduce the following definitions:

$$\mu^0 = -\left(\frac{dW}{d\mathbf{E}}\right)_0 \qquad \alpha = -\left(\frac{d^2W}{d\mathbf{E}^2}\right)_0 \tag{19b}$$

$$\beta = -\left(\frac{d^3W}{d\mathbf{E}^3}\right)_0 \qquad \gamma = -\left(\frac{d^4W}{d\mathbf{E}^4}\right)_0 \tag{19c}$$

These relations constitute the basic support of the derivative techniques. The derivatives can be carried out analytically, as for example, in the coupled perturbed Hartree–Fock technique (CPHF),[32–34] or numerically, as in the finite-field technique (FF).[35,36] These techniques do not provide analysis of the frequency dependence of the molecular polarizabilities. Also, other types of techniques have been developed, such as the time-dependent Hartree–Fock (TDHF) method.[37,38] This is a variational method that solves for the wavefunction of the ground-electronic state in the presence of the external electric field and is used to calculate both static and dynamic linear and nonlinear polarizabilities.

The coupled oscillator technique is yet another type of technique that has been used to evaluate the optical properties of conjugated polymers[39,40] and the nonlinear optical responses of substituted polyenes.[41] In this approach the electrons are treated as harmonically bound particles, for which the resonant frequencies of the coupled oscillators correspond to electronic transition frequencies. Globally, the treatment of such systems perturbed by an external electric field can provide mechanistic insight, in terms of electronic normal modes, associated with the nonlinear processes.

One of the most widely used techniques, the sum-over-state technique (SOS), derives from perturbation theory.[42] It is a powerful technique that allows for the identification of the excited states that play essential roles in the optical nonlinearities.[23,24,43,44] In addition, the SOS technique allows for the calculation of the frequency dependence of the polarizabilities. This particular approach has been applied on the basis of different Hamiltonians but mostly within the INDO (intermediate neglect of differential overlap)[45,46] or CNDO (complete neglect of differential overlap)[47] semiempirical techniques, coupled to single or single and double configuration interaction (CI). As a result of such INDO-CI calculations, one obtains information on (1) the transition energies, E_{gm} (i.e., the energy difference between states m and g); (2) the corresponding transition dipoles, M_{gm} (related to the intensity of the transition, i.e., the oscillator strength, $f \sim E_{gm} \times M_{gm^2}$); and (3) the state dipoles $\mu_g, \mu_m, \ldots$. These results are the inputs for evaluation of the polarizabilities using the SOS expressions.

The SOS expressions for the dominant components of the static first-, second-, and third-order polarizability tensors are (within a Taylor series expansion)[42]

$$\alpha_{xx} = 2 \sum_{m \neq g} \frac{\langle g|\mu_x|m\rangle\langle m|\mu_x|g\rangle}{E_{gm}} \tag{20}$$

$$\beta_{xxx} = 6 \sum_{m \neq g} \sum_{n \neq g} \frac{\langle g|\mu_x|m\rangle\langle m|\bar{\mu}_x|n\rangle\langle n|\mu_x|g\rangle}{E_{gm}E_{gn}} \tag{21}$$

$$\gamma_{xxxx} = 24\left(\sum_{m \neq g} \sum_{n \neq g} \sum_{p \neq g} \frac{\langle g|\mu_x|m\rangle\langle m|\bar{\mu}_x|n\rangle\langle n|\bar{\mu}_x|p\rangle\langle p|\mu_x|g\rangle}{E_{gm}E_{gn}E_{gp}} - \sum_{m \neq g} \sum_{p \neq g} \frac{\langle g|\mu_x|m\rangle\langle m|\mu_x|g\rangle\langle g|\mu_x|p\rangle\langle p|\mu_x|g\rangle}{E_{gm}E_{gp}E_{gp}} \right) \tag{22a}$$

Here x corresponds to the long axis of the molecule; $\langle g|\mu_x|m\rangle$ is the electronic transition moment along the Cartesian x-axis between the reference state described by the ground state $\langle g|$ and excited state $\langle m|$ (denoted simply as M_{gm}); and $\langle m|\bar{\mu}_x|n\rangle$ denotes the dipole difference operator equal to $\langle m|\mu_x|n\rangle - \langle g|\mu_x|g\rangle\delta_{mn}$, where δ_{mn} is the Kronecker delta function (when $n = m$, M_{mm} is simply the state dipole μ_m).

These expressions become slightly modified when the dynamic NLO properties are evaluated (as opposed to the static properties, where $\hbar\omega$ is zero). To calculate the first-order polarizability $\alpha_{xx}(-\omega; \omega)$, the denominator in equation (20) becomes

$$E_{gm} - \hbar\omega - i\Gamma_{gm} \tag{22b}$$

The $\beta_{xxx}(-2\omega; \omega, \omega)$ second-harmonic-generation term is evaluated by changing the denominators in equation (21) to

$$(E_{gm} - 2\hbar\omega - i\Gamma_{gm})(E_{gn} - \hbar\omega - i\Gamma_{gn}) \tag{22c}$$

while in the calculation of the $\gamma_{xxxx}(-3\omega; \omega, \omega, \omega)$ third-harmonic-generation terms, the denominators in equation (22) become

$$(E_{gm} - 3\hbar\omega - i\Gamma_{gm})(E_{gn} - 2\hbar\omega - i\Gamma_{gn})(E_{gp} - \hbar\omega - i\Gamma_{gp}) \tag{22d}$$

and

$$(E_{gm} - 3\hbar\omega - i\Gamma_{gm})(E_{gp} - \hbar\omega - i\Gamma_{gp})(E_{gp} + \hbar\omega - i\Gamma_{gp}) \tag{22e}$$

where ω is the frequency of the perturbing radiation field and Γ_{gm} is the damping factor associated with the excited state $|m\rangle$, which is introduced to treat the nonlinearities on resonance (i.e., when $E_{gm} = \hbar\omega$ or $2\hbar\omega$ or $3\hbar\omega$).

From these denominator expressions it becomes easier to understand what is called the resonance enhancement of the nonlinearities. When the laser energy $\hbar\omega$ (or its second or third harmonic, $2\hbar\omega$ or $3\hbar\omega$) is approaching an electronic transition E_{gm}, the polarizabilities are enhanced since the denominators are reduced considerably.

Simplified versions of these equations have been identified that provide useful model expressions as a two-state model for α and β[48,49] and as a three-term expression for γ.[50-54] Basically, in the two-state models the idea is to focus on the most important excited state (i.e., the one that dominates the SOS analysis). This state is typically strongly coupled to the ground state, and consequently, reduces equations (20) and (21) to

$$\alpha_{xx}[\text{model}] = 2\frac{\langle g|\mu_x|e\rangle\langle e|\mu_x|g\rangle}{E_{ge}} = 2\frac{M_{ge}^2}{E_{ge}} \tag{23}$$

$$\beta_{xxx}[\text{model}] = 6\frac{\langle g|\mu_x|e\rangle(\langle e|\mu_x|e\rangle - \langle g|\mu_x|g\rangle)\langle e|\mu_x|g\rangle}{E_{ge}^2} = 6\frac{M_{ge}^2\,\Delta\mu}{E_{ge}^2} \tag{24}$$

where $\Delta\mu$ is the difference between the dipole moments in the first strongly allowed charge-transfer excited state and the ground state, $\mu_e - \mu_g$. As discussed below, these two-state models are used for the basis of the molecular design strategies for efficient NLO chromophores. In the case of the third-order polarizability γ, a single excited $|e\rangle$ strongly coupled to the ground state $|g\rangle$ is considered again, and the model is now improved by considering a few higher excited states $|e'\rangle$, themselves strongly coupled to the first excited $|e\rangle$.

$$\begin{aligned}\gamma_{xxxx}[\text{model}] &= 24\frac{\langle g|\mu_x|e\rangle(\langle e|\mu_x|e\rangle - \langle g|\mu_x|g\rangle)(\langle e|\mu_x|e\rangle - \langle g|\mu_x|g\rangle)\langle e|\mu_x|g\rangle}{E_{ge}^3} \\ &\quad - 24\frac{\langle g|\mu_x|e\rangle\langle e|\mu_x|g\rangle\langle g|\mu_x|e\rangle\langle e|\mu_x|g\rangle}{E_{ge}^3} \\ &\quad + 24\sum_{e'}\frac{\langle g|\mu_x|e\rangle\langle e|\mu_x|e'\rangle\langle e'|\mu_x|e\rangle\langle e|\mu_x|g\rangle}{E_{ge}^2 E_{ge'}}\end{aligned}$$

$$\gamma_{xxxx}[\text{model}] = 24\left(\frac{M_{ge}^2\,\Delta\mu^2}{E_{ge}^3} - \frac{M_{ge}^4}{E_{ge}^3} + \sum_{e'}\frac{M_{ge}^2 M_{ge'}^2}{E_{ge}^2 E_{ge'}}\right)$$

term notation: D N T (25)

There are three terms:

1. The D (dipolar) term, which is explicitly dependent on $\Delta\mu$, M_{ge}, and E_{ge}, as in the case of β_{xxx}[model].
2. The N (negative) term, which provides a net negative contribution and, as for α_{xx}[model], depends on M_{ge} and E_{ge}. Taken together, these first two terms constitute the two-state model for γ, where only the ground state and the first excited state are thought to contribute to a description of the hyperpolarizability.
3. A third term, the T (two-photon) term, has also to be introduced, that takes into account the participation of higher excited states $|e'\rangle$ associated with two-photon-like processes. This formula is represented pictorially in Fig. 6.6, where we illustrate some of the states and the optical channels (virtual transitions) involved for each of the three terms.

As discussed in the next section, these model expressions include parameters that can be controlled synthetically, in a rational manner, to optimize the NLO response. For instance, it is understood how to design a molecule with a small transition energy; this is a strategy used to enhance the nonlinearities. In fact, although the reduction of the transition energy may result in an increase in the nonlinearity, it corresponds directly to a reduction of the transparency window associated with the materials and therefore the useful range where the molecule can be used without significant absorption of the incident laser beam. Such absorption is quite problematic for most but not all NLO applications. Accordingly, a continuing challenge is to develop molecules that are transparent and still have large nonlinearities.

In practice the dynamic, frequency-dependent nonlinearities are exploited and, since the specific frequencies involved in every process are application

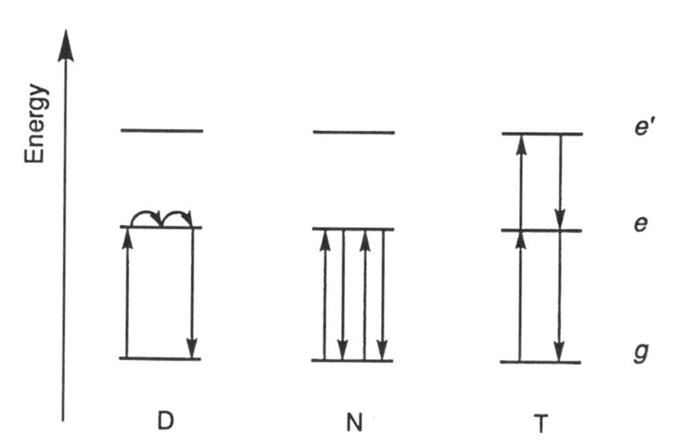

Figure 6.6 Optical channels involved for each of the terms (D, N, and T) of the three-term expression for γ.

dependent, it is not surprising to realize that the optimization of a NLO material (and molecule) is usually specific to a particular application using a particular laser. Nevertheless, for fundamental investigations, the static nonlinearities (i.e., corrected for dispersive enhancement) are usually preferred as a basis for comparison.

6.4 MOLECULAR STRUCTURE–NLO PROPERTIES RELATIONSHIPS

Our ability to exploit the NLO properties of organic materials relies on a fundamental understanding of the relationships between chemical structure and molecular nonlinearities. This is because organic materials are most frequently molecular in nature, in other words, the macroscopic properties depend primarily on the properties of the molecules forming the materials. Early on, organic compounds with conjugated π-electrons were identified as structures that can display large nonlinearities. A cloud of π-electrons appears to be easily polarizable, and such electronic contributions are fast (on the order of femtoseconds). In addition, asymmetric substitution of a conjugated chain with a donor group on one end and an acceptor group on the other end provides the noncentrosymmetry required for a second-order nonlinearity. Here we present the essential features of a model that provides a unified picture of these relationships in organic dyes, including a detailed understanding of the parameters that control the chemical structure.[55–59] Indeed, fine tuning the chemical structure of conjugated chains is a critical key to the rational engineering of NLO materials. The model also provides precise guidelines as to what structures are needed to obtain optimized second- and third-order polarizabilities.

6.4.1 Unified Description for Linear and Nonlinear Polarizabilities[57]

Ground-State Polarization and Bond-Length Alternation In this model, molecular polarizability and hyperpolarizabilities are correlated with the degree of ground-state polarization. The degree of ground-state polarization, or in other words, the degree of charge separation in the ground state, depends primarily on the chemical structure (for example, the structure of the π-conjugated system or the strength of the donor and acceptor substituents) but also on its surroundings (for example, the polarity of the medium). In donor–acceptor polyenes, ground-state polarization is related to a geometrical parameter, the bond-length alternation (BLA), which is defined as the average of the difference in the length between adjacent carbon–carbon bonds in a polymethine [$(CH)_n$] chain. Polyenes have alternating double and single bonds (bond length equal to 1.34 Å and 1.45 Å, respectively) and thus show a high degree of bond-length alternation (BLA = +0.11 Å). To better understand this correlation, it is illustrative to discuss the wavefunction of the ground state in terms of a linear combination of the two limiting resonance structures: a neutral form characterized by a positive BLA and a charge-separated form characterized by a negative BLA (since the double and single bonds pattern is now reversed relative to the neutral form) (Fig. 6.7). For substituted polyenes with weak donors and acceptors, the neutral resonance form dominates the ground-state wavefunction, and the molecule has a high degree of bond-length alternation. With stronger donors and acceptors, the contribution of the charge-separated resonance form to the ground state increases and simultaneously, BLA decreases. When the two resonance forms contribute equally to the ground-state structure, the molecule exhibits essentially no BLA. This zero BLA limit is called cyanine-like, referring to the common structure of a cyanine molecule. Such cyanine molecules are known to be well represented by two degenerate resonance forms, resulting in a structure with virtually no BLA, as represented by the bottom structure in Fig. 6.7. Finally, if the charge-separated form dominates the ground-state wavefunction, the molecule acquires a reversed bond-alternation pattern. BLA is a measurable parameter that is related to the *mixing* of the two resonance forms in the actual ground-state structure of the molecule. Since BLA is related to the degree of ground-state polarization and the mixing of the two resonance forms, it is important to account for the fact that different molecular building blocks (—CH=CH—, —C≡C—, —C_6H_4—) will have different bond-length alternations at a point where there is equal mixing between the two charge-transfer forms.

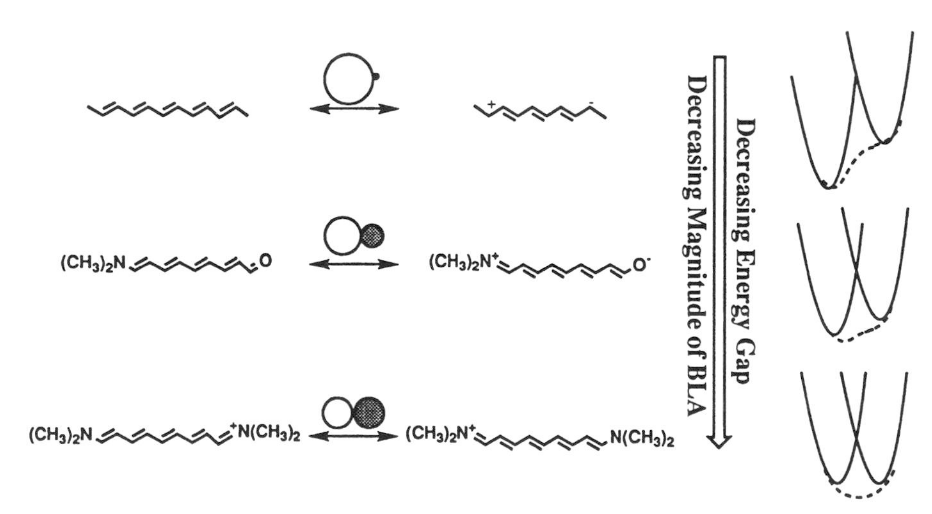

Figure 6.7 (Left) As the two resonance forms contribute more equally to ground-state structure, BLA decreases. The relative contribution of the resonances is represented schematically by the sizes of the balloons over the arrows. (Right) How the energy difference of the potential well for the two resonance forms can be related to the relative contribution of the two resonance structures and therefore to BLA.

***Theoretical Investigations of the Structures*–Properties Relationships**
At second order, a simple four-orbital analysis of the two-state model presented above shows that an optimal combination of donor and acceptor strengths for a given bridge will maximize β.[56] Optimal is the keyword here, since it was a common misbelief earlier that stronger donor and acceptor substituents would always result in larger nonlinearities. To correlate the molecular polarizabilities α, β, and γ with ground-state polarization and consequently with BLA, calculations were performed on the molecule $(CH_3)_2N{-}(CH{=}CH)_4{-}CHO$ including the presence of an external static electric field, **F**, of varying strength. The field **F** is applied such that increasing strength leads to greater ground-state polarization. The first quantum-chemical approach used to investigate these structure–properties relationships involved the AM1 Hamiltonian, with point charges (sparkles) utilized to induce a grading of the ground-state polarization, and using the finite field procedure to evaluate the molecular polarizabilities.[59] A second approach has included calculations performed at the semiempirical intermediate neglect of differential overlap-configuration interaction (INDO-CI) level, with evaluation of the molecular polarizabilities through the sum-over-states formulation at the correlated level (configuration interaction with single and double excitations), as well as by means of more simple two-state and three-term models.[58] In addition, similar studies have been performed using a valence-bond charge-transfer theory,[55] a refined free-electron model,[60] a coupled oscillator approach,[41] or at an *ab initio* level, using an expanded self-consistent reaction-field theory.[31, 61] Although each method differs significantly in its approach to evaluating both the electronic structure and the molecular polarizabilities, they all arrive at essentially the same picture. To illustrate the salient points of these studies, we present a brief summary of the INDO study.

In this study, the ground-state polarization is tuned by applying an external static electric field of varying strength.[58] The range in electric field strength is from about 10^7 to 10^8 V/cm, the same order as the reaction fields determined for common dipolar liquids.[62] It thus appears that the external electric field strengths employed in that approach are similar to what a dipolar molecule may experience upon going from the gas phase to condensed phases, such as polar solvents.[63] The polarizabilities were correlated with the average of the difference in the π-bond order between adjacent carbon–carbon bonds [denoted more simply as the bond-order alternation (BOA), related qualitatively, as is BLA, to *mixing* of the two resonance forms in the actual ground-state structure of the molecule.[58] The π-bond order is related to the amount of π-charge associated with a bond and therefore is a measure of the double-bond character of a given C—C bond (a BOA of about -0.55 is calculated for a donor–acceptor polyene with alternating double and single bonds[64]). The relationships between BOA and the molecular polarizabilities are not straightforward and depend, as we will see, on the order of the response; the results are summarized in Fig. 6.8.

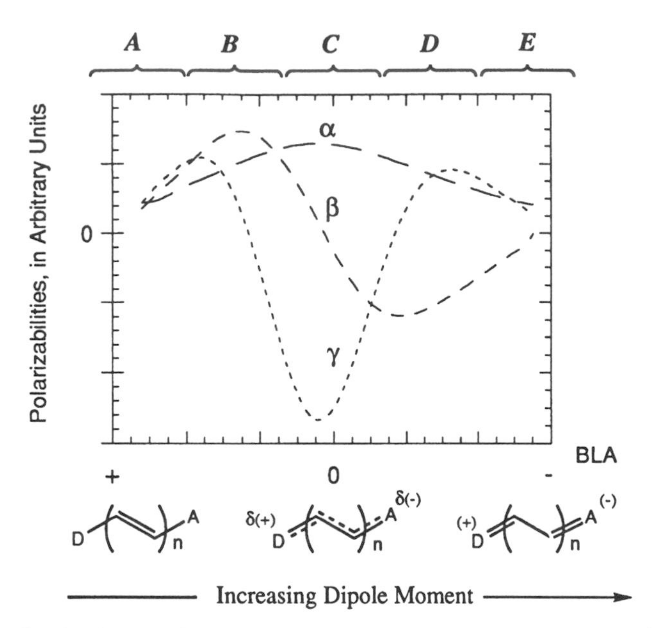

Figure 6.8 Qualitative evolutions of the polarizabilities α, β, and γ (in arbitrary units) as a function of the ground-state polarization, translated in terms of bond-length alternation, BLA, for a donor (D)/acceptor (A) polyene, and on top, definition of regions A through E used as marks on the ground-state polarization axis.

The linear polarizability, α, is maximized when an equal mixing of the two limiting resonance structures takes place, that is, when BOA (or BLA) is zero, referred to as the cyanine limit. The behavior of the first hyperpolarizability, β, is more complex; going from the neutral polyene limit to the cyanine limit, β first increases, peaks in a positive sense for an intermediate structure, decreases, and passes through zero at the cyanine limit. From that limit and going to the charge-separated resonance structure, β continues to decrease and thus becomes negative, peaks in a negative sense, and then decreases again (in absolute value), to become smaller in the charge-separated structure. The second hyperpolarizability, γ, is again more complex and presents two positive peaks associated with intermediate mixing of resonance structures and a large negative value at the cyanine limit. Thus these calculations make specific predictions about what type of molecules of a given length should have optimized nonlinearities.

The behavior of these curves have been investigated in detail, and the two-state model for α and β, as well as the three-term expression for γ, often account quite well for evolution of the nonlinearities, as a function of the ground-state polarization. These model expressions, presented in the preceding section, provide design guidelines for optimization of the properties, since they relate in a first approximation the nonlinearities to other properties, such as the first transition energy and dipole and the ground- and first-excited state dipole.

Experimental Studies To experimentally investigate the structure/nonlinearity relationships, a series of six molecules that spanned a wide range of ground-state polarization and therefore BLA were examined in a variety of solvents.[57, 65, 66] This solvent-dependent study allowed for a fine tuning of the ground-state polarization of each molecule since the more polar the solvent, the larger the contribution of the charge-separated form to the ground state (in other words, the larger is the ground-state polarization).[63] An approximate assessment of degree of mixing between the neutral and charge-separated forms is first made based on data derived from Raman, ultraviolet-visible, ^{1}H nuclear magnetic resonance spectroscopies, and x-ray crystallography,[65–67] and allows the molecules to be associated with a region of BLA according to the definitions of regions A to E shown in Fig. 6.8. The assignments of molecules to regions are shown in Fig. 6.9. The concept of loss and gain of aromaticity, on which the design of such molecules is based, is discussed briefly in Section 6.4.2. Finally, the EFISH, which gave values of $\mu\beta$, and the THG measurements, which gave values of γ, are reported in Table 6.1 (only the dispersion-corrected $\mu\beta(0)$ values are reported.[18, 49]). The qualitative trends observed from these experimental data confirmed the basic shape predicted for β and γ, as shown in Fig. 6.8. For example, for compounds 1.3 and 1.4 (labeled according to their presentation in Table 6.1), which fall in region C, $\mu\beta$ changes sign and γ exhibits a negative peak. For compound 1.5, which falls in region C/D, $\mu\beta$ peaks and γ changes sign from negative to positive.

Derivative Relationships An interesting observation drawn simultaneously from the theoretical and experimental data is the following: As the ground-state polarization increases (in other words, from the theoretical standpoint, as the structure evolves under the influence of the applied field $\mathbf{F}$), the linear polarizability α, the first hyperpolarizability β, and second hyperpolarizability γ are seen to be derivatives, with respect to $\mathbf{F}$, of their next-lower-order polarizability. For example, when α peaks, β changes sign; when β peaks, γ changes sign.

Even if the idea of derivative relationships between various orders of polarizabilities is implied by the definition of the polarization itself, it is only recently that a clarification of how the different polarizabilities are related to one another as the molecular structure changes has been presented.[57] The experimentally observed trends strongly support the validity of an equivalent electric field concept, which relates the various experimental approaches to increasing the ground-state polarization, in a first approximation, to the application of a simple electric field across a polarizable π-conjugated system. These experimental approaches include (1) the tuning of the donor–acceptor strength and (2) the topology of the molecular structure

Compound Number and Resonance Structure	Region	Basis of Assignment
1.1	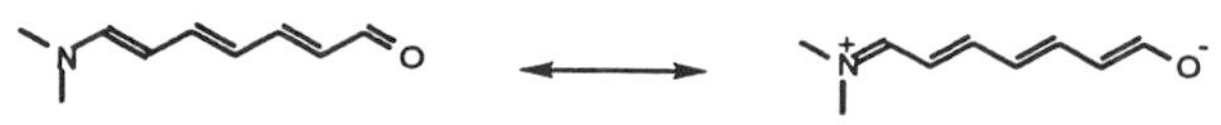 A	X-RAY, RAMAN, NMR, UV
1.2	 B	X-RAY, RAMAN, NMR, UV
1.3	 B/C	^{13}C-NMR, NMR, UV
1.4	 C	X-RAY, NMR, UV
1.5	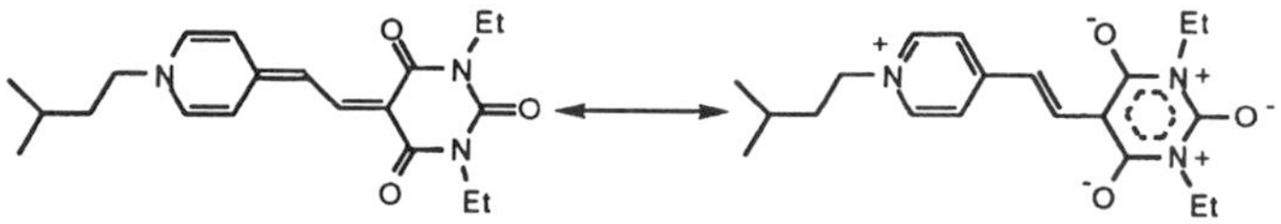 C/D	NMR, UV
1.6	 D/E	NMR, UV

Figure 6.9 Limiting resonance structures and numbering scheme for model donor–acceptor polyene compounds and assignment to regions *A* through *E* (defined in Fig. 6.8); Et, ethyl; Bu, butyl.

TABLE 6.1 $\mu\beta(0)$ and γ Values for the Compounds Presented in Fig. 6.9[a]

Compound	CCl_4 (0.053)	C_6H_6 (0.111)	Dioxane (0.164)	$CHCl_3$ (0.259)	CH_2Cl_2 (0.309)	CH_3CN (0.456)	CH_3NO_2 (0.482)	CH_3OH (0.762)	BOA Region
1.1	236	213	—	247	263	268	327	—	*A*
	—	—	40	95	105	113	113	73	
1.2	245	255	—	281	238	162	136	—	*B*
	40	15	−25	−42	−50	−120	−117	−135	
1.3	272	137	—	133	94	74	−43	—	*B/C*
	—	−20	−100	−135	−145	−205	−220	−166	
1.4	177	168	—	−14	−38	−155	−203	—	*C*
	—	−85	−170	−195	−175	−130	−125	−10	
1.5	—	−117	—	−248	−276	—	−240	—	*C/D*
	—	—	−25	15	30	79	73	150	
1.6	—	—	—	−386	−500	−374	−247	—	*D/E*
	—	—	—	130	167	228	227	291	

[a]Top entries are $\mu\beta(0)$ values (in units of 10^{-48} esu, EFISH measurements at 1.907 μm); bottom entries are γ values (in units of 10^{-36} esu, THG measurements at 1.907 μm). From left to right, the solvent polarity increases, and the corresponding normalized ET_{30} values are given in parentheses.[57]

(i.e., by gain of aromaticity stabilization energy upon charge separation, as discussed below); and (3) the solvent-mediated polarization. These derivative relationships provide a unified picture of linear and nonlinear polarization of organic donor–acceptor compounds and made possible predictions for hyperpolarizabilities of higher-order hyperpolarizabilities (such as δ) as a function of molecular structure (BOA or BLA).

6.4.2 Design of Second-Order NLO Molecules

Prior to the detailed prediction described above in which maximized β is associated with a particular ground-state polarization, molecules for second-order NLO applications were based simply on aromatic π-electron systems unsymmetrically endcapped with electron donor and acceptor groups. Thus, as the electrons interact with the oscillating electric field of light, they will show a preference to shift from donor toward acceptor and a reluctance to shift in the opposite direction. Two prototypical examples of NLO chromophores are *p*-NA and 4-(*N*,*N*-dimethylamino)-4′-nitrostilbene (DANS) (Fig. 6.10). In *p*-NA, the π-system of the benzene ring provides the polarizable electrons, and the amino group acts as donor and the nitro group as acceptor. In DANS, the two benzene rings and the double bond are conjugated, providing a longer π-system.

π-Systems with Reduced Aromaticity With few exceptions, until recently, molecules with sufficiently strong donors and acceptors have not been synthesized to reach the peak of the β curve. If the peak was reached,

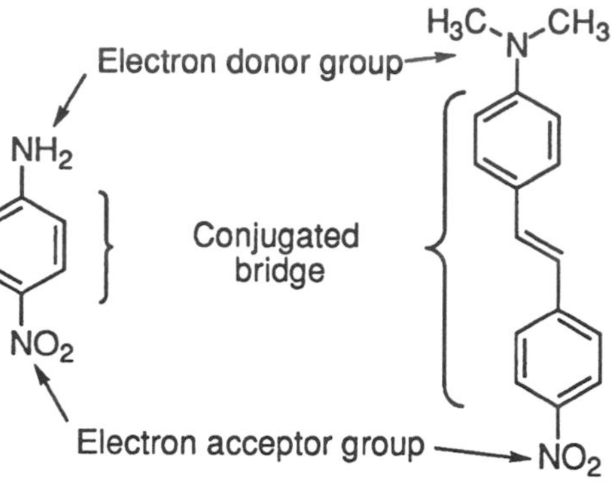

Figure 6.10 Prototypical π-conjugated organic donor–acceptor chromophores for second order: (left) *p*-nitroaniline (*p*-NA); (right) 4-*N*,*N*-dimethylamino-4′-nitrostilbene (DANS).

replacement of one endgroup with an even stronger donor or acceptor would diminish β, and this has been observed only recently.[68] It is therefore necessary to develop synthetic design strategies that will lead to molecules where β is optimized by tuning the ground-state polarization (or simply BLA) of the π-electron system between the donor and acceptor. Such a strategy is based primarily on the realization that molecules with strongly aromatic bridges, such as stilbene, will not have the correct balance of the two limiting resonance forms in the ground state that corresponds to an optimized β value.[56] This is due primarily to the energetic price associated with the loss of aromaticity in charge-separated form (see below). Accordingly, Marder et al. initially examined *N*,*N*-dimethylindoaniline (compound 2.1), which has both an aromatic and a quinonal ring in the canonical neutral and charge-transfer resonance forms. Such conjugated systems are called degenerate π-systems, since the topology of the π-electron bridge no longer leads to an inherent energetic bias in either resonance form. In such bridges the necessary mixing is dictated by the electron affinities of the end groups. A similar strategy has been used to design low-bandgap polymers,[69, 70] and for these materials the mixed aromatic and quinonal character results in enhancements in the linear polarizability. Experimental EFISH β values are reported in Table 6.2, where molecules using this new strategy are compared to other conventional compounds. Even though compound 2.1 is two atoms shorter than compound 2.2 and lacks coplanar rings, $\beta(0)$ is almost a factor of 2 greater. Marder et al. have shown that breaking the degeneracy of the π-system (while keeping the overall length of the molecule constant) in general resulted in a lower hyperpolarizability. Thus ω-(4-*N*,*N*-dimethylaminophenyl)penta-2,4-dien-1-al (compound 2.3), which is structurally similar to compound 2.1 but lacks the double bond critical for the degeneracy of the π-system, has a much lower nonlinearity. Similarly, breaking the aromaticity

TABLE 6.2 Optical Properties (λ_{max}), Dipole Moments (μ), and EFISH Results from Measurements in Chloroform at 1.907 μm [Dispersion-Corrected $\mu\beta(0)$]

Compound	λ_{max} (nm)	μ (D)	$\mu\beta(0)$ (10^{-48} esu)
2.1	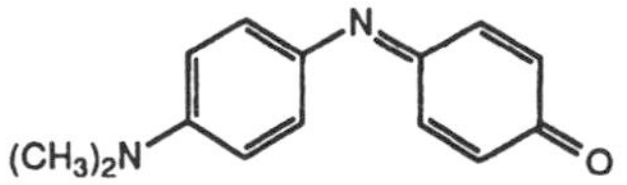 590	4.0	424
2.2	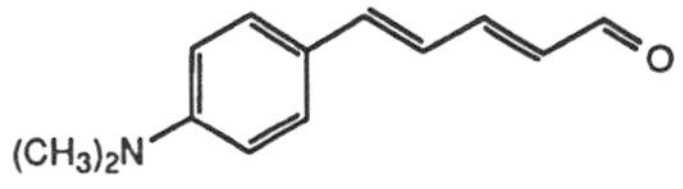 412	6.0	240
2.3	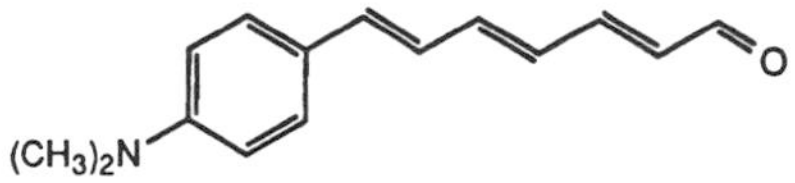 434	6.3	416
2.4	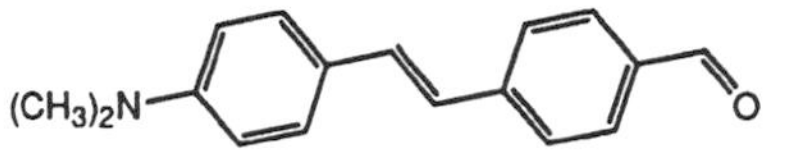 360	3.5	70

in compound 2.4 leads to a π-bridge in compound 2.2 that is closer to being degenerate and consequently, a large enhancement of β. Accordingly, a factor of 3.3 enhancement of β for 6-(4-*N*,*N*-dimethylaminophenyl)-1-nitrohexa-1,3,5-triene, 2.4, vs. 2.2, is observed.[71] These results strongly suggest that the enhanced nonlinearity of compound 2.1 is due largely to the degenerate π-electron bridge.

Subsequently, much greater support for this hypothesis has been published. For example, Dirk et al. examined chromophores using the low aromaticity of a thiazole ring.[72] They noted an enhanced $\beta(0)$ relative to that for DANS and attributed the improved nonlinearity to the lowered aromaticity of the heterocyclic thiazole ring. Cheng et al. have also examined chromophores in which a benzene ring of the bridge was replaced with a heterocyclic furan or thiophene ring.[71] They observed red shifts in the optical absorption spectra of the chromophores and enhanced hyperpolarizabilities compared to the aromatic analogs, as did Dirk. Jen et al. have shown in a similar spirit that replacement of both benzene rings with thiophene rings, in analogy to DANS, results in a twofold increase in $\mu\beta$, where μ is the dipole moment of the molecule.[73]

End Groups and Aromatic Stabilization Other compounds, including types of substituents that gain aromaticity upon charge separation, have been investigated in our group, such as those reported earlier in Table 6.1. It appears that for compounds 1.3 and 1.4, such strong acceptors, coupled to a short polyene, induced too much ground-state polarization to maximize β. Nevertheless, the elongation of the polyene bridge without changing the end groups results in an increase in the nonlinearity that is due, first, to the increasing number of π-electrons in the system, and second, to the fact that the effective ground-state polarization decreases with chain lengthening. This strategy has allowed us to design longer donor–acceptor polyenes, including another efficient acceptor: the 3-phenyl-5-isoxazolone (Table 6.3).[74]

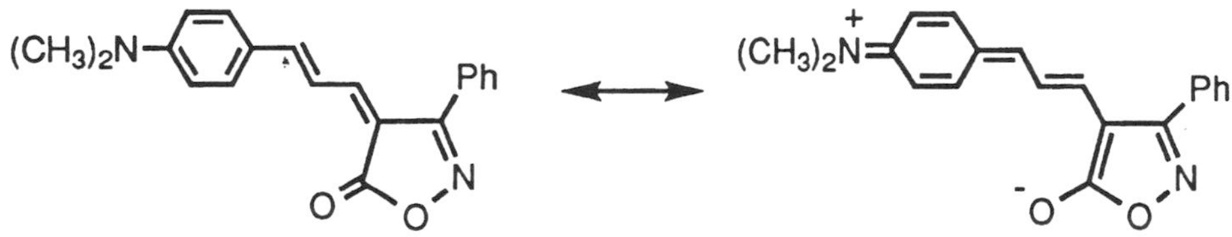

The single-crystal x-ray structures of compounds 3.2 and 3.3 have been determined and provide evidence that these compounds present an appropriate reduced BLA, which can result in a very large nonlinearity β.

Ikeda et al. have reported $\mu\beta$ values for several merocyanines (also known as Brooker dyes[75–77]) with a thiobarbituric acid acceptor.[78, 79] In these cases, where both end groups gain some aromatic stabilization in the charge-separated form, the ground-state polarization is expected to be very large, and accordingly, some of these dyes exhibit negative $\mu\beta(0)$ values. Recently, merocyanines have been examined carefully.[80, 81] The somewhat random or unpredictable behavior noted earlier is explained easily within the framework of the structure–property relationships described above.

Thermal Stability Issue Some earlier rules of thumb about factors that lead to chromophore thermal instability have recently been shown to be inaccurate. For example, it was generally believed that in the common dye, Disperse Red-1, whose structure is similar to DANS, the nitro group was the weak point for thermal decomposition. However, it was recently shown that the alkyl amino functionality was the culprit and that for many chromophores with amino donors, replacement of the alkyl functionality on the nitrogen with an aryl group (such as phenyl) could improve the thermal stability dramatically.[82] In other studies it was shown that for molecules containing a thiophene bridge rather than a simple polyene bridge, it was possible to achieve excellent thermal stability without a major sacrifice of optical nonlinearity.[73, 83, 84] These thermal stability issues are particularly relevant in the context of developing poled polymers materials for electro-optic devices and are discussed in more detail on pages 235–238.

TABLE 6.3 Optical (λ_{max}) and EFISH Results from Measurements in Chloroform at 1.907 μm [Dispersion-Corrected $\mu\beta(0)$]

Compound	λ_{max} (nm)	$\mu\beta(0)$ (10^{-48} esu)
3.1	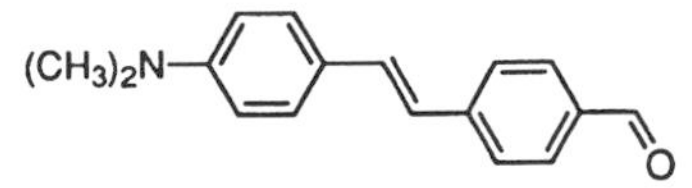 360	70
3.2	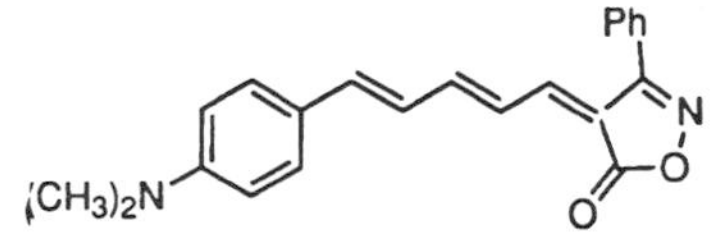 562	1,895
3.3	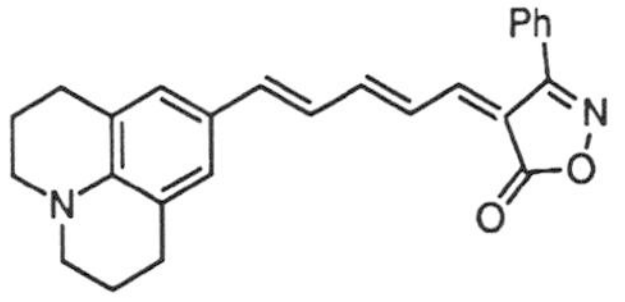 620	3,000
3.4	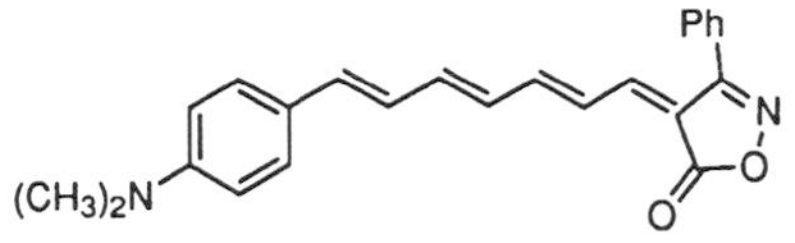 582	4,696
3.5	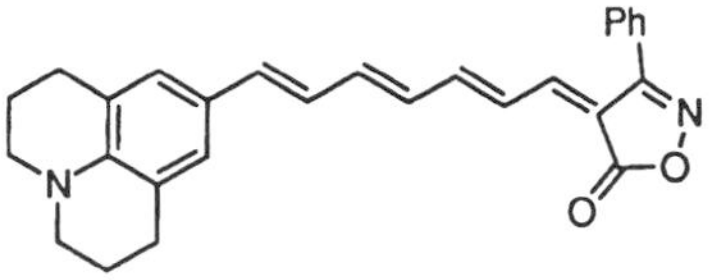 640	4,753
3.6	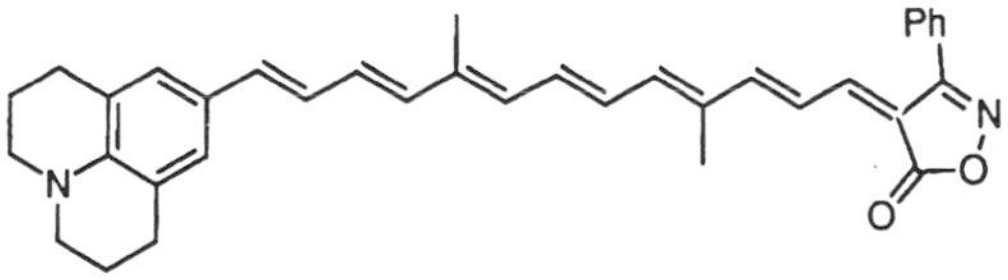 647	13,600

Octupolar Molecules The classic strategy for designing second-order nonlinear optical chromophores is to attach a π-donor and a π-acceptor to a conjugated bridge in an essentially linear geometry, as discussed extensively above. This strategy results in *polar* molecules with hyperpolarizability β. The polarity of such molecules is a disadvantage for obtaining noncentrosymmetric crystals, because the dipole–dipole interaction favors antiparallel alignment, all other factors being equal. Recently, by using an irreducible tensor decomposition, Zyss has shown that β is comprised uniquely by a

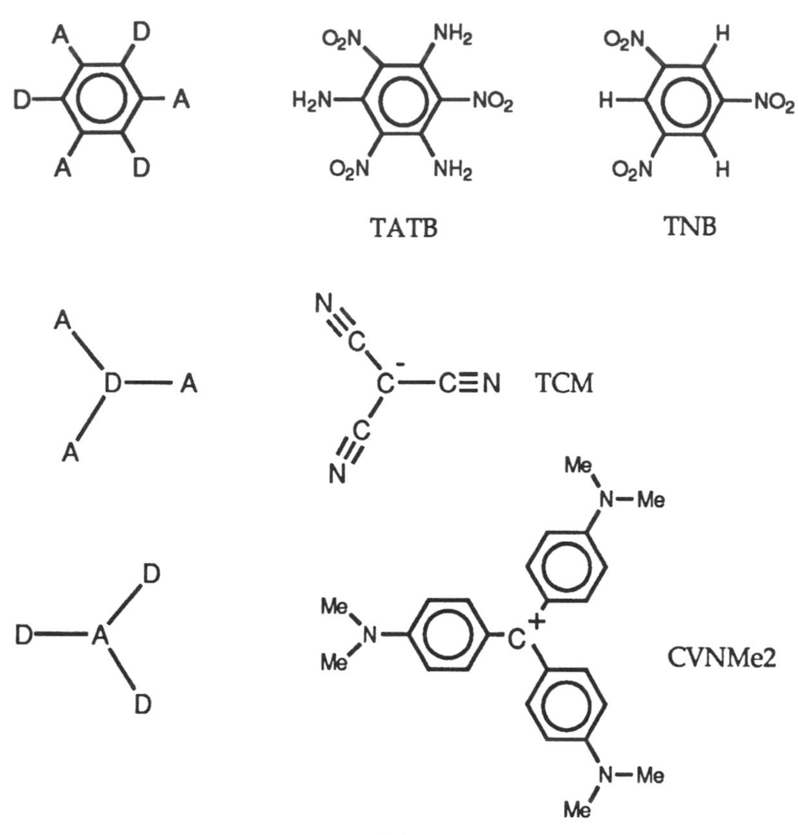

Figure 6.11 Plane template for donor (D) and acceptor (A) in octupolar symmetry, and examples of compounds reported in the literature.

vector part and an octupolar part.[85] Consequently, he has proposed that certain *nonpolar* molecules with threefold rotational symmetry (octupolar molecules) can exhibit nonzero β (Fig. 6.11). Octupolar molecules hold potential for noncentrosymmetric crystallization, reduced aggregation, large off-diagonal tensor components, and improved nonlinearity–transparency trade-offs.

A prototypical octupolar molecule is 1,3,5-triamino-2,4,6-trinitrobenzene (TATB). Despite being nonpolar, this material exhibits a powder SHG efficiency three times that of urea, indicative of a nonzero molecular β.[86] From an experimental point of view, β of a closely related structure (TIATB)[87] has been measured using the hyper Rayleigh scattering (HRS)[88, 89] method. This technique is based on a purely optical nonlinear scattering process and does not require application of an external field or dipole alignment as for EFISH. This study includes comparison to the dipolar analog [i.e., the paranitroaniline (*p*-NA)]. The results presented show that the octupolar molecule possesses a β value two times larger than that of the related dipolar molecule. This is in agreement with the prediction made by Zyss[85] and the theoretical calculations[90] reported for these compounds.

To gain some insight into the structure–property relationships of such octupolar compounds, Joffre et al. have performed some model calculations attempting to compare the influence on the hyperpolarizability β, of dipolar versus octupolar distributions of donors and acceptors around a benzene ring.[91] The results are summarized in Fig. 6.12. The peak function obtained for the dipolar distribution is in qualitative agreement with the relationship discussed earlier, showing again that there is an optimal combination of donor and acceptor strengths required to maximize the hyperpolarizability (in the present work, the strength of the end groups is expressed by the site energy Δ). This conclusion holds for the octupolar distribution, with a maximum observed for a different combination of donor–acceptor (i.e., a different Δ value; the smaller value implies that a combination of weaker donor and/or acceptor should provide a larger octupolar response). Another point illustrated by these results is that the maximum β value is equivalent in both distributions. Finally, the factor 2 in the preceding paragraph referring to the ratio between the β values obtained for TATB and p-NA, is in the context of Joffre's work, expected to belong to the region on the left inside of Fig. 6.12 (i.e., where the systems are substituted with relatively weak donors and acceptors).

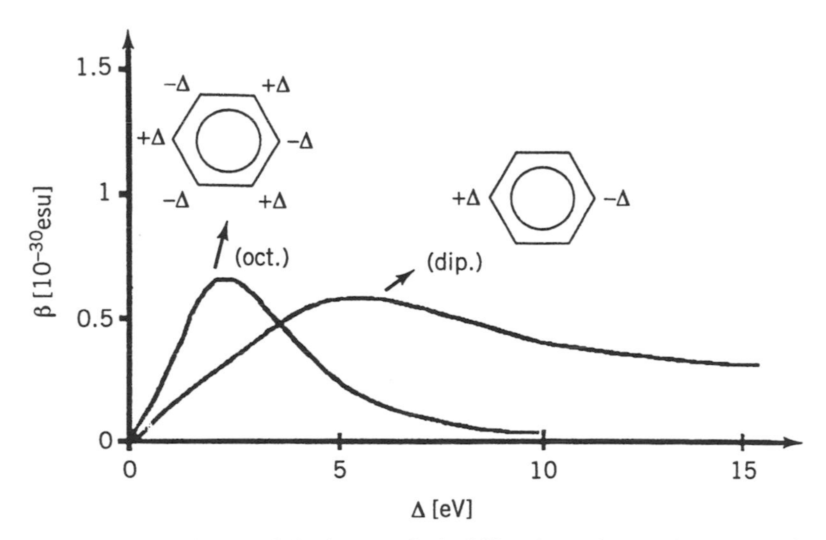

Figure 6.12 Dependence of the hyperpolarizability β on the on-site energy, Δ, and on dipolar and octupolar distribution. $\Delta = 0$ eV corresponds to a nonsubstituted benzene, $+\Delta$ to a donor substituent, and $-\Delta$ to an acceptor; the larger Δ is, the stronger are the end groups. (Adapted from Ref. 90.)

Although several works have reported nonnegligible hyperpolarizability β for octupolar systems, an opportunity for designing new octupolar compounds exists. Besides the work reported by Joffre et al. on an "aromatic" octupolar system, no structure–property relationship has been clearly defined for such nondipolar systems.

Λ-Shaped Molecules The concept of Λ-shaped molecules, presented several years ago, can be seen as an extension of the design strategy of linear donor–acceptor compounds.[92, 93] For the fabrication of a real device, the linear (one-dimensional charge transfer) molecules have to be incorporated into a condensed phase, such as a polymer matrix, a Langmuir–Blodgett (LB) film, or even a crystal. The orientation of the NLO molecules is a decisive point here. Nevertheless, each system has drawbacks that cause a significant decrease in the second-order optical nonlinearity: (1) in polymers, the molecules undergo orientational relaxation; (2) in LB films, they tend to incline from normal to the substrate; and (3) because noncentrosymmetric crystals must be aligned in the phase-matched direction, only 38% of the main component β_{xxx} of the one-dimensional charge transfer is ultimately exploited.

To overcome these losses of activity in NLO materials, Miyata and co-workers have designed two-dimensional charge-transfer molecules, called Λ-shaped molecules.[93] In such molecules, two independent donor–acceptor moities are linked so that the intramolecular charge transfer contributing to the hyperpolarizability has a two-dimensional character (Fig. 6.13). Consequently, the off-diagonal component β_{xyy} becomes nonnegligible and contributes to the macroscopic nonlinearity.

6.4.3 Design of Third-Order NLO Molecules

In early work in this field, the design of third-order materials focused on extended π-conjugated systems.[94, 95] Contrary to the second order, no symmetry requirement restricts the type of structures having potentially high second hyperpolarizability. Oligomers and polymers have thus been widely studied: for example, polyacetylene, polydiacetylene, polyarylenes (such as polythiophene and polyaniline), and poly(arylenevinylenes) [such as poly(*p*-phenylenevinylene)]. The issues of substitution effects, chain-length dependence, and extension of the π-system into a two-dimensional (such as phthalocyanines) and three-dimensional (such as fullerene) framework, are presently under investigation.

Our goal in this section is not to review the literature related to this particular field, but primarily to present through some examples how the molecular structure appears to be, as for the second order, a key parameter that controls the nonlinearity. The structure–property relationships for third-order effects are more complex than for the second-order effects, due to the contribution of multiple excited states and terms entering with

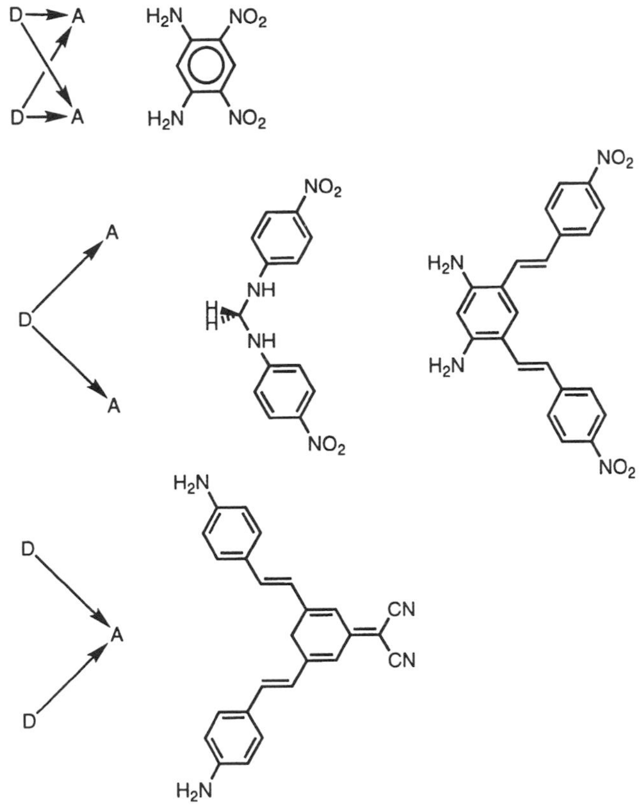

Figure 6.13 Plane template for donor (D) and acceptor (A) Λ-shaped molecules, and examples of two-dimensional charge-transfer compounds.

opposite signs.[50] (See the corresponding SOS, or more simply, the three-term expressions, presented in Section 6.3.)

***Ground-State Polarization and* γ** As discussed briefly in Section 6.4.1, in donor–acceptor-substituted polymethines, γ is related to the ground-state polarization. The γ curve presented in Fig. 6.8 has two positive peaks associated with two intermediate forms that appear (1) between the neutral-polyene and cyanine-like structures and (2) between the cyanine-like and fully charge-separated forms. At the cyanine limit, γ has a large negative peak. The three-term-model expression presented in Section 6.3 accounts for this evolution, and the contribution to the overall response of each individual term is presented in Fig. 6.14.

The origin of the N term can be compared to the polarizability α. In the two-state-model expression, α is proportional to M_{ge}^2/E_{ge}. It is well established that α is maximum at the cyanine limit, where in fact M_{ge} is

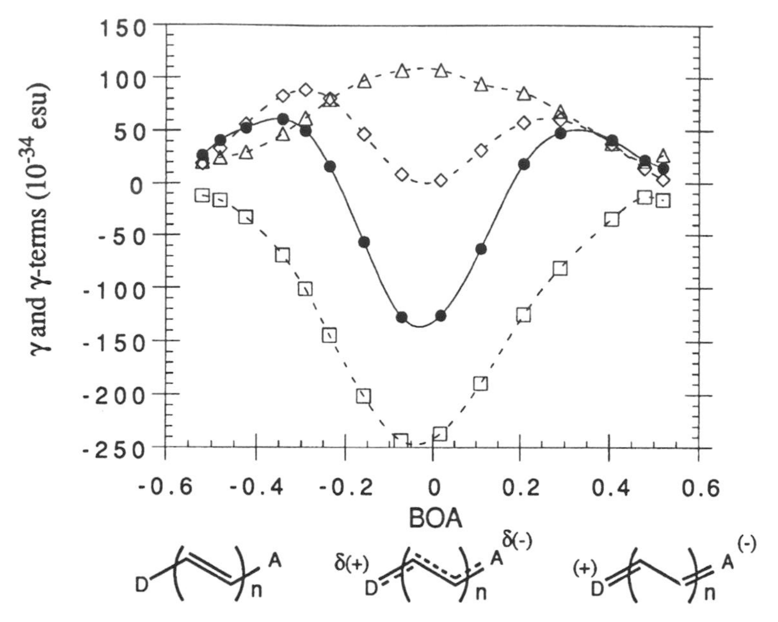

Figure 6.14 Evolution of the three-term-model γ values (full circles), the D (diamonds), N (squares), and T terms (triangles) plotted versus BOA, as calculated for $Me_2N—(CH{=}CH)_4—CH{=}O$.

maximized and where $1/E_{ge}$ is also maximized compared to the polyene limit. Cyanine molecules have been shown experimentally[96] and theoretically[97] to have the largest polarizability displayed by linear polymethine dyes (defined here as dyes with linear arrays of $\leftarrow CH{=}CH \rightarrow_n$, including polyenes, cyanines, merocyanine, etc.). Thus the large (and negative) contribution of the N term to the hyperpolarizability γ, should not be surprising.

The D term is important since it is responsible for the two positive peaks; if it is neglected or if it vanishes (which is the case in a centrosymmetric system), when starting from the polyene limit and increasing the field and thus BOA, γ would immediately start decreasing and then become negative without showing any positive extremum. The D term provides a positive contribution that is maximal in between the polyene and cyanine limits and in between the cyanine and zwitterionic structures; the model is thus able to rationalize the reason why, near the polyene limit, the substitution of a polyene molecule by donor and acceptor groups results in an increase in (positive) γ, as discussed by Garito et al.[52]

Finally, the T term is positive and presents a maximum at the cyanine limit. It partially counteracts the influence of the N term, which also peaks at the cyanine limit. In the strongly bond-alternated structures, the T term is larger in magnitude than the N term, leading to positive γ. With decreasing BOA, the N term increases rapidly in magnitude, reaching a peak at the cyanine limit, where $\Delta\mu$ and therefore the D term vanishes; globally, γ is then negative since the magnitude of the N term is larger than that of the T term, due to the very large values of $1/E_{ge}$ and M_{ge} relative to $1/E'_{ge}$ and M'_{ee}.

Such concepts are relatively new in the field, but experimental evidence such as that presented in Table 6.1 confirms this evolution.[57] These relations are certainly not universal and probably more valid for a one-dimensional system, such as (substituted) polyene or cyanine. The three-term model accounts mainly for the prescriptions proposed by C. Dirk et al., but unifies them in a more general and smoother evolution.[53]

Dipolar Enhancement The enhancement of the third-order response by including a significant D-term contribution to γ (term proportional to $\Delta\mu$), as in noncentrosymmetric and dipolar donor–acceptor polyenes, has been exploited to design substituted carotenoids with very large nonlinearities (Table 6.4).[98, 99] This idea, first proposed by Garito et al.[52] as presented in the model, suggests that coming from the extreme polyene limit and increasing the ground-state polarization by, for instance, using end groups of stronger donor or acceptor strength, can result in enhanced γ. This strategy has guided the investigation (including synthesis and THG measurement) of the compounds reported in Table 6.4, where the relative susceptibilities are reported, comparing values obtained at the peak of the three-photon resonance of each compound. Even considering that each of these values is on resonance, they still should be considered as quite large, since β-carotene itself has a γ value well off resonance of about $11{,}000 \times 10^{-36}$ esu.

Negative-γ Compounds It was thought by many that molecules with negative γ were rare. Aside from cyanine molecules and the very strong donor–acceptor polyenes reported in Table 6.1, few systems with negative γ value have been identified. Squaraine-type molecules have this unusual property.[100-103] This can be explained within the context of the previous model if we simply realize that such molecules can be represented as an equal mixture of resonance forms:

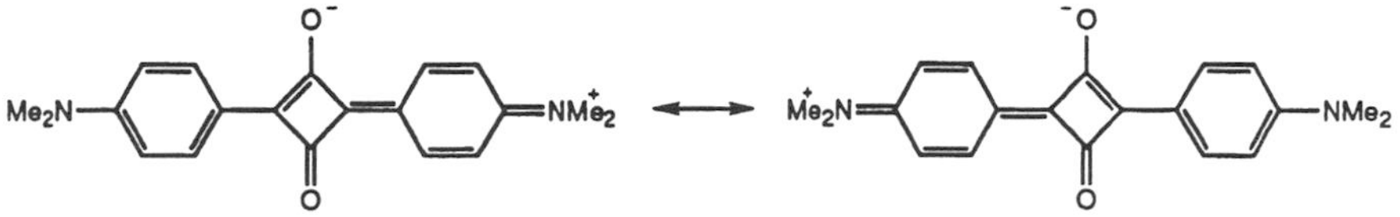

With such a representation it is easy to compare a squaraine-type molecule with a cyanine, having no bond alternation and consequently, a negative γ value. Experimental results confirm the negative sign of γ in squaraine-type

TABLE 6.4 Relative THG Results from Measurements in Chloroform

Compound	Relative Susceptibility [a]
4.1	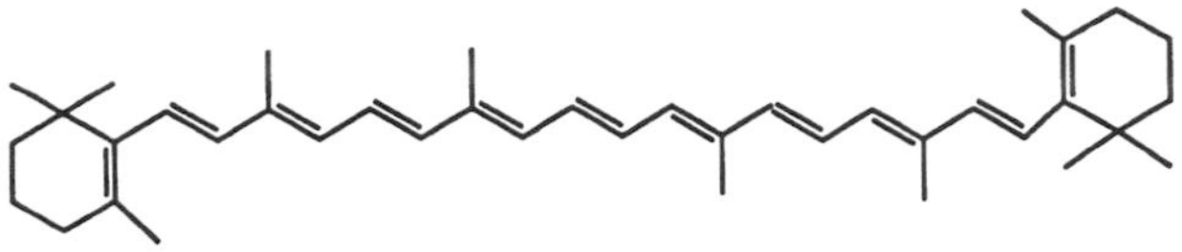 1
4.2	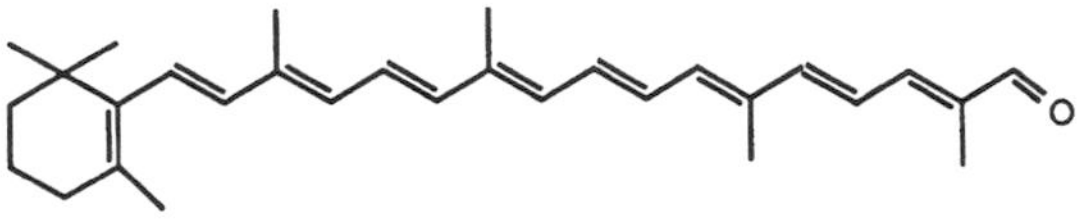 2.2
4.3	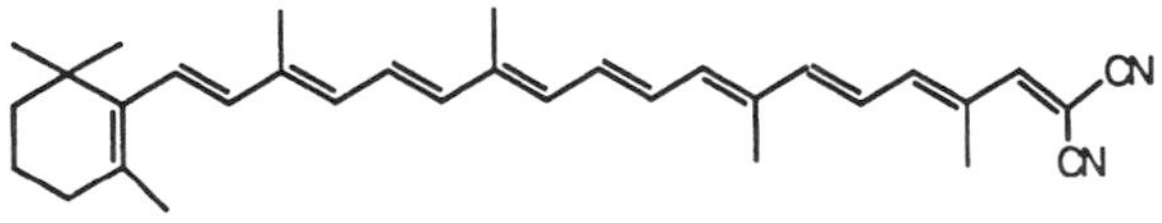 2.4
4.4	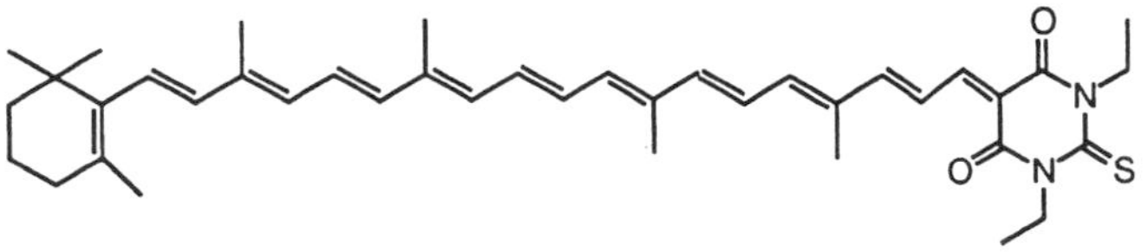 5
4.5	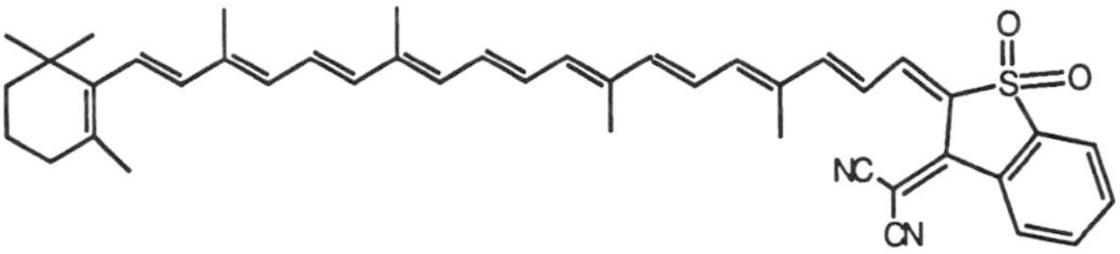 35

[a] The relative susceptibilities compare values obtained at the peak of the three photon resonance of a series of carotenoids substituted with acceptor end group of increasing strength.

molecules. Moreover, it appears that the negative sign comes from a balance mainly between two types of contributions: one corresponding to the N term, and a second including the participation of two-photon excited states (T term).

Chain-Length Dependence Issue Theoretical and experimental studies have demonstrated significant enhancement of hyperpolarizability by extending the length of polyenes. The results are usually discussed in terms of a power law dependence as a function of chain length:

$$\gamma \propto N^{a(N)}, \text{ or in other words, } \log \gamma \propto a(N) \log N, \tag{26}$$

where N is the number of monomer units along the chain (or alternatively, the number of π-electrons, or a length). A series of results provide an exponent, a, which usually ranges between 3 and 5.4 for short chains (the exponent approaching 1 corresponds to a saturation of the nonlinear response). It is worth mentioning that the exponent itself is dependent on the chain length; therefore, care should be taken when comparing different π-systems as well as the same π-systems over various length ranges. The enhancement of γ with chain lengthening varies with the nature of the π-system (e.g., polyene versus polythiophene versus polyparaphenylene),[95] but also with the nature of the endcapping groups (e.g., particularly donor–donor substituents versus acceptor–acceptor versus donor–acceptor).[104-107]

Excited-State Enhancement Another enhancement mechanism for the nonlinear optical processes in π-conjugated systems has been realized through population of the electronic excited states by optical excitation. In other words, the reference state in the sum-over-states expression (22) is no longer the ground state but becomes a low-lying excited state. Garito et al. have shown that by optically exciting α,ω-diphenylhexatriene, at 355 nm (into the $1B_u$ and the slightly lower energy $2A_g$ excited states), the NLO response can be enhanced by more than two orders of magnitude.[108, 109] Moreover, the excited-state absorption at the probe wavelength (DFWM experiment performed at 1.064 μm) was reported to be negligible, and a γ value of $12{,}000 \times 10^{-36}$ esu was obtained, compared to 50×10^{-36} esu, the value measured for the ground state.

By examining the SOS expression, it is easy to understand the origin of such an enhancement when the reference state is no longer g, but e. First, it appears that the energy differences between excited states, E'_{ee}, tend to be smaller than E_{ge}. Second, the number of excited states strongly coupled, and therefore leading to a significant contribution to γ, may be larger (in other words, the number of terms that have significant contributions in the summation is larger).

6.5 MATERIALS DESIGN

6.5.1 Second-Order NLO Materials[110]

As noted earlier, second-order NLO effects occur only in molecules lacking an inversion center. Similarly, a bulk material comprised of nonlinear molecules must also lack an inversion center if the molecular β is to lead to a macroscopic susceptibility $\chi^{(2)}$.[111] Furthermore, the materials must exhibit low optical losses from either absorption or scattering, and they must be orientationally, environmentally, and photochemically stable. In addition, if they are to be incorporated into devices, it is necessary for them to be processible. In this section we examine some of the strategies directed toward synthesizing noncentrosymmetric organic crystals, poled polymers, and self-assembled multilayers.

Organic Crystals As mentioned before, NLO materials for second-order applications should be noncentrosymmetric. Also, it is well known that dipolar molecules (possessing high hyperpolarizability β) tend to crystallize in a centrosymmetric space group. In fact, 75% of all known achiral organic molecules crystallize in centrosymmetric space groups and thus have no net $\chi^{(2)}$.[112] Crystal engineering approaches have been used successfully to induce acentric organization of NLO chromophores in crystals, based on an understanding of the crystals packing forces: electrostatic, hydrogen bond, van der Waals, and π–π^* stacking interactions.

Some crystal engineering strategies that have been investigated are:

1. Use the chirality of sp^3 tetrahedral carbon atom(s) (e.g., MAP;[113] see Fig. 6.15).
2. Design amphiphilic molecules in which a neutral bulky group alternates with the NLO species, to reduce the probability of antiparallel stacking (e.g., COANP[114]).
3. Introduce functionalities (such as OH) that allow for intermolecular hydrogen bonding (e.g., NPP[115] and PNP[116]).
4. Attach an additional polar group (such as CN) to the NLO species, creating a local dipole independent of that due to the intramolecular charge transfer associated to the nonlinearity (NPAN[117]).

In particular, the hydrogen-bonding strategy provides considerable flexibility in the design of noncentrosymmetric crystals and co-crystals of organic materials that can individually be SHG active or inactive. Resulting from various intermolecular interactions between different molecular species, the process of crystal formation can be viewed as a molecular recognition at the surface of the growing crystals. In that context, Etter et al. have examined a series of nitroaniline materials and found a high occurrence of acentric hydrogen-bond aggregates.[118, 119] Another elegant example is the co-crystallization of *p*-aminobenzoic acid and 3,5-dinitroaniline, which forms a two-dimensional acentric lattice.[120] Noncentrosymmetric crystals of *p*NA mixed with a substituted benzene have also been reported by a Japanese group, to be SHG active. Other promising results of molecular aggregation controlled by hydrogen bonding have been reviewed.[121–123]

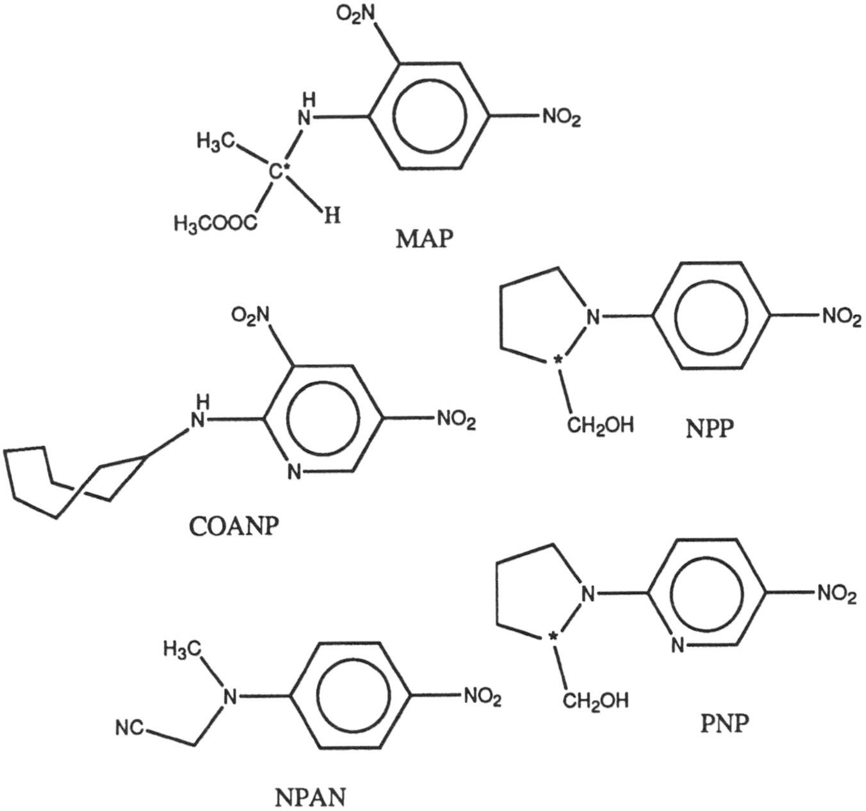

MAP

NPP

COANP

PNP

NPAN

Figure 6.15 Structural formulas of molecular crystals; see the text for references.

Another NLO crystal engineering strategy first introduced by Meredith was to utilize ionic forces in organic crystals.[124] Following this approach, Marder et al.[125–127] and Nakanishi et al.[128] have demonstrated that when crystallized with various anions, cationic chromophores, with large β, often lead to salts with large macroscopic second-order optical nonlinearities. It was found that in a particular class of compounds, 4-*N*-methylstilbazolium salts, the "salt methodology" provides a higher probability of obtaining $\chi^{(2)}$ active crystals, as compared to crystals of conventional neutral dipolar organics. Roughly half of the seventy 4-stilbazolium salts examined by various workers exhibited sizable SHG efficiencies, indicative of a high incidence of noncentrosymmetric packing. More significantly, for most of the nonlinear cations, an anion was found that led to compounds with very efficient SHG. One compound, 4-*N*,*N*-dimethylamino-4′-*N*′-methylstilbazolium toluene-*p*-sulfonate (DAST) (Fig. 6.16), has a powder SHG efficiency larger than any other material to date, 1000 times urea measured at 1.907 μm.[125] For electro-optic applications the largest nonlinear coefficients will occur when

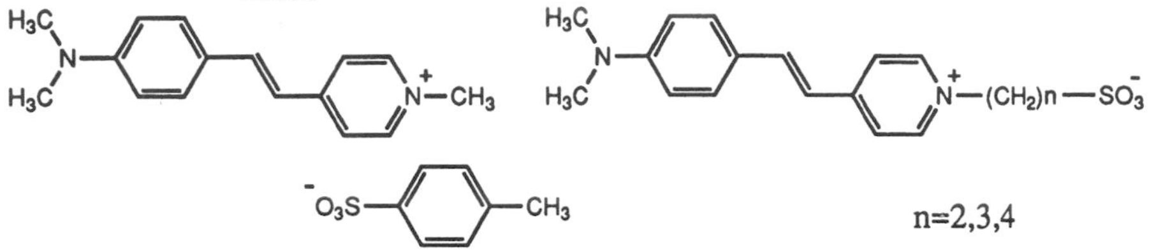

DAST

n=2,3,4

Figure 6.16 Molecular structure of DAST and 4-substituted 4′-(-sulfonato-alkyl)stilbazolium.

all of the molecules are aligned in the same direction. In crystalline DAST, the only deviation from a completely aligned system is the 20° angle between the long axis of the cations and the polar *a*-axis of the crystal. Donor–acceptor, hydrogen-bonding, and π–π stacking interactions often yield polar sheet structures. Marder et al. and Nakanishi et al. have suggested that the formation of alternating cationic and anionic sheets can facilitate the formation of macroscopically polar structures. Thus the exceptional powder SHG efficiencies and the excellent alignment of the chromophore in the crystal suggested that crystals of DAST could have unusually large electro-optic properties. In a preliminary determination, Yakymyshyn et al. reported the electro-optic coefficient, r_{11}, of DAST to be 400 $\pm$ 150 pm/V for frequencies up to at least 100 kHz.[129] More recently, Bosshard et al.[130] have performed measurements on high-quality crystals and have determined an r_{11} of 160 $\pm$ 50 pm/V which is still one of largest values for any organic material. The extremely large coefficients measured for DAST and the high incidence of noncentrosymmetric crystal packing demonstrate the tremendous potential of organic salts crystals for use in applications such as optical signal processing, communications, and interconnect applications requiring highly nonlinear, fast, thermally stable materials.

Another class of stilbazolium derivatives in which the anionic and the cationic part are linked together by a *n*-akyl chain (Fig. 6.16) have been reported to be efficient for SHG.[131] In such an intrinsic zwitterionic system, the anionic and cationic species create a large dipole moment that is independent of that due to the intramolecular charge transfer.

Poled Polymers The method most widely used to achieve noncentrosymmetry artificially, and closest to commercialization, is the poled-polymer approach.[132–134] If dipolar NLO species are dissolved in a polymer and subjected to a large electric field, at or above the temperature at which the polymer becomes rubbery (called its glass transition temperature, T_g), the interaction of the dipole with the field causes the dipolar species to align, to a certain extent, in the direction of the applied field. If the polymer is cooled back to the glassy state, with the field applied, the field-induced noncentrosymmetric alignment can be frozen in place, yielding a material with a second-order optical nonlinearity.[135] However, this thermodynamically unstable alignment quickly decays in polymers such as poly(methylmethacrylate), resulting in a greatly reduced nonlinearity. Thus it is necessary to design systems that would lock the aligned molecules in place more effectively, even at elevated temperatures, and to incorporate more highly nonlinear molecules. In comparison to inorganic crystal-based electro-optic devices, organic polymer devices have the potential for lower-power, higher-speed switching and, in addition, could be less expensive to manufacture since polymeric devices are amenable to automated mass production.

For applications where the electro-optic polymers are used as interconnects on integrated circuits, the polymer must have extremely high thermal stability (since it will have to survive several soldering steps during processing, as well as high operational temperatures) but only modest nonlinearity. In contrast, for applications that demand high sensitivity and low power consumption, such as spatial light modulation, moderate thermal stability is adequate but large nonlinearity is essential. Accordingly, it is possible to distinguish research on poled polymers according to two goals: (1) polymers with large nonlinearities (electro-optic coefficient greater than 30 pm/V) that have long-term thermal stability at 85°C, and (2) polymers with moderate nonlinearity that are stable at high temperatures (at 300°C or greater) for short periods and 105°C for long periods.

Researchers have adopted two approaches to improve the long term stability of molecular alignment in poled polymers. First, several groups have used polymers with very high glass transition temperatures as host matrices (Fig. 6.17). The rationale is that if the T_g of the polymer is roughly 150 to

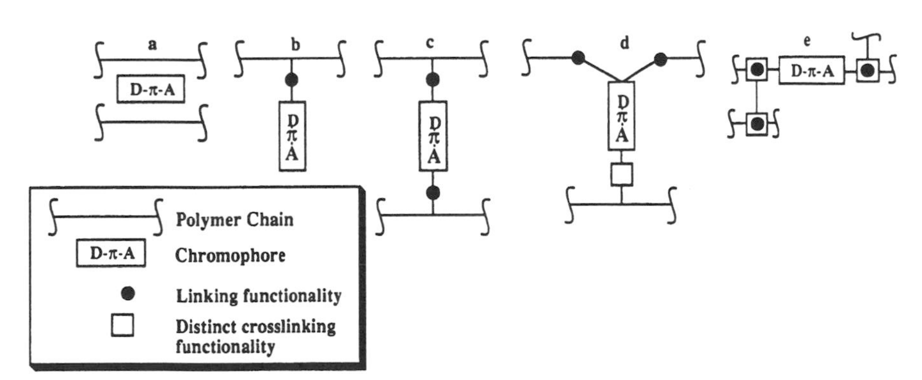

Figure 6.17 Schematic illustration of (*a*) a host–guest polymer; (*b*) a chromophore appended polymer; (*c*) a chromophore cross-linked resin; (*d*) a chromophore appended polymer, subsequently cross-linked with functionality attached to the chromophore; (*e*) a chromophore containing main-chain prepolymer, subsequently cross-linked with an additive, that reacted with functionality in the prepolymer chain. (Adapted from ref. 134.)

TABLE 6.5 Optical (λ_{max}) and EFISH Results from Measurements in Dioxane (for 5.1 and 5.2) or Dichloromethane (for 5.3 and 5.4) at 1.907 μm [Dispersion-Corrected $\mu\beta(0)$]

Compound	λ_{max} (nm)	$\mu\beta(0)$ (10^{-48} esu)
5.1	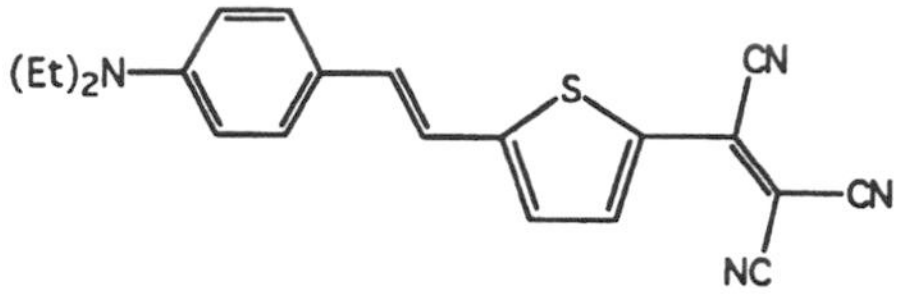 640	3032
5.2	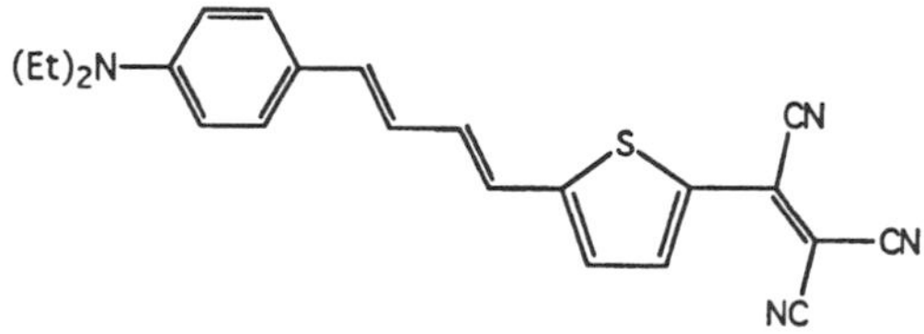 662	4146
5.3	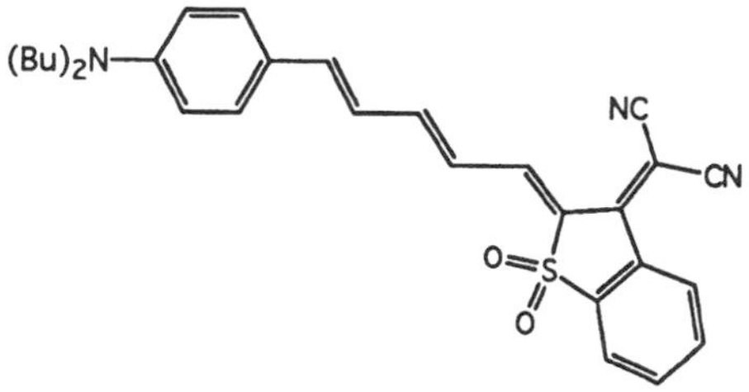 770	3900
5.4	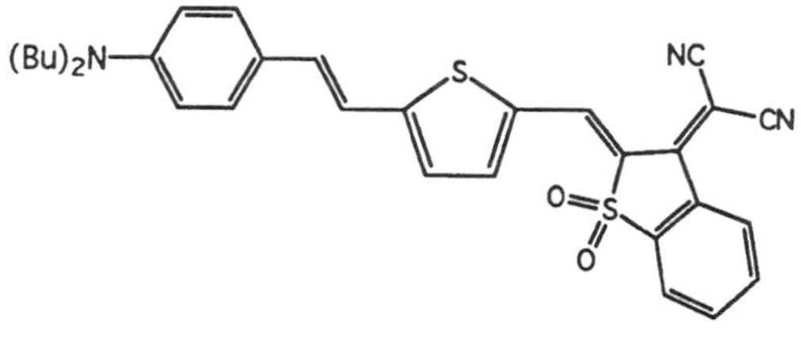 744	5000

200°C above the ultimate operating temperature, decay of the alignment would be negligible over the device lifetime. This was first demonstrated with a commercial polyimide and NLO chromophore,[136] and more recently with highly nonlinear heteroaromatic chromophores (Table 6.5).[137–140] Jen and co-authors have developed classes of chromophores based on the combination of the efficient thiophene conjugated unit and donor and acceptor groups. The chemical stability of the dye under high-temperature processes time and temperature for imidization (30 min at 200°C and for poling: 10 min at 220°C) was monitored by observing the characteristic charge-transfer band in the UV-vis absorption spectrum.[137] The effect of physical aging was also studied. The Jen group has reported long-term stability at the elevated temperatures of 120°C (150°C) for more than 30 days, with the electro-optic coefficient maintained at 80% (60%) of its initial value. Moreover, the semiconductor industry is already accustomed to use polyimide as photoresists for and in packaging of integrated circuitry; thus for integrated optics applications it may be possible for polyimide-based electro-optic polymers to be easily incorporated into devices. The Jen group has recently reported large electro-optic activity and long-term stability of poled guest–host polyquinoline thin films.[139] A large electro-optic coefficient r_{33} of 26 pm/V was retained at 80°C for more than 80 days (by comparison, for lithium niobate r_{33} is 30.8 pm/V[141]).

The motion of chromophores in polymer matrices might be further restricted if the chromophore is covalently attached at one or more sites to the polymer (Fig. 6.17). Accordingly, several groups have explored covalent attachment of chromophores to polyimides in either the main or as a side chain. This approach can lead to polymers with exceptional thermal stability. Jen's group has reported a "facile" two-step synthesis for NLO side-chain aromatic polyimides. The obvious advantages of this procedure include the ease in controlling the loading level of the chromophore and adjusting the polymer backbone rigidity.[142]

However, it appears that the ultimate achievable stability is related to the inherent flexibility of the polymer backbone and the degree of freedom in the linkage between the chromophore and the polymer (i.e., whether the chromophore is attached in the main chain or in a side chain). Thus systems with rigid backbones and short rigid linkers will tend to have the highest T_g and also the best temporal stability. A possible disadvantage of greatly increased backbone or side-chain rigidity is that it may lead to some degree of crystallinity, which can result in significant optical losses due to scattering.

In various systems the cross-linking process can take place at temperatures under 200°C; however, in the high-T_g polymer approach, the minimum poling temperature being considered seriously is 225 to 250°C, and for the integrated optics applications, poling temperatures in excess of 350°C may be necessary.[143–145] The latter thermal requirements are extremely demanding for organic molecules, and for chromophores with extended conjugation, stability even at 200°C is nontrivial. Accordingly, as discussed earlier, one active avenue of research is to identify chromophores that are both highly nonlinear and thermally robust.

The molecular hyperpolarizability/thermal stability trade-off has been studied most systematically by the group from IBM Almaden. In particular, they reported the general observation that replacement of aliphatic diakylamino donor groups with diarylamino groups in NLO chromophores always results in an increase in thermal stability, and sometimes in ample nonlinearities.[82] Recently, they reported the preparation of a soluble NLO-substituted polyimide derivative containing this type of diaryl amino donor head embedded in a polyimide main chain.[146, 147] The resulting polymer is characterized by extraordinary stability of the induced polar order, due partially to the high glass temperature ($T_g \sim 350°C$) of the system.

More recently, dyes that contain a very strong heterocyclic acceptor, the Foron Blue, have also been synthesized (Table 6.5) in our group.[140] The efficiency of a polyene bridge (compound 5.3) versus a thiophene (compound 5.4) has been demonstrated, the latter being more thermally stable. The results on compound 5.4 demonstrate that it is a potent chromophore to be incorporated into high-T_g polymers. The development of functionalized analogs of compound 5.4 for covalent attachment or cross-linkage polymers are under investigation.

Self-Assembled Monolayers Another approach to assembling acentric materials involves the covalent self-assembly of intrinsically acentric multilayers of NLO chromophores on an inorganic oxide substrate. Compared to the more traditional Langmuir–Blodgett films, these self-assembled thin films are thermodynamically stable; also contrary to the poled-polymer technique, no electric field is required to induce the acentric alignment of the chromophores. The preparation of highly ordered films based on trichlorosilane derivatives has been reported.[148, 149]

Marks et al. have shown that layers of oriented chromophores can be assembled sequentially on a glass slide with the chains in a given layer cross-linked.[150, 151] Subsequent layers are covalently attached to the prior layers. Multilayer structures have been built up with a repeating reaction sequence, giving alternating stilbazolium chromophore layers and soft polysiloxane layers that connect the molecules transverse to the stacking direction, thus providing stability.

Katz has developed another approach involving assembly of zirconium phosphonate layers.[152] Silicon and glass substrates were phosphorylated and then reacted with $ZrOCl_2$, followed by reaction with 4-{4-[*N*,*N*-bis(2-hyroxyethyl)amino]phenylazo}phenylphosphonic acid to generate a first monolayer. This layer could then be phosphorylated with $POCl_3$ and the previous reaction sequence starting with $ZrOCl_2$ could then be repeated. Twenty-five layers were readily assembled. Plots of the thickness, absorbance, and square root of second-harmonic intensity versus number of layers all gave straight lines, indicating that subsequent layers were well ordered. The films exhibited good thermal stability, showing a diminution of the second-harmonic intensity only above 150°C, at which point the spectroscopic evidence suggests that the chromophore itself began to decompose.

Photorefractive Polymers and Crystals The demonstration of photorefractivity in polymers by Moerner and co-workers at IBM in 1991 has sparked tremendous interest in the organic NLO community.[153] Photorefractive (PR) effects in inorganic electro-optic crystals such as $BaTiO_3$, $LiNbO_3$, and GaAs have been known for many years. In recent years, photorefractive effects in organic materials were observed first in doped electro-optic organic crystals[154, 155] and then in photoconductive, nonlinear optical polymers.[153] Because of their potential for large photorefractive figures of merit and the processing advantages afforded by polymeric materials, work on photorefractive polymers has taken off. Research in this area was reviewed in detail by Moerner and Silence in 1994.[156] In this section we provide a brief introduction to PR polymers and highlight some of the significant recent advances.

The PR effect is associated with the spatial modulation of the refractive index resulting from light-induced space-charge fields in electro-optic materials.[156, 157] Such light-induced changes in the refractive index can be utilized for optical data storage, holographic recording, processing and amplification of images, and optical limiting. There are two necessary but not sufficient requirements for true photorefractivity: photoconductivity and electro-optic activity. However, even in materials possessing both of these properties, other effects associated with excited-state populations and photochemical processes, for example, may give rise to effects similar in some ways to the PR effect but will not show the clear signature of true photorefractivity: asymmetric power transfer between two laser beams in a two-beam coupling experiment.

In PR materials, a space-charge field is created by photoexcitation of mobile charge carriers that are preferentially transported out of the illuminated regions of the material and trapped. This leads to the formation of an electric field, which causes an electro-optic refractive index change. As illustrated in Fig. 6.18, two laser beams crossing at an angle in a PR material lead to an interference pattern of alternating light and dark regions, with mobile charge carriers produced in the light regions and trapped in the dark regions. The alternating charged regions create a periodic space-charge field and a refractive index grating that is spatially shifted from the intensity grating. The phase of the photorefractive index grating is shifted by $\pi/2$ relative to the intensity grating. Such photorefractive gratings can efficiently diffract a reading light beam and are the basis for various optical processing and data storage schemes. Furthermore, the phase-shifted index grating, and the asymmetric power transfer between beams that it enables, allows for optical gain (i.e., amplification of a weak signal by power transfer from a stronger beam).

Because the composition of polymeric materials can be readily controlled [e.g., the nonlinear optical chromophore and the type and amount of various molecular dopants (photosensitizers, charge-transport agents, and traps)], and because they can be formed into various thin film (or bulk) structures, there is a great deal of interest in photorefractive polymers. The large electro-optic coefficients (tens to 100 pm/V) and low dielectric constants (in the range 3 to 6) of typical electro-optic organic crystals and polymers hold the potential for large values of the photorefractive figure of merit $(n^3 r/\varepsilon)$, perhaps as high as an order of magnitude above that for the best inorganic crystals. Large electro-optic coefficients, efficient photo-carrier generation,

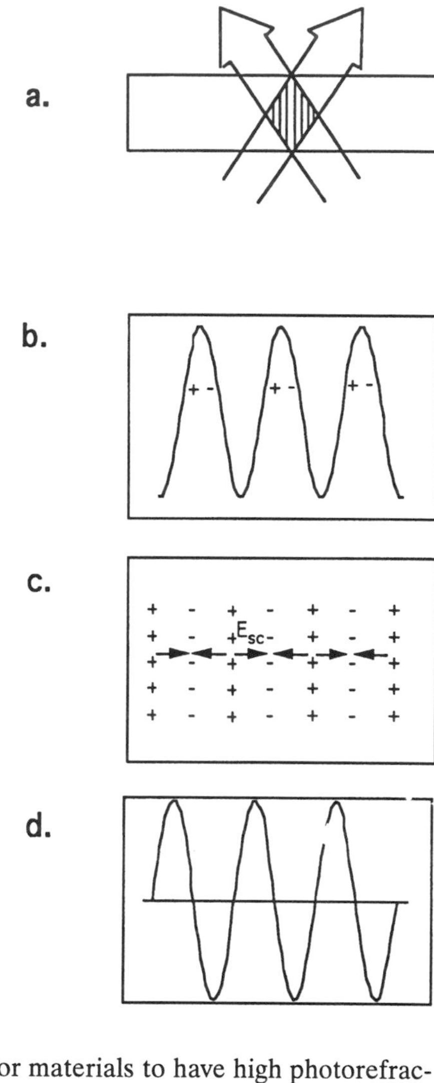

Figure 6.18 Photorefractive effect: (*a*) crossing writing beams create an optical interference pattern; (*b*) photogenerated charge carriers are produced; (*c*) mobile charges are transported and trapped, creating an alternating space-charge field; (*d*) the space-charge field induces an index grating via the electro-optic effect.

and fast charge transport are required for materials to have high photorefractive efficiencies and short response times. These issues are driving much of the current work on photorefractive polymers.

Four approaches have been taken to include the necessary functionalities in photorefractive polymers. The first approach involves a polymer with attached chromophores (either as pendant groups or in the polymer backbone) that possess a finite first hyperpolarizability, which can be doped with charge generating, transporting, and trapping agents.[153] The second involves a charge transporting or photoconductive polymer to which NLO chromophores, charge-generating, and charge-trapping species may be added.[158] The third approach uses an inert polymer binder with sensitizer, charge transport, and NLO chromophores added as molecular dopants.[159] The fourth approach is where the required functionalities are included by covalent attachment to a polymer of all the necessary agents: NLO chromophore, and charge-generating, charge-transporting, and charge-trapping agents.[160] Being synthetically simpler, the first three approaches have been used for most of the PR polymers reported.

The first observation of PR behavior in polymers was reported by Ducharme et al.[153] They used an epoxy polymer, bisA-NPDA, with NPDA being the NLO chromophore, doped with 30% DEH (a charge-transport agent) (Fig. 6.19). The electro-optic response in the polymer was induced by poling with a strong dc electric field. Photocarriers were produced by exciting on the absorption edge of the bisA-NPDA: DEH polymer (i.e., the NLO

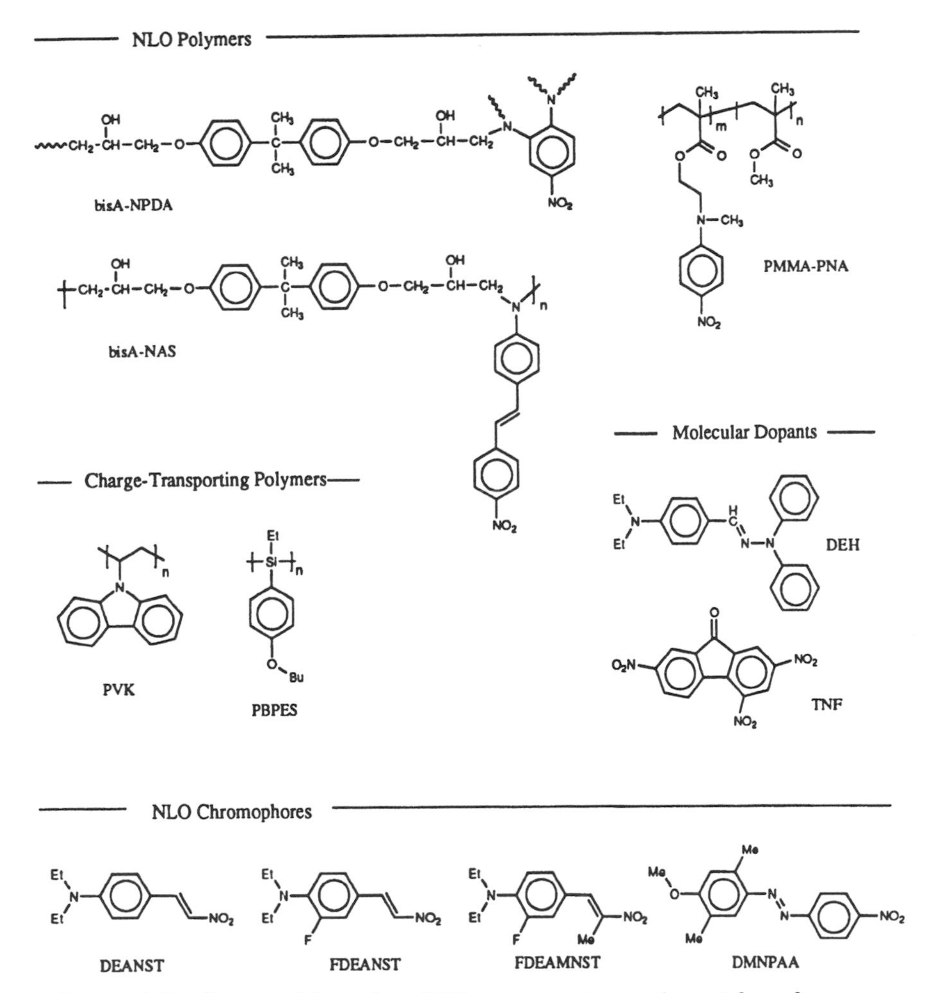

Figure 6.19 Structural formulas of PR components; see the text for references.

chromophore was also the sensitizer). PR gratings with diffraction efficiencies up to 2×10^{-5} (a value of 10^{-3} was reported for a related polymer with a higher electro-optic nonlinearity) were observed with a write beam intensity of 13 W/cm^2 and a poling field of about 120 kV/cm for a 350-μm-thick film. The response time of this PR polymer was about 100 s. Two beam coupling experiments established that the response of the bisA-NPDA : DEH system was truly photorefractive and not due to some other mechanism.[161] Thus this work demonstrated that photorefractive polymers were possible and sparked significant activity in this area.

A wide variety of PR polymers have now been reported, and major advances in both the magnitude and response time of the PR effect in polymers have been made since the original observation. These advances result from several factors, including introduction of sensitizers and use of improved sensitizers, recognition and use of an orientational enhancement mechanism in low-T_g PR polymers, use of chromophores with larger hyperpolarizabilities (and polarizability anisotropies), and maximization of loading with NLO chromophore.

It has been recognized that orientation of the NLO chromophore by the space-charge field and the birefringence that is created can make a major contribution to the photorefractive effects observed, particularly in relatively low-T_g PR polymers, where chromophore orientational mobility is high.[156, 162] This effect may be operative in many PR polymers where chromophores are mobile, whether by design or not. High loading levels of molecular charge transport or NLO chromophore dopants are commonly used, and this can lead to plasticization of a host polymer and thus to high chromophore mobility. As discussed below, intentional incorporation of a plasticization agent to achieve low T_g has resulted in the best PR performance reported to date, and a major contribution to this performance is made by the orientational mechanism.

In the NLO acrylate polymer-based PMMA-PNA : DEH materials (Fig. 6.19), use of sensitizer additives at relatively low (about 0.1 wt %) concentration led to increased diffraction efficiencies and enhanced response times.[163, 164] With C_{60} as the sensitizer, a diffraction efficiency of 4.8×10^{-5} and a response time of 0.25 s was obtained with an applied field of 114 kV/cm. These results were attributed to increased quantum efficiency for carrier generation and an enhanced carrier concentration due to addition of C_{60}. Increased diffraction efficiencies and large PR gain have been obtained in a linear epoxy NLO polymer system. In bisA-NAS : DEH(29%) (with no added sensitizer), improvements in sample fabrication that allowed use of a very strong electric field (600 kV/cm), a photorefractive gain coefficient of up to 56 cm^{-1} has been achieved and a diffraction efficiency of 12% was demonstrated.[165] Although this PR gain was quite high, net gain (when the PR gain exceeds the absorption coefficient) was not demonstrated in this system.

The first PR polymer that used a charge-transporting polymer host was a polyvinylcarbazole (PVK)-based system: PVK : DEANST : C_{60}, where DEANST (32 wt %) was the NLO chromophore and C_{60} (1.9 wt %) was the sensitizer.[158] A diffraction efficiency of 2×10^{-5} was reported for this system. Subsequent work using PVK as a host along with FDEANST as the NLO chromophore and TNF as a sensitizer led to significantly improved performance (a diffraction efficiency of 1% and a 0.1-s response time) and was the first PR polymer to show net two-beam coupling gain (7.2 cm^{-1}).[166] A polysilane system has also been used as a charge-transporting polymer host. The PBPES : FDEAMNST (40 wt %) : C_{60}(0.2 wt %) system showed a very fast grating growth time of 0.04 s and a small net two-beam coupling gain (0.7 cm^{-1}).[167]

In a major breakthrough in PR polymers, Meerholz et al. prepared a PVK-based PR polymer with the composition DMNPAA : PVK : ECZ : TNF (50 : 33 : 16 : 1 wt %), which included a very high loading level of NLO chromophore (DMNPAA) as well as a significant amount of plasticizer (*N*-ethyl carbazole, ECZ) and a small amount of photosensitizer (TNF).[168] This system exhibited by far the highest diffraction efficiency (86%, limited by reflection and absorption loss) and net two-beam coupling gain (over 200 cm^{-1}) ever reported for a PR polymer system. Measurements of the polarization anisotropy ratio of the photorefractive index changes showed that the orientational enhancement mechanism was mostly responsible for the results observed. Thus, in this heavily plasticized, low-T_g system, with an extended chromophore (possessing large polarizability anisotropy, hyperpolarizability, and dipole moment), exceptionally large photorefractive index changes are realized.

With the advances in PR polymer over the past couple of years, the primary performance (diffraction efficiencies and response times) of PR polymers is now competitive with those of leading inorganic photorefractives such as $KNbO_3$ and $BaTiO_3$. These advances have allowed PR polymers to be utilized successfully in device applications such as optical pattern recognition,[169] volume holography,[170] and dynamic holography.[171] Significant issues remain to be addressed, such as long-term stability of the very low-T_g composites, preparation of higher-T_g systems with large photorefractive gain and fast response, and in developing a detailed mechanistic understanding of the phenomena involved in these complex materials. The rapid advances in PR polymers and their use in optical applications over the past few years provides substantial impetus for continued research in this area.

6.5.2 Third-Order NLO Materials

Various classes of organic materials are currently under study as third-order materials: liquids, molecular solids, charge-transfer complexes, π-conjugated polymers, organometallic compounds, organic composites, and liquid crystals have been reviewed.[94, 95] Once again the type of application guides the choice of materials; those possessing large third-order nonlinear optical properties, such as nonlinear refraction and nonlinear absorption, can be used in devices for purely optical switching, modulation, phase conjugation, and limiting.

All-Optical Signal Processing One of the principal applications of the third-order NLO materials is in the area of all-optical signal processing. The particular materials requirements have been reviewed elsewhere.[172] Briefly, a useful figure of merit that defines the optical switching performance of a third-order NLO material is

$$F(\lambda) = \frac{\chi^{(3)}(\lambda)}{\alpha'(\lambda)\tau} \tag{27}$$

where $\chi^{(3)}$ is the third-order susceptibility that determines the response, α' denotes the sum of the linear and nonlinear absorption, and τ is the lifetime of the response. This suggests that the combination of a large magnitude of $\chi^{(3)}$, a small linear and nonlinear absorption, α', and a short lifetime, τ, is desirable for a given device operating at the light wavelength λ. For silica-based optical fibers, the optimal operating wavelength for all-optical switching is 1.5 μm. Therefore, the third-order NLO material should possess no absorption near 1.5 μm (one-photon transition around 0.8 to 0.85 eV), 0.75 μm (two-photon transition at 1.65 eV), and 0.5 μm (three-photon transition at 2.5 eV). The device should also be operational with light intensities less than 1 MW/cm^2, on a path length of about 1 cm, unless the losses can be kept very low, in which case somewhat longer lengths can be used. Consequently, the desired third-order material should possess a $\chi^{(3)}$ on the order of 10^{-8} esu (off-resonance region); such materials have not yet been identified.

Chromophores for Optical Limiting Materials that exhibit optical limiting are currently of interest for the protection of sensors and human eyes from high-intensity laser light as well as for optical pulse shaping and processing applications.[173-176] Optical limiters exhibit high linear transmission at low intensities, and for high intensities the transmission drops to a low value, thus blocking high-intensity pulses and providing protection while allowing good ambient vision. Chromophores that exhibit reverse saturable absorption (RSA) are of great interest for the development of optical limiting materials. RSA occurs when a chromophore has weak-to-moderate ground-state absorption over a given spectral range and strong absorption from optically excited states in the same wavelength range. Accordingly, such chromophores become increasingly strongly absorbing with increasing intensity and can be used in nonlinear absorptive materials, which convert absorbed optical energy to heat. Significant advances have been made in chromophores for

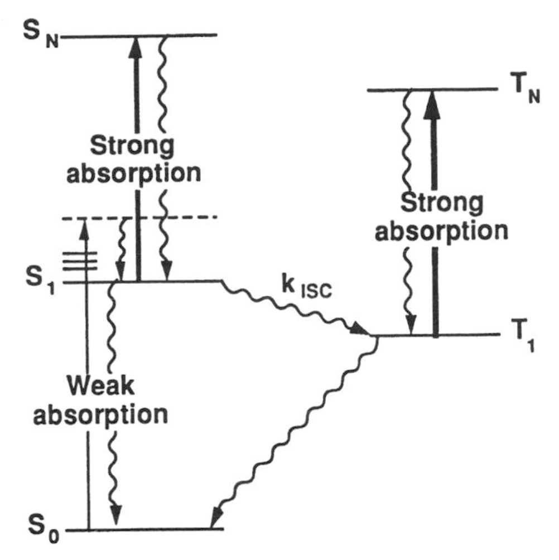

Figure 6.20 Schematic energy-level diagram for a chromophore with a singlet ground state. Parameters are defined in the text.

optical limiting in the last few years, and work in this area has been reviewed.[177, 178] In this section we provide a brief introduction to reverse saturable absorbers and highlight the use of systematic molecular approaches to the optimization of RSA chromophores. Finally, we describe recent efforts toward optimal use of RSA materials in optical limiting devices.

The important states typically involved in the reverse saturable absorption response of chromophores is illustrated in Fig. 6.20. Optical excitation of the chromophore leads to prompt population of the lowest excited singlet state from which excited singlet–singlet absorption can occur. The triplet state, T_1, is populated by a typically slow process: the intersystem crossing from the excited singlet state, S_1. Upon population of T_1, triplet–triplet absorption may occur to higher triplet states. For reverse saturable absorption to occur, the absorption cross section (σ) of the excited singlet state (s) and/or triplet state (t) must be larger than that of the ground state (g) at the excitation wavelength (i.e., $\sigma_s/\sigma_g > 1$ and/or $\sigma_t/\sigma_g > 1$). Furthermore, there must be sufficient ground-state absorption so that a large fraction of the ground-state population can be optically pumped to the excited state. This leads to figures of merit called the saturation intensity ($I_s = h\nu/\sigma_g\tau$, where τ is the excited-state lifetime) or saturation fluence ($F_s = h\nu/\sigma_g\Phi$, where Φ is the quantum yield for formation of the absorbing state). I_s is the appropriate parameter when τ is less than the laser pulse duration and F_s is relevant to the case where τ is greater than the pulse duration. Generally, the value of σ_g must be large enough so that I_s or F_s is much lower than the material's damage intensity or fluence, respectively.

A wide range of chromophores have been shown to exhibit RSA. RSA was first reported for the dyes sulfonated indanthrone and sudanschwartz B, in

1967.[179] However, it was not until 1985 that interest in RSA picked up with the recognition of the potential applications to optical limiters, laser pulse shaping, and spatial light modulation. Since that time, several classes of reverse saturable absorber chromophores, including porphyrins,[180-183] indanthrones,[184] metal cluster compounds,[185, 186] fullerenes,[187] cyanines,[188] phthalocyanines,[189, 190] and naphthalocyanines,[191, 192] have been identified. The optical limiting properties of these materials have been reviewed in some detail.[177, 178]

The vast majority of RSA chromophores have been examined for optical limiting only at a single wavelength (532 nm, the second-harmonic wavelength of Nd : YAG lasers) and pulse duration (typically about 8 ns for Q-switched lasers). Recently, measurements of the wavelength-dependent optical limiting of RSA chromophores have been reported.[193] In general, a variety of measurements are needed to establish the mechanism of optical and photophysical parameters that determine the response of reverse saturable absorbers. A minimum set of measurements for adequate characterization of chromophores is pulse energy-dependent transmission measurements with two different pulse durations (preferably picosecond and nanosecond pulses), measurement of the ground-state absorption cross section, and determination of the S_1 fluorescence lifetime and triplet quantum yield. In the pulse energy-dependent transmission measurements, attention must be paid to the optical geometry such that the total transmitted energy is detected and beam spreading contributions due to nonlinear refraction or thermal effects are avoided. Only through such a set of measurements can the relative contributions of excited singlet–singlet and triplet–triplet absorptions to the optical limiting be assessed. With this knowledge, the suitability of particular chromophores for use with a given laser pulse duration (or range of durations) can be judged and strategies for molecular modifications to optimize the rates of photophysical processes can be developed.

As an example of an approach to the optimization of RSA chromophores, we highlight progress made in our laboratories on phthalocyanine complexes. Phthalocyanine (Pc) metallo-organic complexes have received considerable attention as reverse saturable absorbers for the Nd : YAG second harmonic at 532 nm.[189, 194] As shown in Fig. 6.21, metallophthalocyanines typically exhibit a strong, sharp π–π^* (Q-band) absorption at about 680 nm, followed by a region of relatively weak absorption between 400 and 600 nm. In this same region, Pc exhibit strong excited-state S_1–S_n and T_1–T_n transitions. Reverse saturable absorption and optical limiting with Pc at 532 nm was first reported for chloroaluminum phthalocyanine (ClAlPc).[190] A relatively detailed picture of the photophysics of ClAlPc and other phthalocyanines has been developed. Values for the absorption cross sections for the S_0–S_1, S_1–S_n, and T_1–T_n transitions, the S_1 fluorescence lifetime, and the intersystem crossing rate have been established.[191, 195] Examination of the photophysical properties of ClAlPc revealed that σ_t/σ_g (about 30) is about three times larger than σ_s/σ_g (about 10) but that τ_{isc} (i.e., k_{isc}^{-1}) has a value of 18 ns. On

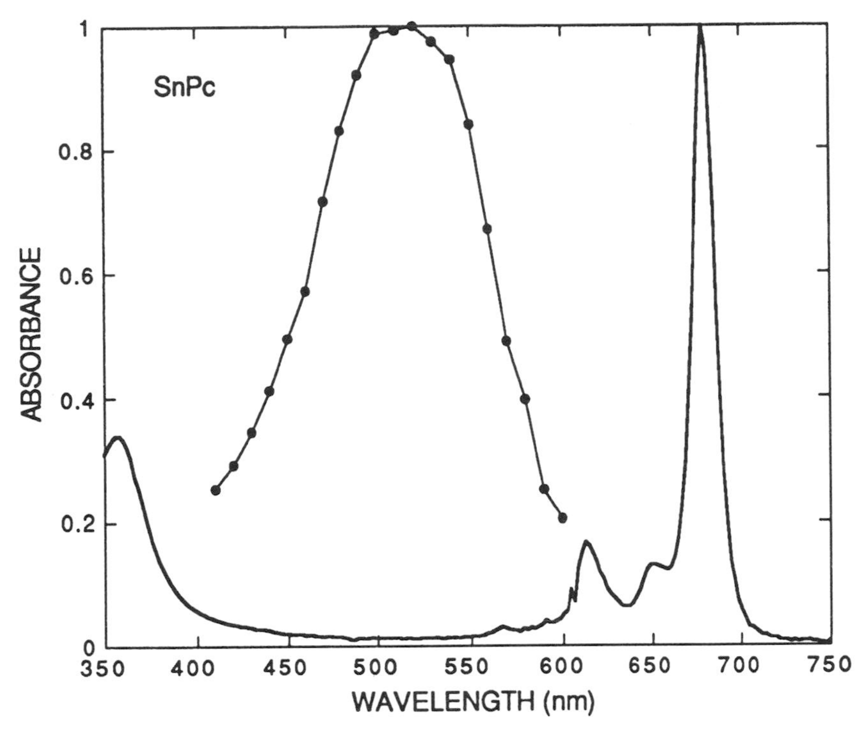

Figure 6.21 Electronic absorption spectrum of bis[tri-(*n*-hexyl)siloxy]SnPc in toluene solution at 7.1×10^{-6} *M*. Also shown (curve with circles) is the transient absorption spectrum (arbitrary units) of SnPc in toluene obtained 100 ns after excitation at 355 nm. (From Ref. 189.)

the time scale of a 1- to 10-ns laser pulse, there would be little buildup of triplet population during the pulse, and the reverse saturable absorption would be dominated by the excited singlet population and σ_s/σ_g. Accordingly, we employed a strategy to increase the intersystem crossing rate and make better use of the strong triplet–triplet absorption.

An approach based on the heavy-atom effect to enhancement of the RSA response of phthalocyanines for 532-nm, 8-ns pulses has been demonstrated.[189] In the heavy-atom effect, attachment of an atom of high atomic number to a chromophore leads to an increase in the intersystem crossing rate due to the increase in the spin-orbit coupling for π-electrons resulting from orbital mixing with the metal. The increase in the intersystem crossing rate provides a means for making more effective use of the triplet–triplet absorption for short pulse durations. In a series of phthalocyanines containing metals and metalloids from groups IIIA (Al, Ga, In) and IVA (Si, Ge, Sn, and Pb), central metal atoms of increased atomic number led to an increase in the effective excited-state absorption cross section and

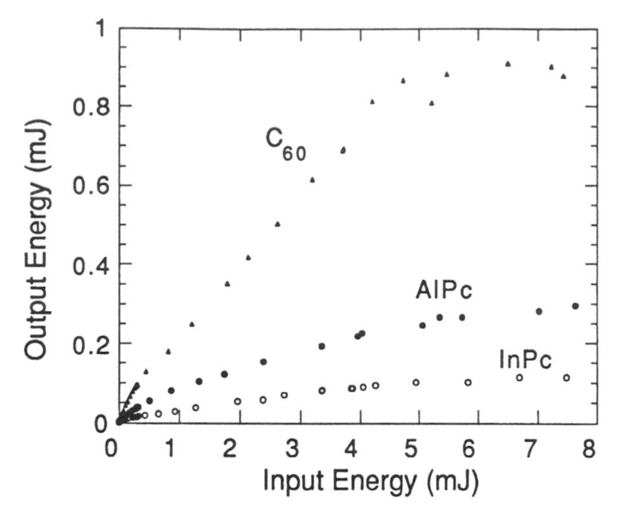

Figure 6.22 Optical limiting response of solutions of C_{60} (triangles), AlPcCl (solid circles), and InPc(*t*-butyl)$_4$Cl (open circles) in an f/5 optical geometry. Data are for 532-nm, 8-ns pulses and sample linear transmittance of 70%. (Adapted from Ref. 198.)

the optical limiting response. In Fig. 6.22 the performance of a heavy-atom phthalocyanine, InPc(*t*-butyl)4Cl, compared to previous RSA chromophores, such as C_{60} and AlPcCl, is illustrated.

While the performance of reverse saturable absorbers has been investigated in a variety of optical limiting geometries, the results have fallen short of the desired performance levels of strong signal suppression of 10^3 to 10^5, combined with a linear transmittance of > 50%. A significant conceptual step in the design of optical limiters utilizing reverse saturable absorbers was the realization that the problem of managing the fluence of a convergent optical beam in an absorbing material leads to a general requirement for a nonhomogeneous concentration profile of the absorber.[178, 196] Miles has recently considered both the management of fluence in the material and maximal utilization of the nonlinear response of the reverse saturable absorber.[197] In reverse saturable absorbers, maximal attenuation of the energy of the optical pulse requires strong saturation of the excited-state population. The key to establishing strong saturation along the optical path in the material is that the fluence along the path must be maintained above (say 2.5 times) the saturation fluence (F_s) and, at the same time, controlled so that it is below the limit imposed by optical damage. Miles showed that this condition can be achieved by using fully saturated absorbers with a nonhomogeneous concentration profile along the propagation direction z:[197]

$$N(z) = \frac{2}{\sigma_e |z|} \tag{28}$$

In a recent experiment, we set out to test this optical limiter design concept by using a practical approximation to the ideal hyperbolic concentration profile, which involved a geometrically expanding array of uniform concentration plates.[198] To implement the limiter design, a series of ClIn(*t*-butyl)$_4$Pc-doped PMMA plates were used. As shown in Fig. 6.23, the results for a three-plate device in an f/5 optical system with 8-ns and, 532-nm pulses demonstrate an order-of-magnitude enhancement in optical limiting response. The strong signal transmission was 0.18%, compared to the linear transmittance of 70%, and this is the highest performance reported to date. This initial experiment shows that there is considerable promise for the use of appropriate concentration gradients of strong reverse saturable absorbers in the development of high-performance optical limiter devices.

6.6 CONCLUSIONS

Our understanding of NLO materials has advanced tremendously over the past several years. For example, there now exist strategies for the optimization of molecular second- and third-order nonlinearities as well as for improving the chemical and thermal stability of NLO chromophores. Poled polymers with nonlinearities as large as state-of-the-art inorganic materials

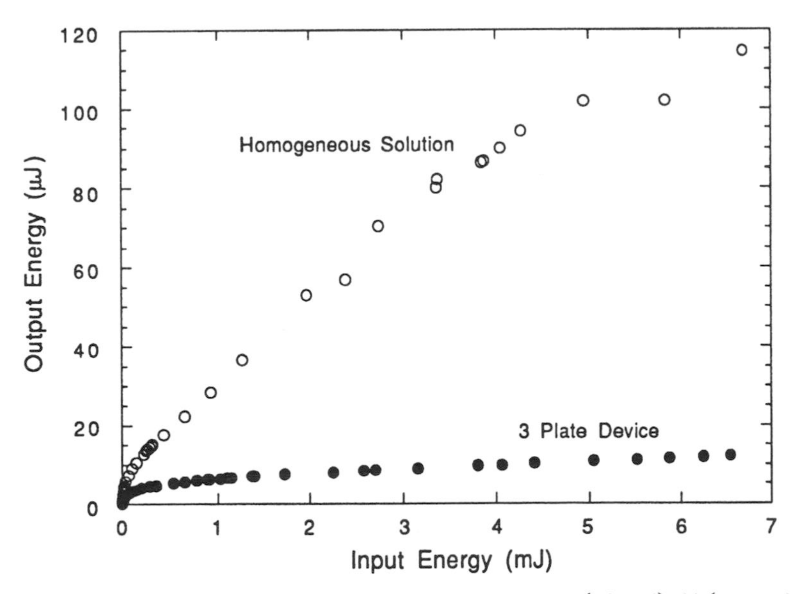

Figure 6.23 Optical limiting response of solutions of InPc(*t*-butyl)$_4$Cl (open circles) and a three-plate "bottleneck" limiter based on InPc(*t*-butyl)$_4$Cl-doped PMMA (solid circles) in an f/5 optical geometry. Data are for 532-nm, 8-ns pulses and sample linear transmittance of 70%. (Adapted from Ref. 198.)

have now been reported, and systematic approaches to locking in this nonlinearity, at high temperatures, have been demonstrated. Optical limiting materials with performance approaching that needed for protection of human eyes from lasers have been demonstrated. Nonlinearities much larger than those realized to date are achievable *if* we can build the "best of the bests" all into one system. If so, the great promise of organic materials for NLO applications will be realized.

REFERENCES

1. *Optical Nonlinearities in Chemistry*, special issue, *Chem. Rev.* **94** (1994).
2. J. L. Bredas, R. R. Chance, Eds., *Conjugated Polymeric Materials: Opportunities in Electronics, Optoelectronics, and Molecular Electronics*, NATO-ARW Series E, Vol. E182, Kluwer, Dordrecht, 1990.
3. D. S. Chemla, J. Zyss, Eds., *Nonlinear Optical Properties of Organic Molecules and Crystals*, Vol. 1 & 2, Academic Press, San Diego, CA, 1987.
4. A. J. Heeger, J. Orenstein, D. R. Ulrich, Eds., *Nonlinear Optical Properties of Polymers*, MRS Symposium Proceedings, Vol. 109, Materials Research Society, Pittsburgh, PA, 1988.
5. S. R. Marder, J. E. Sohn, G. D. Stucky, Eds., *Materials for Nonlinear Optics: Chemical Perspectives*, ACS Symposium Series, Vol. 455, American Chemical Society, Washington, DC, 1991.
6. J. Messier, F. Kajzar, P. N. Prasad, *Organic Molecules for Nonlinear Optics and Photonics*, NATO-ARW Series E, Vol. E194, Kluwer, Dordrecht, 1989.
7. J. Messier, F. Kajzar, P. N. Prasad, Eds., *Organic Molecules for Nonlinear Optics and Photonics*, NATO-ARW Series E, Vol. E194, Kluwer, Dordrecht, 1991.
8. P. N. Prasad, D. J. Williams, *Introduction to Nonlinear Optical Effects in Molecules and Polymers*, Wiley-Interscience, New York, 1991.
9. D. J. Williams, Ed., *Nonlinear Optical Properties of Organics and Polymeric Materials*, ACS Symposium Series, Vol. 233, American Chemical Society, Washington, DC, 1983.
10. G. Zerbi, Ed., *Organic Materials for Photonics: Science and Technology*, Elsevier, Amsterdam, 1993.
11. G. D. Stucky, S. R. Marder, J. E. Sohn, in *Materials for Nonlinear Optics: Chemical Perspectives*, ACS Symposium Series, Vol. 455, S. R. Marder, J. E. Sohn, G. D. Stucky, Eds., American Chemical Society, Washington, DC, 1991, pp. 2–30.
12. A. D. Buckingham, *Adv. Chem. Phys.* **12**, 107 (1967).
13. P. A. Franken, A. E. Hill, C. W. Peters, G. Weinreich, *Phys. Rev. Lett.* **7**, 118 (1961).
14. I. P. Kaminow, *An Introduction to Electro-optic Devices*, Academic Press, San Diego, CA, 1974.
15. V. V. Shkunov, B. Y. Zel'dovich, *Sci. Am.* **253**, 54 (1985).
16. D. M. Pepper, *Sci. Am.* **254**, 74 (1986).

17. B. F. Levine, C. G. Bethea, *Appl. Phys. Lett.* **24**, 445 (1974).
18. B. F. Levine, C. G. Bethea, *J. Chem. Phys.* **63**, 2666 (1975).
19. K. D. Singer, A. F. Garito, *J. Chem. Phys.* **75**, 3572 (1981).
20. L. T. Cheng, W. Tam, S. H. Stevenson, G. R. Meredith, G. Rikken, S. R. Marder, *J. Phys. Chem.* **95**, 10631 (1991).
21. P. D. Maker, R. W. Terhune, *Phys. Rev. A* **137**, 801 (1965).
22. R. W. Hellwarth, *Prog. Quant. Electron.* **5**, 1 (1977).
23. J. L. Bredas, in *Organic Materials for Photonics: Science and Technology*, G. Zerbi, Ed., Elsevier, Amsterdam, 1993, pp. 127–153.
24. D. R. Kanis, M. A. Ratner, T. J. Marks, *Chem. Rev.* **94**, 195 (1994).
25. J. M. Andre, J. Delhalle, J. L. Bredas, *Quantum Chemistry Aided Design of Organic Polymers: An Introduction to the Quantum Chemistry of Polymers and Its Applications*, World Scientific, Singapore, 1991.
26. M. M. Karelson, M. C. Zerner, *J. Phys. Chem.* **96**, 6949 (1992).
27. A. Willetts, J. E. Rice, *J. Chem. Phys.* **99**, 426 (1993).
28. K. V. Mikkelsen, Y. Luo, H. Agren, P. Jorgensen, *J. Chem. Phys.* **100**, 8240 (1994).
29. J. Yu, M. C. Zerner, *J. Chem. Phys.* **100**, 7487 (1994).
30. S. Dibella, T. J. Marks, M. A. Ratner, *J. Am. Chem. Soc.* **116**, 4440 (1994).
31. C. Dehu, F. Meyers, E. Hendrickx, K. Clays, A. Persoons, S. R. Marder, J. L. Bredas, *J. Am. Chem. Soc.* **117**, 10127 (1995).
32. S. P. Karna, P. N. Prasad, M. Dupuis, *J. Chem. Phys.* **94**, 1171 (1991).
33. S. P. Karna, M. Dupuis, E. Perrin, P. N. Prasad, *J. Chem. Phys.* **92**, 7418 (1990).
34. C. E. Dykstra, P. G. Jasein, *Chem. Phys. Lett.* **109**, 388 (1984).
35. H. A. Kurtz, J. J. P. Stewrad, K. M. Dieter, *J. Comput. Chem.* **11**, 82 (1990).
36. H. D. Cohen, C. C. J. Roothaan, *J. Chem. Phys.* **43**, S34 (1965).
37. S. P. Karna, M. Dupuis, *J. Comput. Chem.* **12**, 487 (1991).
38. S. P. Karna, M. Dupuis, *Chem. Phys. Lett.* **171**, 201 (1990).
39. S. Mukamel, H. X. Wang, *Phys. Rev. Lett.* **69**, 65 (1992).
40. A. Takahashi, S. Mukamel, *J. Chem. Phys.* **100**, 2366 (1994).
41. G. H. Chen, S. Mukamel, *J. Am. Chem. Soc.* **117**, 4945 (1995).
42. B. J. Orr, J. F. Ward, *Mol. Phys.* **20**, 513 (1971).
43. B. M. Pierce, *J. Chem. Phys.* **70**, 165 (1989).
44. J. R. Heflin, K. Y. Wong, O. Zamanikhamiri, A. F. Garito, *Phys. Rev. B Condens. Matter* **38**, 1573 (1988).
45. J. A. Pople, D. L. Beveridge, P. A. Dobossh, *J. Chem. Phys.* **47**, 2026 (1967).
46. J. Ridley, M. Zerner, *Theor. Chim. Acta* **32**, 111 (1973).
47. J. A. Pople, D. P. Santry, G. A. Segal, *J. Chem. Phys.* **43**, S129 (1965).
48. J. L. Oudar, *J. Chem. Phys.* **67**, 446 (1977).
49. J. L. Oudar, D. S. Chemla, *J. Chem. Phys.* **66**, 2664 (1977).
50. F. Meyers, S. R. Marder, B. M. Pierce, J. L. Bredas, *Chem. Phys. Lett.* **228**, 171 (1994).

51. A. F. Garito, J. R. Heflin, K. Y. Wong, O. Zamani-Khamiri, *Proc. SPIE Int. Soc. Opt. Eng.* **971**, 2 (1988).
52. A. F. Garito, J. R. Heflin, K. Y. Wong, O. Zamani-Khamiri, in *Organic Materials for Nonlinear Optics*, R. A. Hann and D. Bloor, Eds., Royal Society of Chemistry, London, 1989, p. 16.
53. C. W. Dirk, L. T. Cheng, M. G. Kuzyk, *Int. J. Quantum Chem.* **43**, 27 (1992).
54. B. M. Pierce, *Proc. SPIE Int. Soc. Opt. Eng.* **1560**, 148 (1991).
55. D. Q. Lu, G. H. Chen, J. W. Perry, W. A. Goddard, *J. Am. Chem. Soc.* **116**, 10679 (1994).
56. S. R. Marder, D. N. Beratan, L. T. Cheng, *Science* **252**, 103 (1991).
57. S. R. Marder, C. B. Gorman, F. Meyers, J. W. Perry, G. Bourhill, J. L. Bredas, B. M. Pierce, *Science* **265**, 632 (1994).
58. F. Meyers, S. R. Marder, B. M. Pierce, J. L. Bredas, *J. Am. Chem. Soc.* **116**, 10703 (1994).
59. C. B. Gorman, S. R. Marder, *Proc. Natl. Acad. Sci. USA* **90**, 11297 (1993).
60. C. Kuhn, *Synth. Met.* **41–43**, 3681 (1991).
61. J. L. Rivail, D. Rinaldi, *Chem. Phys.* **18**, 233 (1976).
62. C. J. F. Bottcher, *Theory of Electric Polarization: Dielectrics in Static Fields*, Elsevier, Amsterdam, 1973.
63. C. Reichardt, *Solvents and Solvent Effects in Organic Chemistry*, VCH, Weinheim, Germany, 1988.
64. L. Salem, *The Molecular Orbital Theory of Conjugated Systems*, W. A. Benjamin, Menlo Park, CA, 1966.
65. S. R. Marder, J. W. Perry, G. Bourhill, C. B. Gorman, B. G. Tiemann, K. Mansour, *Science* **261**, 186 (1993).
66. G. Bourhill, J. L. Bredas, L. T. Cheng, S. R. Marder, F. Meyers, J. W. Perry, B. G. Tiemann, *J. Am. Chem. Soc.* **116**, 2619 (1994).
67. G. Bourhill, M. Mansour, B. G. Tiemann, C. B. Gorman, S. Biddle, S. R. Marder, J. W. Perry, *Proc. SPIE Int. Soc. Opt. Eng.* **1853**, 208 (1993).
68. S. R. Marder, C. B. Gorman, B. G. Tiemann, L. T. Cheng, *J. Am. Chem. Soc.* **115**, 3006 (1993).
69. J. L. Bredas, *Synth. Met.* **117**, 115 (1987).
70. J. L. Bredas, A. J. Heeger, F. Wudl, *J. Chem. Phys.* **85**, 4673 (1986).
71. L. T. Cheng, W. Tam, S. R. Marder, A. E. Stiegman, G. Rikken, C. W. Spangler, *J. Phys. Chem.* **95**, 10643 (1991).
72. C. W. Dirk, H. E. Katz, M. L. Schilling, L. A. King, *Chem. Mater.* **2**, 700 (1990).
73. A. K. Y. Jen, V. P. Rao, K. Y. Wong, K. J. Drost, *J. Chem. Soc. Chem. Commun.*, 90 (1993).
74. S. R. Marder, L. T. Cheng, B. G. Tiemann, A. C. Friedli, M. Blanchard-Desce, J. W. Perry, J. Skindhoj, *Science* **263**, 511 (1994).
75. L. G. S. Brooker, U.S. Patent 2,170,807 (1939).
76. L. G. S. Brooker, G. H. Keyes, R. H. Sprague, R. H. VanDyke, E. VanLare, G. VanZandt, F. L. White, H. W. J. Cressman, S. G. Dent, *J. Am. Chem. Soc.* **73**, 5332 (1951).

77. L. G. S. Brooker, G. H. Keyes, R. H. Sprague, R. H. VanDyke, E. VanLare, G. VanZandt, F. L. White, *J. Am. Chem. Soc.* **73**, 5126 (1951).
78. H. Ikeda, Y. Kawabe, T. Sakai, K. Kawasaki, *Chem. Lett.*, 1803 (1989).
79. H. Ikeda, T. Sakai, K. Kawasaki, *Chem. Phys. Lett.* **179**, 551 (1991).
80. R. Ortiz, S. R. Marder, L. T. Cheng, B. G. Tiemann, S. Cavagnero, J. W. Ziller, *J. Chem. Soc. Chem. Commun.*, 2263 (1994).
81. G. U. Bublitz, R. Ortiz, C. Runser, A. Fort, M. Barzoukas, S. R. Marder, S. G. Boxer, *J. Am. Chem. Soc.* **119**, 2311–2312 (1997).
82. C. R. Moylan, R. J. Twieg, V. Y. Lee, S. A. Swanson, K. M. Betterton, R. D. Miller, *J. Am. Chem. Soc.* **115**, 12599 (1993).
83. V. P. Rao, A. K. Y. Jen, K. Y. Wong, K. J. Drost, *J. Chem. Soc. Chem. Commun.*, 1118 (1993).
84. V. P. Rao, Y. M. Cai, A. K. Y. Jen, *J. Chem. Soc. Chem. Commun.*, 1689 (1994).
85. J. Zyss, I. Ledoux, *Chem. Rev.* **94**, 77 (1994).
86. I. Ledoux, J. Zyss, J. S. Siegel, J. Brienne, J. M. Lehn, *Chem. Phys. Lett.* **172**, 440 (1990).
87. T. Verbiest, K. Clays, C. Samyn, J. Wolff, D. Reinhoudt, A. Persoons, *J. Am. Chem. Soc.* **116**, 9320 (1994).
88. K. Clays, A. Persoons, *Phys. Rev. Lett.* **66**, 2980 (1991).
89. K. Clays, A. Persoons, *Rev. Sci. Instrum.* **63**, 3285 (1992).
90. J. L. Bredas, F. Meyers, B. M. Pierce, J. Zyss, *J. Am. Chem. Soc.* **114**, 4928 (1992).
91. M. Joffre, D. Yaron, R. J. Silbey, J. Zyss, *J. Chem. Phys.* **97**, 5607 (1992).
92. H. S. Nalwa, K. Nakajima, T. Watanabe, K. Nakamura, A. Yamada, S. Miyata, *J. Appl. Phys.* **30**, 983 (1991).
93. H. S. Nalwa, T. Watanabe, S. Miyata, *Adv. Mater.* **7**, 754 (1995).
94. H. S. Nalwa, *Adv. Mater.* **3**, 341 (1993).
95. J. L. Bredas, C. Adant, P. Tackx, A. Persoons, B. M. Pierce, *Chem. Rev.* **94**, 243 (1994).
96. S. Dahne, K. D. Nolte, *J. Chem. Soc. Chem. Commun.*, 1042 (1972).
97. V. P. Bodart, J. Delhalle, J. M. Andre, J. Zyss, *Can. J. Chem.*, **63**, 1631 (1985).
98. S. R. Marder, W. E. Torruellas, M. Blanchard-Desce, V. Ricci, G. I. Stegeman, S. Gilmour, J.-L. Brédas, J. Li, G. U. Bublitz, S. G. Boxer, *Science* **276**, 1233 (1997).
99. V. Ricci, W. E. Torruellas, G. I. Stegeman, S. Gilmour, S. R. Marder, M. Blanchard-Desce, in *Organ. Thin Films Photon. Appl. Tech. Dig.* 1995, Vol. 21 (Optical Society of America, Portland, OR, 1995) pp. 46–47.
100. Q. L. Zhou, R. F. Shi, O. Zamani-Khamari, A. F. Garito, *Nonlinear Opt.*, **6**, 145 (1993).
101. Y. Z. Yu, R. F. Shi, A. F. Garito, C. H. Grossman, *Opt. Lett.* **19**, 786 (1994).
102. C. Poga, T. M. Brown, M. G. Kuzyk, C. W. Dirk, *J. Opt. Soc. Am. B. Opt. Phys.* **12**, 531 (1995).

103. C. W. Dirk, W. C. Herndon, F. Cervanteslee, H. Selnau, S. Martinez, P. Kalamegham, A. Tan, G. Campos, M. Velez, J. Zyss, I. Ledoux, L. T. Cheng, *J. Am. Chem. Soc.* **117**, 2214 (1995).
104. M. Blanchard-Desce, J. M. Lehn, M. Barzoukas, I. Ledoux, J. Zyss, *Chem. Phys.* **181**, 281 (1994).
105. C. B. Gorman, S. R. Marder, *Chem. Mater.* **7**, 215 (1995).
106. G. Puccetti, M. Blanchard-Desce, I. Ledoux, J. M. Lehn, J. Zyss, *J. Phys. Chem.* **97**, 9385 (1993).
107. F. Meyers, J. L. Bredas, in *Organic Materials for Nonlinear Optics III*, G. J. Ashweel, D. Bloor, Eds., Royal Society of Chemistry, London, 1993, p. 1.
108. J. R. Heflin, D. C. Rodenberger, R. F. Shi, M. Wu, N. Q. Wang, Y. M. Cai, A. F. Garito, *Phys. Rev. A* **45**, 4233 (1992).
109. D. C. Rodenberger, J. R. Heflin, A. F. Garito, *Nature* **359**, 309 (1992).
110. S. R. Marder, J. W. Perry, *Adv. Mater.* **5**, 804 (1993).
111. D. J. Williams, *Angew. Chem. Int. Ed. Engl.* **23**, 690 (1984).
112. J. F. Nicoud, R. J. Twieg, in *Nonlinear Optical Properties of Organic Molecules and Crystals*, D. S. Chemla, J. Zyss, Eds., Academic Press, San Diego, CA, 1987, pp. 227–296.
113. J. L. Oudar, R. Hierle, *J. Appl. Phys.* **48**, 2699 (1977).
114. P. Gunter, C. Bosshard, K. Sutter, H. Arend, G. Chapuis, R. J. Twieg, D. Dobrowolski, *Appl. Phys. Lett.* **50**, 486 (1987).
115. J. Zyss, J. F. Nicoud, M. Coquillay, *J. Chem. Phys.* **81**, 4160 (1984).
116. R. J. Twieg, C. W. Dirk, *J. Chem. Phys.* **85**, 3539 (1986).
117. P. V. Vidakovic, M. Coquillay, F. Salin, *J. Opt. Soc. Am. B* **4**, 998 (1987).
118. T. W. Panunto, Z. Urbanczyklipkowska, R. Johnson, M. C. Etter, *J. Am. Chem. Soc.* **109**, 7786 (1987).
119. M. C. Etter, K. S. Huang, *Chem. Mater.* **4**, 824 (1992).
120. M. C. Etter, G. M. Frankenbach, *Chem. Mater.* **1**, 10 (1989).
121. M. C. Etter, *Acc. Chem. Res.* **23**, 120 (1990).
122. E. Fan, C. Vicent, S. J. Geib, A. D. Hamilton, *Chem. Mater.* **6**, 1113 (1994).
123. J. M. Lehn, *Angew. Chem. Int. Ed. Engl.* **29**, 1304 (1990).
124. G. R. Meredith, in *Nonlinear Optical Properties of Organic and Polymeric Materials*, ACS Symposium Series, Vol. 233, D. J. Williams, Ed., American Chemical Society, Washington, DC, 1983.
125. S. R. Marder, J. W. Perry, W. P. Schaefer, *Science* **245**, 626 (1989).
126. S. R. Marder, J. W. Perry, W. P. Schaefer, *J. Mater. Chem.* **2**, 985 (1992).
127. S. R. Marder, J. W. Perry, C. P. Yakymyshyn, *Chem. Mater.* **6**, 1137 (1994).
128. S. Okada, A. Masaki, H. Matsuda, H. Nakanishi, M. Kato, R. Muramatsu, M. Otsuka, *Jpn. J. Appl. Phys.* **29**, 1112 (1990).
129. C. P. Yakymyshyn, S. R. Marder, K. R. Stewart, E. P. Boden, J. W. Perry, W. P. Schaefer, in *Organic Materials for Nonlinear Optics II*, Vol. 91, R. A. Hann, D. Bloor, Eds., Royal Society of Chemistry, London, 1991, pp. 108–114.
130. C. Bosshard, K. Sutter, R. Schlesser, P. Gunter, *J. Opt. Soc. Am. B* **10**, 867 (1993).

131. C. Serbutoviez, J. F. Nicoud, J. Fischer, I. Ledoux, J. Zyss, *Chem. Mater.* **6**, 1358 (1994).

132. D. M. Burland, R. D. Miller, C. A. Walsh, *Chem. Rev.* **94**, 31 (1994).

133. L. R. Dalton, A. W. Harper, B. Wu, R. Ghosn, J. Laquindanum, Z. Liang, A. Hubbel, C. Xu, *Adv. Mater.* **7**, 519 (1995).

134. S. R. Marder, J. W. Perry, *Science* **263**, 1706 (1994).

135. K. D. Singer, J. E. Sohn, S. J. Lalama, *Appl. Phys. Lett.* **49**, 248 (1986).

136. J. W. Wu, J. F. Valley, S. Ermer, E. S. Binkley, J. T. Kenney, G. F. Lipscomb R. Lytel, *Appl. Phys. Lett.* **58**, 225 (1991).

137. A. K. Y. Jen, K. Y. Wong, V. P. Rao, K. Drost, Y. M. Cai, *J. Electron. Mater.* **23**, 653 (1994).

138. K. Y. Wong, A. K. Y. Jen, *J. Appl. Phys.* **75**, 3308 (1994).

139. Y. M. Cai, A. K. Y. Jen, *Appl. Phys. Lett.* **67**, 299 (1995).

140. M. Ahlheim, M. Barzoukas, P. V. Bedworth, M. Blanchard-Desce, A. Fort, Z. Y. Hu, S. R. Marder, J. W. Perry, C. Runser, M. Staehelin, B. Zysset, *Science*, **271**, 335–337 (1996).

141. A. Yariv, P. Yeh, *Optical Waves in Crystals*, Wiley, New York, 1984.

142. T. A. Chen, A. K. Y. Jen, Y. M. Cai, *J. Am. Chem. Soc.* **117**, 7295 (1995).

143. C. Xu, B. Wu, L. R. Dalton, P. M. Ranon, Y. Shi, W. H. Steier, *Macromolecules* [illegible] 6716 (1992).

144. C. Xu, B. Wu, L. R. Dalton, Y. Shi, P. M. Ranon, W. H. Steier, *Macromolecules* **25**, 6714 (1992).

145. M. A. Hubbard, T. J. Marks, W. P. Lin, G. K. Wong, *Chem. Mater.* **4**, 965 (1992).

146. T. Verbiest, D. M. Burland, M. C. Jurich, V. Y. Lee, R. D. Miller, W. Volksen, *Science* **268**, 1604 (1995).

147. R. D. Miller, D. M. Burland, M. Jurich, V. Y. Lee, C. R. Moylan, J. I. Thackara, R. J. Twieg, T. Verbiest, W. Volksen, *Macromolecules* **28**, 4970 (1995).

148. N. Tillman, A. Ulman, J. S. Schildkraut, T. L. Penner, *J. Am. Chem. Soc.* **110**, 6136 (1988).

149. N. Tillman, A. Ulman, T. L. Penner, *Langmuir* **5**, 101 (1989).

150. D. Li, T. J. Marks, T. Zhang, G. K. Wong, *Synth. Met.* **41–43**, 3157 (1991).

151. D. Li, M. A. Ratner, T. J. Marks, C. Zhang, K. J. Yang, G. H. Wong, *J. Am. Chem. Soc.* **112**, 7389 (1990).

152. H. E. Katz, G. Scheller, T. M. Putvinski, M. L. Schilling, W. L. Wilson, C. E. D. Chidsey, *Science* **254**, 1485 (1991).

153. S. Ducharme, J. C. Scott, R. J. Twieg, W. E. Moerner, *Phys. Rev. Lett.* **66**, 1846 (1991).

154. K. Sutter, P. Gunter, *J. Opt. Soc. Am. B Opt. Phys.* **7**, 2274 (1990).

155. K. Sutter, J. Hulliger, P. Gunter, *Solid State Commun.* **74**, 867 (1990).

156. W. E. Moerner, S. M. Silence, *Chem. Rev.* **94**, 127 (1994).

157. P. Gunter, J. P. Huignard, Eds., *Photorefractive Materials and Applications I, II*, Vols. 61, 62, Springer-Verlag, Heidelberg, 1988, 1989.

158. Y. Zhang, Y. P. Cui, P. N. Prasad, *Phys. Rev. B. Condens. Matter* **46**, 9900 (1992).

159. S. M. Silence, J. C. Scott, J. J. Stankus, W. E. Moerner, C. R. Moylan, G. C. Bjorklund, R. J. Twieg, *J. Phys. Chem.* **99**, 4096 (1995).

160. Y. M. Chen, Z. H. Peng, W. K. Chan, L. P. Yu, *Appl. Phys. Lett.* **64**, 1195 (1994).

161. C. A. Walsh, W. E. Moerner, *J. Opt. Soc. Am. B. Opt. Phys.* **9**, 1642 (1992).

162. W. E. Moerner, S. M. Silence, F. Hache, G. C. Bjorklund, *J. Opt. Soc. Am. B Opt. Phys.* **11**, 320 (1994).

163. S. M. Silence, F. Hache, M. Donckers, C. A. Walsh, D. M. Burland, G. C. Bjorklund, R. J. Twieg, W. E. Moerner, *Proc. SPIE Int. Soc. Opt. Eng.* **1852**, 253 (1993).

164. S. M. Silence, C. A. Walsh, J. C. Scott, W. E. Moerner, *Appl. Phys. Lett.* **61**, 2967 (1992).

165. M. Liphardt, A. Goonesekera, B. E. Jones, S. Ducharme, J. M. Takacs, L. Zhang, *Science* **263**, 367 (1994).

166. M. C. J. M. Donckers, S. M. Silence, C. A. Walsh, F. Hache, D. M. Burland, W. E. Moerner, R. J. Twieg, *Opt. Lett.* **18**, 1044 (1993).

167. S. M. Silence, J. C. Scott, F. Hache, E. J. Ginsburg, P. K. Jenkner, R. D. Miller, R. J. Twieg, W. E. Moerner, *J. Opt. Soc. Am. B Opt. Phys.* **10**, 2306 (1993).

168. K. Meerholz, B. L. Volodin, K. M. Sandalphon, B. Kippelen, N. Peyghambarian, *Nature* **371**, 497 (1994).

169. C. Halvorson, B. Kraabel, A. J. Heeger, B. L. Volodin, K. Meerholz, K. M. Sandalphon, N. Peyghambarian, *Opt. Lett.* **20** 76 (1995).

170. J. J. Stankus, S. M. Silence, W. E. Moerner, G. C. Bjorklund, *Opt. Lett.* **19**, 1480 (1994).

171. B. L. Volodin, K. M. Sandalphon, K. Meerholz, B. Kippelen, N. V. Kukhtarev, N. Peyghambarian, *Opt. Eng.* **34**, 2213 (1995).

172. B. M. Pierce, in *Molecular and Biomolecular Electronics*, Vol. 240, R. R. Birge, Ed., American Chemical Society, Washington, DC, 1994, pp. 243–302.

173. R. L. Crane, M. Khoshnevisan, K. Lewis, E. W. VanStryland, Eds., *Materials for Optical Limiting*, Vol. 374, Materials Research Society, Pittsburgh, PA, 1995.

174. K. P. J. Reddy, *Curr. Sci.* **61**, 520 (1991).

175. A. Penzkofer, *Appl. Phys. B* **46**, 43 (1988).

176. M. J. Soileau, Ed., *Materials for Optical Switches, Isolators, and Limiters, Proc. SPIE* **1105** (1989).

177. J. W. Perry, in *Nonlinear Optics of Organic Molecular and Polymeric Materials*, H. Nalwa, S. Miyata, Eds., CRC Press, Orlando, FL, 1997, pp. 813–840.

178. L. W. Tutt, T. F. Boggess, *Prog. Quantum Electron.* **17**, 299 (1993).

179. C. R. Giuliano, L. D. Hees, *IEEE J. Quantum Electron.* **3**, 358 (1967).

180. J. H. Si, M. Yang, Y. X. Wang, L. Zhang, C. F. Li, D. Y. Wang, S. M. Dong, W. F. Sun, *Appl. Phys. Lett.* **64**, 3083 (1994).

181. S. Guha, K. Kang, P. Porter, J. F. Roach, D. E. Remy, F. J. Aranda, D. Rao, *Opt. Lett.* **17**, 264 (1992).

182. H. S. Fei, L. Han, X. C. Ai, R. Yin, J. C. Shen, *Chin. Sci. Bull.* **37**, 298 (1992).

183. W. Blau, H. Byrne, W. M. Dennis, J. M. Kelly, *Opt. Commun.* **56**, 25 (1985).

184. R. C. Hoffman, K. A. Stetyick, R. S. Potember, D. G. Mclean, *J. Opt. Soc. Am. B. Opt. Phys.* **6**, 772 (1989).
185. G. R. Allan, D. R. Labergerie, S. J. Rychnovsky, T. F. Bogess, A. L. Smirl, L. Tutt, *J. Phys. Chem.* **96**, 6313 (1992).
186. L. W. Tutt, S. McCahon, *Opt. Lett.* **15**, 700 (1990).
187. L. W. Tutt, A. Kost, *Nature*, **356**, 225 (1992).
188. S. Hughes, G. Spruce, B. S. Wherrett, K. R. Welford, A. D. Lloyd, *Opt. Commun.* **100**, 113 (1993).
189. J. W. Perry, K. Mansour, S. R. Marder, K. J. Perry, D. Alvarez, I. Choong, *Opt. Lett.* **19**, 625 (1994).
190. D. R. Coulter, V. M. Miskowski, J. W. Perry, T. H. Wei, E. W. VanStryland, D. J. Hagan, *Proc. SPIE* **1105**, 42 (1989).
191. T. H. Wei, D. J. Hagan, M. J. Sence, E. W. VanStryland, J. W. Perry, D. R. Coulter, *Appl. Phys. B* **54**, 46 (1992).
192. J. W. Perry, L. R. Khundkar, D. R. Coulter, D. Alvarez, S. R. Marder, T. H. Wei, M. J. Sense, E. W. VanStryland, D. J. Hagan, in *Organic Molecules for Nonlinear Optics and Photonics*, NATO ASI Series E, Vol. 194, J. Messier, F. Kajzar, P. N. Prasad, Eds., Kluwer, Dordrecht, The Netherlands, 1991, pp. 369–382.
193. J. R. Heflin, S. Wang, D. Marciu, C. Figura, R. Yordanov, *Proc. SPIE* **2530**, 176–187 (1995).
194. J. S. Shirk, R. G. S. Pong, F. J. Bartoli, A. W. Snow, *Appl. Phys. Lett.* **63**, 1880 (1993).
195. K. Mansour, D. Alvarez, Jr., K. J. Perry, I. Choong, S. R. Marder, J. W. Perry, *Proc. SPIE* **1853**, 132 (1993).
196. S. W. McCahon, L. W. Tutt, U.S. patent 5,080,469 (1992).
197. P. Miles, *Appl. Opt.* **33**, 6965 (1994).
198. J. W. Perry, K. Mansour, I.-Y. S. Lee, X.-L. Wu, P. V. Bedworth, C.-T. Chen, D. Ng, S. R. Marder, P. Miles, T. Wada, M. Tian, H. Sasabe, *Science*, **273**, 1533–1536 (1996).

CHAPTER 7

Nanoparticles and Nanostructural Materials

KENNETH J. KLABUNDE AND CATHY MOHS
Department of Chemistry, Kansas State University, Manhattan, Kansas

7.1 A WORLD BOUNDED BY CHEMISTRY AND SOLID-STATE PHYSICS

7.1.1 Nanoscale Regime

Chemistry and physics are two fields of science that evolved in an intermingled way and are still inseparable in any practical sense. However, if we view chemistry as the study of atoms and molecules, a realm of matter whose dimensions are generally less than 1 nm, while a major branch of physics deals with solids of essentially an infinite array of bound atoms or molecules of greater than 100 nm, a significant gap exists between the regimes. Figure 7.1 illustrates this gap, which deals with particles of 1 to 100 nm, or approximately from 10 to 10^6 atoms or molecules per particle.[1]

In this nanoscale regime neither quantum chemistry nor classical laws of physics hold.[2] In materials (metals, semiconductors, or insulators) where strong chemical bonding is present, delocalization of valence electrons can be extensive, and the extent of delocalization can vary with size. This effect, coupled with structural changes with size variation, can lead to different chemical and physical properties, *depending on size*. Indeed, it has now been demonstrated that a host of properties depend on the size of such nanoscale particles, including magnetic, optical, melting points, specific heats, and surface reactivity. Furthermore, when such ultrafine particles are consolidated into macroscale solids, these bulk materials sometimes exhibit new properties (e.g., enhanced plasticity).

Chemistry of Advanced Materials: An Overview, Edited by Leonard V. Interrante and Mark J. Hampden-Smith.
ISBN 0-471-18590-6

Chemistry	Nanoscale regime	Solid State Physics

I------------------I--I--------------------------------------I

one atom	10 atom cluster	100 atom cluster	1000 atom cluster	10,000 atom cluster	1×10^6 atom cluster	bulk

0 1 2 3 5 7 10 100

Diameter (nm)

Figure 7.1 Size relationships of chemistry, nanoparticles, and solid-state physics. (From Ref. 1.)

If we contemplate these findings for a moment, it becomes clear that a huge new field of science awaits investigation. Think of the multitude of combinations of two, three, or more elements with particles of varying sizes! Each change in composition or size can lead to different physical and chemical properties! It soon becomes clear that an almost infinite number of possibilities present themselves. And this new field of clusters/nanophase materials, lying between the traditional fields of chemistry and solid-state physics, touches upon disciplines such as electronics, astronomy, mathematics, and engineering.

It becomes obvious that interdisciplinary research is required for progress to be made. The most important aspects are synthesis, physical properties, and chemical properties, but the most important of these at this time is *synthesis*. The nanoparticles under study are almost always prepared in the laboratory (as opposed to naturally occurring), are sometimes reactive with oxygen and water, and are difficult to produce in a monodisperse (one size only) form. Thus creative synthesis schemes that lead to gram or kilogram quantities of pure materials are absolutely essential before this new field of science can be developed for the benefit of humankind.

7.1.2 Gas-Phase Clusters

A question could be posed: What shape do small clusters take on as they grow to 10, 50, 100, or 10,000 atoms? Would irregular defective shapes, or symmetrical, close-packed shapes be formed, and at what size do true metallic properties take hold? In recent years it has become possible to attempt to answer such questions by experimentation. Elegant gas-phase metal cluster growth processes have been developed, such as Smalley's pulsed cluster beam (PCB) apparatus.[3–5] This method depends on pulsed laser vaporization of elements. The resultant atoms or small clusters are injected into a relatively cold helium steam, where cluster growth can take place. A vast array of clusters has been produced in this way and detected by laser ionization and mass spectrometry. Some cluster stoichiometries have been found to be more favorable than others, and these often represent closed-shell clusters and occasionally, MacKay icosohedra (those clusters where 1, 13, 55, 147, 309, 501, 932,... metal atoms are completely encapsulated in close-packed structures).

However, Riley and co-workers,[6,7] using a continuous flow cluster beam (CFCB) apparatus, have shown that both open and closed-shell clusters do form, and that gas-phase annealing can change the reactivity (and supposedly the shape) of certain iron and nickel particles.

These results and those of Jarrold and co-workers[8,9] show that relatively cool small metal clusters can take on unexpected geometrical shapes, presumably of a defective nature with many dangling corners or edges. They also show that geometrical shape can determine reactivities, so chemical reactivity and shape are important parameters in our examination of nanoparticles.[1]

7.1.3 Condensed-Phase Nanoparticles

Supramolecular and Cluster Compounds In Section 7.1.2 it was shown that metal atoms can be grown to small, isolated metal clusters. If we continue this approach, starting with the smallest and building toward larger particles, and also consider molecular growth, the areas of "supramolecular chemistry" (large molecule) and cluster compounds come into focus. Fascinating structures are being synthesized: For example, C_{60},[10-12] clusters of C_{60},[13] a "molecular Ferric wheel,"[14] self-assembled and self-replicating structures,[15] and cluster compounds[16] have been reported. However, since these topics cannot be covered here, and because an extensive review of free atoms, clusters, and nanoscale particles has appeared that covered mainly particles of 1 to 1000 atoms or 1 to 3 nm,[1] we will generally be concerned here with the regime 3 to 10 nm. Furthermore, mainly inorganic nanoparticles will be of interest as opposed to organic and/or van der Waals clusters.

Inorganic Nanoparticles[17]

Metals It is important to distinguish between naked and ligand-stabilized metal clusters/particles. Those considered naked include gas-phase clusters, discussed earlier, and those grown in low-temperature matrices in relatively inert diluents.[1] Then, of course, there are the ligand-stabilized clusters that constitute small metal clusters that have been trapped by strongly bound ligands to the extent that they are now stable, soluble molecules.[16] Unfortunately, such cluster compounds are usually limited to a metal core of about 40 atoms, much smaller than of interest to us here.[18]

One of the unique developments relevant here is the synthesis of a class of giant ligand-stabilized metal clusters by Schmid and co-workers,[19] whose formulated opinion is that colloids can serve as giant clusters. Schmid defines clusters as larger, perhaps up to 100 nm. Usually, it is possible to structurally characterize the smaller metal clusters by x-ray crystallography so that their geometrical shapes are known exactly. Colloids are generally polycrystalline and exist in a distribution of sizes. However, they are ligand or solvent stabilized but in a more dynamic way. An example of these beautiful structures is shown in Fig. 7.2, represented by $Au_{55}(PPh_3)_{12}Cl_6$ and $Pd_{561}(phen)_{36}O_{190-200}$.

Semiconductors Moving from the metals to semiconductors opens another wide array of nanoparticles, including Si, Ge, ZnO, TiO_2, CdS, and CdSe. These are particularly unusual materials because a unique property of semiconductors, the bandgap energy, changes with particle size, which of course has immense interest for solar cells, energy storage, photovoltaic cells, and much more. Figure 7.3, for example, illustrates the expected pattern of molecular orbitals and allowed electronic transitions for a 7-nm CdS nanoparticle.

Unique synthetic approaches are required to prepare and stabilize such binary (predominately ionic) nanoparticles, such as the inverse micelle method

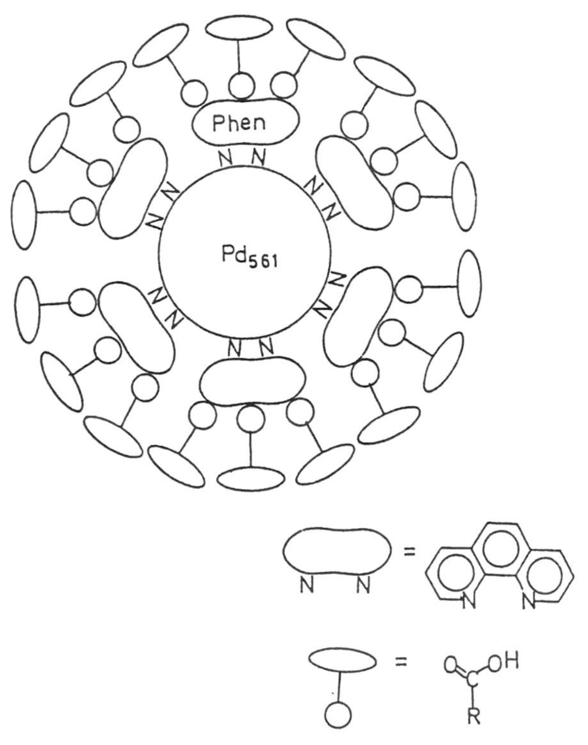

Figure 7.2 Pd_{561} nucleus in Pd_{561} $phen_{36}O_{190-200}$. (After Ref. 20.)

(discussed in Section 7.2.2) and thiol capping:[21]

$Cd(ClO_4)_2$ inside a hydrophilic nanocavity in a sea of organic solvent + $C_6H_5Se - Si(CH_3)_3$ in a sea of organic solvent

↓

C_6H_5
|
CdSe CdSe—C_6H_5
C_6H_5—SeCd SeCd
CdSe CdSe—C_6H_5
C_6H_5—SeCd SeCd
|
C_6H_5

In fact, devices have already been fabricated using nanocrystals of CdS, including thiol-capped, self-assembled monolayers.[22] Needless to say, semiconductor nanocrystals currently hold a special significance in the growing field of nanophase materials.

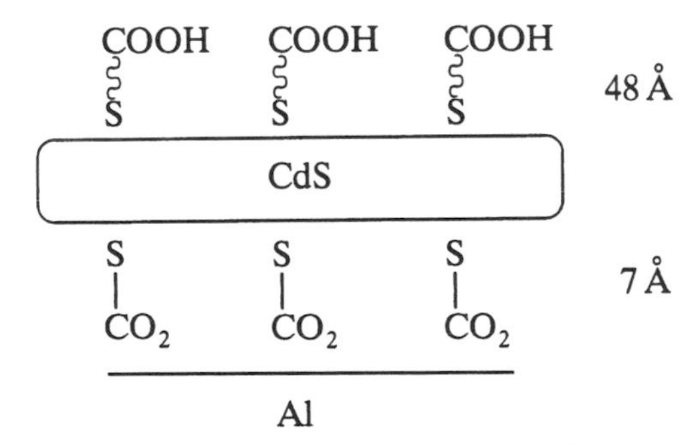

Dielectrics (Insulator Nanoparticles) Can simple, ionic, insulator particles (e.g., NaCl, MgO, CaF_2, or TiO_2) be unique due to particle size? The answer is definitely "yes." If we consider that nanoparticles of crystalline substances have about 10^{19} interfaces/cm^3 and range in surface areas up to 500 or even 800 m^2/g, fascinating possibilities come to mind. For example, if we compact but do not grow the nanocrystals, solids with multitudinous grain boundaries are formed. In the cases of CaF_2[23] and TiO_2,[24] solid samples are obtained which undergo plastic deformation at room temperature, presumably by diffusional creep. It can be proposed that further work in the area of

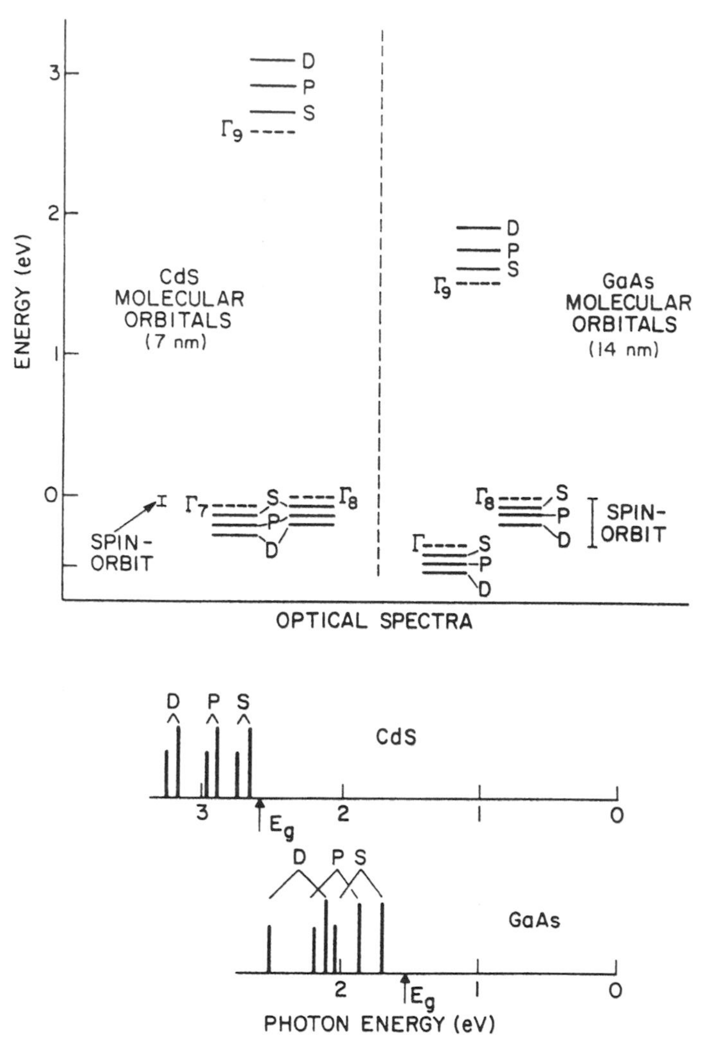

Figure 7.3 Molecular orbitals and allowed discrete electron transitions for spherical crystallites of CdS (7 nm) and GaAs (14 nm). (After Ref. 2.)

consolidated nanophase materials may lead to ceramics with increased flexibility, less brittleness, and perhaps greater strength. It may be possible to form materials with a large fraction of atoms at grain boundaries, perhaps in atomic arrangements that are unique. It may also be possible to produce binary materials of normally immiscible compounds or elements.

Another aspect to consider is that smaller and smaller particles may take on different crystalline forms or at least different morphologies, and this could affect their surface reactivities and/or adsorption properties. For example, Fig. 7.4 shows that 4-nm MgO crystallites have quite a different

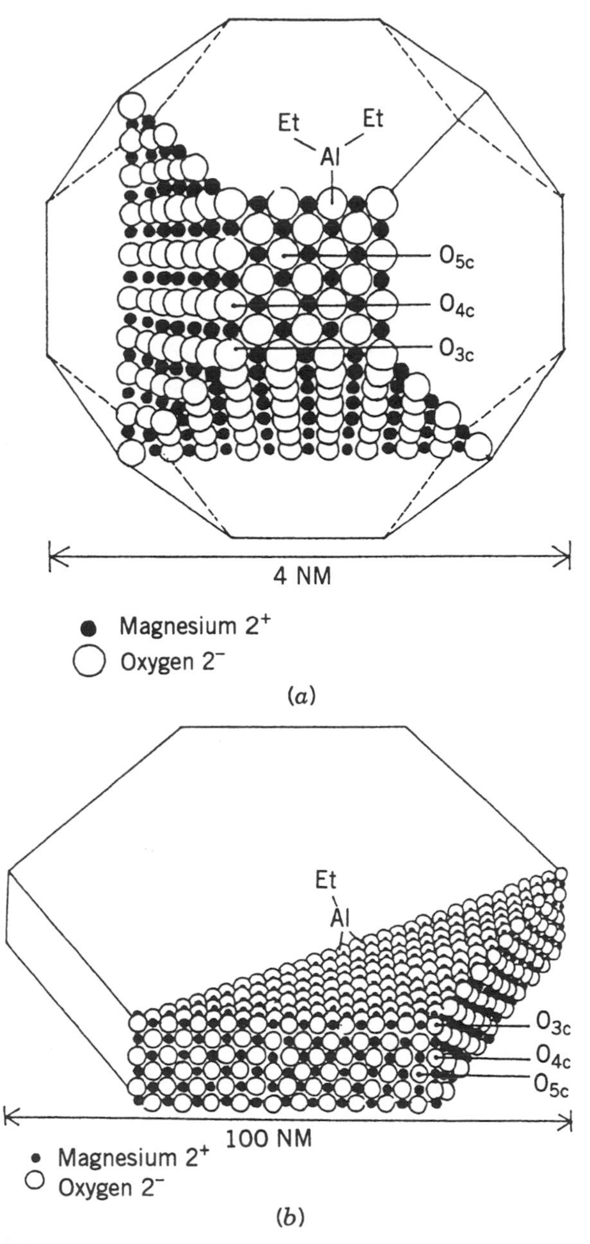

Figure 7.4 Idealized scale drawings of a 4-nm MgO nanocrystal and 9-nm-thick microcrystal. (From Ref. 25.)

morphology than larger crystallites and should possess more edge and corner sites (lower coordination sites) that certainly could affect adsorption properties.[25, 26]

7.1.4 Future Directions

If we consider that the periodic table of the elements "is a puzzle that has been given to us by God"[27] that holds a huge treasure chest of new solid materials, think of what this means in the realm of nanoparticles. Every known solid substance and every material yet to be discovered will yield a new set of properties, dependent on size. Optical properties, magnetic properties, melting points, specific heats, and crystal morphologies can all be influenced because nanophase materials serve as a bridge between the molecular and condensed phases. It would seem that the likelihood of new discoveries and new applications is extremely high. Many scientists look forward to the art and the science of nanophase materials.[27, 28]

7.2 METHODS OF PREPARATION

"Every gemstone, every metallic alloy, every superconductor, every speck of dust, every semiconducting crystal, every piece of biological tissue, absolutely every material that ever was, is, or will be remains latent within the periodic table until someone or something puts its atomic building blocks together."[27] This quote is the essence of the frontiers of materials chemistry, and it takes on a fourth dimension when we add the fact that this infinite array of materials can have different properties with change in particle size! We now examine how chemists have approached the synthesis of these exciting new structures.

7.2.1 Metal Particles

Metal Atom Aggregation in Inert Gases and Cold Matrices Perhaps the most straightforward approach to producing metallic nanoparticles is by controlled aggregation of atoms. It is relatively easy to vaporize metals using resistive heating from crucibles, electron-beam heating, or laser pulses.[1, 29] As mentioned earlier, rather elegant methods have been developed for gas-phase metal cluster growth processes employing pulsed laser evaporation of metals. The atoms are ejected into a cold helium gas stream and clusters up to about 100 atoms are detected by laser ionization and mass spectrometry. However, this approach does not lead to particles large enough for isolation as nanoparticles, and the scale is too small for synthetic purposes. Nonetheless, these studies are teaching us a great deal about formation, stabilities, and reactivities of very small metal clusters.

For synthetic purposes, a similar and much older method has been described, the gas evaporation method (GEM).[1] The GEM refers to vaporization of elements in a static pressure of inert gas, helium or argon. Clustering takes place near the vaporization source, and the particles eventually deposit on the inner walls of the evaporation chamber. This approach has been used for many years on a reasonably large scale.[30-33] A vacuum chamber is charged with 10 to 50-torr helium, and the metal or metals are evaporated by resistive heating from a high-temperature crucible. The ultimate particle size depends on pressure as well as gas temperature. The fine particles can be collected on a cool grid or on the walls of the chamber. A slight variation of this method has been used to prepare soot particles containing a significant portion of the now famous Buckminster fullerenes (C_{60}, C_{70}, etc.).[11] In this adaption the vaporation source is a carbon arc. A further adaption allows a reactive gas to react with the deposited metallic nanocrystals. For example, TiO_2 can be produced in nanoscale by vaporization of Ti in an inert gas. Nanoparticles of Ti were collected on a cold finger at 77 K. Then an oxidizing gas was inlet, and TiO_2 nanoparticles were formed. These were scraped down in a hold chamber (as shown in Fig. 7.5), compacted, and then mechanical properties were determined.[23]

Another approach that has broad appeal is the growth of clusters from atoms in low-temperature matrices. Depending on experimental conditions, the smallest metal clusters (dimers, trimers, etc.) can be trapped in frozen argon, krypton, or xenon and spectroscopically analyzed.[34-36] Much larger clusters can be prepared by allowing further growth, and materials can be isolated in multigram quantities as ultrafine powders.[37,38] This approach, often employing organic solvents as a matrix host material, has been termed the solvated metal atom dispersion (SMAD) method for ultrafine metal powders. In a typical experiment, 1 to 2 g of a metal (almost any metal) is vaporized from a high-temperature crucible and codeposited over 1 to 2 h with 100 g of organic solvent (such as pentane or toluene) in a 3-L vacuum chamber cooled in liquid nitrogen.[1,39] Upon warming form 77 K to room temperature, the ultrafine particles form in a very reactive state and with surface areas ranging from 30 to 80 m^2/g. Attempts have been made to understand the clustering process. Generally, slow warming of the matrix using a large molar excess of solvent to metal and use of more polar (more strongly ligating) solvents aid in the formation of smaller particles.[40] Thus nickel powders have smaller crystallite sizes in the order THF $<$ toluene $<$ pentane $\simeq$ hexane.[41,42]

After extensive study of the SMAD process, the following points can be made:[40]

1. Upon slight warming and matrix softening and meltdown, solvated atoms became labile and begin to oligomerize.
2. Within a certain narrow range of solvent viscosity and temperature, the oligomerization is partially reversible, and competition is established

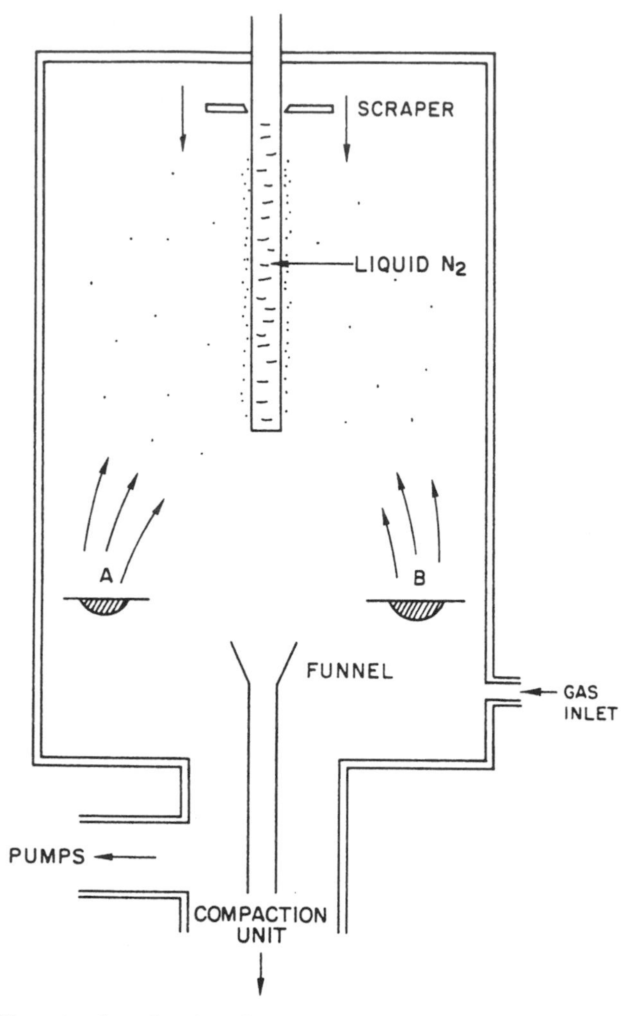

Figure 7.5 Vaporization chamber for nanoparticle formation; a reactive gas can be present during vaporization or can be added later. (After Ref. 2.)

between further cluster growth, declustering, and solvent reaction/ligation. A milder, selective growth process is enabled.

3. If the warming rate is rapid through this critical narrow viscosity range, the cluster growth process dominates and rapid growth occurs. A nonselective growth process is enabled.

4. As the clusters/particles grow, they become less mobile with size increase, and eventually the solvent reaction/ligation rate catches up, and further growth is stopped.

The SMAD method has been used to prepare a wide array of useful nanomaterials:[40] active metals for organic chemistry,[37] catalysts,[43,44] as precursors to metal carbides,[42] nonaqueous colloidal metals useful in metal-film-forming processes,[45] metal particles in polymers,[46-48] and mono- and bimetallic magnetic particles.[49]

Thermal or Sonocative Decomposition of Metal Carbonyls Heating or sonication can also be used on certain inorganic metal complexes to prepare nanoscale metal particles. Iron nanoparticles (5 to 20 nm) have been prepared by thermolysis of $Fe(CO)_5$ in polymer solutions, and sonication has also been used quite successfully.[50] Generally, the metal particles have carbonaceous impurities, and these tend to allow isolation of noncrystalline materials.[51]

Reduction of Metal Ions By reducing metal ions in solution, metal atoms can be formed, and these agglomerate into nanoparticles in much the same way as discussed earlier for metal atoms in warming matrices. However, the process of reduction/agglomeration can be quite complex. Nonetheless, nanoscale particles are usually the result. Some common reduction methods are considered next.

Borohydride Reductions The reduction of metal ions in aqueous solution has been known since the 1950s and derived from classified research during the World War II Manhattan Project.[52] Continued work showed that it was an effective way to produce ultrafine metallic particles or metal boride particles for catalytic purposes.[53] In fact, it was found that the forming metal particles (e.g., Co) were effective catalysts for borohydride hydrolysis in the water medium:

$$BH_4^- + 2H_2O \xrightarrow{\text{catalytic particles}} BO_2^- + 4H_2$$

Over the years, this approach to reduction of aqueous metal ions has been extant and has aided the preparation of magnetic materials.[54-56]

Researchers have continually pushed the limits of the BH_4^-/M^{n+} system, and there is confusion in the literature about what can actually be reduced. The procedure also has a notorious reputation for irreproducibility with regard to crystalline versus noncrystalline ("amorphous"), to final boron content in the particles, the best method of reagent mixing, and to oxidative sensitivity of the fine particles. Due to these problems, a series of investigations have been reported, where some of the complex chemistry of the reduction process was elucidated. It was shown that aqueous and non-aqueous media can yield very different results.[57,58] For example, with Co^{2+} reduction in aqueous solution, nanoparticles of Co_2B was the product, and the complete stoichiometry is shown in the reaction[57]

$$2Co^{2+} + 4BH_4^- + 9H_2O \rightarrow Co_2B + 12.5H_2 + 3B(OH)_3 \qquad (1)$$

An intermediate in this process could be an aquatic dimer species (see Scheme 7.1). On the other hand, in nonaqueous media (dry diglyme: $CH_3OCH_2CH_2OCH_2CH_2OCH_2$), reduction of $CoBr_2$ yielded nanoscale cobalt metal. In this case a diglyme solvated species is believed to be an intermediate:[58]

$$Co^{2+} + BH_4^- \xrightarrow{\text{diglyme}} [(\text{diglyme})_n Co(BH_4)_2] \rightarrow Co_{(s)} + H_2 + B_2H_6$$

Indeed, it was found that progressive addition of small amounts of water caused the reaction product to change progressively from $Co_{(s)}$ to Co_2B. It was also found that the Co_2B and $Co_{(s)}$ nanoparticles can serve as catalysts both for the hydrolysis of BH_4^-, thus depleting the solution of BH_4^- before all of the Co^{2+} was reduced, and for decomposition of B_2H_6:

$$B_2H_6 \xrightarrow[\text{Co catalyst}]{\text{diglyme}} B_{(s)} + 3H_2$$

These findings combined with similar studies of $Fe^{2+/3+}$, Ni^{2+}, and Cu^{2+} led to conclusions about major products that are formed (shown in Table 7.1) and that exposure to air can cause exothermic oxidation processes that can lead to the wrong conclusion regarding what the true primary product is.[60]

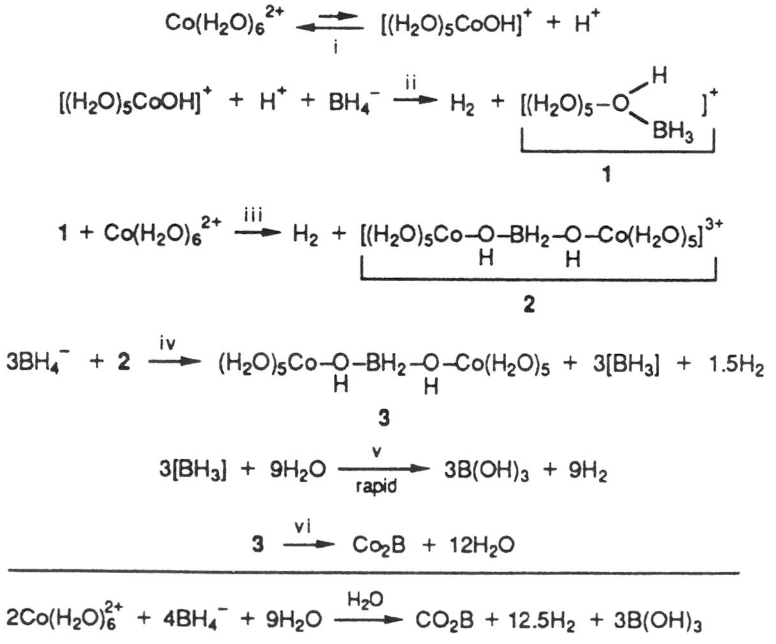

$$Co(H_2O)_6^{2+} \underset{i}{\rightleftharpoons} [(H_2O)_5CoOH]^+ + H^+$$

$$[(H_2O)_5CoOH]^+ + H^+ + BH_4^- \xrightarrow{ii} H_2 + \underbrace{[(H_2O)_5\text{-}O(H)BH_3]^+}_{\mathbf{1}}$$

$$\mathbf{1} + Co(H_2O)_6^{2+} \xrightarrow{iii} H_2 + \underbrace{[(H_2O)_5Co\text{-}O(H)\text{-}BH_2\text{-}O(H)\text{-}Co(H_2O)_5]^{3+}}_{\mathbf{2}}$$

$$3BH_4^- + \mathbf{2} \xrightarrow{iv} \underbrace{(H_2O)_5Co\text{-}O(H)\text{-}BH_2\text{-}O(H)\text{-}Co(H_2O)_5}_{\mathbf{3}} + 3[BH_3] + 1.5H_2$$

$$3[BH_3] + 9H_2O \xrightarrow[\text{rapid}]{v} 3B(OH)_3 + 9H_2$$

$$\mathbf{3} \xrightarrow{vi} Co_2B + 12H_2O$$

$$2Co(H_2O)_6^{2+} + 4BH_4^- + 9H_2O \xrightarrow{H_2O} CO_2B + 12.5H_2 + 3B(OH)_3$$

Scheme 7.1 Proposed chemical reaction steps in aqueous borohydride reduction of cobalt ions. (From Ref. 57.)

TABLE 7.1 Trends in Borohydride Reduction of First-Row Transition-Metal Ions

		Metal-Containing Products of Borohydride Reduction		
Metal Ion	Reduction Potential, E^a	In Water Under Argon	In Diglyme Under Argon	In Water Under Air
Fe^{3+}	-0.036^b	Fe	FeB	Fe, FeO_x
Fe^{2+}	-0.41	Fe	Fe_2B^c	Fe, FeO_x
Co^{2+}	-0.28	Co_2B	Co^d	$Co_3(BO_3)_2$, Co
Ni^{2+}	-0.23	Ni_2B	Ni, Ni_2B, Ni_3B^c	Ni, NiO
Cu^{2+}	$+0.34$	Cu	Cu^e	Cu, Cu_2O

Source: Ref. 59.

[a] Standard condition $M^{2+} + 2e^- \rightarrow M_{(s)}$.
[b] $Fe^{3+} + 3e^- \rightarrow Fe_{(s)}$.
[c] $Ni(BH_4)_2$ formed first; slow decomposition at 25°C yielded a mixture of solid Ni, Ni_2B, and Ni_3B. $Fe(BH_4)_2$(diglyme) and $Fe(BH_4)_3$(diglyme) are stable in diglyme, and upon heating to 70°C the precipitation of FeB or Fe_2B takes place.
[d] $Co(BH_4)_2$ is not stable.
[e] $Cu(BH_4)_2$ formed first; thermal decomposition at 90°C yield $Cu_{(s)}$.

For example, nanoparticle Co_2B reacts with O_2 to yield encapsulated $Co_{(s)}$:

$$4Co_2B + 3O_2 \rightarrow 8Co_{(s)} + B_2O_{3(s)}$$

In summary, BH_4^- reduction of metal ions can be very useful, but to obtain good reproducibility, the following points were made: (1) add the aqueous solutions of the metal ion to a fresh $NaBH_{4(aq)}$ solution by means of a T-joint with rapid mixing; or (2) add the aqueous metal ion solution to solid $NaBH_4$ rapidly with good mixing; (3) always work under airless conditions; (4) the particles formed are nanocrystalline with crystallite size 1 to 3 nm, and with incorporation of some solvent and heat treatment will always cause crystal growth and loss of solvent; (5) if a mixture of nanoparticle metal and metal boride is present before heat treatment, a M_xB_y solid solution can be formed, and the M/B ratio will depend on the composition of the original nanopowder mixture.

It should be possible to use the BH_4^- reduction process for metal ions with reduction/oxidation potentials lower than about 1.0 V. If reaction rates and experimental precautions are taken, it should be possible to prepare bimetallic alloy and/or bimetallic core-shell particles.[61]

Borohydride reductions even work well in reverse micelle systems (see Section 7.2.2 for a description of this technique). For example, Co particles as small as 1.8 nm and as large as 4.4 nm have been prepared, depending on the surfactant used, solution concentrations, and other factors.[62, 63] As would be expected based on the earlier discussion, if enough water is present, Co_2B, Ni_2B, and Ni—Co—B nanoparticles can be prepared.

Similar chemistry of hydroorganoborates also leads to reduction of metal ions to metallic nanoparticles. By the following general reaction, a series of metal particles of size range 6 to 100 nm were prepared:

$$MX_n + nNR_4BEt_3H \rightarrow M_{(s)} + nNR_4X + nBEt_3 + (n/2)H_2 \quad (2)$$

where M = metals of groups 6 to 11; X = Cl, Br; n = 2, 3; R = alkyl groups. In this procedure the boron content was generally low, less than 1.5%. Mixed metal particles such as Fe—Co have also been prepared in this way.[64, 65]

Alkali Metal (Li, Na, K) Reductions Rieke and co-workers[66, 67] have described several procedures for using the high reducing potential of alkali metals for preparing nanoscale *metal* particles under mild conditions. It is imperative that dry solvents (usually diglyme, THF, or other ethers) are used and that all work is carried out under anaerobic conditions. The solvent must at least partially dissolve the metal salt in question. Examples are

$$NiCl_2 + 2K \xrightarrow{\text{diglyme}} Ni_{(s)} + 2KCl \quad (3)$$

$$AlCl_3 + 3K \xrightarrow{\text{xylene}} Al_{(s)} + 3KCl \quad (4)$$

As can be seen by the $AlCl_3$ case, a compound that could not be reduced by the borohydride method, the alkali reduction method can be applied to almost any metal salt, and the product is usually nanocrystalline metal deposited on microcrystalline KCl, NaCl, or LiCl. These products are extremely reactive with oxygen and a variety of organic compounds, and in fact, the main thrust of the Rieke work has been to develop new organic chemistry using these nanoscale "activated" metals.[68, 69]

An important improvement in this preparative method has been developed where naphthalene is added to a diglyme slurry of potassium metal. The naphthalene is soluble in the diglyme but has a high enough electron affinity that the K metal transfers electrons to it, producing potassium naphthalide. In this way the potassium is brought into solution:

$$K + C_{10}H_8 \xrightarrow{\text{diglyme}} [C_{10}H_8]^{\dot{-}} \; K^+ \quad (5)$$

and a strong reducing agent is formed that reacts more rapidly with added metal salts.

A very similar procedure has allowed preparation of "organically solvated magnesium."[70,71] In this instance, activated Mg metal can be complexed and solubilized by adding anthracene:

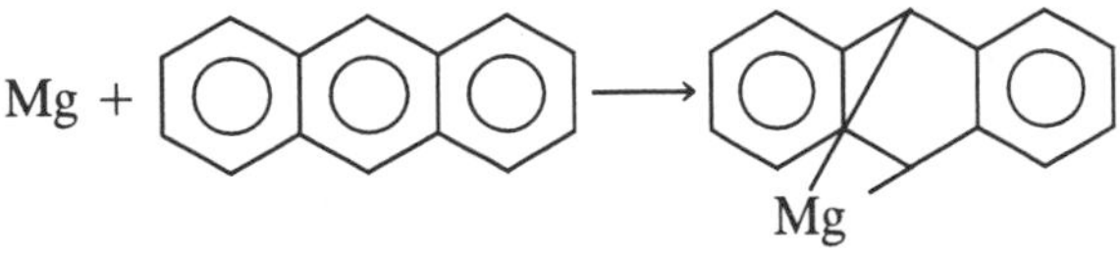

Mg + [anthracene] ⟶ [Mg-anthracene adduct]

This reagent is also excellent for carrying out reductions of transition-metal salts to yield nanoparticles.

Radiolysis Methods Radiation of aqueous solutions of Ag^+ causes the reduction of Ag^+ to Ag° due to the presence of solvated electrons. When colloid stabilizers such as polyphosphates are also present, extremely small silver clusters and cluster cations can be prepared and observed in solution for many hours.[72] In this way elementary steps leading to clusters have been studied. For example, $Ag\circ$ atoms have been found to react with Ag^+ to give Ag_2^+, and dimerization can occur to give Ag_4^{2+}. This species actually is in equilibrium:[73, 74]

$$Ag_4^{2+} \rightleftharpoons Ag^{3+} + Ag^+ \rightleftharpoons Ag_2 + 2Ag^+ \tag{6}$$

Basically, the radiolysis method allows spectroscopic study of the cluster growth process in aqueous or alcohol–water mixtures.

Reduction by Organic Glycols or Hydrazine Inorganic salts can be reduced in polyalcohols such as ethylene glycol simply by extended heating and boiling, and small particles of Co, Ni, Cu, and other metals can be formed.[75] Similarly, bimetallic colloidal particles (e.g., PdCu) have been prepared by thermal decomposition of the acetates in boiling bromobenzene, xylene, or 2-ethoxyethanol.[76, 77]

Finally, the reduction of Cu^{2+} by hydrazine hydrate in boiling methanol leads to Cu metal nanoparticles. The forming particles need not be stabilized by polyalcohols or other stabilizing agents. A range of particle sizes were observed (3 to 30 nm with a mean size of 14 nm), but an interesting finding was that atomic resolution TEM revealed the presence of both multiply twinned and parallel twinned morphologies.[78]

7.2.2 Semiconductor Particles

Most relevant to the nanoparticle theme has been development of methods for precipitating and encapsulating, simultaneously such binary compounds as CdS, CdSe, PdTe, and other metal–nonmetal combinations, and we will now examine some innovations in synthetic methods. Of course, the precipitation of sulfides and selenides has been carried out for a very long time, and certain methods are even included in textbook qualitative analysis schemes. Small particle sizes can be encouraged by using high dilution and low temperatures.[74, 79] However, to improve control of particle sizes in the nanometer size range, a variety of encapsulating/protecting procedures have been reported recently, including use of zeolites, porous glass gels, metaphosphates, vesicles, and micelles.

Zeolites and Other Porous Solids Zeolites contain exchangeable metal cations within their cage framework. Herron has carried out cadmium ion–sodium ion exchange by slurrying several grams of zeolite LZY-52 (sodium zeolite Y from Linde) in an aqueous $Cd(NO_3)_2$ solution. After filtration, washing, drying, and heat treatment under oxygen, the dry Cd-exchanged zeolite was treated with H_2S, and very small $(CdS)_4$ clusters were trapped within the sodalite framework.[80, 81] Similar results have been obtained for Cd^{2+}–zeolite–H_2Se reactions, yielding encapsulated $(CdSe)_x$ clusters[82] and Pb^{2+}–zeolite–H_2S reactions.[83] In a similar way, where nanocavities serve as templates for growth of nanoparticles, a membrane-based synthetic approaches has been devised.[84] Commercially available polymeric filtration membranes, porous aluminas, porous glasses, and zeolites have all been employed to some degree.

Inverse Micelles and Related Systems Other approaches at "arrested precipitation" have been summarized briefly by Steigerwald and Brus.[85] It was pointed out that small clusters always form with internal crystallinity by typical precipitation reactions,

$$Cd^{2+} + Na_2Se \rightarrow (CdSe)_s + 2Na^+ \tag{7}$$

and this implies that nucleation is either a reversible process (dissolution/readsorption until each ion is perfectly placed in the embryonic lattice) or there are severe stereoelectronic requirements that must be met.

To be able to work with more exotic combinations, it became necessary to devise reagents that were soluble in organic as well as aqueous solutions. A typical example is a trimethylsilyl($—Si(CH_3)_3$) reagent:

$$CdCl_2 + Se[Si(CH_3)_3]_2 \rightarrow (CdSe)_s + 2(CH_3)_3Si—Cl \tag{8}$$

Such reagents not only give solubility advantages but also allow avoidance of using toxic gases such as H_2S, H_2Se, and H_2Te. By changing the CH_3

substituent to larger R groups (e.g., pentamethylethyl, shown below), steric bulk can allow reaction rate control.

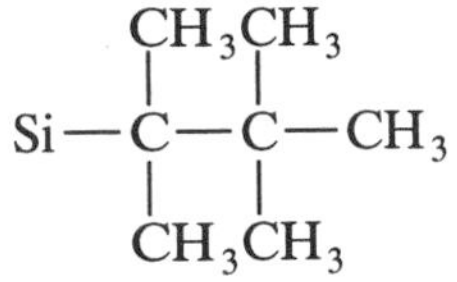

$$
\begin{array}{ccccc}
 & CH_3 & & CH_3 & \\
 & | & & | & \\
Si- & C & - & C & -CH_3 \\
 & | & & | & \\
 & CH_3 & & CH_3 &
\end{array}
$$

Other examples of new organometallic precipitation schemes include the preparation of $(NiTe)_s$, $(PdTe)_s$, and $(NiAs)_s$:[86, 87]

$$Et_3P{-}Te + Ni(COD)_2 \rightarrow (NiTe)_s + Et_3P + 2COD \tag{9}$$

$$2Co_2(CO_8) + 8Et_3P - Te \xrightarrow[-PEt_3]{-CO} Co_6Te_8(PEt_3)_6 \xrightarrow[heat]{-PEt_3} (CoTe)_s \tag{10}$$

These are interesting developments, and the driving force has been to develop new ways to achieve monodisperse nanocrystals of controllable size. Thus the precipitation reactions must be "controllably arrested." This brings us to a discussion of inverse micelles.

Fendler and co-workers described enhanced stability of CdS nanoparticles toward flocculation if the particles were entrapped in nanometer-sized water pockets.[88, 89] Such pockets are formed when surfactant molecules are placed in hydrocarbon solvents, thus causing the polar ends to gather together (Fig. 7.6). Addition of water-soluble salts causes the salt ions to gather in the pockets, and upon addition of a precipitating agent (e.g., Se^{2-} addition to encapsulated Cd^{2+}), a small crystallite of CdSe is grown in the nano-reactor

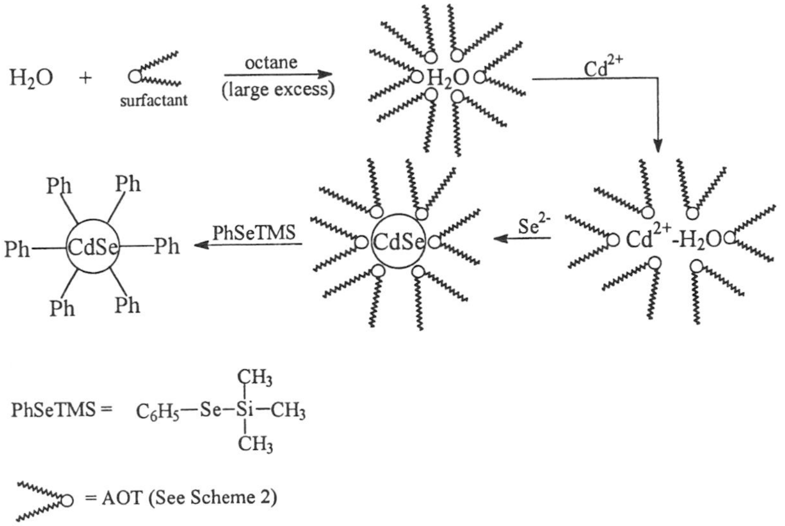

H_2O + surfactant $\xrightarrow[\text{(large excess)}]{\text{octane}}$ H_2O micelle $\xrightarrow{Cd^{2+}}$ Cd^{2+}-H_2O micelle $\xrightarrow{Se^{2-}}$ CdSe micelle $\xrightarrow{\text{PhSeTMS}}$ Ph-capped CdSe

$$PhSeTMS = C_6H_5{-}Se{-}Si(CH_3)_3$$

= AOT (See Scheme 2)

Figure 7.6 Inverse micelle method of preparing nanoscale CdSe particles and capping with phenyl groups.

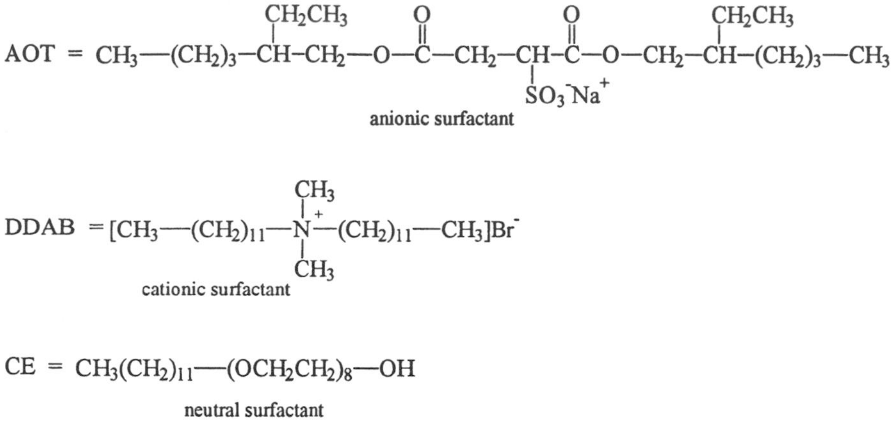

$$\text{AOT} = CH_3{-}(CH_2)_3{-}\underset{}{\overset{CH_2CH_3}{\overset{|}{C}H}}{-}CH_2{-}O{-}\overset{O}{\overset{\|}{C}}{-}CH_2{-}\underset{SO_3^-Na^+}{\underset{|}{C}H}{-}\overset{O}{\overset{\|}{C}}{-}O{-}CH_2{-}\overset{CH_2CH_3}{\overset{|}{C}H}{-}(CH_2)_3{-}CH_3$$

anionic surfactant

$$\text{DDAB} = [CH_3{-}(CH_2)_{11}{-}\underset{CH_3}{\underset{|}{\overset{CH_3}{\overset{|}{N^+}}}}{-}(CH_2)_{11}{-}CH_3]Br^-$$

cationic surfactant

$$\text{CE} = CH_3(CH_2)_{11}{-}(OCH_2CH_2)_8{-}OH$$

neutral surfactant

Scheme 7.2 Typical surfactants used in preparing inverse micelles.

pocket, and the size of the particle may be governed by the size of the pocket.

Steigerwald and Brus[79] have reported a major advance in the technology by "grafting" a micelle encapsulated crystallite with SePh groups. In this way, the CdSe crystallite is grown and is covalently linked to an organic coating. This allows isolation of a "capped molecular" species, which is soluble in organic solvents.

In such reverse micelle systems, the ultimate control of particle size is very much dependent on the molar ratio of water to surfactant (designated as w). Indeed, crystallites of CdSe can range in size from 1.5 to 6.0 nm, with w ranging from 0 to 8.[90,91] Although the exact mechanism of cluster growth is not known, it is clear that the amount of water present can affect micelle pocket size, and ultimately, the size of the CdSe crystallite obtained.

Herron and co-workers have reported similar capping schemes for CdS and have reported cluster empirical formulas. For example, for a crystallite size of about 2.0 nm, elemental analysis yields a formula of $Cd_1S_{0.54}(SC_6H_5)_{0.92}(HSC_6H_5)_{0.04}$, suggesting the presence of mainly grafted $S{-}C_6H_5$ groups but also some intact $H{-}S{-}C_6H_5$ species.[92] In this study, ^{113}Cd NMR confirmed the presence of sphalerite crystalline CdS as the particle core and was able to show an increased percentage of Cd in this core as the $S/S{-}C_6H_5$ ratio increased. The surfactants used are varied, but several typical molecules are shown in Scheme 7.2.

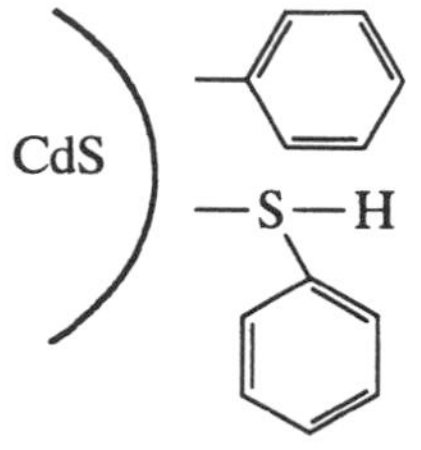

CdS

—S—H

Recently, inverse micelles have been used to prepare a wide variety of nearly monodisperse nanoparticles, including metals such as Pt[93] and Au[94] as well as metal borides[95] and Fe_3O_4.[96, 97]

Gels, Phosphates, and Polymers The inverse micelle approach to encapsulating nanoparticles depends on solvation energies, or in other words, the favorable enthalpy of inverse micelle formation that greatly overcomes the unfavorable entropy term. Another way to form cavities that might be capable of encapsulating nanoparticles is by forming bridged polysilesquioxanes, as reported by Choi and Shea.[98] In this work disilyl species such as $Cl_3Si—CH_2CH_2CH_2CH_2CH_2CH_2SiCl_3$ are hydrolyzed, forming bridged polymers as shown in Scheme 7.3. In the xerogel process for producing these polymeric materials, surface areas of about 530 m^2/g and pore diameters of 3 to 4 nm were determined. The CdS nanoparticles that grew in the cavities and were trapped were on the order of 6 to 9 nm.

Other sol-gel approaches have also been reported,[99] but most of them are covered in Chapter 9. There have been other interesting reports where semiconducting nanocrystals have been grown in certain types of cultured

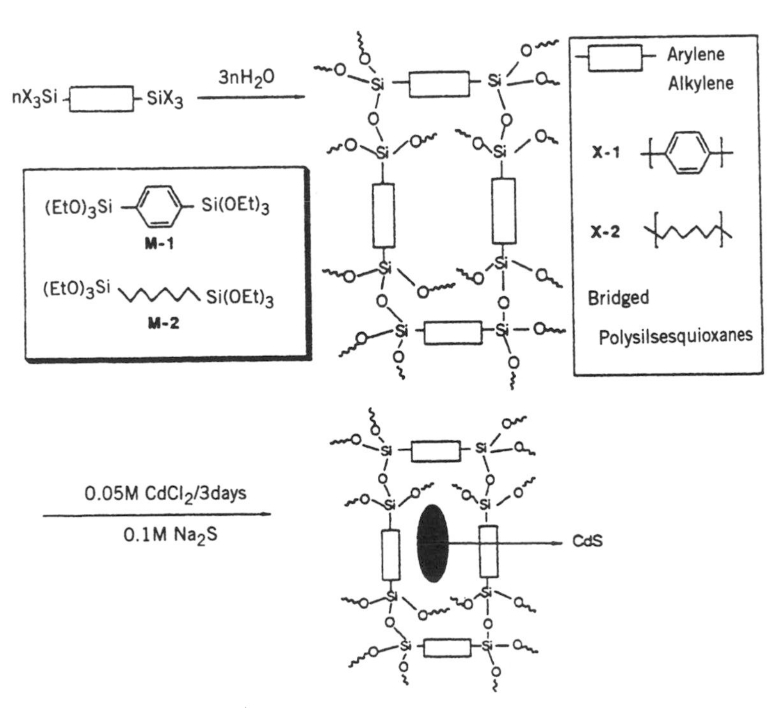

Scheme 7.3 Bridged polysilsesquioxanes as cavities for nanoparticle formation and trapping. (After Ref. 98.)

yeasts.[100] Also, by using newly developed reagents, GaP and GaAs have been prepared.[101, 102]

$$AsCl_3 + Na/K \rightarrow As(Na)_3 \quad (11)$$

$$GaCl_3 + As(Na)_3 \rightarrow GaAs \quad (12)$$

$$GaCl_3 + As(SiMe_3)_3 \rightarrow GaAs \quad (13)$$

7.2.3 Metal Oxides / Sulfides / Nitrides: Ceramic Nanoparticles

The preparation of monodiperse particles of metal oxides and hydroxides with specific shapes is as old as colloid chemistry itself.[103, 104] However, next we review briefly some of the more recent advances that have led to smaller and smaller particles.

Sol-Gel Methods: Aerogels and Xerogels This topic is covered in some detail in Chapter 9 of this volume. Only a very brief discussion is given here.

Direct precipitation of aqueous metal ions by addition of hydroxide can initially cause gel formation and then ultrafine precipitates. However, drying and dehydration usually cause severe sintering, collapse of pore structures, and loss of surface area. This is believed to occur because the solvent, water, causes a vapor–liquid interface within the channels and pores and results in surface tension, creating concave menisci inside the network. As the water is evaporated and the menisci reduce, the buildup of tensile force acting on the walls of the channels and pores causes shrinkage. The resultant product is often hard and glassy (not nanophase) and is called a *xerogel.*

To preserve the texture of the wet gel, the vapor interface must be eliminated during the drying process. Two approaches have been used in combination: (1) replace the water with solvents of higher vapor pressure and less surface tension, such as alcohols and aromatics; and (2) heat the wet gel to above the critical temperature of the solvent in a high-pressure autoclave and then vent gaseous-fluidlike solvent rapidly as a vapor (hypercritical drying), yielding *aerogels*.[105–108]

A pertinent example is the optimization of solvent mix and conditions for preparing $Mg(OH)_2$ (900 m^2/g, 3.3-nm crystallites) and MgO (400 m^2/g, 4.2-nm crystallites), as shown below.[108, 109]

1. $Mg_{(s)} + CH_3OH \xrightarrow[CH_3OH]{\text{reflux}} Mg(OCH_3)_2 + H_2$

2. Toluene is added to give a 20 : 80 by volume mix of CH_3OH : toluene.

3. $Mg(OCH_3)_2 + H_2O \xrightarrow[\text{toluene}]{CH_3OH} \underset{\text{gel}}{Mg(OH)_2} + 2CH_3OH$

4. $\underset{\text{gel}}{Mg(OH)_2} \xrightarrow[\substack{CH_3OH,\ \text{toluene} \\ \text{(hypercritical drying)}}]{265°C} \underset{\substack{\text{ultrafine} \\ \text{particles}}}{Mg(OH)_2}$

5. $Mg(OH_2) \xrightarrow{25-500°C} \underset{\text{nanocrystals}}{MgO} + H_2O$

This general approach has been used to prepare a wide series of nanoparticles of MgO, CaO, Al_2O_3, ZrO_2, TiO_2, SiO_2, and others.

A further improvement in the hypercritical drying step has involved the use of supercritical CO_2 as replacement solvent. Quite often, light, "airy" monoliths can be prepared where the original gel network has been very well preserved rather than just serving as a precursor to ultrafine powders.[106, 107]

Precipitation and Digestion: Some Recent Advances Nanoparticles can sometimes by produced by the age-old direct precipitation methods if reaction conditions and post-treatment conditions are carefully controlled. For example, controlled precipitation of metal hydroxides is possible by careful manipulation of pH. In some cases, particle size can be controlled by a second step, where a metal hydroxide is converted to an oxide by heat treatment in aqueous solution, a process called digestion. In this way, small particles can be prepared, which span a broad size range, yet retain uniformity of composition, crystallinity, and morphology.[110, 111] As an example, manganese ferrite ($MnFe_2O_4$) nanoscale particles ranging from 5 to 180 nm have been prepared from Fe^{3+}, Fe^{2+}, and Mn^{2+} aqueous solutions.[111] Particle-size control was available by adjusting the ratio of total metal ion in solution to hydroxide concentration. Further control was achieved by the digestion procedure. Thus, upon digestion, nanocrystalline material formed, and the final particle size was a maximum at a [metal ion]/[OH^-] ratio of 0.32 or when $x = 9.4$:

$$\begin{aligned} &MnCl_{2(aq)} + 2FeCl_{3(aq)} + x\,NaOH_{(aq)} \\ &\quad \rightarrow MnFe_2O_{4(s)} + (x-8)NaOH_{(aq)} + 4H_2O + 8NaCl_{(aq)} \end{aligned} \tag{14}$$

Precipitation reactions have been used to prepare La, Ba, and Cu oxalates from electrolyte solutions, and after heat treatment, a high-T_c oxide superconductor, $La_{1.85}Ba_{0.15}CuO_4$, was formed.[112] Matijević has presented an excellent review of the literature regarding the precipitation of nanoscale particles.[113]

Aerosol Spray Pyrolysis This ingenious method involves forming tiny droplets of salt solutions by nebulization. The droplets are then entrained in flowing hot gas. As the droplets flow, they are heated higher and higher, causing first solvent evaporation and then consolidation or chemical reaction. A major advantage of this approach includes particle-size control by adjusting salt solution concentration and by the nebulization procedure. Another advantage is that each particle can be multicomponent with little or no segregation.[114, 115]

A typical example can be demonstrated in the preparation of gadolinium iron garnet nanoparticles $Gd_3Fe_5O_{12}$.[116] The preparation of this compound starts with solutions of $Gd(NO_3)_{3(aq)}$ and $Fe(NO_3)_{3(aq)}$ mixed together at atomic ratios of 3 : 5 for Gd : Fe. The concentrations of the solutions could be varied to control final particle sizes.

Aerosol spray pyrolysis apparatus includes a nebulizer, aerosol formation, dryer, and collector. This apparatus allows very high temperatures to be achieved, often up to 800°C or higher. Usually, nitrogen is employed as the flow gas, although, on occasion, oxygen is added if oxidizing power is desired.[117]

Since metal oxides are usually the desired products, in the example at hand, some pyrolytic chemical decomposition reactions must take place during the aerosol heating. A balanced equation would be

$$3Gd(NO_3)_3 + 5Fe(NO_3)_3 \rightarrow Ga_3Fe_3O_5 + 6O_2 + 24NO_2 \qquad (15)$$

Since O_2 and NO_2 are the most likely product gases, the hot environment is highly oxidizing, which would ensure that the highest stable oxidation states of the metal ions will be obtained (even if lower oxidation states such as Fe^{2+} were used). If lower oxidation states or mixed oxidation states are desired, for example, as in manganese ferrite $MnFe_2O_4$, nitrate salts are not the best choice. Indeed, in the case of $MnFeO_4$, only when $FeCl_3/MnCl_2$ salts were used was the synthesis successful, probably by the following reaction:[118]

$$2FeCl_3 + MnCl_2 + 4H_2O_{(g)} \xrightarrow{650°C} MnFe_2O_{4(s)} + 8HCl_{(g)} \qquad (16)$$

It is obvious that the choice of anion is crucial, along with temperature, flow rate, and starting solution concentrations.

Reactive Evaporation A modification of the GEM is to evaporate a metal in the presence of a small amount of oxidizing gas. During the gas-phase clustering step, the atoms/clusters are oxidized, and the oxidized clusters then collect on the walls or on a cold finger.[119] Using this approach (actually with or without oxidizer), an apparatus can be fashioned so that the powder collected can be removed and collected in a holding unit or compaction unit (see Fig. 7.5).[2,120]

Heating of Inorganic and Organometallic Compounds Recent results regarding pyrolysis of new inorganic/organometallic compounds are promising for obtaining particles of refractory compounds such as lanthanide borides and nitrides. Shore and co-workers[121] have prepared several borohydride complexes and characterized them thoroughly: $(CH_3CN)_3Yb(BH_4)_2$; $(C_5H_5N)_4Yb(BH_4)_2$; $(CH_3CN)_2Eu(BH_4)_2$; $(C_5H_5N)_2Eu(BH_4)_2$. The lanthanide metal ions are very resistant to reduction and so are capable of forming stable BH_4^- complexes. Upon pyrolysis, lanthanum borides could be produced:

$$(CH_3CN)_xYb(BH_4)_2 \xrightarrow[-CH_3CN,\,-H_2,\,-Yb]{1000°C,\ \text{vacuum}} YbB_4 \qquad (17)$$

In another example, LaDuca and Wolczanski[122] have prepared lanthanum nitrides by pyrolysis of silyamino complexes:

$$LnCl_3 + 3LiN[Si(CH_3)_3]_2 \xrightarrow[-LiCl]{THF}$$

$$[[(CH_3)_3Si]_2N]_3Ln \xrightarrow[150-180°C]{N_2} \xrightarrow[210°]{NH_3} LnN_{1-x}NH_3 \xrightarrow[heat]{} LnN$$

$$Ln = Y, La, Sm, Eu, Tb, Yb, Pr, Nd, Er \quad (18)$$

Finally, it is of interest again to consider metal vapor deposition approaches to formation of metal carbides. For example, highly reactive nickel particles can be prepared by nickel atom clustering in low-temperature alkanes.[123] During this process some carbonaceous material is incorporated in the particles due to attack of pentane on the highly reactive growing clusters. Thus particles of "active nickel" and "active carbon" coexist at room temperature. However, upon heating to about 200°C, the composite particles were converted to Ni and Ni_3C; in the presence of additional carbon sources (e.g., octene, octane, or methyl methacrylate), Ni_3C was formed cleanly and completely. Similar results were obtained with palladium. These results show that relatively low temperatures (180 to 200°C) will cause this conversion, and these mild conditions are apparently possible due to the small size and reactive nature of the Ni_x—C_yH_z particles.

$$Ni_{vapor} + pentane \xrightarrow{77\,K} \xrightarrow{300\,K} Ni_xC_yH_z \xrightarrow[octane]{180°} Ni_3C \quad (19)$$

7.3 PHYSICAL PROPERTIES

Nanoparticles display new physical properties for two reasons: (1) finite-size effects in which electronic bands give way to molecular orbitals as the size decreases, and (2) surface/interface effects (surface atoms/bulk atoms is about 1 in a 3-nm particle, or 50% surface atoms). The latter circumstance, that small particles represent surface matter in macroscopic quantities, is often not fully appreciated in interpretive schemes.

7.3.1 Metallic Behavior

Harrison and Edwards have given an excellent discussion of electrons in small metallic particles and in cessation of metallic behavior with decrease in size.[124] Of course, it can be recognized that a single metal atom could not behave as a metal. In fact, clusters of 100 atoms may not exhibit such behavior; however, the region of 100 to 10,000 atoms should, according to theory, show quantum size effects and probably a transition from the non-metallic to metallic behavior. That is, there is a gradual evolution of solid-state characteristics, such as quasi-continuous density for electronic states, thus the potential appearance of a conduction band.

Figure 7.7 demonstrates one-electron energy gaps as a function of particle diameters as well as calculated surface fraction. Thus, for a particle containing N atoms, the spacing of adjacent energy levels denoted by Δ will be of the order E_F/N (E_F is the Fermi energy). If this energy gap for electronic excitation is surmountable by thermal or incident radiation, quantum size effects are to be expected. Indeed, for sodium metal the cessation of conducting behavior should occur for particle diameters below 2.0 nm. Furthermore, if the energy gap is larger than the thermal energy, changes in the magnetic properties of the itinerant electrons should be substantial.

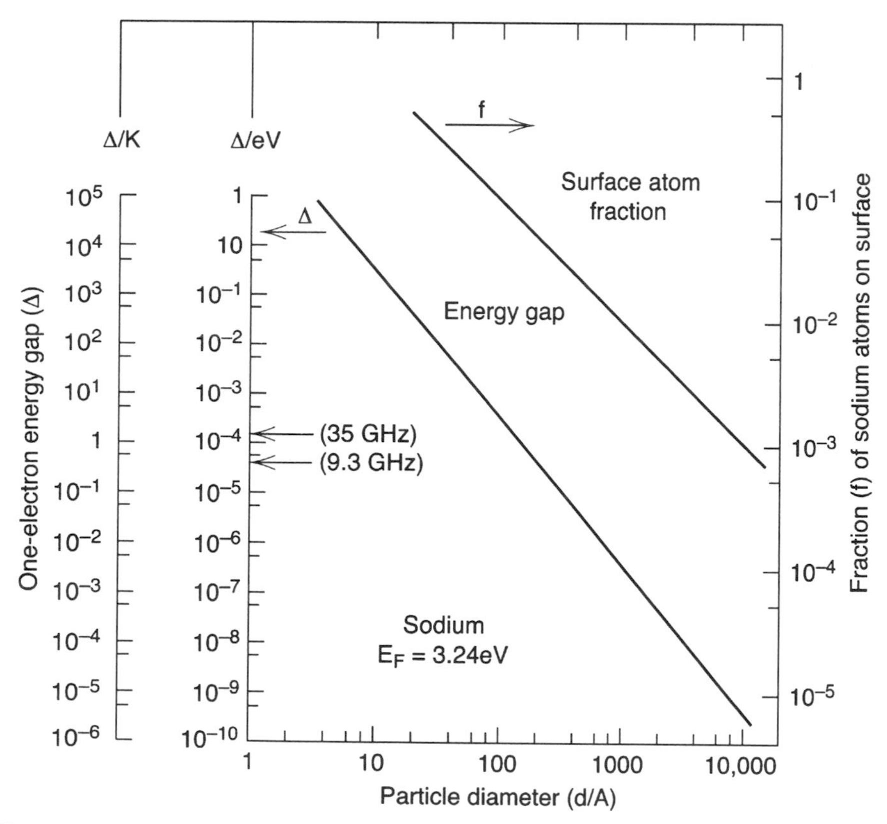

Figure 7.7 Plot of average energy level spacing Δ as a function of particle diameter, *d*, and the calculated surface fraction, *f*, of atoms in a cluster as a function of *d*. (After Ref. 124.)

7.3.2 Magnetic Behavior

A brief introduction to magnetism is needed to understand magnetic nanoparticles. The origin of magnetism in Fe, Co, and Ni (the ferromagnetic metals) is due to collective electrons or itinerant electrons. The following criteria for the existence of ferromagnetism have been established: (1) there must be electrons in partially filled energy bands so that vacant energy levels are available for electrons with unpaired spins to move into; (2) the density of levels in the band must be high so that spin alignment results in only a small increase in energy; and (3) the crystal lattice must have the proper spacing so that *d*-electron spins in one atom can align with the spins in a neighboring atom.[125, 126]

Increasing temperature causes thermal agitation that works against these exchange forces, and above a certain temperature (T_c = Curie temperature) ferromagnetism disappears and ferromagnets become paramagnets. Due to the existence of these forces, it may be expected that all ferromagnetic materials would spontaneously be magnetized to saturation below T_c. However, this is usually not the case, and Weiss predicted the existence of magnetic domains to explain this.[127]

Domains are small regions, about 22 nm in bulk iron metal, within the specimen which are spontaneously magnetized to saturation but are oriented in such a way that the sample as a whole has a reduced magnetization. Domain walls exist between domains made up of atoms of more random spin orientation.

Application of an external magnetic field to a demagnetized sample will cause an increase in the sample's internal magnetic field (domains are caused to align). The maximum internal magnetization is called the saturation magnetization, σ_s. Upon reversing the external field, eventually reaching zero applied field, the sample will often retain a remnance magnetization, and upon applying a negative external field, σ_s is driven to zero. The negative external field (energy) necessary is the coercivity, H_c. These parameters are of great importance in our following discussion of small-particle magnetism.

In 1930, Frenkel and Dorfman[128] predicted that a particle of ferromagnetic material below some critical size would behave as a single magnetic domain. The critical size for single-domain behavior depends on several factors, including particle shape and σ_s. A collection of single-domain particles that are not interacting with each other can reverse their magnetization only by rigid rotation of the magnetization vector of each individual particle. As compared to bulk material, which reverses magnetization by domain-wall displacement at low fields, rigid rotation requires large reverse fields so as to overcome any anisotropies (shape, crystal, stress) that oppose the rotation. Thus, in ultrafine particles it has been possible to observe high coercivities, and this has found great application in the field of information storage.

Carmen[129] reported that for iron nanoparticles of 25 nm diameter, a maximum H_c was observed, whereas at smaller sizes, H_c decreased and superparamagnetism sets in.[130] Superparamagnetism was first reported by Neel[131] and is due to sufficient thermal energy equilibrating the magnetization of an assembly of single-domain particles over a time scale shorter than the time of the experiment. The particles behave similar to paramagnets, but rather than having a small moment of a few Bohr magnetons, the moment is a sum of the moments of all the atoms in the particle.

Superparamagnetic particles obey essentially the same physical laws as paramagnets (μ is the moment for a particle, whereas μ for a paramagnet is the moment for an atom). Therefore, superparamagnetic materials will exhibit no hysteresis and will have temperature-dependent properties such that magnetization curves obtained at different temperatures will superimpose when plotted against H/T.

Below a critical temperature T_B, the blocking temperature, superparamagnetic behavior disappears and ferromagnetism sets in.[132] Because there is an exponential dependence of critical temperatures (T_C, T_N, T_B) on particle volume, there is a well-defined (in theory) nanoparticle size at which transition to stable, single-domain particles occurs. For an ideal spherical iron particle having a diameter of 11.5 nm, a relaxation time of 10^{-1} s at 300 K should be observed, and it will behave as a superparamagnet. Increasing the size to 15.0 nm should yield a relaxation time of 10^9 s at 300 K, so a particle of this size should be stable and behave as a ferromagnet.

Indeed, the temperature dependence of the magnetic behavior of superparamagnetic materials can be used to estimate particle size and to deduce blocking temperatures.[133] However, such calculations must make major assumptions, such as no interparticle interactions, spherical particle shape, good crystallinity, and clean particles. In reality, this is never the case.

We see from this introduction to magnetic particles that the physical effects of magnetism can be extremely informative with regard to quantum size effects. However, to interpret the data, the particles themselves must be well characterized with regard to *shape*, *crystallinity*, and the *chemical state of the surface*. This being the case, nanoparticle magnetism studies must fall more and more into the realm of the materials chemist.

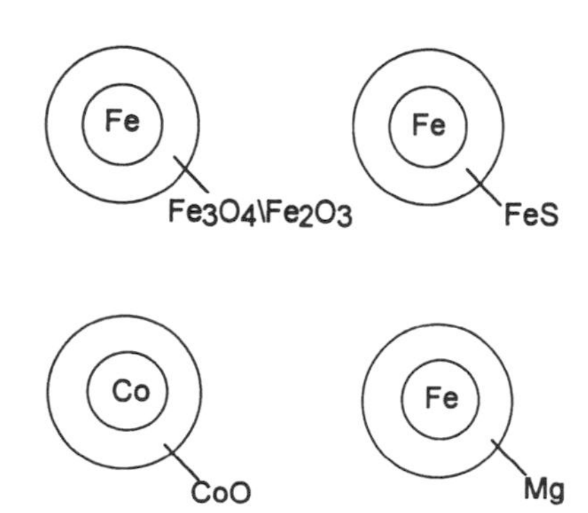

Figure 7.8 Core/shell magnetic particles synthesized and characterized.

Although there are many, many reports concerning nanoparticle magnetism dealing with oxides, mixed oxides, and sulfides as well as metals that could be discussed herein,[134] a few recent examples illustrate how materials chemistry is of some aid in creating core/shell nanoparticles that thereby help to protect the ferromagnetic core from oxidation and to allow examination of surface effects. Shown in Fig. 7.8 are nanoparticle core/shell systems that will be discussed.

It has been clearly shown that a magnetic coating material has a dramatic effect on coercivity H_c (often discussed in terms of exchange anisotropy). Meiklejohn and Bean studied fine particles of Co metal with an outer layer of CoO. An usually large H_c was explained in terms of exchange coupling between the spins of ferromagnetic Co and antiferromagnetic CoO; thus the reversal of the spins of Co in the Co core material was resisted by the strong crystal anisotropy of the CoO.[135]

Analogous results were obtained by Gangopadhyay and co-workers,[136] where H_c was found to increase with a decrease in particle size from 12 nm to 3 nm, while σ_s increased with increasing size. These results were explained by proposing that H_c is strongly affected by the interaction between the Fe oxide shell and the Fe core. The highest H_c obtained at room temperature was 1050 O_e for a particle with a 14.0-nm core, and its value at 10 K was 1425 Oe (just 37% higher). However, smaller particles with 2.5-nm core size went from a negligible H_c at 150 K to 3400 Oe at 10 K, showing the much stronger influence of temperature on the smaller nanoparticles. It was proposed that the smaller Fe core "feels" much more the effect of the Fe oxide shell, due to the higher Fe oxide/Fe ratio. The strong decrease in H_c with temperature increase was explained as due to the superparamagnetic behavior of the Fe oxide shell.

Overall, the interaction between the superparamagnetic shell and the Fe core is responsible for the high H_c values at low T and its drastic T dependence, and the effect is most pronounced in the smallest particles, where fluctuations of the magnetic moment of the shell are large enough to affect the entire particle. This rationale is supported by the onset of superparamagnetism in the shell at about 30 K when measured with a SQUID magnetometer.

The severe dependence of H_c on the shell material begs the question "what about other shell materials, magnetic or nonmagnetic?" (see Fig. 7.9). Experiments with [FeS]Fe nanoparticles have been very interesting in this regard. Again it was found that H_c and σ_s were strongly influenced by the coating material in conjunction with particle size. However, in the case of [Mg]Fe nanoparticles, strikingly different results were found. The nonmagnetic Mg coating allowed H_c to be very small and possessed a small temperature dependence.[137]

An important conclusion is that surface coatings in such small particles can *dominate* H_c and can also control temperature dependence. A reasonable explanation for the behavior of H_c, an extrinsic property, is that the spin-flip

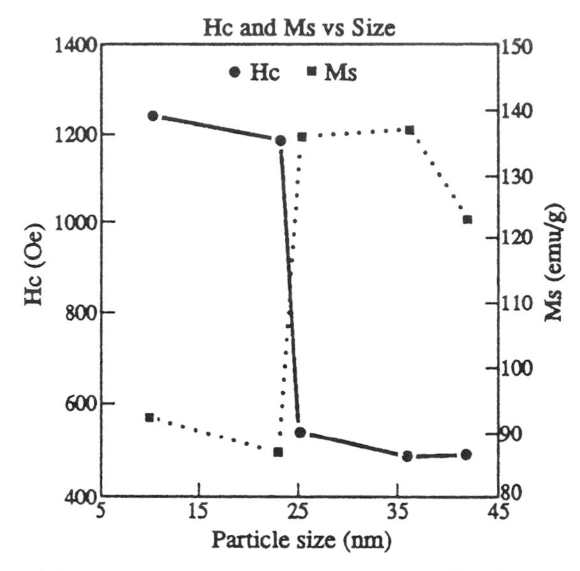

Figure 7.9 Coercivity versus Fe core particle size in the [FeS]Fe system. (From Ref. 125.)

behavior of a magnetic coating can help to pin the total spin of the core material. In the absence of a magnetic coating, no such pinning can occur, so H_c is small. Unfortunately, this rationale does not help in trying to explain the dependence of σ_s, an intrinsic property, on core particle size. Thus the question remains: Does a pristine Fe particle or cluster have a σ_s that is size dependent? Of course, we know that long-range order is necessary for ferromagnetic behavior. However, what is the magnetic moment per atom in $Fe, Fe_2, Fe_3, \ldots, Fe_n$? Recent gas-phase experiments on Fe, Co, and Ni clusters with magnetic moments measured using a Stern–Gerlach apparatus, particles deflected in a magnetic field, show that as metal clusters decrease in size from a few hundred atoms to a few tens of atoms per cluster, Fe, Co, and Ni clusters have all shown about 30% enhancements in magnetic moment/atom.[138-143] This is apparently due to a narrowing of the density of states at the surface enhancing the magnetism.

Although these Stern–Gerlach experiments are "the ultimate" in the study of pristine magnetic metal clusters, the problem encountered is that the clusters are not big enough to approach a single-domain particle, which contains thousands of atoms, and the particles cannot be isolated and studied in depth.

An approach to preparing isolable quantities of "relatively" small (large compared with the Stern–Gerlach gas-phase work) magnetic metal particles is the inverse micelle approach. One investigation has reported that spherical, quite monodisperse Co particles with average diameters controllable between 1.8 and 4.4 nm can be prepared in certain surfactants.[144-145] The surfactant employed, DDAB, is a cationic material with a hydrophilic, positively charged head group close to the Co particle surface. It is proposed that this cationic environment may be able to remove electron density from the surface. Added electron-donating ligands should have the opposite effect. Thus, pyridine and quinuclidine,

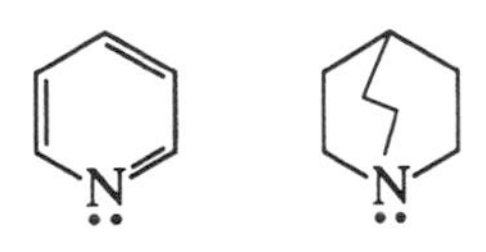

pyridine quinuclidine

can quench this magnetism by as much as 40%, and the effect is reversible. This ligand-modulated magnetism is a remarkable finding that promises to help explain surface effects on magnetism.

In summary, size and surface effects on magnetism are still not clearly understood in nanoparticles, but ongoing experiments are slowly allowing a better understanding to emerge. In the general field of nanoparticle studies, magnetism investigations, due to the many measurement probes that are available, perhaps yield the greatest promise for unraveling complex, cooperative atomic behavior in these ultrasmall entities.

7.3.3 Binding Energies and Melting Points

Theoretical approaches have been used to predict orbital properties and cohesive energies for model clusters of up to several hundred metal atoms.[146] Simple Hückel methods were used for spherically symmetric clusters in simple cubic, body-centered cubic, face-centered cubic, and hexagonal close-packed constrained geometries. The Hückel orbitals (supershells) remained separated even up to 500 to 600-atom cluster sizes.

A classical droplet model provided a good fit to cluster atomization energies and correctly extrapolated to the bulk cohesive energies. In general, these calculations showed that the atomization energy per atom for spherical clusters increased somewhat irregularly on going from 1 to 500 atoms.[146, 147] If the atoms are more weakly bound in the smaller metal clusters, this should translate into lower melting points for the smaller clusters. In fact, there is experimental evidence for this.

Martin and co-workers found that gas-phase sodium clusters formed in a cold He stream of −80°C formed a series of clusters that contained a large portion of geometric shell structures, that is, closed-shell clusters. However, the growth of these clusters in higher-temperature He led progressively to fewer and fewer closed-shell structures. These data, with certain assumptions, were interpreted as evidence for size-dependent melting.[148] Sodium

TABLE 7.2 Melting Temperature of $(Na)_n$ Clusters

Shell	n	r (Å)	T_m (K)	T_m/T_0
6	923	16.0	288 ± 4	0.776
7	1,415	18.4	288 ± 4	0.776
8	2,057	20.9	288 ± 4	0.776
9	2,869	23.3	294 ± 2	0.792
10	3,871	25.8	298 ± 2	0.803
11	5,083	28.2	298 ± 2	0.803
12	6,525	30.7	303 ± 3	0.817
13	8,217	33.1	303 ± 3	0.817
14	10,179	35.6	303 ± 3	0.817

Source: Ref. 148.

Na_n clusters containing 1000 atoms appeared to melt at 288 K, while clusters of 10,000 atoms melted at 303 K, and bulk sodium metal melts at 371 K. Table 7.2 lists some of these data.

Buffat and Borel[149] have reported a somewhat similar size dependence for gold clusters; indeed, comparably sized crystallites of metals, inert gases, and molecular crystals have all been shown to melt at temperatures considerably below that of the bulk substances.[150-156] Melting in semiconductor nanocrystals has also come under study.[157] Using an electron microscope as a tool of observation, it was determined that 3.2-nm (radius) crystallites melted at about 1120 K, 1.8 nm at 1000 K, and 1.2 nm at 600 K, whereas bulk CdS melts at 1680 K. This rather large range, 600 to 1680 K, is remarkable and has several important implications. Annealing of consolidated nanocrystals to remove defects should be possible at relatively low temperature; however, thermal stability of photoelectric devices based on nanocrystals of semiconductors will have lower thermal stabilities than anticipated originally. An advantage may be that fusing of nanocrystals to form films could be done at relatively lower temperatures.

How are these lower melting temperatures rationalized? Melting occurs when the chemical potential of the solid and liquid phases are equal. Parameters that are important in determining the chemical potential are molar latent heat of fusion, surface tension, and density. Upon decrease in size, nanocrystals have a considerably larger surface tension.

As pointed out earlier, lattice parameters, space between atoms or ions, become smaller with decreasing size. Perhaps the best way of picturing these findings is that the atoms or ions in smaller particles "like each other more" and form structures that are less crystalline, thereby easier to melt. On the other hand, atom vaporization energies decrease with decreasing particle sizes because there are fewer neighbors to keep each atom from escaping.

7.3.4 Optical and Electronic Properties

Ultrasmall particles, whether metal or semiconductor, exhibit unique optical properties. This was first noticed by Faraday in his description of the various colors of colloidal gold solutions.[158] After the classic work of Zsigmondy[159] and others, this property of gold nanoparticles is reasonably well understood.[103, 160] Mie[161] developed the theory of plasmon resonance to explain the visible absorption bands exhibited by small metallic particles, which has been discussed thoroughly by Ferrell and co-workers.[162] Basically, when a metallic particle becomes small enough, the atoms behave collectively to develop a "molecular orbital" set within which visible light absorption takes place. This absorption band progressively red-shifts, shifting to lower energies, as the particle size grows as in Au, Au_2, Au_3, Au_n, and eventually gives way to broad bulk metal absorption. Thus the position of the plasmon absorption band can be used to determine the particle size roughly, although solvent effects can also affect this band position. For example, particle sizes of 5 to 6 nm have absorption bands in the range 530 to 570 nm in nonaqueous solvents.[160]

More recent work has dealt with absorption in semiconductor nanoparticles. A prodigious amount of work has been published in recent years, which indicates the special significance and potential applications of such particles. Steigerwald and Brus have reviewed this area briefly.[163] The materials investigated most vigorously have been Si, CdS, and CdSe. Figure 7.10 depicts why semiconductor nanoparticles have progressively varying optical absorption bands, depending on size. Note that in bulk silicon there are bands for valence electrons and an empty conduction band. Light promotes an electron over the bandgap into the conduction band; thus bulk Si is a photoconductor. However, in Si nanoparticles, a set of molecular orbitals with discrete energy levels is developed. Note the larger bandgap compared with the bulk, and this gap increases still further in the smallest "cluster" Si_2. Similar diagrams can be constructed for all nanoparticle semiconductors.

Figure 7.11 illustrates absorption spectra for CdS particles of 1.0-, 2.0-, and 6.0-nm diameters. A sharp band at about 360-nm wavelength is observed for the smallest particle, whereas the 2.0- and 6.0-nm particles show broad, shifted bands, as predicted.[164] Since the absorption spectra are usually very broad, photoluminescence is often a complementary probe. In fact, such studies of indirect bandgap material, AgBr clusters, have been examined in some detail.[165] The effect of small size on indirect gap semiconductors may be twofold. First, the expected blue shift is observed with smaller and smaller sizes. Second, the shape of the bands may change; that is, a normally forbidden transition may become "less forbidden" in a small cluster.

Indeed, as early as 1970, Berry noted anomalous absorption behavior in 5.0-nm crystallites of AgBr.[166] Rossetti and co-workers speculated that small AgBr crystallites might begin to resemble direct bandgap materials with

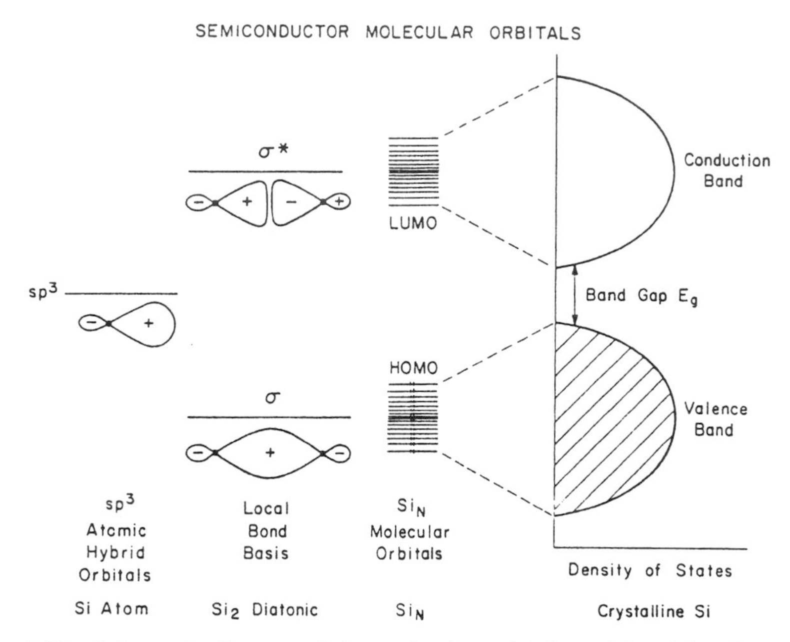

Figure 7.10 Schematic diagram of the molecular orbital model and band structure for silicon. (After Ref. 163.)

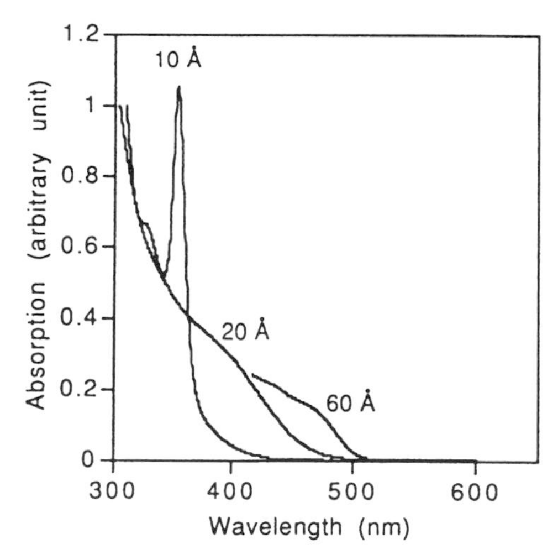

Figure 7.11 Absorption spectra of capped CdS nanocrystals. (After Ref. 164.)

decreasing size, due to the mixing of surface-state wavefunctions into the bulk band structure.[167]

In analogous work it has been reported that dramatic luminescence enhancement takes place by the binding of simple amines to nanoparticles of CdS or Cd_3As_2.[168] Apparently, modification of midbandgap states, probably associated with excess Cd^{2+}, takes place. Such states can dominate radiationless decay, and quantum yields approached unity.

7.3.5 Nonlinear Optical Properties

When semiconductor clusters in the range 1 to 15 nm are doped into organic polymers or glasses, the resultant materials possess nonlinear optical properties. Materials possess third-order optical nonlinearity if their refractive index depends, reversibly, on the intensity of incident light. Under such conditions, the induced polarization of the material contains a term that has a 10^3 dependence on the electric field of the incident light, hence the term *third-order nonlinear optical effect*.[169] In the nanoscale size range, there are discrete electronic states rather than a continuous energy band, and *bound* electron–hole pairs (bound excitons) rather than free electrons and holes.

Due to the large surface area in small particles, the bound exciton is quickly trapped by surface defects, within less than picoseconds, forming a trapped electron–hole pair.[169] Thus we have the situation where electron–hole pairs are on the surface, so adsorbed species might be expected to affect nonlinearity. Indeed, this is the case as shown by absorption of amines, such as NH_3 or dimethyformamide or by embedding in Nafion polymer. These observations support the idea of the local field model (the mechanism of the resonant nonlinearity is attributed to the reduction in absorption strength in the presence of an optically generated trapped electron–hole pair).[169]

Some particularly interesting results have recently came from capped CdS particles in polyacetylene.[170] This new class of nonlinear optical material is unique because the absorption spectra of the capped CdS and of the polyacetylene form a natural transmission window between the band edge of the CdS (around 510 nm) and the absorption peak of the polyacetylene (about 628 nm). The nonlinear refractive index of both components is negative within the regime. Since the nonlinear refractive index changes more slowly with wavelength than does the absorbance, the two reinforce each other. Thus, with such a combination of nanoparticles and polymers, it should be possible to choose a transmission window optimized for a specific device requirement.[171]

It has also been possible to produce core/shell semiconductor nanoparticles by inverse micelle techniques.[172] For example, CdS coated with ZnS exhibits a very different optical absorption spectra than particles of a solid solution of ZnS–CdS, pointing out the versatility of such layered composites and the further possibility of tailoring optical properties.[172, 173] Nanoscale metal particles can also exhibit nonlinear optical properties when embedded in polyacetylene like polymers.[174] Actually, gold clusters and polydiacetylene are each of interest separately for their nonlinear optical properties (NLO). Taken together, they hold special promise.

Using the SMAD metal vapor technique, gold clusters were dispersed in poly(diphenylbutadiyne) and other NLO diyne systems. When 2.0-nm gold particles were present at an estimated gold volume fraction of 7%, a 200-fold enhancement in the third-order NLO coefficient at 1.064 μm was observed. Hence the combination of nanoparticles and NLO polymers appears quite promising.

The ultimate goal of much of this effort is to produce a material for *optical switching*. Up to this point, only glass can be used, even though it exhibits only a weak nonlinear effect. About 100 m of glass is necessary for complete switching.[175] This length needs to be reduced, and many semiconductors and some polymers, which possess much stronger nonlinearity, could conceivably be used. However, to date, none can be grown to even centimeter length with optical loss as low as glass. So this is exactly the area where nanoparticles may be valuable, probably doped in polymers, making composites with low optical loss but high nonlinearity.

7.4 CHEMICAL PROPERTIES OF SURFACES

As mentioned earlier, if nanoparticles are small enough, a very significant portion of the total atoms are on the surface; for example, a 3-nm particle has 50% surface atoms. In addition, we might expect intrinsically different surface chemistry due to unusual morphologies, unusual surface defects perhaps in high concentrations, and unusual electronic states affecting surface chemistry. In the following discussion we consider what is known about surface chemistry of nanoparticles with differences in surface chemistry versus bulk samples as the primary concern.

7.4.1 Metals

We begin this section by considering briefly the differences in chemistry of metal atoms versus the smallest clusters. The question of whether *single* metal atoms or *small clusters* are more chemically reactive has been answered as "it depends on the metal in question and under what conditions."[1,176] In one study, the reactivity of Mg, Mg_2, Mg_3, Mg_4, Ca, Ca_2, Ca_3, and Ca_x with alkyl halides at very low temperatures showed that clusters were more reactive. This was attributed to the energetically favorably formation of cluster Grignard reagents:

$$Mg_2 + CH_3Br \rightarrow (Mg_2^+CH_3Br^-) \rightarrow CH_3Mg_2Br \qquad (20)$$

The lower ionization energy of Mg_2 versus Mg and the possibility of a four-centered transition state were proposed as rationale for this finding where cluster reactivity exceeded atom reactivity.[177] Recent theoretical reports tend to support the idea that RMg_nX should be stable; however, none have been isolated and thoroughly characterized as yet. There is the possibility that these compounds would exist as "Mg ligated species."[178]

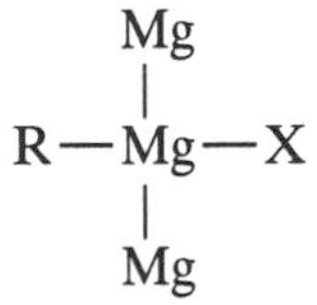

```
      Mg
      |
R—Mg—X
      |
      Mg
```

With the advent of the gas-phase cluster beam apparatus for studying the chemical properties of gas-phase metal clusters,[1] a host of interesting experiments have been reported. Riley and co-workers[179] were the first to report that annealing of Fe_n clusters in the gas phase (followed by cooling) sometimes caused a *lowering* of cluster reactivity with H_2. This finding suggests that the initial cluster formed is defective, thus more reactive, and upon annealing becomes more symmetrical, thus less reactive on a comparative basis.[179]

Cox, Kaldor, and Trevor and co-workers have studied Pt_n gas-phase species extensively. They were able to activate CH_4 for the first time on an unsupported metal cluster.[1, 180] The reaction of CH_4 with Pt_n has a distinct cluster size dependence, where Pt_2–Pt_5 were most reactive and Pt_6–Pt_{24} least reactive. A mild correlation with the availability of low-coordination-number metal atoms was found. Thus these experiments indicate that defect sites encourage CH_4 activation. This idea has also been supported in many earlier studies of metal clusters in low-temperature hydrocarbon matrices.[181]

So indirect evidence for chemical reactivity changes due to morphological changes is strong. However, let us also consider physical evidence. As particle size increases from a few atoms up to thousands and tens of thousands, several interesting things occur. First, the atomic-scale growth mechanisms are variable and sometimes unique.[182] For example, in the case of the microstructure of colloidal silver, there is evidence for a polytetrahedral growth sequence.[183]

If construction of a metal lattice were to take place by the successive addition of individual metal atoms to the simplest stable unit (i.e., the tetrahedron), eventually it becomes clear that with relatively small numbers of atoms a polytetrahedral growth sequence would be preferred over a bulk lattice (hexagonal close packed or cubic close packed). The polytetrahedral growth leads to pentagonal symmetry.

It must be concluded that these polytetrahedral structures eventually change over to the normal hexagonal- or cubic-close-packed crystal morphology, and Edwards and co-workers have presented evidence that this change

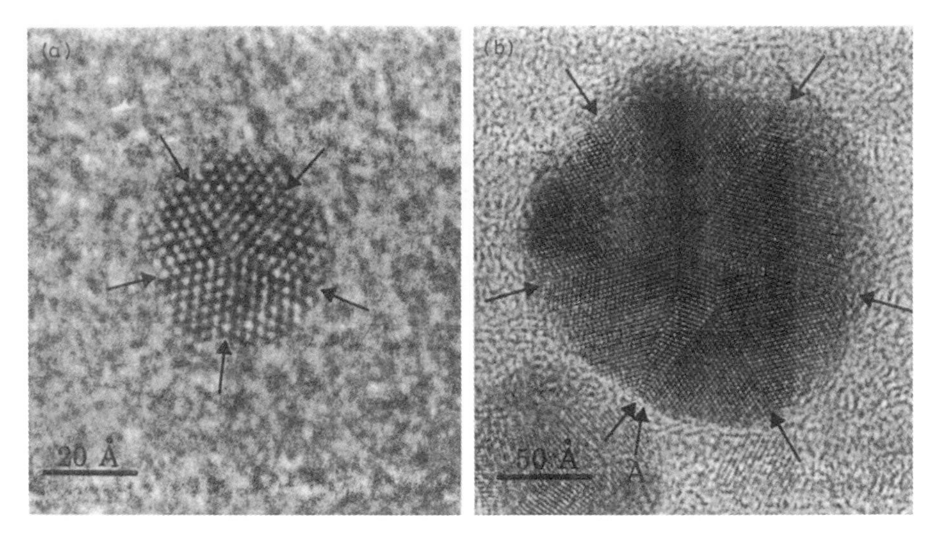

Figure 7.12 TEM photographs of 2.0- and 5.0-nm silver particles showing polytetrahedral symmetry and changeover to bulk crystal form. (After Ref. 183.)

can be observed by electron microscopy at about 5-nm silver particle size (see Fig. 7.12).

Henglein and co-workers have explored the chemistry of nanoscale Ag_n particles in aqueous solution and have reported some rather unexpected results.[184, 185] The unusual reactivity of Ag_n with nucleophilic reagents such as NH_3, PH_3, CN^-, I^-, SH^-, and SR^- is attributed to the coordinative unsaturation of surface silver atoms, and this is more pronounced in nanoparticles. The surface plasmon absorption band for the silver particles was perturbed upon interaction of the nucleophilic reagents and could be used to monitor the reactions.

Small oligomeric silver clusters were found to be very strong reducing reagents, and this property is a function of size (smaller clusters being more highly reducing):

$$Ag_n \xrightleftharpoons{\text{aqueous}} Ag_{n-1} + Ag^+ + e^- \tag{21}$$

The trend is opposite to the size dependence of the ionization potential in vacuo, and this finding is apparently due to the binding energy of a silver atom in the clusters. Generally, the binding becomes greater with increasing n until it reaches the value of 2.6 eV for bulk silver. Thus, for smaller clusters, reaction (21) is more favorable.[186] This is a fascinating discovery, and further comparisons of E_n^0 for solvated and nonsolvated (in vacuo) are needed for other metals.

Based on the previous discussion, it might be expected that the photochemical oxidation of Ag_n would also be size dependent, and indeed it is to a certain extent.[187] Silver particles (about 3 nm) were easily photo-oxidized in aqueous solution in the presence of N_2O, and they can behave as photocatalysts for N_2O decomposition in the presence of a reducing solvent such as 2-propanol. However, in the absence of such a solvent, dissolution of the silver particles takes place:

Photoelectron emission: $$Ag_n + h\nu \rightarrow Ag_n^+ + e^-_{(aq)} \quad (22)$$

Back reaction: $$Ag_n^+ + e^-_{(aq)} \rightarrow Ag_n \quad (23)$$

Electron scavanging: $$e_{(aq)} + N_2O + H_2O \rightarrow N_2 + OH^- + \cdot OH \quad (24)$$

Positive hole injection: $$\cdot OH + Ag_n \rightarrow OH^- + Ag_n^+ \quad (25)$$

Dissolution: $$Ag_n^+ \rightarrow Ag_{n-1} + Ag^+ \quad (26)$$

The reducing potential and photoreducing potential of small silver clusters/particles can be used to reduce a variety of organic materials, including CCl_4 and nitrobenzene, for example:

$$Ag_n + CCl_4 \rightarrow Ag_{n-1} + Ag^+ + Cl^- + \cdot CCl_3 \quad (27)$$

Rao and co-workers have reported analogous results with Ag_n and Ni_n clusters.[188] The dissociation of O_2 on small silver clusters was favored, and the proportion of O_2 molecules cleaved did not depend on temperature, whereas temperature is important as is the case for large clusters and bulk silver. The change in reactivity was greatest at the metal–insulator transition, and this probably implies that a combination of reducing potential and coordinatively unsaturated metal atom availability (molecular orbitals for O_2 complexation) is important.

These examples show that silver nanoparticles do have unique chemical properties. These useful properties are perhaps best represented by Fig. 7.13, where small silver particles serve the role of a catalyst for reduction of water by free radicals.[189] Bradley and co-workers have also investigated particle-size

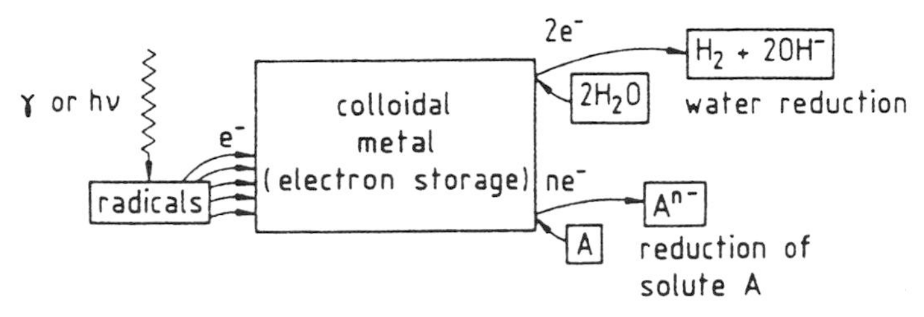

Figure 7.13 Colloidal metal-catalyzed reduction of water by free radicals. (After Ref. 189.)

effects on surface chemistry.[190-192] Both infrared (IR) and nuclear magnetic resonance (NMR) studies of adsorbed CO on colloidal Pt particles (1 to 5 nm generally) indicated that $\nu_{C=0}$ and ^{13}C NMR shifts were found at intermediate values between carbonyl cluster compounds and absorbed CO on bulk metal surfaces.[190] An interesting feature was the absence of a Knight shift in the NMR (when observed, the underlying metal substrate has a conduction band). Accordingly, the particles (about 1 nm) were below the minimum size needed to exhibit such a downfield shift.

Adsorption of CO onto a series of polymer-stabilized Pd colloids (2.0 to 8.0 nm) showed, according to IR, two modes of adsorption, terminal and bridged. The smallest particles gave the highest ratio of terminal/bridged, while the largest showed only bridged.

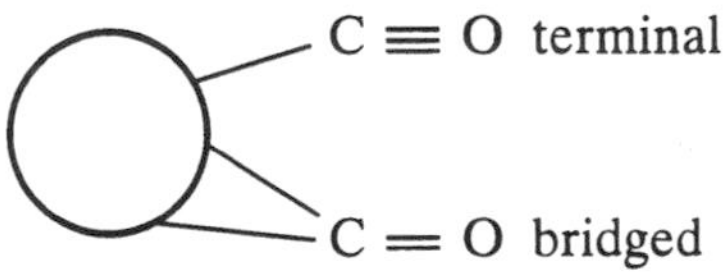

$C \equiv O$ terminal

$C = O$ bridged

This result was interpreted as an indication of the smaller particles exhibiting a larger proportion of low-coordination-number metal atoms, an intuitive relationship argued many years ago by Handeveld and Hartog.[193]

From these studies and many others,[1, 29, 194-197] it has become clear that the chemistry and photochemistry of nanoscale metal particles is dependent on size and morphology and can be modified by adsorbed ligands. Although trends have been established, more quantitative comparisons of solvated versus nonsolvated particles would be welcomed.

Hence it may not come as a surprise that the surface chemical properties of colloidal metals also affect their film-forming properties. "Living" colloidal Pd and Au have been prepared in nonaqueous solvents, and these serve as precursors for the formation of Pd and Au films, sometimes called the chemical liquid deposition method.[198-200] It was determined that atom-by-atom growth of Pd or Au clusters in low-temperature acetone led to negatively charged particles that were ligated by many solvent molecules. Electron scavanging from the solvent took place yielding the negatively charged particles, and in this way were stabilized toward aggregation by both electronic and steric (solvation) effects. Upon film formation by removal of excess solvent, the film surface exhibited a positive shift in the core-level binding energy. This finding and the results of electrical potential biasing experiments were consistent with the surface metal clusters maintaining their negative charge and coated with organic ligated solvent and solvent fragments. However, careful heating caused loss of the organic species and the formation of metal films. The CLD procedure was especially successful for forming gold films on sulfur-containing polymers such as polyphenylene sulfide[201] and silver.[202]

Finally, in dealing with the surface chemistry of metallic nanoparticles, the feature of reduction of adsorbed secondary ions should be discussed. When colloidal silver particles adsorb Cd^{2+} (in the presence of 1-hydroxyethylmethyl radicals), reduction takes place, yielding a Cd° coating.[203] The mechanism of reduction apparently involves first transfer of electrons to the colloidal silver followed by reduction of adsorbed Cd^{2+} ions. Once the Cd atoms are formed on the surface of the silver particle, a blue shift of the surface plasmon absorption of silver is observed. At a coverage of three monolayers, the surface plasmon band is shifted to 260 nm, where pure Cd absorbs. Thus some control of the absorption band is readily possible. In a similar way, lead-coated gold particles have been prepared with a change in optical properties.[204] Deposition of lead adatoms caused the plasmon absorption band of gold to be shifted to shorter wavelengths, which can be rationalized by a Pb → Au electron donation and double-layer charging. The optical spectra observed for various mono-, di-, and trilayer lead on gold were in good agreement with calculated spectra according to Mie theory calculations. In general, this approach of layering Cd and/or Pd or silver or gold can be used to give good control of the position of this optical band.

7.4.2 Semiconductors

The optical properties of semiconductor nanoparticles can also be perturbed significantly by adsorption of solvent molecules on ligands.[205] Recall that nanoparticles are often 30 to 50% surface atoms and molecules, so perturbating of energy levels/chromophores of the surface atoms can affect the entire particle. In an extreme case, the semiconductor surface can undergo reactive chemisorption and optical properties can be drastically quenched, such as when CdS nanoparticles are treated with H_2S:

$$CdS_{(surf)} + H_2S \rightleftharpoons Cd(HS)_{2(surf)} \tag{28}$$

On the other hand, McLendon and co-workers have shown that adsorption of small amounts of tertiary amines on CdS particles caused dramatic (up to 450%) increases in the quantum yield for exciton emission.[206] Both CdS and Cd_3As_2 particles were so affected, and it is believed that Cd^{2+}–ligand interaction filled or eliminated defect sites that would normally have been present for promoting nonradiative electron–hole pair recombination.

Henglein[189] has discussed in some detail the effects of small size and surface adsorption and the electron–hole-pair recombination phenomenon.

As might be anticipated, the lifetime of the hole-pair separation is crucial for the practical use of semiconductor nanoparticles in photochemically driven oxidation/reduction reactions. Indeed, solvent effects, ligand effects, and added metallic catalysts have all come under intense study. We mention just a few of the hundreds of papers dealing with these issues.

Ellis and co-workers have noted photoluminescent changes in single-crystal CdS and CdSe when alkenes were adsorbed.[207] Generally, the magnitude of the enhancement correlated with alkene basicity (the more electron donation to CdS, the more enhancement) and was related to photoionization potentials. This ligand–CdS electronic coupling is also evident in light-driven catalytic processes. For example, Shiragami and co-workers have found that visible light can photocatalyze the reduction of ketones and alkenes in the presence of colloidal CdS,[208] and enhanced reduction in the presence of 2 to 5-nm nanoparticles compared with bulk CdS was attributed to a size quantization effect. It was proposed that the nanoparticles can induce sequential electron transfer by Cd° formation on the surfaces and then by efficient two-electron transfer to the adsorbed organic. Bulk CdS, with fewer defect sites and a shifted absorption band, was not capable of the key step of Cd° formation, apparently. This study serves as the first example of visible-light-induced reductive carbon–carbon bond formation on semiconductor particles.

Hoffman and co-workers have demonstrated that oxidation processes can also be driven by photocatalysis on quantum-sized semiconductor particles: for example, the product of H_2O_2 and organic peroxides.[209] Indeed, it would seem reasonable that such oxidation processes may be even more favorable than reduction since it is believed that during electron–hole pair formation, the hole (h^+) as well as the electron migrate to the surface of the nanoparticles, where it is available to oxidize adsorbed species.

When the electron is the species of interest, often better results are obtained when a small patch of a metal is also present on the surface of the semiconductor nanoparticle. The electron migrates to the metal, and then the electron transfer to the adsorbed species is catalyzed by the metal. The metal can also serve to catalyze H_2 formation from adsorbed $H\cdot$ species, for example, when H_2O or H_2S are being reduced. Thus Rufus and co-workers have found that Rh on CdS serves well for H_2S reduction:[210]

$$H_2S \xrightarrow[\text{CdS Rh}]{h\nu} H_2 + S \tag{29}$$

Still another unique finding was reported by Horvath and Fendler, where it was shown that CdS particles could serve to mediate transmembrane photoelectron transfer in surfactant vesicles. This system provided a well-behaved matrix for compartmentalization of the components of the electron transfer process confined within the aqueous pools of the vesicles.[211]

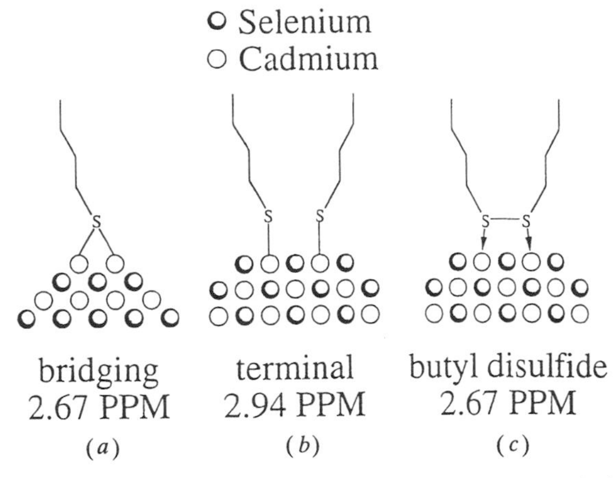

Figure 7.14 (*a*) Butanethiolate ligand bridging two surface cadmium ions; (*b*) forming terminal bonds to single cadmium ions; (*c*) attached butyl disulfide datively bound to the surface and easily desorbed. (After Ref. 212.)

Although many applications of semiconductor particles have been explored, very few detailed studies dealing with surface characterization have been reported. Since it is clear that surface effects can be dominant in determining photochemistry/photophysics, such studies are certainly warranted. Recently, one such investigation dealt with ^{1}H NMR characterization of CdSe nanoparticle surfaces.[212] It was demonstrated that butanethiolate (BuS) could bind the CdSe nanocrystals in three ways (Fig. 7.14). Quantitative results concerning the amounts of attached ligands could be assessed by NMR and related to the average number of attached groups per nanocrystallite.

According to TEM, the particles averaged 3.5 ± 0.4 nm or an average of 811 atoms plus surface ligands. Although results were somewhat variable, a coverage of 133% consisting of combined bonding modes (Fig. 7.14) was found, suggesting the presence of some free BuSSBu. In comparing the results with those for CdSe single crystals, it appears that nanocrystallites of CdSe have far more defect sites than commercially available single crystals, but the small particles show no evidence of faceting. Their overall spherical shape suggests the presence of numerous steps along the surface.

7.4.3 Metal Oxides (Insulators and Semiconductors)

Due to the refractory nature of most of the metal oxides, the formation of ultrasmall particles is actually facilitated. The highly ionic nature of some materials, especially MgO, Al_2O_3, ZrO_2, and TiQ_2, allows the formation of many types of stable defect sites, including edges, corners, and anion/cation vacancies. Of course, some metal oxides are semiconductors, and common examples are TiO_2 and ZnO. The bandgap of TiO_2 lies in the UV range, and again a blue shift is observed as particle size decreases.

Numerous reports have appeared describing the photochemical decomposition of water and organics using TiO_2 particles as catalysts, sometimes including metals as catalysts (e.g., patches of Pd° or Pt° on TiO_2 particles).[213-217] Since all of this photochemistry depends on the lifetime of the electron–hole pair in the nanoparticles, it could be particularly interesting to know if there is size dependence on the photochemical reduction of water or oxidation of hydrocarbons or chlorocarbons.

Riegel and Bolton have proposed that variations in the rate constants for hole–electron recombination are most important, as opposed to subsequent oxidation or reduction reactions.[218]

$$\text{UV light} \xrightarrow{TiO_2} e^-h^+ \xrightarrow[H_2O]{O_2,\, OH^-} \cdot O_2^- + \cdot OH \tag{30}$$

Similar reasoning was expressed when different surface area TiO_2 particles were employed as photocatalysts for degradation of chlorophenols.[219] When the surface/volume ratio became too high (i.e., when the surface area of the TiO_2 approached 150 m^2/g), the number of surface defect sites available to facilitate the e^-h^+ recombinations become too great.

One system that lends itself well to further analysis of nanoparticles size effects is the photo-oxidative degradation of $CHCl_3$ with TiO_2 in water.[220] In our laboratory two samples of TiO_2 particles were prepared (Table 7.3) by aerogel or conventional means and compared with a commercial sample of Degussa P-25.[221]

$$CHCl_3 + 2H_2O \xrightarrow[h\nu]{TiO_2} CO_2 + 6HCl \tag{31}$$

The results of this study seemed to confirm the concept that when the nanoparticles become too small, the e^-h^+ recombination rate is so fast as to

TABLE 7.3 Physical Properties of TiO_2 Nanoparticles Employed in Photocatalytic Degradation of $CHCl_3$

Sample	Preparation Method	Surface Area (m^2/g)	Crystallite Size (nm)	Crystalline Form
AP–TiO_2	Aerogel	150	8.9	Anatase
CP–TiO_2	Conventional	10	20	Anatase
Degussa P–25	—	55	18.2	80% Anatase/20% rutile

detrimentally override the expected increased efficiency in the initial light absorption step. This enhanced recombination efficiency is due to two things: (1) the small volume of the particles, thus limiting the distance that e^- and h^+ can get apart from each other, and (2) the presence of a high concentration of defect sites on the smaller nanoparticles (Fig. 7.15).[222, 223]

It can be concluded that in the application of nanoparticles for photodegradation of chlorocarbons, an optimum size is needed and that "smaller is not always better." It is becoming clearer and clearer that surface defect sites on nanoparticles are of crucial importance in many aspects of their

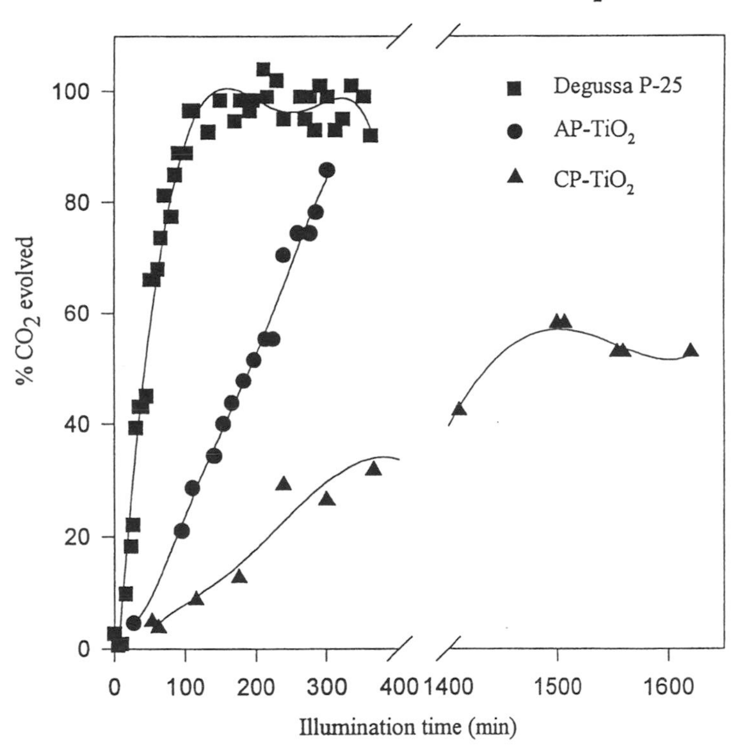

Figure 7.15 Photodegradation of $CHCl_3$ in water with different sources of TiO_2. AP, nanoscale TiO_2; CP, microscale. Degussa is commercially available TiO_2. (From Ref. 222.)

surface chemistry. If we turn to the insulator metal oxides, we note that there have been numerous studies of their surfaces in an attempt to clarify the types of defect sites that can exist.[224-227]

Although there are a large number of defect sites, the most common are coordinatively unsaturated ions due to planes, edges, corners, anion/cation vacancies, and electron excess centers, for example, for MgO:

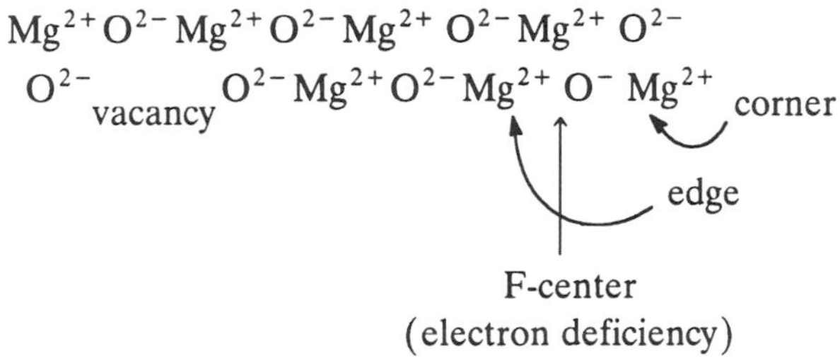

$$\begin{array}{l} Mg^{2+}O^{2-}Mg^{2+}O^{2-}Mg^{2+}\ O^{2-}Mg^{2+}\ O^{2-} \\ O^{2-}{}_{\text{vacancy}}\ O^{2-}Mg^{2+}O^{2-}Mg^{2+}\ O^{-}\ Mg^{2+} \end{array}$$

corner

edge

F-center
(electron deficiency)

Such sites are often invoked as the reactive sites for many interesting and useful reactions, including methane activation,[228] D_2–CH exchange,[229] CO oligomerization,[230] oxygen exchange in CO_2,[231] and H_2O.[232]

Magnesium oxide and calcium oxide are particularly interesting in nanoparticle form. It has been possible to prepare MgO in very high surface area (500 m^2/g) with average crystallite sizes of about 4 nm.[233] The high surface areas and the intrinsically high surface reactivity allows these materials to be especially effective as adsorbents; in fact, they have been called "destructive adsorbents" due to their tendency to adsorb and simultaneously destroy, by bond-breaking processes, a series of toxic chemicals. For example, organophosphorus compounds (mimics of poisonous nerve agents) are destroyed at relatively low temperature and in high capacity. These nanoparticles are also effective for destruction of organochlorine compounds, and they adsorb large quantities of SO_2, CO_2, HCl, HBr, and other gases.[234-237]

Nanoscale MgO and CaO can be considered as stoichiometric chemical reagents. Their high surface areas dictate that 30 to 40% of the MgO(CaO) moieties are on the surface, thus allowing surface-gas reactions to occur approaching a stoichiometric range:

$$CCl_4 + 2CaO \longrightarrow CO_2 + 2CaCl_2 \tag{32}$$

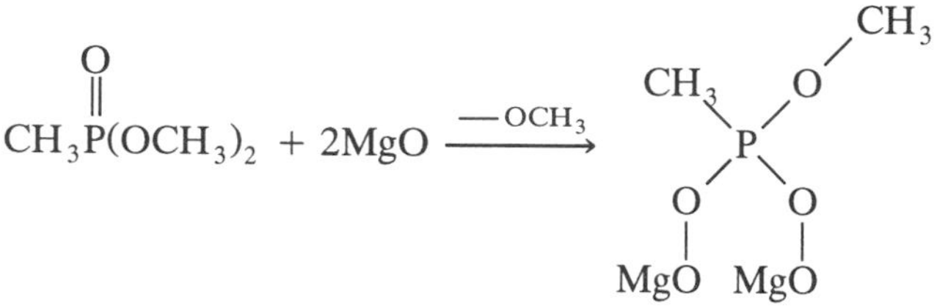

$$CH_3\overset{\overset{\displaystyle O}{\|}}{P}(OCH_3)_2 + 2MgO \xrightarrow{-OCH_3} \begin{array}{c} CH_3 \quad O{-}CH_3 \\ \diagdown\ \diagup \\ P \\ \diagup\ \diagdown \\ O \qquad O \\ | \qquad | \\ MgO \quad MgO \end{array} \tag{33}$$

The surface reactivities are generally much higher for the nanoparticles than for larger crystallites. This appears to be due to the presence of large surface concentrations of edges, corner, and other defects.[238] Figure 7.4 illustrates a model of a 4-nm MgO nanocrystal. Under the electron microscope the particles appear spherical; however, x-ray diffraction confirms a cubic internal structure. It is likely that cleavage planes exist by loss of corners, which would expose (110) and (111) faces as well as the normal (100) faces.

Some of the inherent reactivity may be due to the presence of isolated —OH groups, as determined by $AlEt_3$ surface reactions, which indicated that mostly terminal O-$AlEt_2$ species formed.

$$\text{(surface)}\!-\!(OH)_3 + 2AlEt_3 \longrightarrow \text{(surface)}\!-\!O\!-\!AlEt\,;\ \text{(surface)}\!<\!(O)_2\!>\!AlEt + 3EtH \qquad (34)$$

This suggests that their —OH groups were well separated and isolated from one another. In addition, it was determined that there are inherent differences in the Brønsted acidity of their surface —OH groups compared with larger crystallites. Overall, the nanoparticles of MgO are definitely intrinsically different in chemical reactivity. This coupled with their high surface areas promises to lead to new surface-adsorbate chemistry.

Recent evidence has added an additional dimension to this new surface chemistry. It has been found that layered (core/shell) particles (e.g., $[Fe_2O_3]MgO$ nanoparticles) exhibit even higher reactivity and reaction capacity in chlorocarbon destructive adsorption.[239] This enhancement appears to be due to a facile Cl^-/O^{2-} exchange within the nanoparticle (Fig. 7.16). These results support the contention that nanoparticle metal oxides are new and very special chemical reagents, and continuing investigations should be fruitful.

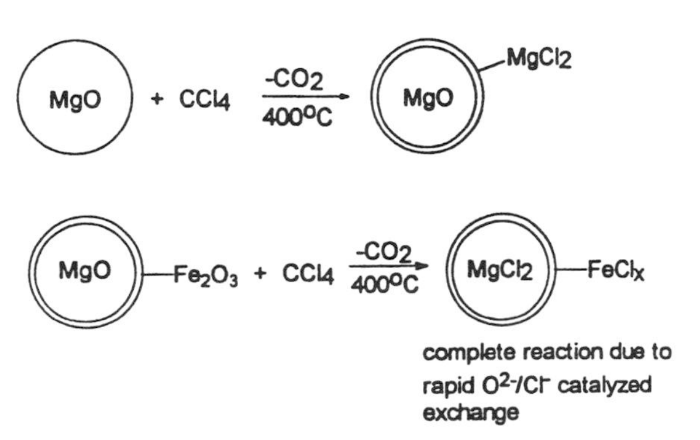

Figure 7.16 Catalytic effect of Fe_2O_3 coating on MgO to $MgCl_2$ conversion with CCl_4 at 400°C. (From Ref. 239.)

7.5 POSSIBLE APPLICATIONS OF NANOPARTICLES

7.5.1 Destructive Adsorbents

The intrinsic surface reactivities coupled with high surface areas allow nanoparticle metal oxides to be considered a new family of adsorbents that strongly chemisorb the incoming species usually by dissociative processes. Thus the term *destructive adsorbent* is appropriate. Two possible applications come to mind:

1. Use in air-purification cartridges for buildings, military vehicles, airplanes, and other enclosed areas.
2. As an alternative to incineration of toxic substances, such as chlorocarbons, PCBs, military agents, and other toxic chemicals. Temperatures are much lower than incineration temperatures, and large volumes of hot, flowing gases are not necessary.[240]

7.5.2 Water Purification

Nanoparticles of metals such as Fe, Zn, and Sn have demonstrated high reactivities for chlorocarbons in an aqueous environment.[241] These results imply that membranes or flow-through chambers could be constructed that might be used for groundwater decontamination:

$$4Zn + CCl_4 + 4H_2O \rightarrow CH_4 + 2Zn(OH)_2 + 2ZnCl_2 \qquad (35)$$

7.5.3 New Catalysts

Within the field of heterogeneous catalysis was the first true application of nanoparticles, especially supported metal particles on catalyst supports. In a sense, research on nanoparticle catalysis has been going on for decades and has led to an industry that is vital to the country's economy to the extent of 20% of the GDP.[242]

Thus nanoparticles are already extremely important in commerce. Nonetheless, new catalytic applications are still very probable. This has been shown to be especially true for ultrafine particulates produced by the solvated metal atom dispersion (SMAD) method.[243] It was shown that dehydrogenation, hydroformylation, hydrogenation, methanation, isomerization, Fischer–Tropsch reactions, and other catalytic processes can be enhanced by the use of new nanoparticles and bimetallic nanoparticles. With the availability of effective reducing reagents, for treating transition metal salts in either aqueous or nonaqueous media, nanoscale catalytic metal particles are becoming readily available to all scientists.[56–60, 244]

7.5.4 Information Storage

In this area nanoparticles have also already found their way into important commerce.[245] All modern audio and video tapes depend on the magnetic properties of fine particles. However, we are now learning how to produce ever smaller, more highly magnetic materials with better control of coercivity. Further work should allow much enhanced clarity for lower costs, and possibly applications not yet imagined.

7.5.5 Refrigeration

The promise of refrigerators and air conditioners that do not need refrigeration fluids (Freons, HFCs, etc.) is emboldened in new research on magnetic nanoparticles. It has already been demonstrated on a small scale that an entropic advantage can be gained in magnetic particle field reversal. That is, upon application of a magnetic field, the entropy of a magnetic species changes, and if adiabatic conditions are maintained, the field application will result in a temperature change. This ΔT is called the magnetocaloric effect, and the magnitude depends on the size of the magnetic moment, heat capacity, and temperature dependence of magnetization. With nanoparticles these parameters are quite different from consolidated bulk materials, and they hold promise for the construction of better magnetocaloric refrigeration units than have been possible to date.[246] If further research on the synthesis of better magnetic nanoparticles with engineering advantages is successful, it is hard to overestimate the tremendous improvements this could mean for society and the environment.

7.5.6 Solar Cells and Environmental Cleanup

Semiconductor nanoparticles hold the potential for more efficient solar cells for both photovoltaic (electricity production) and water splitting (hydrogen production). In addition, these particles show promise for decontamination of water through photo-oxidation and photoreduction of contaminants.

7.5.7 Optical Computers

Another dream of scientists in this field concerns the use of quantum confined semiconductor systems for optical transitions, spatial light modulators, and other devices depending on nonlinear optical properties.[247]

7.5.8 Improved Ceramics and Insulators

It has been shown that nanoparticles of ceramic materials can be compressed at relatively low temperatures into solids that possess better flexibility and malleability.[2] After further development the use of ceramics as a replacement for metals may be possible. It should also be noted that aerogel-prepared materials generally have very low densities, can be translucent or transparent, and have low thermal conductivities and unusual acoustic properties. They have found various applications, including detectors for radiation, superinsulators, solar concentrators, coatings, glass precursors, catalysts, insecticides, and destructive adsorbents.

7.5.9 Cosmetics

Ultrafine particles should prove useful in the cosmetics industry, where lightness, texture, and coverage are so important.

7.5.10 Thin Films

The discovery of stable nonaqueous metallic colloidal solutions has given us a new precursor material for the formation of thin metallic films.[45, 248]

ACKNOWLEDGMENTS

The skillful work of many students is gratefully acknowledged, especially D. G. Park, E. B. Zuckerman, M. Franklin, O. Koper, J. Stark, D. Zhang, S. Decker, Y. Jiang, T. Boronia, A. Khaleel, S. Mahashwari, Y. Jiang, E. Lucas, W. Li, F. Tian, N. Sun, A. Bedilo, P. Hooker, I. Lagadic, G. Glavee, and G. Youngers. We also thank Jan Vaughan and Jada Heiman for their help in preparation of the manuscript. The generous support of the Army Research Office and the National Science Foundation is also gratefully acknowledged.

REFERENCES

1. K. J. Klabunde, *Free Atoms, Clusters and Nanoscale Particles*, Academic Press, San Diego, CA, 1994, pp. 2, 36.
2. R. P. Andres, R. S. Averback, W. L. Brown, L. E. Brus, W. A. Goddard III, A. Kaldor, S. G. Louie, M. Moskovits, P. S. Peercy, S. J. Riley, R. W. Siegel, F. Spaepen, Y. Wang, *J. Mater. Res.* **4**, 704 (1989).
3. T. G. Dietz, M. A. Duncan, D. E. Powers, R. E. Smalley, *J. Chem. Phys.* **74**, 6511 (1981).
4. M. D. Morse, M. E. Geusic, J. R. Heath, R. E. Smalley, *J. Chem. Phys.* **83**, 2293 (1985).
5. S. Marayama, L. R. Anderson, R. E. Smalley, *Rev. Sci. Instrum.* **61**, 3686 (1990).
6. E. K. Parks, G. C. Nieman, L. G. Pobo, S. J. Riley, *J. Chem. Phys.* **88**, 6260 (1988).
7. S. J. Riley, *Z. Phys. D Atom. Mol. Clusters* **12**, 537 (1989).

8. M. F. Jarrold, J. E. Bower, *J. Chem. Phys.* **87**, 5728 (1987).
9. M. F. Jarrold, J. E. Bower, *J. Am. Chem. Soc.* **110**, 70 (1988).
10. R. F. Curl, R. E. Smalley, *Science* **242**, 1017 (1988).
11. W. Kratschmer, L. D. Lamb, K. Fostiropoulos, D. R. Huffman, *Nature* **347**, 354 (1990).
12. H. Kroto, *Science* **242**, 1139 (1988).
13. D. J. Whales, *J. Chem. Soc. Faraday Trans.* **90**, 1061 (1994).
14. K. L. Taft, C. D. Delfs, G. C. Papaefthymiou, S. Foner, D. Gatteschi, S. J. Lippard, *J. Am. Chem. Soc.* **116**, 823 (1994).
15. E. A. Wintner, M. M. Conn, J. Rebek, Jr., *J. Am. Chem. Soc.* **116**, 8877 (1994).
16. D. M. P. Mingos, D. J. Whales, *Introduction to Cluster Chemistry*, Prentice Hall, Upper Saddle River, NJ, 1990.
17. E. Matijević, *Langmuir* **10**, 8 (1994).
18. A. W. Olsen, K. J. Klabunde, *Encyclopedia of Physical Science*, Academic Press, San Diego, CA, 1987, Vol. 8, pp. 110–146.
19. G. Schmid, *Endeavor N.S.* **4**, 172 (1990).
20. G. Schmid, in *Aspects of Homogeneous Catalysis*, Vol. 7, R. Ugo, Ed., Kluwer, Dordrecht, The Netherlands, 1990, pp. 1–36.
21. M. L. Steigerwald, A. P. Alivisatos, J. M. Gibson, T. D. Harris, R. Kortan, A. J. Muller, A. M. Thayer, T. M. Duncan, D. C. Douglass, L. E. Brus, *J. Am. Chem. Soc.* **110**, 3046 (1988).
22. V. L. Colvin, A. N. Goldstein, A. P. Alivisatos, *J. Am. Chem. Soc.* **114**, 5221 (1992).
23. J. Karch, R. Birringer, H. Gleiter, *Nature* **330**, 556 (1987).
24. R. W. Siegel, S. Ramasamy, H. Hahn, L. Zonqquan, L. Ting, R. Gronsky, *J. Mater. Res.* **3**, 1367 (1988).
25. H. Itoh, S. Utamapanya, J. V. Stark, K. J. Klabunde, J. R. Schlup, *Chem. Mater.* **5**, 71 (1993).
26. J. V. Stark, M. S. thesis, Kansas State University, 1995.
27. I. Amato, *Science*, **252**, 644 (1991), writing about materials synthesis in the laboratories of F. Desalvo and J. D. Corbitt.
28. E. Matijević, *Langmuir* **2**, 12 (1986).
29. K. J. Klabunde, *Chemistry of Free Atoms and Particles*, Academic Press, San Diego, CA, 1980.
30. C. Hayashi, *Phys. Today*, 44 (Dec. 1987); *J. Vac. Sci.* **A5**, 1375 (1987).
31. G. A. Niklasson, *J. Appl. Phys.* **62**, 258 (1987), and references therein.
32. K. Kirmura, S. Bandow, *J. Chem. Soc. Jpn.* **56**, 3578 (1983).
33. S. Yatsuya, S. Katsukabe, R. Uyeda, *Jpn. J. Appl. Phys.* **12**, 1675 (1973).
34. M. P. Andrews, G. A. Ozin, *J. Phys. Chem.* **90**, 3353 (1986).
35. R. J. VanZee, W. Weltner, Jr., *J. Chem. Phys.* **92**, 6976 (1990).
36. D. M. Lindsay, G. A. Thompson, Y. Wang, *J. Phys. Chem.* **91**, 2630 (1987).
37. K. J. Klabunde, T. O. Murdock, *J. Organ. Chem.* **44**, 3901 (1979).
38. S. C. Davis, K. J. Klabunde, *Chem. Rev.* **82**, 153 (1982).

39. K. J. Klabunde, P. L. Timms, P. S. Skell, S. Ittel, *Inorg. Syn.* **19**, 59 (1979).
40. K. J. Klabunde, G. Cardenas-Trivino, in *Active Metals*, A. Fürstner, Ed., VCH, Weinheim, Germany, Chapter 6, pp. 237–275.
41. S. C. Davis, S. C. Severson, K. J. Klabunde, *J. Am. Chem. Soc.* **103**, 3024 (1981).
42. P. Hooker, B. J. Tan, K. J. Klabunde, S. Suib, *Chem. Mater.* **3**, 947 (1991).
43. K. J. Klabunde, S. C. Davis, H. Hattori, Y. Tanaka, *J. Catal.* **54**, 254 (1978).
44. K. J. Klabunde, Y.-X. Li, B. J. Tan, *Chem. Mater.* **3**, 30 (1991).
45. K. J. Klabunde, G. Youngers, E. J. Zuckerman, B. J. Tan, S. Antrim, P. M. A. Sherwood, *Eur. J. Inorg. Solid State Chem.* **29**, 227 (1992).
46. G. Cardenas-Trivino, C. C. Retamal, *Polymer Bull.* **26**, 611 (1991).
47. A. W. Olsen, Z. H. Kafafi, *J. Am. Chem. Soc.* **113**, 7758 (1991).
48. G. Sergeev, V. Zagorsky, M. Petrukhina, *J. Mater. Chem.* **5**, 31 (1995).
49. K. J. Klabunde, D. Zhang, G. N. Glavee, C. M. Sorensen, G. C. Hadjipanayis, *Chem. Mater.* **6**, 784 (1994).
50. (a) C. H. Griffiths, M. P. O'Horo, T. W. Smith, *J. Appl. Phys.* **50**, 7108 (1979). (b) K. S. Suslick, *High Energy Processes in Organometallic Chemistry*, ACS Symposium Series, Vol. 333, American Chemical Society, Washington, DC, 1987, pp. 191–208.
51. J. Van Wonterghem, S. Morup, S. W. Charles, S. Wells, J. Villadsen, *Phys. Rev. Lett.* **55**, 410 (1985).
52. H. I. Schlesinger, H. C. Brown, A. E. Finholt, J. R. Gilbreath, H. R. Hockstra, E. K. Hyde, *J. Am. Chem. Soc.* **75**, 215 (1953).
53. H. C. Brown, C. A. Brown, *J. Am. Chem. Soc.* **84**, 1493, 1494 (1962).
54. J. Van Wonterghem, S. Morup, C. J. W. Koch, S. W. Charles, S. Wells, *Nature* **322**, 622 (1986).
55. A. Corrias, G. Ennas, G. C. Licheri, G. Marongiu, G. Paschina, *Chem. Mater.* **2**, 363 (1990).
56. L. Yiping, G. C. Hadjipanaysis, C. M. Sorensen, K. J. Klabunde, *J. Appl. Phys.* **69**, 5141 (1991).
57. G. N. Glavee, K. J. Klabunde, C. M. Sorensen, G. C. Hadjipanaysis, *Langmuir* **9**, 162 (1993).
58. G. N. Glavee, K. J. Klabunde, C. M. Sorensen, G. C. Hadjipanaysis, *Inorg. Chem.* **32**, 474 (1993).
59. G. N. Glavee, K. J. Klabunde, C. M. Sorensen, G. C. Hadjipanaysis, *Langmuir* **10**, 4726 (1994).
60. G. N. Glavee, K. J. Klabunde, C. M. Sorensen, G. C. Hadjipanaysis, *Inorg. Chem.* **34**, 28 (1995).
61. L. Yiping, Z. X. Tang, G. C. Hadjipanaysis, K. J. Klabunde, C. M. Sorensen, C. M. Chow, T. Ambrose, J. Xiao, F. Kaatz, A. Ervin, *Nanostruct. Mater.* **2**, 131 (1993).
62. J. P. Chen, K. M. Lee, C. M. Sorensen, K. J. Klabunde, G. C. Hadjipanaysis, *J. Appl. Phys.* **75**, 5876 (1994).
63. J. B. Nagy, I. Bodart-Ravet, E. G. Derouane, *J. Chem. Soc. Faraday Trans.* **87**, 189 (1989).

64. H. Bönnemann, W. Brijoux, T. Joussen, *Angew. Chem. Int. Ed. Engl.* **29**, 273 (1990).
65. D. Zeng, M. J. Hampden-Smith, *Chem. Mater.* **5**, 681 (1993).
66. R. D. Rieke, *Acc. Chem. Res.* **10**, 301 (1977).
67. G. L. Rochfort, R. D. Rieke, *Inorg. Chem.* **25**, 348 (1986).
68. K. Tsai, J. L. Dye, *Chem. Mater.* **5**, 540 (1993).
69. R. D. Rieke, *CRC Crit. Rev. Surf. Chem.* **1**, 131 (1991).
70. H. Bönnemann, B. Bogdanović, *Angew. Chem. Int. Ed. Engl.* **22**, 728 (1983).
71. H. Bönnemann, B. Bogdanović, R. Brinkmann, B. Spliethoff, D. W. He, *J. Organomet. Chem.* **451**, 23 (1993).
72. P. Mulvaney, A. Henglein, *J. Phys. Chem.* **94**, 4182 (1990).
73. T. Linnert, P. Mulvaney, A. Henglein, H. Weller, *J. Am. Chem. Soc.* **112**, 4657 (1990).
74. A. Henglein, *Chem. Rev.* **89**, 1861 (1989).
75. F. Fievet, J. P. Lapier, M. Figlarz, *MRS Bull.*, 29 (Dec. 1989).
76. (a) K. Esumi, T. Tano, K. Torigoe, K. Meguro, *Chem. Mater.* **2**, 564 (1990).
(b) Y. Wang, N. Toshima, *J. Phys. Chem.* **101**, 5301 (1997).
77. J. S. Bradley, E. W. Hill, C. Klein, B. Chaudret, A. Duteil, *Chem. Mater.* **5**, 254 (1993).
78. A. C. Curtis, D. G. Duff, P. P. Edwards, D. A. Jefferson, B. F. G. Johnson, A. I. Kirkland, A. S. Wallace, *J. Phys. Chem.* **92**, 2270 (1988).
79. M. L. Steigerwald, L. E. Brus, *Acc. Chem. Res.* **23**, 183 (1990).
80. Y. Wang, N. Herron, *J. Phys. Chem.* **92**, 4988 (1988).
81. N. Herron, in *Materials for Nonlinear Optics: Chemical Perspectives*, ACS Symposium Series, Vol. 455, S. R. Marder, J. E. Sohn, G. D. Stucky, Eds., American Chemical Society, Washington, DC, 1991, 582.
82. K. Moller, M. M. Eddy, G. D. Stucky, N. Herron, T. Bein, *J. Am. Chem. Soc.* **111**, 2564 (1989).
83. K. Moller, T. Bein, N. Herron, W. Mahler, Y. Wang, *Inorg. Chem.* **28**, 2914 (1989).
84. C. R. Martin, *Science* **266**, 1961 (1994).
85. M. L. Steigerwald, L. E. Brus, *Annu. Rev. Mater. Sci.* **19**, 471 (1989).
86. J. G. Brennan, T. Siegrist, S. M. Stuczynski, M. L. Steigerwald, *J. Am. Chem. Soc.* **111**, 9240 (1989).
87. M. L. Steigerwald, T. Siegrist, S. M. Stuczynski, *Inorg. Chem.* **30**, 2256 (1991).
88. M. Meyer, C. Wallberg, K. Kurihara, J. H. Fendler, *J. Chem. Soc. Chem. Commun.*, 90 (1984).
89. J. H. Fendler, *Chem. Rev.* **87**, 877 (1987).
90. H. Watzke, J. H. Fendler, *J. Phys. Chem.* **91**, 854 (1987).
91. Y. Wang, A. Suna, W. Mahler, R. Kasowski, *J. Chem. Phys.* **87**, 7315 (1987).
92. N. Herron, Y. Wang, H. Eckert, *J. Am. Chem. Soc.* **112**, 1322 (1990).
93. M. Boutonnet, J. Kizling, P. Stenius, G. Maire, *Colloids Surf.* **5**, 209 (1982).
94. K. Kurihara, J. Kizling, P. Stenius, J. H. Fendler, *J. Am. Chem. Soc.* **105**, 2574 (1983).

95. N. Lufimadio, J. B. Nagy, E. G. Derouane, in *Surfactants in Solution*, Vol. 3, K. J. Mittal, B. Lindman, Eds., Plenum Press, New York, 1984, p. 1483.
96. M. Gobe, K. Kon-No, K. Kandori, A. Kitahara, *J. Colloid. Interface Sci.* **93**, 293 (1983).
97. K. M. Lee, C. M. Sorensen, K. J. Klabunde, G. C. Hadjipanayis, *IEEE Trans. Magn.* **28**, 3180 (1992).
98. K. M. Choi, K. J. Shea, *J. Phys. Chem.* **98**, 3207 (1994).
99. Y. Zhang, N. Raman, J. K. Bailey, J. Brinker, R. M. Crooks, *J. Phys. Chem.* **96**, 9098 (1992).
100. C. T. Dameron, R. N. Reese, R. K. Mehra, A. R. Kortan, P. J. Carroll, M. L. Steigerwald, L. E. Brus, D. R. Winge, *Nature* **338**, 596 (1989).
101. S. S. Kher, R. L. Wells, *Chem. Mater.* **6**, 2056 (1994).
102. (a) R. L. Wells, C. G. Pitt, A. T. McPhail, A. P. Purdy, S. Shafieezad, R. B. Hallock, *Chem. Mater.* **1**, 4 (1989). (b) M. A. Olshavsky, A. N. Goldstein, A. P. Alivisatos, *J. Am. Chem. Soc.* **112**, 9438 (1990).
103. B. Jirgensons, M. E. Straumanis, *Colloid Chemistry*, Macmillan, New York, 1962, pp. 119, 130, 258, 306.
104. D. J. Shaw, *Introduction to Colloid and Surface Chemistry*, 2nd ed., Butterworth, London, 1970.
105. S. S. Kistler, *J. Phys. Chem.* **36**, 52 (1932).
106. S. J. Teichner, G. A. Nicolaon, M. A. Vicarini, G. E. E. Gardes, *Adv. Colloid Interface Sci.* **5**, 245 (1976).
107. H. D. Gesser, P. C. Goswami, *Chem. Rev.* **89**, 765 (1989).
108. S. Utamapanya, K. J. Klabunde, J. R. Schlup, *Chem. Mater.* **3**, 175 (1991).
109. H. Itoh, S. Utamapanya, J. V. Stark, K. J. Klabunde, J. R. Schlup, *Chem. Mater.* **5**, 71 (1993).
110. T. Sugimoto, E. Matijevic, *J. Colloid Interface Sci.* **74**, 227 (1980).
111. Z. X. Tang, C. M. Sorensen, K. J. Klabunde, G. C. Hadjipanayis, *J. Colloid Interface Sci.* **146**, 38 (1991).
112. J. D. Jorgensen, H. B. Schuttler, D. G. Hinks, D. W. Capone II, K. Zhang, M. B. Brodsky, D. J. Scalapino, *Phys. Rev. Lett.* **58**, 1024 (1987).
113. E. Matijević, *Acc. Chem. Res.* **14**, 22 (1981).
114. G. L. Messing, T. J. Gardner, R. R. Ciminelli, *Sci. Ceram.* **12**, 117 (1983).
115. K. Nonaka, S. Hayaski, K. Okada, N. Otsuka, T. Yano, *J. Mater. Res.* **6**, 1750 (1991).
116. H. K. Xu, C. M. Sorensen, K. J. Klabunde, G. C. Hadjipanayis, *J. Mater. Res.* **7**, 712 (1992).
117. Z. X. Tang, S. Nafis, C. M. Sorensen, G. C. Hadjipanayis, K. J. Klabunde, *IEEE Trans. Magn.* **25**, 4236 (1989).
118. Q. Li, C. M. Sorensen, K. J. Klabunde, G. C. Hadjipanayis, *Aerosol Sci. Technol.* **19**, 453 (1993).
119. G. Lassaletta, A. Fernández, J. P. Espinos, R. González-Elipe, *J. Phys. Chem.* **99**, 1484 (1995).

120. R. W. Siegel, H. Hahn, in *Current Trends in the Physics of Materials*, M. Yussouff, Ed., World Scientific, Singapore, 1987, p. 403.
121. J. P. White III, H. Deng, S. G. Shore, *Inorg. Chem.* **30**, 2337 (1991).
122. R. L. LaDuca, P. T. Wolczanski, *Inorg. Chem.* **31**, 1311 (1992).
123. P. Hooker, B. J. Tan, K. J. Klabunde, S. Suib, *Chem. Mater.* **3**, 947 (1991).
124. M. R. Harrison, P. P. Edwards, in *The Metallic and Nonmetallic States of Matter*, P. O. Edwards, C. N. R. Rao, Eds., Taylor & Francis, London, 1985, p. 389.
125. K. A. Easom, K. J. Klabunde, C. M. Sorensen, G. C. Hadjipanayis, *Polyhedron* **13**, 1197 (1994).
126. B. D. Cullity, *Introduction to Magnetic Materials*, Addison-Wesley, Reading, MA, 1972.
127. P. Weiss, *Compt. Rend.* **143**, 1136 (1906).
128. J. Frenkel, J. Dorfman, *Nature* **126**, 274 (1930).
129. E. E. Carmen, *Powder Met.* **4**, 1 (1959).
130. F. E. Luborsky, T. O. Paine, *J. Appl. Phys.* **31**, 66s (1960).
131. L. Neel, *C.R. Acad. Sci.* **228**, 664 (1949).
132. C. P. Bean, J. D. Livingston, *J. Appl. Phys.* **30**, 120s (1959).
133. C. P. Bean, I. S. Jacobs, *J. Appl. Phys.* **27**, 1448 (1956).
134. (a) A. H. Morrish, K. Haneda, X. Z. Zhou, in *Nanophase Materials: Synthesis, Properties, Applications*, G. C. Hadjipanayis, R. W. Siegel, Eds., Kluwer, London, 1994, pp. 515–536. (b) D. J. Sellmeyer, ibid., pp. 537–554. (c) C. L. Chien, ibid., pp. 555–569. (d) A. E. Berkowitz, F. E. Spada, F. T. Parker, ibid., pp. 587–594. (e) S. Morup, S. Linderoth, ibid., pp. 595–612. (f) E. P. Giannelis, V. Mehrotra, J. K. Vassiliou, R. D. Shull, R. D. MacMichael, R. F. Ziolo, ibid., pp. 617–624. (g) A. S. Arrott, T. L. Templeton, Y. Yoshida, ibid., pp. 663–674.
135. W. H. Meikeljohn, C. P. Bean, *Phys. Rev.* **105**, 904 (1957).
136. S. Gangopadhyay, G. C. Hadjipanayis, B. Dale, C. M. Sorensen, K. J. Klabunde, V. Papaefthymiou, A. Kostikas, *Phys. Rev. B* **45**, 9778 (1992).
137. K. J. Klabunde, D. Zhang, G. N. Glavee, C. M. Sorensen, G. C. Hadjipanayis, *Chem. Mater.* **6**, 784 (1994).
138. W. A. de Heer, P. Melani, A. Chatelain, *Phys. Rev. Lett.* **65**, 488 (1990).
139. I. M. L. Billas, A. Chatelain, W. A. de Heer, *Science* **265**, 1682 (1994).
140. J. Merikoski, J. Timonen, M. Manninen, P. Jena, *Phys. Rev. Lett.* **66**, 938 (1991).
141. J. P. Bucher, D. C. Douglass, L. A. Bloomfield, *Phys. Lett.* **66**, 3052 (1991).
142. S. N. Khanna, S. Linderoth, *Phys. Rev. Lett.* **67**, 742 (1991).
143. D. C. Douglass, A. J. Cox, J. P. Bucher, L. A. Bloomfield, *Phys. Rev. B* **47**, 12874 (1993).
144. J. P. Chen, C. M. Sorensen, K. J. Klabunde, G. C. Hadjipanayis, *J. Appl. Phys.* **75**, 5876 (1994).
145. J. P. Chen, C. M. Sorensen, K. J. Klabunde, G. C. Hadjipanayis, *Phys. Rev. B* **51**, 11527 (1995).
146. D. M. Lindsay, Y. Wang, T. F. George, *J. Cluster Sci.* **1**, 107 (1990).
147. C. E. Klots, *J. Phys. Chem.* **92**, 5864 (1988), also discusses evaporation of small particles.

148. T. P. Martin, U. Näher, H. Schaber, U. Zimmerman, *J. Chem. Phys.* **100**, 2322 (1994).

149. Ph. Buffat, J. P. Borel, *Phys. Rev. A* **13**, 2287 (1976).

150. S. J. Peppiat, J. R. Sambles, *Proc. R. Soc. London A* **345**, 387 (1975).

151. M. Y. Hahn, R. L. Whetten, *Phys. Rev. Lett.* **61**, 1190 (1988).

152. C. Solliard, *Surf. Sci.* **106**, 58 (1981); C. L. Briant, J. J. Burton, *J. Chem. Phys.* **63**, 2045 (1975).

153. W. D. Kristensen, E. J. Jensen, R. M. J. Cotterill, *J. Chem. Phys.* **60**, 4161 (1974).

154. J. D. Honeycutt, H. C. Andersen, *J. Phys. Chem.* **91**, 4950 (1987).

155. J. Waltersdorf, A. S. Nepijko, E. Pippel, *Surf. Sci.* **106**, 64 (1981).

156. C. Solliard, M. Flueli, *Surf. Sci.* **156**, 487 (1985).

157. A. N. Goldstein, C. M. Echer, A. P. Alivisatos, *Science* **256**, 1425 (1992).

158. M. Faraday, *Philos. Trans. R. Soc. London Ser. B* **147**, 145 (1857).

159. (a) R. Zsigmondy, P. A. Thiessen, *Das Kollide Gold*, Barth, Leipzig, 1925, p. 159. (b) See also J. Turkevich, P. C. Stevenson, J. Hillier, *J. Phys. Chem.* **57**, 670 (1953).

160. For historical reviews and studies of non-aqueous systems see: (a) M. T. Franklin, K. J. Klabunde, in ACS Symposium Series, Vol. 333, K. S. Suslick, Ed., American Chemical Society, Washington, DC, 1987, p. 246. (b) S. T. Lin, M. T. Franklin, K. J. Klabunde, *Langmuir* **2**, 259 (1986). (c) M. T. Franklin, M.S. thesis, Kansas State University, 1987.

161. C. Mie, *Ann. Phys. (Leipzig)* **25**, 377 (1908).

162. T. L. Ferrell, T. A. Callcott, R. J. Warmack, *Am. Sci.* **73**, 344 (1985).

163. M. L. Steigerwald, L. E. Brus, *Acc. Chem. Res.* **23**, 183 (1990).

164. Y. Wang, N. Herron, *Phys. Rev. B* **42**, 7253 (1990).

165. K. R. Johansson, G. McLendon, A. P. Marchetti, *Chem. Phys. Lett.* **179**, 321 (1991).

166. C. J. Berry, *Photogr. Sci.* **18**, 169 (1970).

167. R. Rossetti, R. Hull, J. M. Gibson, L. E. Brus, *J. Chem. Phys.* **83**, 1406 (1985).

168. T. Dannhauser, M. O'Neil, K. Johansson, D. Whitten, G. McLendon, *J. Phys. Chem.* **90**, 6074 (1986).

169. Y. Wang, *Acc. Chem. Res.* **24**, 133 (1991).

170. Y. Wang, N. Herron, W. Mahler, A. Suna, *J. Opt. Soc. Am. B* **6**(4), 808 (1989).

171. R. E. Schwerzel, *Naval Research News*, Vol. 46(4), Office of Naval Research, Arlington, VA, 1994, pp. 24–29.

172. A. R. Kortan, R. Hull, R. L. Opila, M. G. Bawendi, M. L. Steigerwald, P. J. Carroll, L. E. Brus, *J. Am. Chem. Soc.* **112**, 1327 (1990).

173. See also: (a) R. Vogel, P. Hoyer, H. Weller, *J. Phys. Chem.* **98**, 3183 (1994), for tailored optical properties by supporting semiconductor particles of sulfides on nanoporous oxides; and (b) G. Lassaletta, A. Fernández, J. P. Espinós, A. R. González-Elipe, *J. Phys. Chem.* **99**, 1484 (1995), for blue-shifted TiO_2 absorptions due to surface effects of a SiO_2 support.

174. A. W. Olsen, Z. H. Kafafi, *J. Am. Chem. Soc.* **113**, 7758 (1991).

175. L. T. Chang, N. Herron, Y. Wang, *J. Appl. Phys.* **66**, 3417 (1989).
176. K. J. Klabunde, *Chemistry of Free Atoms and Particles*, Academic Press, San Diego, CA, 1980.
177. K. J. Klabunde, A. Whetten, *J. Am. Chem. Soc.* **108**, 6529 (1986).
178. A. V. Nemukhin, I. A. Topol, F. Weinhold, *Inorg. Chem.* **34**, 2980 (1995).
179. E. K. Parks, B. H. Weiller, P. S. Bechthold, W. F. Hoffman, G. C. Niemann, L. P. Pobo, S. J. Riley, *J. Chem. Phys.* **88**, 1622 (1988).
180. D. J. Trevor, D. M. Cox, A. Kaldor, *J. Am. Chem. Soc.* **112**, 3742 (1990).
181. K. J. Klabunde, G. H. Jeong, A. W. Olsen, in *Selective Hydrocarbon Activation: Principles and Progress*, J. A. Davies, P. L. Watson, A. Greenberg, J. F. Liebman, Eds., VCH, New York, 1990, Chapter 13, p. 433.
182. S.-L. Chang, P. A. Thiel, *Crit. Rev. Surf. Chem.* **3**, 239 (1994).
183. D. G. Duff, A. C. Curtis, P. P. Edwards, D. A. Jefferson, B. F. G. Johnson, D. E. Logan, *J. Chem. Soc. Chem. Commun.*, 1264 (1987).
184. F. Strelow, A. Fojtik, A. Henglein, *J. Phys. Chem.* **98**, 3032 (1994).
185. T. Linnert, P. Mulvaney, A. Henglein, *J. Phys. Chem.* **97**, 679 (1993).
186. A. Henglein, *Ber. Bunsenges. Phys. Chem.* **94**, 600 (1990).
187. T. Linnert, P. Mulvaney, A. Henglein, *Ber. Bunsenges. Phys. Chem.* **95**, 838 (1991).
188. C. N. R. Rao, V. Vijayakrishnan, A. K. Santra, M. W. J. Prins, *Angew. Chem. Int. Ed. Engl.* **31**, 1062 (1992).
189. A. Henglein, *Top. Cur. Chem.* **143**, 115 (1988).
190. J. S. Bradley, J. M. Millar, E. W. Hill, S. Behal, *J. Catal.* **129**, 530 (1991).
191. J. S. Bradley, J. M. Millar, E. W. Hill, *J. Am. Chem. Soc.* **113**, 4016 (1991).
192. J. S. Bradley, E. W. Hill, S. Behal, C. Klein, *Chem. Mater.* **4**, 1234 (1991).
193. R. Van Hardeveld, F. Hartog, *Surf. Sci.* **15**, 189 (1969).
194. H. P. Kaukonen, U. Landman, C. L. Cleveland, *J. Chem. Phys.* **95**, 4997 (1991).
195. D. Meisel, *J. Am. Chem. Soc.* **101**, 6133 (1979).
196. V. VuKović, J. M. Nedeljković, *Langmuir* **9**, 980 (1993).
197. L. N. Lewis, R. J. Uriante, N. Lewis, *J. Mol. Catal.* **66**, 105 (1991).
198. G. Cardenas-T., K. J. Klabunde, E. B. Dale, *Langmuir* **3**, 986 (1987).
199. G. Cardenas-T., R. Oliva-C., K. J. Klabunde, *Eur. J. Solid State Inorg. Chem.* **33**, 1135–1147 (1996).
200. B. J. Tan, P. M. A. Sherwood, K. J. Klabunde, *Langmuir* **6**, 105 (1990).
201. K. J. Klabunde, G. Youngers, E. B. Zuckerman, B. J. Tan, S. Antrim, P. M. Sherwood, *Eur. J. Inorgan. Solid State Chem.* **29**, 227 (1992).
202. B. Rabinovitch, R. H. Horner, K. J. Klabunde, P. Hooker, *Metalsmith* **14**, 41 (1994).
203. A. Henglein, P. Mulvaney, T. Linnert, A. Holzworth, *J. Phys. Chem.* **96**, 2411 (1992).
204. P. Mulvaney, M. Giersig, A. Henglein, *J. Phys. Chem.* **96**, 10419 (1992).
205. Y. Wang, N. Herron, *J. Phys. Chem.* **91**, 5005 (1987).

206. T. Kannhauser, M. O'Neil, K. Johansson, D. Whitten, G. McLendon, *J. Phys. Chem.* **90**, 6074 (1986).
207. G. J. Meyer, L. K. Leung, J. C. Yu, G. C. Lisensky, A. B. Ellis, *J. Am. Chem. Soc.* **111**, 5146 (1989).
208. T. Shiragami, H. Ankyu, S. Fukami, C. Pac, S. Yanagida, H. Mori, H. Fujita, *J. Chem. Soc. Faraday Trans.*, 1055 (1992).
209. A. J. Hoffman, E. R. Carraway, M. R. Hoffman, *Environ. Sci. Technol.* **28**, 776 (1994).
210. I. B. Rufus, V. Ramakrishnan, B. Viswanathan, J. C. Kuriacose, *Langmuir* **6**, 565 (1990).
211. O. Horvath, J. H. Fendler, *J. Phys. Chem.* **96**, 9591 (1992).
212. S. A. Majetich, A. C. Carter, J. Belot, R. D. McCollough, *J. Phys. Chem.* **98**, 13705 (1994).
213. J. Kiwi, M. Grätzel, *Angew. Chem. Int. Ed. Engl.* **18**, 624 (1979).
214. A. Kay, R. Humphrey-Baker, M. Grätzel, *J. Phys. Chem.* **98**, 952 (1994).
215. J. Fan, J. T. Yates, Jr., *J. Phys. Chem.* **98**, 10621 (1994).
216. G. Dagan, M. Tomkiewicz, *J. Phys. Chem.* **97**, 12651 (1993).
217. Z. Goren, I. William, A. J. Nelson, A. J. Frank, *J. Phys. Chem.* **94**, 3784 (1990).
218. G. Riegel, J. R. Bolton, *J. Phys. Chem.* **99**, 4215 (1995).
219. P. Pichat, C. Guillard, C. Maillard, L. Amalric, J. C. D'Oliveira, in *Photocatalytic Purfication and Treatment of Water and Air*, D. F. Ollis, H. Al-Ekabi, Eds., Elsevier, Amsterdam, 1993, pp. 207–223.
220. C. Y. Hsiao, C. L. Lee, D. F. Ollis, *J. Catal.* **82**, 418 (1983).
221. C. Mohs, K. J. Klabunde, in *Nanophase Materials*, G. C. Hadjipanayis, R. W. Siegel, Eds., Kluwer, Dordrecht, The Netherlands, 1994, pp. 121–124.
222. C. Mohs, unpublished results, in Ph.D. thesis, Kansas State University, 1996.
223. See K. M. Schindler, M. Kunst, *J. Phys. Chem.* **94**, 8222 (1990), for a further discussion of surface effects.
224. H. Liu, L. Feng, X. Zhang, O. Xue, *J. Phys. Chem.* **99**, 332 (1995).
225. K. Tanabe, *Solid Acids and Bases*, Academic Press, San Diego, CA, 1970.
226. M. Utiyama, H. Hattori, K. Tanabe, *J. Catal.* **53**, 237 (1978).
227. R. M. Morris, K. J. Klabunde, *Inorg. Chem.* **22**, 682 (1983).
228. D. J. Driscoll, W. Martin, J. X. Wang, J. H. Lunsford, *J. Am. Chem. Soc.* **107** (1985).
229. M. F. Hoq, K. J. Klabunde, *J. Am. Chem. Soc.* **108**, 2114 (1986).
230. (a) R. M. Morris, K. J. Klabunde, *J. Am. Chem. Soc.* **105**, 2633 (1983). (b) M. A. Nygren, L. G. M. Pettersson, Z. Barandiaran, L. Seijo, *J. Chem. Phys.* **100**, 2010 (1994). (c) T. Tashiro, J. Ito, R. B. Sim, K. Miyazawa, E. Hamada, K. Toi, H. Kobayashi, T. Ito, *J. Phys. Chem.* **99**, 6115 (1995).
231. H. Tsuji, T. Shishido, A. Okamura, Y. Gao, H. Hattori, H. Kita, *J. Chem. Soc. Faraday Trans.*, 803 (1994).
232. Y. X. Li, K. J. Klabunde, *Chem. Mater.* **4**, 611 (1992).
233. S. Utamapanya, K. J. Klabunde, J. R. Schlup. *Chem. Mater.* **3**, 175 (1991).
234. Y. X. Li, K. J. Klabunde, *Langmuir* **7**, 1388 (1991).

235. O. Koper, Y. X. Li, K. J. Klabunde, *Chem. Mater.* **5**, 500 (1993).
236. Y. X. Li, J. R. Schlup, K. J. Klabunde, *Langmuir* **7**, 1394 (1991).
237. J. Stark, unpublished results.
238. H. Itoh, S. Utamapanya, J. Stark, K. J. Klabunde, J. R. Schlup, *Chem. Mater.* **5**, 71 (1993).
239. K. J. Klabunde, A. Khaleel, D. Park, *High Temp. Sci.* **33**, 99 (1995).
240. Y. X. Li, K. J. Klabunde, *Langmuir* **7**, 1388 (1991).
241. T. Boronia, K. J. Klabunde, G. B. Sergeev, *Environ. Sci. Technol.* **29**, 1511–1517 (1995).
242. V. Haensel, R. Burwell, *Sci. Am.* **225**, 46 (1971).
243. K. J. Klabunde, Y. X. Li, B. J. Tan, *Chem. Mater.* **3**, 30 (1991).
244. (a) H. Bönnemann, B. Korall, *Angew. Chem. Int. Ed. Engl.* **31**, 1490 (1992). (b) H. Bönnemann, W. Brijoux, T. Joussen, *Angew. Chem. Int. Ed. Engl.* **29**, 273 (1990).
245. (a) E. Matejivic, *Annu. Rev. Mater. Sci.* **15**, 483 (1985). (b) E. Matejivic, *MRS Bull.* 18 (Dec. 1989).
246. (a) R. D. Shull, R. D. McMichael, L. J. Swartzendruber, L. H. Bennett, in *Studies of Magnetic Properties of Fine Particles and Their Relevance to Materials Science*, J. J. Dormann, D. Fiorani, Eds., Elsevier, Amsterdam, 1992, pp. 161–169. (b) R. D. McMichael, R. D. Shull, L. J. Swartzendruber, L. H. Bennett, *J. Magn. Mater.* **111**, 29 (1992).
247. N. Herron, *Chem. Technol.*, 542 (1989).
248. G. Cardenas-T., K. J. Klabunde, *Bol. Soc. Chil. Quim.* **33**, 163 (1988).

CHAPTER 8

Nanoporous Materials

PETER T. TANEV, JEAN-RÉMI BUTRUILLE, and THOMAS J. PINNAVAIA
Department of Chemistry, Michigan State University, East Lansing, Michigan

8.1 INTRODUCTION

Porous materials created by nature or by synthetic design have found great utility in all aspects of human activity. Their pore structure is usually formed in the stages of crystallization or subsequent treatment and consists of isolated or interconnected pores that may have similar or different shapes and sizes.[1] The pore shape can be roughly approximated to cylindrical, slit shaped, or ink-bottle or cone shaped. Depending on the predominant pore size, the solid materials are classified by IUPAC as: (1) *microporous*, having pore sizes below 2.0 nm; (2) *macroporous*, with pore sizes exceeding 50.0 nm; and (3) *mesoporous*, with intermediate pore sizes between 2.0 and 50.0 nm.[2] The use of macroporous materials as adsorbents and catalysts is relatively limited due to their low surface area and large nonuniform pores. Microporous and mesoporous materials, or more generally *nanoporous materials*, however, are widely used in adsorption, separation technology, and catalysis.[3,4] Nanoporous materials can be structurally amorphous, paracrystalline, or crystalline. Amorphous materials, such as silica gel or alumina gel, do not possess long-range order, whereas paracrystalline solids, such as γ- or η-Al_2O_3 are quasi-ordered, as evidenced by the broad peaks on their x-ray diffraction patterns.[5] Both classes of materials exhibit broad distribution of nanopores that limits their shape selectivity and effectiveness as adsorbents, ion exchangers, and catalysts. The only class of nanoporous materials possessing *uniform or nearly uniform pore sizes* includes crystalline *zeolites*, related *molecular sieves*, and *pillared lamellar solids*. Because of their *uniform nanopore size*, these materials can discriminate or "sieve" molecules by shape and size.

Chemistry of Advanced Materials: An Overview, Edited by Leonard V. Interrante and Mark J. Hampden-Smith.
ISBN 0-471-18590-6 © 1998 Wiley-VCH, Inc.

For instance, molecules with kinetic diameters smaller than the uniform nanopores will be allowed to diffuse, adsorb, and transform catalytically in the pores, whereas those with slightly larger kinetic diameters will not.[6] In addition, the uniform pore network of these materials can play an important shape-selective role in ruling out undesired reaction pathways by favoring specific active complex and reaction products. These important properties of nanoporous materials have made them very suitable for size-specific industrial applications in adsorption, separation (ion exchange) and shape-selective catalytic processes. In this chapter we describe some recent developments on synthesis and catalytic applications of nanoporous zeolites, molecular sieves, and pillared lamellar solids.

8.2 ZEOLITES AND MOLECULAR SIEVES

Zeolites are naturally occurring or synthetic highly crystalline aluminosilicates with the general chemical formula $[M_{x/n}(AlO_2)]_x \cdot [SiO_2]_y \cdot mH_2O$. Their frameworks are negatively charged, due to the replacement of Si^{4+} by Al^{3+}. Here n represents the formal positive charge on the cation. In natural zeolites the charge is balanced by M monovalent or polyvalent metal cations, such as Na^+ or Ca^{2+}. In synthetic zeolites the charge could also be balanced by an organic cation: for example, tetrapropylammonium ion. According to their Si/Al framework ratio, the zeolites are classified as high-silica zeolites, $Si/Al > 5$; intermediate-silica zeolites, $5 \geq Si/Al > 2$; and low-silica zeolites, $1 \leq Si/Al \leq 2$.[7] As the Si/Al ratio increases the zeolite framework becomes more hydrophobic. Thus high-silica zeolites exhibit high adsorption affinity for organic moieties, whereas low-silica zeolites (because of their Lewis and Brønsted acidity) adsorb predominantly water.

Figure 8.1 illustrates the formation of a typical zeolite lattice from elementary framework building blocks. The straight lines here represent oxygen bridges, and the vertices are occupied by the Al or Si atoms. The blocks are usually polyhedron units that are formed during synthesis by cross-linking of the TO_4 tetrahedra (where T is Al and Si). The spatial arrangement of these framework units is orchestrated by the cationic template. For example, formation of the sodalite cage occurs by cross-linking of the AlO_2^- and SiO_2 tetrahedra around the cationic template. This gives rise to the corresponding sodalite truncated octahedron building block (Fig. 8.1). As shown in the figure, further cross-linking between these sodalite cages (by sharing the apical oxygen atoms) can generate different three-dimensional zeolite frameworks, such as that of sodalite, zeolite A, or faujasite (zeolite X or Y). However, the formation of zeolite A and faujasite is accomplished through the participation of smaller building blocks, denoted by zeolite chemists as double four-ring (D4R) or double six-ring (D6R) building blocks.[8] All known zeolites and molecular sieves are formed in a similar fashion by putting together polyhedron units with different shapes and sizes.[9-11] The pore

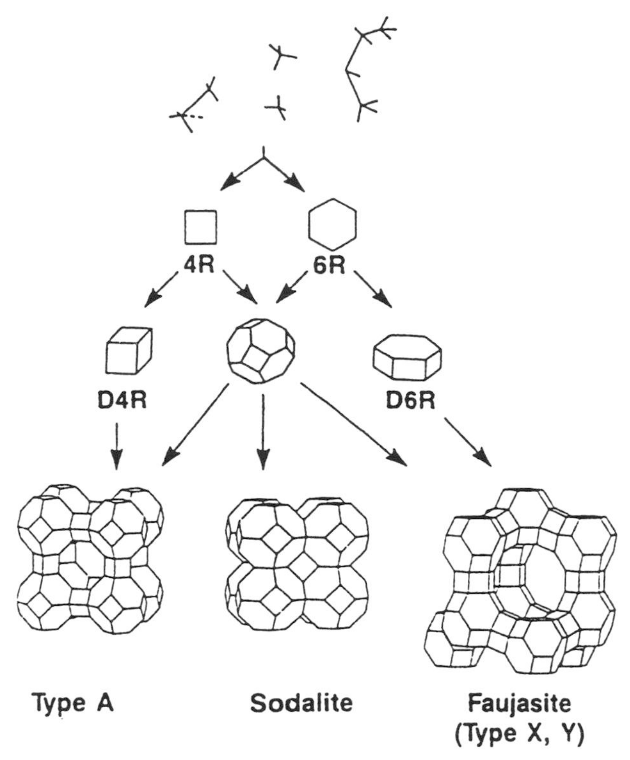

Figure 8.1 Schematic representation of the formation of the aluminosilicate members of sodalite family of zeolites. The different frameworks are obtained by putting together sodalite cages and other polyhedron units in a different arrangement. (Reproduced from D. E. W. Vaughan, *Chem. Eng. Prog.* **85**, 25 (1988) with permission of the American Institute of Chemical Engineers. Copyright © 1988 AIChE. All rights reserved.)

network of a typical zeolite, which is confined by the solid framework, consists of cavities and connecting windows of uniform size occupied by cations and water molecules. Because of their *uniform pore size* and ability to "sieve" molecules by shape and size, zeolites are considered as a subclass of molecular sieves.

Molecular sieves are also crystalline framework materials but with nonaluminosilicate nature. The tetrahedral positions in the framework of molecular sieves are occupied solely by Si or by combinations of other T atoms, such as Al, Ga, Ti, V, and P. The pure SiO_2 molecular sieves are neutral, due to the charge compensation between the $+4$ silicon and the four -2 O atoms (each of these O atoms is shared between two Si tetrahedra). In the aluminophosphate-based molecular sieves, for example, the ratio of Al/P is kept at 1 and the framework is again neutral, due to charge compensation between the AlO_4^- and PO_4^+ tetrahedra. There are many natural zeolites, few natural molecular sieves, and large and continuously growing number of synthetic zeolites and molecular sieves with or without natural counterparts.[11]

8.2.1 Templating Approaches to Microporous Molecular Sieves

Synthetic zeolites and molecular sieves are prepared under hydrothermal conditions from aluminosilicate or metallophosphate gels. The synthesis of the first zeolite without synthetic counterpart was accomplished by Barrer in 1948.[12] This date marks the beginning of an extensive use of alkali metal aluminosilicate gels to crystallize zeolite structures. It has been postulated that the hydrated alkali metal cations play a *structure directing role*. The first synthesis of a zeolite (zeolite A) using an organic template, in particular the trimethylammonium cation, was reported by Barrer and Denny in 1961.[13] Since that pioneering work, a myriad of different organic directing agents (with one or more functional groups) were put to test in the search of new and more exotic zeolite structures. It did not take very long to realize that the proper selection of template is of extreme importance for the preparation of a particular framework and pore network. Thus, today, we have hundreds of molecular sieves with more than 85 different frameworks that are classified and well systematized by Meier and Olson in an *Atlas of Zeolite Structure Types*.[11] Excellent up-to-date reviews of the use of various organic templates and their corresponding structures, as well as the mechanism of structure directing are given elsewhere.[14–18] Here we briefly illustrate some of the conventional preparation approaches to high-silica zeolites and pure silica molecular sieves.

The synthesis of high-silica zeolites usually involves the addition of charged organic molecules (mostly quaternary ammonium cations) to the aluminosilicate or silica gel. The obtained reaction mixture is then placed in an autoclave and heated at high temperature (from 100 to 200°C) for a prolonged period of time (1 to 180 days) to afford the crystalline product. According to Davis,[16] the role of the organic molecules in the synthesis of high-silica zeolites is that of (1) space-filling species, (2) structure-directing agents, or (3) templates. A typical example of *space-filling* function is illustrated by the fact that 22 and 13 different organic molecules can be used to synthesize zeolites ZSM-5 and ZSM-48, respectively.[19] When a specific structure is prepared in the presence of a particular organic molecule, we talk about a *structure-directing* role of the organic molecule. For example, the preparation of the high-silica zeolite SSZ-26 was accomplished by the structure-directing role of an exotic organic ion *N,N,N,N′,N′,N′*-hexamethyl-8,11-[4.3.3.0]dodecane diammonium cation.[20] The *templating* function also involves crystallization of the molecular sieve in the presence of single organic species. However, the absence of a free rotation of the organic guest in the pore network of the inorganic host is considered to be specific for a templating function. For example, the preparation of ZSM-18 was accomplished in the presence of a triquat ammonium cation ($C_{18}H_{36}N^+$). The cage of this molecular sieve was found to have the same trifold rotational symmetry as the organic guest. The lack of rotational freedom of the triquat molecule in the cavity of zeolite ZSM-18 was attributed[16] to the true templating role of this organic guest. The preparation of high-silica zeolites such as ZSM-5, ZSM-11, ZSM-12, and ZSM-23 is usually performed in the presence of small quaternary ammonium "templates" such as tetrapropyl, tetrabutyl, or tetraethyl ammonium ions and di- trimethyl- diammonium heptane, respectively.[15, 17, 18]

Davis et al. elaborated on the mechanism of formation of high-silica microporous zeolites, in particular ZSM-5 (Fig. 8.2).[21] According to the proposed mechanism, the organic cations tend to organize water molecules in their vicinity. The free energy of this templating solution is minimized by hydrophobic hydration of *single organic cations*. Upon the addition of a silicate precursor, the negatively charged silicate species (at high pH the silica species are negatively charged[6]) begin to condense and polymerize on the surface of the single organic cations, giving rise to a long-range ordered zeolite framework. Thus, the geometry of the *single organic cation* is transferred into the particular ZSM-5 inorganic framework. It is important to emphasize that the cation arranges the TO_4 tetrahedra in the framework walls of the microporous molecular sieves in very ordered fashion. This is evidenced by the multiple reflections observed on the corresponding x-ray diffraction patterns.[22]

Primary amines and diamines such as propylamine, *i*-propylamine,[23] diaminopentane, diaminohexane, and diaminododecane[24] were also found to direct the synthesis of the ZSM-5 structure. However, Hearmon et al. have pointed out[25] that the foregoing framework assembly process (under the specific reaction conditions) is directed by the protonated forms of these amines. On the other hand, the use of neutral primary amines of short alkyl chain lengths afforded predominantly microporous molecular sieves.

In summary, most of the existing microporous high-silica zeolites and molecular sieves were assembled using quaternary ammonium cations or protonated forms of amines or diamines, prolonged reaction times, and hydrothermal reaction conditions. The few cases involving the use of neutral primary amines as templates also afforded microporous high-silica molecular sieves. The microporous character of these molecular sieves could be attributed to the crystallization of the solid framework around *single organic species*.

8.2.2 Applications of Zeolites and Molecular Sieves

The industrial application of zeolites was launched in the late 1950s following the remarkable accomplishment made by Milton with the preparation of zeolite A[26] and zeolite X.[27] The following years witnessed rapid growth in the field of zeolite and molecular sieve synthesis. Currently, thousands of patents and publications in a number of respected scientific magazines, such as

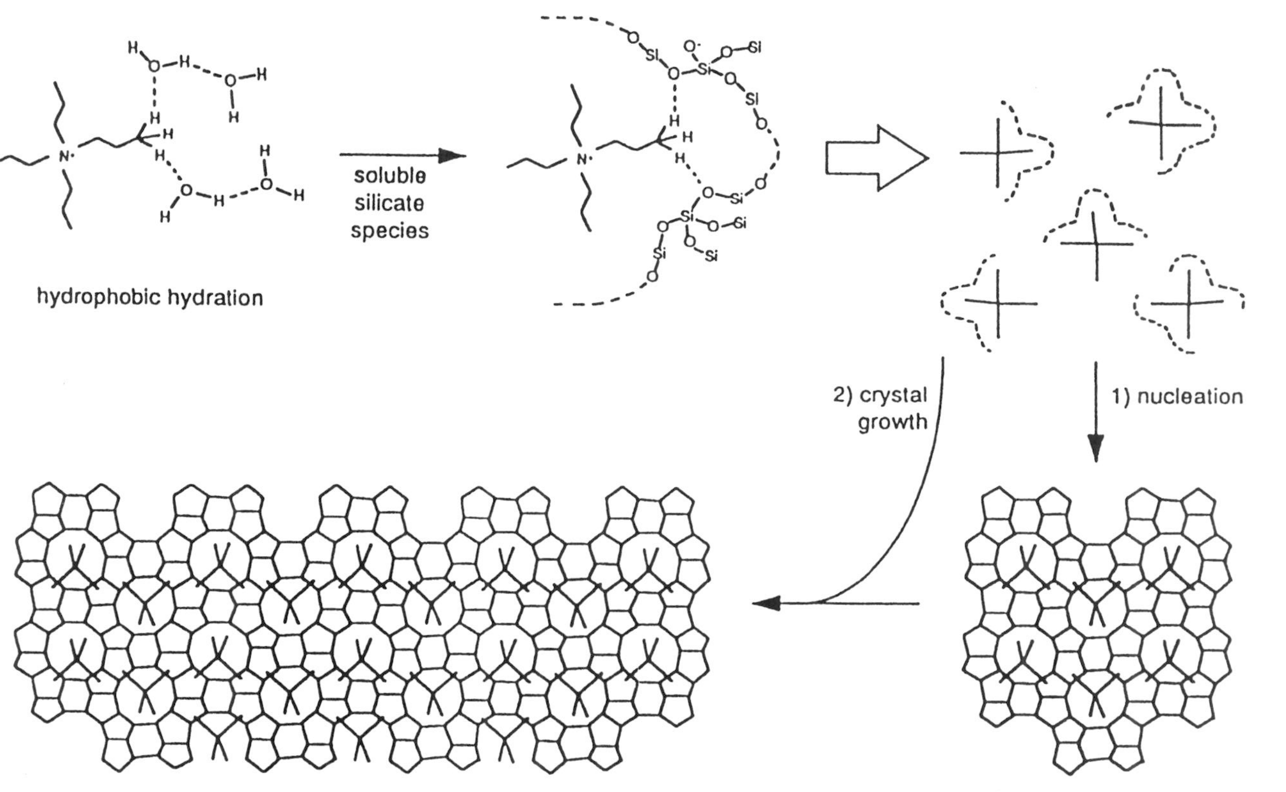

Figure 8.2 Mechanism of formation of high-silica microporous zeolite ZSM-5. (Reprinted with permission from M. E. Davis, C.-Y. Chen, S. L. Burkett, and R. F. Lobo, in *Better Ceramics Through Chemistry VI*, MRS Symposium Proceedings, Vol. 346, A. K. Cheetham, Ed., Materials Research Society, Pittsburgh, PA, 1994, p. 831. Copyright 1994, Materials Research Society.)

Zeolites, Microporous Materials, Advanced Materials, Chemistry of Materials, Applied Catalysis, Catalysis Letters, Journal of Catalysis, Journal of the Chemical Society, Chemical Communications, and even *Nature* and *Science*, are devoted (per year) to the synthesis and the properties of existing and newly discovered zeolites and molecular sieves.

Because of their uniform pore size and significant pore volume, zeolites, and molecular sieves are widely used in adsorption, separation technology (ion exchange), and catalysis.[3–5, 27] Depending on their Si/Al framework ratio zeolites are classified as low-silica ($1 \leq Si/Al \leq 2$), intermediate-silica ($2 < Si/Al \leq 5$), and high-silica zeolites ($Si/Al > 5$).[7] The low- and intermediate-silica zeolites are polar, and therefore, extremely suitable as adsorbents for removing water and polar molecules from valuable industrial gases.[3, 27] On the other hand, these zeolites exhibit high ion exchange capacity and are used extensively as ion exchangers. For example, zeolite A is used on a very large scale as water softener in detergents.[4, 30] Almost two-thirds of the world demand for zeolites is based on the need for detergent builders. Very recently, the low-silica-containing zeolites Y and scolecite have been used to separate fructose–glucose mixtures[28] and even anitbiotics, [29] respectively.

The high-silica zeolites, such as ZSM-5 or its pure silica analog silicalite-1, exhibit high affinity toward organic molecules and the ability to selectively adsorb organic pollutants from wastewaters or rivers.[7] However, the small pore size of these molecular sieves (approximately 0.57 nm) precludes the possibility for adsorption and separation of the toxic polyaromatic chlorohydrocarbons from the waste or drinking waters. Thus, high-silica zeolites and molecular sieves with larger uniform pore size are extremely desirable.

There is little doubt that the enormous growth of the field of zeolites and molecular sieves was due to the discovery of their catalytic potential in fluid catalytic cracking (FCC) of heavy petroleum fractions and other refining processes.[30, 31] The acidic aluminosilicate zeolite Y (H^+ form) was found to be effective FCC catalyst for conversion of the "middle distillates" to gasoline. Currently, this particular zeolite is used on a very large scale in the oil-refining industry. However, due to the small pore size of this microporous framework (about 0.74 nm), the large hydrocarbon molecules from the "bottom of the barrel" cannot penetrate the pore volume and hence cannot be converted to gasoline.

Zeolites and molecular sieves are also very useful as shape-selective catalysts and catalytic supports. The first example of shape-selective catalysis, demonstrated by Weisz et al.,[32] showed that both primary and secondary alcohols can diffuse and undergo dehydration in the large-pore (about 0.74 nm) zeolite X, whereas only primary alcohols were accommodated and dehydrated in the small-pore-size zeolite A (about 0.43 nm). Later, it became clear that the uniform micropore size not only limits the shape and size of the reactants that could penetrate the framework but also influences the reaction selectivity by ruling out the shape and size of the corresponding reaction products. These unusual properties of zeolites were quickly realized

TABLE 8.1 Major Commercial Zeolite Processes

Process	Zeolite	Product	Dollars/ton[a]
Catalytic cracking	Faujasite	Gasoline, fuel oil	1.5–3000
Hydrocracking	Faujasite	Kerosene, jet fuel, benzene, toluene, xylene	12,000 Pt[b]
Hydroisomerization	Mordenite	*i*-Hexane, heptane (octane enhancer)	12,000 Pt[b]
iso/*n*-Paraffin separation	Ca-A	Pure *n*-paraffins	5,000
Dewaxing	ZSM-5	Low-pour-point	60,000 Pt[b]
	Mordenite	lubes	14,000 Pt[b]
Olefin drying	K-A	Polyolefin feed	4,000
Benzene alkylation	ZSM-5	Styrene	60,000
Xylene isomerization	ZSM-5	Paraxylene	60,000 [b]

Source: Reproduced from D. E. W. Vaughan, *Chem. Eng. Prog.* **85**, 25 (1988) with permission of the American Institute of Chemical Engineers.

[a] Pt does not include the price of the recoverable Pt and Pd component, which may vary between 100 and 300 troy ounces per ton.

[b] Costs are difficult to determine because of combinations with other licensing services.

and a number of important catalytic processes using zeolites were developed in the last three decades. Some of the commercial catalytic processes involving zeolites are summarized in Table 8.1. It did not take long to realize that the potential of these microporous molecular sieves is strictly limited by the small pore size of their frameworks.

8.2.3 Recent Advances in the Field of Microporous Molecular Sieves

Because of the limitations noted above, a myriad of new organic directing agents, aluminosilicate compositions, and reaction variables were put to test in the last three decades in attempts to expand the uniform micropore size and prepare new and stable frameworks with useful properties. Despite this considerable effort, until 1988 the larger pore size available in zeolites and molecular sieves (see Fig. 8.3) was still that of the synthetic faujasite analogs, zeolites X and Y (prepared as early as 1959).[33]

Replacement of the aluminosilicate gels with aluminophosphate gels led in 1982 to the advent of aluminophosphate molecular sieves (AlPOs) by Wilson and co-workers.[34] However, the first AlPOs, AlPO-5 and AlPO-11, exhibited even smaller pore size than zeolite Y. Nevertheless, attention shifted toward the preparation of aluminorphosphates, and in 1988, Davis and co-workers[35] disclosed the first 18-membered-ring aluminophosphate molecular sieve (denoted VPI-5) with an hexagonal arrangement of one-dimensional channels and uniform pore size of approximately 1.2 nm (see Fig. 8.3). Three years after that, Estermann and colleagues discovered[36] a 20-membered-ring gallophosphate molecular sieve, cloverite, with a three-dimensional channel

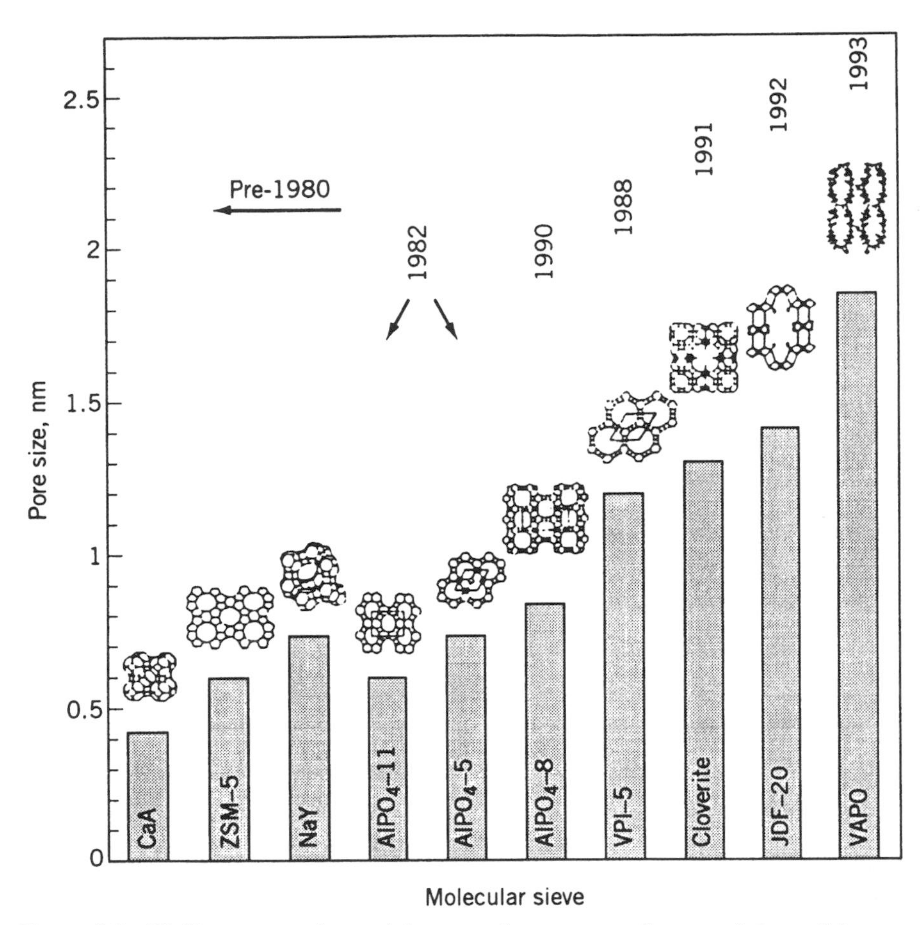

Figure 8.3 Uniform pore size and framework structure of some of the well-known microporous zeolites and recently developed molecular sieves.

system and a uniform pore size of 1.3 nm. In 1992, Thomas and his co-workers reported[37] yet another 20-membered-ring aluminophosphate molecular sieve, denoted JDF-20, having a uniform pore size of 1.4 nm. Very recently, preparation of a vanadium phosphate molecular sieve (VAPO) with a 1.8-nm lattice cavity was announced by Haushalter and colleagues.[38] The actual pore size of the latter two materials is still to be determined, since sorption data are lacking. The thermal stability of VPI-5 and cloverite seems to be much lower than that of the high-silica zeolites and molecular sieves,[39, 40] whereas that of the novel JDF-20 and the VAPO is still to be determined.

All of these newly developed microporous structures were again prepared using *single cationic species* (primary ammonium, diammonium, tertiary, and quaternary ammonium ions) as framework templates. In addition, the eventual processes that could be performed over these materials would still be restricted to small molecules of equivalent "micropore" size.

8.2.4 Mesoporous Molecular Sieves

Assemblies of Surfactant Molecules as Templates: A Milestone to the Synthesis of Mesoporous Molecular Sieves The considerable synthetic effort toward expanding the uniform micropore size available in zeolites and molecular sieves met with limited success until 1991. Simultaneously, clay scientists studied the behavior of intercalated *assemblies* of surfactant molecules in the galleries of their lamellar hosts.[41] This fact had an important impact on the discovery of the mesoporous molecular sieves. Thus, in 1990, while studying the ion-exchange reaction between a single-layered silicic acid host (kanemite, $NaHSi_2O_5 \cdot 3H_2O$) and long-chain quaternary ammonium cations, Kuroda and co-workers reported[42] that the layers seems to distort and cross-link around the cations to form a new mesoporous structure. The driving force for this layer folding process is most likely the ion pairing between the positively charged assemblies of quaternary ammonium cations (S^+) and the negatively charged layered host (I^-). Upon calcination of this organic–inorganic complex, the surfactant was burned off and the one-dimensional mesopore network was made available to molecules. The adsorption studies revealed that the cross-linked material possesses a nearly uniform mesopore size centered at about 3.0 nm, and the ^{29}Si MAS NMR confirmed the hypothesis for significant cross-linking of the structure because of the large increase in the Q^4/Q^3 ratio.[42] The resulting material can be described as having a "ribbon-candy"-like structure, as depicted in Fig. 8.4.

In 1991, scientists at Mobil Oil Research and Development disclosed in a lengthy series of patent applications the preparation and some of the most important properties of the first family of mesoporous molecular sieves (denoted M41S).[43,44] A detailed summary of the Mobil patents on mesoporous molecular sieves was provided very recently by Casci.[45]

Mobil's M41S Family of Mesoporous Molecular Sieves According to Mobil's technology, long-chain quaternary ammonium surfactants minimize their energy in solution by assembling into micelles (Fig. 8.5*A*).[46] Under certain conditions these micelles can adopt a rodlike shape and even organize into long-range hexagonal arrays with the charged head groups pointing toward the solution and the long hydrocarbon chains (hydrophobic) pointing toward the center of the micelles. The ability of the long-chain quaternary ammonium cations to form rodlike micelles and long-range hexagonal arrays in aqueous solutions (with rod diameters in the mesopore range 2.0 to 4.0 nm) has been known for a very long time after the pioneering work of Luzzati (1968).[47] The formation of the micellar rods and their organization into hexagonal arrays is strongly dependent on the surfactant's alkyl chain length, concentration, the nature of the halide counterion, and temperature of the solution.[47–49] Upon the addition of a silicate precursor, for example, sodium silicate, the negatively charged silica species (I^-) condense and polymerize

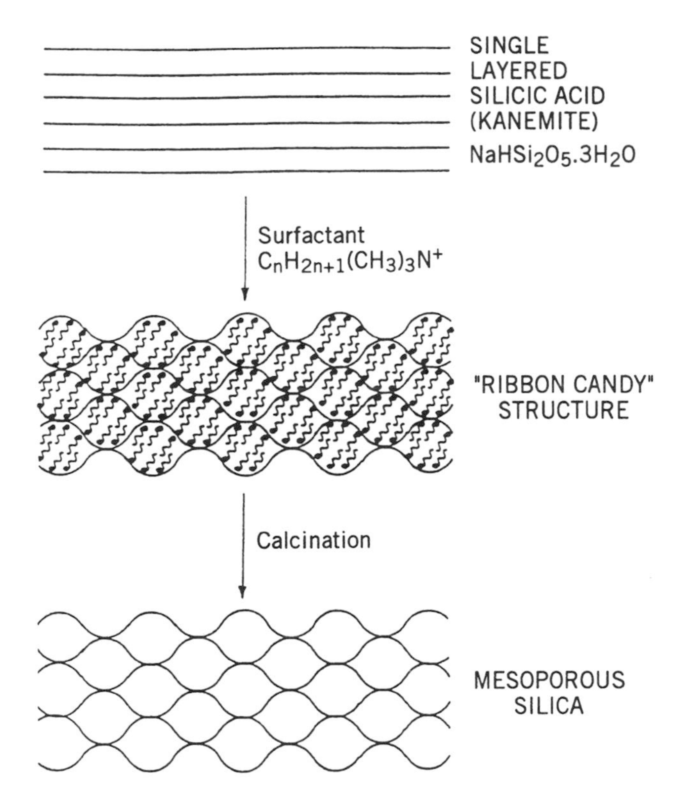

Figure 8.4 Schematic illustration of the rearrangement of a layered silicic acid host (kanemite) into a "ribbon-candy"-like structure in the presence of assemblies of long-chain quaternary ammonium cations as templates.

on the surface of the positively charged micelles (S^+), giving rise to the corresponding hexagonal S^+I^- organic–inorganic complex (Fig. 8.5*A*). The calcination of the complex revealed the hexagonal solid framework of this particular mesoporous molecular sieve, denoted MCM-41.

Another important contribution of Mobil researchers is the disclosure that the uniform mesopore size of MCM-41 can be varied in the range 1.3 to 10.0 nm simply by varying the surfactant alkyl chain length (from 6 to 16 carbon atoms) or by adding an auxiliary organic (e.g., trimethylbenzene) into the internal hydrophobic region of the micelles. Thus a hexagonal MCM-41 silicas with *uniform mesopore size* of 2.0, 4.0, 6.5, and even 10.0 nm have been reported (Fig. 8.5*B*). A typical Mobil preparation of MCM-41 involves the use of an aqueous solution of cationic surfactant, such as cetyltrimethyl ammonium chloride, partially exchanged for $OH^-(C_{16}H_{33}(CH_3)_3NOH/Cl)$ over an ion-exchange resin. The surfactant concentration of this solution was selected to be approximately 25 wt %, which is larger than the critical micelle concentration necessary for the formation of rodlike micelles (about 11 to

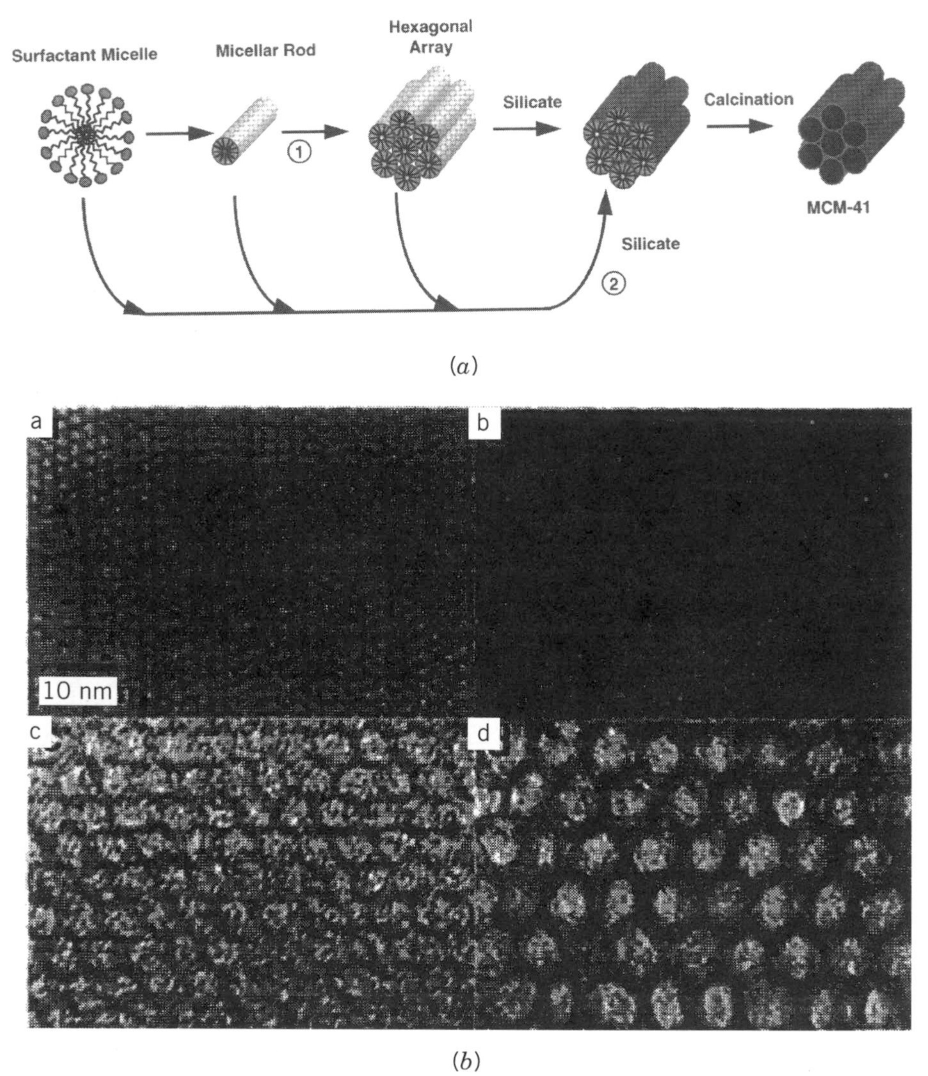

(*a*)

(*b*)

Figure 8.5 (*A*) Mobil S^+I^- mechanistic routes for the formation of the hexagonal MCM-41: route (1), liquid crystal phase initiated, and route (2), silicate anion initiated. (*B*) Transmission electron micrographs of MCM-41 samples with uniform mesopore sizes of (*a*) 2.0, (*b*) 4.0, (*c*) 6.5, and (*d*) 10.0 nm. (Reprinted with permission from J. S. Beck et al., *J. Am. Chem. Soc.* **114**, 10834 (1992). Copyright 1992, American Chemical Society.)

20.5 wt %).[50] The sources of silica were varied from sodium silicate, amorphous fumed silica, colloidal silica, or tetraethyl orthosilicate (TEOS). In some cases aluminum-substituted analogs were prepared using alumina sources selected from the group of sodium aluminate, aluminum sulfate, and pseudoboehmite.

The following is a typical reaction mixture composition[43] for the preparation of MCM-41 (expressed in moles):

$$1.0SiO_2 : 0.03Al_2O_3 : 0.007Na_2O : 0.183(CTMA)_2O : 0.156(TMA)_2O : 23.5H_2O$$

The addition of tetramethylammonium hydroxide (TMAOH) was probably dictated by the need for a base capable of dissolving the amorphous silica source. The preparation comprises mixing of the foregoing inorganic precursors with the solution of template and autoclaving of the corresponding reaction mixture at temperatures from 100 to 150°C for a period of time from 4 to 144 h. The crystalline product was recovered by filtration, air dried, and subjected to heating in N_2 atmosphere at 550°C for 1 h followed by calcination in air at the same temperature for 6 h.[46]

The use of different surfactant/silica ratios afforded, in addition to the hexagonal MCM-41 phase, cubic MCM-48 and lamellar M41S phases. Figure 8.6 illustrates the XRD patterns of these materials. This entire family of mesoporous molecular sieve phases prepared by S^+I^- templating was denoted as M41S.[46] Surfactant/silica ratios lower than 1 were found to give the hexagonal MCM-41 material (pattern *A*), whereas ratios larger than 1 afforded the cubic (MCM-48) counterpart (pattern *B*). In addition, surfactant/silica ratios much larger than 1 afforded the lamellar M41S mesophase (pattern *C*). However, the lamellar phase was found to be unstable and collapsed upon calcination at elevated temperatures.[46]

Templating Pathways to Mesoporous Molecular Sieves Mobil's mechanism of the formation of M41S materials involves strong electrostatic interactions and ion pairing between quaternary ammonium cations (S^+), as structure directing agents, and anionic silicate oligomer species (I^-). The recently reported[48] preparation of related hexagonal mesoporous structures by rearrangement of a layered silicate host (kanemite) can also be considered a derivative of the electrostatic approach to mesoporous molecular sieves. Originally, two possible variations of this S^+I^- electrostatic mechanism were proposed by Beck et al. (see Fig. 8.5*A*).[46] According to route 1, the surfactant molecules in solution organize spontaneously into micellar liquid crystals. The addition of the silicate source leads to condensation and polymerization of the anionic silicate species on the surface of the positively charged micelles by a mechanism of ion pairing. Further condensation of the inorganic precursor in the continuous solvent (water) region leads to the formation of the inorganic framework walls. The second mechanistic direction proposed by Mobil (route 2 in Fig. 8.5*A*) suggests that the addition of anionic silicate

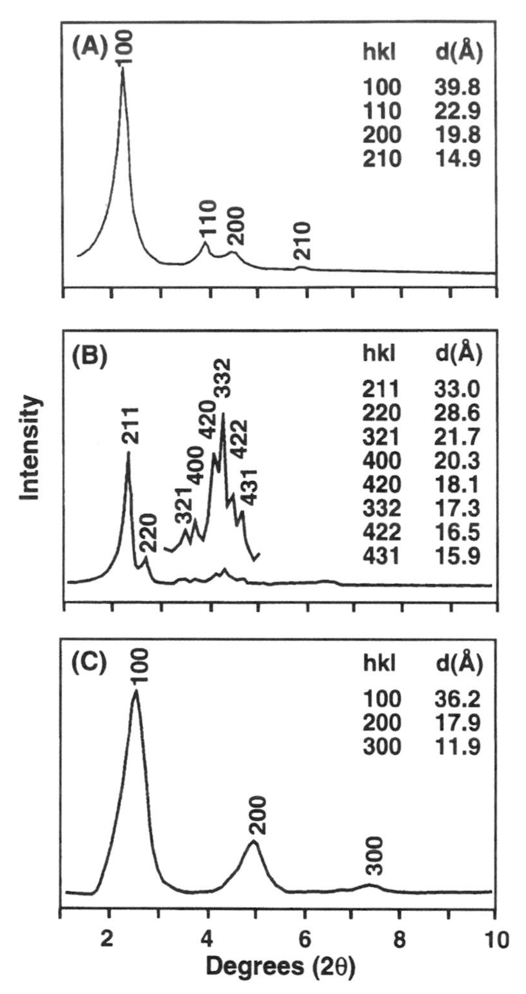

Figure 8.6 Powder x-ray diffraction patterns of calcined (*A*) hexagonal MCM-41, (*B*) cubic *Ia3d* MCM-48 phase, and (*C*) as-synthesized lamellar M41S phase. (Reprinted with permission from J. S. Beck et al., *J. Am. Chem. Soc.* **114**, 10834 (1992). Copyright 1992, American Chemical Society.)

to the templating solution triggers formation of the hexagonal liquid-crystalline phase.

Stucky and his co-workers also elaborated on the S^+I^- mechanism and identified three closely coupled phenomena for the formation of the surfactant–silica mesophases.[49] These are (1) multidentate binding of silicate oligomers to the surfactant headgroups, (2) preferred polymerization of silicate oligomers at the surfactant–silicate interface, and (3) charge-density matching across the surfactant–silicate interface. According to these workers, at high pH, the reaction mixture contains small silica oligomers such as single or double three- and four-membered rings of different negative charges (see Fig. 8.1). These oligomeric anions act as multidentate ligands for the cationic headgroups of the surfactant, leading to a strongly interacting surfactant–silicate interface. Perhaps the most important implication of this work is that the framework wall thickness of the MCM-41 materials is *limited*. The number of anionic silicate oligomers that can be incorporated into the walls of MCM-41 materials by S^+I^- templating is *restricted* by the strong electrostatic repulsions between the I^- silicate species and the S^+I^- charge matching interactions. Thus the framework wall growth of MCM-41 is terminated when charge compensation is achieved. This hypothesis is supported by the observation that the framework wall thickness (0.8 to 0.9 nm) of MCM-41 did not vary over a wide range of reaction conditions and surfactant chain lengths.[49]

Davis and co-workers have also elaborated on the S^+I^- templating mechanism.[50] Based on ^{14}N NMR analyses of their quaternary ammonium surfactant solutions and corresponding MCM-41 products, these authors concluded that the formation mechanism does not involve the liquid crystalline phase as suggested originally by Mobil (route 1). Instead, they proposed that randomly ordered rodlike micelles form initially and interact via electrostatic interactions with the anionic silicate species to give polymerized monolayers of silica on the surface of the micelles (Fig. 8.7). These silica-coated micelles pack spontaneously into long-range hexagonal array. The driving force for this process was believed to be the further condensation of the silicate species. Thus, this work provided a refinement of the route 2 originally proposed by Beck et al. (see Fig. 8.5*B*).[46] An important finding supported by ^{29}Si MAS NMR and elemental analysis data is that complete condensation of the anionic silicate species in the walls of the electrostatically templated MCM-41 is not possible because framework $(SiO)_{4-x}(SiO)_x^{x-}$ species are necessary for charge compensation of the occluded quaternary ammonium template.[50] This implies that the cationic template in typical S^+I^- MCM-41 preparations is strongly bound to the negatively charged framework and difficult to recover by nondestructive methods.

Recently, Stucky and colleagues further extended[51] the electrostatic assembly approach to mesoporous molecular sieves by proposing four complementary synthesis pathways (Fig. 8.8). Pathway 1 involved the direct co-condensation of anionic inorganic species (I^-) with a cationic surfactant (S^+) to give assembled ion pairs (S^+I^-), the original synthesis of M41S silicas being the prime example.[46] In the charge-reversed situation (pathway 2) an anionic template (S^-) was used to direct the self-assembly of cationic inorganic species (I^+) via S^-I^+ ion pairs. Pathway 2 has been found to give a hexagonal iron and lead oxide and different lamellar lead and aluminum oxide phases. Pathways 3 and 4 involved counterion (X^- or M^+)-mediated

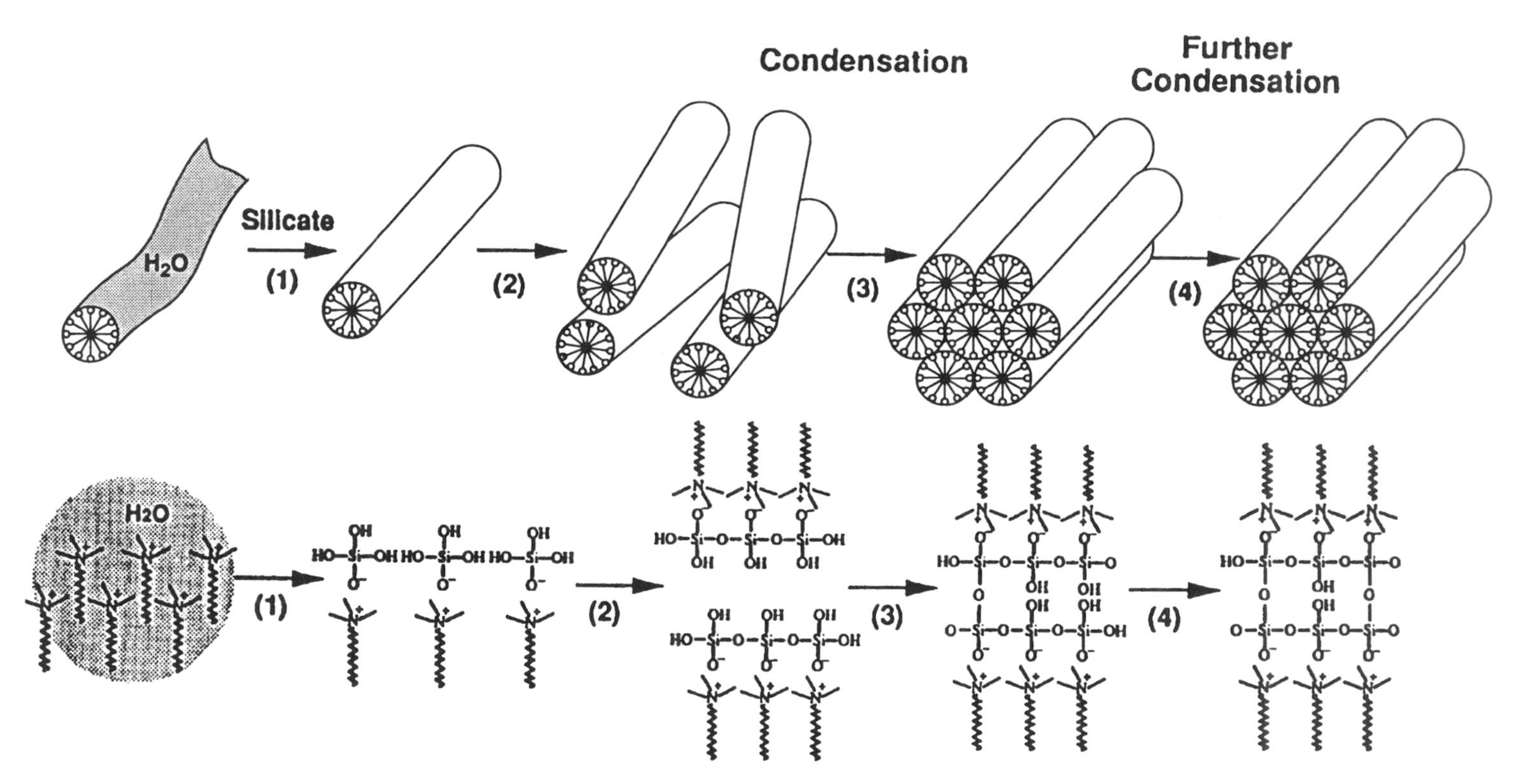

Figure 8.7 Mechanism for the S^+I^- formation of MCM-41 proposed by Davis et al. (Reprinted with permission from M. E. Davis, C.-Y. Chen, S. L. Burkett, and R. F. Lobo, in *Better Ceramics Through Chemistry VI*, MRS Symposium Proceedings, Vol. 346, Pittsburgh, PA, 1994, p. 831. Copyright 1994, Materials Research Society.)

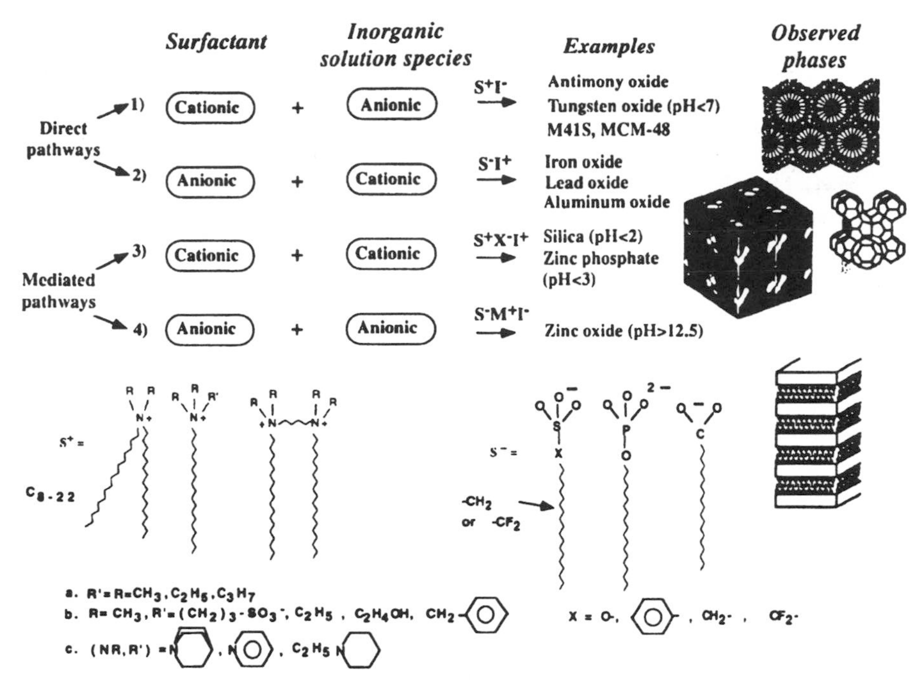

Figure 8.8 Schematic representation of the four complementary electrostatic templating pathways to ordered mesostructures. (Reprinted with permission from Q. Huo et al., *Nature* **368**, 317 (1994). Copyright 1994, Macmillan Magazines Ltd.)

assemblies of surfactants and inorganic species of similar charge. These counterion-mediated pathways afforded assembled solution species of type $S^+X^-I^+$ (where $X^- = Cl^-$ or Br^-) or, $S^-M^+I^-$ (where $M^+ = Na^+$ or K^+), respectively. The viability of pathway 3 was demonstrated by the synthesis of a hexagonal MCM-41 using a quaternary ammonium cation template and strongly acidic conditions (5 to 10 *M* HCl or HBr) to generate and assemble positively charged framework precursors.

In another example, provided by the same researchers, a condensation of anionic aluminate species was accomplished by alkali cation–mediated (Na^+,K^+) ion pairing with an anionic template ($C_{12}H_{25}OPO_3^-$). The preparation of the corresponding $Al(OH)_3$ phase in this case has been attributed to the fourth pathway ($S^-M^+I^-$).[51] The contributions of Stucky and co-workers for the preparation of mesoporous molecular sieves with nonsilicate composition, especially transition-metal oxides, cannot be overstated. Transition-metal oxide mesoporous molecular sieves could be very important in a number of catalytic processes, such as methane oxidation and photocatalytic decomposition of large organic pollutants. Unfortunately, all of Stucky's templated mesoporous metal oxides, including the lamellar alumina phase, were unstable to template removal by calcination or other methods.[51] Thus a synthetic approach that will generate stable transition-metal mesoporous phases is highly desired.

Pathway 3 ($S^+X^-I^+$) afforded not only the preparation of hexagonal MCM-41 but also a *Pm3N* cubic and a lamellar phase.[51] A typical preparation involved the addition of TEOS to a strongly acidic solution (5 to 10 *M* HCl or HBr) of quaternary ammonium surfactant and aging of the reaction mixture at ambient temperature for more than 30 min.[51] It has been postulated that the interactions between the cationic silica species [$\equiv Si(OH_2)^+$] and halide-cationic surfactant headgroups are mediated by the large excess of halide ions (X^-). However, the main driving force for mesostructure formation is also electrostatic. The need for large excess of corrosive acidic reagent will require special reactor equipment and waste disposal considerations in potential industrial-scale preparation.

Very recently, we have reported[52] a neutral ($S^\circ I^\circ$) templating route (denoted as pathway 5) to hexagonal mesoporous molecular sieves (HMSs). They postulated that the formation of HMS mesostructures occurs through the organization of neutral primary amine surfactant molecules (S°) into neutral rodlike micelles (see Fig. 8.9). The addition of neutral inorganic precursor, for example TEOS, to the solution of template affords the hydrolyzed intermediate $Si(OC_2H_5)_{4-x}(OH)_x$ species. These species participate in H-bonding interactions with the lone pairs on the surfactant headgroups, affording surfactant–inorganic complexes in which the surfactant part could be viewed as a hydrophobic tail and the inorganic precursor as a bulky headgroup. This significantly changes the packing of the surfactant–inorganic complexes obtained and probably triggers the formation of rodlike micelles in solutions of neutral primary amine surfactants. Further hydrolysis and condensation of the silanol groups on the micelle–solution interface afford short-range hexagonal packing of the micelles and framework wall formation. We also cannot preclude the possibility of having a slightly different pathway involving preorganized spherical micelles of surfactant molecules or even surfactant bilayer arrays in the initial ethanol–water solutions of template. However, we think that a rearrangement into rodlike micelles takes place upon addition of inorganic precursor that is again triggered by H-bonding interactions between the lone pairs on the surfactant headgroups and the intermediate silica precursor species.

A typical preparation of HMS materials involves the addition of tetraethyl orthosilicate (TEOS) (1.0 mol) to a solution of neutral primary amine (0.27 mol) in ethanol (9.09 mol) and deionized water (29.6 mol). The reaction mixture is aged at ambient temperature for 18 h, and the resulting HMS silica is recovered by filtration or air drying. In contrast to the electrostatic templating pathways, the neutral templating approach allows for a cost reduction in large-scale preparations by employing mild reaction conditions

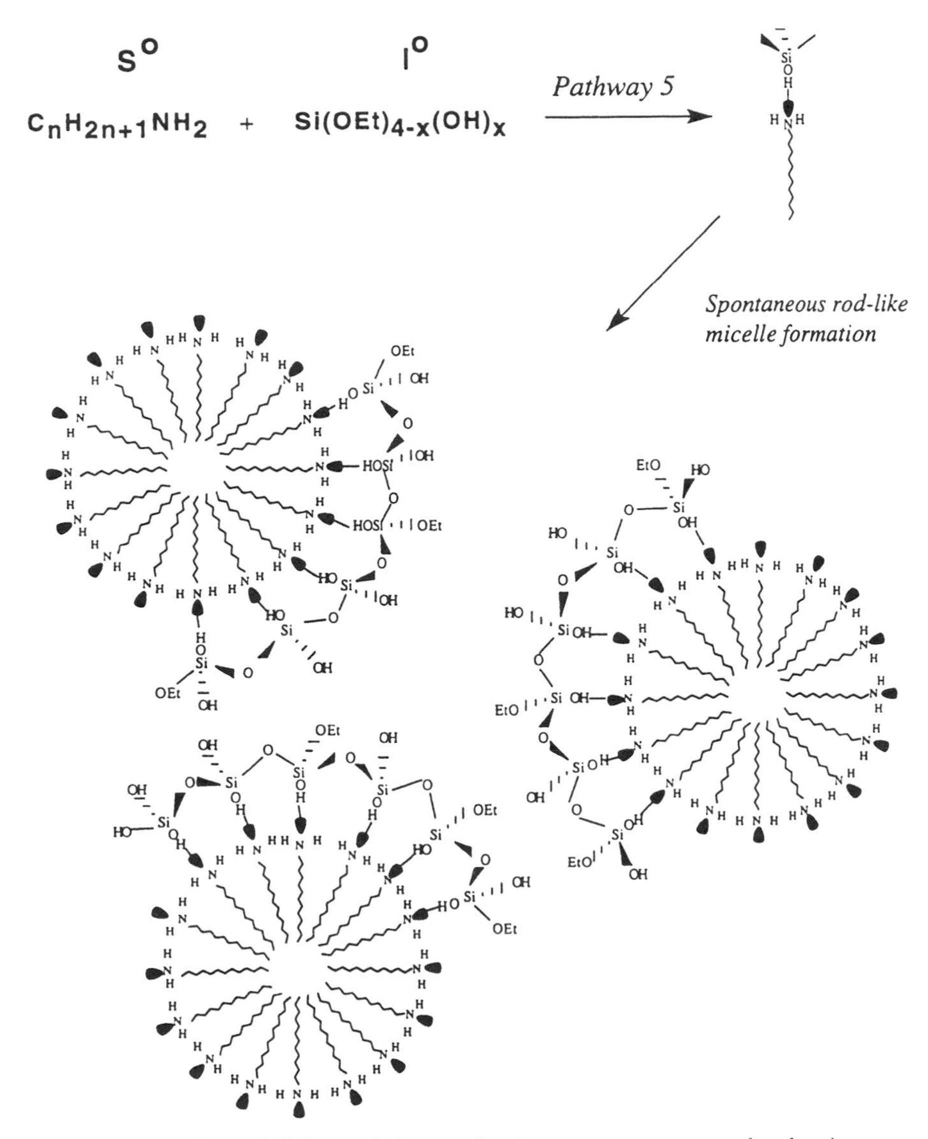

Figure 8.9 Neutral S°I° templating mechanism to mesoporous molecular sieves.

and cheap primary amine surfactants that can easily be recovered and recycled by simple solvent extraction.

Properties of the Mesoporous Molecular Sieves Prepared by Different Templating Approaches The x-ray diffraction patterns of M41S materials prepared by the S^+I^- templating pathway are presented in Fig. 8.6.[46] The hexagonal MCM-41 phase exhibited just a few *hk*0 reflections (100, 110, 200, and 210) and lacked reflections with nonzero *l* component. This has been attributed to the absence of local order in the framework walls of the hexagonal MCM-41. The *hk*0 reflections observed were assigned to scattering arising from the uniform hexagonally arranged mesopores. Very recently, Fröba et al., performed SiK XANES measurements on a series of M41S samples, related high-silica zeolites, lamellar silicate hosts, amorphous glasses, and high-temperature silica phases.[53] Comparison of the results revealed that M41S samples exhibit intermediate local wall order between the high-temperature silica phases and amorphous glasses, similar to that of the lamellar silicates. This led the authors to propose that the walls of M41S silicas closely resemble curved silicate sheets. However, the comparison of the ^{29}Si NMR data reveals that the MCM-41 materials exhibit much broader and overlapping Q^3 and Q^4 peaks, whereas those of a crystalline lamellar silicate such as magadite are much sharper and well resolved.[55] This suggests larger variety of Si—O—Si bond angles in MCM-41 silicas than in the lamellar silicates.

The XRD patterns of the MCM-41 prepared by the acidic $S^+X^-I^+$ templating (pathway 3) also possessed the *hk*0 reflections characteristic of the hexagonal phase.[51] In contrast, the XRD patterns of HMS materials prepared by $S°I°$ templating exhibited *single* d_{100} reflections accompanied with diffuse scattering centered at approximately 5° 2Θ.[52] Higher-order Bragg reflections of the hexagonal structure were not resolved in the patterns of HMS materials. However, we and others have demonstrated[43, 52, 54] that similar "single-reflection" MCM-41-type products still possess short-range hexagonal symmetry. Therefore, the diffuse scattering at about 5° is attributable to broadening of the remaining *hk*0 reflections of the hexagonal phase. We believe that future studies aimed at optimizing the reaction conditions for our $S°I°$ templating approach would ultimately afford long-range-ordered hexagonal, cubic, and lamellar mesoporous materials.

The framework wall thickness of MCM-41 materials prepared by the S^+I^- pathway have been found to be *limited* by the charge compensation between the positively charged micelles and the negatively charged inorganic species.[49] According to the original report by Beck and co-workers,[46] the framework wall thickness varies from 0.8 to 1.3 nm, with variation of the surfactant chain length. Stucky and co-workers observed even smaller variations of the framework wall thickness (0.8 to 0.9 nm) for MCM-41 silicas prepared by S^+I^- templating under a wide range of reaction conditions and surfactant chain lengths.[49] This difference in wall thickness reported by the two groups could be attributed to the use of Cl^- or Br^- salts of the quaternary ammonium ions as surfactants in the work of Stucky, whereas Beck et al. used partially exchanged Cl^-/OH^- solutions of surfactants. Davis et al. also found that the framework wall thickness of their S^+I^- MCM-41 materials was in the range of 1.0 nm.[55] The framework wall thickness of MCM-41 samples prepared by the acidic $S^+X^-I^+$ pathway also seems to be similar, 1.0 to 1.2 nm.[51, 56]

In general, the small framework thicknesses of the electrostatically templated MCM-41 materials could seriously limit their thermal and hydrothermal stability. In an attempt to improve the thermal stability of the S^+I^- MCM-41 aluminosilicates, Fajula and co-workers varied the amount of S^+ surfactant, sodium aluminate, and water in their reaction mixtures and observed the effect of this variation on the framework wall thickness.[57] They stated that the alkalinity of the solution had the most influence on pore wall thickness. In particular, six out of 10 samples showed framework wall thicknesses of less than 1.0 nm, one showed about 1.3 nm, and only three exhibited thicknesses of from 1.5 to 1.6 nm. However, given the range of variation of the framework wall thickness observed by the foregoing groups[46, 49, 55] for the same class of materials (from 0.8 to 1.3 nm) it is unlikely that the variation of wall thickness observed by Fajula et al. is due exclusively to differences in alkalinity. An important finding of this work was that MCM-41 samples with thicker pore walls exhibit higher thermal stability. This was nicely illustrated by a comparison of the degree of pore volume contraction exhibited by thin- and thicker-walled MCM-41 samples after heat treatment in air at 550°C.[57]

We have reported[52] that the framework wall thickness of HMS molecular sieves prepared by $S^\circ I^\circ$ templating with primary amine surfactants is much larger (from 1.7 to 3.0 nm) than that exhibited by their electrostatic counterparts. The much thicker pore walls of HMS materials could be attributed to the absence of electrostatic repulsions between the neutral intermediate-silicate species and to the lack of charge-matching interactions between the neutral surfactant and the neutral inorganic precursor.[52, 56] As illustrated by Fajula and co-workers,[57] thicker pore walls are highly desired as improving the thermal and hydrothermal stability of the mesoporous framework. In this sense, neutral templating may provide for the preparation of mesoporous molecular sieves with much improved thermal stability.

Since the S^+I^- electrostatic templating pathway is based on charge matching between a cationic surfactant and an anionic inorganic reagent, the template is strongly bonded to the charged framework and difficult to recover. In the approach originally disclosed,[46] the template was not recovered but simply burned off by calcination at elevated temperatures. Recently, the scientists at Mobil came up with a better idea: to ion-exchange the ionic surfactant with an acidic cation donor solution.[58] Also, Stucky and co-authors have demonstrated that much of the template–halide ion pairs in MCM-41 prepared by the acidic $S^+X^-I^+$ method can be displayed by ethanol extraction.[51] Thus ionic template recovery is possible provided that *exchange ions* or *ion pairs* are present in the extraction process. In contrast, the $S^\circ I^\circ$ templating approach to mesoporous molecular sieves allowed for the facile and environmentally benign recovery of the neutral template by simple solvent extraction.[52] Owing to the absence of counterions, the neutral template recovered can easily be recycled and reused by simple evaporation of the solvent.

Scientists at Mobil reported the ^{29}Si MAS NMR of as-synthesized M41S samples prepared by the electrostatic S^+I^- pathway.[59] Using deconvolution techniques, they resolved three broad and overlapping peaks in the spectra of MCM-41 and MCM-48 and two peaks in the spectrum of the thermally unstable lamellar phase. The spectra of both hexagonal MCM-41 and cubic MCM-48 were essentially the same as for amorphous silica. This is not surprising given the amorphous character of the framework walls and the expected wide range of Si—O—Si bond angles. Interestingly, all S^+I^- phases exhibited more noncondensed $Si(SiO)_x(OH)_{4-x}$ framework units (where $x = 2$ or 3, respectively, for Q^2 and Q^3) than fully condensed ($x = 4$) units, as evidenced by the low ratios of $Q^4/Q^2 + Q^3$ (on average around 0.67). This result is extremely significant and implies a relatively *low degree of cross-linking* of the "electrostatic" molecular sieves. The low degree of cross-linking (a ratio of 0.77) is also evident from the data of Davis et al.,[50,55] who showed that cationic template removal by ion exchange with HCl–ethanol solutions, as well as by calcination, resulted in further cross-linking of the Q^2 and Q^3 units into Q^4 units. The fact that the HCl–ethanol treatment causes some condensation of the silanol groups is not surprising since the acid-catalyzed condensation of silanols is well known. However, HCl–ethanol treatment causes significant lowering of the d_{100} values, which translates into thinner pore walls or smaller mesopore sizes. In this connection, Davis and co-workers made a very important observation, namely, that 100% Q^4 state or full cross-linking of the electrostatic M41S framework cannot be reached because SiO^- groups are needed for charge compensation of the quaternary ammonium cationic template (S^+). This fact suggests that the ionic template in a typical M41S preparation is strongly bound to the negatively charged framework and difficult to recover.

The ^{29}Si MAS NMR spectra of the as-synthesized M41S phases prepared by the acidic $S^+X^-I^+$ pathway generally also exhibit low Q^4/Q^3 ratios < 1.0.[51,56] However, the positive charge of the surfactant is balanced in part by the excess of halide counterions, which allowed for displacement of 85 wt % of the template–halide ion pairs by solvent extraction.

In contrast to the electrostatic templating pathways, our neutral $S°I°$ pathway affords much more cross-linked HMS frameworks, as judged by the ^{29}Si MAS NMR spectra.[56] This is evidenced by the very high Q^4/Q^3 ratios of our as-synthesized and solvent-extracted samples (2.70 and 3.12, respectively). This observation is not surprising, due to the lack of electrostatic repulsions and charge matching and the corresponding thicker pore walls of our HMS materials.

The adsorption properties of electrostatically templated M41S materials have been the subject of intensive studies from the first days of their discovery. Beck et al. demonstrated[46] that both benzene and N_2 adsorption isotherms of their MCM-41 samples exhibit a sharp adsorption uptake in the low-relative-pressure region indicative of capillary condensation in framework-confined mesopores. The BET specific surface area[2] was estimated to be around 1000 m^2/g and the total adsorbed volume in the range 0.7 to 1.2 cm^3/g. The mesopore size distribution of S^+I^- MCM-41 samples was calculated by the method of Horvath–Kawazoe.[60] This method assumes a slit-shaped pore model and was originally designed for determination of the micropore size distribution. Surprisingly, the mesopore sizes determined by this method were found to be in a good agreement with these determined from transmission electron micrographs. The mesopore size of MCM-41 samples prepared by S^+I^- templating was found to be in the range 1.8 to 3.7 nm, depending on the surfactant chain length.[46] Figure 8.10*A* shows typical N_2 adsorption–desorption isotherms for amorphous silica, zeolite NaY, and MCM-41 prepared by the S^+I^- pathway. It is obvious that the isotherm for MCM-41 differs quite dramatically from those for the amorphous silica and zeolite NaY. The presence of a large hysteresis loop at $P_i/P_o > 0.5$ in the isotherm of the amorphous silica is indicative of nonuniform textural or interparticle mesopores. The isotherm of zeolite NaY (pore size about 0.74 nm) does not exhibit a significant hysteresis loop and is characterized by strong adsorption uptake at very low relative pressures, owing to adsorption in micropores.[2] Due to the absence of framework-confined mesopores, NaY and amorphous silica lack the sharp adsorption feature observed on the isotherm of MCM-41.

Davis and co-workers measured cyclohexane and water adsorption isotherms for a S^+I^- pure-silica MCM-41 and aluminosilicate MCM-41 samples.[55] Both samples exhibited very high cyclohexane adsorption capacities (0.4 g/g dry solid) and very low water adsorption capacities (0.05 g/g). However, the aluminosilicate showed slightly higher adsorption affinity toward water. This result shows that pure-silica MCM-41 samples are hydrophobic and that the Al substitution could introduce partial hydrophilicity. This feature of the pure-silica MCM-41 samples makes them very attractive as potential adsorbents for large organic pollutants from waste or drinking waters.

The N_2 adsorption–desorption isotherms of MCM-41 samples prepared by the acidic $S^+X^-I^+$ templating route are very similar to those exhibited by the S^+I^- counterparts.[52, 56] However, both electrostatic templating routes afford MCM-41 samples that lack appreciable textural mesoporosity. This is evidenced by the absence of a significant hysteresis loop in their N_2 adsorption–desorption isotherms in the region $P_i/P_o > 0.4$. The lack of textural mesoporosity for the electrostatically templated MCM-41 materials could impose serious limitations on their use in diffusion-controlled processes.

Recently, we have demonstrated that $S^\circ I^\circ$ templating affords HMS materials with very small scattering domain sizes (less than 17.0 nm) and complementary framework-confined and textural mesoporosity (Fig. 8.10*B*).[52, 56] The

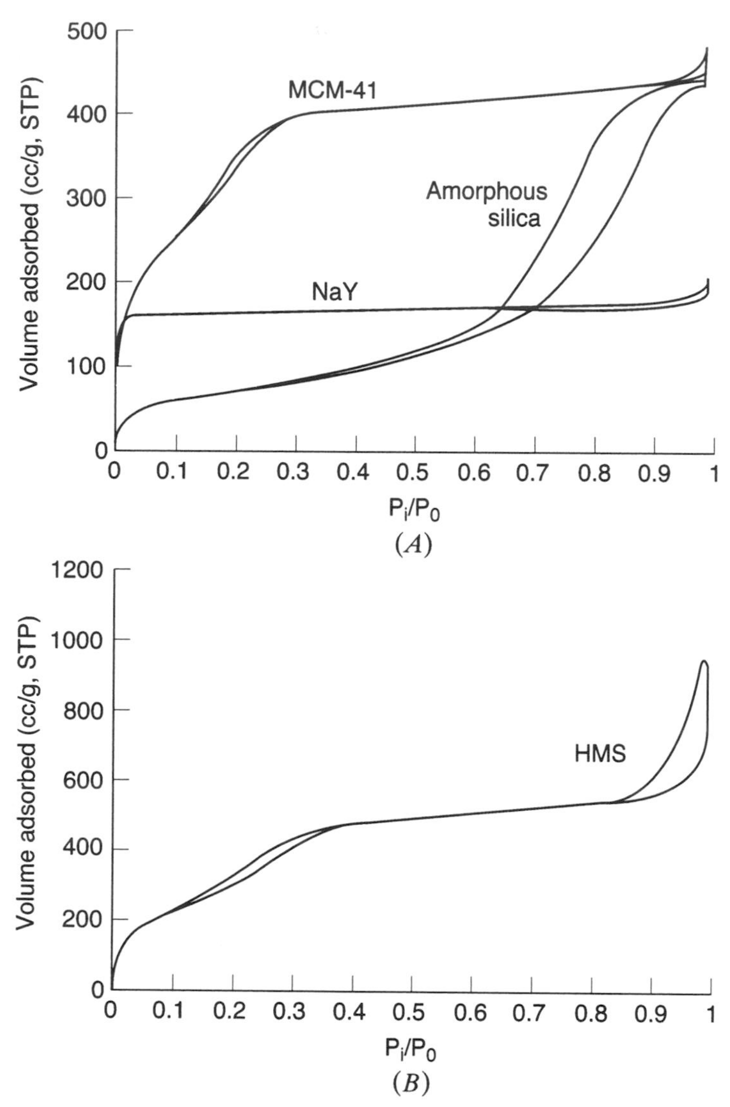

Figure 8.10 N_2 adsorption–desorption isotherms for (*A*) amorphous silica, zeolite NaY, MCM-41 prepared by the S^+I^- pathway, and (*B*) HMS prepared by the neutrals S°I° templating route.

small crystallite size and substantial textural mesoporosity are very desirable for accessing the framework-confined mesopores and for improving the performance as adsorbents and catalysts.[61, 62]

In summary, the electrostatic templating pathways to ordered mesostructures used assemblies of charged surfactant ions (S^+ or S^-) as templates to organize an inorganic framework from charged inorganic oxide precursors

(I^- or I^+). These charged templates are usually expensive, strongly bonded to the charged inorganic framework, and difficult to recover. In general, the electrostatically bonded templates are removed from the framework of S^+I^- MCM-41 materials by either a combustion process or by an ion-exchange reaction with an ion donor solution. Also, ion pairs are necessary to displace the template from the framework of $S^+X^-I^+$ MCM-41 materials. Electrostatic templating affords as-synthesized MCM-41 materials with a relatively low degree of framework cross-linking and small framework wall thickness (from 0.8 to 1.2 nm). This could seriously influence the thermal and hydrothermal stability of these molecular sieves, especially in applications that require severe operating conditions. In addition, MCM-41 samples prepared by electrostatic templating seem to lack appreciable textural mesoporosity. The lack of textural mesoporosity could lead to serious diffusion limitations in many potential applications.

Neutral templating ($S^\circ I^\circ$) allows for the preparation of mesoporous molecular sieves with more completely cross-linked framework, large framework wall thickness, small particle sizes, and complementary framework-confined and textural mesoporosity. In addition, the $S^\circ I^\circ$ approach allows for cost reduction by employing less expensive reagents and mild reaction conditions while providing for the effective and environmentally benign recovery and recyclability of the template.

8.2.5 Important Catalytic Applications of Microporous Molecular Sieves

The isomorphous substitution of synthetic zeolites and molecular sieves with metal atoms capable of performing different chemical (mostly catalytic) tasks is quickly emerging as an important aspect of today's approach to the design of heterogeneous catalysts. For the purpose of our discussion, we briefly review the catalytic applications of some well-known and relatively new industrially important microporous zeolites: Y, ZSM-5, and TS-1.

The industrially important aluminum containing zeolites X, Y, and high-silica ZSM-5 are used on a very large scale as cracking, alkylation, and isomerization catalysts. Because of their uniform pore size and shape-selective properties, they are much more active and specific than the amorphous alumina–silica catalysts. The substitution of Al in the frameworks of these zeolites requires protons for balancing the negative charge. These protons are localized at one of the four oxygen atoms of the AlO_4 tetrahedra. The AlOHSi Brønsted acidity sites are responsible for the high activity of the zeolites exhibited in a broad range of catalytic conversions of alkanes, alkenes, and alcohols. There is increasing demand in recent years for treating heavier feeds and for shape-selective catalytic alkylation or isomerization involving large organic molecules. However, the small uniform pore size of these industrially important catalysts (≤ 0.74 nm) severely limits their catalytic potential to molecules of small kinetic diameters.

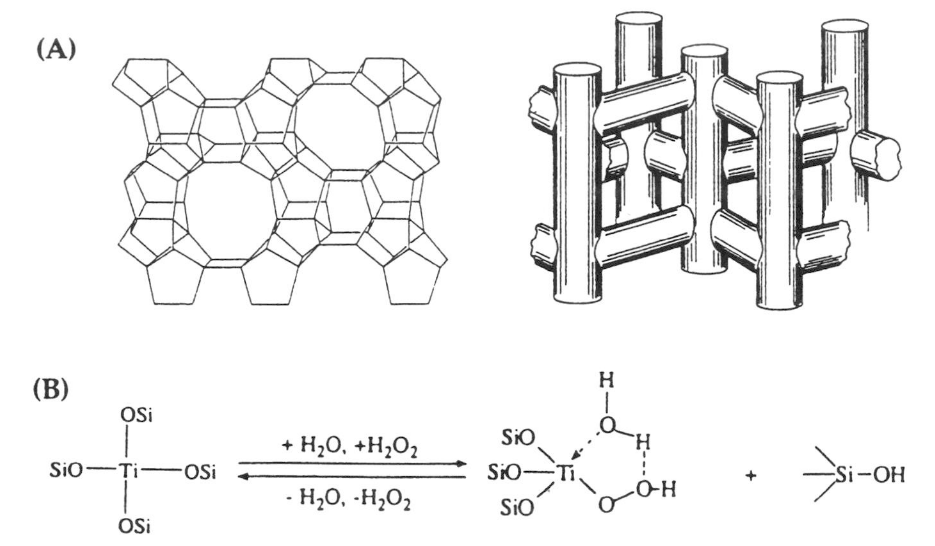

Figure 8.11 (*A*) Schematic representation of the TS-1 solid framework (left) and corresponding pore network (right). (Reprinted with permission from G. T. Kokotailo, S. L. Lawton, D. H. Olson, *Nature* **272**, 437 (1978). Copyright 1978, Macmillan Magazines Ltd.) (*B*) Formation of an active Ti-peroxocomplex upon addition of aqueous H_2O_2 to site-isolated tetrahedral Ti atoms in the framework of TS-1. (Reprinted with permission from C. B. Khow et al., *J. Catal.* **149**, 195 (1995). Copyright 1995, Academic Press, Inc.)

Titanium silicalite-1 (denoted TS-1), with a ZSM-5 framework and pore size of about 0.6 nm, is emerging as a valuable industrial catalyst, due to its ability to oxidize organic molecules under mild reaction conditions. The hydrothermal synthesis of TS-1 was first accomplished by Taramasso et al. in 1983.[63] The three-dimensional framework of TS-1, shown in Fig. 8.11*A*, confines a micropore network of intersecting 10-membered, parallel, elliptical (0.51×0.57 nm) channels along [100], and zigzag, nearly circular (0.54 ± 0.02 nm) channels along [010].[64] Therefore, the size of the framework-confined micropores of TS-1 is equal to the size of these intersecting channels (about 0.6 nm).

Titanium silicalite was found to be an effective liquid-phase oxidation catalyst for a variety of organic molecules in the presence of H_2O_2 as oxidant. The broad spectrum of TS-1 catalyzed reactions includes oxidation of alkanes,[65] oxidation of primary alcohols to aldehydes and secondary alcohols to ketones,[66] epoxidation of olefins,[67] hydroxylation of aromatic compounds,[68] and oxidation of aniline.[69] The production of catechol and hydroquinone from phenol over TS-1 is now an industrially established process.[70] There is a widespread notion that the exceptional catalytic activity of TS-1 is due to the presence of site-isolated titanium atoms in the micropores of the silicalite host (Fig. 8.11*B*). In addition, the ability of these Ti sites to undergo coordination change easily in the presence of H_2O and H_2O_2 and to form a very active titanium peroxocomplex is believed to be of primary importance for the activity observed. However, because of the small pore size of the inorganic framework, the number of organic compounds that can be oxidized by TS-1 is strongly limited to molecules having kinetic diameters equal to or less than about 0.6 nm. Another titanium silicalite, TS-2, with MEL structure, was reported to exhibit similar oxidation properties.[71] The similar catalytic behavior of TS-2 is not surprising in view of the nearly identical size of the silicalite-2 framework-confined micropore channels (about 0.53 nm). Very recently, a Ti-substituted analog of yet another zeolite (zeolite β) with slightly larger micropore size has been reported by Corma et al.[72] The main incentive for preparing Ti-substituted analog of zeolite β was to be able to take advantage of its slightly larger-micropore-size network, composed by intersecting 12-membered ring (0.76 × 0.64 nm) channels along [001] and 12-membered channels (0.55 nm) along [100]. However, the catalytic oxidation chemistry of Ti-substituted zeolite β, with the exception of the slightly higher conversion of cyclododecane relative to TS-1, was again confined to the well-known small substrates subjectable to catalytic oxidation over TS-1 and TS-2 molecular sieves. In addition, the presence of Al^{3+} in zeolite β affords a partially hydrophilic framework exhibiting much higher acidity than TS-1 and TS-2. This could limit the selectivity in catalytic oxidations by bulky alkyl-substituted aromatics or phenols, due to dealkylation of the alkyl groups or isomerization. The small micropore size of two recently discovered Ti-substituted molecular sieves, namely ETS-10[73] and Ti-ZSM-48,[74] would probably again confine their catalytic oxidation chemistry to substrates with small kinetic diameters. Vanadium-substituted silicalite 1 and 2 (denoted VS-1 and VS-2) were also reported,[75] but due to the embedding of V in the same silicalite microporous framework, the catalytic oxidation activity of these molecular sieves is again limited to small organic substrates with kinetic diameters of less than 0.6 nm.

Therefore, there is a need for new metal-substituted mesoporous molecular sieves capable of transforming organic species with kinetic diameters greater than 0.6 nm, especially bulky aromatics. Such metal-substituted mesoporous molecular sieves would greatly complement and extend the catalytic chemistry of the foregoing microporous materials toward much larger organic molecules.

8.2.6 Catalytic Applications of Mesoporous Molecular Sieves

Both mesoporous molecular sieves MCM-41 and HMS offer exciting opportunity for the preparation of large-pore analogs of the above industrially important catalysts. The first preparation of Ti-substituted MCM-41 was demonstrated simultaneously by us[61] and by Corma et al.[76] However, the latter workers used the Mobil's S^+I^- templating route (pathway 1) and prolonged hydrothermal synthesis conditions to prepare their Ti-MCM-41 analog. The crystallinity of this particular material was very poor and the claim for a hexagonal Ti-MCM-41 material is doubtful giving the reported *d*-spacing of 2.8 nm. Thus the XRD pattern of Corma's Ti-MCM-41 probably corresponds to a lamellar rather than a hexagonal phase. The Ti-site isolation in this framework was studied by diffuse reflectance UV-visible spectroscopy and IR spectroscopy. The sample exhibited a UV absorbance at 210 to 230 nm and an IR band at 960 cm^{-1}, which were assigned to site-isolated Ti atoms in tetrahedral (210 nm) and octahedral (230 nm) coordination and to $Si{-}O^{\delta-} \cdots Ti^{\delta+}$ group stretching vibrations, respectively. However, we[77] and others[78] have observed this IR band for Ti-free HMS and MCM-41, and therefore it cannot be considered as evidence of Ti site isolation. Finally, the catalytic activity of this Ti-MCM-41 sample was illustrated by the epoxidation of rather small organic molecules such as hex-1-ene and norbornene in the presence of H_2O_2 and *t*-butylhydroperoxide (THP) as oxidants.

Simultaneously with the report of Corma et al., we reported[61] the preparation of a hexagonal mesoporous silica (HMS) molecular sieve and a Ti-substituted analog (Ti-HMS) by the acid-catalyzed hydrolysis of inorganic alkoxide precursors in the presence of a partially protonated primary amine surfactants (S°/S^+). In the same article we demonstrated the first ambient-temperature preparation of Ti-MCM-41 molecular sieve using the acidic $S^+X^-I^+$ templating route (pathway 3). We also reported that both Ti-HMS and Ti-MCM-41 exhibit remarkable catalytic activity for peroxide oxidation of very large aromatic substrates such as 2,6-di-*t*-butylphenol to the corresponding quinone.[61] Due to its complementary textural and framework-confined mesoporosity, Ti-HMS showed superior catalytic activity for the oxidation of this large organic substrate.

Recently, Sayari and colleagues reported[79] the preparation and catalytic activity of V-MCM-41 for peroxide oxidation of 1-naphthol and cyclododecane. However, the preparation of this molecular sieve again involved an electrostatic S^+I^- templating pathway and hydrothermal treatment at 100°C for 6 days. In addition, the reaction mixture contained significant amounts of Na^+ ions, which are known to be an unwanted impurity, significantly lowering the catalytic activity of the microporous TS-1.[80]

Corma et al. have more recently compared[81] the catalytic activity of Ti-MCM-41 and Ti-zeolite β for epoxidation of α-terpineol and norbornene. Due to its large uniform mesopore size, Ti-MCM-41 exhibited catalytic activity superior to that of Ti-zeolite β for oxidation of these bulky olefins. We have also reported a neutral $S^\circ I^\circ$ templating pathway to mesoporous molecular sieves.[52] We have used this neutral templating route to prepare Ti-HMS molecular sieves with different Ti loadings.[82] The $S^\circ I^\circ$ templating strategy allowed for the effective and environmentally benign recovery and recycling of the neutral primary amine template from Ti-HMS by simple solvent extraction. Ti-HMS showed superior catalytic activity for oxidation of the bulky 2,6-DTBP relative to the Ti-MCM-41 counterpart at all nominal Ti

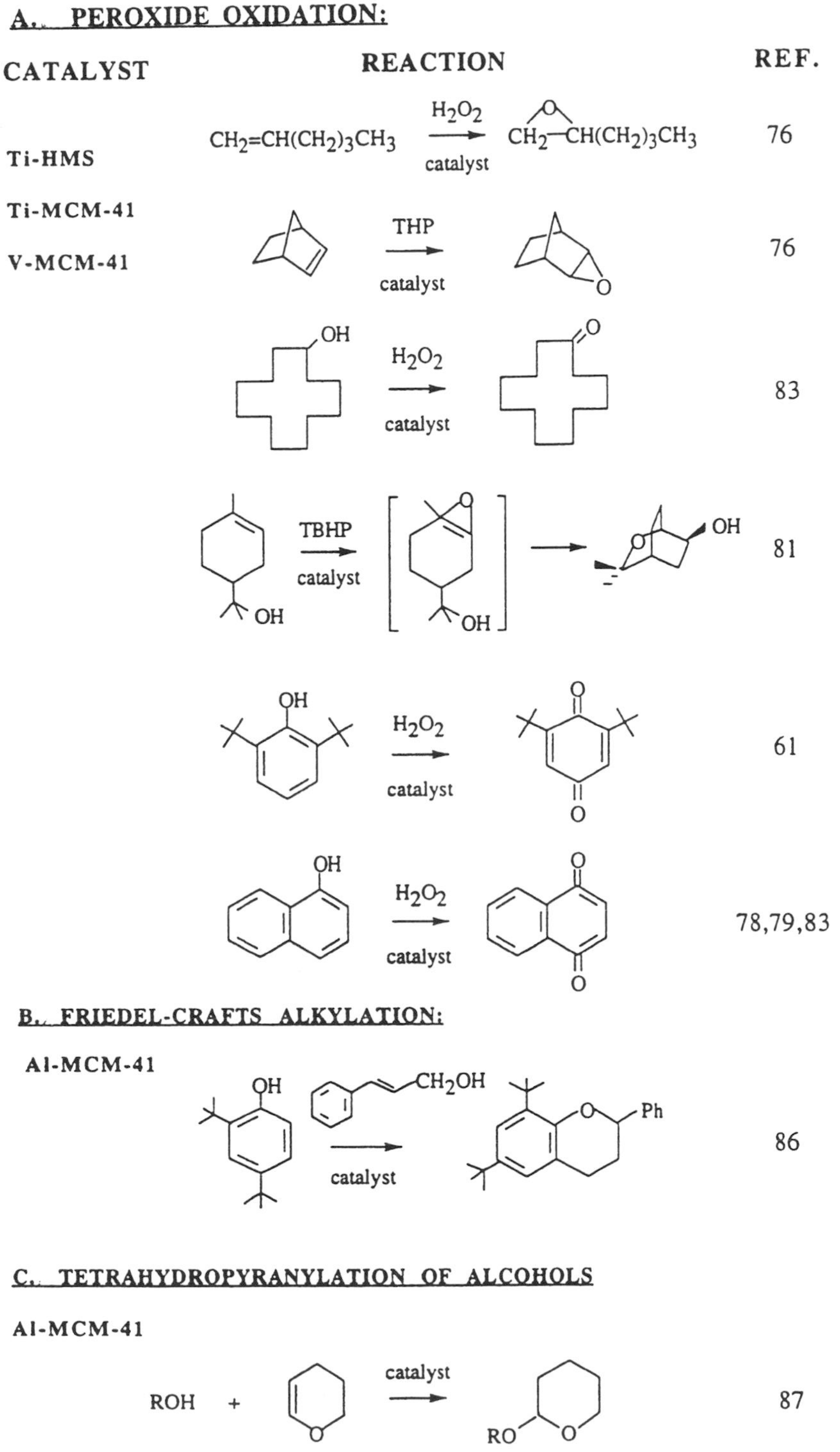

A. PEROXIDE OXIDATION:

CATALYST	REACTION	REF.
Ti-HMS	$CH_2=CH(CH_2)_3CH_3$ → (H_2O_2, catalyst) epoxide $CH_2-CH(CH_2)_3CH_3$	76
Ti-MCM-41	norbornene → (THP, catalyst) norbornene epoxide	76
V-MCM-41	cyclododecanol → (H_2O_2, catalyst) cyclododecanone	83
	(TBHP, catalyst) → [epoxide] → OH	81
	2,6-di-tert-butylphenol → (H_2O_2, catalyst) 2,6-di-tert-butyl-p-benzoquinone	61
	1-naphthol → (H_2O_2, catalyst) 1,4-naphthoquinone	78,79,83

B. FRIEDEL-CRAFTS ALKYLATION:

CATALYST	REACTION	REF.
Al-MCM-41	2,4-di-tert-butylphenol + cinnamyl alcohol (PhCH=CHCH_2OH) → (catalyst) chroman (Ph)	86

C. TETRAHYDROPYRANYLATION OF ALCOHOLS

CATALYST	REACTION	REF.
Al-MCM-41	ROH + dihydropyran → (catalyst) RO-tetrahydropyranyl ether	87

Figure 8.12 Reactions involving large organic molecules catalyzed by metal-substituted MCM-41 and HMS as reported to date.

loadings in the range 1 to 10 mol %. This has been attributed to the complementary textural mesoporosity of Ti-HMS, which facilitates access to the framework-confined mesopores.

In another development Sayari et al. compared the catalytic activity of Ti-HMS and Ti-MCM-41 for peroxide oxidation of 2,6-DTBP, 1-naphthol, and norbornylene.[83] Ti-HMS was found to exhibit better H_2O_2 selectivities as compared to Ti-MCM-41. The decomposition of H_2O_2 over Ti-HMS was found to be very limited, whereas in the case of Ti-MCM-41 all remaining H_2O_2 was decomposed. Thus these authors confirmed our observations that both Ti-HMS and Ti-MCM-41 are very promising catalysts for the oxidation of large organic substrates. Figure 8.12 summarizes most of the catalytic reactions reported to date performed over metal-substituted MCM-41 and HMS materials.

The catalytic expectations regarding Al-substituted MCM-41 derivative are very high. Perhaps, a convincing evidence for that is the large number of patents on MCM-41 applications granted to Mobil Oil during 1991–1992.[84] Most of these applications deal with oligomerization and isomerization of C_3 to C_{10} olefins over Al-MCM-41. There are also patents describing catalytic cracking and dealkylation of branched aromatics.[85] Very recently, Corma and co-workers reported that acidic Al-MCM-41 can catalyze the alkylation of electron reach bulky aromatics, such as 2,4-DTBP with cinnamyl alcohol.[86] This reaction did not proceed with any significant conversion over microporous zeolite Y but meoporous Al-MCM-41 afforded a good yield of the corresponding Friedel–Crafts products.

In another development, Kloetstra and van Bekuum[87] demonstrated that MCM-41 aluminosilicate could catalyze the tetrahydropyranylation of alcohols and phenols. The large mesopore size of MCM-41 allowed for the accommodation and selective catalytic transformation of cholesterol to the corresponding tetrahydropyranyl ether (see Fig. 8.12).

In conclusion, it is obvious that the invention of mesoporous molecular sieves will have an important impact on industrial heterogeneous catalysis. Myriads of new and exciting opportunities for the catalytic transformations of large organic molecules are now open for exploration. We have no doubt that the day is not far off when their industrial application will be considered a common practice.

8.3 POROUS LAMELLAR SOLIDS

Solids with layered structures possess basal planes of atoms that are tightly bonded within the planes but relatively weakly bonded in the direction perpendicular to the planes. The asymmetric bonding interactions translate into greatly different physical properties for the material in the in-plane and out-of-plane directions. The weakly interacting region between the stacked units is usually referred to as the interlayer or gallery region. When the layers are electrically neutral, as in graphite or FeOCl, the galleries are empty and the basal planes of adjacent layers are in van der Waals contact. Neutral guest molecules can often be incorporated between the host layers to form regularly intercalated derivatives. The incorporation of neutral species into the van der Waals gap typically is accompanied by electron transfer reaction between the molecular guest and the layered host.[88] The free-energy change associated with the electron transfer step provides much of the driving force for the intercalation reaction.

There also exists several classes of lamellar solids in which the layered units carry a net electrical charge.[89] These include smectite clays, layered double hydroxides, group IV metal phosphates, among many others. To achieve an electrically neutral structure, counterions, unusually solvated by water or other polar molecules, occupy the gallery region between layers. Thus ionic lamellar solids qualitatively resemble the conventional intercalation compounds formed by electron transfer reactions between neutral guest and layered host precursors. The difference, however, is that in ionic lamellar solids, charge separation between gallery ions and the layers is complete, whereas in conventional intercalates the extent of charge transfer between guest and layered host is seldom complete. Consequently, ionic lamellar compounds can justifiably be described as intercalation compounds, although in practice they are not formed by electron transfer reactions. Instead, they simply crystallize, with complete charge separation between the gallery species and the host layers being a distinguishing feature of their structure.

Owing to their nanoscale periodicity, ionic lamellar solids give rise to very large intracrystalline surface areas of several hundred square meters per gram or more. However, in most cases the gallery surfaces are accessible only to water and other small polar molecules that are capable of solvating the gallery counterions and the charged-layer surfaces. Removing the solvating molecules by outgassing at elevated temperatures results in the recollapse of the galleries, especially if the intercalated counterions are small relative to interstices occupied by the ions on the gallery surfaces. However, if the counterions are relatively large, they can function as molecular props or "pillars" and thereby prevent the galleries from collapsing completely when the solvating medium is removed.[90, 91] The gallery space might then be accessible to other small molecules the size of H_2O (e.g., N_2, CO_2, or NH_3). But simply facilitating the adsorption of small molecules is relatively uninteresting. Ideally, one would like to tailor the gallery structure on a length scale that would allow the accommodation of organic and inorganic molecules for molecular assembly and, perhaps, catalytic chemical conversions. Pillaring reactions of a lamellar host are an important route to achieving these desired structural modifications.

8.3.1 Pillared Layered Materials

Pillared lamellar solids are best described as intercalation compounds that meet three important criteria. These are illustrated schematically in Fig. 8.13.

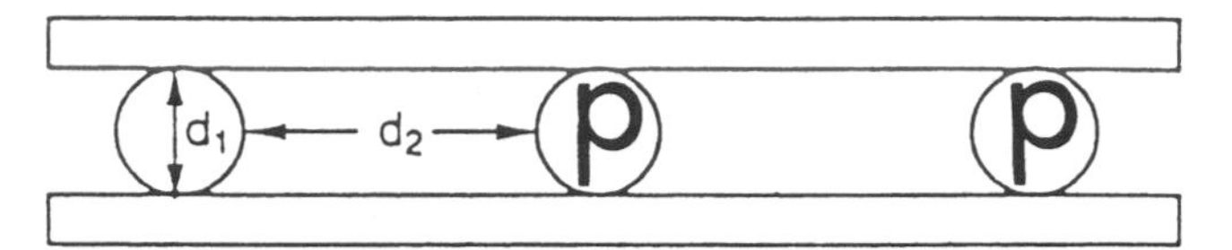

Figure 8.13 Schematic illustration of a pillared lamellar solid in which the pores are defined by the gallery height (d_1) and the lateral free separation (d_2) between pillaring species P.

First, the gallery species must be *sufficiently robust* to provide vertical expansion of the galleries and prevent gallery collapse upon dehydration. Second, the pillars must be *laterally spaced* to allow for interpillar access by molecules as least as large as nitrogen. Simply expanding the galleries to molecular dimensions by intercalation of pillars has no significance for intracrystal adsorption and catalysis if the galleries are effectively stuffed full with the pillars nearly in van der Waals contact. Third, the host layers must be sufficiently *rigid* to sustain the desired lateral separation of the pillars.[92] Flexible or floppy host layers (e.g., graphite) would be of little use in designing galleries that are nanoporous because the layers would simply fold around the pillars and close off the space between the pillars.

To illustrate the use of pillaring reactions in the design of nanoporous solids, we consider primarily two layered structure types that currently are under investigation in several laboratories: smectite clays and layered double hydroxides. In addition, we briefly identify a few other layered systems of particular interest for pillaring.

Smectite Clays Smectite clays are a family of complex layered oxides with 2:1 layer lattice structures, analogous to muscovite, phlogopite, and other mica minerals.[93, 94] Figure 8.14*A* illustrates the 2:1 structure in which a central $MO_4(OH)_2$ octahedral sheet is symmetrically cross-linked to two tetrahedral MO_4 sheets. Aluminum, iron, magnesium, and sometimes lithium occupy the octahedral interstices, whereas silicon and, in part, aluminum occupy tetrahedral sites. Various cations, but especially Na^+ and Ca^{2+}, may occupy the gallery surfaces. The nature of the cations filling the tetrahedral and octahedral sites in the 2:1 layers distinguishes the various members of this mineralogical family. Idealized anhydrous unit cell formulas for representative clays are provided in Table 8.2. A major difference between a smectite and a mica is the layer charge density. Smectites typically have charge densities between 0.45 and 1.2 e^- per $O_{20}(OH)_4$ unit cell, whereas the micas are much more highly charged, with 2.0 e^- per $O_{20}(OH)_4$ unit. The difference in charge density is in part responsible for the fact that certain metal ion-exchange forms of smectite undergo swelling and ion exchange in

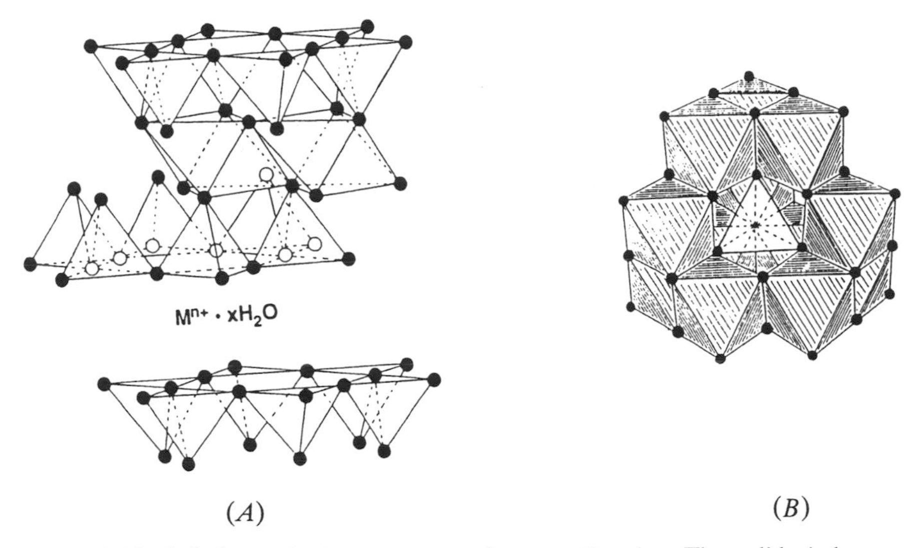

Figure 8.14 (*A*) Layer lattice structure of a smectite clay. The solid circles are oxygen and hydroxyl groups that define two tetrahedral sheets and a central octahedral sheet. The $M^{n+} \cdot nH_2O$ exchange cations occupy positions on the gallery surfaces. (*B*) Polyhedral structure of $Al_{13}O_4(OH)_{24}(H_2O)_{12}{}^{7+}$ used as a pillaring agent for smectite clays.

TABLE 8.2 Anhydrous Unit Cell Formulas for Typical Smectite Clay Minerals

Mineral	Typical Unit Cell Formula
Montmorillonite	$Ca_{0.35}[Mg_{0.70}Al_{3.30}](Si_{8.0})O_{20}(OH)_4$
Beidellite	$Na_{0.7}[Al_{4.0}](Si_{7.3}Al_{0.7})O_{20}(OH)_4$
Hectorite	$Na_{0.6}[Li_{0.6}Mg_{5.4}](Si_{8.0})O_{20}(OH)_4$
Saponite	$Na_{0.9}[Mg_{6.0}](Si_{7.1}Al_{0.9})O_{20}(OH)_4$

water, whereas micas are not swelled by water and are very difficult to ion-exchange.

The concept of pillaring a lamellar solid as a means of forming microporous derivatives was first demonstrated in 1955 by Barrer and MacLeod using smectite clays.[95] In this seminal work, the alkali-metal and alkaline-earth exchange cations in montmorillonite were replaced by quaternary ammonium ions such as Me_4N^+, among others. Subsequent work, reviewed by Barrer,[96] has led to the pillaring of other smectite clays by a variety of onium ions and metal complex ions of different sizes. By varying the charge on the host layer and hence the lateral separation of the pillars, one can tailor the pore structure to differentiate adsorbates on the basis of molecular size. The molecular sieving properties of a pillared clay are sufficient to distinguish between molecules differing in size by a few tenths of an angstrom unit.

Metal oxide pillared clays are chemically and thermally more robust than their microporous organoclay counterparts. These derivatives are prepared by first replacing the gallery Na^+ and Ca^{2+} ions in the native mineral with a robust polycation of high charge. For example, aluminum chlrohydrate,[97, 98] $Al_{13}O_4(OH)_{24}(H_2O)_{12}{}^{7+}$, has been used extensively as a pillaring reagent. The polyhedral structure of this ion is shown in Fig. 8.14*B*. Twelve aluminum ions occupy octahedral interstices, and one is positioned at a tetrahedral site at the center of the polyhedron. Water, hydroxyl groups, and bridging oxygens occupy vertices. Thermal dehydration/dehydroxylation of the intercalated polycation converts the ion to a small, metal oxide–like particle in the gallery region:

$$Al_{13}O_4(OH)_{24}(H_2O)_{12}{}^{7+} \rightarrow 6.5\text{“}Al_2O_3\text{”} + 7H^+$$

Upon heating, protons are released which in part balance the negative charge of the host clay layers. A number of review articles have appeared that summarize the synthesis and physical properties of metal oxide pillared clays derived from the intercalation of polyoxocations of aluminum, zirconium, chromium, and many other metals.[90, 91, 99, 100] The Lewis acid sites provided by coordinatively unsaturated metal ions on the pillar and the Brønsted acidity formed upon thermolysis imparts novel chemical catalytic properties.[101–105] Moreover, the pores between metal oxide pillars often are larger than those found in conventional zeolites.

A schematic illustration of a pillared clay aggregate is provided in Fig. 8.15*A*. Some defects are introduced due to layer folding and irregular platelet sizes. Nevertheless, most of the void volume is in the nanoporous range. It is noteworthy that if the clay platelet size is very small, a new family of nanoporous materials known as delaminated clays can be formed.[106] These are derivatives in which edge-to-face layer aggregation competes with face-to-face layer aggregation. The edge-to face aggregation of layers is facilitated by very small particle size (< 500 Å) or by clays with lath-like layer morphology. Some face-to-face aggregation allows for the presence of micropores as in conventional pillared clays, but the edge-to-face aggregation mechanism is extensive, and this causes a card house structure to form. The competitive edge-to-face and face-to-face aggregation mechanisms in a delaminated clay are illustrated qualitatively in Fig. 8.15*B*.

The extent of face-to-face aggregation is apparently limited to domains less than 70 Å, because these materials are typically x-ray amorphous. In contrast, a pillared clay will exhibit one to several orders of 00*l* x-ray reflections. The purpose of pillaring layered materials is to introduce regular porosity and to enable facile access to the intracrystal gallery space. The porous properties of alumina pillared clays are thus of primary importance. N_2 surface areas of more than 300 m^2/g are typically observed for alumina

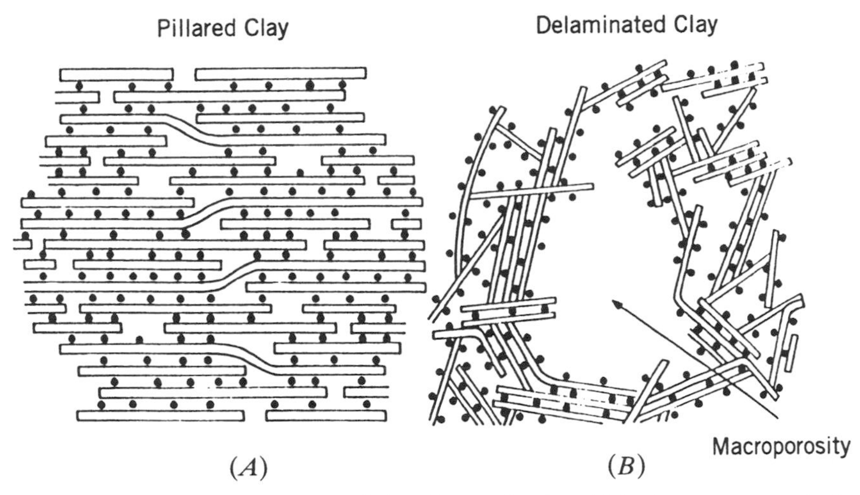

Figure 8.15 Layer aggregation in (*A*) pillared clay in which the layers are stacked primarily face to face and (*B*) delaminated clay in which edge-to-face aggregation competes with face-to-face aggregation. The slabs represent the 1.0-nm-thick clay layers and the filled circles represent the pillaring cation.

pillared clays. The contribution of microporous surface area to the total surface area is usually higher than 75%. Processing conditions during synthesis, drying, and calcining of the as-synthesized material can greatly affect the surface area, pore structure, and thermal stability of the finished product.[107-110] Air-dried materials usually exhibit substantially higher microporous surface area than do freeze-dried materials, the difference being attributed to the extent of face-to-face arrangement of the layers.[111] The particle size of the clay is also an important factor for the pore structure of the pillared product. Thus clays with small particle size, such as laponite[112] or certain saponites,[113, 114] tend to yield delaminated rather than truly pillared products upon reaction with Al_{13}. This behavior is related to the aggregation mechanism of clay tactoids.

The size of pillared clay pores is relatively difficult to estimate. The height of the micropores, around 0.9 nm, is easily obtained from x-ray diffraction. However, the interpillar distance is difficult to evaluate because pillared clays are not crystallographically ordered in the in-plane direction. Adsorption of molecules of various diameters shows that alumina pillared clays exhibit a molecular sieving effect for molecules in the range 0.8 to 1.0 nm.[115] N_2 adsorption[116] and ^{129}Xe NMR[117] studies of alumina pillared clays provide pore size estimates comparable to the gallery height, but the values of the pore sizes of metal oxide pillared clays generally *do not* parallel the gallery height. This was nicely illustrated in a comparison of pore sizes obtained by modeling the adsorption isotherms of chromia, zirconia, alumina, iron oxide, and titania pillared clays.[118, 119] As illustrated in Fig. 8.16, the fitting of the N_2

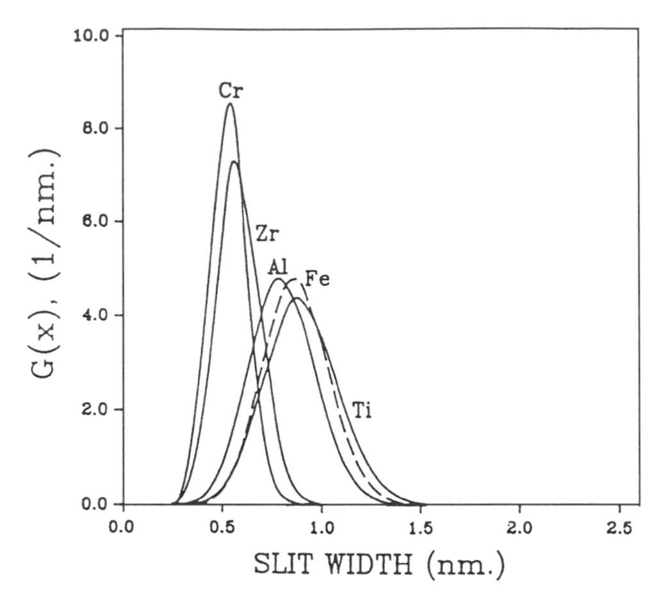

Figure 8.16 JC micropore size distribution for pillared clays.

adsorption data to a Jaroniec–Choma (JC) adsorption model gave pore size maxima in the range 0.5 to 1.0 nm.

A Horvath–Kawazoe (HK) treatment of the adsorption data gave similar results, except that the pore distributions generally tended to be skewed toward larger widths.[119] In the case of alumina pillared montmorillonite, the micropore distribution was found to be bimodal.[120, 121] Since the increasing order of pore maxima did not parallel the order of increasing gallery height for the various pillared clays, it was possible to conclude that the pore size was determined by the lateral separation of the pillars, which, in general, is substantially smaller than the gallery height.

X-ray diffraction peaks (in transmission mode) corresponding to the lateral free distance between cations in clays pillared with organometallic complexes were resolved by Tsvetkov and White.[122] This analysis provided an interpillar distance in the range 1.3 to 1.5 nm for montmorillonite and 0.95 to 1.08 nm for a highly charged fluorohectorite. In an x-ray diffraction and thermodynamic study of Xe and Kr adsorption in pillared montmorillonite, an interpillar distance of 1.8 nm was indicated.[123] These measurements, based on x-ray scattering, probably overestimate the true pillar separation.

The porosity of alumina pillared clays can be modified by incorporating a polymer or a surfactant during the synthesis. Thus Suzuki et al. reported[124] that the interlayer distance of alumina pillared fluorohectorite could be increased to 2.0 nm by adding polyvinyl alcohol to the Al_{13} pillaring solution, even after calcination at 500°C. The co-intercalation of a nonionic surfactant of general formula $C_{12-14}H_{25-29}O(CH_2CH_2O)_5$ and Al_{13} has been reported

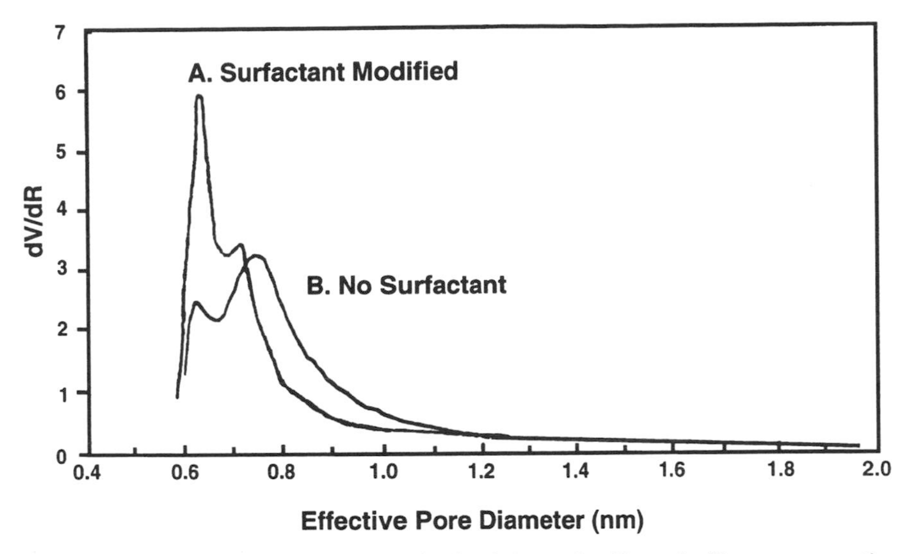

Figure 8.17 Micropore distribution obtained from the Horvath–Kawazoe equation for (*A*) surfactant-modified alumina pillared montmorillonite and (*B*) alumina pillared montmorillonite (no-surfactant).

to yield pillared materials exhibiting an enhanced crystallographic ordering along the layer stacking direction.[125] The calcined alumina pillared product obtained by surfactant-modified synthesis also exhibited a more uniform micropore distribution than that of the material produced without surfactant modification. The micropore distribution obtained from the Horvath–Kawazoe equation is presented in Fig. 8.17.

In most of the pillared lamellar solids reported to date, the gallery height is comparable to the van der Waals thickness of the host layers (about 1.0 nm). That is, the gallery free space constitutes the *minor* component by volume. It should be possible through the proper choice of pillaring agent to molecularly engineer pillared derivatives in which the gallery height is substantially larger than the thickness of the host layers. The term *supergallery* has been proposed to describe derivatives in which the gallery height is two or more times as large as the thickness of the host layers.[126] If the lateral separation between the pillars can be made comparable to the size of the pillars, materials with nanopores should be attainable.

Metal oxide pillared clays derived from intercalated polycation precursors contain oxide aggregates of more of less uniform size, as judged by the fact that they exhibit several orders of 00*l* x-ray reflections. Several investigators have recognized the possibility of preparing pillared clays by direct intercalation of metal oxide sols.[126–129] The latter approach could afford supergallery derivatives. Aqueous metal oxide sols are commercially available with average particle sizes in the range 2.0 to 4.0 nm and a particle size distribution on the order of ± 0.5 nm. Such particles could function as pillars for the formation of supergallery pillared clays, provided that the particle size was not altered upon intercalation by competitive hydrolysis reactions.

The tubular aluminosilicate imogolite, with the empirical formula $SiAl_2O_3(OH)_4$, is an unusual naturally occurring sol particle with a tunnel-like or tubular structure. This material represents a *molecularly regular* sol particle. The external and internal diameters of the tube are approximately 2.4 and about 0.8 nm, respectively, and the tube length can range to several hundred nanometers.[130] Molecular sieving studies have shown that the intratube channel is indeed available for adsorption of molecules with kinetic diameters smaller than 1.0 nm.[131]

Imogolite has been shown to intercalate as a regular monolayer into smectite clays such as Na^+-montmorillonite.[132] These new tubular silicate-layered silicate (TSLS) nanocomposites with a basal spacing of 3.4 ± 0.1 nm can be viewed as a new type of pillared clay in which the pillars themselves are microporous. Figure 8.18 illustrates the structure of a TSLS complex. Although the tubes are aggregated over limited domains in the lateral

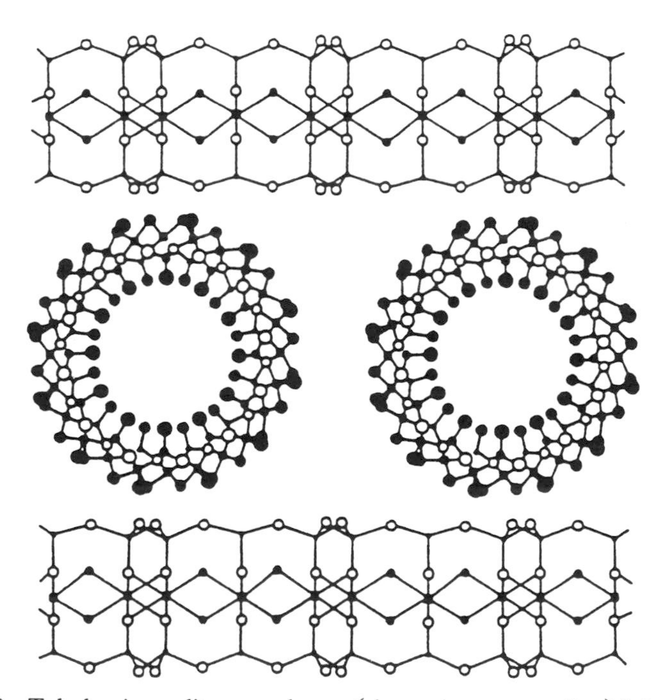

Figure 8.18 Tubular imogolite monolayers (shown in cross section) intercalated in the galleries of layered Na^+-montmorillonite. The composition of the imogolite tubes is $(HO)SiO_3Al_2(OH)_3$, as read from the inner tiers of atomic planes constituting the tube walls.

direction, the packing of the gallery is incomplete, much like a logjam structure. Thus intratube free space is available, but the intertube pores are accessible only through the intratube channels.

A BET surface area of 580 m^2/g and a liquid microporous volume of 0.205 cm^3/g has been observed for the TSLS species.[133] A bimodal microporous behavior was indicated by the nitrogen adsorption behavior, suggesting that nitrogen was accessing both intra- and intertube nanopores. Significantly, the intercalated tubes were thermally stable to about 450°C, whereas the pristine tubes collapse above about 250°C.

Layered Double Hydroxides Layered double hydroxides (LDHs) are complementary to smectite clays insofar as the charge on the layers and the gallery ions is reversed; that is, the host layers of an LDH are two-dimensional polyhydroxyl cations and the gallery species are hydrated anions. The compositions of LDHs are represented by the formula $[M^{II}_{1-x}M^{III}_{x}(OH)_2][A^{n-}]_{x/n} \cdot zH_2O$, where A^{n-} is the gallery anion, and M^{II} and M^{III} are divalent and trivalent cations that occupy the interstices of edge-shared $M(OH)_6$ octahedral sheets. A large number of compositions are possible,[134-138] depending on the choice of M^{II}, M^{III}, A^{n-}, and the layer cation stoichiometry, which typically is in the range $x = 0.17$ to 0.33. As shown in Fig. 8.19*A*, LDH structures consist of $Mg(OH)_2$-like sheets separated by galleries of hydrated A^{n-} ions.

Despite their complementary structural relationship to smectite clays, LDHs are not easily pillared. Owing to their relatively high layer charge density (about 4.0 e^+/nm^2 for LDHs versus about 1.0 e^-/nm^2 for smectites), the galleries of LDHs tend to be filled by the pillaring ions themselves. However, polyoxometalate anions (POMs)[139] with high charge can be effective reagents for the pillaring of LDHs. These ions have structures consisting of several layers of space-filling oxygen atoms. The first crystalline forms of POM-pillared LDHs were reported for Zn_2Al, Zn_2Cr, and Ni_3Al derivatives using $V_{10}O_{28}^{6-}$ as the pillaring reagent.[140, 141] The basal spacings of the pillared products (1.19 nm) corresponded to gallery heights of 0.71 nm (three oxygen planes) and to an orientation in which the C_2-axis of the pillaring anion is parallel to the host layers. Adsorption–desorption of N_2 indicated the presence of both small micropores (< 1.0 nm) and nanopores with a pore maximum near 2.0 nm.

Keggin ions of the type α-$[XM_{12}O_{40}]^{n-}$ shown, in Fig. 8.19*B*, also are efficient reagents for the pillaring of Zn_2Al-LDH structures.[142] The species $[Zn_2Al(OH)_6]NO_3 \cdot 2H_3O$ was found to undergo facile and complete intercalative ion-exchange reaction with aqueous solutions of $[H_2W_{12}O_{40}]^{6-}$ and $[SiV_3W_9O_{40}]^{7-}$ Keggin ions. Interestingly, no ion exchange occurred with $[PW_{12}O_{40}]^{3-}$ or $[SiW_{12}O_{40}]^{4-}$, and only partial exchange was observed with the Keggin-like species $[PCuW_{11}O_{39}(H_2O)]^{5-}$. The latter results suggest that the accessibility of the LDH galleries depends in part on the charge on the

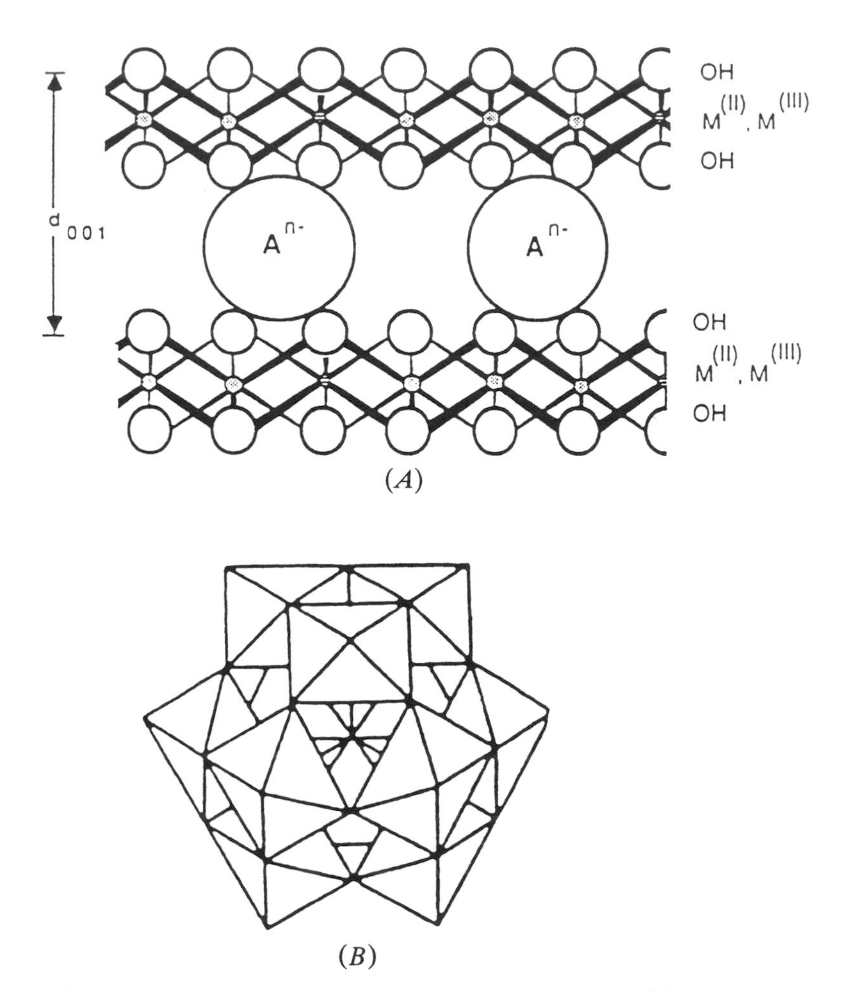

Figure 8.19 (*A*) Structure of layered double hydroxides (LDHs) of the type $[M^{II}_{1-x}M^{III}_{x}(OH)_2][A^{n}]_{x/n} \cdot zH_2O$. Gallery water is not shown. (*B*) Structure of a α-Keggin ion of the type α-$[XM_{12}O_{40}]^{n-}$.

POM. The area needed to accommodate a Keggin ion of diameter is 0.98 nm is 0.849 nm^2, and since the area per unit charge in a Zn_2Al-LDH is only about 0.166 nm^2, a -5 charge on the Keggin ion is needed to spatially accommodate a monolayer of the ion in the LDH gallery. The lack of intercalative ion exchange for -3 and -4 Keggin ions is consistent with these simple geometric considerations. The partial replacement of NO_3- ions by -3 and -4 Keggin ions would require mixing in the same gallery two ions of very different size and charge, and such mixing appears to be thermodynamically unfavorable. Commensurate relationships between the triangular faces of the Keggin ion and those of the gallery surface may also be important in determining gallery access by ion exchange.[143] For instance, the β-isomer of $[SiV_3W_9O_{40}]^{7-}$ is much slower than the α-isomer to intercalate. This difference is lability suggested that the α-isomer prefers to orient in the galleries with the C_2-axis perpendicular to the basal surfaces.

The pillaring of LDHs by POMs is further complicated by the fact that most LDHs are basic, whereas the POM anions are acidic. Hydrolysis reactions of the LDH and POM can result in products that are poorly ordered or that contain multicrystalline phases. A promising route to pillared forms of basic LDHs is based on the exchange reaction of the desired POM with an expanded LDH precursor intercalated by a large organic anion such as *p*-toluenesulfonate or terephthalate.[141] The large organic anion is very readily replaced by the POM, and competing side reactions are minimized. Thus it is possible to prepare pillared forms of highly basic Mg_2Al-LDH hosts with acidic POM pillars such as $[V_{10}O_{28}]^{6-}$ and $[Mo_7O_{24}]^{6-}$. Well-ordered LDHs interlayered by large organic anions can be relatively difficult to prepare, but a more general and reliable route to such precursors has been described.[144] The latter method utilizes as a precursor a hydroxide-exchanged form of the LDH, such as synthetic meixnerite, $[Mg_3Al(OH)_8]OH \cdot 2H_2O$. These derivatives swell in polar solvents such as glycerol or glycerol–water mixtures. The swelling greatly enhances the accessibility of the gallery hydroxide ions for reaction with organic acids. A variety of LDH intercalates containing long-chain alkyl carboxylates and α,ω-dicarboxylates have been prepared by this route. Thus it is now possible in principle to match the gallery height of the organic LDH precursor to any desired POM pillaring agent.

Disk-shaped metallomacrocyclic anions such as cobalt(II) phthalocyanine tetrasulfonate, $[Co(PcTs)]^{4-}$, are currently being investigated as potential pillaring agents for LDHs.[145] In the case of $[Co(PcTs)]^{4-}$ intercalation into a $Zn_{1-x}Al_x$-LDH with $x = 0.33$, the anion intercalates with the plane of the macrocycle perpendicular to the LDH layers, as illustrated in Fig. 8.20. At a layer charge density of $x = 0.33$, the gallery anions are too closely spaced to function as true pillaring agents. Lowering the layer charge density may increase the lateral separation of the anions. However, even at $x = 0.33$, the Co(II) centers at the edge of the LDH layers are available for catalytic reaction. LDH-intercalated $[Co(PcTs)]^{4-}$ anions, for instance, catalyze the O_2 oxidation of mercaptides to disulfides under ambient conditions. Such transformation using easily recyclable catalysts are of interest for possible applications in the remediation of polluted groundwaters and industrial effluents. Interestingly, immobilization of the macrocyclic anion of LDH surfaces greatly improves the longevity of the complex. Several hundred catalyst turnovers can be achieved using the LDH-immobilized catalyst, whereas only about 25 turnovers can be achieved with the homogeneous complex, owing to competing side reactions which lead to its deactivation. It is also noteworthy that Keggin ions intercalated in LDH galleries are effective oxidation catalysts for olefins using peroxide as the oxidizing agent.[146]

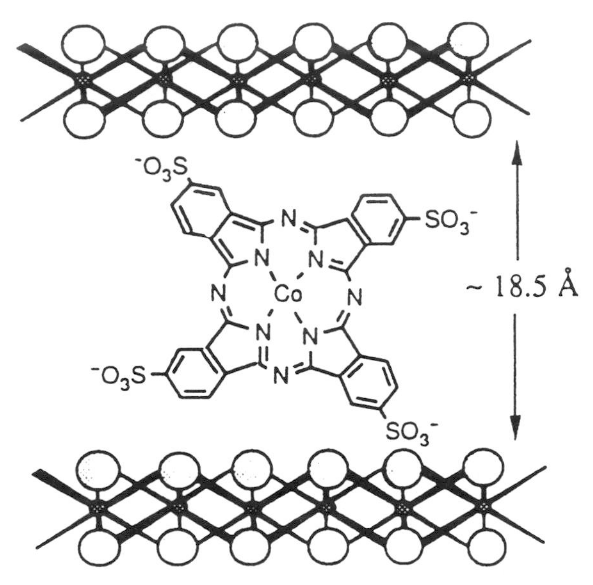

Figure 8.20 Edge-on intercalation of $[Co(PcTs)]^{4-}$ in the galleries of an LDH.

Other Layered Systems Smectite clays and layered double hydroxides represent only two classes of layered materials that can be chemically modified to form nanoporous derivatives. Several other layered compounds are suitable for pillaring reactions, including layered titanates,[147] phosphates and phosphonates,[148-154] silicates,[155] and niobates.[156] The group IV metal phosphates are particularly interesting because, unlike smectite clays, the gallery surfaces are chemically functional. As shown in Fig. 8.21*A*, the phosphate groups in $Zr(HPO_4)_2$ are monoprotonated. The protons can be replaced by acid-base reaction with metal cations, which in turn can be replaced with pillaring cations such as $Al_{13}O_4(OH)_{24}(H_2O)_{12}{}^{7+}$.[148] However, the latter ion completely fills the interlayers and the stuffed galleries are unaccessible for adsorption. A much more promising approach, first demonstrated by Dines et al.,[157] is to replace some of the P—OH groups with rigid difunctional phosphonate groups to form cross-linking bridges between adjacent layers. This is illustrated in Fig. 8.21*B* for a mixed zirconium phosphate/phosphonate containing $P—C_6H_5—C_6H_5—P$ groups as the bridging units. By regulating the relative amounts of phosphate and phosphonate groups in the structure, one can in principle vary the average distance between randomly separated cross-linking units.

8.3.2 Porous Clay Heterostructures

A relatively unexplored approach to designing nanoporous layered materials is to form the pillaring species directly within the galleries of the host rather

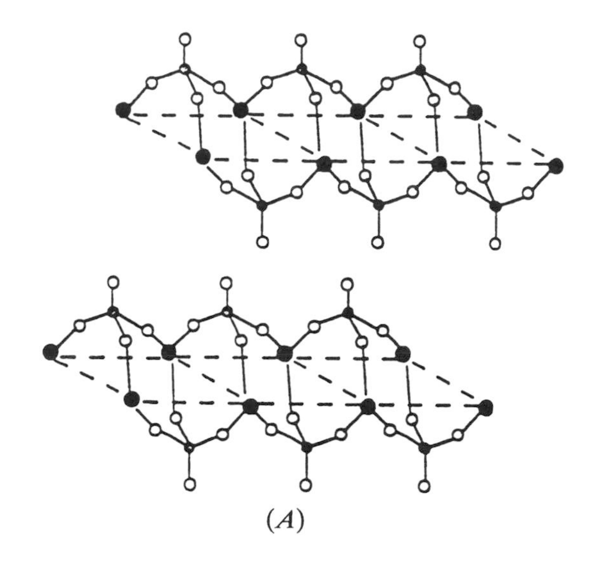

(*A*)

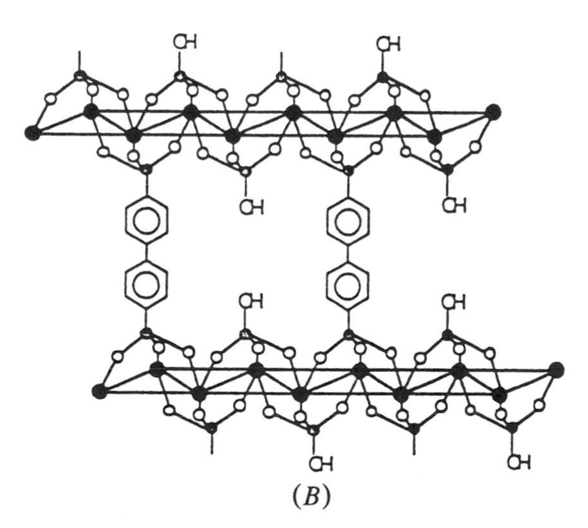

(*B*)

Figure 8.21 Structures of (*A*) $Zr(HPO_4)_2$ and (*B*) a nanoporous mixed diphosphonate/phosphate pillared derivative. In part *A* the closed large and small circles represent zirconium and phosphorus and the open circles represent oxygen.

than exchanging the preformed pillar or pillar precursor as in the examples above. This concept of in situ pillar formation was first demonstrated for the hydrolysis and polymerization of tetraethylorthosilicate (TEOS) in the galleries of layered titanates and sheet silicates (e.g., magadiite) interlayered by alkyl ammonium ions.[158] The gallery polymerization process under aqueous conditions was shown to be sufficiently regular to form crystallographically well-ordered pillared derivatives interlayered by nanoscopic domains of silica. Analogous intragallery polymerization reactions of TEOS can be carried out using neat amines as the gallery swelling agent.[159] This general approach to preparing nanoporous layered materials relies on the formation of pillaring inorganic oxide particles in the space between the onium exchange cations. Thus the size of the micropores depends on the charge density of the layer host, which determines the spatial separation between the onium ions.

More recently, intragallery polymerizations of metal alkoxides have been developed that give rise to a new type of porous structure denoted porous clay heterostructures or PCHs.[160] In this new approach to intragallery synthesis the galleries of a smectite clay host are first intercalated with a mixture of cationic and neutral amine-based surfactants, typically a quaternary alkyl ammonium ion and a primary amine. When a metal alkoxide such as TEOS is added to the surfactant–clay complex in the presence of water, an intragallery mesostructure similar to that of MCM-41 is formed. Thus upon removal of the surfactant, the resulting intercalate is a porous, regularly ordered hybrid composite or heterostructure of two end-member bulk phases, a smectite clay and a mesostructured metal oxide.

Figure 8.22 illustrates the gallery templated synthesis of a PCH formed from TEOS and fluorohectorite. Step (A) in the process involves the formation of an amine-solvated bilayer structure with thickness equivalent to the length of the quaternary cation (filled head groups) and the neutral amine (open headgroups) surfactants. Intercalation of TEOS in step (B) occurs by partial displacement of the neutral amine. In crucial step (C) the templated heterostructure is formed in which a two-dimensional hydrated silica is organized around micellar assemblies of quaternary onium ion and clay heterostructure with a two-dimensional framework of porous silica intercalated between the clay layers. The specific surface area of PCH materials is in the range 470 to 750 m^2/g, and the pore volume is 0.31 to 0.48 cm^3/g.

The choice of surfactant is important in forming the PCH. If the interactions of the onium ion, amine, and TEOS do not lead to an intragallery mesostructure, only a pillared clay derivative will be formed. PCHs containing mesostructured intragallery silica typically exhibit pore distributions with maxima in the supermicropore to small mesopore range (1.7 to 2.5 nm). In contrast, smectites interlayered with silica are pillared with the gallery aggregates closely spaced so as to give pore distributions in the micropore range 0.5 to 1.0 nm.[158–161] Thus PCH materials offer promising opportunities for the adsorption and immobilization of large biomolecules (enzymes) and the catalytic conversion of high-molecular-weight organics, such as those found in low-grade crude oil.

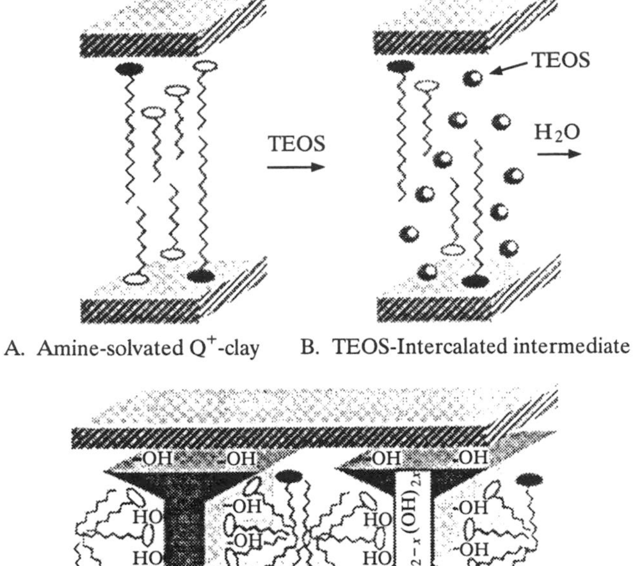

Figure 8.22 Proposed mechanism for the formation of a PCH by gallery-templated synthesis.

8.3.3 Catalytic Properties of Alumina Pillared Clays

Gas Oil Cracking The first results of catalytic cracking with alumina pillared clays were reported by Vaughan et al.[162] and were quite encouraging since the conversion and gasoline yields obtained with pillared clays were competitive with those of faujasite and higher than those of amorphous silica-alumina. These results,[163] with similar results obtained by Occelli and Tindwa,[164] are gathered in Table 8.3. Also, Occelli noted that alumina

TABLE 8.3 Catalytic Cracking Performances of Alumina Pillared Clays

Conditions		Alumina Pillared Clay	Amorphous Silica-Aluminas	REY	Ref.
Heavy oil, 510°C	Conversion (vol. %)	71.0	55.5	73.5	163
	C_5 gasoline (vol %)	51.5	38.0	58.0	
Light oil 480°C	Conversion (vol %)	74.2	—	70.6[a]	164
	C_5 gasoline (vol %)	44.0	—	47.7[a]	

[a]Commercial cracking catalyst.

pillared clays yielded more light cycle gas oil than did conventional cracking catalysts.[165] However, as observed in two reviews of the topic,[166, 167] two major problems have been encountered and have up to now deferred commercialization of these materials. Most pillared clays are selective for coke formation and their hydrothermal stability is low. However, Guan and her co-workers have demonstrated[168] that alumina pillared derivatives of the mixed-layer clay rectorite are hydrothermally more stable than the ultrastable zeolite Y and that the clay is more effective than a zeolite-based catalyst for the cracking of heavy oil fractions. The parent Mg^{2+}-rectorite is inactive for gas oil cracking. A schematic representation of rectorite, showing the smectite- and mica-type galleries of this mixed-layer clay, is shown in Fig. 8.23. The alumina pillared product, however, exhibits selectivity similar to pillared smectite clays and zeolitic FCC catalysts reported previously. Interestingly, the bulk density of pillared rectorite is less than 50% that of pillared montmorillonite. Thus for a given catalyst/oil ratio under MAT reactor conditions, the contact time is twice as long for pillared rectorite as for pillared montomorillonite. Thus the reactivity of pillared rectorite with a

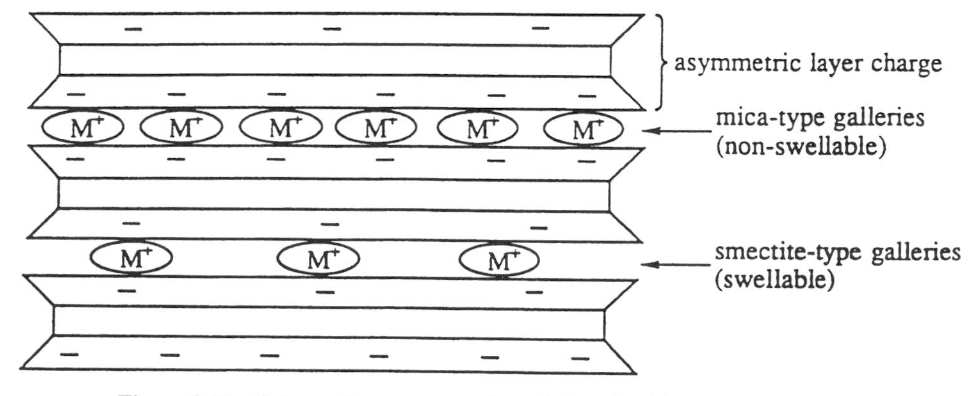

Figure 8.23 Schematic representation of the mixed-layer clay rectorite.

surface area of 160 to 190 m^2/g is comparable to that of a pillared montmorillonite with a surface area near 300 m^2/g. However, it was observed that the low particle density of pillared rectorite could be undesirable for certain commercial applications.

As with other pillared clays, pillared rectorite gives higher light cycle oil yields by cracking a greater fraction of heavy slurry oil. High coke yields have been attributed to the Lewis acidity of the pillared rectorite and to the high iron content. The coke/conversion ratio for pillared rectorite (0.12 w/w) is almost twice that of a zeolite catalyst (0.076). Pillared rectorite is easily regenerated following coking. Calcination at 760°C for 1 h removes carbonaceous deposits without loss of surface area. Occelli has verified that natural rectorites pillared by aluminum polycations exhibits thermal stability and hydrothermal stability far superior to those of montmorillonite catalysts prepared similarly.[169] Pyridine chemisorption studies indicate that Lewis acidity dominates under cracking reaction conditions. Pillared rectorites exposed to steam (760°C for 5 h) have cracking activity comparable to that of steam-aged commercial FCC catalysts containing 35% rare earth type Y zeolite. However, coke selectivity is inferior to that of zeolite-based cracking catalysts. The high-iron content, 0.7 to 1.0 wt % Fe_2O_3, of the pillared clay may have contributed to coke formation.

An improvement in thermal stability as well as in the basal spacing regularity of pillared clays derived from tridecameric cations, has been achieved by replacing the central tetrahedral aluminum ion in the $Al_{13}{}^{7+}$ oligomer with gallium(III) to afford $GaO_4Al_{12}(OH)_{24}(H_2O)_{12}{}^{7+}$, henceforth abbreviated $GaAl_{12}{}^{7+}$.[170, 171] The $Ga_{13}{}^{7+}$ analog, however, affords pillared clays that are less stable than those derived from $Al_{13}{}^{7+}$. NMR studies[172, 173] indicate that for $GaAl_{12}{}^{7+}$ the gallium occupies the central tetrahedral site of the polycation. Thus the improved stability for $GaAl_{12}{}^{7+}$ probably arises from a better size match between Ga^{3+} and the tetrahedral cavity. The advantage of Ga substitution into the pillar is illustrated by the following results. Upon heating at 700°C, the BET surface area of the Al_{13} pillared montmorillonite decreased to 47% of its value at 200°C (from 240 to 114 m^2/g), while that of the Ga_{13} clay decreased to 37% (from 231 to 86 m^2/g). However, the $GaAl_{12}$ clay heated at 700°C retained 71% of its surface area at 200°C (from 277 to 196 m^2/g). Thus the incorporation of Ga into the tetrahedral position of the Keggin-like ion clearly results in a significantly more stable structure. Another group[174] has observed an improvement in surface area (but not necessarily in the thermal stability) of Ga_{13} pillared clays over the temperature range 200 to 500°C. The improved layer stacking order for $GaAl_{12}$ clays relative to that of Al_{13} derivatives was evident from the fact that the $00l$ reflection remained much sharper for the $GaAl_{12}$ clay than for the Al_{13} analog when heated to 700°C.

Shape-Selective Reactions Shape selectivity can occur in a porous catalyst when the reactants, transition state, intermediates or products, and the pores of the catalyst are of similar size. The approximate 0.75-nm gallery

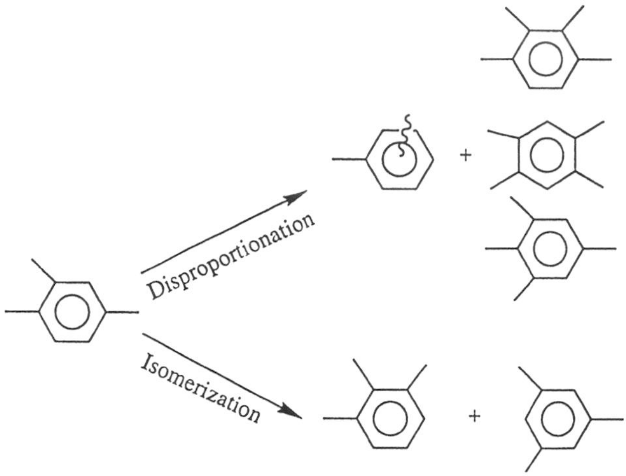

Figure 8.24 Disproportionation and isomerization reactions for 1,2,4-trimethylbenzene.

height of pillared clays makes them potentially suitable materials for shape-selective effects on relatively large molecules. Kikuchi and Matsuda have reviewed different reactions for which shape selectivity have been proposed as an explanation of product distribution.[175] Reactions such as oligomerization, esterification, or cracking have been used to test shape selectivity. These systems are, however, often chemically complicated and do not usually give solid evidence for shape-selective effects. On the other hand, reactions involving aromatic compounds are specially favorable for investigating shape selectivity, because they involve mechanistically similar pathways yielding isomers of different sizes.

The disproportionation of 1,2,4-trimethylbenzene has been used as a test reaction for alumina pillared clays. As shown in Fig. 8.24, 1,2,4-trimethylbenzene (abbreviated 1,2,4-TrMB) can undergo two reactions over an acid catalyst: disproportionation and isomerization. Kikuchi et al.[176] observed that the disproportionation products 1,2,4,5-tetramethylbenzene (abbreviated 1,2,4,5-TeMB) and *o*-xylene were formed in excess of thermodynamical levels over alumina pillared montmorillonite. They proposed that a cage effect in favor of the smaller bimolecular transition state leading to 1,2,4,5-TeMB occurs in the micropores. More recent studies[177] showed that certain nonpillared clays gave even higher 1,2,4,5-TeMB selectivity than pillared clays, thus casting doubt on any shape-selective effect. Increased selectivity in 1,2,4,5-TeMB was always correlated with decreased selectivity for isomerization, suggesting that 1,2,4,5-TeMB selectivity is mainly governed by the isomerization to disproportionation ratio, not by pore size.

Table 8.4 summarizes the selectivity toward 1,2,4,5-TeMB for the disproportionation of 1,2,4-trimethylbenzene over various alumina pillared clays and an ammonium-exchanged montmorillonite. Increasing the calcination

TABLE 8.4 Conversion of 1,2,4-Trimethylbenzene over Various Alumina-Pillared Clays

Clay	Calcination Temperature (°C)	Conversion (%)	Disproportionation	1,2,4,5-TeMB	Ref.
Montmorillonite	400	19.4	90.8	70.7	177
Montmorillonite	650	5.9	96.8	81.0	177
Saponite	400	15.2	80.3	57.1	178
Laponite	400	3.2	90.6	70.3	178
Montmorillonite	500	17	87.9	59	179
Beidellite	500	5.8	90.2	68	179
NH_4-montmorillonite	—	3.3	94.1	72	179

[a]Ammonium-exchanged montmorillonite, included for comparison purposes.

temperature of the alumina pillared montmorillonite resulted in increased 1,2,4,5-TeMB selectivity and decreased isomerization/disproportionation ratio. Higher calcination temperatures mainly affect the Brønsted acidity, which is known to catalyze isomerization reactions. Thus the 1,2,4,5-TeMB selectivity might be in fact mainly governed by the Brønsted/Lewis acidity ratio.

Shape-selective effects for alumina pillared montmorillonite have also been reported for *m*-xylene conversion over alumina pillared clays.[180] By reducing the charge density of the host layer it was possible to vary the lateral separation of the pillars. Because the conversion of *m*-xylene and the isomerization to disproportionation ratio increase with increasing pillar density, it was suggested that isomerization, being a monomolecular process, may be occurring in smaller pores than disproportionation, which is a bimolecular process. However, as we have seen in the previous example, several factors other than pore size can affect the isomerization/disproportionation ratio.

Alkylation of toluene by methanol has also been used as a test reaction for pillared clays. Urabe et al.[181] reported that the yield of *p*-xylene obtained with alumina pillared saponite was significantly higher than the value expected for thermodynamic equilibrium. Alumina pillared montmorillonite gave even higher *p*-xylene yields. The selectivity enhancement for alumina pillared montmorillonite was explained by a pore size effect, alumina pillared montmorillonite having a more regular microporous structure.[182] However, deviations from thermodynamical equilibrium might also contribute to the kinetic control of the alkylation reaction in the pillared clay.[183]

Propene Alkylation of Biphenyl The examples presented above deal with mononuclear aromatics that are probably too small to be subjected to any cage effect in these large micropores. Therefore, we can anticipate that the use of larger probe molecules might provide evidence for shape selectivity. The exceptional shape selectivity exhibited by the relatively large pore zeolite, mordenite, in propene alkylation of liquid phase biphenyl[184] makes

TABLE 8.5 Dependence of Catalytic Activity of Alumina Pillared Wyoming Montmorillonite (APWM) and Alumina Delaminated Laponite (ADL) on Particle Size for the Propene Alkylation of Biphenyl[a]

		Particle Size			
		150–75 μm	75–50 μm	50–45 μm	< 45 μm
APWM (0.1 g)	Conversion (%)	26	36	60	98
	Monoalkyl	86	79	64	8
	Dialkyl	14	18	28	31
	Trialkyl	0	2	8	41
	Tetraalkyl	0	0	0	20
ADL (0.09 g)	Conversion (%)	36	65	79	85
	Monoalkyl	77	56	41	35
	Dialkyl	21	33	37	38
	Trialkyl	2	10	16	19
	Tetraalkyl	0	1	6	8

[a] 250°C and 140 psi.

the reaction very attractive as a test reaction for pillared clays.[185] We have studied clays of various charge densities, charge localization, and extent of layer stacking, such as montmorillonite, beidellite, and laponite. The dependence of the catalytic activity on the catalyst particle size was large, as shown in Table 8.5. Under the liquid-phase conditions employed, the reaction was under diffusion control and the catalytic activity was governed primarily by

TABLE 8.6 Relationship Between Biphenyl Alkylation Activity and Porosity of Alumina Pillared Clays

	Catalyst		
	Delaminated Laponite	Pillared Fluorohectorite	Pillared Montmorillonite
Porosity			
Surface area (m^2/g)	387	214	334
Pore volume (cm^3/g)	0.35	0.27	0.22
Micropore volume	0.03	0.06	0.11
Mesopore volume	0.32	0.21	0.11
Activity			
Catalyst weight (%)	0.66	0.66	0.66
Conversion (%)	89	98	64
Selectivity			
Catalyst weight (%)	0.33	0.2	0.66
Conversion (%)	63	50	64
ortho	32	23	15
meta	30	34	42
para	37	43	43

textural (interparticle) mesoporosity. A distinction was made between alumina pillared montmorillonite, which exhibits less textural mesoporosity, and alumina pillared beidellite and fluorochectorite, for which a significant contribution of the pore volume is due to textural mesopores (Table 8.6).

The product selectivity toward the three possible monoalkylated isomers was studied in the range of 50 to 65% conversion. A relationship between the mesoporosity of the catalyst and the yield of ortho alkylated product was observed. Thus the catalysts with significant textural mesoporosity, such as delaminated laponite, afforded approximately 30% of the ortho isomer, whereas the microporous pillared montmorillonite gave only 15% of ortho-alkylated product. The larger size of the ortho isomer, due to a large twist angle between the phenyl rings, may explain this shape-selective effect. Pillared clays with different basal spacings[186, 187] were prepared to test for a shape-selective effect, but no difference in selectivity was observed upon varying the interlayer spacing from 0.6 to 1.5 nm. A MNDO/PM3 calculation showed that the thermodynamic equilibrium for this system is 8% ortho, 45% meta, and 47% para. Based on the energies calculated for the transition states, the proposed kinetic product distribution was found to be 70% ortho, 20% para, and 10% meta.[188] The catalysts with significant textural mesoporosity yielded a product distribution relatively close to the expected distribution for kinetic control. In contrast, the microporous catalysts exhibited a product distribution closer to that calculated for thermodynamic equilibrium, in which the contact time was long.

8.4 SUMMARY AND CONCLUSIONS

Because of their uniform pore size and ability to discriminate and adsorb molecules on the basis of their size and shape, microporous zeolites and molecular sieves are used in a number of important adsorption, ion-exchange, and catalytic processes. However, the small pore size of these inorganic frameworks severely limits their applications to small molecules of equivalent micropore sizes. Owing to these limitations, a myriad of new organic directing agents, aluminosilicate compositions, and reaction variables were put to test in the last three decades in attempts to expand the uniform micropore size and to be able to process and transform much larger molecules. This effort met with limited success until 1991, due mainly to the use of single guest species, such as alkali metal cations, quaternary ammonium cations, or primary amines as framework templates. The recently discovered hexagonal mesoporous molecular sieves MCM-41 (with uniform pore size in the range 13 to 100 Å) by the scientists at Mobil Oil R & D provide a unique opportunity for a logical extension of these applications toward processing and transforming of much larger molecules. The preparation of MCM-41 was originally accomplished by a templating mechanism involving electrostatic interactions between assemblies of positively charged quaternary ammonium cations (S^+) and anionic inorganic species (I^-) as framework precursors. This electrostatic approach has been extended by providing three complementary templating pathways: the charge-reversed S^-I^+ and the counterion (X^- or M^+)-mediated $S^+X^-I^+$ or $S^-M^+I^-$. However, the electrostatic templating pathways afforded mesoporous molecular sieves with limited framework wall thickness, low-degree-of-framework cross-linking (as determined by ^{29}Si MAS NMR), and little or no textural mesoporosity, which does not contribute to improving thermal stability and to accessing the framework-confined mesopores. In addition, due to the strong electrostatic interactions and charge matching, the cationic template is strongly bound to the framework and difficult to recover.

The recently proposed neutral ($S^\circ I^\circ$) templating approach offers distinct advantages as allowing for the preparation of hexagonal mesoporous molecular sieves (HMS) with (1) thicker pore walls, (2) more completely cross-linked framework for improved thermal stability, and (3) small scattering domain size or complementary textural mesoporosity for better access of the framework-confined mesopores, while providing for the effective and environmentally benign recovery and recycling of the neutral template.

Both mesoporous molecular sieves MCM-41 and HMS offer exciting opportunity for the preparation of large-pore analogs of the industrially important microporous catalysts ZSM-5 and TS-1. Just months after the advent of MCM-41 and HMS, we are witnessing the growing number of applications in which the metal-substituted MCM-41 and HMS molecular sieves are undoubtedly to play an important role. Thus the Al-, Ti-, and V-substituted MCM-41 and HMS were shown to exhibit the unique ability to accommodate and perform catalytic alkylation and oxidation of much larger organic molecules, such as 2,4-di-*t*-butylphenol, cholesterol, 2,6-di-*t*-butylphenol, and 1-naphthol. It is obvious that the invention of mesoporous molecular sieves will have an important impact on industrial heterogeneous catalysis. Myriads of new and exciting opportunities for catalytic transformations of large organic molecules are now open for exploration. We have no doubt that the day is not far off when their industrial application will be considered a common practice.

Lamellar solids, particularly layered silicates and, especially, smectite clays, can readily be transformed into nanoporous derivatives by intercalating robust cations that function as molecular pillars between the host layers. Cations that act as precursors to molecular metal oxide pillars are especially promising materials for adsorption and chemical catalysis, owing to their thermal stability and Brønsted and Lewis acidity. Access to the gallery surfaces of most of the pillared lamellar solids prepared to date depends mainly on the lateral separation of the pillaring aggregate, not the gallery height. It appears that there is as yet no unequivocal example of a cage effect for a metal oxide pillared clay. This is not surprising, because cage effects have been difficult to prove even with crystalline zeolites. The pore distribution is much broader in pillared clays than in zeolites, and this will certainly make it even more difficult to provide unequivocal evidence for shape selectivity. Nevertheless, pillared clays are very reactive solid acid catalysts with pore structures that are readily accessible to molecules that are too large to be accommodated by conventional microporous zeolites and molecular sieves. Gallery templated synthesis of layered heterostructures is a newly discovered alternative approach to the design of nanoporous materials based on lamellar precursors. Pillared clays and their PCH relatives are complementary to mesostructured metal oxides in terms of thermal stability, acidity, and pore size distribution. Thus these distinguishable classes of materials are likely to meet different materials needs.

Future studies of both mesostructured metal oxides and porous lamellar solids are likely to focus in part on improving thermal stability for high-temperature catalytic applications. The advances made in pillared rectorites illustrates the type of improved thermal performance that can be achieved by structural modifications. These clays have the reactivity and selectivity needed to improve the conversion of heavy crude oils to higher-value petroleum products. The immobilization of biological molecules, especially enzymes and metallomacrocyclic complexes, on nanoporous solids is another area that should attract increasing attention. We may also look to the possible use of these materials in environmental applications, as adsorbents and separations membranes, for example. The long-range possibilities for the creative use of nanoporous solids in many other materials areas is most promising, encompassing such diverse applications as sensors, solar energy conversions, and composite design.

ACKNOWLEDGMENTS

The support of much of the research presented in this review by NSF Grants CHE-9224102 and DMR-8903579 and by NIEHS Grants ESO4911B and ESO4911C is gratefully acknowledged.

REFERENCES

1. J. C. P. Broekhoff, B. G. Linsen, in *Physical and Chemical Aspects of Adsorbents and Catalysts*, B. G. Linsen, Ed., Academic Press, London, 1970, p. 1.
2. K. S. W. Sing et al., *Pure Appl. Chem.* **57**, 603 (1985).
3. D. M. Ruthven, *Chem. Eng. Prog.* **2**, 42 (1988).
4. H. G. Karge, J. Weitkamp, Eds., *Zeolites as Catalysts, Sorbents and Detergents Builders*, Studies in Surface Science and Catalysis, Vol. 46, Elsevier, Amsterdam, 1989; M. E. Davis, *Chem. Ind.* **1992**, 137 (1992).
5. H. E. Bergna, Ed., *The Colloid Chemistry of Silica*, Advances in Chemistry Series, Vol. 234, American Chemical Society, Washington, DC, 1994; K. Wefers, C. Misra, *Oxides and Hydroxides of Aluminum*, Alcoa Technical Paper 19, Alcoa Laboratories, Pittsburgh, PA, 1987, p. 52.

6. C. Roger, M. J. Hampden-Smith, D. Schaefer, G. B. Beaucage, *J. Sol-Gel Sci. Technol.* **2**, 67 (1994).
7. E. M. Flanigen, in *Proceedings of the 5th International Conference on Zeolites*, L. V. Rees, Ed., Heyden, London, 1980, p. 760; E. M. Flanigen et al., *Nature* **271**, 512 (1978).
8. D. E. W. Vaughan, *Chem. Eng. Prog.* **85**, 25 (1988).
9. R. M. Barrer, *Zeolites and Clay Minerals as Sorbents and Molecular Sieves*, Academic Press, London, 1978, p. 34.
10. D. W. Breck, *Zeolite Molecular Sieves: Structure, Chemistry and Use*, Wiley, Chichester, West Sussex, England, 1974.
11. W. M. Meier, D. H. Olson, *Atlas of Zeolite Structure Types*, 3rd ed., Butterworth-Heinemann, London, 1992.
12. R. M. Barrer, *J. Chem. Soc.* **1948**, 127 (1948).
13. R. M. Barrer, P. J. Denny, *J. Chem. Soc.* **1961**, 971 (1961).
14. R. M. Barrer, *Zeolites* **1**, 130 (1981).
15. B. M. Lok, T. R. Cannan, C. A. Messina, *Zeolites* **3**, 282 (1983).
16. M. E. Davis, R. F. Lobo, *Chem. Mater.* **4**, 756 (1992); R. L. Lobo, S. I. Zones, M. E. Davis, *J. Incl. Phenom. Mol. Recog. Chem.* **21**, 47 (1995).
17. H. Gies, B. Marler, *Zeolites* **12**, 42 (1992).
18. P. A. Jacobs, J. A. Martens, *Synthesis of High-Silica Aluminosilicate Zeolites*, Studies in Surface Science and Catalysis, Vol. 33, B. Delmon, J. T. Yates, Eds., Elsevier, Amsterdam, 1987.
19. K. R. Franklin, B. M. Lowe, in *Zeolites: Facts, Figures, Future*, Studies in Surface Science and Catalysis, Vol. 49, P. A. Jacobs, R. A. van Santen, Eds., Elsevier, Amsterdam, 1989, p. 174.
20. S. I. Zones, M. M. Olmstead, D. S. Santilli, *J. Am. Chem. Soc.* **114**, 4195 (1992).
21. S. L. Burkett, M. E. Davis, *J. Phys. Chem.* **98**, 4647 (1994).
22. R. von Balmos, J. B. Higgins, *Zeolites* **10**, 313 (1990).
23. M. K. Rubin, E. J. Rosinski, C. J. Plank, U.S. patent 4,151,189 (1979).
24. L. D. Rollmann, E. W. Valyocsik, U.S. patent 4,108,881 (1978).
25. R. A. Hearmon, A. Stewart, *Zeolites* **10**, 608 (1990).
26. R. M. Milton, U.S. patent 2,882,243 (1959).
27. R. M. Milton, U.S. patent 2,882,244 (1959).
28. C. Ho, C. B. Ching, D. M. Ruthven, *Ind. Eng. Chem. Res.* **26**, 1407 (1987).
29. P. K. Shrivastava, R. Prakosh, *J. Sci. Res.* **11**, 13 (1989).
30. E. Roland, in *Zeolites as Catalysts, Sorbents and Detergents Builders*, Studies in Surface Science and Catalysis, Vol. 46, H. G. Karge, J. Weitkamp, Eds., Elsevier, Amsterdam, 1989, pp. 645–659.
31. A. Corma, A. Martinez, *Adv. Mater.* **7**, 137 (1995).
32. P. B. Weisz, V. J. Frillette, R. W. Maatman, E. B. Mower, *J. Catal.* **1**, 307 (1962).
33. M. E. Davis, *Acc. Chem. Res.* **26**, 111 (1993).
34. S. T. Wilson et al., *J. Am. Chem. Soc.* **104**, 1146 (1982).
35. M. E. Davis et al., *Nature* **331**, 698 (1988).

36. M. Estermann et al., *Nature* **352**, 320 (1991).
37. Q. Huo et al., *J. Chem. Soc. Chem. Commun.* **1992**, 875 (1992).
38. V. Soghmonian, Ch. Qin, R. Haushalter, J. Zubieta, *Angew. Chem. Int. Ed. Engl.* **32**, 610 (1993).
39. M. J. Annen et al., *J. Phys. Chem.* **95**, 1380 (1991).
40. A. Merrouche et al., *Zeolites* **12**, 226 (1992).
41. G. Lagaly, K. Beneke, A. Weiss, *Am. Mineral.* **60**, 642 (1975); K. Beneke, G. Lagaly, *Am. Mineral.* **68**, 818 (1983).
42. T. Yanagisawa, T Shimizu, K. Kuroda, C. Kato, *Bull. Chem. Soc. Jpn.* **63**, 988 (1990).
43. J. S. Beck et al., WO patent 91/11390 (1991).
44. J. S. Beck , U.S. patent 5,057,296 (1991).
45. J. L. Casci, in *Advanced Zeolite Science and Applications*, Studies in Surface Science and Catalysis, Vol. 85, J. C. Jansen, M. Stöcker, H. G. Karge, J. Weitkamp, Eds., Elsevier, Amsterdam, 1994, pp. 329–355.
46. J. S. Beck et al., *J. Am. Chem. Soc.* **114**, 10834 (1992); C. T. Kresge et al., *Nature* **359**, 710 (1992).
47. V. Luzzati, in *Biological Membranes*, D. Chapman, Ed., Academic Press, San Diego, CA, 1968, pp. 71–123.
48. S. Inagaki, Y. Fukushima, K. Kuroda, *J. Chem. Soc. Chem. Commun.* **1993**, 680 (1993).
49. A. Monnier et al., *Science* **261**, 1299 (1993).
50. C.-Y. Chen, S. L. Burkett, H.-X. Li, M. E. Davis, *Microporous Mater.* **2**, 27 (1993); M. E. Davis, C.-Y. Chen, S. L. Burkett, R. F. Lobo, in *Better Ceramics Through Chemistry VI*, A. K. Cheetham, Ed., MRS Symposium Proceedings, Vol. 346, Materials Research Society, Pittsburgh, PA, 1994, p. 831.
51. Q. Huo et al., *Nature* **368**, 317 (1994).
52. P. T. Tanev, T. J. Pinnavaia, *Science* **267**, 865 (1995).
53. M. Fröba et al., in *Advances in Porous Materials*, Sh. Komarneni, J. S. Beck, D. M. Smith, Eds., MRS Symposium Proceedings, Vol. 371, Materials Research Society, Pittsburgh, PA, 1995, pp. 99–104.
54. R. Schmidt, D. Akporiaye, M. Stöcker, O. H. Ellestad, in *Zeolites and Related Microporous Materials: State of the Art 1994*, J. Weitkamp, H. G. Karge, H. Pfeifer, and W. Hölderich, Eds., Elsevier, Amsterdam, 1994, pp. 61–68.
55. C.-Y. Chen, H.-X. Li, M. E. Davis, *Microporous Mater.* **2**, 17 (1993).
56. P. T. Tanev, T. J. Pinnavaia, *Chem. Mater.* **8**, 2068 (1996).
57. N. Coustel, F. Di Renzo, F. Fajula, *J. Chem. Soc. Chem. Commun.* **1994**, 967 (1994).
58. D. D. Whitehurst, U.S. patent 5,143,879 (1992).
59. J. C. Vartuli et al., *Chem. Mater.* **6**, 2317 (1994).
60. G. Horvath, K. J. Kawazoe, *J. Chem. Eng. Jpn.* **16**, 470 (1983).
61. P. T. Tanev, M. Chibwe, T. J. Pinnavaia, *Nature* **368**, 321 (1994).
62. N. S. Gnepr et al., *C.R. Acad. Sci. Ser. 2* **309**, 1743 (1989); B. Chauvin et al., *J. Catal.* **111**, 94 (1988).

63. M. Taramasso, G. Perego, B. Notari, U.S. patent 4,410,501 (1983).
64. G. T. Kokotailo, S. L. Lawton, D. H. Olson, *Nature* **272**, 437 (1978).
65. D. R. Huybrechts, L. De Bruycker, P. A. Jacobs, *Nature* **345**, 240 (1990).
66. A. Esposito, C. Neri, F. Buonomo, U.S. patent 4,480,135 (1984).
67. C. Neri, A. Esposito, B. Anfossi, F. Buonomo, Eur. patent 100,119 (1984).
68. A. Esposito, M. Taramasso, C. Neri, F. Buonomo, Brit. patent 2,116,974 (1985); A. Tangaraj, A. Kumar, P. Ratnasamy, *Appl. Catal.* **57**, L1 (1990).
69. H. R. Sonawane et al., *J. Chem. Soc. Chem. Commun.* **1994**, 1215 (1994); S. Gontier, A. Tuel, *Appl. Catal. A* **118**, 173 (1994).
70. G. Bellussi et al., Eur. patent 200,663 (1986).
71. J. S. Reddy, R. Kumar, *J. Catal.* **130**, 440 (1991).
72. M. A. Camblor, A. Corma, A. Martinez, J. Perez-Pariente, *J. Chem. Soc. Chem. Commun.* **1992**, 589 (1992).
73. M. W. Anderson et al., *Nature* **367**, 347 (1994).
74. D. P. Serrano, H.-X. Li, M. E. Davis, *J. Chem. Soc. Chem. Commun.* **1992**, 745 (1992).
75. M. S. Rigutto, H. van Bekkum, *Appl. Catal.* **68**, L1 (1991); P. R. H. Rao, A. V. Ramaswamy, P. Ratnasamy, *J. Catal.* **137**, 225 (1992).
76. A. Corma, M. T. Navarro, J. Perez Pariente, *J. Chem. Soc. Chem. Commun.* **1994**, 147 (1994).
77. P. T. Tanev, T. J. Pinnavaia, unpublished results.
78. A. Sayari, V. R. Karra, J. S. Reddy, I. L. Moudrakovski, in *Advances in Porous Materials*, MRS Symposium Proceedings, Vol. 371, Sh. Komarneni, J. S. Beck, D. M. Smith, Eds., Materials Research Society, Pittsburgh, PA, 1995.
79. K. M. Reddy, I. Moudrakovski, A. Sayari, *J. Chem. Soc. Chem. Commun.* **1994**, 1059 (1994).
80. C. B. Khouw, M. E. Davis, *J. Catal.* **151**, 77 (1995).
81. A. Corma, M. T. Navarro, J. P. Pariente, F. Sanchez, in *Zeolites and Related Microporous Materials: State of the Art 1994*, Studies in Surface Science and Catalysis, Vol. 84, J. Weitkamp, H. G. Karge, H. Pfeifer, W. Hölderich, Eds., Elsevier, Amsterdam, 1994, pp. 69–75.
82. T. J. Pinnavaia, P. T. Tanev, W. Jialiang, W. Zhang, in *Advances in Porous Materials*, MRS Symposium Proceedings, Wol. 371, Sh. Komarneni, J. S. Beck, D. M. Smith, Eds., Materials Research Society, Pittsburgh, PA, 1995, pp. 53–62.
83. J. S. Reddy, A. Dicko, A. Sayari, in *Proceedings of the Division of Petroleum Chemistry*, Anheim, CA, 1995, in press.
84. N. A. Bhore, Q. N. Le, G. H. Yokomizo, U.S. patent 5,134,243 (1992); Q. N. Le, R. T. Thomson, U.S. patent 5,191,144 (1993).
85. Q. N. Le, R. T. Thomson, U.S. patent 5,232,580 (1993).
86. E. Armengol et al., *J. Chem. Soc. Chem. Commun.* **1995**, 519 (1995).
87. K. R. Kloetstra, H. van Bekkum, *J. Chem. Res. (S)* **1995**, 26 (1995).
88. M. S. Whittingham, A. J. Jacobson, Eds., *Intercalation Chemistry*, Academic Press, New York, 1982.
89. G. Lagaly, *Solid State Ion.* **22**, 43 (1986).

90. T. J. Pinnavaia, *Science* **220**, 365 (1983).
91. D. E. W. Vaughan, *ACS Symp. Ser.* **368**, 308 (1988).
92. H. Kim et al., *Phys. Rev. Lett.* **60**, 2168 (1988).
93. R. M. Barrer, *Zeolites and Clay Minerals as Sorbents and Molecular Sieves*, Academic Press, San Diego, CA, 1978, pp. 407–483.
94. G. W. Brindley, G. Brown, Eds., *Crystal Structures of Clay Minerals and Their X-Ray Identification*, Mineralogical Society, London, 1980.
95. R. M. Barrer, D. M. MacLeod, *Trans. Faraday Soc.* **51**, 1290 (1955).
96. R. M. Barrer, *Pure Appl. Chem.* **61**, 1903 (1989).
97. G. Johansson, *Acta Chem. Scand.* **14**, 771 (1960).
98. J. Y. Bottero, J. M. Cases, F. Flessinger, J. E. Poirier, *J. Phys. Chem.* **84**, 2933 (1980).
99. I. V. Mitchell, Ed., *Pillared Lamellar Structures*, Elsevier, New York, 1990.
100. R. Burch, *Catal. Today* **2**, 185 (1988).
101. J. M. Adams, *Appl. Clay Sci.* **2**, 309 (1987).
102. F. Figueras, *Catal. Rev. Sci. Eng.* **30**, 457 (1988).
103. D. Plee, A. Schutz, G. Ponelet, J. J. Fripiat, in *Catalysis by Acids and Bases*, Studies in Surface Science and Catalysis, Vol. 20, B. Imelik, C. Naccache, G. Goudurier, Y. Ben Taarit, J. C. Vedrine, Eds., (Proc. Int. Symp., Villeubanne, Elsevier, Amsterdam, 1985, p. 343.
104. P. Laszlo, *Science* **235**, 1473 (1987).
105. M. L. Occelli, in *Keynotes in Energy-Related Catalysis*, Studies in Surface Science and Catalysis, Vol. 35, S. Kaliaguine, Ed., Elsevier, Amsterdam, 1988, p. 101.
106. M. L. Occelli, S. D. Landau, T. J. Pinnavaia, *J. Catal.* **90**, 256 (1987).
107. D. Plee, F. Borg, L. Gatineau, J. J. Fripiat, *J. Am. Chem. Soc.* **107**, 2362 (1985).
108. R. A. Schoonheydt, H. Leeman, *Clay Miner.* **27**, 253 (1992).
109. C. Doblin, P. Corrigan, T. W. Turney, in *Proceedings of the 10th International Clay Conference*, Australia, in press.
110. R. Molina, A. Vieira-Coelho, G. Poncelet, *Clays Clay Miner.* **40**, 480 (1992).
111. T. J. Pinnavaia, M.-S. Tzou, S. D. Landau, R. H. Raythatha, *J. Mol. Catal.* **27**, 195 (1984).
112. M. L. Occelli, S. D. Landau, T. J. Pinnavaia, *J. Catal.* **104**, 331 (1987).
113. T. Matsuda, H. Nagashima, E. Kikuchi, *Appl. Catal.* **45**, 171 (1988).
114. K. Urabe, H. Sakurai, Y. Izumi, in *Proceedings of the 9th International Congress on Catalysis*, M. J. Philips, M. Ternam, Eds., Chemistry Institute of Canada, Ottawa, 1988, p. 1858.
115. R. A. Schoonheydt, *Introduction to Zeolite Science and Practice*, Studies in Surface Science and Catalysis, H. van Bekkum, E. M. Flanigen, J. C. Jansen, Eds., Elsevier, Amsterdam, 1991, p. 201.
116. D. E. W. Vaughan, in *Perspective in Molecular Sieve Science*, W. H. Flank, T. E. Whyte, Eds., American Chemical Society, Washington, DC, 1988, p. 308.
117. G. Fetter, D. Tichit, L. C. De Ménorval, F. Figuerars, *Appl. Catal.* **65**, L1 (1990).
118. M. S. Bash, E. S. Kikkinides, R. T. Yang, *Ind. Eng. Chem. Res.* **31**, 2181 (1992).

119. M. S. Bash, R. T. Yang, *A.I.Ch.E.J.* **37**, 1357 (1992).
120. A. Gil, M. Montes, *Langmuir* **10**, 291 (1994).
121. Z. Ge, D. Li, T. J. Pinnavaia, *Microporous Mater.* **3**, 165 (1994).
122. F. Tsvetkov, J. White, *J. Am. Chem. Soc.* **110**, 3183 (1988).
123. I. Gameson, W. J. Stead, T. Rayment, *J. Phys. Chem.* **95**, 1727 (1991).
124. K. Suzuki, M. Horio, H. Masuda, T. Mori, *J. Chem. Soc. Chem. Commun.* **1991**, 873 (1991).
125. L. J. Michot, T. J. Pinnavaia, *Chem. Mater.* **4**, 1433 (1992).
126. A. Moini, T. J. Pinnavaia, *Solid State Ion.* **26**, 119 (1988).
127. R. M. Lewis, R. A. Van Santen, U.S. patent 4,637,992 (1987).
128. S. Yamanaka, N. Tatsoo, M. Hattori, *Mater. Chem. Phys.* **17**, 87 (1987).
129. P. B. Malla, S. Komarneni, *Clays Clay Miner.* **41**, 472 (1993).
130. V. C. Farmer, M. J. Adams, A. R. Fraser, F. Palmieri, *Clay Miner.* **18**, 459 (1983).
131. M. J. Adams, *J. Chromatogr.* **188**, 97 (1980).
132. I. D. Johnson, T. A. Werpy, T. J. Pinnavaia, *J. Am. Chem. Soc.* **110**, 8545 (1988).
133. T. A. Werpy, L. J. Michot, T. J. Pinnavaia, *ACS Symp. Ser.* **437**, 120 (1990).
134. S. Miyata, *Clays Clay Miner.* **28**, 50 (1980).
135. W. T. Reichle, *Chemtech* **1986**, 58 (1986).
136. A. deRoy, C. Forano, K. El Malki, J.-P. Besse, in *Synthesis of Microporous Materials*, M. L. Occelli, H. Robson, Eds., Van Nostrand Reinhold, New York, 1992, Vol. 2, pp. 108–169.
137. K. A. Carrado, A. Kostapapas, S. L. Suib, *Solid State Ion.* **26**, 77 (1988).
138. I. J. Park, K. Kuroda, C. Kato, *Solid State Ion.* **42**, 197 (1990).
139. M. T. Pope, *Heteropoly and Isopoly Oxometalates*, Spring-Verlag, New York, 1983.
140. T. Kwon, G. A. Tsigdinos, T. J. Pinnavaia, *J. Am. Chem. Soc.* **110**, 3653 (1988).
141. M. A. Drezdzon, *Inorg. Chem.* **27**, 4628 (1988).
142. T. Kwon, T. J. Pinnavaia, *Chem. Mater.* **1**, 381 (1989).
143. T. Kwon, T. J. Pinnavaia, *J. Mol. Catal.* **74**, 23 (1992).
144. E. Dimotakis, T. J. Pinnavaia, *Inorg. Chem.* **29** 2393 (1990).
145. M. E. Perez, R. Ruano, T. J. Pinnavaia, *Catal. Lett.* **11**, 51 (1991).
146. T. Tatsumi, K. Yamamoto, H. Tajima, H. Tominaga, *Chem. Lett.* **1992**, 815 (1992).
147. R. G. Anthony, R. G. Dosch, in *Preparation of Catalysts V*, Studies in Surface Science and Catalysis, Vol. 63, G. Poncelet, P. A. Jacobs, P. Grange, B. Delmon, Eds., Elsevier, Amsterdam, 1991, p. 637.
148. A. Clearfield, *Comments Inorg. Chem.* **10**, 89 (1990).
149. J. W. Johnson et al., *J. Am. Chem. Soc.* **111**, 381 (1991).
150. A. A. G. Tomlinson, in *Pillared Layered Structures*, I. V. Mitchell, Ed., Elsevier, New York, 1990, p. 91.
151. G. Alberti et al., in *Pillared Layered Structures*, I. V. Mitchell, Ed., Elsevier, New York, 1990, p. 119.
152. G. Cao, T. E. Mallouk, *Inorg. Chem.* **30**, 1434 (1991).

153. S. Yamanaka, M. Hattori, *Inorg. Chem.* **20**, 1929 (1981).
154. D. A. Burwell, M. E. Thompson, *Chem. Mater.* **3**, 730 (1991).
155. W. Schwieger, B. Heidemann, *Rev. Chim. Miner.* **22**, 639 (1985).
156. M. M. Treacy, S. B. Rice, A. J. Jacobson, J. T. Lewandowski, *Chem. Mater.* **2**, 279 (1990).
157. M. B. Dines, R. E. Cooksey, P. C. Griffith, R. H. Lane, *Inorg. Chem.* **22**, 1003 (1983).
158. M. E. Landis et al., *J. Am. Chem. Soc.* **113**, 3189 (1991).
159. J. M. Dailey, T. J. Pinnavaia, *Chem. Mater.* **4**, 855 (1992).
160. A. Galarneau, A. Barodawalla, T. J. Pinnavaia, *Nature* **374**, 529 (1995).
161. G. Fetter et al., *Clays Clay Miner.* **42**, 161 (1994).
162. D. E. W. Vaughan, R. J. Lussier, J. S. Magee, U.S. patents 4,176,090 (1979); 4,248,739 (1981); 4,271,043 (1981).
163. R. J. Lussier, J. S. Magee, D. E. W. Vaughan, *Preprints, 7th Canadian Symposium on Catalysis*, 1980, p. 112.
164. M. L. Occelli, R. M. Tindwa, *Clays Clay Miner.* **31**, 22 (1983).
165. M. L. Occelli, *Ind. Eng. Chem. Prod. Res. Dev.* **22**, 553 (1983).
166. F. Figueras, *Catal. Rev. Sci. Eng.* **30**, 457 (1988).
167. M. Occelli, in *Keynotes in Energy-Related Catalysis*, Studies in Surface Science and Catalysis, Vol. 35, S. Kaliaguine, Ed., Elsevier, Amsterdam, 1988, p. 101.
168. J. Guan, E. Min, Z. Yu, in *Proceedings of the 9th International Congress on Catalysis*, M. J. Philips, M. Ternam, Eds., Chemistry Institute of Canada, Ottawa, 1988, p. 104.
169. M. L. Occelli, in *Preparation of Catalysis V*, Studies in Surface Sciences and Catalysis, Vol. 63, G. Poncelet, P. A. Jacobs, P. Grange, B. Delmon, Eds., Elsevier, Amsterdam, 1991, p. 287.
170. S. M. Bradley, R. A. Kydd, R. Yamdagni, C. A. Fyfe, in *Synthesis of Microporous Materials*, Vol. 2, M. L. Occelli, H. E. Robson, Eds., Van Nostrand Reinhold, New York, 1992, p. 13.
171. F. Gonzàlez et al., *Inorg. Chem.* **31**, 727 (1992).
172. B. Thomas, H. Görz, S. Schönherr, *Z. Chem.* **27**, 183 (1987).
173. S. M. Bradley, R. A. Kydd, R. Yamdagni, *Magn. Reson. Chem.* **28**, 746 (1990).
174. A. Bellaloui, D. Plee, P. Meriaudeau, *Appl. Catal.* **63**, L7 (1990).
175. E. Kikuchi, T. Matsuda, *Catal. Today* **2**, 297 (1988).
176. E. Kikuchi, T. Matsuda, H. Fujiki, Y. Morita, *Appl. Catal.* **11**, 331 (1984).
177. M. Kojima, R. Hartford, C. T. O'Connor, *J. Catal.* **128**, 487 (1991).
178. A. Schutz et al., in *Proceedings of the International Clay Conference*, Denver, 1985, L. G. Schults, H. van Olphen, F. A. Mumpton, Eds., Clay Minerals Society, Bloomington, IN, 1987, p. 305.
179. B. M. Lok, B. K. Marcus, C. L. Angell, *Zeolites* **6**, 185 (1986).
180. T. Mori, K. Suzuki, *Chem. Lett.* **1989**, 2165 (1989).
181. K. Urabe, H. Sakurai, Y. Izumi, *J. Chem. Soc. Chem. Commun.* **1986**, 1074 (1986).

182. K. Urabe, H. Sakurai, Y. Izumi, in *Proceedings of the 9th International Congress on Catalysis*, M. J. Philips, M. Ternam, Eds., Chemistry Institute of Canada, Ottawa, 1988, p. 1858.
183. M. Horio, K. Suzuki, H. Masuda, T. Mori, *Appl. Catal.* **72**, 109 (1991).
184. G. S. Lee, J. J. Mag, S. C. Rocke, J. M. Garcés, *Catal. Lett.* **2**, 197 (1989).
185. J.-R. Butruille, T. J. Pinnavaia, *Catal. Today* **14**, 141 (1992).
186. L. J. Michot, T. J. Pinnavaia, *Chem. Mater.* **4**, 1433 (1992).
187. J. R. McCauley, Internat. patent application PCT/US88/00567 (1988).
188. J.-R. Butruille, Ph.D. thesis, Michigan State University, 1992.

CHAPTER 9

Molecular Precursor Routes to Inorganic Solids

JACQUES LIVAGE, C. SANCHEZ, and F. BABONNEAU
Chemie de la Matière Condensé, Université Pierre et Marie Curie
75252 Paris Cedex 05, France

9.1 INTRODUCTION

Ceramic materials are currently synthesized via the direct reaction, in the solid state, of a mixture of powders. For the reaction to proceed at a reasonable rate, high mobility of the reactants and maximum contact surface between the reacting particles are desirable. The rates of these processes are enhanced by higher temperature and finer particle size. Both thermodynamic and kinetic factors are important in the fabrication of materials from powders. Metastable phases are difficult to obtain. Solid-state reactions do not allow one to control the size and morphology of solid particles. They usually lead to polydispersed powders more than 10 μm in diameter.

New synthesis methods, based on molecular precursors, have been developed during the last few decades. The sol-gel process typically involves the hydrolysis and condensation of metal alkoxides $M(OR)_z$. This leads to the formation of a hydrous oxide network. Thermal decomposition yields a mixture of finely divided and intimately mixed metal oxides. Long-range diffusion is no longer required, products may form at much lower temperatures, sometimes even at room temperature. Moreover, metastable phases can be prepared which remain kinetically stable to quite high temperatures.

Wet chemistry has already been used for a long time for the preparation of oxide powders. The first synthesis of silica from silicon alkoxide was described by Ebelmen more than 150 years ago.[1] The commercial production of sol-gel coatings onto flat glass began in the early 1960s.[2] Wet chemistry

Chemistry of Advanced Materials: An Overview, Edited by Leonard V. Interrante and Mark J. Hampden-Smith.
ISBN 0-471-18590-6

exhibits many advantages for the production of metastable phases or monodispersed powders. Sol-gel chemistry allows the powderless processing of glasses and ceramics. Thin films or fibers can be produced directly from the solution by techniques such as dip coating or spin drawing. More recently, hybrid organic–inorganic materials have been made via the sol-gel route.[3–5] They open a wide range of possibilities and have led to the development of new fields in materials science such as sol-gel optics and molecular composites. Polymeric precursors were proposed by Yajima about 20 years ago for the production of SiC fibers.[6] This process is now being extended to other nonoxide ceramics, such as silicon nitride and boron nitride.

9.2 SOL-GEL CHEMISTRY OF METAL ALKOXIDES

9.2.1 Synthesis of Metal Alkoxides

The sol-gel synthesis of oxide materials is based on the hydrolysis and condensation of metal alkoxides $M(OR)_z$, where M is a metal atom having the oxidation state z and R usually an alkyl group ($R = CH_3, C_2H_5, \ldots$). The chemistry of metal alkoxides has been described intensively in the book by Bradley et al.[7] Many metal alkoxides ($M = Si, Al, Ti, \ldots$) are now commercially available. They are usually synthesized via the reaction of metal chlorides with alcohol, but due to the high reactivity of most alkoxides toward hydrolysis, absolutely anhydrous conditions are required for the successful preparation of metal alkoxides. An exothermic reaction occurs when some metal chlorides are mixed with aliphatic alcohols:

$$MCl_z + hROH \longleftrightarrow MCl_{z-h}(OR)_h + hHCl \tag{1}$$

Such a reaction, called *solvolysis* or *alcoholysis*, could be described as a nucleophilic substitution. The process is initiated by solvation that weakens the O—H bond, followed by the deprotonation of the ROH molecule. Chloroalkoxides and finally, alkoxides $M(OR)_z$ are obtained when the reaction is driven to completion:

$$\overset{\diagup Cl}{M} + \overset{H \diagdown}{}RO \longrightarrow \overset{\diagup Cl \;\; H \diagdown}{M \cdots\cdots\cdots OR} \longrightarrow M{-}OR + HCl \tag{2}$$

The proton is associated with a leaving Cl^- group, and the stability of metal chlorides in alcoholic solutions mainly depends on the polarity of the M—Cl bond. The extent of alcoholysis reactions mainly depend on the electronegativity of the metal atom and the steric hindrance of the alkoxy group.

In the case of less electronegative elements, the metal chloride is not dissociated. Only solvated species such as $LaCl_3 \cdot 3i\text{-}PrOH$, $CeCl_3 \cdot 3i\text{-}PrOH$, or $ThCl_4 \cdot 4EtOH$ are formed and can often be recrystallized from the solution.

For the more electronegative elements such as boron, silicon, or phosphorus, alcoholysis proceeds to completion. This reaction is currently used for the industrial preparation of silicon alkoxides such as tetraethylorthosilicate (TEOS) or tetramethylorthosilicate (TMOS).

$$SiCl_4 + 4EtOH \longrightarrow Si(OEt)_4 + 4HCl \quad (3)$$

The steric hindrance of bulky alkoxy groups can prevent the formation of metal alkoxides. The reaction of $SiCl_4$ with primary and secondary alcohols gives alkoxides, whereas with tertiary alcohols, orthosilisic acid and alkyl chlorides are formed almost quantitatively, as follows:

$$SiCl_4 + 4t\text{-BuOH} \longrightarrow Si(OH)_4 + 4t\text{-BuCl} \quad (4)$$

The chemical reactivity of metal chlorides toward alcohol decreases with the electronegativity of the metal. Most transition-metal chlorides therefore undergo only partial alcoholysis, leading to chloroalkoxides $MCl_{z-h}(OR)_h$. A gradation in reactivity is observed with the tetravalent halides, $SiCl_4 >$ $TiCl_4 > ZrCl_4 > ThCl_4$. The reaction goes to completion with $SiCl_4$, whereas other MCl_4 chlorides lead to solvates such as $TiCl_2(OEt)_2 \cdot EtOH$, $ZrCl_3(OEt) \cdot EtOH$, and $ThCl_4 \cdot 4EtOH$. Therefore, the reaction has to be driven to completion by adding a base such as ammonia, pyridine, or sodium alkoxide:

$$TiCl_4 + 4EtOH + 4NH_3 \longrightarrow Ti(OEt)_4 + 4NH_4Cl \quad (5)$$

$$TiCl_4 + 4NaOEt \longrightarrow Ti(OEt)_4 + 4NaCl \quad (6)$$

NH_4Cl or NaCl precipitates from the solution and can be removed by filtration.

The reaction of metal chloride with alcohols is not the only way to synthesize alkoxides. Other inorganic compounds can be used as precursors for the synthesis of metal alkoxides. Oxides and hydroxides of the nonmetals and high-valent metal ions such as B^{III}, Ge^{IV}, P^{V}, or V^{V} behave as oxyacids and react with alcohols to form esters and water:

$$M(OH)_z + zROH \longleftrightarrow M(OR)_z + zH_2O \quad (7)$$

$$MO_x + 2xROH \longleftrightarrow M(OR)_{2x} + xH_2O \quad (8)$$

Water has to be removed by adding organic solvents such as benzene, which form azeotropes with water. These azeotropes are progressively eliminated upon reflux by using a Dean–Stark apparatus or simply by distillation of the solvent. When heated under reflux, P_2O_5 reacts with EtOH to give a mixture of mono- and diesters:

$$P_2O_5 + 3EtOH \longleftrightarrow PO(OH)(OEt)_2 + PO(OH)_2(OEt) \qquad (9)$$

Vanadium alkoxides are currently synthesized from V_2O_5, or even better from vanadates, as follows:

$$V_2O_5 + 6ROH \longleftrightarrow 2VO(OR)_3 + 3H_2O \qquad (10)$$

The reaction is performed in a solvent such as benzene. Water is eliminated by azeotropic distillation and the alkoxide is purified by distillation under reduced pressure.

Alkaline and alkaline-earth metals are oxidized directly by alcohols:

$$Na + MeOH \longrightarrow NaOMe + \tfrac{1}{2}H_2 \qquad (11)$$

This reaction can be used for the in situ synthesis of such alkoxides directly in the precursor solution for the synthesis of heterometallic alkoxides.

Russian scientists have developed the electrochemical synthesis of metal alkoxides. An anode of a given metal is electrochemically dissolved into an alcohol solvent in the presence of an electroconductive additive.[8] The process involves the following reactions:

$$M \longrightarrow M^{z+} + ze^- \quad \text{(anode)} \qquad (12)$$

$$zROH + ze^- \longrightarrow zOR^- + zH^+ \quad \text{(cathode)} \qquad (13)$$

$$M^{z+} + zOR^- \longrightarrow M(OR)_z \qquad (14)$$

Alkali metal chlorides are added to the batch to get yields as high as 95%. A pilot plant has been built in Tula (Russia) to produce about 2 tons/yr, and many alkoxides (Ti, Zr, Nb, Ta, Y, Sn, Cu, Ge, Mo, W) are now commercially available.

9.2.2 Molecular Structure of Metal Alkoxides

The oxidation state Z of metal ions is often smaller than their usual coordination number N in oxide compounds. Therefore, most metal alkoxides are not coordinatively saturated, and coordination expansion is a general tendency of the sol-gel chemistry of metal alkoxides $M(OR)_z$. Positively charged metal atoms $M^{\delta+}$ tend to increase their coordination number by using their vacant orbitals to accept electrons from nucleophilic ligands, leading to the formation of oligomers or solvates.[7]

Oligomerization The electronic structure of the highest occupied molecular orbitals of the OR^- group is $\ldots \sigma^2, \pi^4$ (where the π-orbitals are nonbonding). Alkoxy groups then behave as σ and π donors, allowing the

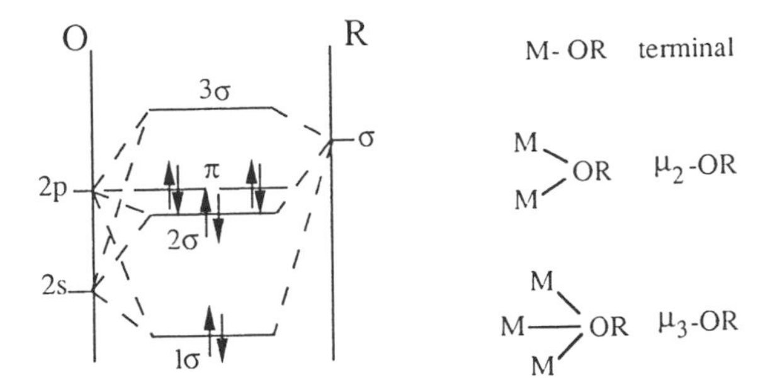

Figure 9.1 Molecular orbitals and electronic structure of the OR^- radical.

formation of μ_2-OR or μ_3-OR bridges (Fig. 9.1). In neat alkoxides or in nonpolar solvents, coordination expansion occurs via the nucleophilic addition of other alkoxide molecules and the formation of OR bridges. Most metal alkoxides therefore exhibit oligomeric molecular structures, $[M(OR)_z]_n$. As a general rule, metal alkoxides usually adopt the smallest possible structural unit consistent with all metal atoms attaining a higher coordination number. A mixture of several oligomers is often observed, depending on parameters such as concentration, temperature, nature of the solvent, oxidation state of the metal atom, or the steric hindrance of alkoxide groups.

The degree of oligomerization of metal alkoxides increases with the size of the metal atom, its electropositive character, and the $N - Z$ difference. It decreases when the electronegativity of the metal atom increases. Some data deduced from cryoscopic and ebullioscopic measurements are reported in Table 9.1.

The molecular complexity of metal alkoxides also depends on the steric hindrance of the alkoxide groups. Most methanolates $M(OCH_3)_z$ are insoluble polymers, whereas bulky secondary or tertiary alkoxy groups tend to prevent oligomerization. Oligomeric species $[Ti(OEt)_4]_n$ have been evidenced for titanium ethoxide, both in the solid state ($n \approx 4$) and in solution ($n \approx 2, 3$), whereas titanium isopropoxide Ti(O-*i*-Pr)$_4$ remains monomeric (Fig. 9.2).

Most alkoxides of monovalent (Li^+, Na^+, K^+) and bivalent (Cu^{2+}, $Zn^{2+}, \ldots$) metals give insoluble polymeric derivatives. Soluble low-valent

TABLE 9.1 Average Degree of Oligomerization *n* of Neat Metal Alkoxides

	Alkoxide			
	$Si(OEt)_4$	$Ti(OEt)_4$	$Zr(OEt)_4$	$Th(OEt)_4$
N	4	6	8	8
$N - Z$	0	2	4	4
n	1	2.9	3.6	6

Source: Ref. 7.

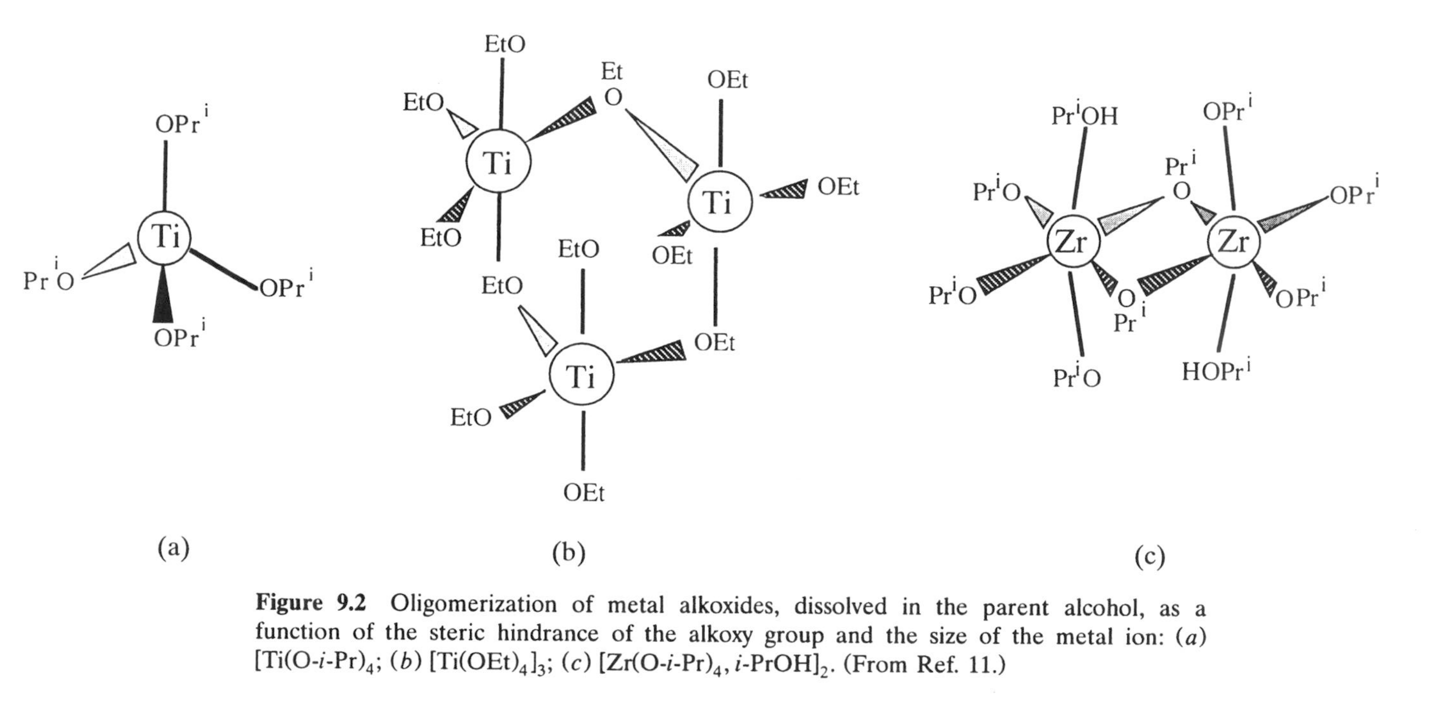

Figure 9.2 Oligomerization of metal alkoxides, dissolved in the parent alcohol, as a function of the steric hindrance of the alkoxy group and the size of the metal ion: (*a*) $[Ti(O\text{-}i\text{-}Pr)_4$; (*b*) $[Ti(OEt)_4]_3$; (*c*) $[Zr(O\text{-}i\text{-}Pr)_4, i\text{-}PrOH]_2$. (From Ref. 11.)

metal alkoxides can be synthesized by replacing the simple alkoxy group OR by polydentate ligands such as 2-methoxyethanol $CH_3OC_2H_4OH$. Such a solvent is used currently to prevent polymerization and obtain soluble molecular precursors, especially with low-valent metal cations such as Cu^{2+}. They have been widely used for the sol-gel synthesis of advanced ceramics such as the high-T_c superconductors $YBa_2Cu_3O_{7-x}$.[9]

Solvation Metal alkoxides are not miscible with water and have to be dissolved in a cosolvent prior to hydrolysis. Coordination expansion could therefore also occur via solvation. Solvate formation is often observed when alkoxides are dissolved in their parent alcohol. The stability of such solvates increases with the size and decreases with the electronegativity of the metal. At room temperature, monomeric $Ti(O\text{-}i\text{-}Pr)_4$ is not solvated, whereas solvated dimeric species $[Zr(O\text{-}i\text{-}Pr)_4(i\text{-}PrOH)]_2$ are formed in *i*-PrOH solutions and $[Zr(O\text{-}n\text{-}Pr)_4]_n$ oligomers ($n \leq 4$) are obtained in nonpolar solvents such as cyclohexane.

Despite their thermodynamic stability, M—OR bonds are rather labile. Alcohol interchange reactions occur when metal alkoxides are dissolved in a nonparent alcohol. These reactions lead to the formation of a whole range of mixed alkoxides. Five different tetrahedral $Si(OEt)_{4-x}(O\text{-}i\text{-}Pr)_x$ species ($0 \leq x \leq 4$), for instance, are clearly evidenced by ^{29}Si NMR when $Si(OEt)_4$ is mixed with *i*-PrOH. This reaction is quite slow with silicon alkoxides; it takes few hours and is generally catalyzed by protons. It is much faster with other alkoxides, such as $Ti(O\text{-}i\text{-}Pr)_4$, and can occur within few minutes in the presence of protons. The choice of the solvent is therefore an important chemical parameter for controlling the kinetics of hydrolysis and condensation reactions. High oligomers in which the metal atom already has a high coordination number are much less reactive than monomeric alkoxides.

Oxoalkoxides The self-condensation of metal alkoxides can occur via ether elimination between two alkoxy groups, leading to the formation of oxo-bridges:[10]

$$\text{M—OR} + \text{RO—M} \longrightarrow \text{M—O—M} + \text{ROR} \tag{15}$$

Such a reaction requires more activation energy than the formation of μ-OR bridges. It can occur upon heating, and μ-oxo complexes are often formed during the purification of metal alkoxides via distillation. Condensation via ether elimination is favored by the smaller size of the μ-oxo ligands and their ability to exhibit higher coordination numbers, up to six, allowing the formation of more compact oligomers (Fig. 9.3).

Oxoalkoxides can also be synthesized from metal chlorides or carboxylates:

$$\text{M—OR} + \text{M—Cl} \longrightarrow \text{M—O—M} + \text{RCl} \tag{16}$$

$$\text{M—OR} + \text{M—OOCR}' \longrightarrow \text{M—O—M} + \text{R}'\text{COOR} \tag{17}$$

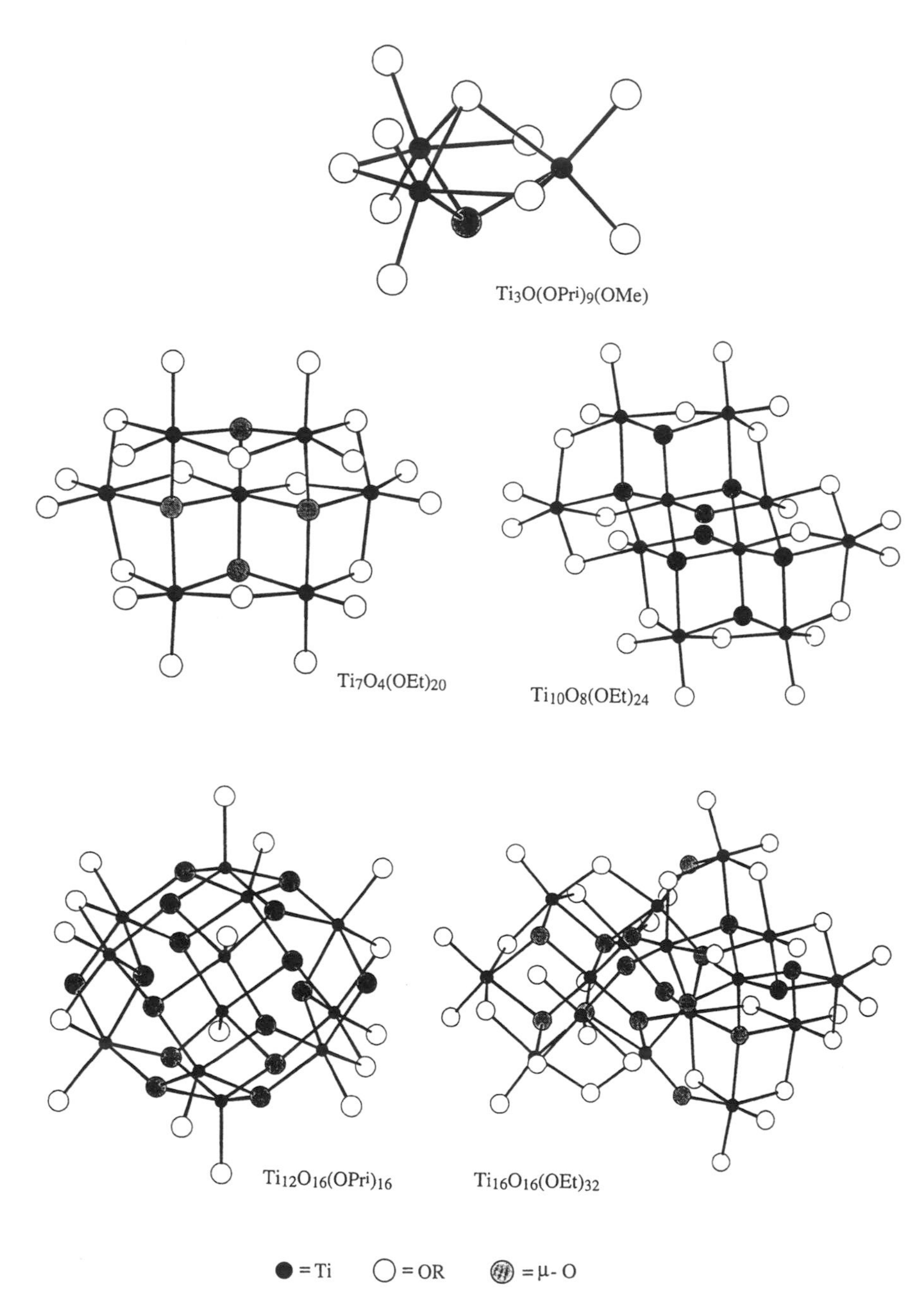

Figure 9.3 Molecular structure of some titanium oxoalkoxides.

The oxo bridge is quite strong and oxoalkoxides are more stable and less reactive than the corresponding alkoxides. They are formed primarily with large and electropositive metals such as $Pb_4O(OEt)_6$, $Pb_6O_4(OEt)_4$, $Bi_4O_2(OEt)_8$, or $Y_5O(O\text{-}i\text{-}Pr)_{13}$. Liquid $NbO(OEt)_3$, for instance, crystallizes slowly in a dry atmosphere, giving $Nb_8O_{10}(OEt)_{20}$ crystals.

9.2.3 Hydrolysis and Condensation of Metal Alkoxides

Oligomerization reactions usually don't lead to the formation of very large polymers or solid phases. This is due primarily to the steric hindrance of the alkoxy group. Condensation reactions then have to be initiated by small and highly reactive OH groups, and sol-gel chemistry is based on the hydrolysis and condensation of metal alkoxides $M(OR)_z$. These nucleophilic reactions can be described as the substitution of alkoxy ligands by hydroxylated species (XOH) as follows:[11]

$$M(OR)_z + xXOH \longleftrightarrow [M(OR)_{z-x}(OX)_x] + xROH \qquad (18)$$

where X stands for hydrogen (hydrolysis), a metal atom (condensation), or even an organic or inorganic ligand (complexation).

These reactions follow an associative S_N2 mechanism:

$$\begin{array}{c}\text{H}\\ \backslash\\ \text{O}^{\delta-}\\ /\\ \text{X}\end{array} + M^{\delta+}—O^{\delta-}—R \rightleftharpoons \begin{array}{c}\text{H}^{\delta+}\\ \backslash\\ \text{O}\\ /\\ \text{X}\end{array}\!\!—M—O^{\delta-}—R \rightleftharpoons XO—M—\begin{array}{c}\text{H}^{\delta+}\\ /\\ \text{O}\\ \backslash\\ \text{R}\end{array}$$

$$\rightleftharpoons XO—M + ROH \qquad (19)$$

The first step of these reactions is the nucleophilic addition of negatively charged $HO^{\delta-}$ groups onto the positively charged metal atom $M^{\delta+}$, leading to an increase of the coordination number of the metal atom in the transition state. The positively charged proton is then transferred toward an alkoxy group and the protonated ROH ligand is finally removed.

The chemical reactivity of metal alkoxides toward hydrolysis and condensation depends mainly on the electronegativity of the metal atom and its ability to increase its coordination number N (i.e., on its size) (Table 9.2). Silicon alkoxides are rather unreactive, while titanium alkoxides are very sensitive to moisture. The hydrolysis rate of $Ti(OEt)_4$ ($k_h \approx 10^{-3}$ M^{-1} s) is about five orders of magnitude greater than that of $Si(OEt)_4$ ($k_h \approx 5.10^{-9}$ M^{-1} s).[2b] Gelation times of silicon alkoxides are on the order of days, whereas titanium alkoxides give gels within a few seconds or minutes.

Despite the fact that the electronegativity of Sn is larger than that of Si, $Sn(O\text{-}i\text{-}Pr)_4$ hydrolyzes very fast because its coordination number is typically six. Most alkoxides are very sensitive to moisture and must be handled with care under a dry atmosphere; otherwise, precipitation occurs as soon as

TABLE 9.2 Hydrolysis Rate of Metal Alkoxides as a Function of the Electronegativity χ, Ionic Radius r, and Maximum Coordination Number N of the Metal

Alkoxide	χ_M	r (Å)	N	Hydrolysis Rate
$Si(O\text{-}i\text{-}Pr)_4$	1.74	0.40	4	Slow
$Ti(O\text{-}i\text{-}Pr)_4$	1.32	0.64	6	Fast
$Sn(O\text{-}i\text{-}Pr)_4$	1.89	0.69	6	Fast
$PO(O\text{-}i\text{-}Pr)_3$	2.11	0.34	4	No hydrolysis
$VO(O\text{-}i\text{-}Pr)_3$	1.56	0.59	6	Fast

TABLE 9.3 Gelation Time for Some $Si(OR)_4$ Precursors

Alkoxide	Gelation TIme (h)
$Si(OMe)_4$	44
$Si(OEt)_4$	242
$Si(OBu)_4$	550

water is present. Alkoxides of highly electronegative elements such as $PO(OEt)_3$ cannot be hydrolyzed under ambient conditions, whereas the corresponding vanadium derivatives $VO(OEt)_3$ are readily hydrolyzed into vanadium pentoxide gels in the presence of a large excess of water.

The sensitivity of metal alkoxides toward hydrolysis and condensation depends on the steric hindrance of the alkoxy groups. It decreases when the size of the OR group increases, as shown in Table 9.3 for silicon alkoxides. The choice of the solvent is therefore very important for the sol-gel synthesis of metal oxides. Even with a given precursor, different solvents can lead to different products. The specific surface area of silica gels prepared from $Si(OMe)_4$ and heated to 600°C, for instance, increases from 170 m^2/g to 300 m^2/g and the mean pore diameter decreases from 36 Å to 29 Å when EtOH is used as a solvent instead of MeOH. This is due mainly to alcohol interchange reactions.[12]

Hydrolysis and condensation rates also depend on the molecular structure of the metal alkoxides. In the case of titanium alkoxides, for instance, monomeric precursors such as $Ti(OPr^i)_4$, in which Ti is only four-coordinate, react very quickly with water, leading to the uncontrolled precipitation of polydispersed TiO_2. The reaction is much slower with oligomeric precursors such as $[Ti(OEt)_4]_n$, in which Ti has a higher coordination number. In 1982, a very important paper from Barringer and Bowen showed that spherical monodispersed TiO_2 powders could be produced via the controlled hydrolysis of diluted solutions of $Ti(OEt)_4$ in EtOH.[13]

$Ti(OBu^n)_4$ is currently preferred as a precursor to TiO_2. It does not react too fast with water and can be handled without too much care. This is because O-n-Bu, with four carbon atoms, is the largest alkoxy group that does not prevent oligomerization. It gives mainly trimeric species, $[Ti(O\text{-}n\text{-}Bu)_4]_3$, when dissolved in benzene and dimeric species, $[Ti(O\text{-}n\text{-}Bu)_4, BuOH]_2$, in its parent alcohol. On the other hand, monomeric precursors would be more convenient for the sol-gel synthesis of multicomponent oxides. The perovskite phase $BaTiO_3$ is formed upon heating around 800°C when $[Ti(OEt)_4]_n$ is used as the Ti precursor. This temperature decreases down to 600°C with the monomeric Ti precursor $Ti(O\text{-}i\text{-}Pr)_4$. The barium precursor is the corresponding alkoxide $Ba(OR)_2$ obtained via the reaction of Ba metal with the alcohol.

The formation of condensed species also depends on the hydrolysis ratio $h = H_2O/M$:

1. Molecular clusters are formed when a very small amount of water is used ($h < 1$). Condensation is then governed mainly by the formation of μ-OR and μ-oxo bridges. Molecular oxo-alkoxides such as $Ti_7O_4(OEt)_{20}$ ($h = 0.6$), $Ti_{10}O_8(OEt)_{24}$ ($h = 0.8$), and even $Ti_{16}O_{16}(OEt)_{32}$ ($h = 1$) have been obtained via the controlled hydrolysis of $Ti(OEt)_4$. These compounds have been characterized by x-ray diffraction and ^{17}O NMR. Many examples of such molecular clusters have been reported in the literature.[14]
2. For very large hydrolysis ratios ($h \gg 10$), all alkoxy groups are removed and an oxide network is formed. Moreover, the high dielectric constant of the aqueous medium leads to the acid or base dissociation of the surface OH groups, and the resulting oxide gels are very similar to those obtained from aqueous solutions.
3. Between these two extremes, alkoxy groups are not completely hydrolyzed. They remain bonded to the growing oxide network giving rise to oxopolymers. Such polymeric species are often very suitable precursors for the deposition of thin films. Good adhesion with oxide substrates is obtained via the condensation of the remaining OR ligands with the hydroxyl groups at the surface of the substrate. Organics are then removed upon calcination at around 300°C and a transparent oxide thin film is finally obtained.

Vanadium alkoxides $VO(OR)_3$ exhibit typical behavior. Their stoichiometric hydrolysis would correspond to $h = 3$. Such an hydrolysis ratio leads to orange transparent gels with $VO(O\text{-}n\text{-}Pr)_3$, while polymeric sols are formed with $VO(O\text{-}t\text{-}Am)_3$. In both cases, the vanadium coordination is square pyramidal as in aqueous gels. It appears that as hydrolysis proceeds, the alkoxy ligands become poor leaving groups, so that they cannot be completely removed. The orange gels should then be described as oxo-polymers $[V_2O_{5-x}(O\text{-}n\text{-}Pr)_x]_n$ ($x \approx 0.4$). Alkoxy ligands remain bonded to the oxide network and are only removed upon heating at around 300°C. SAXS experiments suggest that these orange gels are made up of highly branched polymeric species.[15] Red gels are obtained for large hydrolysis ratios ($h \approx 50$). All alkoxy groups are then removed, and these gels are very similar to the V_2O_5, nH_2O gels obtained from aqueous solutions. They are made of ribbon-like particles and the acidic dissociation of adsorbed water molecules occurs, giving rise to molecular decavanadic species that have been evidenced by ^{51}V NMR.

9.2.4 Chemically Controlled Condensation

Catalyzed Hydrolysis and Condensation Si^{IV} remains fourfold coordinated ($N = Z = 4$) all the way from the alkoxide precursor $Si(OR)_4$ to the oxide SiO_2. Coordination expansion does not occur and silicon alkoxides $Si(OR)_4$ are always monomeric. The electronegativity of Si is rather high ($\chi = 1.74$) and silicon alkoxides are not very sensitive toward hydrolysis and condensation. As shown before, gelation takes several days when water is added to an alcoholic solution of $Si(OR)_4$. Therefore, the hydrolysis and condensation rates of silicon alkoxides are enhanced by acid or base catalysis (Table 9.4).

Acids protonate negatively charged alkoxide ligands and increase the reaction kinetics by producing better leaving groups and eliminating the requirement for proton transfer within the transition state. Basic catalysis provides better nucleophilic OH^- groups for hydrolysis, whereas deprotonated silanol groups $Si{-}O^-$ enhance condensation rates. Both hydrolysis and condensation rates are increased via catalysis; however, acid catalysis mainly increases hydrolysis rates, whereas basic catalysis enhances condensation. In the presence of an excess of acid, condensation can even be prevented, as shown by ^{29}Si NMR spectra, which exhibit mainly Q^0 species (Q^n, n gives the number of $Si{-}O{-}Si$ bridges), corresponding to fully hydrolyzed $Si(OH)_4$ monomers. Basic catalysis favors condensation and ^{29}Si NMR spectra show the rapid formation of Q^4 species, which are linked to four other $[SiO_4]$ tetrahedra. This can also be seen when gelation is performed in two steps. Acid water is first added, leading to a clear and

TABLE 9.4 Gelation Time T_g of a Solution of $Si(OEt)_4/EtOH$ (0.5 M)

Catalyst	Initial pH	T_g (h)
No catalyst	5	1000
HF	1.9	12
HCl	0.05	92
HNO_3	0.05	100
AcOH	3.7	72
NH_3	9.9	107

Source: Ref. 16.

stable solution of hydrolyzed species. Increasing the pH above pH ≈ 4 leads to gelation within a few minutes.

Catalysis not only increases reaction rates but, as in aqueous solutions, leads to polymeric species of different shapes. The electron-donating ability of the ligands decreases as OR > OH > O. Therefore, basic catalysis is directed toward the central units of oligomeric species (Q^2 or Q^3 Si atoms) rather than the end groups (Q^1 Si atom), leading to highly branched species. Upon gelation, compact species are formed, giving rise to dense spherical colloids or porous gels (average pore diameter in the range 50 to 100 Å). This process was developed by Stöber for the industrial production of monodispersed silica particles, 0.05 to 2 μm in diameter.[17] Acid-catalyzed condensation is directed preferentially toward the end groups of oligomeric species. Chain polymers are then formed, in agreement with the observation that acid catalysis combined with a low hydrolysis ratio leads to spinnable sols or monolithic gels of low porosity (average pore diameter < 50 Å).

The chemical reactivity of silicon alkoxides can also be increased via nucleophilic activation by chemical species such as dimethylaminopyridine or F^-, which behave as Lewis bases. A pentavalent intermediate is formed with the nucleophilic atoms (N, O, F) that stretches and weakens the Si—OR bonds. The positive charge δ_{Si} increases and the silicon atom becomes more prone toward further nucleophilic attack by H_2O or Si—OH. This nucleophilic activation increases both hydrolysis and condensation rates.

Acid catalysis has been shown to increase hydrolysis and condensation rates. However, it has to be pointed out that for high acid concentrations, the reverse phenomenon is observed: Acids behave as inhibitors toward condensation. The polymerization of $Ti(O\text{-}n\text{-}Bu)_4$ can be controlled kinetically by using acid inhibitors. A TiO_2 precipitate is formed when a mixture of water and n-BuOH is added to the alkoxide, while a gel is obtained in the presence of a large amount of HCl. The net effect of adding an acid is to protonate Ti—OH groups. This slows down the reaction so that long polymeric chains can form.[18]

Chemical Modification of Alkoxide Precursors Because of coordination expansion, most metal alkoxides other than silicon are highly reactive toward hydrolysis and condensation. They must be handled in a dry atmosphere; otherwise, hydrolysis occurs. Therefore, their chemical reactivity has to be decreased to avoid uncontrolled precipitation. This can conveniently be performed via the chemical modification (or complexation) of the molecular precursor.

Metal alkoxides react with hydroxylated compounds XOH such as carboxylic acids or β-diketones, leading to the departure of alkoxy groups as follows:

$$M(OR)_z + x\text{XOH} \longleftrightarrow M(OR)_{z-x}(OX)_x + x\text{ROH} \qquad (20)$$

Monovalent complexing XO^- species (acetates, β-diketonates) often behave as bidentate (bridging or chelating) ligands, giving rise to an increase of the coordination number N. They are more strongly bonded and less hydrolyzable than the alkoxy groups. The new precursor therefore exhibits a higher coordination of the metal atom, a different molecular structure, and a reduced functionality. Its chemical reactivity is strongly modified, and complexed alkoxides are usually much less sensitive toward hydrolysis and condensation. Nucleophilic chemical additives are currently employed to stabilize highly reactive metal alkoxides and control the formation of condensed species.[19]

Titanium *iso*-propoxide reacts with acetic acid ($AcOH = CH_3COOH$). For a 1 : 1 ratio, a slightly exothermic reaction takes place, leading to a clear solution:

$$Ti(O\text{-}i\text{-}Pr)_4 + AcOH \longrightarrow Ti(OAc)(O\text{-}i\text{-}Pr)_3 + i\text{-}PrOH \qquad (21)$$

Acetate behaves as a bridging ligand. The coordination of titanium increases from four to six and oligomeric $[Ti(O\text{-}i\text{-}Pr)_3(OAc)]_n$ species are formed ($n = 2$ or 3) (Fig. 9.4).

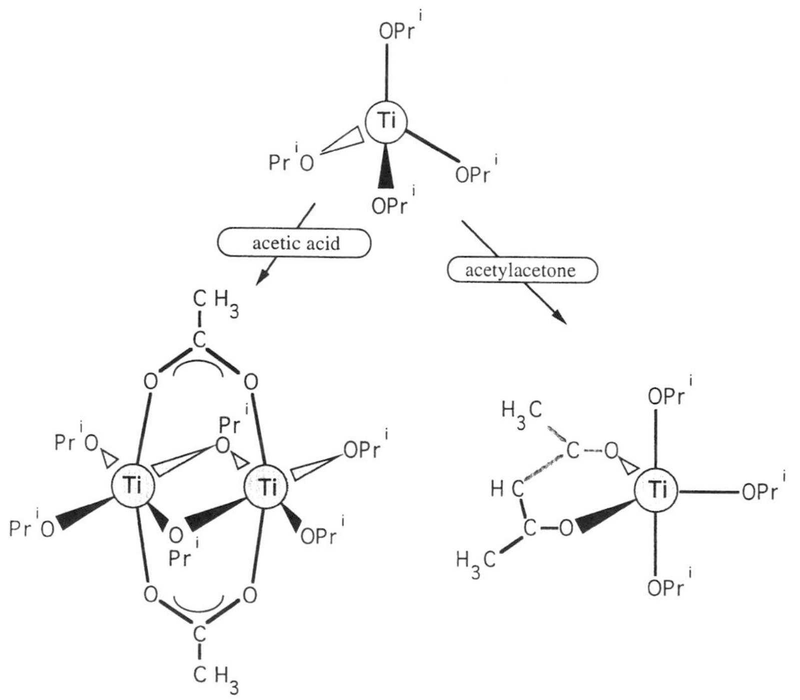

Figure 9.4 Chemical modification of $Ti(O\text{-}i\text{-}Pr)_4$ by acetic acid (CH_3COOH) and acetylacetone ($CH_3{-}CO{-}CH_2{-}CO{-}CH_3$) leads to the formation of $[Ti(O\text{-}i\text{-}Pr)_3(OAc)]_n$ and $[Ti(O\text{-}i\text{-}Pr)_3(acac)]$.

Esterification occurs when more than 1 mol of AcOH is added. Acetic acid in excess reacts with alcohol molecules released during complexation providing the in situ generation of water and giving rise to more condensed species. Only molecular clusters are formed under these conditions. Hexameric species $[Ti_6O_4(O\text{-}i\text{-}Pr)_{12}(OAc)_4]$ are actually obtained upon aging an equimolar mixture of AcOH and $Ti(O\text{-}i\text{-}Pr)_4$ in a closed vessel (Fig. 9.5). Only two-thirds (x = OAc/Ti = 4/6) of the added acetic acid has been used for complexation. Some water is then provided via esterification reactions arising from acetic acid in excess. It leads to the slow hydrolysis of alkoxy groups which are replaced by oxo bridges. In the presence of an excess of water, all organic groups are removed and clear transparent titanium dioxide gels or sols are obtained.

Acetylacetone (acacH = $CH_3—CO—CH_2—CO—CH_3$) is currently used for the stabilization of alkoxide solutions. It reacts readily with $Ti(O\text{-}i\text{-}Pr)_4$, giving a yellow solution of $Ti(acac)(O\text{-}i\text{-}Pr)_3$. However, as acetylacetone behaves as a chelating ligand, only monomeric species are formed, in which Ti is fivefold coordinated (Fig. 9.4). Condensation is observed when water is added. Alkoxy groups are hydrolyzed first. Strongly bonded complexing acac groups are not hydrolyzed, even with an excess of water (unless pH < 2).

$[TiO(acac)]_2$ can be formed upon the hydrolysis of $Ti(acac)_2(OR)_2$. Single crystals have been isolated and x-ray diffraction experiments show dimers with sixfold-coordinated Ti atoms linked through oxygen bridges (Fig. 9.5). Strongly complexing acac ligands cannot be hydrolyzed easily and condensation does not go any further. Larger molecular species can be obtained for smaller amounts of acetylacetone. Single crystals of $Ti_{18}O_{22}(O\text{-}n\text{-}Bu)_{26}(acac)_2$ have been obtained ($x = 0.1$, $h = 1.2$). They are made of a core of 18 $[TiO_6]$ octahedra sharing edges or corners while complexing acac ligands and nonhydrolyzed alkoxy groups remain outside the $[Ti_{18}O_{22}]$ core of the molecule (Fig. 9.5).

Clear sols are obtained when $Ti(O\text{-}n\text{-}Bu)_4$ is hydrolyzed in the presence of acetylacetone. The mean hydrodynamic diameter of the colloidal particles, measured by quasi-elastic light scattering, increases from 2 nm to 40 nm as the hydrolysis ratio increases from $h = 1$ to $h = 4$ (for $x = 0.3$). It decreases from 40 nm to 3 nm when the amount of acetylacetone increases from $x = 0.3$ to $x = 1$ (for $h = 4$).

The hydrolysis of metal alkoxides gives reactive M—OH bonds which lead to condensation and favor the formation of larger species. Complexation leads to nonreactive M—OX bonds which act as polymerization lockers, preventing condensation. A large variety of species are obtained upon hydrolysis and condensation. Molecular clusters, colloidal particles, gels, or precipitates can be formed, depending on the relative amount of hydrolysis ($h = H_2O/M$) and complexation (x = X/M).

All alkoxy groups are not removed when $h \approx z$. Molecular clusters or oxopolymers are formed. The dielectric constant of the organic solvent is

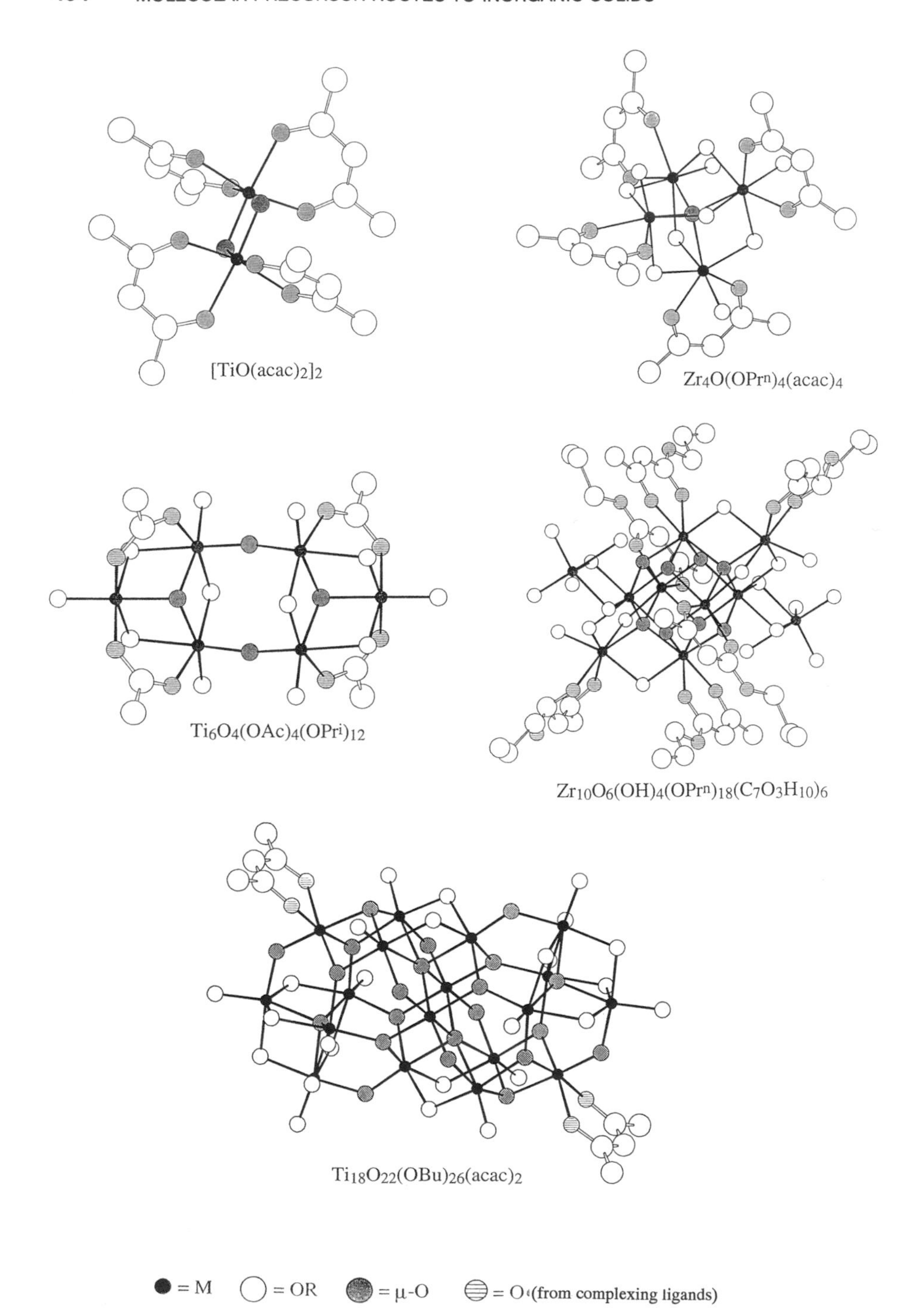

Figure 9.5 Molecular structure of some clusters formed via the partial hydrolysis of chemically modified titanium or zirconium alkoxides.

rather low and the remaining Ti—OH groups are not ionized. The surface charge of polymeric particles is very small and van der Waals interactions prevail. These precursors can be employed advantageously to draw fibers or for the deposition of amorphous thin films and coatings. Remaining organic groups are then calcined upon heating in air and an oxide material is obtained above about 400°C. In the presence of a large excess of water ($h \gg z$), all alkoxy groups are removed and highly cross-linked tridimensional polymers are formed. They lead to hydrous oxides $MO_{z/2}, nH_2O$. The adsorption and dissociation of water molecules at the oxide–solvent interface lead to the formation of charged oxide particles similar to those synthesized from aqueous solutions.

Zirconia powders can readily be obtained via the hydrolysis and condensation of zirconium alkoxides. However, such precursors are highly reactive. This arises from the electrophilicity of Zr^{IV} and its tendency to increase its coordination number up to 8, as in tetragonal zirconia ZrO_2. Coordination expansion occurs as soon as water is added to the alkoxide solution, leading to the uncontrolled precipitation of polydispersed powders. Intermediate molecular compounds corresponding to the early stages of hydrolysis are very difficult to observe. Only very unstable crystals of $Zr_{13}O_8(OCH_3)_{36}$ have been reported in the literature.[20]

Therefore, complexing ligands such as β-diketones are currently added to the alkoxide solution prior to hydrolysis, allowing better control over the formation of oxide particles. A slightly exothermic reaction occurs when acetylacetone is added to a solution of zirconium *n*-propoxide in *n*-propanol. The solution turns pale yellow, suggesting that acetylacetone is bonded to zirconium (charge-transfer coloration) as follows:

$$Zr(O\text{-}n\text{-}Pr)_4 + x\text{acacH} \longrightarrow Zr(O\text{-}n\text{-}Pr)_{4-x}(\text{acac})_x + xn\text{-PrOH} \quad (22)$$

The modified precursor, $Zr(O\text{-}n\text{-}Pr)_{4-x}(\text{acac})_x$, can then be hydrolyzed by adding a given amount of water to the clear solution. The hydrolysis of alkoxy ligands gives reactive Zr—OH groups, allowing condensation to proceed and leading to different oligomeric species, depending on the respective values of x and h.

A gelatinous precipitate of amorphous hydrous zirconia ZrO_2, nH_2O is formed for a very low complexation ratio ($x \leq 0.1$) and large hydrolysis ratio, whereas crystalline $Zr(\text{acac})_4$ precipitates beyond $x \geq 4$. For smaller complexation ratios ($x \approx 1$), few alkoxy groups are removed when a small amount of water is added ($h < 1$). This leads to oligomeric species such as $Zr_4O(O\text{-}n\text{-}Pr)_{10}(\text{acac})_4$ or $Zr_{10}O_6(OH)_4(\text{-}n\text{-}Pr)_{18}(\text{AAA})_6$ (AAA = allylacetoacetate). Hydrolyzed alkoxy groups give oxo or hydroxo bridges, while complexing ligands remain bonded to the metal atom. These organic ligands are located mainly outside the oxo-rich core of the molecular species and act as terminal ligands, preventing further condensation (Fig. 9.5).

Amorphous colloidal particles are obtained when more water is added to the modified precursor solution. Their mean diameter increases when h increases and x decreases but their size remains rather polydisperse. Crystalline zirconia cannot be obtained when hydrolysis is performed at room temperature. Two conditions are required for crystallization to occur: (1) the solution must be heated to favor ion diffusion and increase crystallization rates, as for the thermohydrolysis of aqueous solutions of zirconium oxychloride, and (2) complexing acac ligands and nonhydrolyzed OR groups have to be removed to allow the formation of Zr—O—Zr bonds. These conditions can be obtained by refluxing the colloidal solution in the presence of an organic acid such as *p*-toluene sulfonic acid (PTSA), β-Diketonate ligands become more labile in an acidic medium and can be released in the solution. Zr-acac bonds are then dissociated, allowing the formation of Zr—O—Zr bonds and the crystallization of tetragonal zirconia colloids. NMR and infrared experiments show that organic groups (acac and PTSA) are bonded to the surface of the growing cluster. They prevent aggregation via steric hindrance effects, and stable transparent sols are obtained. These sols are made of monodisperse colloidal particles with a mean hydrodynamic diameter around 3 nm. Aggregation actually occurs rapidly if surface organic groups are removed via dialysis.[21]

These stable ZrO_2 colloids can be used for the production of microporous ceramic membranes. Such membranes are made of randomly packed primary particles. The size of the pores formed by the voids between these particles depends on the size of the particles. The preparation of ceramic membranes with connected micropores requires particles smaller than 10 nm in the final oxide layer and no aggregation of these particles in the sol state. These conditions have been fulfilled with the previous synthesis of zirconia sols. Microporous zirconia membranes have recently been made via the deposition of such sols (x = [acac]/[Zr] ≈ 0.5) onto zirconia ultrafiltration tubes. A thin layer, about 0.2 μm in thickness, made of nonaggregated nanosized particles is obtained after calcination at 500°C. A very narrow pore size distribution is observed with a pore radius around 3 nm. The pore size of these membranes actually decreases as the amount of acetylacetone added to the zirconium alkoxide precursor increases.[22]

9.2.5 Sol-Gel Synthesis of Mixed Oxides

Most advanced ceramics are multicomponent systems having two or more types of cations in the lattice. It is therefore necessary to prepare gels of high homogeneity in which cations of various kinds are distributed uniformly at an atomic scale through M—O—M′ bridges. A major problem in forming homogeneous multicomponent gels is the unequal hydrolysis and condensation rates of the metal alkoxides. This may result in chemical inhomogeneities, leading to higher crystallization temperatures or even undesired crystalline phases. Several approaches have been attempted to overcome this problem, including partial prehydrolysis of less reactive precursors, matching of hydrolysis rates by chemical modification with chelating ligands, and synthesis of heterometallic alkoxides.[23]

Heterometallic Alkoxides The formation of oligomeric species via the nucleophilic addition of alkoxide molecules can be observed when two different metal alkoxides are mixed together. Bimetallic alkoxides were first recognized in 1929 by Meerwein and Bersin during the titration of acidic and basic alkoxides. Bimetallic alkoxides, or oxoalkoxides, are typically obtained upon heating a mixture of two alkoxides, especially when the two metal atoms have a large electronegativity difference.[24] The driving force for the formation of bimetallic alkoxides involves Lewis acid-base reactions in which metal atoms of different electronegativities are involved. Therefore, most bimetallic alkoxides are formed with alkali or alkaline-earth metals. The formation of $Mg[Al(O\text{-}i\text{-}Pr)_4]_2$, for instance, can be described as the chelation of Mg^{2+} (Lewis acid) by $[Al(O\text{-}i\text{-}Pr)_4]^-$ ligands (Lewis base). Bimetallic alkoxides can be formed simply by refluxing a mixture of alkoxides. They are more often synthesized via the reaction of halides or acetates with alkoxides, leading to the formation of esters rather than ethers:

$$M(OR)_n + M'(X)_{n'} \longrightarrow (OR)_{n-1}\, M{-}O{-}M'(OX)_{n'-1} + RX \qquad (X = OAc, Cl) \qquad (23)$$

Heterometallic alkoxides are often more soluble than their parent alkoxides, a property that can be used advantageously for the sol-gel chemistry of nonsoluble alkoxides. Lead acetate $Pb(OAc)_2$, for instance, is not soluble in toluene but can easily be dissolved at room temperature in the presence of $Ti(OEt)_4$, suggesting the formation of bimetallic species. $Pb_4O(OEt)_6$ undergoes complete dissolution in ethanol when $Nb(OEt)_5$ is added, giving rise to $[Pb_6O_4(OEt)_4][Nb(OEt)_5]_4$.

Heterometallic alkoxides can be described as molecular precursors or building blocks for the synthesis of multicomponent oxides. They provide homogeneity at a molecular level and could offer a way to overcome mismatched hydrolysis rates. Their main advantage would be to form heterometallic $M{-}O{-}M'$ bonds in the solution, allowing the low-temperature synthesis of multicomponent materials. $LiNbO_3$ and $MgAl_2O_4$ have been synthesized via the hydrolysis of the bimetallic alkoxides $[LiNb(OEt)_6]$ and $[MgAl_2(OR)_6]$, respectively (Fig. 9.6). Barium titanate, $BaTiO_3$, is generally synthesized by calcining a mixture of $BaCO_3$ and TiO_2 between 1000 and 1200°C. The perovskite $BaTiO_3$ phase synthesized via the sol-gel route is formed around 600°C when $Ti(OPr^i)_4$ and $Ba(OPr^i)_2$ are used as precursors. This crystallization temperature decreases to 50°C when the mixture of alkoxides is refluxed prior to hydrolysis. This is due to the in situ formation of bimetallic oxoalkoxides such as $[BaTiO(O\text{-}i\text{-}Pr)_4(i\text{-}PrOH)]_4$,

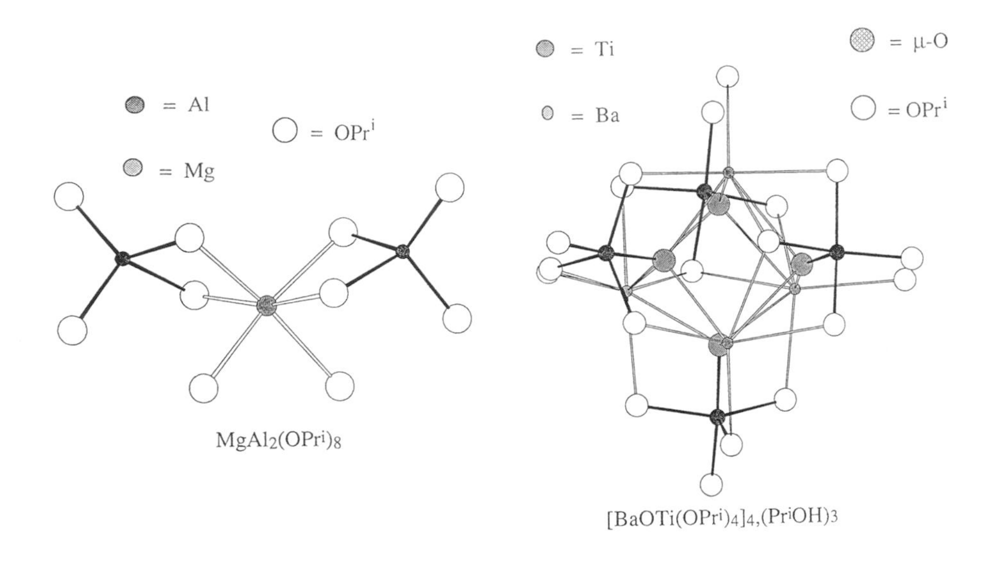

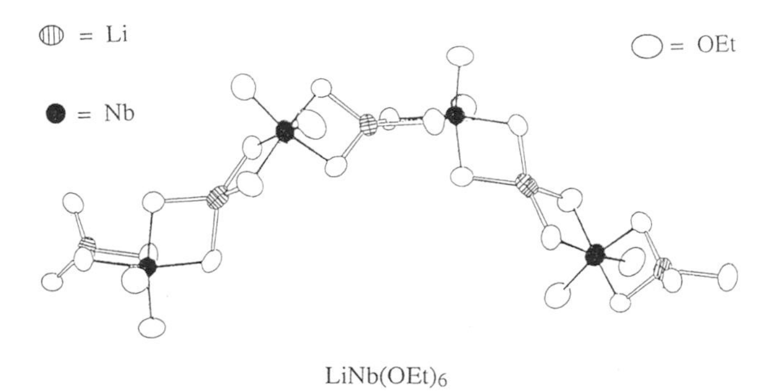

Figure 9.6 Molecular structure of some heterometallic alkoxides.

$[Ba_4Ti_2O(OEt)_{14}, (EtOH)_8]$, $[BaTi_2(OEt)_{10}, (EtOH)_5]$, $[BaTi_4(OEt)_{18}]$, or $[Ba_4Ti_{13}O_{18}(OCH_2CH_2OCH_3)_{24}]$, which have been characterized by x-ray diffraction (Fig. 9.6).

The sol-gel process has also been suggested for the treatment of nuclear wastes.[25] The problem is to immobilize radioactive elements within a stable oxide matrix. Glasses are usually chosen for their ability to dissolve most elements. However, they can be leached out via lixiviation. Ceramics would therefore offer a more stable matrix, but it is quite difficult to incorporate a large variety of different elements within a single compound. Titanate-based ceramics, known as "synrocs," have been developed in Australia for immobilizing high-level nuclear waste from nuclear power reactor fuel reprocessing.

They are made by hydrolyzing a mixture of Al, Ti, and Zr alkoxides with an aqueous slurry of radioactive fission products and Ba and Ca hydroxides. Multioxide ceramic phases such as zirconolite $CaZrTi_2O_7$ are formed after drying and calcination that are more stable than the glasses used for the storage of nuclear wastes.

Silica-Based Mixed Oxides Most ceramics and glasses are based on silica. The problem is then to form Si—O—M bonds in the solution and keep them during the whole hydrolysis and condensation process. However, silicon has a rather high electronegativity ($\chi_{Si} = 1.74$) and does not show any tendency toward coordination expansion. Its chemical reactivity is quite low and very few heterometallic alkoxides or oxo-alkoxides are formed with silicon. It is therefore difficult to prepare homogeneous multicomponent gels in silicate systems because of the large difference in hydrolysis and condensation rates between silicon alkoxide and other alkoxides.

The key point, for the sol-gel synthesis of homogeneous $SiO_2—MO_x$ materials, is to match hydrolysis rates. This can be done either via the partial prehydrolysis of the silicon alkoxide or by slowing down the hydrolysis rate of the other metal alkoxides via complexation.

By far the most used synthetic approach is to partially prehydrolyze silicon tetraethoxide (TEOS) with a small amount of acidic water to form $Si(OEt)_{4-x}(OH)_x$ species ($x \neq 0$). The purpose of this first hydrolysis step is to generate as many Si—OH groups as possible and to consume all the water. Thus the precipitation of oxide particles can be avoided when the more reactive $M(OR)_z$ alkoxide is added. The condensation between silicon species being very slow at low pH, condensation occurs preferentially between Si—OH and M—OR groups leading to the formation of Si—O—M bonds. Such bonds have been evidenced in solution by ^{29}Si and ^{17}O NMR.

Complexation can also be used to slow down the condensation of the more reactive alkoxide. In the case of the $SiO_2—TiO_2$ system, for instance, the reaction of $Ti(O\text{-}i\text{-}Pr)_2(acac)_2$ with silicic acid first provides solute polytitanosiloxane species, followed by further condensation to form cross-linked polymer gels. Heat treatment then leads to transparent monolithic $SiO_2—TiO_2$ glasses. To decrease the reactivity of titanium precursors without using chelating agents, alkoxides with bulky tertiary groups such as $Ti(O\text{-}t\text{-}Am)_4$ can also be used. It has been shown that around pH ≈ 11, its hydrolysis rate is close to that of silicon tetramethoxide.

9.3 HYBRID ORGANIC–INORGANIC COMPOUNDS

Hybrid organic-inorganic compounds are novel multifunctional materials that offer a wide range of interesting properties. This new field of materials science is expanding rapidly. Several international meetings are now devoted to this promising area of research and many review papers have been published during the last three years.[3-5, 26-35]

Several routes, such as intercalation, electrocrystallization, polymer reinforcement, or sol-gel processing, can be used for the synthesis of hybrid compounds. The mild conditions involved in the sol-gel synthesis of metal oxides provide a versatile access to hybrid organic–inorganic materials. The intimate mixing of molecular precursors in organic solvents allows organic and inorganic components to be associated at the molecular level, and the sol-gel route appears to be very convenient for the synthesis of organic–inorganic nanocomposites. However, the reactivity of both organic and inorganic precursors are usually quite different, and phase separation tends to occur. The properties of hybrid materials do not depend only on organic and inorganic components but also on the interface between both phases. The general tendency is therefore to increase interfacial interactions by creating an intimate mixing, or interpenetration, between organic and inorganic networks. Moreover, the formation of chemical bonds between organic and inorganic species would prevent phase separation, allowing the synthesis of molecular composites or organic–inorganic copolymers.

Therefore, hybrid materials will be divided into two classes:[5]

1. Class I corresponds to hybrid systems in which weak interactions such as van der Waals forces and hydrogen bonds or electrostatic interactions are created between organic and inorganic phases. This class involves mainly small organic species embedded within an oxide matrix.
2. Class II corresponds to hybrid compounds where both organic and inorganic components are bonded through strong covalent chemical bonds.

9.3.1 Organic Species in Sol-Gel Matrices (Class I)

The sol-gel synthesis of metal oxides usually leads to highly porous materials. These porous matrices can then be impregnated with a solution of organic species. Organic molecules can also be mixed with metal alkoxide precursors in a common solvent. Hydrolysis and condensation then lead to the formation of an oxide network around the organic species. Both routes have been developed extensively by either inorganic sol-gel or polymer chemists. The resulting materials are amorphous composites belonging to class I. Organic components bring some new physical or chemical properties to the oxide materials. The microstructure and spatial distribution of organic and inorganic components are governed mainly by weak interactions (van der Waals, hydrogen bonds, hydrophilic–hydrophobic behavior) (Fig. 9.7). The porosity of the oxide matrix can be tailored by controlling the gelation rate and the phase separation between organic and inorganic components.

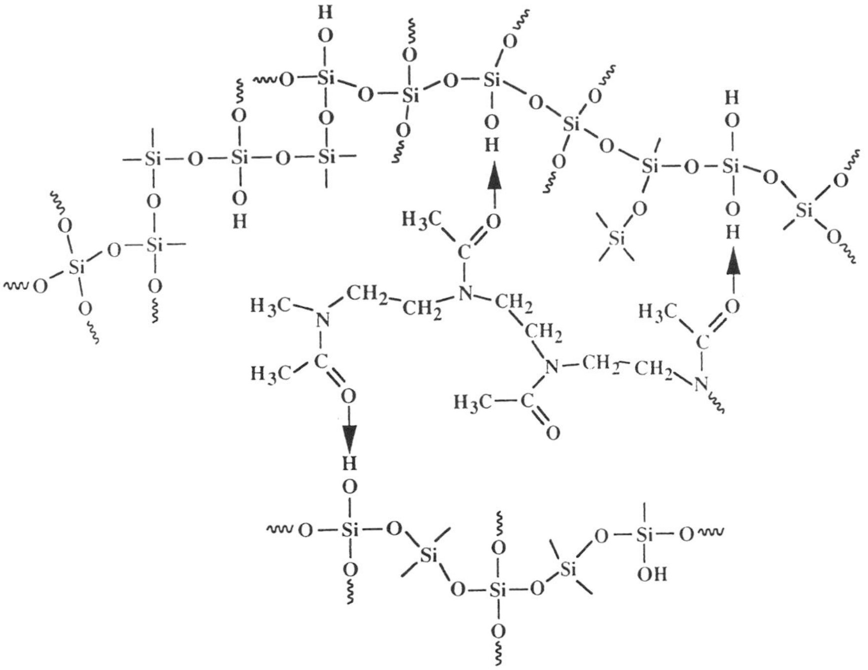

Figure 9.7 Schematic representation of polyoxazoline–silica gel hybrids (class I).

Impregnation The sol-gel synthesis of metal oxides leads to the formation of porous xerogels. It is therefore possible to impregnate the porous oxide with solute organic species. There are, of course, no chemical limitations in this process, and a wide range of organic dyes have been impregnated within porous silica by dipping the porous xerogel into the organic solution. Fluorescent silica microspheres, laser materials, or chemical sensors have been made that way. Impregnation is easy to perform, but organic molecules are just adsorbed at the surface of the pores of the oxide network and they are often leached out by washing.

The open porosity of oxide gels can also be filled with a solution of polymerizable monomers such as methyl methacrylate (MMA). The in situ organic polymerization is then initiated by UV irradiation or thermal treatment. Such PMMA-silica composites are highly transparent. Their hardness is intermediate between those of glasses and polymers. They can be polished, allowing the fabrication of large transparent optical pieces. One of the main advantages of this process is that the inorganic xerogel can be densified at high temperature, prior to impregnation, to control its porosity. Conjugated polymers such as polypyrrole, polyaniline, or poly(*p*-phenylene) have also been impregnated within silica sol-gel matrices. These hybrids exhibit large third-order nonlinear optical (NLO) coefficients.

Similar approaches have been used by polymer chemists to improve the mechanical properties of organic polymers by adding inorganic fillers. Inorganic particles are then mixed with the polymer before it becomes too viscous. However, inorganic particles tend to agglomerate, leading to the formation of heterogeneous macrocomposites. Sol-gel precursors such as metal alkoxides can advantageously be used to get an homogeneous distribution of oxide particles within a polymeric organic network. Two different approaches have been used to prevent phase separation and form nanocomposite materials.

1. A metal-oxo cross-linked polydimethylsiloxane (PDMS) elastomer network was used by Mark[36] and Wilkes.[37] The polymer is swollen with an alcoholic solution of metal alkoxide $M(OR)_z$ (M = Si, Ti, Zr, Al). The solute alkoxide diffuses within the polymeric network. Hydrolysis and condensation then lead to the in situ formation of small, nonaggregated, oxide particles, which reinforce the mechanical properties of the organic polymer.
2. Interpenetrating networks have been synthesized by Novak[4] via the simultaneous polymerization of both organic and inorganic components. Silica-based hybrids have been made with organic polymers obtained by ring-opening metathesis polymerization of cyclic alkenols or free-radical-initiated polyaddition. Such a method appears to be specially suitable for the production of nanocomposites, with minimal shrinkage and a wide range of silica/polymer ratios (Fig. 9.8).

Organic Dyes in Sol-Gel Matrices Organic molecules can be dissolved together with metal alkoxides in a common solvent. The hydrolysis and condensation of the alkoxide precursor lead to the formation of an oxide network in which nonreactive organic species remain embedded. Such hybrid materials have been used primarily to trap organic dyes within oxide matrices, giving rise to a rapidly expanding field known as sol-gel optics.[38-40]

Organic dyes exhibit a wide range of optical properties and can be used for the fabrication of optical systems such as dye lasers or photochromic, fluorescent, or NLO devices.[29-31] Their poor thermal instability precluded their incorporation into inorganic oxide matrices until the use of sol-gel-derived glasses started about 10 years ago. The primary advantages of sol-gel matrices are the following:

1. Oxide matrices offer better protection than polymers toward heat, UV irradiation, and mechanical stresses.
2. Aggregation processes such as dimerization can be minimized within sol-gel matrices, and larger amounts of organic dyes can usually be dissolved in the presence of sol-gel precursors.

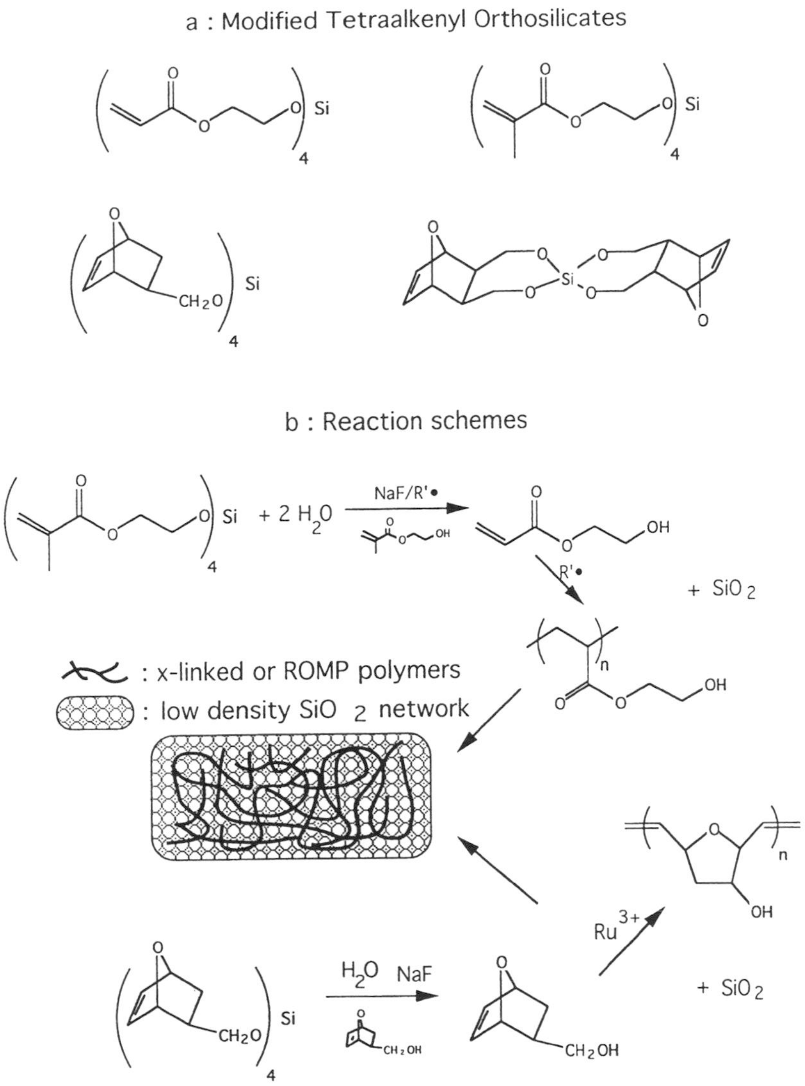

a : Modified Tetraalkenyl Orthosilicates

b : Reaction schemes

$NaF/R'\bullet$

$+ 2 H_2O$

$+ SiO_2$

: x-linked or ROMP polymers

: low density SiO_2 network

Ru^{3+}

H_2O NaF

CH_2OH

Figure 9.8 Modified silicon alkoxides and reaction pathway for the formation of organic–inorganic interpenetrating networks.

3. Sol-gel matrices can be designed in a large variety of shapes, such as thin films or planar waveguides, and coatings onto optical fibers or microspheres.

Many organic dyes, such as rhodamines, perylenes, spiro-oxazines, and porphyrins, have been incorporated into sol-gel oxide matrices. They exhibit a wide range of optical properties and can be used for fluorescence, photochromism, photochemical hole burning, or nonlinear optics. Up to now, most of the work has been done on laser materials.

Tunable solid-state lasers are based on transition-metal ions such as Cr^{III} or Ti^{III}, but their emission is restricted to the infrared. Organic dyes are able to give tunable laser emission in the visible. This would be very important for technological applications. Therefore, much work has been carried out to use organic dyes in solid matrices. This was, of course, limited to organic polymers, but the mild conditions of the sol-gel route allow "fragile" organic dyes to be associated with inorganic networks. This was demonstrated simultaneously by two groups, in Israel and Japan, who embedded several organic laser dyes (rhodamines, perylenes, etc.) within silica xerogels. They were followed by an increasing number of researchers, and the sol-gel route now opens a land of opportunities for the processing of solid-state tunable lasers in the visible.

All this work pointed out the role of dye–matrix interactions in the photo response of these hybrid materials. It is well known now that sol-gel matrices prevent the dimerization of organic dyes such as rhodamine. This is an obvious advantage, as dimerization quenches fluorescence. This suggests that associations between organic dyes are weaker than interactions with the oxide network. Rhodamine molecules should be adsorbed at the surface of the micropores so that dimerization remains negligible even for high concentrations ($c \approx 10^{-2}$ M). Hydrogen bonds are presumably formed between the nucleophilic heteroatoms (N, O) of the dye and the hydroxyl groups Si—OH of the silica matrix (c.f., Fig. 9.7).

The development of solid-state dye lasers is now a rapidly expanding field.[40] Up to now, the best results concerning solid tunable dye lasers in the visible range have been obtained with perylene red and pyrromethene dyes embedded within hybrid matrices. These matrices are made via the hydrolysis condensation of methyltriethoxysilane (MTEOS) or vinyltriethoxysilane (VTEOS) precursors with amino functionalized alkoxysilanes such as 3-aminopropyltriethoxysilane. An excellent efficiency of 63% was obtained with a pyrromethene 597 doped matrix. A 6-mJ output energy was obtained for 10-mJ pump energy (2 J/cm^2 fluence). These materials exhibit good thermal and photochemical stability. Laser emission can be tuned over a range of more than 50 nm. For a pump energy of 0.7 mJ at a 1-Hz repetition rate, the perylene red dye–doped matrix still exhibits an output energy of 50% of the initial value (0.1 mJ) after more than 10^5 shots. These hybrid materials are close to application, but some progress has still to be made to improve the stability of organic dyes and to prevent excessive heating upon laser impulsions.

Photochromic hybrids are very promising materials for smart windows, display devices, or solar glasses. Photochromic organic dyes such as 2,3-diphenylindenone oxide, spiropyranes, and spyroxazine derivatives have been incorporated within sol-gel matrices. Homogeneous dispersion, strong absorption of the colored form, good photo and thermal stability, and a controlled thermal fading rate are some major requirements for photochromic applications. Direct and reverse photochromic effects have been observed when spiropyranes are embedded within sol-gel matrices synthesized from $Si(OMe)_4$ and $EtSi(OEt)_3$, respectively. Interactions between the dye and the matrix must be responsible for such different photochromic effects.[42]

Chemical Sensors and Biosensors In sol-gel-derived optical devices, organic dyes have to be protected against the environment in order to avoid degradation or leaching. Therefore, dense matrices are used and best results are obtained with nonporous hybrid matrices (class II). The problem is quite different when chemical sensors have to be made. Organic dyes are then embedded within porous sol-gel matrices, so that small molecules can diffuse through the oxide network. Trapped organic dyes are then able to react reversibly with analytes from the outside medium. This was first demonstrated by research scientists from Jerusalem, who showed that porous sol-gel glasses could be used for the preparation of chemical-sensing materials. Since then many tests have been performed for the detection of metal cations, pH, anions, and even organic or inorganic molecules. Detection is usually performed via the color modification of an organic dye. One of the main advantages of the sol-gel process is that thin films, rods, disks, microspheres, or plates of any desired shape can be molded. MacCraith et al.[41] coated optical fibers as waveguides for the evanescent wave excitation of immobilized dyes. The optical fiber is coated with a microporous sol-gel film containing an organic dye such as fluorescein. Several pH indicators can be trapped in the same sensor, thus increasing the range of detection. Leaching of the reagents from the sol-gel matrix is currently the main drawback, limiting their application to low cost disposable devices. However, this drawback could be overcome if the organic dyes were chemically bonded to the matrix (see below).

The chemical conditions provided by sol-gel chemistry are mild enough to allow the incorporation of biomolecules. Their biochemical activity is preserved, and they can react with smaller molecules that diffuse through the pores of the xerogel. The protective action of the sol-gel cage prevents leaching and enhances their stability significantly. Moreover, the transparency of sol-gel matrices makes them suitable for the optical detection of colored reactions. Enzymes, antibodies, and even whole cells have been entrapped within sol-gel matrices for the realization of biocatalysts and biosensors.[43-48]

Biological species are very sensitive to alcohol and pH. The usual sol-gel procedure has then to be slightly modified to meet these requirements. Ethanol is strictly prohibited, to avoid the denaturation of the proteins, but small amounts of methanol can be tolerated. Tetramethoxysilane (TMOS) is therefore used as a molecular precursor rather than TEOS. The neat alkoxide is prehydrolyzed under acid-catalyzed conditions before biomolecules are added to the solution. No alcohol can be added so that the mixture of nonmiscible TMOS and water is usually sonicated in order to have intimate mixing. Methanol is released during this hydrolysis reaction. It behaves as a co-solvent, and a transparent solution of hydrolyzed precursors is obtained after a few minutes. This solution is then mixed with a buffered solution (pH $\approx$ 7) containing the biomolecules. Condensation reactions are quite fast at this pH, and a gel network rapidly grows in which the biomolecules remain embedded. Some enzymes are quite fragile and entrapment has sometimes to be performed at around 0°C.

Many enzymes have been trapped within sol-gel matrices for the realization of biosensors. The most studied example is glucose oxidase (GOD).[43-47] This enzyme is known to catalyze the oxidation of D-glucose by O_2 to give D-gluconic acid and hydrogen peroxide. The hydrogen peroxide formed during this reaction can be detected via another enzymatic reaction using horseradish peroxidase (HRP) to catalyze the oxidation of an organic dye by H_2O_2. The presence of glucose is measured quantitatively via optical absorbance. The oxidation of glucose can also be detected electrochemically using an electrode sensitive to the concentration of O_2 in the solution. Redox mediators such as ferrocene have also been used to detect the redox reactions occurring at the active center of GOD.[47] The reduced mediator is regenerated by the anode and the exchange Faradaic current is proportional to the amount of converted glucose. The electrochemical detection of enzymatic reactions could lead to the development of sol-gel biosensors, as it does not require the synthesis of transparent gels.

The sol-gel route has recently been extended to the realization of immunoassays in order to detect specific antigen–antibody reactions. However, the open porosity of the gel is usually not very large (pore size < 50 Å). This could be a real drawback when large species such as antibodies have to diffuse through the pores of the gel. Therefore, up to now, sol-gel-derived immunoassays have been performed with large antibodies trapped within a sol-gel matrix or covalently bonded at the surface of a sol-gel film. The immobilization of antibodies on silica substrates has already been widely described. The silica surface is usually functionalized with an organosilane such as γ-aminopropyltriethoxysilane, containing an alkoxysilyl group at one end and an organic functional group at the other end. A similar method was used recently with sol-gel films synthesized from functionalized alkoxysilane precursors containing active groups (SH, NH_2, etc.). Immunoglobulins are covalently bound at the surface of the silica film via a coupling agent such as glutaraldehyde. Specific antigen–antibody reactions were then detected by surface plasmon resonance.

Synthetic antibodies (antifluorescein, antinitroaromatics) have also been encapsulated within sol-gel matrices. Small molecular haptens (fluorescein, nitroaromatics) then diffuse through the pores of the sol-gel matrix, where they have been shown to bind selectively to the encapsulated antibodies.

Such assays can be used for the detection of chemical contaminants, but they are not really useful for medical purposes. Immunoassays have to be performed with real parasites (virus, bacteria, or protozoa), and human sera, rather than purified monoclonal antibodies, has to be used. This was realized recently via the sol-gel entrapment of the parasitic protozoan *Leishmania*. Electron microscopy shows that the cellular organization and integrity of the plasma membrane of entrapped parasites are preserved. Moreover, they retain their antigenic activity and react specifically with the corresponding antibodies. Sol-gel matrices have also been used for the realization of enzyme-linked immunosorbent assays (ELISA) directly with the sera of infected patients.[48]

9.3.2 Chemically Bonded Organic–Inorganic Hybrids (Class II)

Class II corresponds to hybrid organic–inorganic compounds where both the organic and inorganic components are bonded through covalent bonds (Fig. 9.9). These new materials have been developed simultaneously by sol-gel chemists using mainly organosilane precursors $R'_xSi(OR)_{4-x}$ and polymer chemists via the co-condensation of functionalized polymers (PDMS, PTMO, etc.) with TEOS. These hybrid materials are known as ormosils (organically modified silicates), ormocers (organically modified ceramics), ceramers (ceramic polymers), or polycerams (polymeric ceramics).[3–5, 26–35]

Two general methods have been developed for the synthesis of these hybrid organic–inorganic materials:

1. The second network is formed within the matrix of an already formed polymeric network. The first network must be functionalized to form chemical bonds with the second component. Both inorganically functionalized organic macromonomers or organically functionalized inorganic metal-oxo cores can be used.
2. The hydrolysis and condensation of organically functionalized alkoxide precursors such as an alkoxysilane, $R'_xSi(OR)_{4-x}$. The organic group R′ behaves as either a network former or a network modifier, depending on whether it gives rise to organic polymerization or not.

Hybrid Molecular Precursors The chemistry of hybrid organic–inorganic gels has been developed mainly with silica-based materials by Schmidt at Würzburg.[3, 28] Currently, the most common way to introduce an organic group into a silica network is to use organoalkoxysilane precursors, $R'_xSi(OR)_{4-x}$. Under most sol-gel conditions the Si—C bond is stable toward hydrolysis and the organic R′ group remains bonded to the inorganic network. Methyl-substituted silicon alkoxides ($R' = CH_3$) are the simplest precursors for the sol-gel synthesis of hybrid gels. They are at the borderline between two kinds of materials. Depending on x, they can be used as

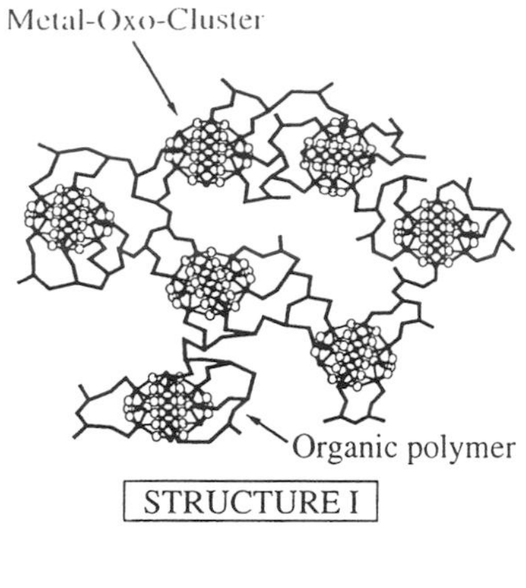

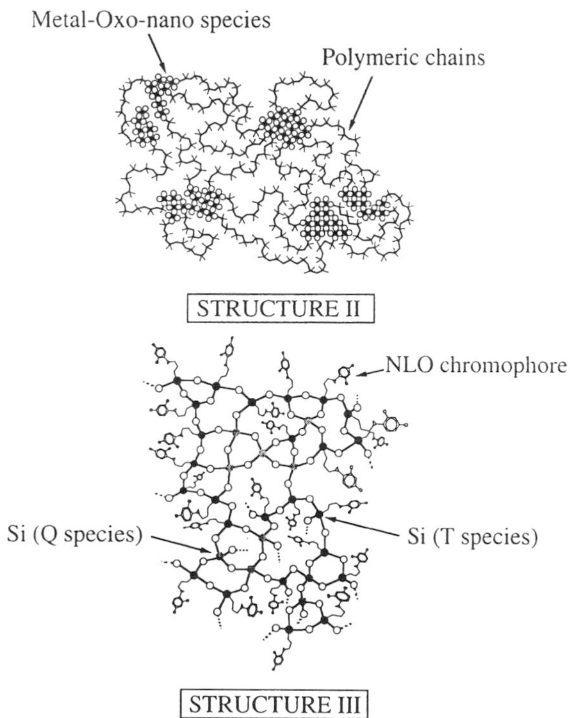

Figure 9.9 Schematic structure of some organic–inorganic hybrids (class II): (I) well-defined inorganic molecular clusters cross-linking organic polymers, (II) polydispersed oxide nanoparticles linked by organic polymers (PDMS, PTMO, ...), (III) hybrid copolymers made of short $(T{-}O{-}T)_a$ and $(Q{-}O{-}Q)_b$ units.

precursors for silicones ($x = 2$ and 3) or silica ($x = 4$). Their chemistry has therefore been studied extensively in both fields. In acidic conditions, the hydrolysis rate of organosilanes increases significantly when x decreases. The hydrolysis rate of $Si(OEt)_4$ is three orders of magnitude smaller than that of $(CH_3)_2Si(OEt)_2$. The reverse is observed in basic medium. $Si(OR)_4$ species are hydrolyzed, whereas the hydrolysis rate of $(CH_3)_2Si(OEt)_2$ is close to zero and remains quite small for $CH_3Si(OEt)_3$. The chemical reactivity of each precursor toward hydrolysis and condensation can then, in principle, be tailored to fit with the others. The reactivity of methyl-substituted silicon alkoxides being very low in a basic medium, in contrast to pure silicon alkoxides, the synthesis of hybrid systems made from alkoxysilane and metal alkoxides is usually performed in an acidic medium, to match the chemical reactivity of all precursors.

Trifunctional Alkoxysilanes, $R'Si(OR)_3$ A wide range of trifunctional alkoxysilanes, $R'Si(OR)_3$, are commercially available. They have been widely used as coupling reagents for the functionalization of silica surfaces. Used in the sol-gel process, they offer many advantages for basic or applied research in such fields as optics, catalysis, selective membranes, molecular recognition, sensors, or molecular imprinting.

The organic group R′ can have different functions and generally acts as a network modifier. In that case, R′ groups can be simple organic groups such as alkyl or aryl groups. $CH_3Si(OR)_3$ is often used to provide some hydrophobic character to a silica matrix or to modify its mechanical properties. After hydrolysis and condensation, each Si remains bonded to a methyl group, and thus the gel network is less rigid since the degree of cross-linking of the oxide network decreases. Highly porous coatings have been prepared recently at ambient pressure without the need of supercritical drying. The silica surface was covered with methyl groups by using CH_3SiCl_3 gas. These surface groups make shrinkage on drying reversible. As the solvent is withdrawn, the gel springs back to a porous state. Aerogel films with 98.5% porosity have been made by using this approach.

R′ can also be a functional group used to bring some new property to the oxide matrix and trifunctional alkoxysilanes $R'Si(OR)_3$ (R′ being an organic chromophore) are now currently used to bring optical properties into the hybrid materials (solid-state dye lasers, photochromic devices, photochemical hole burning, nonlinear optics). Several studies show that the optical response increases significantly when the organic dye is covalently bonded to a class II hybrid sol-gel matrix rather than simply embedded within an oxide network. The stability of organic dyes on laser irradiation is improved when they are grafted to the sol-gel matrix. The best results with photochromics are obtained with spiroxazine derivatives covalently bonded to hybrid matrices made from methyltriethoxysilane and/or glycidyloxypropyl-trimethoxysilane precursors. Their color-fading speed is similar to that observed in ethanol, while their photostability is much better. Photochemical hole burning (PHB) can be observed at higher temperature. Trialkoxy-functionalized porphyrins grafted to hybrid organic–inorganic matrices exhibit a sharp PHB effect at 77 K, whereas this effect cannot be observed above 4 K when the porphyrin is not bonded to the silica network. This behavior is probably related to an increase of the dye–matrix rigidity that should decrease the thermal blurring of the photochemically burned hole occurring via vibrations.[49]

Quadratic nonlinear optics (NLO) is a quite new domain of sol-gel optics, where the optical properties are strongly dependent on and improved by the covalent bonding of the organic dye to the matrix. In this field, class II materials are required. It is well known that second-order effects cannot be observed in centrosymmetric systems. The random orientation of organic dyes in sol-gel matrices rules out second-harmonic generation (SHG) and most work has been performed with third-order processes. Second-order nonlinearities can only be observed after poling to orient all organic dyes within the sol-gel matrix. However, because of thermal relaxation processes, organic dyes tend to lose their common orientation with time and SHG disappears rapidly. Better results are obtained when NLO dyes are grafted to the oxide matrix via covalent bonds. A nonresonant second-order nonlinearity of about 80 pm/V has recently been reported.[50] These high values are now very competitive with those reported for inorganic ($LiNbO_3$) and polymeric materials (Fig. 9.10).

Second-order NLO response can be improved by increasing the rigidity of the amorphous sol-gel matrix (i.e., the glass transition temperature T_g) and the number of covalent bonds formed by each dye. The best orientational stability is obtained when both ends of the chromophore are anchored so that the chromophore principal axis is collinear with the polymer backbone. It has recently been shown that the covalent incorporation of an alkoxy trifunctionalized diazobenzene derivative as a main-chain constituent of a sol-gel matrix leads to a large second-harmonic coefficient (27 pm/V at 1.32 μm) and a high thermal stability at 100°C, close to that of polyimide/sol-gel interpenetrating networks. The processing conditions are also very important. Care must be taken so that the hybrid matrix does not become too hard before a reasonable degree of alignment is obtained by electric field poling.

Polymerizable organic groups R′ can also be used with trifunctional silicon alkoxides $R'Si(OR)_3$. They then behave as network formers so that an organic network is formed together with the oxide network. Most common polymerizable functions introduced so far into hybrid gels are vinyl, epoxy, or methacrylate groups. Organic polymerization is induced by photochemical or thermal curing. Patternable coatings have been prepared from a mixture of TEOS with organosilane precursors such as γ-glycidyloxypropyltrimethoxysilane, γ-methacryloxypropyltrimethoxysilane, and vinyltrimethoxysilane. The mixture is refluxed for several hours and then spin-coated onto a substrate. Patterning can be realized by photolithographic methods or even direct laser writing. Protective coatings can be deposited onto glass or metallic substrates.

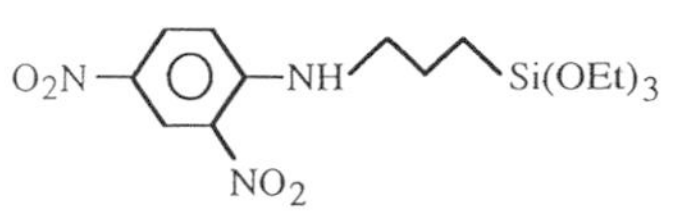

O_2N–[C₆H₃(NO₂)]–NH–$(CH_2)_3$–$Si(OEt)_3$

NO_2

TSDP

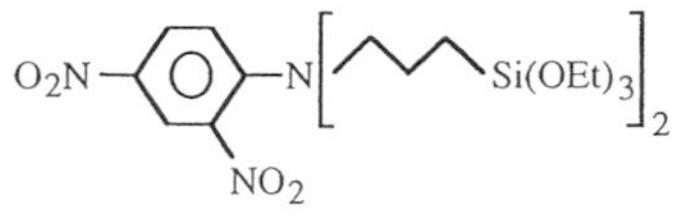

O_2N–[C₆H₃(NO₂)]–N[$(CH_2)_3$–$Si(OEt)_3$]$_2$

NO_2

BTPDNA

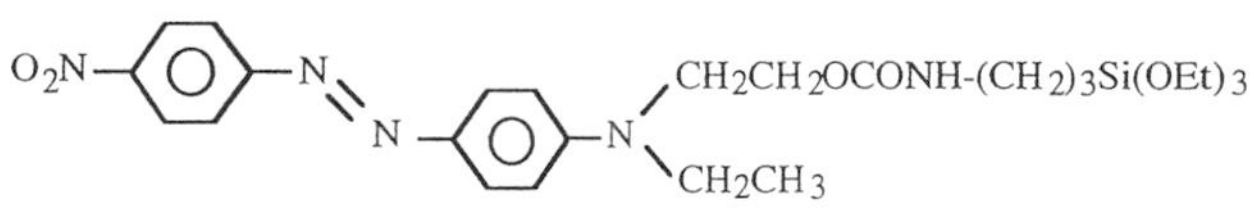

O_2N–C_6H_4–N=N–C_6H_4–N(CH_2CH_2OCONH-$(CH_2)_3Si(OEt)_3$)(CH_2CH_3)

DR1UPTEOS

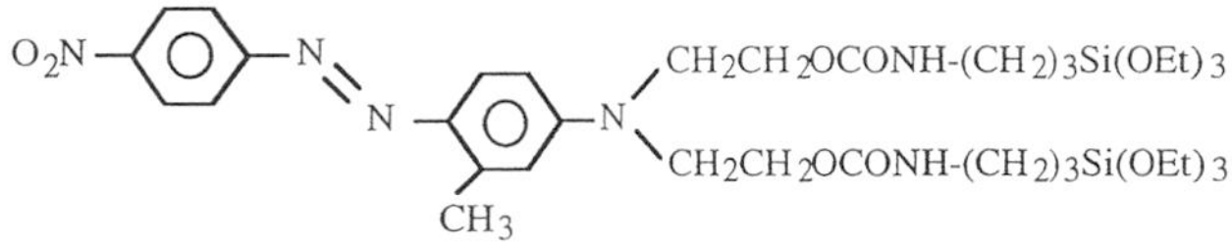

O_2N–C_6H_4–N=N–$C_6H_3(CH_3)$–N(CH_2CH_2OCONH-$(CH_2)_3Si(OEt)_3$)$_2$

CH_3

ICTES-Red 17

$(EtO)_3Si(CH_2)_3$-$HNCOO(CH_2)_2$–$S(O_2)$–C_6H_4–N=N–C_6H_4–N(CH_2CH_2OCONH-$(CH_2)_3Si(OEt)_3$)$_2$

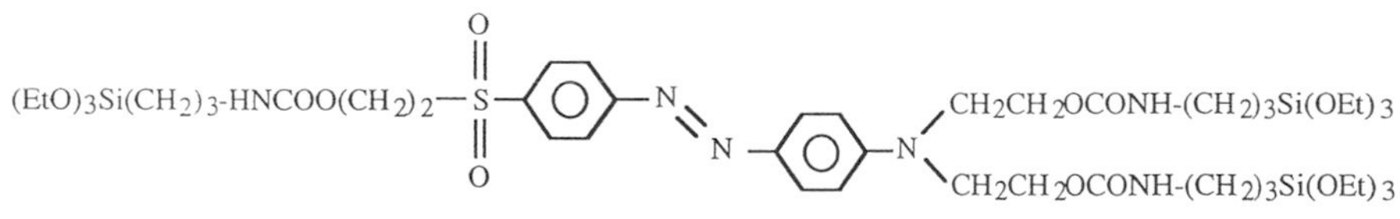

ICTES-ADSAB

Figure 9.10 Examples of organic NLO chromophores that can be grafted onto silica matrices.

Organic polymers can also be released by chemical reactions or calcination, leaving highly porous silica materials.

Difunctional Alkoxysilanes, $R'R''Si(OR)_2$ The most common R′ (or R″) groups of $R'R''Si(OR)_2$, are methyl or phenyl groups. Powders obtained via the hydrolysis of $(CH_3)_2Si(OR)_2$ can be used as soft abrasives for dermatologic applications. Upon hydrolysis and condensation these difunctional precursors lead to cyclic species or short chains rather than high polymers. Precursors of higher functionality have to be added as cross-linking reagents. As the amount of cross-linking reagent increases, viscous liquids are first obtained, then viscoelastic gels and brittle glasses. Both methyl and phenyl groups make the network hydrophobic, a property that could be interesting for protective coatings. Polar solvents such as water or alcohols are not retained in the gel network. They do not have to be removed upon drying or heating, avoiding shrinking processes. Thick coatings can then be obtained without cracks. Phenyl siloxanes are sometimes preferred to methyl siloxanes, to increase the thermal stability, glass transition temperature, or refractive index.

Under acidic conditions, TEOS reacts quite rapidly with $(CH_3)_2Si(OEt)_2$ (DEDMS), preventing the formation of long PDMS chains. If the amount of TEOS is low (DEDMS/TEOS ≈ 9), tetrafunctional units remain rather isolated and serve to cross link PDMS chains. An interconnected network of difunctional (D) and tetrafunctional (Q) units is formed when the amount of TEOS increases (DEDMS/TEOS ≤ 2).

Transition-metal alkoxides (Ti, Zr) do not act only as cross-linking agents between difunctional units. They are also able to catalyze the condensation of siloxane units into long PDMS chains. This catalytic role together with the tendency of transition-metal alkoxides to enhance phase separation are quite interesting for film formation. The presence of long chains cross-linked by segregated metal-oxo nano species in the film makes the material more flexible, allowing the deposition of crack-free transparent coatings several microns thick.

During the synthesis of hybrid networks via the hydrolysis and co-condensation of alkoxysilanes and metal alkoxides, nano or macrophase separation tends to occur as a result of thermodynamic incompatibilities between organic and inorganic components. This process has to be carefully controlled. Many properties of the hybrid materials depend on this parameter (transparency, porosity, film-forming ability). Phase separation is a consequence of the segregation between hydrophilic oxide species and lypophilic organic moieties. The extent of phase separation can be tuned via the control of the polymerization rates of both components and by the formation of chemical bonds between organic and inorganic components R—Si—O—M (M = Si, Ti, Zr, Al, V, etc.). However, heterometallic oxo bonds are not very stable toward hydrolysis: Their stability depends mainly on the water content, pH, the nature of the metal M, and the number of organic R groups carried by the silicon atom.

Transition-Metal Hybrid Precursors Most metal atoms are less electronegative than silicon. Therefore, M—C bonds are more ionic and are rapidly cleaved by water. Complexing organic ligands have then to be used and very few hybrid gels have been synthesized with metals other than silicon. The role of complexing ligands such as carboxylic acids and β-diketones as network modifiers was described in the first part of this chapter. Polymerizable carboxylic acids have recently been used for the sol-gel synthesis of zirconium oxide–based hybrid gels. Methacrylic acid is first used to complex zirconium alkoxide precursors, followed by UV copolymerization of methacrylate-functionalized zirconium oxide sols, γ-methacryloxypropyltrimethoxysilane, and organic monomers. However, as complexing ligands, carboxylates are less strong than β-diketones. They can be partly removed upon hydrolysis, and many chemical bonds between organic and inorganic components are broken.

Chelating ligands such as acetyl acetoxy ethyl methacrylate (AAEM) having both a strong chelating group and a polymerizable methacrylate function have also been used.[33,51] Both organic and inorganic polymerizations were run simultaneously by hydrolyzing the AAEM-modified precursor solutions in the presence of a radical initiator. Both inorganic polycondensation and organic polymerization occur and the chemical bonds between organic and inorganic moieties are preserved. These hybrids can be pictured as zirconium oxo species chemically bonded to polymeric methacrylate chains via β-diketo complexing functions. For high complexation ratios, zirconium-oxo clusters are connected through long polymethacrylate chains, while for lower complexation ratios, short organic chains cross-link larger zirconium oxo particles. These hybrid copolymers could be better described as highly porous micro gels made of inorganic blocks embedded in a hybrid matrix.

The principle used for the synthesis of these hybrid zirconium-oxo-poly-AAEM copolymers can be extended to many other systems. It rests on the competition between inorganic (oxo, hydroxo) and organic ligands (alkoxo, complexing ligands) toward the transition-metal atom. Following the same principles, titanium alkoxides complexed with unsaturated organic groups (isogeunol, itaconic succinic anhydride, 2-acrylamido-2-methylpropanesulfonic acid) have been used as heterofunctional precursors[34] for the synthesis of TiO_2-based hybrid copolymers.

Functionalized Polymers and Oligomers

Polymeric Bridges For many years metal–organic precursors (silicon alkoxides, titanium alkoxides, titanium acetylacetonates, triethanolamine titanium chelates) have been used for the reinforcement of natural polymers such as polysaccharides, cellulosic materials, or vegetable oil derivatives. Hydrolyzed oxo species behave as cross-linking reagents with the hydroxyl groups of the organic polymers. These cross-linking reactions modify the rheological behavior of the polymer and enable better processing.

In sol-gel chemistry, covalent bonds between organic and inorganic components can be introduced conveniently via the functionalization of organic preformed polymers with hydroxyl —CH_2—OH, silanol Si—OH, or trialkoxysilyl —Si—$(OR)_3$ moieties. These reactive groups undergo condensation reactions with metal alkoxides $M(OR)_z$. A wide range of polymers have been used in the sol-gel literature, but most papers deal with polydimethylsiloxane (PDMS), polyimides, and polyamides using $Si(OR)_4$, $Ti(OR)_4$, or $Zr(OR)_4$ as cross-linking alkoxides.

These hybrids are interesting for the synthesis of high-performance or multifunctional polymeric materials. Their mechanical properties depend strongly on the degree of phase dispersion that can be reached in the nanocomposite materials. A real mastery of the process involves the control of many parameters, such as the ratio between organic and inorganic components, the nature and amount of catalysts, the hydrolysis ratio, the temperature, the chemical nature and molecular weight of functionalized macromonomers, the number of anchoring groups, the cross-linking alkoxides, or the solvent. Small changes in the experimental procedure lead to a wide variety of hybrid organic–inorganic materials.

Almost 10 years ago Mark[36] and Wilkes[37] reported the synthesis of hybrid materials from PDMS and TEOS. These nanocomposites, containing both ceramic and polymeric components, were called ceramers. They showed that the condensation of $Si(OEt)_4$ with silanol or alkoxysilyl groups at the end of telechelic PDMS polymers increases their modulus, toughness, and density. Organic and inorganic components are actually formed of Si—O—Si bonds, so that they are highly compatible and phase separation can be avoided. TEM experiments show that the silica particles in PDMS are less than 100 Å in diameter, with a relatively narrow size distribution and very little aggregation. More homogeneous hybrids are formed as the molecular weight of PDMS decreases but the hardness decreases, too. Oligomers with higher functionality are obtained when siloxane oligomers are end-capped with triethoxysilane rather than silanol groups. The chance of reacting with hydrolyzed TEOS increases, leading to more homogeneous products.

Homogeneity depends on the hydrolysis conditions. The hydrolysis rate of the alkoxide has to be fast enough for co-condensation to occur. Otherwise, the self-condensation of the functionalized PDMS leads to the formation of larger polymers and phase separation between organic and inorganic components. Therefore, acid catalysis is currently used to promote the hydrolysis of TEOS. The structure of PDMS-TEOS hybrids depends strongly on catalysts. Small, well-defined silica particles are formed with base catalysis, while acid catalysis leads to poorly defined, "fuzzy" particles. This is consistent with the well-known hydrolysis–condensation behavior of TEOS. Acid catalysts gives polymeric structures that are less branched and less compact than those obtained from a basic catalyst.

The role of acid catalysts are actually more complex. They not only change the hydrolysis–condensation rates of TEOS but may also change the siloxane distribution via the hydrolysis of Si—O—Si bonds. As a consequence, even for a given concentration of PDMS and TEOS, very different materials can be obtained, depending on the amount of acid catalyst. This led to the discovery of "rubbery ormosils" by Mackenzie.[52] These materials exhibit properties similar to those of organic rubbers even with an amount of silica close to 70 wt %. Hydrolysis is performed with a high acid concentration ($H^+/Si = 0.3$). The resulting materials should consist of PDMS chains of medium length and small porous silica particles. These pores provide the free volume needed for chain motion. As a consequence, PDMS chains can curl and uncurl in the presence of external stress and the hybrid materials exhibits rubbery elasticity.

The mechanical properties of hybrid materials, made of ethoxysilyl terminated PDMS-TEOS or ethoxysilyl-terminated PTMO-TEOS, can be improved by adding transition-metal oxides such as TiO_2 or ZrO_2. However, $Ti(OR)_4$ or $Zr(OR)_4$ alkoxides cannot be added directly; otherwise, the fast hydrolysis of these alkoxides would lead to the precipitation of TiO_2 or ZrO_2 particles. Therefore, care has to be taken to match the reactivity of both components. As for mixed oxides, this can be done via the pre hydrolysis of TEOS or by chemical modification of $Ti(OR)_4$.

Organic Molecular Building Blocks Organic molecular building blocks can be used as bridging units for the synthesis of hybrid materials. Inorganic components arise from the hydrolysis and condensation of silicon alkoxides. Organic bridges are used to modify the properties of the oxide materials. For example, porosity can be created by inserting organic spacers into a silicate network. Hexafunctional alkoxysilane precursors, in which trialkoxysilane groups are separated by rigid phenylene spacers, have been developed by Loy and Shea.[53] After drying, these precursors lead to highly porous materials with specific surface areas of around 1000 m^2/g. These materials show promising potential applications in the field of chromatographic supports, membranes (pore template strategies), catalyst and catalyst supports, optical devices (NLO), protective coatings, and as encapsulants (e.g., to trap quantum dots) (Fig. 9.11).

More recently, *star gels* have been described by Sharp at DuPont.[54] Precursor molecules are made of an atomic, linear, or cyclic core, with a number of flexible radial arms terminated by $Si(OR)_3$ groups. When hydrolyzed, these precursors form network structures whose elements are interconnected by flexible cross-links. Clear, compliant glasses and thick films have been made from star gels.

Inorganic Building Blocks Oxide particles are usually polydispersed and randomly distributed within the organic matrix. More regular structures can be obtained via the functionalization of already preformed oxo-clusters such as $[Si_8O_{20}]^{8-}$ (Fig. 9.12). Hydrosilylation reactions are performed to

$Si(OMe)_3$

$(MeO)_3Si$ $Si(OMe)_3$

$(MeO)_3Si-C{\equiv}C-Si(OMe)_3$

$(MeO)_3Si-C{\equiv}C-C{\equiv}C-Si(OMe)_3$

$(MeO)_3Si$ $Si(OMe)_3$

$(MeO)_3Si$ $Si(OMe)_3$ n

n = 1, 2, 3

$(MeO)_3Si$ S $Si(OMe)_3$

a : Trimethoxysilylated organometallic gel precursors

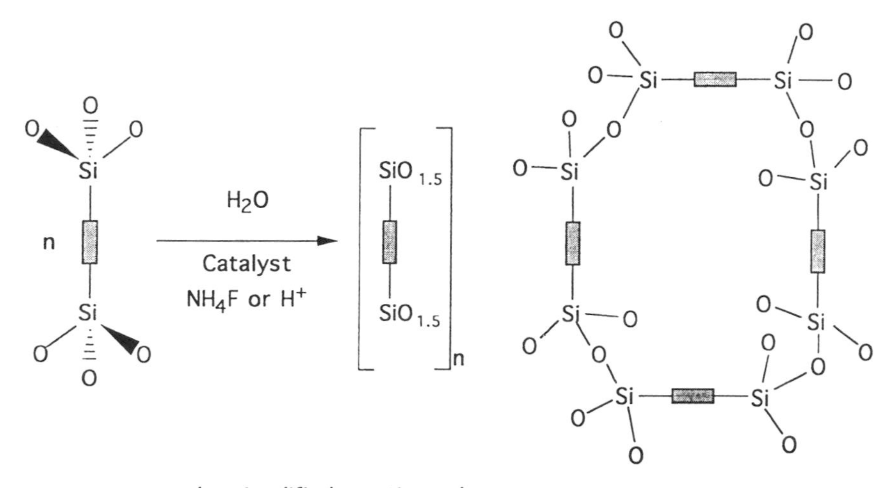

b : simplified reaction scheme

Figure 9.11 Molecular structure of polyfunctional alkoxysilane and formation of an open hybrid network.

bind similar inorganic building blocks ($[CH_3)_2HSi]_8Si_8O_{20}$ and $[CH_2{=}CH(CH_3)_2Si]_8Si_8O_{20}$) with other reactive siloxane species such as divinyl tetramethyldisiloxane, tetramethyl cyclotetrasiloxane, or polymethyl hydrogenosiloxane.[55, 56] The synthesis of silsesquioxane-siloxane copolymers from hydroxy functionalized polyhedral silsesquioxanes has been reported.[57] A silsesquioxane monomer $Si_8O_{11}R_8(OH)_2$ ($R = c\text{-}C_6H_{11}$) with a well defined structure including two reactive hydroxyl groups that can be converted into additional functionalities was used. This precursor reacts with a variety of difunctional silanes such as X—R—X [X = NMe; R = $SiMe_2$, $SiMe_2(CH_2{=}CH)$, $SiMe_2(CH_2)_2SiMe_2$]. These copolymers are highly soluble in $CHCl_3$ and THF. They look like clear plastics when hot pressed or deposited from solution as film. They decompose at about 500°C, a tempera-

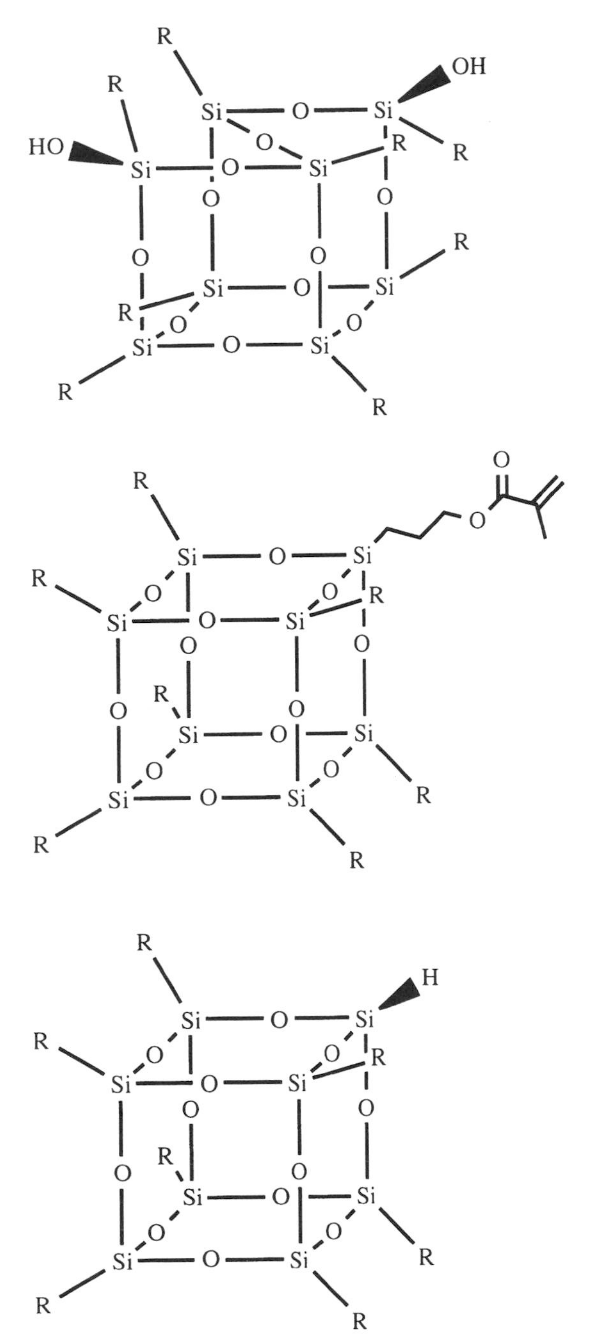

Figure 9.12 Functionalized polysilsesquioxane building blocks. (R = c-C_6H_{11}, c-C_5H_9.)

ture significantly higher than that reported for PDMS (350°C). This work was recently extended to a large family of oligomeric silsesquioxanes that were used as building blocks for the synthesis of silsesquioxane-based polymers and hybrid materials.[58]

Non-silica-based oxo-clusters can also be used. The hydrolysis of $RSn(OR')_3$ (R = butyl, butenyl, OR′ = O-*i*-Pr, O-*t*-Am) leads to $[(RSn)_{12}O_{14}(OH)_6](OH)_4$ molecular clusters.[59] The tin oxohydroxo core is a macrocation associated with two exchangeable hydroxy counter anions. This cagelike cluster is surrounded by 12 *n*-butyl chains, which prevent further condensation. Several strategies can be proposed to make hybrid macromolecular networks:

1. Linkage of the $[(RSn)_{12}O_{14}(OH)_6]^{2+}$ macrocations through electrostatic interaction and hydrogen bonds, by difunctional carboxylates (adipate, dodecanoate, terephthalate) via the substitution of the hydroxy counter anions.[60]
2. Functionalization of the macrocations with polymerizable carboxylates (such as methacrylates) through electrostatic interactions, followed by the organic polyaddition of the double bonds of the methacrylate function with methacrylate monomers.
3. Organic polymerization of the double bond carried by the polymerizable organic function covalently bonded to tin.

Forming a chemical bond between organic species and transition-metal oxide clusters is more difficult. M—C bonds are more polar than Si—C or Sn—C and would be broken in the presence of water. Complexing polymerizable organic ligands such as methacrylic acid and allyl acetoacetate have then to be used. Molecular clusters such as $Nb_4O_4(O\text{-}i\text{-}Pr)_8(OMc)_4$, $Ti_6O_4(OEt)_8(OMc)_8$, and $Zr_{10}O_6(OH)_4(O\text{-}n\text{-}Pr)_{18}(AAA)_6$ (OMc = methacrylate, AAA = allylacetoacetate) are formed via the controlled hydrolysis of metal alkoxides in the presence of these complexing agents.[5] They are made of a metal oxo core surrounded by polymerizable complexing ligands. Further hydrolysis has to be avoided to preserve these clusters. Therefore, organic polymerization must be performed in anhydrous organic solvents without nucleophilic species.

Organic functionalization can also be obtained via M—O—Si—C bonds as reported for some organometallic derivatives of polyoxometallates.[61] These organically functionalized compounds were synthesized by reacting a trichlorosilane, $RSiCl_3$ (R = vinyl, allyl, methacryl, styryl), with the lacunar Keggin polyanion $K_4SiW_{11}O_{39}$, giving rise to $SiW_{11}O_{40}(SiR)_2$. Organic polymerization is then performed with a radical initiator, leading to the formation of polymethacrylate or polystyrene chains. As a result, a hybrid macromolecular network made of silicon-modified polyoxotungstates cross-linked via polymeric chains is obtained.

9.4 POLYMERIC ROUTE TO NONOXIDE CERAMICS

The sol-gel process is a chemical route for preparing oxide glasses and ceramics. In a similar way, preceramic polymers can be designed for use as precursors for nonoxide ceramics.[62-64] The principle of such a synthetic approach is summarized below:

$$\boxed{\text{Polymer}} \xrightarrow{\Delta} \boxed{\text{Ceramics}}$$

$$—\text{M}—\text{X}—\text{M}—\text{X}—\text{M}— \qquad \text{M}_m\text{X}_x$$

(M = Si, Al, B etc. and X = C, N

This process is an extension to the well-known polymer pyrolysis route to carbon materials, which consists of subjecting carbon-based polymers to controlled pyrolysis to convert them to carbon bodies. Most of the commercial carbon materials (e.g., carbon–carbon composites, glassy carbon foams, graphite cloth, fibers) depend significantly or exclusively on this synthetic approach. The importance of this processing route for making carbon bodies is derived not only from its versatility and reasonable cost, but also from the basic limitations of other processing methods, such as melt processing or sintering.

Similar processing limitations for other ceramic materials have focused increased attention on pyrolysis of polymers containing Si, C, N, and B. This interest stems from the importance of ceramic products such as SiC, Si_3N_4, B_4C, AlN, and BN and from the potentially versatile and practical processing of these important ceramics that polymer pyrolysis allows. As for carbon, such a synthetic route introduces the important potential of making fibers as well as foams, coatings, and bulk bodies. The latter may involve the use of the polymer as a binder in normal ceramic powder processing or in the formation of ceramic matrix composites (CMCs). The main example in this field is Nicalon fiber, commercialized in Japan and prepared by the Yajima route, which consists of three main processing steps:[66] (1) spinning of a preceramic polymer, (2) cross-linking of the preceramic polymer fiber (curing step), and (3) thermal conversion of the preceramic polymer fiber into SiC at 1000 to 1200°C (pyrolysis step).

The yield of the ceramic residue, as well as characteristics ranging from shape retention of the original polymeric body to the details of the microstructure, is affected significantly by the choice of the precursor material, its characteristics, and the pyrolysis conditions. These variables provide a substantial challenge in precursor selection and in understanding the pyrolysis process, but lead to product versatility and a variety of technological applications.

9.4.1 Criteria for Useful Preceramics

A polymer must fulfill several criteria to be useful as a ceramic precursor:[65]

1. *Rheology*. The type of shaping method used (drawing or extrusion for fibers, impregnation for composites, etc.) engenders constraints on what is considered useful polymer rheology. For fiber applications it is commonly desirable to have non-Newtonian viscoelasticity such that during spinning, the polymer will flow readily without necking. Non-Newtonian viscoelastic behavior can be obtained either by using high-molecular-weight, chain-entangled linear polymers or highly branched oligomers with a gel-like nature. All spinnable preceramics developed to date are low-molecular-weight, highly branched, gel-like oligomers {e.g., polycarbosilane (PCS), $[MeHSiCH_2]_x$, Me = CH_3}.
2. *Latent reactivity*. Once shaped, the precursor fiber must retain some chemical reactivity that allows it to be rendered infusible. Otherwise, during pyrolysis, the integrity of the object will be lost (e.g., by melting or creep) long before the chemical constituents are transformed into ceramic. Infusibility is commonly obtained through reactions that provide extensive cross-linking. The Yajima polycarbosilanes are normally cross-linked by air oxidation.
3. *Pyrolytic degradation*. In current-generation preceramic polymers, the rates and mechanisms by which extraneous organic ligands are removed as gaseous products during the pyrolytic transformation of the precursor body into the ceramic body must be carefully controlled to ensure uniform densification and to prevent retention of impurities and/or the creation of gas generated flaws (e.g., pores). For example, if the pyrolytic degradation of $[MeHSiCH_2]_x$ was to proceed optimally, the overall process would be

$$[MeHSiCH_2]_x \xrightarrow{\Delta} xCH_4 + xH_2 + xSiC \qquad (24)$$

   Unfortunately, almost half of the excess carbon [e.g., every other $Me(CH_3)$ group] is retained in the final ceramic fiber.
4. *Ceramic yield and density changes*. The volume changes associated with pyrolytic conversion to ceramic materials must also be minimized, to maximize control of the final body dimensions, porosity, and densification-induced stresses. Therefore, it is important to formulate preceramics that contain the minimal amounts of extraneous ligands that permit satisfaction of the foregoing criteria, yet provide high weight percent conversions (ceramic yield) and maximum densification during conversion to ceramic product. Consequently, those preceramic polymers that require extraneous carbon containing ligands for spinnability, stability, or because of monomer availability (cost) are likely to contain the minimum necessary, typically as methyl groups.

5. *Selectivity and microstructure.* The architecture of the preceramic monomer unit can control selectivity to ceramic products and to some extent microstructure. For example, pyrolysis of $[MeHSiNH]_x$ leads to SiC_xN_y ceramics at temperatures of > 600°C, whereas pyrolysis of the isostructural $[H_2SiNMe]_x$ gives Si_3N_4 and free C.[66] To date, only limited studies on structure–product correlations for SiC precursors have appeared. The major concern in choosing a precursor system for a target ceramic is the facility with which synthesis permits manipulation of the polymer architecture, independent of the chemical composition. This in turn will permit chemically similar precursors to be made if fine tuning is necessary, or grossly different precursors if the first product is far from the expected product.

The general criteria noted above serve as a basis for the selection of potential organometallic precursors for processing both oxide and nonoxide ceramics. For specific materials, additional criteria can also play a role, including ease of synthesis, purification, and stability toward air and moisture.

9.4.2 Yajima Route to SiC Fibers

Most of the research in the field of nonoxide preceramic polymers has been devoted to silicon carbide precursors.[65, 67] The SiC preceramic polymers belong to the following families:

Polysilanes: $\left[\overset{|}{\underset{|}{Si}} \right]_n$

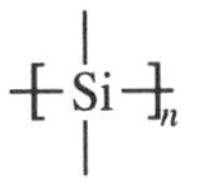

Polycarbosilanes: $\left[\overset{|}{\underset{|}{Si}}-CH_2 \right]_n$ $\left[\overset{|}{\underset{|}{Si}}-C-C \right]_n$

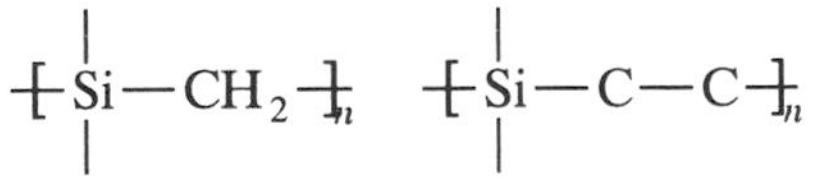

Due to the instability of compounds containing (Si=Si) and (Si=C) π-bonds, organosilicon polymers cannot be obtained using methods similar to the classical olefin polymerization. The following basic synthetic routes to polysilanes and polycarbosilanes have been used:

Dehalocoupling:

$$MeRSiX_2 + M \longrightarrow \left[MeRSi \right]_x + MX \tag{25}$$

Dehydrocoupling:

$$RSiH_3 \xrightarrow{\text{catalyst}} \left[RSiH \right]_x + H_2 \tag{26}$$

Catalytic disproportionation of halo-organodisilanes:

$$[RSiX_2]_2 \rightarrow RX_2Si\text{-}[RXSi]_x\text{-}SiX_2R + RSiX_3 \qquad (27)$$

Ring-opening polymerization:

$$[R_2SiCH_2]_2 \xrightarrow{\text{catalyst}} \text{-}[R_2SiCH_2]_x\text{-} \qquad (28)$$

Hydrosilylation:

$$CH_2{=}CH{-}SiR_2H \xrightarrow{\text{catalyst}}$$

$$H\text{-}[SiR_2CH_2CH_2]_x\text{-}[SiR_2CH(CH_3)]_y\text{-}CH{=}CH_2 \qquad (29)$$

The dehalocoupling reactions have received special attention, especially after the pioneering contribution of Yajima and his co-workers during the 1970s, who demonstrated that high-strength, high-modulus silicon carbide fibers could be prepared starting from a polycarbosilane.[68]

Precursor Chemistry PCS chemistry, as developed by Yajima, relies on two well-studied chemical reactions:

1. *Alkali-metal dehalocoupling of chlorosilanes.* Polydimethylsilane is obtained from dimethyldichlorosilane:

$$x(CH_3)_2SiCl_2 + 2xM \xrightarrow{\text{refluxing toluene}} \text{-}[(CH_3)_2Si]_x\text{-} + 2xMCl \qquad (30)$$

$(M = Li, Na)$

2. *Kumada rearrangement.* Around 400 to 450°C, the polysilane is converted into a polycarbosilane after insertion of methylene groups into the Si—Si backbone.

$$\text{—}\underset{|}{\overset{CH_3\,|}{Si}}\text{—}\underset{|}{\overset{|}{Si}} \xrightarrow{400^\circ C} \text{—}\underset{|}{\overset{H\,|}{Si}}\text{—}CH_2\text{—}\underset{|}{\overset{|}{Si}}\text{—} \qquad (31)$$

Yajima's earliest papers in 1975 describe the preparation of PCS, $\text{-}[MeHSiCH_2]_x\text{-}$, from dodecamethylcyclohexasilane, $\text{-}[(CH_3)_2Si]_6\text{-}$.[68] This cyclic oligosilane was prepared via reaction (30) using lithium metal. This synthetic approach to PCS using Li was technically difficult. It required fractionation steps that were exceedingly time consuming to get the final polymer. However, this polymer could be either melt spun or dry spun from benzene solutions to form 10 to 20-μm-diameter precursor fibers. Pyrolysis of these fibers provided the first SiC fibers prepared from preceramic polymers. Unfortunately, there are several disadvantages to this preparative method, as mentioned previously. In 1976, Yajima et al. published a new, simple, and more economical method for preparing PCS wherein molten sodium (in refluxing toluene) is used instead of Li.[69] The resulting polydimethylsilane (PDMS) is obtained in yields of over 80%.

The Kumada rearrangement [reaction (31)] is the critical step in the generation of PCS with superior processing properties. Consequently, considerable effort has been invested in optimizing the PDMS/PCS conversion process. In the most common method, PDMS is heated to 320°C in flowing Ar, melted, refluxed for 5 h, and then heated to 470°C to remove volatile materials. The resulting PCS (PC-470) is obtained in a 50 to 55% yield and has a number average molecular weight, $Mn = 1500$ Da.

Efforts to improve the synthesis have focused on (1) increasing the yield, (2) lowering the reaction temperature, (3) lowering the reaction time, and (4) eliminating the need for an autoclave. Two alternative approaches have been explored. The first involves heating to 470°C (rather than 320°C) in an autoclave (PC-470). However, use of an autoclave makes the process technically difficult and increases the cost. Alternatively, several groups explored the use of Lewis acid catalysts such as polyborodiphenylsiloxane (BDPSO), prepared by reaction of boric acid with Ph_2SiCl_2, and normal pressures (PC-B). Polymer molecular weights are in the range $Mn = 1000$ to 2000 Da. Both PC-470 and PC-B are produced on an industrial scale.

The formation of PCS (PC-470) from PDMS was proposed by Yajima et al. to occur via the Kumada rearrangement. The transformation of the polysilane backbone into a polycarbosilane backbone is assumed to occur via radical reactions.[70] The primary reaction involves pyrolytic cleavage of Si—Si bonds, whose energy is lowest (222 kJ/mol), compared to Si—C (318 kJ/mol) and C—H (414 kJ/mol). It leads to the formation of silapropylene- and Si—H-containing moieties, which subsequently react to give $Si{-}CH_2{-}Si$ linkages. Ideally, total conversion should lead to the formation of polysilapropylene or PCS, $\left[MeHSiCH_2 \right]_x$.

$$-Si(CH_3)_2-Si(CH_3)_2-Si(CH_3)_2- \xrightarrow{\Delta} -Si(CH_3)_2-Si^{\circ}(CH_3)(^{\circ}CH_2)- + H-Si(CH_3)_2- \longrightarrow$$

$$-Si(CH_3)_2-SiH(CH_3)-CH_2-Si(CH_3)_2- \xrightarrow{\Delta} -Si^{\circ}(CH_3)(^{\circ}CH_2)- + H-SiH(CH_3)-CH_2-Si(CH_3)_2- \longrightarrow$$

$$-SiH(CH_3)-CH_2-SiH(CH_3)-CH_2-Si(CH_3)_2- \xrightarrow{\Delta} \left(-SiH(CH_3)-CH_2- \right)_n$$

The insertion of CH_2 groups into the Si—Si backbone coincident with the formation of Si—H bonds is demonstrated clearly in the IR spectrum of the resulting PC-470 by the appearance of new bands at 1020 and 1355 cm^{-1} ($Si—CH_2—Si$) and at 2100 cm^{-1} (ν_{Si-H}) and by UV spectra that indicate the disappearance of the σ–σ^* absorptions of the polysilane backbone. However, the typical chemical analysis for PC-470 ($SiC_{1.77}O_{0.03}H_{3.7}$) is not that expected for $\left[MeHSiCH_2 \right]_x$, ($SiC_2H_{6.10}$).

^{29}Si NMR experiments in solution and on powders using magic angle spinning (MAS) techniques indicate the presence of two different Si units: one corresponding to SiC_3H, where the Si atom is bonded to three C atoms and one H atom as in $\left[MeHSiCH_2 \right]_x$ and one to SiC_4 ($\delta \approx 0$ ppm), where the Si atom is surrounded by four C atoms. PC-470 is thus a copolymer consisting of two types of monomer units:

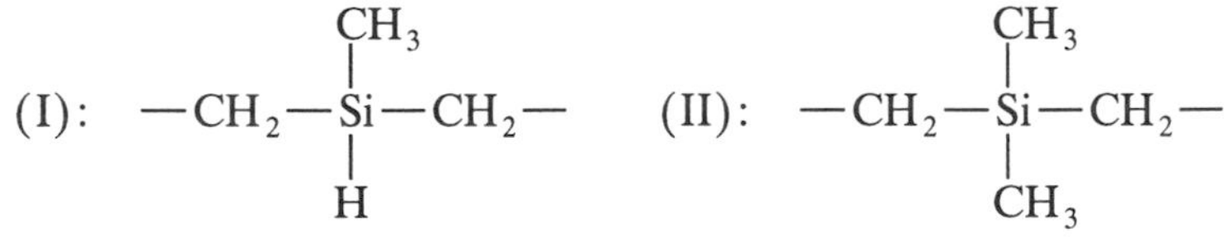

$$\text{(I):}\quad -CH_2-\underset{\displaystyle H}{\overset{\displaystyle CH_3}{\underset{|}{\overset{|}{Si}}}}-CH_2- \qquad \text{(II):}\quad -CH_2-\underset{\displaystyle CH_3}{\overset{\displaystyle CH_3}{\underset{|}{\overset{|}{Si}}}}-CH_2-$$

The presence of type II units suggests the existence of extensive cross-linking between the chains, leading to a broad distribution of Si and C sites in agreement with the first structural model proposed:[71]

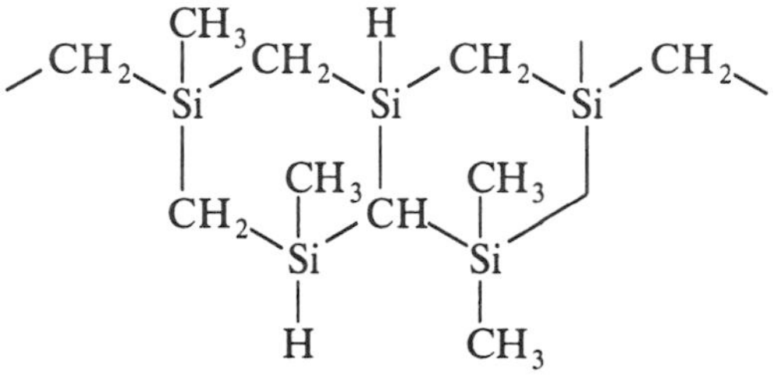

```
        CH3         H
 /CH2\   |   /CH2\  |  /CH2\   |   /CH2\
        Si          Si        Si
         |          |          |
         |   CH3    |   CH3   /
        CH2\  |  /CH\  |  /
              Si        Si
              |         |
              H         CH3
```

Proposed structural model for polycarbosilane PC-470.

Pyrolysis Chemistry Pyrolysis of PC-470 in argon gives typical ceramic yields of about 60%. According to TG-DTA and gas evolution curves, the pyrolysis process can be divided into six stages:[65]

1. Below 400°C, low-molecular-weight polycarbosilanes are removed.
2. Between 400 and 550°C, the molecular weight of PC-470 increases: the weakest bonds in the polymer, the Si—H bonds, are consumed as shown by IR and ^{29}Si MAS-NMR, with no evidence for the formation of Si—Si bonds. Thus it was concluded that Si—H bonds condense with $Si—CH_3$ groups, leading to $Si—CH_2—Si$ linkages.
3. The third stage, from 550 to 850°C, corresponds to the transformation into an inorganic material with the elimination of Si—H, $Si—CH_3$,

and Si—CH_2—Si moieties, as characterized by a loss of H_2 and CH_4. The changes, observed in the ^{29}Si and ^{13}C MAS-NMR spectra, are associated with the cross-linking reactions that occur around the C atoms. The increase in linewidth is due to an increase in the Si site distributions; almost all Si sites are surrounded by four C atoms, but the nature of the C atoms changes as the decomposition proceeds with increasing time and temperature. The conversion of the polymeric network into an inorganic SiC network can be illustrated as follows:

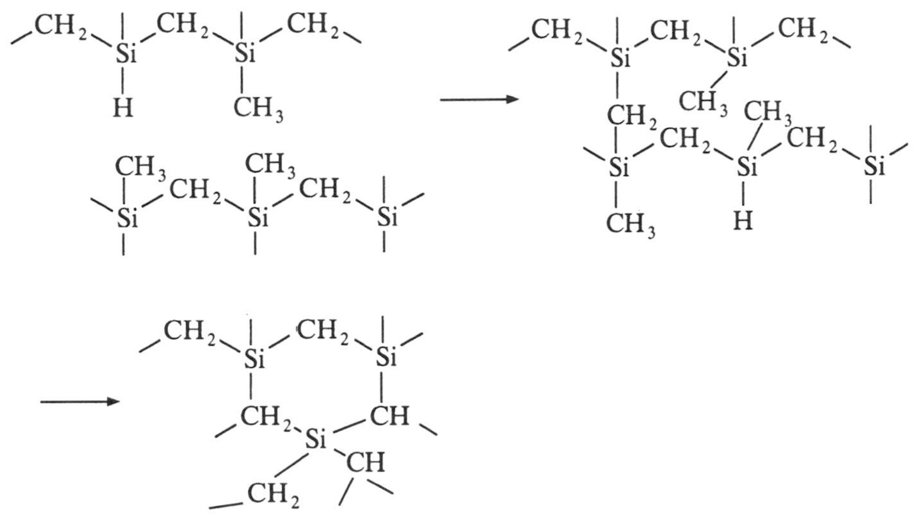

```
/CH2\ | /CH2\ | /CH2\             /CH2\ | /CH2\ | /CH2\
     Si       Si                       Si       Si
     |        |                        |       /
     H        CH3          ---->       CH2  CH3   CH3
                                       | /CH2\  / /CH2\ |
  CH3      CH3                        \Si       Si      Si/
  | /CH2\  | /CH2\ |                   |        |       |
 \Si       Si      Si/                 CH3      H
  |        |       |

        /CH2\ | /CH2\ |
             Si       Si/
             |        |
 ---->      /CH2\   /CH \
                 Si
               /   \CH
            _CH2   / \
```

Suggested mechanism for conversion of PC-470 to ceramic product.

In this temperature range, additional competing decomposition processes lead to the formation of C=C double bonds as a prelude to the formation of a free-carbon phase. At 800°C, the material can be described as an hydrogenated SiC with an average of two CH groups per Si atom and excess carbon.

4. The fourth stage of the pyrolysis occurs between 850 and 1000°C, characterized by minimal weight loss and no gas evolution. This stage is considered as transitional.
5. The fifth stage, between 1000 and 1200°C, marks the onset of crystallization. At 1000°C the XRD pattern is indicative of β-SiC, with crystallites around 2 nm in size as seen by TEM. However, the sample appears heterogeneous as some amorphous (about 20%) SiC-like particles are also observed, along with the presence of aromatic carbon layers, two or three piled up to form basic structural units (BSU). At 1200°C, TEM shows the disappearance of the amorphous phase and the association of BSUs edge by edge to form oriented rims of free graphitic carbon around the SiC crystals.

6. The last stage of the pyrolysis process occurs above 1200°C and corresponds to a crystal growth process. However, even at 1500°C, the XRD patterns are not well defined, probably as a consequence of the presence of excess carbon and grain sizes ranging from 5 to 13 nm, depending on the sample.

Properties of Fibers PCS precursor fibers can be processed by melt-spinning the polymer at 300 to 350°C. To avoid further melting during pyrolytic transformation to a ceramic fiber, a curing step is performed wherein the fiber is heated in air. The cross-linking reactions are based on the formation of siloxane units:

$$2\text{—}\overset{|}{\underset{|}{\text{Si}}}\text{—H} + O_2 \longrightarrow \text{—}\overset{|}{\underset{|}{\text{Si}}}\text{—O—}\overset{|}{\underset{|}{\text{Si}}}\text{—} + H_2O \qquad (32)$$

$$2\text{—}\overset{|}{\underset{|}{\text{Si}}}\text{—OH} \longrightarrow \text{—}\overset{|}{\underset{|}{\text{Si}}}\text{—O—}\overset{|}{\underset{|}{\text{Si}}}\text{—} + H_2O \qquad (33)$$

The oxygen content in the polymeric fiber depends on the curing temperature and time. Pyrolysis of the cured PCS fibers provides ceramic yields of 80 to 85%, 15 to 20% higher than for uncured PCS, which correlates directly with the presence of the oxygen cross-links. The precursor fibers undergo transformation from polymer to ceramic at 500 to 800°C. Oxygen is retained during the polymer-to-ceramic transformation, as shown by the persistence of peaks due to SiC_xO_{4-x} species in XPS and ^{29}Si MAS-NMR experiments. The oxygen content in the final fiber can range from 10 to 15 wt %. The commercial Nicalon fiber is thus a Si—C—O fiber rather than a Si—C fiber. Structural models have been proposed based on numerous characterization investigations, which describe the system as "a glassy silicon oxycarbide," in which extremely small SiC crystallites (< 3 nm) and free, graphitic carbon particles of less than 1 nm in size exist as minor phases in a continuous Si—C—O major phase that represents 40 to 50% of the molar fraction of Si atoms.[72]

Nicalon fibers with diameters of 10 to 20 μm exhibit tensile strengths $\geq$ 2.3 GPa and elastic moduli of $\approx$ 200 GPa.[73] The mechanical properties dramatically decrease when the fibers are heat-treated above 1200°C (tensile strength 0.56 GPa, Young's modulus $\approx$ 100 GPa at > 1500°C). These changes can be ascribed to the extensive structural modifications that occur. Indeed, above 1200°C, loss of CO and possibly SiO occurs due to carbothermal reduction reactions:

$$\begin{aligned} SiO_2 + 3C &\longrightarrow SiC + 2CO \\ SiO_2 + C &\longrightarrow SiO + CO \end{aligned} \qquad (34)$$

At 1400°C, rapid growth of β-SiC crystallites occurs simultaneously (crystallite sizes $\approx$ 7 nm). All of these changes have detrimental effects on the fiber properties.

Advantages and Drawbacks of the Process The main advantages of the Yajima polymer can be summarized as follows:

- Good spinnability
- High ceramic yield
- High mechanical properties of the ceramic fibers.

Despite these advantages, this process has some drawbacks:

- *Multistep synthetic procedure*. The elaboration of PCS relies first on the preparation of polydimethylsilane (PDMS), which is infusible, insoluble, and therefore intractable, a key problem in processing. Moreover, dehalocoupling is a biphasic reaction (metal + chlorosilane) with highly reactive alkali metals, which requires purification steps to remove by-product salts. Then the PDMS-PCS conversion requires use of an autoclave at 470°C unless a Lewis acid catalyst is introduced that allows the reaction to occur at normal pressures. The overall synthetic procedure is thus technically quite difficult.
- *Undefined polymer structure*. Despite extensive structural investigations, the structure of this branched polymer is not yet well defined and is quite dependent on the detailed procedure, thereby introducing potential batch-to-batch variability.
- *Oxidative curing step*. The presence of oxygen in the ceramic fiber is detrimental to the mechanical properties at high temperature.
- *Presence of free carbon*. Excess carbon in the fibers should not have a good influence with regard to oxidation resistance. Excess carbon comes from the pyrolytic degradation of organic groups present in the starting PCS, which is characterized by a high C/Si ratio (> 2).

9.4.3 Alternative Routes to SiC

Linear Polymers Linear well-defined polycarbosilanes can be prepared via ring-opening polymerization of various derivatives of 1,3-disilacyclobutane:[65]

$$\text{cyclo-}[R_2Si(CH_2)_2SiR_2] \xrightarrow{\text{catalyst}} \left[\underset{R}{\overset{R}{|}}\!\!\text{Si}-CH_2 \right]_n \tag{35}$$

- *Poly(dimethylsilaethylene)*. High-molecular-weight poly(dimethylsilaethylene) can be prepared from 1,1,3,3-tetramethylsilacyclobutane. This polymer gives very low ceramic yields, due primarily to its linear structure and the lack of reactive groups. Pyrolysis causes depolymerization reactions that lead to the formation of volatile cyclic species rather than branches and cross-links. For linear polymers, latent functionality is absolutely necessary to allow cross-linking between chains during the first pyrolysis steps.

- *Poly(silapropylene).* Two synthesis have been developed for poly(silapropylene): ring-opening polymerization of 1,3-dichloro-1, 3-dimethyl-1,3-disilacyclobutane or through chlorination and reduction with $LiAlH_4$ of poly(dimethylsilaethylene). The as-prepared polymers give poor ceramic yields (5 to 20%), but a preliminary heat treatment under nitrogen (400°C) favors cross-linking reactions, leading to higher ceramic yields (60 to 70%).
- *Poly(silaethylene).* A patent claimed the preparation of linear poly(silaethylene) (R=H) that will have the stoichiometric C/Si ratio for SiC. High ceramic yield (85%) and formation of pure SiC at 900°C were reported.

Right C / Si Stoichiometry Excess carbon in the ceramic fiber can lower its resistance toward oxidation at high temperature. One way to minimize this carbon phase is to start with a preceramic polymer with a C/Si molar ratio close to 1. One example of such a polymer is polymethylsilane (PMS). Two synthesis have been developed:

- *Dehalocoupling of $MeHSiCl_2$ with K or Na.* Polymers with compositions $(MeSiH)_x(MeSi)_{1-x}$ can be obtained which are soluble for $x = 0.65$ to 0.85. The two main problems related to these polymers are low ceramic yields (15 to 20%) and the presence of excess Si in the final ceramic. They can be partially solved by modifying the polymers via hydrosilylation with organic or organosilicon compounds.
- *Dehydrocoupling of silanes.* PMS can be prepared in one step synthesis according to

$$x\mathrm{RSiH_3} \xrightarrow{\text{catalyst}} x\mathrm{H_2} + \left[\mathrm{RSiH}\right]_x \tag{36}$$

The usual catalysts are dicyclopentadienyl dimethyl metallocenes: Cp_2MMe_2 (M = Ti, Zr). The polymers provide good ceramic yields (70 to 75%, 900°C) and lead to relatively pure SiC. A drawback in this synthesis is that $MeSiH_3$ is flammable, as well as the resulting polymer with catalyst.

Nonoxidative Curing Process Once spun, the preceramic polymer fiber has to be rendered infusible before pyrolysis. In the case of the Yajima-PCS, the reactivity of the Si—H bonds is used in the oxygen curing step to cross-link via the formation of siloxane bonds. The introduction of oxygen is, as mentioned previously, detrimental for the properties of the fiber at high temperature. The most important oxygen-free curing methods are electron-beam radiation or γ-ray treatment under inert atmosphere. Radiation curing was applied to Yajima fibers, and new low-oxygen-content fibers called Hi-Nicalon are now available, which have improved thermal stability, higher elastic modulus, and creep resistance.[74]

Introduction of a Heteroelement To improve the high-temperature properties of Si—C fibers, heteroelements have been introduced into the preceramic polymer, to help in retaining the amorphous state of the ceramic at a higher temperature. Indeed, rapid SiC crystal growth seems detrimental to good mechanical properties.

The best example is the preparation of poly(titanocarbosilane) (PTC) by introducing titanium into a PCS-based polymer.[75] The titanium precursor is a titanium alkoxide that reacts with PCS. Fibers prepared from PTC with 1.5 to 4 wt % Ti are marketed by Ube Ind. under the trade name Tyranno. The characteristic of the fibers are the following: narrow distribution of diameters around 8.5 μm, elastic modulus $\approx$ 170 GPa, and tensile strength $\geq$ 3 GPa. However, heat treatment at 1300°C leads to severe degradation of the mechanical properties and a thermal stability inferior even to that of Nicalon.

9.4.4 Preceramic Precursors for Nitrides

Silicon Nitride Precursors The precursors for silicon nitride are polysilazanes (PSZ):[67]

$$\left[-\overset{|}{\underset{|}{\mathrm{Si}}}-\overset{|}{\mathrm{N}}- \right]_n$$

They can be prepared directly by ammonolysis or aminolysis of chlorosilanes:

$$\equiv\mathrm{Si}-\mathrm{Cl} + -\mathrm{RNH} \longrightarrow \equiv\mathrm{Si}-\mathrm{NR}- + -\mathrm{RNH}\cdot\mathrm{HCl} \qquad (37)$$

If R = H, perhydropolysilazanes are formed, precursors for pure Si_3N_4. If R contains carbon, PSZs will lead to Si—C—N ceramics if pyrolyzed under argon and to Si—N ceramics if ammonia is used.

- *Perhydropolysilazanes*. Spinnable PSZ with the right stoichiometry (Si/N = 1) can be prepared from ammonolysis of the complex $H_2SiCl_2 \cdot$ 2-pyridine. With such an approach, Tonen Corporation developed silicon nitride fibers of small diameter ($\approx$ 10 μm) which exhibit high tensile strength (2.5 GPa) and high tensile modulus (250 GPa). Excess Si can be present in the final ceramics, due to the loss of nitrogen-rich volatile species, but the stoichiometry can be adjusted by pyrolyzing under ammonia or hydrazine. Behavior similar to that of Nicalon fibers is observed at $T \geq 1300$°C, with a drop in the mechanical properties related to the crystallization of the fibers.
- *Poly(N-methylsilazanes)*. Polysilazanes in which various hydrocarbon groups are present on silicon and nitrogen have been developed for

Si—C—N ceramics. Presence of a second phase (silicon carbide or silicon carbo-nitride) could delay crystallization and thus give ceramics with improved high-temperature properties. Several syntheses have been developed; for example:

Aminolysis of dichlorosilane:

$$H_2SiCl_2 \xrightarrow{CH_3NH_2} [H_2Si{-}N(CH_3)]_4 + H{-}N(CH_3){-}\left[SiH_2{-}N(CH_3)\right]_n{-}H \qquad (38)$$

Transition-metal-catalyzed dehydrocoupling:

$$R_2SiH_2 + R'NH_2 \xrightarrow{\text{catalyst}} H_2 + \left[SiR_2{-}NR'\right]_n \qquad (39)$$

- *Nitridation of PS and PCS*. Pyrolysis of cross-linked PCS and PS under ammonia or hydrazine constitutes a general route to highly pure Si_3N_4. Nitridation and carbon removal occurs between 400 and 500°C according to

$$\equiv Si{-}CH_3 + NH_3 \longrightarrow \equiv Si{-}NH_2 + CH_4 \qquad (40)$$

Boron Nitride Precursors Boron nitride fibers, films, and composites are usually prepared by classical methods. For example, boron nitride fibers are prepared industrially by nitridation of boron oxide fibers. After ammonia treatment they are subjected to hot stretching at 2200°C to get almost pure BN fibers, but this step makes this method quite expensive. Some preceramic routes to boron nitride have thus been explored.[75] Most of the preceramic polymers for boron nitride are derived from borazines or boranes.

- *Borazine*. Borazine $[HB{-}NH]_3$ is isostructural with benzene.

$$(HBNH)_3$$

N- and/or B-substituted borazines can also be prepared and poly(borazines) can be obtained either by thermolysis or by cross-linking. The thermal polymerization of the borazine$[HB{-}NH]_3$ and its conversion to hBN has been studied.[76] A large variety of polymers have also been prepared by cross-linking borazines and several reactions can be used:[77]

$$=B{-}Cl \xrightarrow{Na} =B{-}B= + NaCl \qquad (41)$$

$$=B{-}Cl + H_2NR \longrightarrow =B{-}NR{-}B= \qquad (42)$$

$$=B{-}Cl + (Me)_3SiNRR' \longrightarrow =B{-}NRR' + (Me)_3SiCl \qquad (43)$$

One example using equation (43) is the reaction of $[ClB{-}NH]_3$ with hexamethyldisilazane

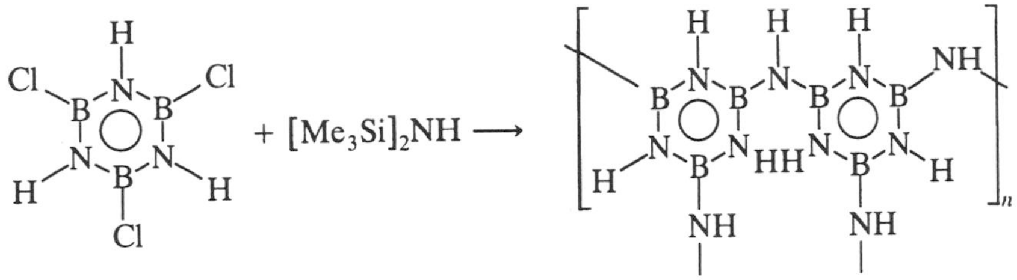

$$[ClB{-}NH]_3 + [Me_3Si]_2NH \longrightarrow [\ldots]_n \qquad (44)$$

- *Boranes*. Polymers can be prepared by reacting boranes with functional amines or ammonia. Boranes can be functionalized with unsaturated organic groups to ensure cross-linking of the polymer. Usually, these polymers are deficient in nitrogen and/or contain carbon, so that pyrolysis under ammonia is required to prepare boron nitride materials.

9.5 CONCLUSION

Chemistry has been employed for the processing of glasses and ceramics since the very beginning of these technologies several thousands of years ago. However, these strong connections with chemistry have been largely forgotten by practitioners of this field, and chemistry is currently only a very small component of glass and ceramic processing. The molecular precursor routes recast this long-standing history leading to the development of novel materials and new technologies.

The history of the sol-gel process is already very long.[78] The synthesis of silicon alkoxides was reported by Ebelmen in the middle of the nineteenth century.[1] The first application of these molecular precursors for materials science was for the preservation of stone around 1930, but sol-gel technology really began in 1939 when Geffcken and Berger of Jena Glaswerk Schott patented a novel treatment of glass surfaces based on silicon alkoxide and aluminum acetylacetonate.[79] These coatings, densified at low temperature, were used for their antireflection and scratch-resistant properties. In the early 1960s, the sol-gel process was applied for the production of ThO_2 and

UO_2 microspheres in the nuclear industry.[80] The development of sol-gel science began much later and the first International Workshop on Glasses and Ceramics from Gels was held in 1982.[81] Since then, hundreds of scientists have become involved in this field and sol-gel science is now reaching its maturity.

In the beginning, molecular precursors were designed mainly for the production of the usual glasses and ceramics. The aim was to develop cheaper processing routes and to improve the properties of these materials. The main applications were for the production of fine powders, coatings, and fibers.[2] In the field of nonoxide ceramics, the discovery of Yajima in 1975 that high-strength, high-modulus SiC fibers can be prepared from polycarbosilane was really the starting point of numerous investigations in the field of preceramic polymers.[6] The successful development and utilization of such polymer pyrolysis processing requires significant contributions from synthetic chemists and their close interaction with ceramists.

The molecular precursor route illustrates the important role of chemistry in materials science, but good chemistry does not necessarily mean good ceramics. Rheological properties should be suitable for film deposition or spinning, ceramic yields should be high, and drying and pyrolytic degradation should be controlled. This point is illustrated clearly by the Yajima process: the starting polycarbosilane presents several disadvantages, such as a multistep synthetic process and poorly defined structure. However, it is the only precursor that has led to the commercialization of SiC fibers.

One main question, still under investigation, is the relationship between the molecular structure of the precursor and the microstructure of the ceramic. The chemical composition of the precursor solution has a direct influence on the chemical composition of the ceramics, but does the molecular architecture also influence the organization of the ceramics? This is a field in which materials chemists should bring answers.

One of the most attractive prospects for application of chemical routes to ceramics should be the synthesis of novel materials that cannot be prepared via the classical solid-state reactions. Hybrid organic–inorganic or even bioinorganic materials open a land of opportunities for materials science. These nanocomposites bridge high-temperature materials such as glasses and ceramics with very fragile species such as organic compounds or biomolecules. This field is still in its infancy but there is little doubt that great advances are still to come.

REFERENCES

1. J. J. Ebelmen, *C. R. Acad. Sci.*, **19**, 398 (1844).
2. (a) L. Klein, Ed., *Sol-Gel Technology*, Noyes Publications, Park Ridge, NJ, 1988. (b) C. J. Brinker, G. W. Sherer, *Sol-Gel Science*, Academic Press, San Diego, CA, 1990.
3. H. Schmidt, B. Seiferling, *Mater. Res. Soc. Symp. Proc.* **73**, 739 (1986).

4. B. M. Novak, *Adv. Mater.* **5**, 422 (1993).
5. C. Sanchez, F. Ribot, *New J. Chem.* **18**, 1007 (1994).
6. S. Yajima, *Handbook of Composites*, Vol. 1, *Strong Fibres*, W. Watt, B. V. Perov, Eds., Elsevier, Amsterdam, 1985.
7. D. C. Bradley, R. C. Mehrotra, D. P. Gaur, *Metal Alkoxides*, Academic Press, London, 1978.
8. E. P. Kovsman, S. I. Andruseva, L. I. Solovjeva, V. I. Fedyaev, M. N. Adamova, T. V. Rogova, *J. Sol-Gel Sci. Technol.* **2**, 61 (1994).
9. M. Kakihana, *J. Sol-Gel Sci. Technol.*, **6**, 7 (1996).
10. R. C. Mehrotra, A. Singh, *Chem. Soc. Rev.* **25**, 1 (1996).
11. J. Livage, M. Henry, C. Sanchez, *Prog. Solid State Chem.* **18**, 259 (1988).
12. K. C. Chen, T. Tsuchiya, J. D. Mackenzie, *J. Non-Cryst. Solids* **81**, 227 (1986).
13. E. A. Barringer, H. K. Bowen, *J. Am. Ceram. Soc.* **65**, C-199 (1982).
14. (a) S. Doeuff, Y. Dromzee, F. Taulelle, C. Sanchez, *Inorg. Chem.* **28**, 4439 (1989). (b) P. Toledano, M. In, C. Sanchez, *C. R. Acad. Sci. Paris Ser. II* **313**, 1247 (1991).
15. M. Nabavi, C. Sanchez, J. Livage, *Eur. J. Solid State Inorg. Chem.* **28**, 1173 (1991).
16. E. J. A. Pope, J. D. Mackenzie, *J. Non-Cryst. Solids* **87**, 185 (1986).
17. W. Stöber, A. Fink, E. Bohm, *J. Colloid Interface Sci.* **26**, 62 (1968).
18. M. R. Mueller, J. D. Britt, *J. Chem. Ed.*, 890 (1994).
19. J. Livage, C. Sanchez, *J. Non-Cryst. Solids* **145**, 11 (1992).
20. B. Morrison, *Cryst. Sec. B* **33**, 303 (1977).
21. M. Chatry, M. Henry, M. In, C. Sanchez, J. Livage, *J. Sol-Gel Sci. Technol.*, **1**, 233 (1994).
22. C. Guizard, A. Julbe, A. Larbot, L. Cot, in *Chemical Processing of Ceramics*, B. I. Lee, E. J. A. Pope, Eds., Marcel Dekker, New York, 1994, p. 501.
23. C. D. Chandler, C. Roger, M. J. Hampden-Smith, *Chem. Rev.* **93**, 1205 (1993).
24. K. G. Caulton, L. G. Hubert-Pfalzgraf, *Chem. Rev.* **90**, 969 (1990).
25. E. R. Vance, *MRS Bull.* **19**, 28 (1994).
26. Y. Chujo, T. Saegusa, *Adv. Polym. Sci.* **100**, 12 (1992).
27. D. E. Rodrigues, G. L. Wilkes, *J. Inorg. Organomet. Polym.* **3**, 197 (1993).
28. H. Schmidt, *Structure Bonding* **77**, 119 (1992).
29. R. Reisfeld, C. K. Jorgensen, *Structure Bonding* **77**, 207 (1992).
30. B. Dunn, J. Zink, *J. Mater. Chem.* **1**, 903 (1991).
31. D. Avnir, *Acc. Chem. Res.* **28**, 328 (1995).
32. M. R. Landry, B. K. Coltrain, C. J. T. Landry, J. M. O'Reilly, *J. Polym. Sci. B* **33**, 637 (1995).
33. C. Sanchez, M. In, *J. Non-Cryst. Solids* **147–148**, 1 (1992).
34. U. Schubert, N. Husing, A. Lorenz, *Chem. Mater.* **7**, 2010 (1995).
35. P. Judeinstein, C. Sanchez, *J. Mater. Chem.* **6**, 1 (1996).
36. G.-S. Sur, J. E. Mark, *Eur. Polym. J.* **21**, 1051 (1985).
37. G. L. Wilkes, B. Orler, H. H. Huang, *Polym. Prepr.* **26**, 300 (1985).
38. J. D. Mackenzie, D. R. Ulrich, Eds., *Sol-Gel Optics I*, *Proc. SPIE*, 1328 (1990).

39. J. D. Mackenzie, Ed., *Sol-Gel Optics II*, *Proc. SPIE*, 1758 (1992).
40. J. D. Mackenzie, Ed., *Sol-Gel Optics III*, *Proc. SPIE*, 2288 (1994).
41. L. C. Klein, Ed., *Sol-Gel Optics: Processing and Applications*, Kluwer, Boston, 1993.
42. D. Levy, D. Avnir, *J. Phys. Chem.* **92**, 734 (1988).
43. S. A. Yamanaka, F. Nishida, L. M. Ellerby, C. R. Nishida, B. Dunn, J. S. Valentine, J. I. Zink, *Chem. Mater.* **4**, 495 (1992).
44. D. Avnir, S. Braun, O. Lev, M. Ottolenghi, *Sol-Gel Optics II*, *Proc. SPIE*, 1758 (1992).
45. D. Avnir, S. Braun, O. Lev, M. Ottolenghi, *Chem. Mater.* **6**, 1605 (1994).
46. C. Dave, B. Dunn, J. S. Valentine, J. I. Zink, *Anal. Chem.* **66**, 1120A (1994).
47. P. Audebert, C. Demaille, C. Sanchez, *Mater. Chem.* **5**, 911 (1993).
48. J. Livage, C. Roux, J. M. da Costa, I. Desportes, J. F. Quinson, *J. Sol-Gel Sci. Technol.* **7**, 45 (1996).
49. A. V. Veret-Lemarinier, J.-P. Galaup, A. Ranger, F. Chaput, J.-P. Boilot, *J. Lumin.* **64**, 223 (1995).
50. B. Lebeau, C. Sanchez, S. Brasselet, J. Zyss, G. Froc, M. Dumont, *New J. Chem.* **20**, 13 (1996).
51. M. In, C. Gérardin, J. Lambard, C. Sanchez, *J. Sol-Gel Sci. Technol.* **5**, 101 (1995).
52. J. D. Mackenzie, Y. J. Chung, Y. Hu, *J. Non-Cryst. Solids* **147–148**, 271 (1992).
53. D. A. Loy, K. J. Shea, *Chem. Rev.* **95**, 1431 (1995).
54. K. Sharp, *J. Sol-Gel Sci. Technol.*, **8**, 541 (1997).
55. D. Hoebbel, I. Pitsch, D. Heidemann, H. Jancke, W. Hiller, *Z. Anorg. Allg. Chem.* **583**, 133 (1990).
56. D. Hoebbel, I. Pitsch, D. Heidemann, *Z. Anorg. Allg. Chem.* **592**, 207 (1991).
57. J. D. Lichtenham, N. Q. Vu, J. A. Carter, J. W. Gilman, F. J. Feher, *Macromolecules* **26** 2141 (1993).
58. J. D. Lichtenham, Y. A. Otonari, M. C. Carr, *Macromolecules* **28**, 8435 (1995).
59. F. Ribot, F. Banse, F. Diter, C. Sanchez, *New J. Chem.* **19**, 1145 (1995).
60. F. Banse, F. Ribot, P. Toledano, J. Maquet, C. Sanchez, *Inorg. Chem.* **34**, 6371 (1995).
61. P. Judenstein, *Chem. Mater.*, **4**, 4 (1991).
62. K. J. Wynne, R. W. Rice, *Annu. Rev. Mater. Sci.* **14**, 297 (1984).
63. J. Lipowitz, *Ceram. Bull.* **70**, 1888 (1991).
64. J. Bill, F. Aldinger, *Adv. Mater.* **7**, 9 (1995).
65. R. M. Laine, F. Babonneau, *Chem. Mater.* **5**, 260 (1993).
66. R. M. Laine, F. Babonneau, K. Y. Blowhowiak, R. A. Kennish, J. A. Rahn, G. J. Exarhos, K. Waldner, *J. Am. Ceram. Soc.* **78**, 137 (1995).
67. M. Birot, J. P. Pillot, J. Dunogues, *Chem. Rev.* **95**, 1443 (1995).
68. S. Yajima, J. Hayashi, M. Omori, *Chem. Lett.*, 931 (1975).
69. S. Yajima, J. Hayashi, M. Omori, K. Okamura, *Nature* **261** 683 (1976).
70. S. Yajima, Y. Hasegawa, J. Hayashi, M. Iimura, *J. Mater. Sci.* **13**, 2569 (1978).
71. Y. Hasegawa, K. Okamura, *J. Mater. Sci.* **21**, 321 (1986).

72. C. Laffon, A. M. Flank, P. Lagarde, M. Laridjani, R. Hagege, P. Olry, J. Cotteret, J. Dixmier, J. L. Miquel, H. Hommel, A. P. Legrand, *J. Mater. Sci.* **21** 1503 (1989).
73. K. Okamura, *Composites* **18**, 107 (1987).
74. M. Takeda, Y. Imai, H. Ichikawa, T. Ishikawa, T. Seguchi, K. Okamura, *Ceram. Eng. Sci. Proc.* **12**, 1007 (1991).
75. S. Yajima, T. Iwai, T. Yamamura, K. Okamura, Y. Hasegawa, *J. Mater. Sci.* **16**, 1349 (1981).
76. P. J. Fazen, E. E. Remsen, J. S. Beck, P. J. Carroll, A. R. McGhie, L. G. Sneddon, *Chem. Mater.* **7**, 1942 (1995).
77. R. T. Paine, C. K. Narula, *Chem. Rev.* **90**, 73 (1990).
78. T. E. Wood, H. Dislich, *Sol-Gel Sci. Technol., Ceram. Trans.* **55**, 3 (1995).
79. W. Geffcken, E. Berger, U.S. patent 2,366,516 (1945).
80. F. T. Fitch, M. G. Sanchez, M. C. Vanik, Br. patent 905,919 (1962).
81. J. Gottardi, Ed., *J. Non-Cryst. Solids*, 48 (1982).

GENERAL BIBLIOGRAPHY

Main Books on Sol-Gel Processing

- *Sol-Gel Technology for Thin Films, Fibers, Preforms, Electronics and Specialty Shapes*, L. C. Klein, Ed., Noyes Publications, Park Ridge, NJ, 1988.
- *Sol-Gel Science*, C. J. Brinker, G. W. Scherer, Academic Press, San Diego, CA, 1990.
- *Sol-Gel Optics: Processing and Applications*, L. Klein, Ed., Kluwer, Boston, 1993.
- *Metal Alkoxides*, D. C. Bradley, R. C. Mehrotra, D. P. Gaur, Academic Press, London, 1978.
- *Ceramic Precursor Technology and Its Applications*, C. K. Narula, Marcel Dekker, New York, 1995.
- *Chemical Synthesis of Advanced Ceramic Materials*, D. Segal, Cambridge University Press, Cambridge, 1989.

Proceedings and Special Issues

- *International Workshop on Glasses and Ceramics from Gels*
  J. Gottardi, Ed., *Journal of Non-Crystalline Solids* **48** (1982).
  H. Scholze, Ed., *Journal of Non-Crystalline Solids* **63** (1984).
  J. Zarzycki, Ed., *Journal of Non-Crystalline Solids* **82** (1986).
  S. Sakka, Ed., *Journal of Non-Crystalline Solids* **100** (1988).
  M. Aegerter, Ed., *Journal of Non-Crystalline Solids* **121** (1990).

L. Esquivias, Ed., *Journal of Non-Crystalline Solids* **147–148** (1992).

J. Livage, Ed., *Journal of Sol-Gel Science and Technology* **2** (1994).

R. Almeida, Ed., *Journal of Sol-Gel Science and Technology* **8** (1997).

- *Better Ceramics Through Chemistry*, in "Materials Research Society Symposium Proceedings"

  C. J. Brinker, D. E. Clark, D. R. Ulrich, Eds., Vol. 32 (1984).

  C. J. Brinker, D. E. Clark, D. R. Ulrich, Eds., Vol. 73 (1986).

  C. J. Brinker, D. E. Clark, D. R. Ulrich, Eds., Vol. 121 (1988).

  B. J. J. Zelinski, C. J. Brinker, D. E. Clark, D. R. Ulrich, Eds., Vol. 180 (1990).

  M. J. Hampden-Smith, W. G. Klemperer, C. J. Brinker, Eds., Vol. 271 (1992).

  A. K. Cheetham, C. J. Brinker, M. L. Mecartney, C. Sanchez, Eds., Vol. 346 (1994).

  B. Coltrain, D. Schaefer, C. Sanchez, G. Wilkes, Eds., Vol. 435 (1996).

- *Science of Ceramics Chemical Processing*, H. L. Hench, D. R. Ulrich, Eds., Wiley, New York, 1986.
- *Ultrastructure Processing of Advanced Ceramics*, J. D. Mackenzie, D. R. Ulrich, Eds., Wiley, New York, 1988.
- *Ultrastructure Processing of Advanced Materials*, D. R. Uhlmann, D. R. Ulrich, Eds., Wiley, New York, 1992.
- *Sol-Gel Optics*

  J. D. Mackenzie, D. R. Ulrich, Eds., *Sol-Gel Optics*, *SPIE Proc.* **1328** (1990)

  J. D. Mackenzie, Ed., *Sol-Gel Optics II*, *SPIE Proc.* **1758** (1992)

  J. D. Mackenzie, Ed., *Sol-Gel Optics III*, *SPIE Proc.* **2288** (1994)

- *Chemistry, Spectroscopy, and Applications of Sol-Gel Glasses: Structure and Bonding*, Vol. 77, R. Reisfeld, C. K. Jorgensen, Eds., Springer-Verlag, Berlin, 1992.
- *Hybrid Organic Inorganic Composites*, J. E. Mark, Y. C. Lee, P. A. Bianconi, Eds., ACS Symposium Series, Vol. 585, American Chemical Society, Washington, DC, 1995.
- *Sol-gel Derived Materials*, J. Ying, ed., special issue of *Chem. Mater.*, November, 1997.

Review Articles

- Sol-gel chemistry of transition metal oxides, J. Livage, M. Henry, C. Sanchez, *Progress in Solid State Chemistry* **18**, 259 (1988).
- Chemical aspects of solution routes to Perovskite-phase mixed-metal oxides from metal-organic precursors, C. D. Chandler, C. Roger, M. J. Hampden-Smith, *Chem. Rev.* **93**, 1205 (1993).

- Sol-gel processing of complex oxide films, G. Yi, M. Sayer, *Ceramic Bulletin* **70**, 1173 (1991).
- Sol-gel processing of transition-metal alkoxides for electronics, G. R. Lee, J. A. Crayton, *Advanced Materials* **5**, 43442 (1993).
- Hybrid nanocomposite materials, between inorganic glasses and organic polymers, B. M. Novak, *Advanced Materials* **5**, 422 (1993).
- The sol-gel process, L. L. Hench, J. K. West, *Chemical Review* **90**, 33 (1990).
- Ceramics via polymer pyrolysis, K. J. Wynne, R. W. Rice, *Annual Review of Materials Science* **14**, 297 (1984).
- Polymer-derived ceramic fibers, J. Lipowitz, *Ceramic Bulletin* **70**, 1888 (1991).
- Preceramic polymer routes to silicon carbide, R. M. Laine, F. Babonneau, *Chemistry of Materials* **5**, 260 (1993).
- Comprehensive chemistry of polycarbosilanes, polysilazanes, and polycarbosilazanes as precursors of ceramics, M. Birot, J. P. Pillot, J. Dunogues, *Chemical Review* **95**, 1443 (1995).
- Hybrid organic–inorganic materials, C. Sanchez and F. Ribot, Guest Eds., special issue of the *New Journal of Chemistry* **18** (1994).
- Organic–Inorganic Hybrid Materials, L. C. Klein, C. Sanchez, Guest Eds., special issue of the *Journal of Sol-Gel Science and Technology* **5** (1995).
- Molecular design of hybrid organic–inorganic materials synthesized through sol-gel chemistry, C. Sanchez, F. Ribot, *New Journal of Chemistry* **18**, 1007 (1994).
- Organic chemistry within ceramic matrices: doped sol-gel materials, D. Avnir, *Accounts of Chemical Research* **28**, 328 (1995).
- Optical properties of sol-gel glasses doped with organic molecules, B. Dunn, J. Zink, *Journal of Materials Chemistry* **1** 903 (1991).
- Chemistry and applications of inorganic–organic polymers (organically modified silicates, H. Schmidt, B. Seiferling, *Materials Research Society Symposium Proceedings* **73**, 739 (1986).
- Organic polymer hybrids with silica gel formed by means of the sol-gel method, Y. Chujo, T. Saegusa, *Advances in Polymer Science* **100**, 12 (1992).
- Small-angle x-ray scattering investigations of fractal dimensions of hybrid inorganic–organic network (ceramers) based on tetraethylorthosilicate and polytetramethylene oxide, D. E. Rodrigues, G. L. Wilkes, *Journal of Inorganic and Organometallic Polymers* **3**, 197 (1993).
- Structural models for homogeneous organic–inorganic hybrid materials: simulations of small-angle x-ray scattering profiles, M. R. Landry, B. K. Coltrain, C. J. T. Landry, J. M. O'Reilly, *Journal of Polymer Science, B, Polymer Physics* **33**, 637 (1995).

- Hybrid nanocomposites materials: between inorganic glasses and organic polymers, B. M. Novak, *Advanced Materials* **5**, 422 (1993).
- Thin films, the chemical processing up to gelation, H. Schmidt, *Structure and Bonding* **77**, 119 (1992).
- Optical properties of colorants or luminescent species, R. Reisfeld, C. K. Jorgensen, *Structure and Bonding* **77**, 207 (1992).
- Synthesis, structural principles and reactivity of heterometallic alkoxides, *K. G. Caulton, L. Hubert-Pfalzgraf*, Chemical Review **90**, 969 (1990).
- Hetrometallic alkoxides and oxoalkoxides as intermediates in chemical routes to mixed metal oxides, L. G. Hubert-Pfalzgraf, *Polyhedron* **13**, 1181 (1994).
- An investigation of group(IV) alkoxides as property controlling reagents in the synthesis of ceramic materials, T. J. Boyle, R. W. Schwartz, *Comments in Inorganic Chemistry* **16**, 243 (1994).
- Enzymes and other proteins entrapped in sol-gel materials, D. Avnir, S. Braun, O. Lev, M. Ottolenghi, *Chemistry of Materials* **6**, 1605 (1994).
- Sol-gel encapsulation methods for biosensors, C. Dave, B. Dunn, J. S. Valentine, J. I. Zink, *Analytical Chemistry* **66**, 1120A (1994).
- Bridged polysilsesquioxanes: highly porous hybrid organic–inorganic materials, *D. A. Loy, and K. J. Shea*, Chemical Review **95**, 143 (1995).
- Silsesquioxanes, R. H. Baney, M. Itoh, A. Sakakibara, T. Suzuki, *Chem. Rev.* **95**, 1409 (1995).
- Hybrid organic–inorganic materials: a land of multidisciplinarity, P. Judeinstein, C. Sanchez, *Journal of Materials Chemistry* **6**, 1 (1996).
- Bioactivity in sol-gel glasses, J. Livage, *Comptes-Rendus de l'Académie des Sciences, Paris* **322**, 417 (1996).

CHAPTER 10

Layered Transition Metal Oxides and Chalcogenides

PAUL A. SALVADOR and THOMAS O. MASON

Department of Materials Science and Engineering, Northwestern University, Evanston, Illinois

MICHAEL E. HAGERMAN* and KENNETH R. POEPPELMEIER

Department of Chemistry, Northwestern University, Evanston, Illinois

10.1 INTRODUCTION

The extraordinary diversity of today's advanced materials is based on an enhanced knowledge of how to attain novel structures displaying new properties that lead to improved performance. Similarly, the major stimulus for efforts to improve existing materials and synthesize novel ones has been the critical need in the marketplace for entirely new levels of performance. A key realization affecting the development of modern materials has been the coupling of a material's external properties to its internal structure.[1] All materials possess an inner architecture, that is, a hierarchy of successive structural levels. The materials chemist's understanding of this architecture and control over its structure is of paramount importance in the attainment of novel materials and the delineation of materials properties. In our review we highlight the chemical factors that control the inner architecture of solids and can be exploited to design materials rationally. In particular, we focus on two classes of advanced materials, layered oxides and chalcogenides, providing a broad overview of the scope, importance, and relevance of layering in oxides and chalcogenides. In addition, we review important concepts to this

* Present address: Department of Chemistry, Northern Arizona University, Flagstaff, Arizona 86011

Chemistry of Advanced Materials: An Overview, Edited by Leonard V. Interrante and Mark J. Hampden-Smith.
ISBN 0-471-18590-6 © 1998 Wiley-VCH, Inc.

field, and materials chemistry in general, including dimensionality, anisotropy, and topochemistry.

10.1.1 Significance of Layered Oxides and Chalcogenides

Inorganic materials that have layered structures, or that contain tunnels or channels in their crystalline structure, have a rich chemistry and have been studied extensively not only for their chemical and physical characteristics but for their potential use in many materials applications.[2-8] Areas in materials science and technology such as microelectronics, optics, chemical sensing, catalysis, and superconductivity have benefited from the ability to control structures, properties, and stabilities of layered oxides and chalcogenides.[9] Moreover, host–guest compounds, which have been realized through the intercalation of guest molecules into inorganic layered hosts, represent a new class of premier advanced materials.[10] Potential applications include catalysts,[11, 12] semiconductors,[13, 14] superconductors,[15] photoactive,[16] linear[17] and nonlinear optical materials,[18] chemical sensors,[19] supramolecules,[20] bioactive materials,[21] ionic conductors,[22] conductive polymer assemblies,[23-25] and nanomaterials.[26, 27] Not only do these layered materials exhibit a wide variety of technologically important properties, they also offer a unique opportunity to exploit crystalline architecture and achieve desired material properties.[28, 29]

Low-dimensional solids have been considered extensively during the past years for their intriguing physical properties (superconductivity, low-dimensional magnetism, and charge density waves). Interest in layered oxides has been renewed with the discovery of high-T_c superconductivity in copper-based materials, and the structure, crystal chemistry, and properties of layered cuprates have been investigated intensely.[30, 31] Other layered transition-metal oxides have attracted considerable attention for their electrical, magnetic, and optical properties.[26, 30-32] Layered chalcogenides, which have been reviewed by Rouxel et al.,[33-35] exhibit a remarkable chemical reactivity and are able to react with atoms or molecules without significant changes in their structure.[36] Low-dimensional solids afford unique opportunities to illustrate the link between coordination chemistry and the solid state.[34] The remarkable similarities of structures, orbital descriptions, and unusual physical properties are a manifestation of this close relationship. General considerations of the relationship between the electronic properties and structural arrangements and stoichiometries in layered and one-dimensional compounds have been described in detail.[35, 37]

10.1.2 Dimensionality in Advanced Materials

Dimensionality, in particular reduced dimensionality, is playing an ever-increasing role in the fields of materials science and materials chemistry because of the diverse properties exhibited by low-dimensional materials. Low-dimensional materials result from a reduction in the number of isotropic directions for a particular property (e.g., electrical conductivity). Nye[38] points out that all crystals are anisotropic for some properties. For instance, a cubic crystal is often anisotropic for elasticity, photoelasticity, and other properties, yet is sometimes misconceived to be isotropic for all properties.[38] Understanding the relation between the anisotropies of a property and the inner architecture is essential when considering low-dimensional materials. Therefore, when we are defining layered, or two-dimensional materials, we stress the property of interest and its relation to the inner architecture, in particular the bonding present in the material.

A three-dimensional material is characterized by a structure in which anisotropy is minimal, as reflected in both the bonding and resultant properties. It follows that a two-dimensional material can therefore be characterized by a structure where two directions are nearly isotropic with respect to bonding or properties, while the third is highly anisotropic. A one-dimensional material also has an anisotropy that is manifested strongly in one dimension, while the other two need not be isotropic. The distinguishing feature between one- and two-dimensional materials is that the direction of interest for a one-dimensional material is in the direction of the anisotropy, while for a two-dimensional material the focus is in the isotropic directions. Materials properties and underlying anisotropies are closely related to their inner architecture, which is of primary interest to the materials chemist. Although advanced materials based on layered (two-dimensional) structures can seem quite unrelated from their structural aspects or properties of interest (i.e., the intercalation compound Li_xTiS_2 and the superconductor $YBa_2Cu_3O_{7-x}$), the chemistry that controls these features is strikingly similar. Transition-metal oxides and chalcogenides are exemplary and elegant cases where the materials chemist can indeed control the underlying chemistry and the ultimate dimensionality of materials. Research in this area, with the aim of understanding and tailoring the resultant anisotropic properties, remains a fruitful area of research.

Classic layered compounds include graphite,[39] clays,[40] and aluminates,[41] in which the weak van der Waals bonding between adjacent layers differs from the largely covalent intralayer bonding, resulting in an anisotropy that can be exploited in, among other applications, intercalation compounds and fast ionic conductors.[42] Although the demarcation of layers in these classic examples is based on distinct bonding differences in the direction perpendicular to the layers, advanced layered materials have been realized by acknowledging that dimensionality can owe its origins to subtler structural variations. Thus, layering in cuprates can be attributed to cooperative Jahn–Teller distortions where the elongated axis is perpendicular to the layers, resulting in strongly two-dimensional electronic properties and superconductivity.[31,43–48] To clarify the nature of layering in oxides and chalcogenides, with respect to structure and properties, in Section 10.2.2 we address structural dimensionality in these materials with an emphasis on the chemical differences between two- and three-dimensional structures.

The scope of structural inorganic chemistry of solids is vast and continues to expand. Burdett highlights the important structural features of solids and depicts how the materials chemist makes progress in understanding the rapidly growing collection of observations concerning these crystalline architectures.[49] Burdett concludes that if we cannot describe, organize, and categorize solid-state structures geometrically, we are at an immediate disadvantage in appreciating their electronic stabilities and materials properties. The traditional way of visualizing solid structures containing extended arrays is to build them up from cation-centered polyhedra, which may be linked together by vertices, edges, faces, or various combinations of these modes. While this provides a useful method of visualizing the network of bonding in solids, it is often misleading because it places emphasis on the coordination geometry of the cation rather than that of the anion. Another method of looking at these structures is to describe them in terms of the cation occupancy of the interstices generated by a pseudo-close-packed anionic array. The anion arrangement of many structures approximates the close packing of spheres, of which the cubic close-packed and hexagonal close-packed are the best known.

Figure 10.1*a*–*c* show three different ways to occupy the octahedral interstices of a hexagonal close-packed anionic array with the stoichiometry MX_2. Overall, because the metal sits in the octahedral holes formed by the close-packed anions, and there exists one octahedral hole for every close-packed sphere, in order to match the given ratio of cation to anion (one to

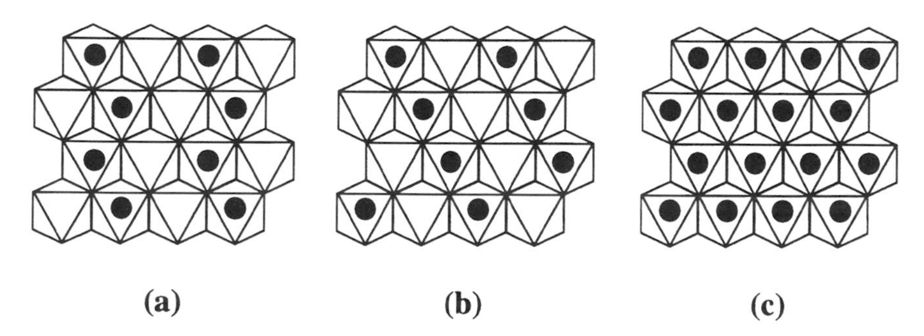

Figure 10.1 Three possible methods of filling octahedral interstices of a hexagonal close-packed anionic array of stoichiometry MX_2. Dark spheres represent M cations filling the interstices above the X layer, and empty triangles represent vacant interstices above the layer. The filled/empty interstices below the X layer are opposite to that above the layer. Thus (*a*) and (*b*) have adjacent sheets (above/below) which are half occupied, while (*c*) has alternating filled and empty sheets. (Adapted from Ref. 49.)

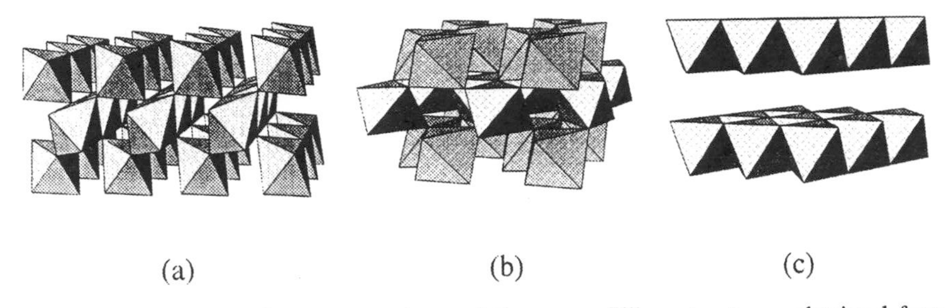

Figure 10.2 Polyhedral representations of the space-filling structures obtained from stacking layers depicted in Fig. 10.1: (*a*) rutile structure; (*b*) α-PbO_2 structure; (*c*) CdI_2 structure. (Adapted from Ref. 49.)

two, respectively), half of the holes must be occupied. Filled spheres in the figure denote M cations in the interstices above the close-packed layer, while empty triangles represent vacant interstices above the anionic layer. The filling of the interstices below the anionic layer is opposite to that above the layer (i.e., a filled interstice above the layer has a corresponding unfilled interstice below). Thus, in Fig. 10.1*a* and *b*, the set of interstices between every adjacent pair of anionic sheets is half occupied. However, in Fig. 10.1*c*, every other set is completely empty. The three structure types that result are rutile (TiO_2), α-PbO_2, and CdI_2, respectively. Polyhedral diagrams for these structures are given in Fig. 10.2*a–c*. Notice that although the polyhedra share edges with each other in all three structures, (*a*) and (*b*) are framework structures, whereas (*c*) is a layered structure. The layered structure shown in (*c*) is generally called the cadmium halide type and has no extended cation–anion bonding in the [001] direction. An understanding of these types of structural differences, especially in advanced materials, is invaluable in appreciating the variance observed in materials properties and electronic stabilities. Detailed accounts of the electronic structures of layered materials have been given previously.[50-53]

10.1.3 Design of Layered Materials

We are only at the beginning of a more extended design of chemical interactions in one, two, and three dimensions.[9] The rational synthesis of materials requires knowledge of crystal chemistry, phase equilibria, and reaction kinetics. Chemical methods of synthesis play a crucial role in the design and discovery of novel materials. Rao has recently provided a detailed summary of the approaches to the synthesis of inorganic materials, including many layered oxides and chalcogenides.[54] It is important to note that in this review we intend to focus our attention not on exact methods of synthesis but on the methodology of achieving a desired chemical architecture (in our case, layering in advanced materials) through rational materials design. Although the choice of the synthetic pathway or technique is often important in achieving desired materials,[55, 56] our primary emphasis will be on controlling the chemical architecture of advanced materials through various chemical considerations (detailed in Section 10.2.1). Subtle control over materials properties can be achieved by variation of the crystalline architecture as well as the chemical composition.[10] Moreover, some reaction processes, such as topochemical transformations and intercalation reactions, can be mediated by the underlying chemical architecture and are briefly discussed below.

An understanding of topochemical processes is essential in a discourse on layered compounds. A solid-state reaction is said to be topochemical when its reactivity is controlled by the crystal structure. Therefore, the chemical nature of the reaction constituents controls the reactivity indirectly, through its influence on the crystalline architecture. The reduction of ABO_3 perovskites to yield $A_2B_2O_5$ and such defect oxides as $ABO_{3-\delta}$ is found to be topochemical.[54] The topochemical transformation that occurs in such reactions involves the reduction of metal–oxygen octahedra, thus forming metal–oxygen square pyramids (e.g., MnO_5), tetrahedra (e.g., FeO_4), or square-planar units (NiO_4), as discussed in Section 10.2.2. The reduction of the high-temperature superconductor $YBa_2Cu_3O_7$ to $YBa_2Cu_3O_6$ is also a topochemical process. In this reaction the CuO_4 units in the CuO chains along the *b* direction transform to O—Cu—O linear units.[54]

Another important topochemical process, which has found widespread use in a variety of applications, is intercalation. The term *intercalation* literally refers to the act of inserting into a calendar some extra interval of time, such as February 29 in a leap year.[57] Intercalation reactions of solids involve the insertion of a guest species (ion or molecule) into a solid host lattice without any major rearrangement of the host framework.[54] Intercalation compounds have proven important for a variety of commercial applications, including petroleum refining,[58, 59] chemical production,[60-62] energy storage,[57, 63] and electrochromic displays.[57] In general terms, an intercalation reaction can be described by the following equation:

$$x(\text{guest}) + \{\ \}_x[\text{host}] \longrightarrow (\text{guest})_x[\text{host}] \qquad (1)$$

where $\{\ \}_x$ represents a vacant site. Three intercalation reactions that have proven important for advanced applications are[57]

$$x\text{Li} + \text{TiS}_2 \longrightarrow \text{Li}_x\text{TiS}_2 \qquad \text{(batteries)} \qquad (2)$$

$$x\text{H} + \text{WO}_3 \longrightarrow \text{H}_x\text{WO}_3 \qquad \text{(electrochromic displays)} \qquad (3)$$

$$(x/2)\text{O}_2 + \text{La}_2\text{CuO}_4 \longrightarrow \text{La}_2\text{CuO}_{4+\delta} \qquad \text{(superconductivity)} \qquad (4)$$

The wide array of materials that can be realized through the intercalation of one guest species is remarkable, as demonstrated by the intercalation compounds of lithium.[54] A critical review of the metal intercalates of the transition-metal dichalcogenides and related host materials has been provided by Whittingham.[64]

10.2 MATERIALS CHEMIST'S APPROACH

Materials chemistry is chemistry driven by the needs and opportunities of modern materials.[65] It creates opportunities and provides society with new ways to address problems such as the scarcity of resources, the maintenance of economic growth and productivity, and the generation of capital and competitiveness in the global marketplace.[66-69] Correlations between the microscopic structure of advanced materials and observed macroscopic properties have proven important in the design of new materials and the fabrication of novel device structures.[70] The principal characteristics of materials chemistry include a focus on solids with useful properties, an emphasis on synthesis, and a long-term goal of understanding structure–bonding–property relationships.[65] In this section we focus on the chemical nature of advanced materials and its influence on materials design.

Materials science is the science of the relations between processing, structure, and properties of materials.[71] The materials chemist is concerned with the synthesis of advanced materials and the description of their compositional, structural, and morphological features, knowledge of the principles of their chemical bonding, and characterization of their physical properties.[72] The understanding of any of these relationships is far from complete, thus putting the materials chemist in the position of both investigating the underlying chemistry of materials and using this knowledge to design new compounds.

10.2.1 Realization of Layered Compounds: Chemical Considerations

Roy has previously addressed rational materials design and presented two intriguing questions (paraphrased below) which are central to the themes of this review and should be considered as we attempt to describe the realization of layered materials:[73] (1) To what extent can we design and make a new material with a particular desired property? (2) What, then, are the knowledge bases in materials science that would guide the practitioner in attempting to synthesize a new material with particular properties? In the following sections we address both questions by examining how one can foster layering in oxides and chalcogenides as a result of an enhanced understanding of materials chemistry. The following chemical considerations are essential for control over layering in oxides and chalcogenides by materials chemists: (1) stoichiometry, (2) coordination chemistry, (3) ionic size and charge, (4) acid-base chemistry, and (5) oxidation-reduction capability. As will be seen below (Sections 10.2.2 and 10.3), some of these considerations are often inseparable in real materials. However, it is essential to recognize the relevant contributions from each of these considerations if rational materials design is to be realized. Ultimately, it is our limited understanding of how the inner architecture is controlled by these chemical factors that hinders our ability to design advanced materials rationally.

10.2.2 Layered Oxides and Chalcogenides: Important Examples

To develop the importance of the chemical considerations cited above, we focus on three exemplary classes of layered materials: dichalcogenides, perovskite-related oxides, and bronzes. As mentioned in Section 10.1.2, a fundamental approach to the description of inorganic structures is the pseudo close packing of anionic layers. Two-dimensional (layered) materials are easily envisioned using this description, owing to the layered nature of the packing itself. The layered dichalcogenides, MX_2 (M = metal cation, X = chalcogen),[33-35] lend themselves well to this layering scheme, with the anisotropic direction perpendicular to the close-packed layers. Figure 10.3 illustrates the nature of layering in these materials, which results from the transition-metal cation residing only between every other close-packed anionic layer, giving XMX-type layers. These layers are held together by weak van der Waals forces between the chalcogens of adjacent layers, as illustrated in Fig. 10.3*a*, which gives rise to a variety of technologically important properties (i.e., chemical sensing, catalysis, electrical and optical data storage). Although no transition metal–chalcogen bonds extend in the direction perpendicular to the layers, intralayer chalcogen–metal bonding plays an important role in the structure and properties of these materials.

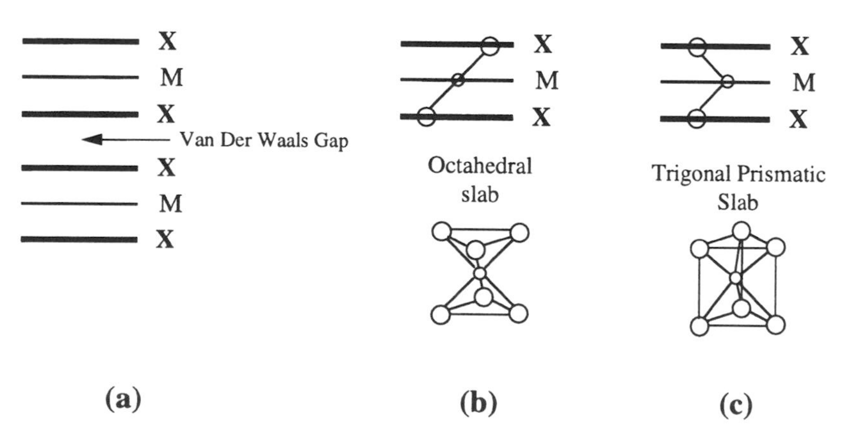

Figure 10.3 Simplified view of the XMX slabs in layered dichalcogenides: (*a*) two XMX slabs and the van der Waals gap between them; (*b*) octahedral XMX slab; (*c*) trigonal prismatic XMX slab. (Adapted from Ref. 33.)

The electronic properties of the layered dichalcogenides are dominated by the orbital interactions between the transition metal *d*-orbitals and the *s*- and *p*-orbitals of the chalcogen.[33, 34, 74–76] A schematic band diagram, corresponding to the atomic orbitals for the MX_2 compounds, is given in Fig. 10.4*a*. Modifications in the stacking sequences of close-packed anionic layers can lead to a variety of structural motifs, which can be understood with respect to the electronic configuration of the transition metal and the ionicity of the cation–anion bonds. Figure 10.3*b* and *c* illustrate the two basic types of coordination for transition-metal dichalcogenides of the formula MX_2. The transition metal is octahedrally coordinated in Fig. 10.3*b*, owing to the AbC stacking sequence of the layers (where anionic and cationic layers are denoted by upper- and lowercase letters, respectively, and A and C indicate different layer orientations). The AbA stacking sequence results in trigonal prismatic coordination of the transition metal, as shown in Fig. 10.3*c*. The preferred structure of the transition-metal dichalcogenides depends on the position and occupancy of the *d*-orbitals on the transition metal with respect to the chalcogen *s*- and *p*-orbitals. Group IV transition metals have no valence electrons and prefer octahedral coordination, with the Fermi level lying between the filled chalcogen bands and the empty metal states, resulting in semiconductors with large bandgaps.[33, 76] Group V transition metals, however, with one electron in the *d*-orbitals, prefer prismatic coordination, as this lowers the energy of the half-filled metal band.[76] These concepts are illustrated in the simplified band models for the transition-metal dichalcogenides given in Fig. 10.4*b* and *c*. Thus both the physical properties (metal–insulator transition) and the structure (octahedral or prismatic stacking) of these layered materials can be understood in terms of the electronic configuration of the transition metal.

Transition-metal oxides, however, do not immediately lend themselves to the layering scheme presented above, which depends on van der Waals

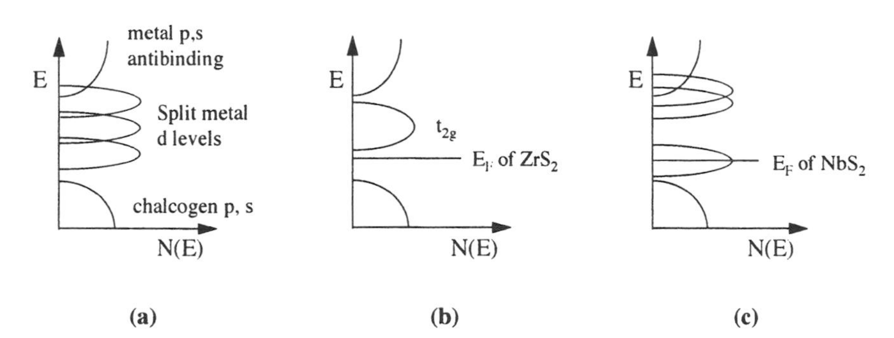

Figure 10.4 Simplified band models for layered dichalcogenides: (*a*) general scheme of metal and chalcogen bands; (*b*) octahedral slabs as in ZrS_2 with an empty *d*-band; (*c*) trigonal prismatic slabs as in NbS_2 with a half-filled metal *d*-band. (Adapted from Ref. 33.)

bonding between anions in the anisotropic direction. The main difference between the chalcogenides and oxides arises from the higher electronegativity and lower polarizability of oxygen compared to the chalcogens, leading to more ionic metal–oxygen bonds than metal–chalcogen bonds.[33,34,74] The anionic repulsion between close-packed oxygen layers is stronger than the weak van der Waals attraction, thereby destabilizing the layered MX_2 structure type for nearly all transition metal oxides.[33] Figure 10.5 illustrates the difference between TiS_2 (*a*) and TiO_2 (*b*), where TiO_2 has the rutile structure. The rutile structure, typically found for MO_2 oxides, can be described as a close-packed oxygen lattice, where the M cations occupy octahedral sites between all layers, as opposed to alternating layers in TiS_2, resulting in a highly covalent network structure having similar bonding between cations and anions in all directions. We see below that differences in this bonding along a unique direction can result in low-dimensional oxides.

Layering in complex oxides can still be understood using the close-packed approach, although the direction of anisotropy is not necessarily perpendicular to the close-packed layers as it was in the layered dichalcogenides discussed above. Perovskite compounds with the general stoichiometry ABO_3 lend themselves particularly well to this description.[77] Katz and Ward[78] and Wells[79] originally applied the AO_3 construction to the description of the stoichiometric perovskite structure. The perovskite structure can be described as a cubic close-packed arrangement of AO_3 layers in which the B cations reside in octahedral sites between these layers. Figure 10.6*a* illustrates one of these AO_3 layers and how identical layers are stacked to form the three-dimensional solid. Successive layers are stacked *ABC ABC* (as for cubic close packing), where layers *A*, *B*, and *C* are denoted by triangles in the figure. Each layer is stacked such that these triangles (which connect oxygen) form trigonal antiprisms. The resulting structure is illustrated in Fig. 10.6*b*, with the layered direction, [111], highlighted (in addition, other

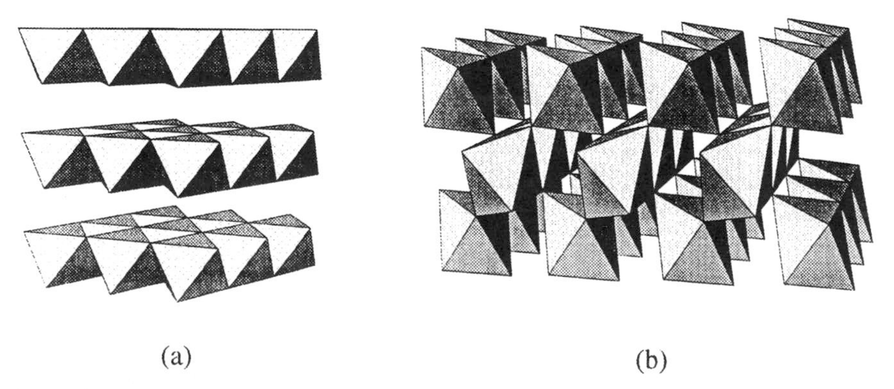

Figure 10.5 Polyhedral representations of the (*a*) TiS_2 and (*b*) TiO_2 (rutile) structures. Note that TiS_2 is layered, whereas TiO_2 is a framework structure.

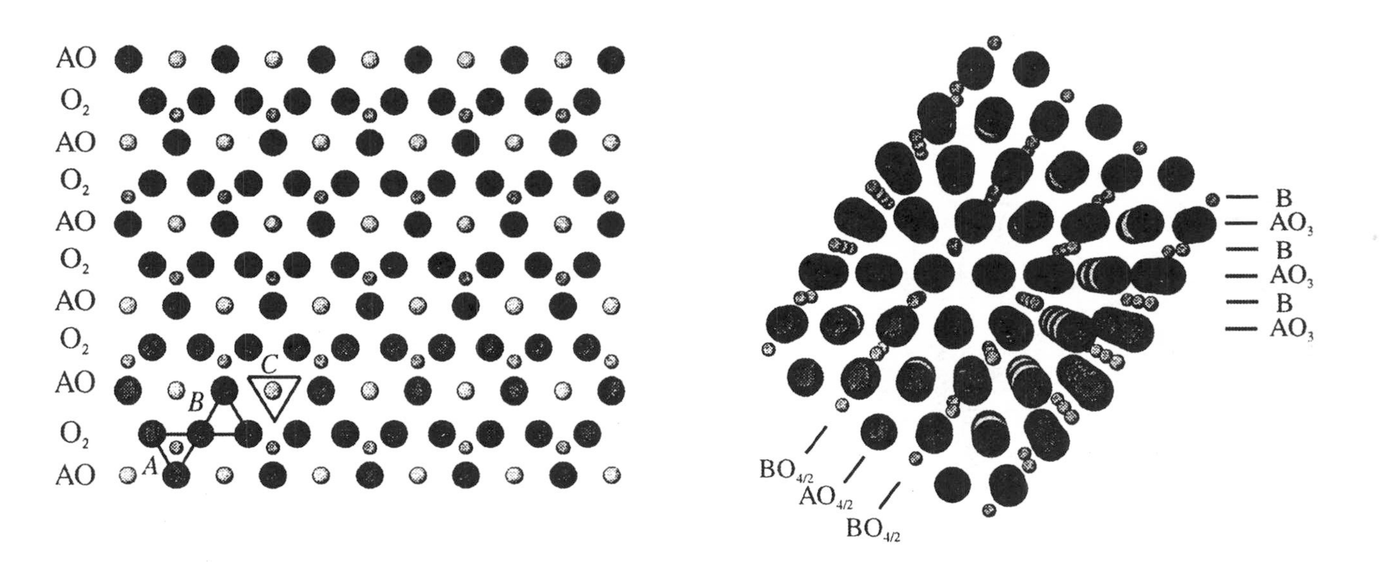

Figure 10.6 (*a*) AO_3 slice of an ideal cubic perovskite with important atomic rows (AO and O_2) denoted. The triangles *A*, *B*, and *C* stack to form trigonal antiprisms, and thus generate the perovskite structure. (*b*) Perovskite structure shown as cubic close-packed lattice with important planes (AO_3, $BO_{4/2}$, and $AO_{4/4}$) denoted. The (100) is in the plane of the page. In both (*a*) and (*b*), large dark spheres denote oxygen, intermediate-size shaded spheres denote A-cations, and small shaded spheres denote B-cations. (Adapted from Ref. 77.)

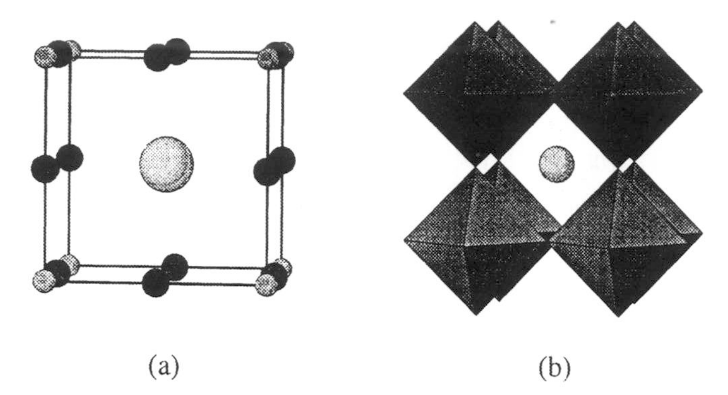

Figure 10.7 Two depictions of the ideal cubic perovskite lattice. (*a*) Transition metal–oxygen connectivity is highlighted along with important atomic planes. Dark spheres denote oxygen, large shaded spheres denote A-cations, and small shaded spheres denote B-cations. (*b*) Polyhedral representation showing the corner-sharing BO_6 octahedra. The large sphere is the A-cation. (Adapted from Ref. 77.)

important planes of atoms are denoted, such as the $BO_{4/2}$ and $AO_{4/4}$ planes). Alternatively, the perovskite lattice can be described as a $BO_{6/2}$ framework of corner-shared octahedra that contains A cations with 12-coordinate sites, as illustrated in Fig. 10.7. The AO_3 layers emphasized in Fig. 10.6*b* correspond to the {111} family of planes in the depiction of the unit cell given in Fig. 10.7, while the (001) and the (002) planes in Fig. 10.7 correspond to the $BO_{4/2}$ and $AO_{4/4}$ planes denoted in Fig. 10.6*b*. The perovskite cell serves as a structural block in a wide variety of complex structures, both layered and three-dimensional, with the simple cubic perovskite containing three-dimensional connectivity of the transition-metal oxygen framework. We consider both descriptions of perovskite-related compounds to better understand the realization of layering in this fascinating class of materials.

Anderson et al.[80] have reviewed stoichiometric double perovskites of the formula $A'A''B'B''O_6$ (primes indicate the possibility of different cations with respect to the ordering on the B-cation sublattice). Three distinct B-cation sublattices, shown in Fig. 10.8 with respect to the B-cation polyhedra, were found to exist. The first (Fig. 10.8*a*) is a random arrangement, where the B′ and B″ cations are distributed statistically over the same site. The second (Fig. 10.8*b*) is a rock salt arrangement, where the B′ and B″ cations alternate along any crystallographic axis. The third arrangement is a layered one (Fig. 10.8*c*), in which the B′ and B″ octahedra order in alternate layers. This layered arrangement has only been found for La_2CuSnO_6.[80–82] It was demonstrated that the difference in ionic size and charge of the constituent cations had the most dramatic effect on the ultimate structure of these compounds, while the coordination preferences played an important role in

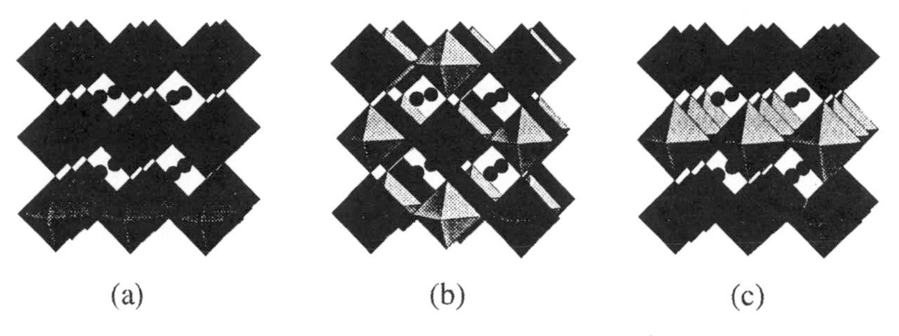

Figure 10.8 (*a*) Random, (*b*) rock salt, and (*c*) layered arrangement of the B-cations (BO_6 octahedra) in double perovskites $A'A''B'B''O_6$. In (*b*) and (*c*), dark octahedra represent $B'O_6$ octahedra and light octahedra represent $B''O_6$ octahedra. Spheres represent A-cations. (After Ref. 80.)

the crossover between the random and rock-salt structures, as well as in realizing a layered structure.[80] All three sublattice types have been observed in the Ln_2CuMO_6 series (Ln = lanthanide, M = tetravalent cation), where the charge difference is 2 and the size difference can be tailored by the choice of Ln and M cations.

The Goldschmidt tolerance factor,[83] which estimates the three-dimensional mismatch between the A—O bonds and the B—O bonds in the perovskite structure, is defined as

$$t = \frac{r_A + r_O}{\sqrt{2}\,(r_B + r_O)} \tag{5}$$

where r_A, r_B, and r_O are the ionic radii[84] of the respective ions. We can write the tolerance factor for a double perovskite ($A'A''B'B''O_6$) with respect to the average ionic radii of the cations on each site:[80]

$$t = \frac{[(r_{A'} + r_{A''})/2] + r_O}{\sqrt{2}\{[(r_{B'} + r_{B''})/2] + r_O\}} \tag{6}$$

where $r_{A'}$, $r_{A''}$, $r_{B'}$, $r_{B''}$, and r_O are the ionic radii of the respective ions. While the tolerance factor is a measure of the framework mismatch between the A and B cations, a measure of the two-dimensional mismatch caused by layering is the difference in the B-cation ionic radii. Both the B-cation sublattice and the phase stability are affected by these size differences, as illustrated in Fig. 10.9. In this figure, the known phases in the Ln_2CuMO_6 series (Ln = lanthanide, M = metal) are plotted with respect to both the Goldschmidt tolerance factor and the B-cation size difference.[80] The random perovskite arrangement is observed when the B-cation size difference (x axis) is less than +0.01 Å (with the exception of La_2CuMnO_6). When this

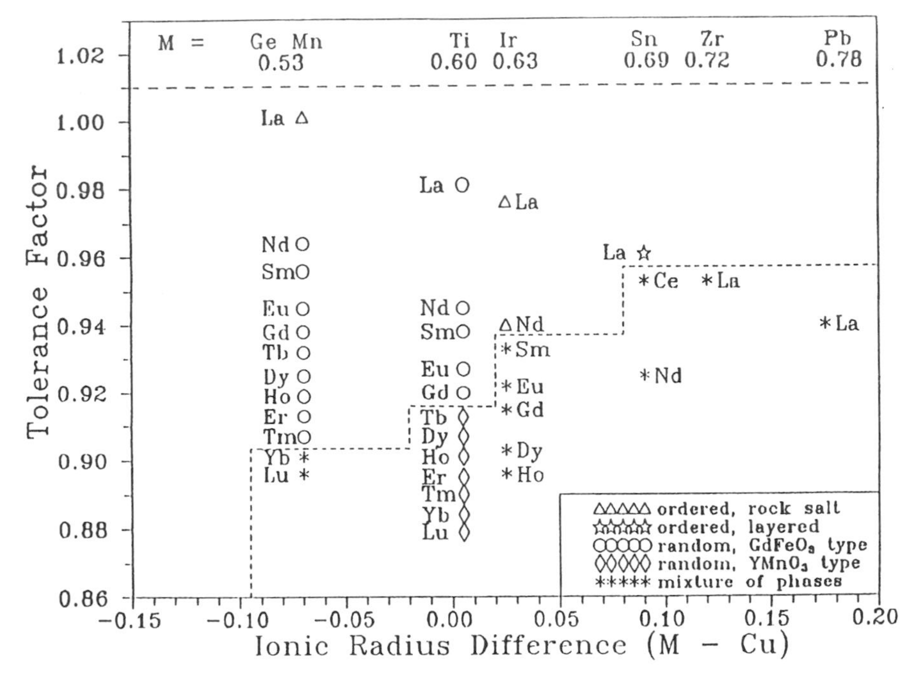

Figure 10.9 Phase field of the double perovskites Ln_2CuMO_6 (Ln = lanthanide, Y) as a function of the tolerance factor, t, and B-cation ionic radius differences. (Adapted from Ref. 80.)

difference is between +0.01 and +0.08 Å, the rock-salt perovskite structure is observed. The layered perovskite structure is observed only when the size difference is greater than +0.08 Å. The perovskite structure becomes unstable for small values of t (y-axis). As t is decreased, either a phase boundary is reached or the structure converts to nonperovskite $YMnO_3$ structure. In addition to the appropriate size and charge differences, the electronic configuration of Cu^{2+} and its propensity to adopt Jahn–Teller distorted polyhedra provides the driving force for layering in La_2CuSnO_6.[81, 82] We will see below that additional chemical considerations affect the oxygen-deficient perovskites, $ABO_{3-\delta}$.

Before moving to the oxygen-deficient perovskites, it is worth examining the double perovskite structures with respect to the close-packed AO_3 layers in order to understand the nature of dimensionality in materials. Figure 10.10 illustrates the B-cation arrangements in double perovskites with respect to the close-packed layers. This representation corresponds to the view of the perovskite in Fig. 10.6*b*. The close-packed layers are denoted *A*, *B*, and *C*. The B-cation pattern on the (110) face of the cubic perovskite is shown. The random arrangement (Fig. 10.10*a*) has a single B site and a three-layer repeat as described above (*ABC* stacking of the AO_3 layers). Both the

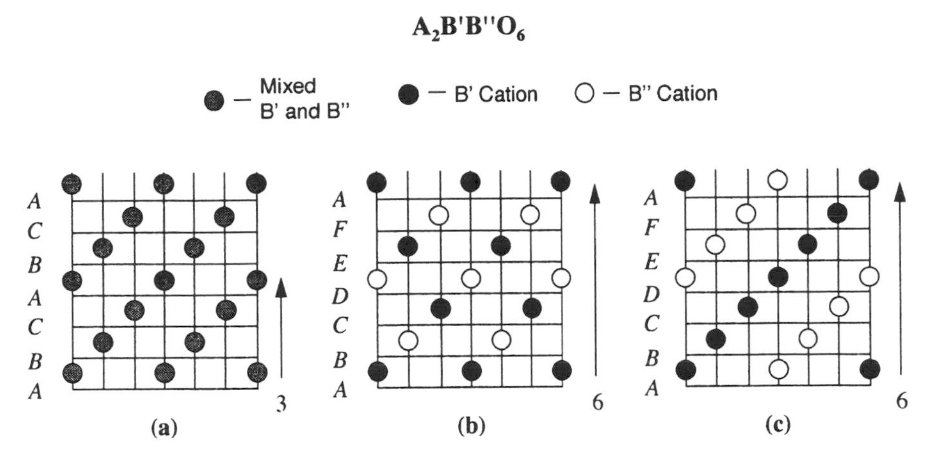

Figure 10.10 The B-cation arrangement perpendicular to the AO_3 layers for (*a*) random, (*b*) rock-salt ordered, and (*c*) layered double perovskites.

rock-salt and layered sublattices have a six-layer repeat, but they differ with respect to the B-cation ordering. The rock-salt sublattice (Fig. 10.10*b*) results from an alternation of B cations in the octahedral interstices in every other layer, while the layered sublattice (Fig. 10.10*c*) results from an ordering of the B-cations within these (111) layers. The structural similarities of these compounds are borne out in this description, and it is clear that two-dimensional, or layered, features result from intricate aspects of these materials, such as the electronic structure, which is dependent on the B′—O orbital interaction. In the layered case, B′—O—B′ interactions (and B″—O—B″) occur in two dimensions while B′—O—B″ interactions occur in the anisotropic direction (see Fig. 10.8*c*). In the rock-salt case B′—O—B″ interactions extend in all three crystallographic directions and therefore exhibit three-dimensional electronic properties (see Fig. 10.8*b*) despite the segregation of the B-cations into alternate (111) planes (or layers).

The majority of layered perovskite oxides are oxygen deficient, having the general formula $ABO_{3-\delta}$. Given the large number of layered oxygen-deficient perovskites and the dirth of examples when stoichiometric, it is clear that oxygen vacancies exert an important structure-directing role. Oxygen vacancies can either be distributed randomly over the oxygen sublattice or ordered on specific crystallographic sites, with the latter resulting in both framework (three-dimensional) and layered (two-dimensional) structures. Examples of these framework structures are $Ca_2Mn_2O_5$,[85, 86] $Ca_2Co_2O_5$,[87] and $La_4BaCu_5O_{13}$.[88] The structure of $Ca_2Mn_2O_5$ is illustrated in Fig. 10.11*a*. The three-dimensional connectivity of the MnO_5 square pyramids is immediately apparent in this depiction. This is in contrast to the structure of $YBaCuFeO_5$,[89] shown in Fig. 10.11*b*, in which the square pyramids of CuO_5 and FeO_5 separate into distinct double layers, and transition metal–oxygen

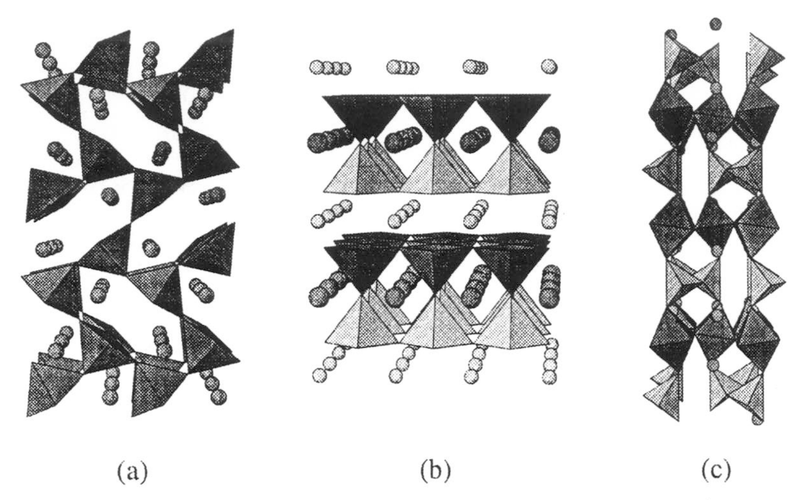

Figure 10.11 Polyhedral representations of the idealized structures of (*a*) $Ca_2Mn_2O_5$, (*b*) $YBaCuFeO_5$, and (*c*) Ca_2FeAlO_5. In (*a*) square pyramids are MnO_5 units and spheres represent calcium ions. In (*b*) small, light spheres represent yttrium, large, dark spheres represent barium, and the light and dark square pyramids are FeO_5 and CuO_5 polyhedra, respectively. In (*c*) the spheres represent calcium and the polyhedra represent $(Fe/Al)O_6$ and $(Fe/Al)O_4$ units. [After Refs. 85 and 86 (*a*); 89 (*b*); 90 and 91 (*c*).]

bonding, in the direction perpendicular to the layers, terminates over a short distance. There are also materials that contain elements of both framework and layered structures. The structure of Ca_2FeAlO_5[90, 91] is illustrated in Fig. 10.11*c* and can be described as a two-dimensional structure, owing to the alternating layers of octahedra and tetrahedra. However, this structure can also be described as a framework structure because of the three-dimensional metal–oxygen bonding.

We have already encountered this mixed structural motif with the layered double perovskite, La_2CuSnO_6, where two-dimensional characteristics are a result of the B-cation order over two octahedral sites. However, in Ca_2FeAlO_5, the similar coordination preferences of Al^{3+} and Fe^{3+} result in cation mixing over the two crystallographically distinct sites. Other examples of mixed structures include $La_2Ni_2O_5$,[92, 93] $Ca_2Fe_2O_5$,[94, 95] and $LaSr_2Fe_3O_8$.[96–99] These three structures can be described as having layers of alternating polyhedra (octahedra/square planes, octahedra/tetrahedra, and octahedra/octahedra/tetrahedra, respectively) surrounding a single B-cation type. Thus they cannot be viewed as layered as a result of B-cation order, nor as a result of termination of metal–oxygen bonding in one unique direction. However, two-dimensional properties have been observed in these layered polyhedral structures,[96–99] illustrating that advanced materials with subtle differences in their structures can lead to novel properties and deserve a great deal more theoretical and experimental attention. In Section 10.3.1 we address multiple B-site cations, focusing on cuprates which order in separate polyhedral layers owing to their coordination preferences. The occurrence of two-dimensional $CuO_{4/2}{}^{2-}$ planes in many cuprates leads to superconductivity as a result of the layered polyhedral units. Several excellent reviews of oxygen deficient perovskites, and their ordered superstructures, have emphasized the importance of the B-cation coordination preferences and electronic structure in determining the oxygen vacancy order in these materials.[100, 101] Moreover, in addition to the chemical considerations mentioned above for the stoichiometric perovskites (ionic size and charge, electronic configuration), other factors, such as the degree of nonstoichiometry, coordination preference of the A cations, and oxidation-reduction of the cations, have important roles in determining the dimensionality of layered perovskite oxides.[77]

A preeminent example of these factors working in concert to stabilize a layered oxide is the compound $YBa_2Cu_3O_{7-x}$ (hereafter referred to as Y123), where x $(0 \leq x \leq 1)$ denotes the capability of variable oxygen stoichiometry. Polyhedral representations of $YBa_2Cu_3O_6$[102-104] and $YBa_2Cu_3O_7$[105, 106] are displayed in Fig. 10.12*a* and *b*, respectively, and are based on a ordered, triple, oxygen-deficient perovskite cell of the general formula $A_3B_3O_{9-\delta}$. A-site, B-site, and oxygen vacancy order results in bilayers of $CuO_2{}^{2-}$ stoichiometry and two-dimensional electronic properties, including superconductivity with a T_c value of 92 K.[105] Oxygen has been completely removed (in both compounds) from every third $(001)_c AO_{4/4}$ plane of the perovskite subcell, and Y occupies the A-site in this plane of ordered oxygen vacancies, owing to the preference of Y^{3+} to occupy 8-coordinate sites. Barium occupies the two higher-coordinated sites in this triple perovskite structure (see Fig. 10.12). No extended copper–oxygen bonding

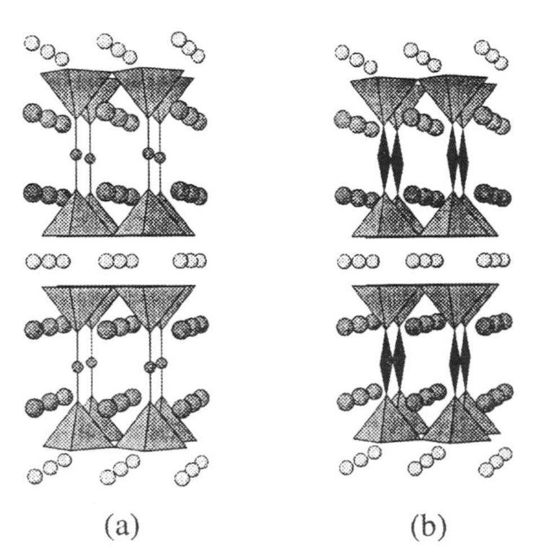

Figure 10.12 Polyhedral representations of the idealized structures of (*a*) $YBa_2Cu_3O_6$ and (*b*) $YBa_2Cu_3O_7$. In (*a*), the square pyramids are CuO_5, the smallest spheres are copper, the intermediate spheres are yttrium, and the largest spheres are barium. In (*b*), the square pyramids are CuO_5, the square planes are CuO_4, the smaller spheres are yttrium, and the larger spheres are barium. [After Refs. 103 (*a*) and 106 (*c*).]

occurs in the *c*-direction in these compounds, as was seen in $YBaCuFeO_5$. Additional oxygen vacancy order results in the formation of two distinct types of copper polyhedra in both compounds, which alternate parallel to the *c*-axis. In $YBa_2Cu_3O_6$ (Fig. 10.12*a*) this additional oxygen vacancy order [removing all of the oxygen from every third $(001)_c$ $BO_{4/2}$ layer] causes the coordination around copper to alternate as SLS SLS parallel to the *c*-axis (where S denotes square pyramidal and L denotes linear). In $YBa_2Cu_3O_7$, however, the additional oxygen vacancy order (removing the oxygen from every [100] row in every third $(001)_c BO_{4/2}$ layer) causes the coordination around copper to alternate as SPS SPS (where P denotes planar), as shown in Fig. 10.12*b*. This illustrates the effects of the coordination preference and electronic structure of the B-cation, d^9 Cu^{2+} (square pyramidal), d^8 Cu^{3+} (square planar), and d^{10} Cu^{1+} (linear). However, A-cation size effects cannot be neglected for the stabilization of layered structures in the oxygen-deficient cuprate-perovskites. For instance, although $LaBa_2Cu_3O_{7-x}$[107–110] (La123) is layered and isostructural with $YBa_2Cu_3O_{7-x}$, no layered phases have been found in the $(La, Sr)_mCu_mO_{3m-x}$ system. Instead, framework structures, such as $(La, Sr)_4Cu_4O_{10}$,[111] $(La, Sr)_5Cu_5O_{13}$,[112, 113] and $(La, Sr)_8Cu_8O_{20-\delta}$,[114–117] are formed when strontium replaces barium as the divalent A-site cation in these oxygen-deficient cuprate-perovskites. A complicated balance, which at present is only partially understood, exists between the chemical effects of the constituent cations in these materials and the structure observed.

Another important class of layered transition-metal oxides are the Ruddlesden–Popper phases $(A_{n+1}B_nO_{3n+1})$,[118, 119] which can be described as an intergrowth of a single rock-salt type AO block and *n* (*n* is any integer) perovskite ABO_3 units. These phases have generated interest owing to their two-dimensional magnetic properties[120, 121] and, more recently, their two-dimensional electronic properties and the occurrence of superconductivity.[122, 123] The general formula for these intergrowth phases can be described as $(AO)_m(ABO_{3-x})_n$, where *m* and *n* are integers denoting the number of rock-salt and perovskite blocks, respectively. The ability of the perovskite structure to tolerate oxygen deficiency leads to the general formula given above, where *x* represents the oxygen deficiency in the perovskite blocks. Figure 10.13*a–c* illustrate three members of this series: La_2CuO_4[124] ($m = 1$, $n = 1$, $x = 0$), $La_2CaCu_2O_6$[125] ($m = 1$, $n = 2$, $x = 0.5$), and $Sr_3Ti_2O_7$[119] ($m = 1$, $n = 2$, $x = 0$), respectively. The rock-salt layer provides the anisotropy in the bonding along the *c*-direction in these materials, resulting in two-dimensional properties. Thus the perovskite transition metal–oxygen bonding framework extends in two directions ad infinitum, while it is spatially confined in the third direction. Materials adopting these structures have been reviewed by Müller-Buschbaum[32] and Raveau et al.,[30] with emphasis on the superconducting cuprates.

Cooperative Jahn–Teller distortions, described before for the layered cuprate perovskite, are also seen in layered oxides with the K_2NiF_4 structure (or distorted variations). As a result of this distortion, the transition-metal

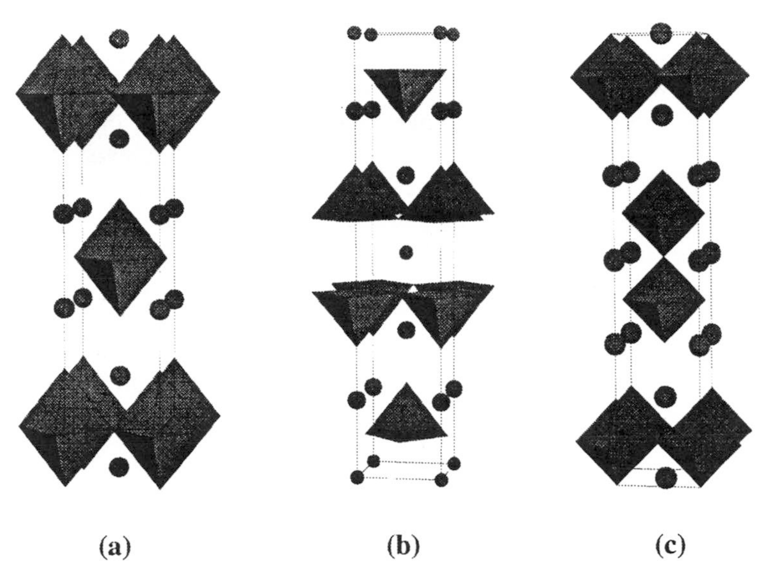

Figure 10.13 Polyhedral representations of the idealized structures for (*a*) La_2CuO_4, (*b*) $La_2CaCu_2O_6$, and (*c*) $Sr_3Ti_2O_7$. In (*a*), spheres are lanthanum and octahedra are CuO_6. In (*b*), large spheres are lanthanum, smaller spheres are calcium, and square pyramids are CuO_5. In (*c*) spheres are strontium and octahedra are TiO_6. [After Refs. 124 (*a*), 125 (*b*), and 119 (*c*).]

octahedra (see Fig. 10.13*a*) are elongated along the [001] direction, as in La_2CuO_4,[124] $SrLaNiO_4$,[126] and $SrNdMnO_4$.[127] The coordination preference of the B-cation affects the structures of these materials, especially with respect to oxygen vacancy ordering, for instance $Ca_2MnO_{3.5}$[85] or $La_2CaCu_2O_6$,[111] which both have square pyramidal coordination about Mn^{3+} (d^4) and Cu^{2+} (d^9), respectively. In addition, A-site effects dramatically influence the dimensionality of the metal–oxygen connectivity. In $La_2CaCu_2O_6$ (Fig. 10.13*b*), two-dimensional $CuO_{4/2}$ nets exist, whereas for $NdSr_2Cu_2O_{6-x}$[128] (not shown) copper–oxygen networks terminate over short range in all directions, owing to the coordination preferences of Nd^{3+} and Sr^{2+} as well as the reduction in the oxygen stoichiometry to compensate for the charge differences on the A-site. As in the layered perovskite oxides, extended two-dimensional metal–oxygen bonding, in the oxygen-deficient Ruddlesden–Popper oxides ($A_{n+1}B_nO_{3n+1-\delta}$), is dependent on several factors, such as size, charge, coordination preference, electronic configuration, and stoichiometry of the A- and B-cations. In addition, each of these factors affects the amount of oxygen deficiency a material can tolerate and thus leads to a complex dependency of the structure and dimensionality of these advanced materials on the underlying chemistry of the constituent ions.

Finally, another fascinating class of layered compounds are low-dimensional bronzes. The term *bronze* applies to a variety of crystalline phases of transition-metal oxides, typically ternary compounds of the type $A_xM_zO_y$. Greenblatt has provided an excellent review of molybdenum oxide bronzes with quasi-low dimensional properties.[17] These compounds have gained considerable interest owing to novel materials properties including highly anisotropic transport characteristics, metal-to-semiconductor transitions driven by charge-density wave states, and superconductivity. The electrical properties are, as with the other oxides discussed herein, determined by the overlap of the *d*-levels of the transition metal (molybdenum) with, and the *p*-levels of, the oxygen.[129-131] The A-cations donate electrons into an empty conduction band of primarily *d*-character in the molybdenum bronzes, and the extent of delocalization of the *d*-electrons determines the optical and transport properties of these materials.[17] The coordination preference of the B-cation determines the intralayer bonding patterns, which is typically edge- and corner-sharing units of BO_6 octahedra in the bronzes, while the stoichiometry and coordination preference of the A-cation determines the interlayer structure and electronic properties.

The red bronzes, of stoichiometry $A_{0.33}MoO_3$, have layered crystal structures as illustrated in Fig. 10.14. The A-cations hold the sheets together, as shown in Fig. 10.14*a*. The layers consist of clusters of six edge-sharing MoO_6 octahedra, which corner share to create two-dimensional sheets, as shown in Fig. 10.14*b*. Two electrons are donated to the layers (by the A-cations) per cluster of MoO_6 octahedra, yet these materials exhibit semiconducting properties, owing to the localization of carriers, which results from long Mo—O bond distances in the *b*-direction.[132] However, the Mo—O bond length is decreased in the blue bronzes, $A_{0.3}MoO_3$ (A = K, Rb, Tl), resulting in metal-

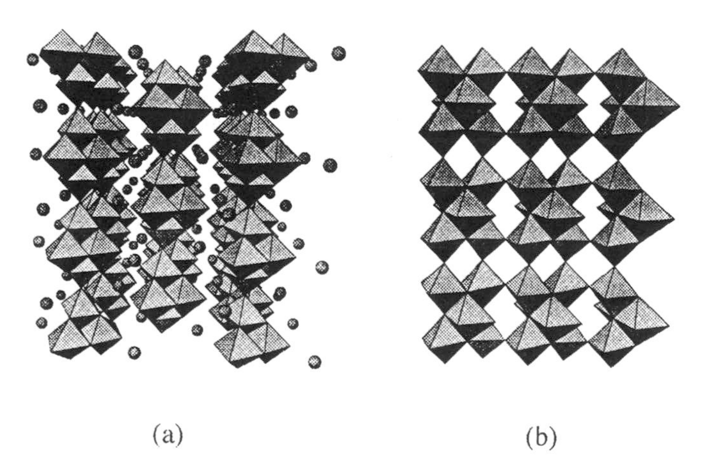

Figure 10.14 Crystal structure of the $A_{0.33}MoO_3$ red bronzes (*a*) emphasizing the layers of Mo—O octahedra and (*b*) the connectivity of the infinite sheets of MO_6 octahedra. The view in (*b*) is of one of the three layers seen in (*a*), rotated 90°, with the A cations removed for clarity.

lic properties.[17] The crystal structure of $A_{0.3}MoO_3$ blue bronze is similar to that of the red bronzes, but the layers consist of clusters of 10 edge/corner-sharing MoO_6 octahedra, which corner share to form the infinite layers.[133, 134] Three distinct Mo positions occur in the layer, with two involved in infinite chains along the *b*-direction, which are responsible for the anisotropic, one-dimensional metallic nature of these materials and the observed metal-to-semiconductor transition (driven by charge-density wave formation).[17] The low-dimensional electronic, magnetic, and optical properties are a direct consequence of the anisotropy in the structure.[17]

Although transition-metal oxides and chalcogenides can seem unrelated when one cursorily inspects their structures or properties, the underlying chemical factors that control the inner architecture of either class of advanced materials is strikingly similar. We have used important examples of layered oxides and chalcogenides to introduce these chemical factors and how they control dimensionality and properties by modifications in the chemical architecture. However, we have focused primarily on cationic effects. The distinction between the two classes of materials can also be understood from the underlying chemistry of the anions. As mentioned previously, the higher electronegativity and lower polarizability of oxygen, compared to those of the chalcogenides, leads to more ionic metal–oxygen bonds than metal–chalcogenide bonds.[33, 34, 74] Thus small, highly charged, strongly polarizing cations can lead to oxides with structures closely related to the chalcogenides.[74] This is evidenced in the bronzes, which contain cations such as Mo^{6+} and have a variety of low-dimensional structures, including van der Waals bonded layer structures.[17] Fortunately, a more complete understanding of the interplay of anionic and cationic factors is evolving. The ability to predict the structure of new materials, as well as precise control over the inner architecture of known compounds, is in its early stages. Continuing research efforts, as discussed in the following sections, which focus on a more complete understanding of these chemical factors, are helping in the realization of a rational materials design approach to chemical synthesis.

10.3 LAYERING IN ADVANCED MATERIALS

In the following section, the chemical considerations introduced in Section 10.2, which can be exploited to control inner architecture of advanced materials, are emphasized. Although these chemical considerations can be applied to several classes of advanced materials, we focus on two types of layered oxides and chalcogenides, layered cuprates and intercalation compounds, to demonstrate the scientific and technological interest in layered materials and to promote an enhanced understanding of rational materials design.

10.3.1 Layered Cuprates

With the discovery of high-T_c superconductivity in the layered cuprate $La_{2-x}Ba_xCuO_4$[122] (see Section 10.2.2 for structural description) came a windfall of research in transition-metal oxides with layered structures containing copper, and to a lesser extent, those transition metals that have easily accessible mixed oxidation states (e.g., iron, cobalt, and nickel). Structurally, all of the known cuprate superconductors can be built up from the layering of perovskite-like units (containing $CuO_{4/2}$) with various other structural elements (i.e., perovskite, rock-salt, or fluorite layers).[30, 32, 135] This scheme was introduced in Section 10.2.2 for $YBa_2Cu_3O_{7-x}$, containing layers of different perovskite-like blocks, and La_2CuO_4, containing perovskite and rock-salt blocks. It has also been shown that the layering in these compounds can be described through the close packing of anion-deficient AO_{3-x} layers, as discussed in Section 10.2.2, where the anisotropic layers are at an angle to the close-packed direction.[77, 136] The essence of layering in these cuprates is in the Jahn–Teller behavior of the Cu^{2+} cations, which cooperatively distort, placing the elongated axis perpendicular to the CuO_2^{2-} layers. However, both three- and two-dimensional structures can exist for compounds with similar stoichiometries and underlying chemistries (as seen in Section 10.2.2 between $LaBa_2Cu_3O_7$ and $(La, Sr)_mCu_mO_{3m-x}$ or between $La_2CaCu_2O_6$ and $NdSr_2Cu_2O_6$). Through careful consideration of the chemical aspects of perovskite-based materials, new layered cuprates have been discovered, as discussed below. In addition, these new compounds help further our understanding of the chemistry that controls advanced materials. Despite the wide number of layered cuprates (both superconductors and nonsuperconductors) known to date, new materials that have particular advantages over known ones (such as better stability, more refractory natures, control over defects, and ease of processibility) will continue to be of interest. In this section we outline some of the novel layered cuprates and how one can control the ordering in these complex oxides. For an in-depth introduction to the crystal chemistry of layered cuprates, the reader is encouraged to consult the excellent review by Raveau et al.[30] For an introduction to the complex electronic nature of these two-dimensional systems, the insightful reviews of Yu and Freeman,[45] Burdett,[46–48] and Goodenough[31, 44] are suggested.

We have already discussed the double perovskite La_2CuSnO_6 (Section 10.2.2) with respect to the chemical factors that stabilize the layered structure. There are many more known oxygen-deficient double perovskites with layered structures, including $YBaCuFeO_5$,[89, 137–140] $LaSrCuAlO_5$,[141, 142] $LaSrCuGaO_5$,[143, 144] and Ba_2CuHgO_4.[145, 146] The important structural feature that gives these materials two-dimensional electronic properties is the $CuO_{4/2}^{2-}$ network. In the stoichiometric compounds, copper is divalent, and the overall oxygen content is related to the net cationic charge. Thus, while $YBaCuFeO_5$, $LaSrCuAlO_5$, and $LaSrCuGaO_5$, all have one-sixth of the normal perovskite oxygen sites vacant, owing to constituent cationic charges,

Ba_2CuHgO_4 has one-third of the normal perovskite oxygen sites vacant. This reduction in the overall oxygen content is a result of replacing the trivalent lanthanide and tetravalent B″-cation (in $Ln_2CuB''O_6$) with divalent alkaline earths (Ba^{2+}, Sr^{2+}) and lower-charged B″-cations (Fe^{3+}, Al^{3+}, Ga^{3+}, and Hg^{2+}). However, such a large degree of nonstoichiometry on the oxygen sublattice is accommodated by the coordination preference of the substituted B-cations. These coordination preferences are illustrated in Fig. 10.15, which shows polyhedral representations of the structures of $LaSrCuGaO_5$ (*a*), $LaSrCuAlO_5$ (*b*), and Ba_2CuHgO_4 (*c*) ($YBaCuFeO_5$ is shown in Fig. 10.11*b*). Note that the copper polyhedra and the B″ polyhedra have separated into distinct layers in all these materials. The concerted order of the B-cations and the oxygen vacancies results in the alternating polyhedral motifs of octahedra/tetrahedra, square pyramids/tetrahedra, and octahedra/linear units, for $LaSrCuGaO_5$, $LaSrCuAlO_5$, and Ba_2CuHgO_4, respectively.

The coordination preferences of Ga^{3+} and Al^{3+} for tetrahedral environments, and that of Hg^{2+} for linear environments, result in the structures described above. Although the copper coordination is best described as square pyramidal in $LaSrCuAlO_5$, there is another oxygen atom, making it pseudo-six coordinate, slightly outside the first coordination sphere of copper.[141, 142] Since this oxygen is outside the coordination environment of

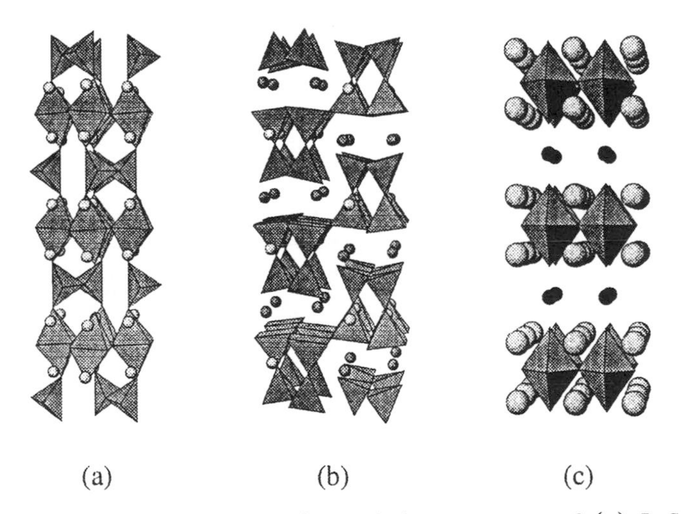

Figure 10.15 Polyhedral representations of the structures of (*a*) $LaSrCuGaO_5$, (*b*) $LaSrCuAlO_5$, and (*c*) Ba_2CuHgO_4. In (*a*) the spheres represent lanthanum/strontium and the polyhedra are CuO_6 octahedra and GaO_4 tetrahedra. In (*b*) the dark spheres represent strontium/lanthanum (where the first cation listed preferentially occupies the site), the light spheres represent lanthanum/strontium, and the polyhedra are CuO_5 square pyramids and AlO_4 tetrahedra. In (*c*) the large spheres represent barium, the small spheres represent mercury, and the polyhedra are CuO_6 octahedra. [After Refs. 143 and 144 (*a*); 141 and 142 (*b*); 146 (*c*).]

copper, termination of the B—O bonding occurs on this oxygen atom in the aluminum tetrahedron. In contrast, the tetrahedra in the gallium analog are arranged such that this oxygen atom bridges the copper and aluminum, maintaining sixfold coordination around the copper.[143, 144] Whereas the gallium material crystallizes in the Brownmillerite structure, with rows of oxygen vacancies removed from alternate [001] rows in every other (100) $BO_{4/2}$ plane, the aluminum material crystallizes with a novel structure closely related to Brownmillerite, but the rows of vacant oxygen atoms are stacked on top of one another in every other $BO_{4/2}$ plane.[141-144] While the coordination requirement of the smaller Al^{3+} cation (coupled with the flexible coordination capabilities of Cu^{2+}), as compared to Ga^{3+}, is the driving force for the structural variation between these two materials, it is accompanied by a preferential ordering of the two A-cations as well.

The A-site coordination in a stoichiometric perovskite is 12 and can be described in terms of the four oxygen atoms in the $BO_{4/2}$ layers above and below the $AO_{4/4}$ layer, and the four oxygen atoms in the latter layer. The coordination can be written, with respect to the coordination in each of these layers, as 4 + 4 + 4 (the middle number referring to the $AO_{4/4}$ layer). In $LaSrCuGaO_5$, there is a single eight-fold A-site that lanthanum and strontium statistically occupy, which can be described as a 4 + 3 + 1 coordination site, with the 1 referring to the coordinated oxygen atom in the $GaO_{4/2}$ layer.[143, 144] The oxygen vacancy order in $LaSrCuAlO_5$ results in two A-sites, one of which is a 4 + 3 + 1 site and is preferentially occupied by lanthanum, and the other which is a 4 + 3 + 2 site and is preferentially occupied by strontium.[141, 142] The slightly larger strontium prefers the higher coordinated site, but the order is not complete, owing to the similarity between the sizes of the two cations.

The cation order in the aluminate and gallate materials is contrasted by that in $YBaCuFeO_5$, which has complete A-site order but B-site order that is not completely understood.[89, 137-140] There are two A-sites in this compound, one of which is occupied exclusively by yttrium and can be described as a 4 + 0 + 4 site, and the other occupied exclusively by barium and described as a 4 + 4 + 4 site. The larger barium occupies the normal perovskite-type site (12-fold), while the small yttrium prefers lower coordination numbers and occupies the eightfold site.[89, 137] In this material it is the similarity of the B-cation coordination preferences that makes the order over the two square-pyramidal sites difficult to distinguish. While some investigators report a mixing of Cu^{2+} and Fe^{3+} over the two square-pyramidal sites,[137] others have reported segregation of the two cations into layers, as illustrated in Fig. 10.11*b*.[89, 138, 140] It appears that the bidimensional structural characteristics of this compound are affected most by the A-site cation, coupled with the propensity of the B-cations to adopt square-pyramidal coordination environments. Further investigations are necessary to gain a complete understanding of the cation ordering in this material.

Oxidation-reduction behavior in these materials is governed predominantly by the ability of copper to exist as both Cu^{2+} and Cu^{3+} in the $CuO_{4/2}^{2-}$ planes. This phenomenon is most dramatically illustrated by the material $Ba_2CuHgO_{4+\delta}$, in which copper is oxidized by excess oxygen in the $Hg\square_{4/2}$ layer (where $\square$ denotes an oxygen vacancy with respect to the normal perovskite oxygen site).[145, 146] Oxidation of the copper in this layered material results in a superconducting transition temperature, T_c, of 94 K.[145, 146] Although oxidation of copper has been achieved in $La_{1-x}Sr_{1+x}CuGaO_5$ by increasing the ratio of Sr^{2+} to La^{3+}, superconductivity has not been observed and is probably a result of the limited range of the solid solution ($x \leq 0.15$).[143, 144, 147, 148] Further research on related materials is necessary to understand the role of doping on the structure and superconducting properties of these layered cuprates, based on oxygen-deficient double-perovskite structures.

The chemical factors that were described above for the double perovskites can be extended to the oxygen-deficient triple perovskites, such as $YSr_2Cu_2GaO_7$,[149, 150] $YSr_2Cu_2AlO_7$,[151] and $LaBa_2Cu_2MO_8$ (M = Nb, Ta).[152-154] The overall oxygen stoichiometry is reflective of the total cationic charge (assuming divalent copper) and the coordination preferences of the B cations. Polyhedral representations of $YSr_2Cu_2GaO_7$ and $LaBa_2Cu_2TaO_8$ are given in Fig. 10.16. Note the similarity of these structures to those of $YBa_2Cu_3O_6$ and $YBa_2Cu_3O_7$, shown in Fig. 10.12*a* and *b*, respectively. In all these compounds, double layers of CuO_5 square pyramids are found, also seen in $La_2CaCu_2O_6$ (Fig. 10.13*a*). The latter compound, $YBa_2Cu_3O_7$, and $YSr_2Cu_2GaO_7$ can be made to superconduct with transition temperatures of 60,[123] 93,[105] and 70 K,[155] respectively.

The linear and planar copper units in the 123 compounds (Fig. 10.12) are replaced by the gallium tetrahedral and niobium octahedral units in $YSr_2Cu_2GaO_7$[149, 150] and $LaBa_2Cu_2TaO_8$,[152, 153] respectively (Fig. 10.16). The coordination preference of Ga^{3+} for tetrahedral environments results in a gallium coordination similar to that in $LaSrCuGaO_5$, but these $Ga(O\square)_{4/2}$ ($\square$ denotes a vacant oxygen site) layers are separated by two layers of CuO_5 square pyramids in $YSr_2Cu_2GaO_7$ rather than the single CuO_6 octahedra seen in $LaSrCuGaO_5$. The coordination preferences of the constituent A-cations are satisfied by the two sites in this compound, with Y^{3+} exclusively occupying the 4 + 0 + 4 eightfold site and the larger strontium occupying the more distorted 4 + 3 + 1 eightfold site (as in $LaSrCuGaO_5$[143, 144]).[149, 150] Although complete ordering of the oxygen vacancies occurs in the $Ga(O\square)_{4/2}$ layer, as described above for $YSr_2Cu_2GaO_7$, the smaller aluminum cation cannot support this completely ordered framework in $YSr_2Cu_2AlO_7$ and has thus been modeled with a disordered vacancy arrangement in the $Al(O\square)_{4/2}$ layer.[151] Despite this, the A-cations retain an ordered distribution over the two sites, with the disordered oxygen vacancy in the coordination sphere of the larger Sr^{2+}.[151] A similar, disordered vacancy arrangement has been

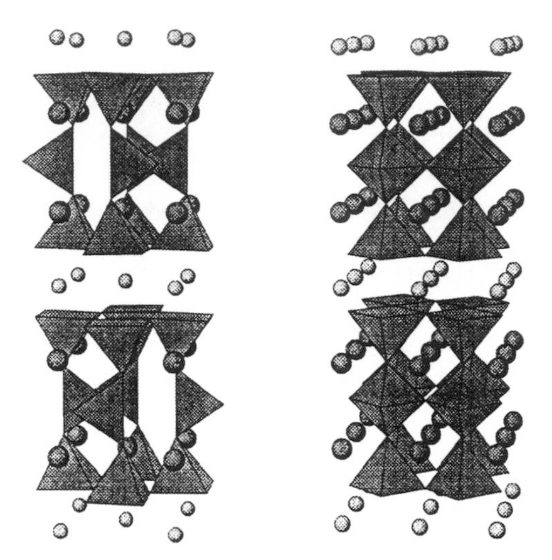

Figure 10.16 Polyhedral representations of the (*a*) $YSr_2Cu_2GaO_7$ and (*b*) $LaBa_2Cu_2TaO_8$ structures. In (*a*) the small spheres represent yttrium, the large spheres represent strontium, and the polyhedra are CuO_5 square pyramids and GaO_4 tetrahedra. In (*b*) the light spheres represent lanthanum, the dark spheres represent barium, and the polyhedra are CuO_5 square pyramids and TaO_6 octahedra. [After Refs. 149 (*a*) and 154 (*b*).]

postulated for the cobalt analog, $YSr_2Cu_2CoO_7$, where the intermediately sized Co^{3+} can support short-range order as proposed for the aluminum compound, but not complete order as in the gallium compound.[156, 157] The disordered model presented above describes the average, long-range structure of these compounds, yet the true local structure has not been determined. A complete understanding of the effect on the oxygen vacancy arrangement of the size of the tetrahedral cation in these triple perovskites has not yet been attained and warrants further investigation.

The higher charge on Ta^{5+} compared to that on Ga^{3+} results in an increase in the oxygen content which is accommodated in the $TaO_{4/2}$ layers, resulting in the octahedral units in $LaBa_2Cu_2TaO_8$ (and the niobium analog).[152-154] Similar to the compounds discussed above, A-site order is observed in $LaBa_2Cu_2TaO_8$, with the smaller La^{3+} occupying the $4 + 0 + 4$ eightfold site, and the larger Ba^{2+} occupying the $4 + 4 + 4$ twelvefold site.[152, 153] In addition, the coordination preferences of the B cations are also satisfied by the order of Cu^{2+} and Ta^{5+} in square pyramidal and octahedral coordination, respectively.

Oxidation of these triple perovskites has been successful in the 123 compounds, as discussed in Section 10.2.2, as well as in the $Y_{1-x}Ca_xSr_2Cu_2GaO_7$.[147, 149] Calcium substitution of yttrium in the latter compound results in a superconducting material when synthesized under high oxygen pressure.[149] Superconductivity has not been observed in the niobium and tantalum compounds, owing presumably to the long Cu—O bonds (about 2 Å) which are approximately 0.5 Å larger than known hole doped superconductors.[152–154] However, superconductivity has been observed in related materials containing blocks of this structure (copper square pyramids connected by niobium octahedra) intergrown with fluorite-like blocks,[158, 159] with a transition temperature of 28 K.[160, 161]

The stabilization of layered materials in the oxygen-deficient quadruple perovskites depends on the chemical factors detailed above for the other layered perovskites, as illustrated in the $Ln'Ln''Ba_2Cu_2M_2O_{11}$ (Ln = lanthanide and Y, M = Ti and Sn) materials.[162–169] Figure 10.17 illustrates the polyhedral representation of the quadruple perovskite structure, based on the structure of $La_2Ba_2Cu_2Sn_2O_{11}$.[162] In the known materials that adopt this structure, $\frac{1}{12}$ of the normal perovskite oxygen sites are vacant (as reflected in the stoichiometry), resulting in a layered polyhedral motif of *SOOS SOOS* (where *S* denotes square pyramids and *O* denotes octahedra),

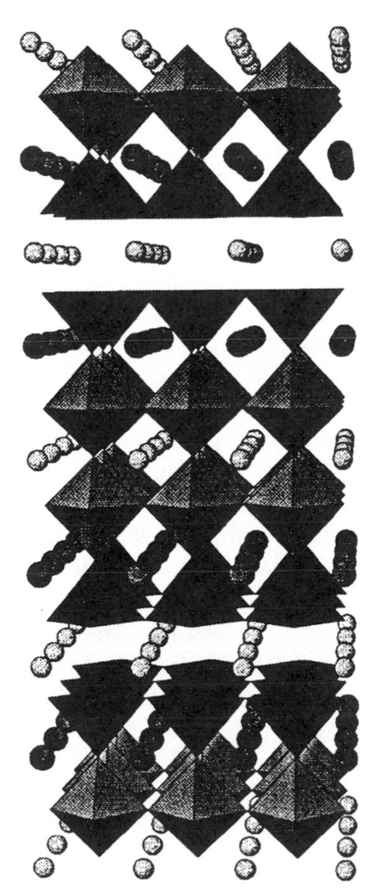

Figure 10.17 Polyhedral representation of the quadruple perovskite structure, $La_2Ba_2Cu_2$-Sn_2O_{11}. The light spheres represent lanthanum, the dark spheres represent barium, and the polyhedra are CuO_5 square pyramids and SnO_6 octahedra. (After Ref. 162.)

as shown in Fig. 10.17. Note that the ordered oxygen vacancy (oxygen removed from every fourth $AO_{4/4}$ layer) is similar to that seen in the layered cuprates discussed previously. This type of order is promoted by the Cu^{2+} cation and an A-site ion whose coordination preference is satisfied by a 4 + 0 + 4 environment. There are two more A sites in this structure, both of which are 4 + 4 + 4 coordination sites. The larger of the two sites is occupied by Ba^{2+} and is between the dissimilar $BO_{4/2}$ layers, while the smaller of the two sites is between the $MO_{4/2}$ layers (M = Sn, Ti) and is occupied by the lanthanide cation.[162, 163, 165, 166, 169]

In $La_2Ba_2Cu_2Sn_2O_{11}$, the large B-cation framework is only stabilized with the A-cation combination of lanthanum and barium.[162] However, simultaneous replacement of tin by titanium and lanthanum by a smaller lanthanide (e.g., $La_{2-x}Ln_xBa_2Cu_2Sn_{2-y}Ti_yO_{11}$) allows the quadruple-perovskite structure to be retained and has been attributed to a tolerance factor effect.[163, 164] Complete substitution of tin by smaller titanium has yielded a series of materials, $Ln_2Ba_2Cu_2Ti_2O_{11}$ (Ln = Nd, Sm, Eu, Gd), which adopt the quadruple-perovskite structure and have a single lanthanide residing on both the 8- and 12-fold lanthanide sites.[163, 165–168] The size of the lanthanide is important in determining the long-range order. In the tin system $La_2Ba_2Cu_2Sn_2O_{11}$, lanthanum occupies the 4 + 0 + 4 site and the quadruple-perovskite structure is retained, whereas when the smaller Ti^{4+} cation occupies the octahedral sites, the quadruple-perovskite structure is destabilized by a large cation on the eightfold site.[166, 167, 169] Thus $La_2Ba_2Cu_2Ti_2O_{11}$ exhibits only short-range order, while the $Ln_2Ba_2Cu_2Ti_2O_{11}$ materials form completely ordered structures when Ln = Nd, Sm, Eu, and Gd. When half the lanthanum are replaced by a smaller cation in $La_2Ba_2Cu_2Ti_2O_{11}$ [e.g., $LaLn'Ba_2Cu_2Ti_2O_{11}$ (Ln′ = Y, Ho, Er)], long-range order of the quadruple-perovskite structure is observed, and the smaller cation preferentially occupies the eightfold site.[169] Thus layering in this complex perovskite material is closely related to the relative sizes of the constituent cations over the multiple sites.

In contrast to the double and triple perovskites, superconductivity has not been reported to date for the layered quadruple-perovskite cuprates. It was recently demonstrated that $La_2Ba_2Cu_2Sn_2O_{11}$, $La_2Ba_2Cu_2Ti_2O_{11}$, and $Eu_2Ba_2Cu_2Ti_2O_{11}$ exhibit high-temperature electrical properties similar to those of known superconducting cuprates (including the single- and double-perovskite cuprates); however, the concentration of electronic carriers is well below that necessary for superconductivity.[147, 170–174] In addition, the electronic structure has recently been calculated for $Eu_2Ba_2Cu_2Ti_2O_{11}$, demonstrating that it has a two-dimensional band structure characteristic of the layered cuprate superconductors.[175] However, doping studies have not been successful in increasing the number of electronic carriers in these materials.[176, 177] As with other layered cuprates, the introduction of electronic species (through the oxidation or reduction of copper) is in competition with ionic compensation mechanisms (such as the production of oxygen vacancies or interstitials) and solid solution boundaries.[147, 170–174, 178, 179] Strict control over the type and amounts of defects, which control the oxidation state of copper and affect the CuO_2^{2-} planes, that are present in these complex oxides is necessary if superconductivity is to be realized in these materials. Crystal chemical modifications to the structure, such as the substitution of yttrium for lanthanum in $La_2Ba_2Cu_2Ti_2O_{11}$, can affect the relative energies of defect formation and therefore favor the generation of electronic carriers as the predominant compensation mechanism of the introduction of chemical dopants.[176] Further research in this area is necessary to understand how low-temperature properties such as superconductivity are related to the equilibrium defect populations during synthesis or subsequent anneals as well as for the rational design of novel superconductors.

10.3.2 Intercalation Compounds

Intercalation chemistry provides a fruitful approach to the rational synthesis of inorganic compounds.[180] As mentioned previously, intercalation describes the reversible insertion of guest species into a lamellar host, with maintenance of the structural features of the host.[42] Intercalated guests include neutral, cationic, and anionic species. The intercalation of charged species typically involves an oxidation-reduction process between the guest and the host network, as described in Section 10.1.3. For example, the insertion of an A^+ cation results from ionization of the electropositive guest element A, and a concordant reduction in the host matrix from Z to Z^-. An example is the intercalation of Li in TiS_2 wherein the Li^+ cation induces a reduction of the Ti metal center from Ti^{4+} to Ti^{3+} (see Fig. 10.18 for the structure of $LiTiS_2$ as compared to Fig. 10.5*a* for that of TiS_2). Similarly, an electronegative atom B can be reduced to B^-, thus resulting in an oxidation of the host matrix to Z^+. The redox potential of the host therefore controls which species are potential intercalants. Hagenmuller[74, 75] points out that layered oxides and chalcogenides generally play the role of acceptors in intercalation reactions, which is in contrast to the amphoteric nature of graphite in intercalation reactions. Yet intercalation of oxygen in the rock-salt blocks of La_2CuO_4 results in donor behavior by the Cu^{2+} ion, illustrating the capability of these materials to intercalate anionic species as well. The wide range of both host–guest combinations and applications for intercalation compounds has led to several somewhat diverse general classification schemes.

Grenier et al. have classified several kinds of intercalation compounds according to their host composition:[181] (1) graphite and related compounds; (2) dichalcogenides of transition metals; (3) layered oxides (MoO_3), V_2O_5, β-alumina, and clays; (4) layered oxychlorides (FeOCl, $FeMoO_4Cl$); (5) halogenides (ZrCl and $ZrCl_2$); and (6) metallic alloys ($LaNi_5$). This categorization illustrates the remarkable diversity and rich chemistry of intercalation materials, of which transition-metal oxides and chalcogenides (discussed herein) are important examples.

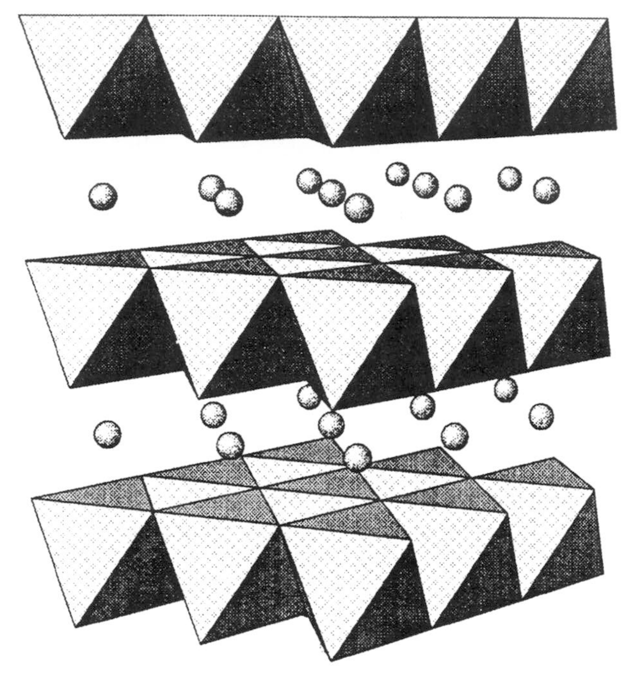

Figure 10.18 Polyhedral representation of $LiTiS_2$. Spheres are lithium cations residing in the van der Waals gap between octahedral XMX slabs.

Li and Whittingham offer a scheme that distinguishes materials according to the relevant intercalative redox and ion-exchange reactions:[180]

1. The single-sheet structures of graphite and boron nitride.
2. Transition-metal oxides and dichalcogenides, MX_2, which have readily reducible cations and have found extensive use as electrodes for energy storage and in electrochromic displays.
3. β-Aluminas, naturally occurring minerals (such as mica, vermiculite, and montmorillonite), and synthetic transition-metal phosphates, arsenates, or silicates, which are less likely to undergo redox behavior but exhibit complex ion-exchange and solvation chemistry (these materials have been investigated extensively as potential solid electrolytes and for ion exchange).

This classification is useful in that it introduces the relationships between types of layering and the materials properties and applications observed.

Because of the diversity in structural types as well as in the properties and applications that can be exhibited by intercalation compounds, as attested to by the difficulty in their classification, we cannot address the entirety of interesting materials in this brief discourse. We will attempt to introduce several of the materials categories that were listed above, in both classification schemes, with a focus on the chemical considerations that are relevant to both the stabilization of these materials and their effects on the ultimate properties or applications of intercalation compounds. It is important to note that several of the chemical considerations apply to each of the materials discussed below. However, for the purpose of emphasizing control over inner architecture, we focus our attention on particular chemical considerations for each class of material. Although an attempt has been made to survey various classes of intercalation compounds with unique materials applications, this list is by no means exhaustive.

We have already introduced the structure and chemical considerations that control the chemistry of the dichalcogenides and related oxide materials. The intercalation properties of these materials have been reviewed extensively.[33-35, 57, 74-76, 182-185] One of the interesting chemical features of the intercalation compounds of layered oxide materials is that the dichalcogenide structures, which are unstable in the pure form for oxides, are stabilized when intercalants are introduced, owing to the screening of the anionic repulsion of neighboring O_2^{2-} anionic layers by the cationic species between them. A variety of structural transitions occur in A_xMX_2 ($A = Li^+$, Na^+, K^+, Rb^+, or Cs^+; M = 3*d* or 4*d* element; $X = O^{2-}$, S^{2-}, or Se^{2-}) intercalation compounds, which have been studied for their energy storage and electrochemical applications, owing to the range of x values possible ($0 \leq x \leq 2$). The coordination preference of the intercalant affects the displacement of adjacent layers relative to one another as a function of the ionic size of the alkali metal, the amount of intercalation (x value), and the nature of the host slabs (discussed in Section 10.2.2).[33]

In the typical layered MX_2 materials, there are empty octahedral interstices between the slabs, as described in Section 10.1.1, which can be occupied by the intercalated A^+ cations. The filling of these sites marks a progressive structural change from a CdI_2 (Fig. 10.1*c*) structure to a NiAs-type structure, as for Li_xTiS_2 materials (for structure of $x = 0$, see Fig. 10.5*a*, and for $x = 1$, see Fig. 10.18). However, prismatic coordination of the intercalated species is possible if the layers displace themselves with respect to one another. Transitions are observed from octahedral coordination of the intercalant to prismatic coordination as a result of the three factors listed above. Octahedral coordination is favored for small alkali cations (Li and Na), small interlayer separations, and higher values of electron transfer (large x) to the host lattice (MX_2^{x-}), owing to the ability of the octahedral environment to accommodate higher anionic charges than the trigonal prism.[33] The structural transitions across any given series can be understood within this conceptual framework. Thus, while octahedral coordination is seen for $A = Li^+$, in A_xTiS_2, $0 \leq x \leq 1$, a transition from prismatic to octahedral coordination is observed for $A = Na^+$ with increasing x, and $A = K^+$, Rb^+, and Cs^+ are found in prismatic coordination for all x. Similar structural distortions are observed in other intercalated dichalcogenides and oxides.[33–35, 57, 74–76, 182–185]

Weigers et al.[186] have reported a new class of layered compounds, which have the stoichiometry $(MS)_nTS_2$ (M = Sn, Pb, Bi, or rare-earth elements; T = Nb or Ta; n = 1.08 to 1.19) and adopt structures that belong to the class of materials known as "misfit"-layer compounds.[187] Although the formula of these compounds is often abbreviated MTS_3, the earlier formula is indicative of the layered nature of these compounds, which consists of alternating layers of TS_2 layers, as described previously for the dichalcogenides (Section 10.2.2) and MS layers. The MS layer can be described as a distorted NaCl-type layer with the M cation in a distorted square pyramidal coordination by sulfur. Both the coordination preference of the transition-metal cation and the overall stoichiometry strongly influence the observed structure and properties of these materials. As in the dichalcogenide layered materials, octahedral or trigonal prismatic coordination of the transition metal can occur, depending on the electronic structure and coordination preference of the particular cation (see Section 10.2.2 and Figs. 10.3 and 10.4). Thus octahedral coordination is observed for T = Ti, Cr, and V, while trigonal prismatic coordination is observed when T = Nb or Ta.[188] The posttransition main-group or rare-earth elements, M, prefer a rock-salt type of lattice for the given overall stoichiometry MX, and occupy a distorted square pyramidal coordination environment resulting from the packing of two rock-salt-type layers, as illustrated in Fig. 10.19.

These two layers (MS and TS_2) have slightly different a-lattice constants (in the plane of the layers), resulting in a layer misfit and a stoichiometry that deviates from the 1 : 1 ratio of M to T for the ideal $(MS)TS_2$ compound (n = 1). The value n in the chemical formula denotes the mismatch between

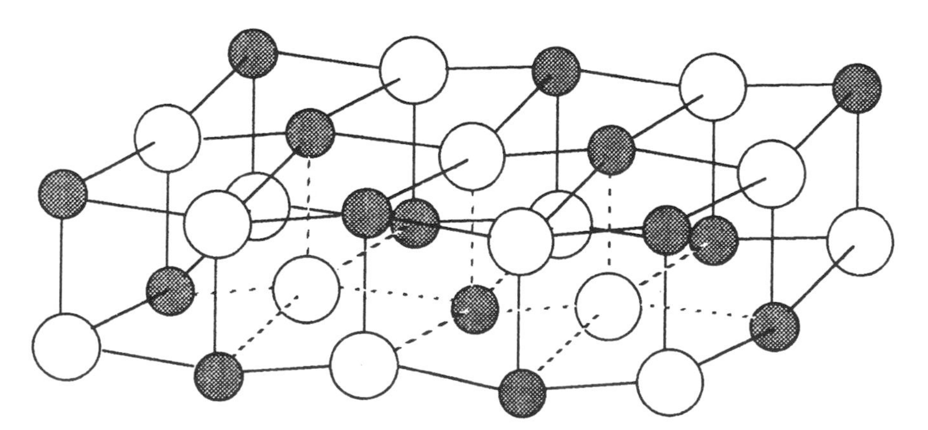

Figure 10.19 Idealized structure of MX unit in the $MX(TX_2)$ misfit layer compounds. The shaded spheres represent M cations and the unshaded spheres represent the X anions. (After Ref. 186.)

the two layers as it is reflected in the overall stoichiometry of the material. Experimentally, n is found to range from 1.08 to 1.28 for known compounds.[188] In addition, stoichiometry plays another important role in determining the properties of these compounds, owing to the ability of the MS layers to support a high concentration of cation vacancies. Rouxel et al.[188] have shown that in $(GdS)_{1.27}CrS$, cationic vacancies occur at the gadolinium site, resulting in the more exact formula $(Gd_{1.16}\square_{0.11}S_{1.27})CrS_2$, where $\square$ denotes a vacancy on the gadolinium site. The observed vacancies help one explain the semiconducting properties of these misfit layers, as they compensate the excess electrons introduced to the CrS_2 layer by the $(GdS)_n$ layer. If no vacancies were present, one would expect metallic behavior in the dichalcogenide layer, owing to the extra electrons occupying the Cr—S conduction band, but the introduction of cation vacancies removes these excess electrons and results in semiconducting behavior.

Layered structures that contain an ideal ratio of 1 : 2 of the number of rock-salt layers and dichalcogenide layers, $MX(TX_2)_2$ (M = Sn, Pb, Bi; T = Ti, Nb, Ta; X = S, Se), have been synthesized and characterized.[189–193] These materials are structurally similar to the compounds introduced above, with the exception that the rock-salt misfit layer is sandwiched between two layers of the dichalcogenide block (i.e., X—T—X X—T—X). As in the compounds above, the lattice mismatch between the rock-salt layer and the dichalcogenide layer results in a stoichiometry of $MX_{1+y}(TX_2)_2$, where $1.12 \leq y \leq 1.18$ has been observed. The introduction of the van der Waals gap in the double layer of the dichalcogenide structure results in similar intercalation behavior to the parent structures.[194–197] Preferential intercalation of cobaltocene in the van der Waals gap between the two X—T—X layers has recently been demonstrated in $(PbS)_{1.18}(TiS_2)_2$, $(PbS)_{1.14}(TaS_2)_2$, $(PbSe)_{1.12}(NbSe_2)_2$.[193] Thus these complex-layered materials display important intercalation properties, owing to the similarity of their structural features to those of the parent intercalation materials.

Intercalation compounds based on layered oxide materials have attracted considerable attention recently, and this interest has yielded new materials related to the Ruddlesden–Popper phases. The structures of several Ruddlesden–Popper materials were introduced in Section 10.2.2 and are illustrated in Fig. 10.13; however, they exhibit little interlayer reaction chemistry. This has been attributed to the high interlayer charge density of these oxides.[198] Dion et al.[199] successfully synthesized layered perovskite materials of the formula $MCa_2Nb_3O_{10}$ (M = K, Rb, Cs), where the interlayer cation is a monovalent alkali metal and has an interlayer cation density that is half of the related Ruddlesden–Popper phases. Further research in this area demonstrated that a variety of structures and stoichiometries [with the general formula $M(A_{n-1}B_nO_{3n+2})$] could be stabilized that undergo facile ion-exchange and intercalation reactions in this interlayer region.[198, 200–206]

The perovskite blocks of the $M(A_{n-1}B_nO_{3n+2})$ series of compounds (M = alkali metal, Tl, NH_4) are similar to the $A_2(A_{n-1}B_nO_{3n+2})$ Ruddles-

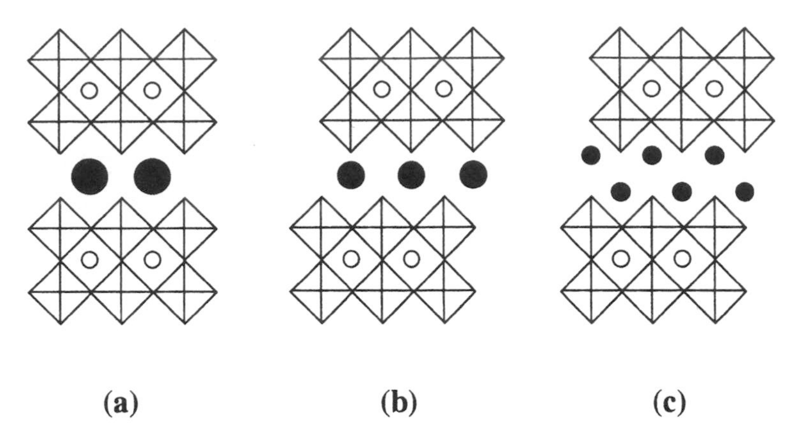

Figure 10.20 (100) projections of the idealized structures of $MLnNb_2O_7$ compounds (*a*) M = Rb, (*b*) M = K, and (*c*) M = Na. The octahedra are NbO_6 units and the small unfilled spheres are La ions. The dark filled spheres represent the M cation positions. In the case of M = Na, the sites are only 50% occupied. (After Ref. 204.)

den-Popper phases where the constituent cations reported to stabilize the former materials have been A = rare earth, Ca, Sr, Ba and B = Nb, Ti. The relative displacement of the perovskite blocks from layer to layer has been shown to be affected by the M cation. In the Ruddlesden–Popper phases the displacement is $(a + b)/2$, where a and b are the unit cell dimensions in the basal plane of the layers, resulting in the displacements shown in Fig. 10.13. In addition, the coordination around the interlayer A cation is a distorted rock-salt type. The Ruddlesden–Popper structure is retained in the $M(A_{n-1}B_nO_{3n+2})$ system when $M = Li^+$ and Na^+, as in $LiCa_2Nb_3O_{10}$,[199] $NaCa_2Nb_3O_{10}$,[199] and $NaLaNb_2O_7$,[204] but only half of the interlayer sites are occupied in a random fashion. When the cation size is increased, the alkali metal occupies a single interlayer site (between the two O layers which terminate the perovskite slabs) which no longer retains the rock-salt coordination of the Ruddlesden-Popper phases, and thus affects the relative displacements of the layers. The largest M cations, $M = Rb^+$ and Cs^+, occupy an eight-fold site between perovskite layers which are not displaced relative to each other, while intermediate size monovalent cations stabilize layers which displace such that trigonal prismatic coordination of the M^+ cation occurs.[198–204, 206] Figure 10.20 illustrates the difference between the three structural types as the M^+ cation size is decreased in $MLaNb_2O_7$ corresponding to (*a*) M = Rb, (*b*) M = K, (*c*) M = Na.[204]

Ion exchange of these materials can lead to the protonated form $H(A_{n-1}B_nO_{3n+1})$, which are solid acids and react with organic bases to form intercalated alkyl-amine compounds, among others that exhibit large interlayer expansions.[198, 200–206] Recently, pillaring studies on four layered oxides, $KCa_2Nb_3O_{10}$, $KCa_3Nb_3TiO_{13}$, $KCa_4Nb_3Ti_2O_{16}$, and $KCaSr_{0.5}Nb_3Ti_{0.5}O_{11.5}$, possessing the above-described structures, by Ram and Clearfield[207] demonstrated that the aluminum keggin ion, $[Al_{13}O_4(OH)_{24}(H_2O)_{12}]^{7+}$, could also be intercalated between the perovskitelike sheets of these layered titanoniobates. These intercalated inorganic polymers arrange themselves so as to give large porosity for sorption and catalytic activity at high temperature.[207] The similarity of the above-described structures to the Ruddlesden–Popper phases have led to the formation of $M_2Ln_2Ti_3O_{10}$ phases (M = alkali metal, Ln = La or rare earth) that possess the $n = 3$ Ruddlesden–Popper structure.[203, 205] These materials, which contain the monovalent interlayer cation, also exhibit facile ion-exchange reactions and stable protonated forms $H_2Ln_2Ti_3O_{10}$, as well as hydrated forms.[203, 205] However, the protonated Ruddlesden–Popper phases do not undergo intercalation reactions with organic bases, indicating a difference between the Brønsted acidity of these phases and the structurally related phases described above.[208] This has been attributed to the difference in the interlayer cation arrangements between these two structure types.[208] Intercalation of organic bases into inorganic solid acid materials is discussed further below.

Other examples of inorganic hosts that allow intercalation of organic guests include classes of materials with anionic species in addition to oxygen, such as the layered phosphates and layered oxyhalides. The realization of these composite layered phosphates is a result of the acid-base chemistry of the inorganic host, specifically that of the interlayer cations, owing to the resistance of the framework to undergo extensive redox behavior (in contrast to the dichalcogenides). Transition-metal phosphates represent an intriguing class of materials with applications in chemical sensing and as ion-exchange agents. Several layered phosphates of the group IV transition metals have been reported with compositions based on $Zr(HPO_4)_2 \cdot H_2O$ (α-ZrP) and $Zr(PO_4)(H_2PO_4) \cdot 2H_2O$ (β-ZrP) formulas.[209] By replacing the OH group of the *o*-phosphoric acid with organic species, intercalated derivatives may be realized.[210–212] Li and Whittingham[180] have reported the preparation and intercalation of a new layered titanium phosphate, $TiO(OH)(H_2PO_4) \cdot 2H_2O$. An understanding of the acid-base properties of both host and guest was crucial in achieving the desired inner architecture, evidenced through careful control of the synthesis conditions, particularly the pH, which yield the layered product. The resultant solid acid, much like the protonated Ruddlesden–Popper oxides discussed above, exhibits ion exchange of the interlayer cation and reaction with organic bases which increases the interlayer separation. Clearfield and Ortiz-Avila[20] have also investigated polyether and polyamine derivative layered zirconium phosphates as a new class of supramolecules. Recently, Liu and Kanatzidis[213] achieved the topotactic polymerization of intercalated polyaniline in several layered phosphates. These materials are important for nanochemical applications, discussed further in Section 10.4.

Recognizing the similarity of various organic molecular cations to the highly electropositive atoms, led Chatakondu et al.[214] to explore the preparation of organometallic counterparts to the molybdenum bronzes discussed in Section 10.2.2. They describe the preparation and solid-state properties of compounds formed by the intercalation of redox-active cubane clusters into MoO_3 or FeOCl. In comparison with the compounds discussed above, which are not prone to reduction, these materials exhibit redox behavior. This work illustrates that electron-rich organometallic compounds (such as cubane clusters and ferrocene) can be intercalated into MX_2 (M = Ti, Zr, Hf, Nb, Ta; X = S, Se) and related host materials. Kanatzidis et al.[24, 25] have investigated crystalline inorganic hosts as media for the synthesis of conductive polymers. The method demonstrated involves the insertion of polymer chains (such as polypyrole, polythiophene, and polyaniline) into inorganic host frameworks (such as FeOCl) through in situ intercalative polymerization of a monomer (contrasting the postintercalative polymerization discussed above for the phosphates) using the host itself as an oxidant. FeOCl serves as a convenient redox-intercalation host, as shown in Fig. 10.21, which shows

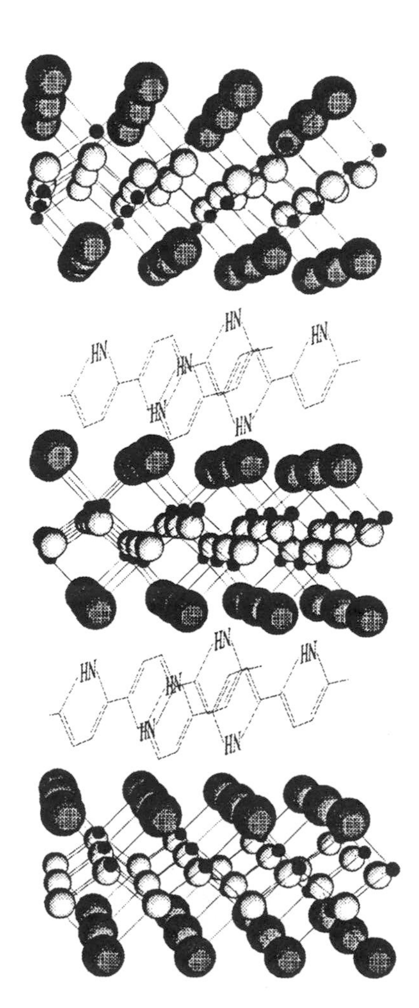

Figure 10.21 Polypyrole intercalation between layers in FeOCl. Small dark spheres are iron, intermediate-sized light spheres are oxygen, and largest shaded spheres are chlorine ions. (Adapted from Ref. 24.)

polypyrole trapped between layers of FeOCl, yielding a highly conductive inorganic–organic polymer microlaminate.[23] Intercalation compounds derived from the foregoing strategy, such as that shown in Fig. 10.21, are composed of alternating monolayers of positively charged conducting polymer chains and negatively charged FeOCl layers. The redox behavior of FeOCl has also been investigated with respect to Li^+ intercalation, and it is seen that reduction of the metal center from Fe^{3+} to Fe^{2+} results in the number of Fe^{2+} centers being equal to x in Li_xFeOCl (and thus the upper limit of intercalation is $x = 1$).[215–217] Materials with polymer chains in well-defined environments such as inorganic host structures represent a new class of materials with the potential for novel or improved properties, and research in this area is only in its early stages. In both of the examples of organic guest intercalation into inorganic host materials, an understanding of the oxidation-reduction chemistry of the host plays a crucial role in the realization of the desired composite.

As our final example of an advanced intercalation compound, we consider the nanoporous tin(IV) chalcogenides, which represent a new class of advanced materials with open frameworks and high potential for nanochemical applications (i.e., chemical sensing).[13, 14, 19] In contrast to the relatively dense layers of the oxide and chalcogenide materials discussed above, whose layers can be derived from the close packing of ions, these materials have open framework structures, as illustrated in Fig. 10.22. Although these materials are structurally layered (or two-dimensional), the open-framework nature of each layer can result in lower dimensional properties, such as quantum electronic properties.[14] Through careful selection of the guest species, a range of crystal structures and pore shapes can be obtained.[19] The compositional tuning of the band properties of isostructural TMA–SnS_xSe_{1-x}–1 (where TMA refers to tetraethylammonium and the 1 refers to a structure type) has been investigated by Ahari et al.[14] While the particular guest species controls the final structure type, the overall reaction stoichiometry (x value in the formula above) within a given structural type strongly influences the resulting semiconductor behavior and quantum size effects. In fact, these properties can be tuned according to Vegard's law between the two end members TMA–SnS–1 and TMA–SnSe–1.[14]

As mentioned above, the nanoporous tin chalcogenide materials exhibit quantum electronic properties but have a layered macroscopic structure.[14] Similarly, the low-dimensional bronzes, discussed in Section 10.2.2, can have one- or two-dimensional electronic properties, which are determined by the anisotropies of metal–oxygen bonding within a structural layer.[17] As discussed in Section 10.1.2, layering in complex materials (or low-dimensional properties in general) result from structural anisotropies as reflected in particular properties. Therefore, the framework structures of various transition-metal oxides (discussed above) may have two-dimensional electronic or magnetic properties, owing to cation ordering or anisotropies in the metal–oxygen bonding along a particular direction. The materials chemist, in general, is faced with the challenge of understanding the relationship be-

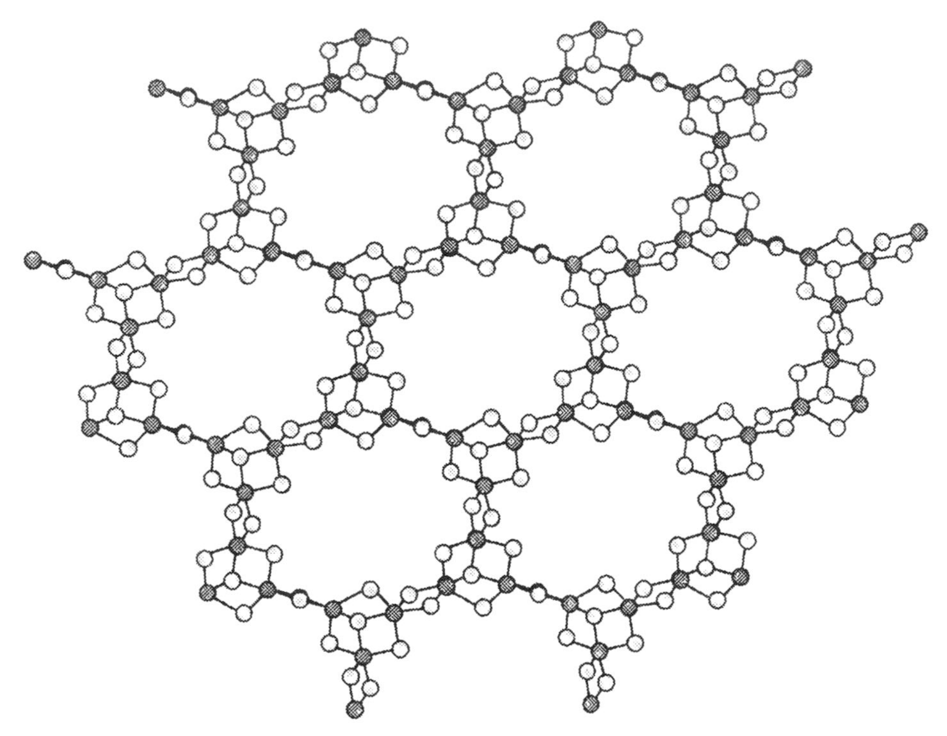

Figure 10.22 Single layer of the nanoporous TMA–SnSe–1 material, illustrating the 24-ring hexagonal-shaped pores. TMA is Me_4N^+ and 1 indicates the structure type. The dark spheres represent Sn atoms and the light spheres represent Se atoms. (Adapted from Ref. 19.)

tween the structure, composition, and properties of materials, and to use this knowledge to design novel materials with desired properties (i.e., layered structures and two-dimensional properties). We have used specific examples of layered cuprates and intercalation compounds to emphasize the chemical factors that affect the structure and properties of advanced materials in general, and to demonstrate how researchers have exploited this understanding to design novel materials. Future directions in materials chemistry will continue toward the attainment of a complete understanding of these chemical factors and their effects on the aforementioned relationship.

10.4 SUMMARY AND FUTURE PROSPECTS

Many challenges remain in predicting the structure, composition, and properties of advanced materials. DiSalvo has noted:[218] "The synthesis of novel solids is as much an art as a science." Moreover, it has become increasingly apparent that progress in inorganic materials chemistry relies upon an enhanced understanding of methods to synthesize and characterize new solids that possess unique structures and properties. The current trend is toward rational design of solids based on an accumulated knowledge of crystal chemistry, thermodynamics, and reactivity, as well as the fundamental relationship between crystal structure and properties.[55] In this review we have highlighted this important relationship and chose as our examples layered cuprates and intercalation compounds. Research in both of these areas will continue to yield novel and interesting materials. In this section we address current research that focuses our attention on the rational design of materials. We also conclude, and broaden our scope somewhat to discuss future advanced materials such as inorganic–organic composites and nanomaterials.

10.4.1 Chemical Architecture and Rational Design of Advanced Materials

In this review we have stressed important chemical considerations that influence the inner architecture of advanced materials. Specifically, we have focused on five chemical factors that control the structure, dimensionality, and resultant properties of materials: (1) stoichiometry, (2) coordination chemistry, (3) ionic size and charge, (4) acid-base chemistry, and (5) oxidation-reduction capability. Various examples of layered oxide and chalcogenide materials, with particular emphasis on layered cuprates and intercalation compounds, were used to illustrate the effect of these chemical factors on the structure of layered compounds. Although we lack a complete understanding of these chemical considerations and their relative contributions to the stabilization of layered materials, great progress has been made in the design of layered oxides and chalcogenides (among other classes of materials), the synthesis of novel materials, and the optimization of known ones.

In this chapter we have focused on the crystal chemistry of layered materials while discussing synthetic techniques only briefly. However, to realize rational materials design, we must not only appreciate the foregoing chemical considerations, but also comprehend synthetic conditions that promote the appropriate chemical considerations required to yield specific inner architectures. Gopalakrishnan has identified two critical steps in the "tailormaking" of solid-state materials:[55] The first is to identify the ideal crystal structure and probable chemical composition that would give rise to the desired materials property; the next is to develop and implement appropriate procedures for the synthesis and characterization of the advanced material.

Roy has also addressed a logical approach to synthesizing new materials to specification.[73] Three of his criteria overlap with our discussion of advanced layered materials: (1) define the structure or the element of the structure that is thought to be critically correlated with the desired property, (2) define the compositional boundary conditions correlated with the desired property, and (3) combine into possible structure–composition areas using crystal chemistry structure field maps; the remaining criteria emphasize the synthetic and empirical conditions that are useful to materials design. The reader is encouraged to consult the preceding references for additional approaches to rational materials design. Future strategies for the realization of novel advanced materials will undoubtedly rely on the combination of factors that control the inner architecture, as well as synthesis methods, which lead to new compounds. In addition to controlling the crystal architecture of advanced materials, strict control of anisotropies is also required at the microscopic and macroscopic levels.

10.4.2 Higher Levels of Structure

In Section 10.1 we stated that all materials possess an inner architecture, that is, a hierarchy of successive structural levels. This review has focused on the crystal structure level and the issues of dimensionality, anisotropy, topochemistry, and to some extent, how these relate to physical properties. Anisotropy, however, also manifests itself in the morphology of crystallites on the micrometer scale. Anisotropic crystals are often acicular (needlelike) or platey in morphology. Large numbers of these crystallites must be assembled into the microstructure of "real" materials, which are almost always polycrystalline in nature. This is a challenging proposition from a processing point of view. Furthermore, as opposed to polycrystalline materials composed of isotropic, equiaxed crystallites (equal size in all three dimensions), grain boundaries in highly anisotropic materials present severe behavioral discontinuities, especially with respect to transport properties. A prime example of this is the well-known "weak link" behavior at grain boundaries in high-T_c superconducting oxides.

There are a number of strategies to overcome the grain boundary limitations in layered anisotropic materials. These usually involve (1) minimizing the mismatch between the individual planes and grain-to-grain misorientation angles by "texturing," and (2) eliminating grain boundaries altogether. The first strategy can be accomplished "cold" (in the so-called "green" or prefired state) or "hot," by a number of melt-forming or forging procedures. One way to actualize the second strategy is by epitaxial deposition of single-crystal thin films. Deposition techniques include laser ablation, metal–organic chemical vapor deposition (MOCVD), and metal–organic molecular beam epitaxy (MOMBE). Research on these topics is currently, and will continue to be, an important area of materials chemistry and materials science and is extremely important to the realization of technological applications of advanced materials.

10.4.3 Prospects for the Future

Throughout our discourse on layered materials we have attempted to address current research trends and future directions. In addition to those opportunities discussed above, other important and emerging areas in oxide and chalcogenide research include the overlap between organic and inorganic materials and mixed anionic materials (such as the layered pnictide oxides[219]), both of which have a large potential for yielding novel materials with unique structures and properties. Cheetham has recently discussed the future of advanced materials and examined the interface between materials science and inorganic chemistry.[10] He concluded that a greater emphasis will be placed on combined inorganic–organic materials, including composites and nanomaterials. Nanochemistry is an open frontier for novel materials and device design, with a focus on the synthesis of nanometer size pieces of matter in one, two, and three dimensions. Nanoscale objects have gained considerable attention, as they exhibit novel physical properties that are a direct consequence of their size. Ozin has provided a review that emphasizes host–guest inclusion chemistry aimed at the synthesis of new hosts and the organization and assembly of a range of guest precursors on the internal surfaces of the nanoscale-dimensional channels, layers, and cavity spaces of a variety of hosts, including layered oxides and chalcogenides.[26] Nanochemistry has generated novel composite materials with high potential for electronic, magnetic, optical, and photonic device applications.

More specifically, new directions in the field of layered oxides and chalcogenides have employed nanochemical strategies in the development of inorganic–organic composite materials. Composite materials offer versatility in structural tunability and enhanced chemical and thermal stabilities. Three interesting composite materials, each with unique materials applications, are (1) organic macromolecules in layered inorganic hosts (fast ionic conductors,[24] nonlinear and electro-optical materials,[220] solid-state batteries[27]), (2) layered oxide semiconductors (electron transport, light absorption, multielectron redox reactions),[221] and (3) molecular recognition in layered metal phosphates[11,222] and nanoporous tin(IV) chalcogenides[13,19] (chemical sensing). These layered host systems, along with their respective intercalated guests, exhibit intriguing structural features, such as staging, stacking, order–disorder transitions, and incommensurate phases, each of which affords important technological advantages. Moreover, these materials offer enhanced materials design capabilities, as they are amenable to systematic structural design by modulation of both the host framework and the guest constituents. The intercalation of polymers between the sheets of layered ceramic hosts (e.g., PEO in homoionic montmorillonite layer silicate) provides access to new nanocomposites with novel physical and mechanical properties, owing to the synergism between the individual components (see Fig. 10.23).[24,223]

Finally, research that focuses at furthering our understanding of the relationships between the structure, properties, and foregoing chemical considerations, is essential for advancement in the field of materials chemistry. Layered transition-metal oxides and chalcogenides have not only proven themselves to be bellwether examples with important technological applications, but research on these compounds has been extremely beneficial to

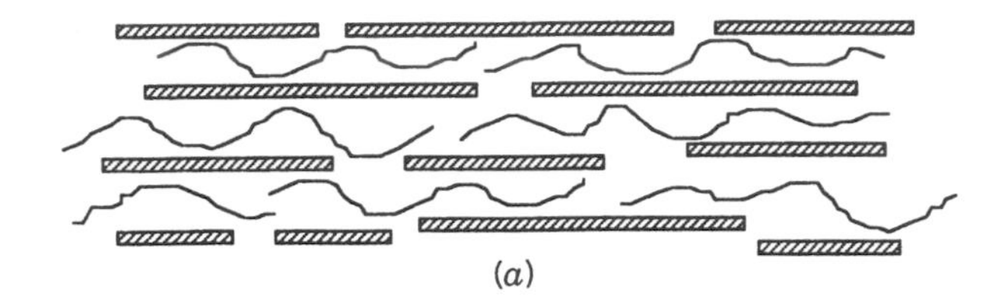

(a)

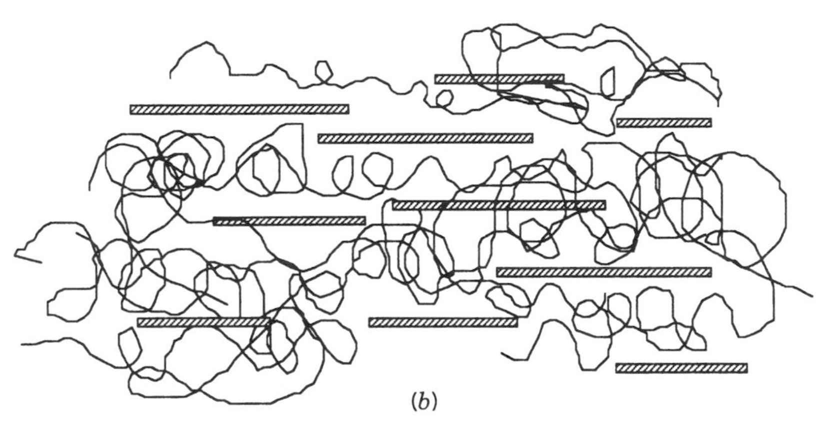

(b)

Figure 10.23 Structural representation of polymer–ceramic composites: (*a*) composite structure of a single polymer layer intercalated between ceramic layers; (*b*) composite structure of delaminated ceramic layers dispersed in a continuous polymer matrix. (Adapted from Ref. 26.)

furthering our general understanding of the chemistry of materials. We have attempted to stress these ideas in this review of layered oxides and chalcogenides. We would like to emphasize that this chapter has not been an exhaustive review of all such materials and their properties, owing to the size and breadth of topics that such a document would require. We have attempted to guide the interested reader to more in-depth reviews on particular topics, while providing an introduction to the materials chemistry of an extremely diverse class of materials. In particular, we hope that our contribution to this volume has conveyed the relevance, excitement, and importance of layered oxides and chalcogenides.

ACKNOWLEDGMENTS

The authors would like to thank the following people for their contributions: J. W. Koenitzer, K. B. Greenwood, H. Fogel, and S. J. Ford. This work was supported by the National Science Foundation through the Science and Technology Center for Superconductivity (DMR-912000) and the MRL Program at the Materials Research Center of Northwestern University (DMF-9120521).

REFERENCES

1. G. L. Liedl, *Sci. Am.* **255**, 127 (1986).
2. R. Brec, *Solid State Ion* **22**, 3 (1986).
3. R. Steffen, R. Schöllhorn, *Solid State Ion.* **22**, 31 (1986).
4. P. J. Squattrito, S. A. Sunshine, J. A. Ibers, *Solid State Ion.* **22**, 53 (1986).
5. P. G. Dickens, S. J. Hibble, *Solid State Ion.* **22**, 69 (1986).
6. C. B. Roxlo, M. Daage, D. P. Leta, K. S. Liang, S. Rice, A. F. Ruppert, R. R. Chianelli, *Solid State Ion.* **22**, 97 (1986).
7. K. A. Carrado, A. Kustapapas, S. L. Suib, *Solid State Ion.* **22**, 117 (1986).
8. M. Balkanski, *Mater. Sci. Eng.* **B3**, 1 (1989).
9. T. Bien, in *Supramolecular Architecture: Synthetic Control in Thin Films and Solids*, T. Bien, Ed., American Chemical Society, Washington, DC, 1992, p. 1.
10. A. K. Cheetham, *Science* **264**, 794 (1994).
11. A. Clearfield, in *Supramolecular Architecture: Synthetic Control in Thin Films and Solids*, T. Bien, Ed., American Chemical Society, Washington, DC, 1992, p. 128.
12. K. A. Carrado, K. B. Anderson, P. S. Grutkoski, in *Supramolecular Architecture: Synthetic Control in Thin Films and Solids*, T. Bien, Ed., American Chemical Society, Washington, DC, 1992, p. 155.
13. T. Jiang, G. A. Ozin, R. L. Bedard, *Adv. Mater.* **7**, 166 (1995).
14. H. Ahari, G. Ozin, R. L. Bedard, S. Petrov, D. Young, *Adv. Mater.* **7**, 370 (1995).
15. T. Ohtani, Y. Sano, Y. Yokota, *J. Solid State Chem.* **103**, 504 (1993).
16. J.-D. Guo, K. P. Reis, M. S. Whittingham, *Solid State Ion.* **53–56**, 305 (1992).
17. M. Greenblatt, *Chem. Rev.* **88**, 31 (1988).
18. G. S. Ferguson, E. R. Kleinfield, *Adv. Mater.* **7**, 414 (1995).
19. H. Ahari, C. L. Bowes, T. Jiang, A. Lough, G. A. Ozin, R. L. Bedard, S. Petrov, D. Young, *Adv. Mater.* **7**, 375 (1995).
20. A. Clearfield, C. Y. Ortiz-Avila, in *Supramolecular Architecture: Synthetic Control in Thin Films and Solids*, T. Bien, Ed., American Chemical Society, Washington, DC, 1992, p. 178.
21. D. Avnir, S. Braun, M. Ottolenghi, in *Supramolecular Architecture: Synthetic Control in Thin Films and Solids*, T. Bien, Ed., American Chemical Society, Washington, DC, 1992, p. 384.
22. D. Jones, R. E. Mejjad, J. Rozière, in *Supramolecular Architecture: Synthetic Control in Thin Films and Solids*, T. Bien, Ed., American Chemical Society, Washington, DC, 1992, p. 220.
23. D. A. Burwell, M. E. Thompson, in *Supramolecular Architecture: Synthetic Control in Thin Films and Solids*, T. Bien, Ed., American Chemical Society, Washington, DC, 1992, p. 166.
24. M. G. Kanatzidis, H. O. Marcy, W. J. McCarthy, C. R. Kannewurf, T. J. Marks, *Solid State Ion.* **32–33**, 594 (1989).
25. M. G. Kanatzidis, C.-G. Wu, H. O. Marcy, D. C. DeGroot, J. L. Schindler, C. R. Kannewurf, M. Benz, E. LeGoff, in *Supramolecular Architecture: Synthetic Control in Thin Films and Solids*, T. Bien, Ed., American Chemical Society, Washington, DC, 1992, p. 194.

26. G. A. Ozin, *Adv. Mater.* **4**, 612 (1992).
27. C. Huber, M. Sadoqi, T. Huber, D. Chako, *Adv. Mater.* **7**, 163 (1995).
28. R. Robson, B. F. Abrahams, S. R. Batten, R. W. Gable, B. F. Hoskins, J. Liu, in *Supramolecular Architecture: Synthetic Control in Thin Films and Solids*, T. Bien, Ed., American Chemical Society, Washington, DC, 1992, p. 256.
29. J. Rouxel, in *Supramolecular Architecture: Synthetic Control in Thin Films and Solids*, T. Bien, Ed., American Chemical Society, Washington, DC, 1992, p. 88.
30. B. Raveau, C. Michel, M. Hervieu, D. Groult, in *Crystal Chemistry of High-T_c Superconducting Copper Oxides*, Springer Series in Materials Science, Vol. 15, U. Gonser, A. Mooadian, R. M. Osgood, M. B. Panish, H. Sakaki, Eds., Springer-Verlag, Berlin, 1991.
31. J. B. Goodenough, A. Manthiram, *J. Solid State Chem.* **88**, 115 (1990).
32. H. Müller-Buschbaum, *Angew. Chem. Int. Ed. Engl.* **28**, 1472 (1989).
33. J. Rouxel, *Chem. Scri.* **28**, 33 (1988).
34. J. Rouxel, *Acc. Chem. Res.* **25**, 328 (1992).
35. J. Rouxel, A. Meerschaut, P. Gressier, *Synth. Met.* **34**, 597 (1989).
36. M. S. Whittingham, *Solid State Ion.* **22**, 1 (1986).
37. R. Hoffman, *Solids and Surfaces: A Chemist's View on Bonding in Extended Structures*, VCH, New York, 1988.
38. J. F. Nye, *Physical Properties of Crystals*, Oxford University Press, Oxford, 1985.
39. N. Bartlett, B. W. McQuillan, in *Intercalation Chemistry*, M. S. Whittingham, A. J. Jacobson, Eds., Academic Press, San Diego, CA, 1982, p. 19.
40. T. J. Pinnavaia, *Science* **220**, 365 (1983).
41. B. C. Tofield, in *Intercalation Chemistry*, M. S. Whittingham, A. J. Jacobson, Eds., Academic Press, San Diego, CA, 1982, p. 181.
42. M. S. Whittingham, in *Intercalation Chemistry*, M. S. Whittingham, A. J. Jacobson, Eds., Academic Press, San Diego, CA, 1982, p. 1.
43. J. B. Goodenough, *Mater. Res. Bull.* **8**, 423 (1973).
44. F. Devaux, A. Manthiram, J. B. Goodenough, *Phys. Rev. B* **41**, 8723 (1990).
45. J. Yu, A. J. Freeman, *J. Electron. Spectrosc. Relat. Phenom.* **66**, 281 (1994).
46. J. K. Burdett, *Adv. Chem. Phys.* **83**, 207 (1992).
47. J. K. Burdett, *Inorg. Chem.* **32**, 3915 (1993).
48. J. K. Burdett, *Acc. Chem. Res.* **28**, 227 (1995).
49. J. K. Burdett, *Chem. Rev.* **88**, 3 (1988).
50. E. Doni, R. Girlanda, in *Electronic Structure and Electronic Transitions in Layered Materials*, V. Grasso, Ed., D. Riedel, Boston, 1986, p. 1.
51. N. Suzuki, S. Tomishima, K. Motizuki, in *New Horizons in Low-Dimensional Electron Systems*, H. Aoki, M. Tsukada, M. Schlüter, F. Lévy, Eds., Kluwer, Boston, 1992, p. 56.
52. A. Koma, in *New Horizons in Low-Dimensional Electron Systems*, H. Aoki, M. Tsukada, M. Schlüter, F. Lévy, Eds., Kluwer, Boston, 1992, p. 85.
53. M. S. Hybertsen, M. Schlüter, in *New Horizons in Low-Dimensional Electron Systems*, H. Aoki, M. Tsukada, M. Schlüter, F. Lévy, Eds., Kluwer, Boston, 1992, p. 229.

54. C. N. R. Rao, *Chemical Approaches to the Synthesis of Inorganic Materials*, Wiley, New York, 1994.
55. J. Gopalakrishnan, *Chem. Mater.* **7**, 1265 (1995).
56. J. D. Corbett, in *Solid State Chemistry: Techniques*, A. K. Cheetham, P. Day, Eds., Clarendon Press, Oxford, 1987, p. 1.
57. M. S. Whittingham, R. R. Chianelli, *J. Chem. Ed.* **57**, 569 (1980).
58. J. S. Dailey, T. J. Pinnavaia, *Chem. Mater.* **4**, 855 (1992).
59. M. Holtz, T.-R. Park, J. Amarasekera, S. A. Solin, T. J. Pinnavaia, *J. Chem. Phys.* **100**, 3346 (1994).
60. J.-R. Butruille, L. J. Michot, O. Barrès, T. J. Pinnavaia, *J. Catal.* **139**, 664 (1993).
61. E. G. Rightor, M. S. Tzou, T. J. Pinnavaia, *J. Catal.* **130**, 29 (1991).
62. E. P. Giannelis, E. G. Rightor, T. J. Pinnavaia, *J. Am. Chem. Soc.* **110**, 3880 (1988).
63. D. Petridis, P. Falaras, T. J. Pinnavaia, *Inorg. Chem.* **31**, 3530 (1992).
64. M. S. Whittingham, *Prog. Solid State Chem.* **12**, 41 (1978).
65. R. A. Laudise, in *Materials for Nonlinear Optics: Chemical Perspectives*, S. R. Marder, J. E. Sohn, G. D. Stucky, Eds., American Chemical Society, Washington, DC, 1991, p. 410.
66. J. P. Clark, M. C. Flemings, *Sci. Am.* **255**, 51 (1986).
67. J. S. Mayo, *Sci. Am.* **255**, 51 (1986).
68. P. Chaudhari, *Sci. Am.* **255**, 137 (1986).
69. J. M. Rowell, *Sci. Am.* **255**, 147 (1986).
70. G. D. Stucky, M. L. F. Phillips, T. E. Gier, *Chem. Mater.* **1**, 492 (1989).
71. E. S. Machlin, *Materials Science in Microelectronics*, Giro Press, New York, 1995, p. 1.
72. H. R. Oswald, A. Reller, *Pure Appl. Chem.* **61**, 1323 (1989).
73. R. Roy, *Solid State Ion.* **32–33**, 3 (1989).
74. P. Hagenmuller, *Solid State Ion.* **40–41**, 3 (1990).
75. P. Hagenmuller, *Mater. Sci. Eng.* **B3**, 253 (1989).
76. W. Y. Liang, *Mater. Sci. Eng.* **B3**, 139 (1989).
77. M. T. Anderson, J. T. Vaughey, K. R. Poeppelmeier, *Chem. Mater.* **5**, 151 (1993).
78. L. Katz, R. Ward, *Inorg. Chem.* **3**, 205 (1964).
79. A. F. Wells, *Structural Inorganic Chemistry*, Oxford University Press, New York, 1984.
80. M. T. Anderson, K. B. Greenwood, G. A. Taylor, K. R. Poeppelmeier, *Prog. Solid State Chem.* **22**, 197 (1993).
81. M. T. Anderson, K. R. Poeppelmeier, *Chem. Mater.* **3**, 476 (1991).
82. M. T. Anderson, K. R. Poeppelmeier, S. A. Gramsch, J. K. Burdett, *J. Solid State Chem.* **102**, 164 (1993).
83. V. M. Goldschmidt, *Str. Nor. Vidensk-Akad. Oslo* **1**, 1 (1926).
84. R. D. Shannon, *Acta Crystallogr.* **A32**, 751 (1976).
85. K. R. Poeppelmeier, M. E. Leonowicz, J. M. Longo, *J. Solid State Chem.* **44**, 89 (1982).

86. K. R. Poeppelmeier, M. E. Leonowicz, J. C. Scanlon, J. M. Longo, *J. Solid State Chem.* **45**, 71 (1982).
87. K. Vidyasagar, J. Gopalakrishnan, C. N. R. Rao, *Inorg. Chem.* **23**, 1206 (1984).
88. C. Michel, L. E. Rakho, M. Hervieu, J. Pannetier, B. Raveau, *J. Solid State Chem.* **68**, 143 (1987).
89. J. T. Vaughey, K. R. Poeppelmeier, in *Proceedings of the International Electronic Ceramics Conference*, Special Publication 804, National Institute of Standards and Technology, Washington DC, 1991, p. 419.
90. W. C. Hansen, L. T. Brownmiller, R. H. Bogue, *J. Am. Chem. Soc.* **50**, 396 (1928).
91. A. A. Colville, S. Geller, *Acta Crystallogr.* **B27**, 2311 (1971).
92. C. N. R. Rao, J. Gopalakrishnan, K. Vidyasagar, A. K. Ganguli, A. Ramanan, L. Ganapathi, *J. Mater. Res.* **1**, 280 (1985).
93. K. Vidyasagar, A. Reller, J. Gopalakrishnan, C. N. R. Rao, *J. Chem. Soc. Chem. Commun.*, 7 (1985).
94. E. F. Bertaut, P. Blum, A. Sagnieres, *C. R. Acad. Sci.* (*Paris*) **244**, 2944 (1957).
95. A. A. Colville, *Acta Crystallogr.* **B26**, 1469 (1970).
96. P. D. Battle, T. C. Gibb, P. Lightfoot, *J. Solid State Chem.* **84**, 237 (1990).
97. P. D. Battle, T. C. Gibb, S. Nixon, *J. Solid State Chem.* **79**, 86 (1989).
98. P. D. Battle, T. C. Gibb, S. Nixon, *J. Solid State Chem.* **79**, 75 (1989).
99. P. D. Battle, T. C. Gibb, S. Nixon, *J. Solid State Chem.* **77**, 124 (1988).
100. C. N. R. Rao, J. Gopalakrishnan, K. Vidyasagar, *Ind. J. Chem.* **23A**, 265 (1984).
101. P. Hagenmuller, M. Pouchard, J. C. Grenier, *J. Mater. Educ.* **12**, 297 (1990).
102. A. Santoro, S. Miraglia, F. Beech, S. A. Sunshine, D. W. Murphy, L. F. Schneemeyer, J. V. Waszczak, *Mater. Res. Bull.* **22**, 1007 (1986).
103. P. Bordet, C. Chaillout, J. J. Capponi, J. Chenavas, M. Marezio, *Nature* **327**, 687 (1987).
104. G. Roth, B. Renker, G. Heger, M. Hervieu, B. Domenges, B. Raveau, *Z. Phys. B Condens. Matter* **69**, 53 (1987).
105. M. K. Wu, J. R. Ashburn, C. J. Torng, P. H. Hor, R. L. Meng, L. Gao, Z. J. Huang, Y. Q. Wang, C. W. Chu, *Phys. Rev. Lett.* **58**, 908 (1987).
106. M. A. Beno, L. Soderholm, I. D. W. Capone, D. G. Hinks, J. D. Jorgenson, J. D. Grace, I. K. Schuller, C. U. Segre, K. Zhang, *Appl. Phys. Lett.* **51**, 57 (1987).
107. A. Maeda, T. Yabe, K. Uchinokura, S. Tanaka, *Jpn. J. Appl. Phys.* **26**, L1368 (1987).
108. C. Michel, F. Deslandes, J. Provost, P. Lejay, R. Tournier, M. Hervieu, B. Raveau, *C. R. Acad. Sci. Paris Ser. II* **304**, 1169 (1987).
109. M. Hervieu, N. Nguyen, C. Michel, F. Deslandes, B. Raveau, *C. R. Acad. Sci. Paris Ser. II* **305**, 1063 (1987).
110. A. Maeda, T. Noda, H. Matsumoto, T. Wada, M. Izumi, T. Yabe, K. Uchinokura, S. Tanaka, *J. Appl. Phys.* **64**, 4095 (1988).
111. K. Otzschi, Y. Ueda, *J. Solid State Chem.* **107**, 149 (1993).
112. C. Michel, L. Er-Rakho, M. Hervieu, J. Pannetier, B. Raveau, *J. Solid State Chem.* **68**, 143 (1987).

113. K. Otzschi, A. Hayashi, Y. Fujiwara, Y. Ueda, *J. Solid State Chem.* **105**, 573 (1993).
114. L. Er-Rakho, C. Michel, B. Raveau, *J. Solid State Chem.* **73**, 514 (1988).
115. W. T. Fu, Q. Xu, A. A. Verheijin, J. M. von Ruitenbeeck, H. W. Zandbergen, L. J. de Jongh, *Solid State Commun.* **73**, 291 (1990).
116. H. Fujishita, M. Sera, M. Sato, *Physica C* **175**, 165 (1991).
117. K. Otzschi, K. Koga, Y. Ueda, *J. Solid State Chem.* **115**, 490 (1995).
118. S. N. Ruddlesden, P. Popper, *Acta Crystallogr.* **10**, 538 (1957).
119. S. N. Ruddlesden, P. Popper, *Acta Crystallogr.* **11**, 54 (1958).
120. J.-C. Bouloux, J.-L. Soubeyroux, G. L. Flem, P. Hagenmuller, *J. Solid State Chem.* **38**, 34 (1981).
121. G. L. Flem, G. Demazeau, P. Hagenmuller, *J. Solid State Chem.* **44**, 82 (1982).
122. J. G. Bednorz, K. A. Müller, *Z. Phys. B* **64**, 189 (1986).
123. R. J. Cava, B. Batlogg, R. B. von Dover, J. J. Krajewski, J. V. Waszczak, R. M. Fleming, W. F. Peck Jr., L. W. Rump Jr., P. Marsh, A. C. W. P. James, L. F. Schneemeyer, *Nature* **345**, 602 (1990).
124. J. M. Longo, P. M. Raccah, *J. Solid State Chem.* **6**, 526 (1973).
125. N. Nguyen, L. Er-Rakho, C. Michel, J. Choisnet, B. Raveau, *Mater. Res. Bull.* **15**, 891 (1980).
126. H. Pausch, H. Müller-Buschbaum, *Z. Naturforsch.* **34B**, 328 (1979).
127. K. Sander, U. Lehmann, H. Müller-Buschbaum, *Z. Anorg. Allg. Chem.* **480**, 153 (1981).
128. J. R. Grasmeder, M. T. Weller, *J. Solid State Chem.* **85**, 88 (1990).
129. J. B. Goodenough, *Bull. Soc. Chim. Fr.*, 1200 (1965).
130. J. B. Goodenough, *Czech, J. Phys.* **B17**, 304 (1967).
131. P. G. Dickens, D. J. Nield, *Trans. Faraday Soc.* **64**, 13 (1968).
132. G. Travaglini, P. Wachter, *Solid State Commun.* **47**, 217 (1983).
133. J. Graham, A. D. Wadsley, *Acta Crystallogr.* **20**, 93 (1966).
134. M. Ghedira, J. Chenavas, M. Marezio, J. Marcus, *J. Solid State Chem.* **57** (1985).
135. Y. K. Tao, Y. Y. Sun, P. H. Hor, C. W. Chu, *J. Solid State Chem.* **105**, 171 (1993).
136. P. Ganguly, N. Shah, *Physica C* **208**, 307 (1993).
137. L. Er-Rakho, C. Michel, P. Lacorre, B. Raveau, *J. Solid State Chem.* **73**, 531 (1988).
138. C. Meyer, F. Hartmann-Boutron, Y. Gros, P. Strobel, *Solid State Commun.* **76**, 163 (1990).
139. M. Pissas, C. Mitros, G. Kallias, V. Psycharis, D. Niarchos, A. Simopoulus, A. Kostikas, C. Christides, K. Prassidies, *Physica C* **185–189**, 553 (1991).
140. M. Pissas, C. Mitros, G. Kallias, V. Psycharis, A. Simopoulus, A. Kostikas, D. Niarchos, *Physica C* **192**, 35 (1992).
141. J. B. Wiley, L. M. Markham, J. T. Vaughey, T. J. McCarthy, M. Sabat, S.-J. Hwu, S. N. Song, J. B. Ketterson, K. R. Poeppelmeier, in *Chemistry of High-Temperature Superconductors II*, D. L. Nelson, T. F. George, Eds., American Chemical Society, Washington DC, 1988, p. 304.

142. J. B. Wiley, M. Sabat, S.-J. Hwu, K. R. Poeppelmeier, *J. Solid State Chem.* **87**, 250 (1990).

143. J. T. Vaughey, R. Shumaker, S. N. Song, J. B. Ketterson, K. R. Poeppelmeier, *Mol. Cryst. Liq. Cryst.* **184**, 335 (1990).

144. J. T. Vaughey, J. B. Wiley, K. R. Poeppelmeier, *Z. Anorg. Allg. Chem.* **598–599**, 327 (1991).

145. S. Putilin, E. V. Antipov, O. Chmaissem, M. Marezio, *Nature* **362**, 226 (1993).

146. J. L. Wagner, P. G. Radaelli, D. G. Hinks, J. D. Jorgensen, J. F. Mitchell, B. Dabrowski, G. S. Knapp, M. A. Beno, *Physica C* **210**, 447 (1993).

147. G. W. Tomlins, N.-L. Jeon, T. O. Mason, D. A. Groenke, J. T. Vaughey, K. R. Poeppelmeier, *J. Solid State Chem.* **109**, 338 (1994).

148. J.-S. Zhou, J. B. Goodenough, *Phys. Rev. B* **51**, 3104 (1995).

149. J. T. Vaughey, J. P. Thiel, D. A. Groenke, C. L. Stern, K. R. Poeppelmeier, B. Dabrowski, D. G. Hinks, A. W. Mitchell, *Chem. Mater.* **3**, 935 (1991).

150. G. Roth, P. Adelmann, G. Hegar, R. Knitter, T. Wolf, *J. Phys.* **1** (1991).

151. Q. Huang, S. A. Sunshine, R. J. Cava, A. Santoro, *J. Solid State Chem.* **102**, 534 (1993).

152. N. Murayama, E. Sudo, K. Kani, A. Tsuzuki, S. Kawakami, M. Awano, Y. Torii, *Jpn. J. Appl. Phys.* **27**, L1623 (1988).

153. C. Greaves, P. R. Slater, *Physica C* **161**, 245 (1989).

154. M.-J. Rey, P. Dehaudt, J. Joubert, A. W. Hewat, *Physica C* **167**, 162 (1990).

155. B. Dabrowski, P. Radaelli, D. G. Hinks, A. W. Mitchell, J. T. Vaughey, D. A. Groenke, K. R. Poeppelmeier, *Physica C* **193**, 63 (1992).

156. Q. Huang, R. J. Cava, A. Santoro, J. J. Krajewski, W. F. Peck, *Physica C* **193**, 196 (1992).

157. T. G. N. Babu, J. D. Kilgour, P. R. Slater, C. Greaves, *J. Solid State Chem.* **103**, 472 (1993).

158. L. Rukang, Z. Yingjie, Q. Yitai, C. Zuyao, *Physica C* **176**, 19 (1991).

159. L. Rukang, Z. Yingjie, X. Cheng, C. Zuyao, Q. Yitai, F. Chengao, *J. Solid State Chem.* **94**, 206 (1991).

160. R. J. Cava, J. J. Krajewski, H. Takagi, H. W. Zandbergen, R. B. V. Dover, W. F. Peck Jr., B. Hessen, *Physica C* **191**, 237 (1992).

161. T. J. Goodwin, H. B. Radousky, R. N. Shelton, *Physica C* **204**, 212 (1992).

162. M. T. Anderson, J. P. Zhang, K. R. Poeppelmeier, L. D. Marks, *Chem. Mater.* **4**, 1305 (1992).

163. A. Gormezano, M. T. Weller, *J. Mater. Chem.* **3**, 771 (1993).

164. A. Gormezano, M. T. Weller, *J. Mater. Chem.* **3**, 979 (1993).

165. K. B. Greenwood, M. T. Anderson, K. R. Poeppelmeier, D. L. Novikov, A. J. Freeman, B. Dabrowski, S. A. Gramsch, J. K. Burdett, *Physica C* **235–240**, 349 (1994).

166. P. Gómez-Romero, M. R. Palacín, J. Rodríguez-Carvajal, *Chem. Mater.* **6**, 2118 (1994).

167. M. R. Palacín, A. Fuertes, N. Casañ-Pastor, P. Gómez-Romero, *Adv. Mater.* **6**, 54 (1994).

168. R. A. Jennings, C. Greaves, *Physica C* **235–240**, 989 (1994).
169. K. B. Greenwood, G. M. Sarjeant, K. R. Poeppelmeier, P. A. Salvador, T. O. Mason, B. Dabrowski, K. Rogacki, Z. Chen, *Chem. Mater.* **7**, 1355 (1995).
170. P. A. Salvador, L. Shen, T. O. Mason, K. B. Greenwood, K. R. Poeppelmeier, *J. Solid State Chem.*, **119**, 80 (1995).
171. M.-Y. Su, C. E. Elsbernd, T. O. Mason, *J. Am. Ceram. Soc.* **73**, 415 (1990).
172. M.-Y. Su, C. E. Elsbernd, T. O. Mason, *Physica C* **160**, 114 (1989).
173. M.-Y. Su, S. E. Dorris, T. O. Mason, *J. Solid State Chem.* **75**, 381 (1988).
174. T. O. Mason, in *Electronic Ceramic Materials*, J. Nowotny, Ed., Trans Tech, Zurich, 1992, p. 503.
175. D. L. Novikov, A. J. Freeman, K. R. Poeppelmeier, *Phys. Rev. B*, **43**, 9448 (1996).
176. P. A. Salvador, T. O. Mason, K. Otzschi, K. B. Greenwood, K. R. Poeppelmeier, B. Dabrowski, *J. Am. Chem. Soc.* **119**, 3756 (1997).
177. A. Gormezano, M. T. Weller, *Physica C* **235–240**, 999 (1994).
178. L. Shen, P. A. Salvador, T. O. Mason, *J. Am. Ceram. Soc.* **77**, 81 (1994).
179. B.-S. Hong, T. O. Mason, in *Superconductivity and Ceramic Superconductors II*, K. M. Nair, U. Balachandran, Y.-M. Chiang, A. S. Bhalla, Eds., American Ceramic Society, Westerville, OH, 1991, p. 95.
180. Y. J. Li, M. S. Whittingham, *Solid State Ionics* **63–65**, 391 (1993).
181. J.-C. Grenier, A. Wattiaux, J.-P. Doumerc, P. Dordor, F. Fournes, J.-P. Chaminade, M. Pouchard, *J. Solid State Chem.* **96**, 20 (1992).
182. M. S. Whittingham, *J. Solid State Chem.* **29**, 303 (1979).
183. C. Delmas, *Mater. Sci. Eng.* **B3**, 97 (1986).
184. E. Sandre, R. Brec, J. Rouxel, *J. Solid State Chem.* **88**, 269 (1990).
185. J. P. Kemp, P. A. Cox, *J. Phys. Chem. Solids* **51**, 575 (1990).
186. G. A. Wiegers, A. Meetsma, S. von Smaalen, R. J. Haange, J. Wulff, T. Zeinstra, J. L. deBoer, S. Kuypers, G. V. Tendeloo, J. V. Landuyt, S. Amelinckx, A. Meerschaut, P. Rabu, J. Rouxel, *Solid State Commun.* **70**, 409 (1989).
187. E. Mackovicki, B. G. Hyde, in *Structure and Bonding*, Vol. 46, Springer-Verlag, Heidelberg, 1981, p. 101.
188. J. Rouxel, Y. Moëlo, A. Lafond, F. J. DiSalvo, A. Meerschaut, R. Roesky, *Inorg. Chem.* **33**, 3358 (1994).
189. G. A. Wiegers, A. Meerschaut, in *Non-commensurated Layered Compounds*, A. Meerschaut, Ed., Trans Tech, Zurich, 1992.
190. A. Meerschaut, L. Guemas, C. Auriel, J. Rouxel, *Eur. J. Solid State Inorg. Chem.* **27**, 557 (1990).
191. A. Meerschaut, C. Auriel, J. Rouxel, *J. Alloys Compounds* **183**, 129 (1992).
192. C. Auriel, A. Meerschaut, R. Roesky, J. Rouxel, *Eur. J. Solid State Inorg. Chem.* **29**, 1079 (1992).
193. L. Hernán, J. Morales, L. Sánchez, J. L. Tirado, J. P. Espinós, A. R. G. Elipe, *Chem. Mater.* **7**, 1576 (1995).
194. C. Auriel, A. Meerschaut, P. Deniard, J. Rouxel, *C. R. Acad. Sci. Paris Ser. II* **313**, 1255 (1991).

195. L. Hernán, P. Lavela, J. Morales, J. Pattanayak, J. L. Tirado, *Mater. Res. Bull.* **26**, 1211 (1991).
196. C. Barriga, P. Lavela, J. Morales, J. Pattanayak, J. L. Tirado, *Chem. Mater.* **4**, 1021 (1992).
197. L. Hernán, J. Morales, L. Sánchez, J. L. Tirado, *Chem. Mater.* **5**, 1167 (1993).
198. A. J. Jacobson, J. W. Johnson, J. T. Lewandowski, *Inorg. Chem.* **24**, 3727 (1985).
199. M. Dion, M. Ganne, M. Tournoux, *Mater. Res. Bull.* **16**, 1429 (1981).
200. M. Dion, M. Ganne, M. Tournoux, J. Ravez, *Rev. Chim. Miner.* **21**, 92 (1984).
201. M. Dion, M. Ganne, M. Tournoux, J. Ravez, *Rev. Chim. Miner.* **23**, 61 (1986).
202. A. J. Jacobson, J. T. Lewandowski, J. W. Johnson, *J. Less Common Met.* **116**, 137 (1986).
203. M. Gondrand, J. C. Joubert, *Rev. Chim. Miner.* **24**, 33 (1987).
204. J. Gopalakrishnan, V. Bhat, B. Raveau, *Mat. Res. Bull.* **22**, 413 (1987).
205. J. Gopalakrishnan, V. Bhat, *Inorg. Chem.* **26**, 4299 (1987).
206. R. A. M. Ram, A. Clearfield, *J. Solid State Chem.* **94**, 45 (1991).
207. R. A. M. Ram, A. Clearfield, *J. Solid State Chem.* **112**, 288 (1994).
208. S. Uma, A. R. Raju, J. Gopalakrishnan, *J. Mater. Chem.* **3**, 709 (1993).
209. A. Clearfield, *Comments Inorg. Chem.* **10**, 89 (1990).
210. G. Alberti, U. Constantino, S. Allulli, N. J. Tomassini, *J. Inorg. Nucl. Chem.* **40**, 1113 (1978).
211. M. B. Dines, P. M. DiGiacomo, *Inorg. Chem.* **20**, 92 (1981).
212. D. A. Burwell, M. E. Thompson, *Chem. Mater.* **3**, 730 (1991).
213. Y.-J. Liu, M. G. Kanatzidis, *Chem. Mater.* **7**, 1525 (1995).
214. K. Chatakondu, M. L. H. Green, J. Qin, M. E. Thompson, P. J. Wiscman, *J. Chem. Soc. Chem. Commun.*, 223 (1988).
215. R. H. Herber, Y. Maeda, *Inorg. Chem.* **19**, 3411 (1980).
216. J. Rouxel, P. Palvadeau, *Rev. Chim. Miner.* **19**, 317 (1982).
217. G. A. Fatseas, P. Palvadeau, J. P. Venien, *J. Solid State Chem.* **51**, 17 (1984).
218. F. J. DiSalvo, *Science* **247**, 649 (1990).
219. S. L. Brock, S. M. Kauzlarich, *Chemtech* **25**, 18 (1995).
220. J. S. O. Evans, S. Barlow, H.-V. Wong, D. O'Hare, *Adv. Mater.* **7**, 163 (1995).
221. D. Rong, Y. I. Kim, H.-G. Hong, J. S. Kruger, J. E. Mayer, T. E. Mallouk, *Coord. Chem. Rev.* **97**, 237 (1990).
222. G. Cao, T. E. Mallouk, *Inorg. Chem.* **30**, 1434 (1991).
223. E. R. Hitzky, P. Arando, *Adv. Mater.* **2**, 545 (1990).

CHAPTER 11

Biomaterials

CAROLE C. PERRY

Department of Chemistry and Physics, The Nottingham Trent University, Clifton Lane, Nottingham NG11 8NS, Nottinghamshire, England

11.1 INTRODUCTION

The normal description of a *biomaterial* is "a substance that is used in prostheses or in medical devices designed for contact with the living body for an intended application and for an intended time period." There are also solid-state biomaterials produced by living organisms such as bones, teeth, spines, and shells that are produced with specific mechanical and structural functionality and which show us the beauty and yet complexity of composite biological materials. It is important to understand the structural and mechanical characteristics of these materials, how they are constructed and how their structures are regulated in order to produce new materials with improved biocompatibility and potential usage in the biomaterials field. In addition, research on biomolecular and solid-state biological materials may, alternatively, lead to the genesis of materials for a range of other applications, including optics, optoelectronics and catalysis. All three principal areas are discussed in this chapter.

11.1.1 Raison d'ètre for Biomaterials Research

Table 11.1 lists the number of yearly implant procedures in the United States alone. The development of biomaterials for use in medical devices has occurred in response to the growing number of patients afflicted with traumatic and nontraumatic conditions. As the population grows older, there is increased need for medical devices to replace damaged or worn tissues.

Chemistry of Advanced Materials: An Overview, Edited by Leonard V. Interrante and Mark J. Hampden-Smith.
ISBN 0-471-18590-6 © 1998 Wiley-VCH, Inc.

TABLE 11.1 Implant Procedures per Year in the United States

Procedures	Number Performed (millions)
Bed sores	8.0
Breast prostheses	0.24
Burns	0.10
Corneal grafts	0.04
Dental implants	0.05
Diabetic skin wounds	3.7
Fixation devices	0.10
Heart valves	0.43
Hip and knee implants	0.25
Intraocular lenses	1.20
Skin excision	0.59
Tendon and ligament replacement	0.15
Total hip	0.17
Total knee	0.10
Vascular bypass grafts	0.19
Vascular grafts	0.20

Source: Ref. 1.

The annual value of the implant market is $1 billion and requires the skills of clinicians, surgeons, engineers, chemists, physicists, and materials scientists to work cooperatively in the development of materials for clinical use.

11.1.2 Medical Biomaterials

There are five classes of biomaterials of this type: metals, ceramics, composites, biological materials/polymers, and synthetic polymers. The choice of synthetic to replace tissues is largely governed by the physical properties of the natural tissue and the properties of the proposed replacement. In general, natural and synthetic polymers are used to replace skin, tendon, ligament, breast, eye, vascular system, and facial tissues, and metals, ceramics, and composites are used to replace or reinforce bone and dentin. These tissues clearly require materials of different strength, as shown in Table 11.2, where natural materials and replacement materials are compared with respect to their strength. The materials are also represented pictorially with respect to their Young's modulus in Section 11.5.1. The strengths and the elastic moduli of the materials listed cover a wide range. It is often the case that the strengths of the materials used to replace natural components are stronger and/or stiffer, which can lead to problems of compatibility both in respect to the mechanical behavior of the host implant and in terms of the biologic response. Applications for the various classes of materials are given in Tables 11.3 to 11.6 and are discussed in more detail in later sections of this chapter.

TABLE 11.2 Physical Properties of Tissues and Materials Used in Their Replacement

Tissue or Material	Ultimate Strength (MPa)	Modulus (MPa)
Soft tissue		
Arterial wall	0.5–1.72	1.0
Hyaline cartilage	1.3–1.8	0.4–19
Skin	2.5–16	6–40
Tendon/ligament	30–300	65–2500
Hard tissue (bone)		
Cortical	30–211	16–20 (GPa)
Cancellous (porous)	51–193	4.6–15 (GPa)
Polymers		
Synthetic rubber	10–12	4
Glassy	25–100	1.6–2.6 (GPa)
Crystalline	22–40	(0.015–1) (GPa)
Metal alloys		
Steel	480–655	193 (GPa)
Cobalt	655–1400	195 (GPa)
Platinum	152–485	147 (GPa)
Titanium	550–680	100–105 (GPa)
Ceramics		
Oxides	90–380 (GPa)	160–4000 (GPa)
Hydroxyapatite	600	19 (GPa)
Composites		
Fibers	0.9–4.5 (GPa)	62–577 (GPa)
Matrices	41–106	0.3–3.1

Source: Ref. 1.

Although these materials find many uses in the clinical environment, they were not originally engineered for biomaterials applications. In crude terms, they became biomaterials when, by a series of trial-and-error experimentation, they were implanted in the human body in a variety of forms and found to work. Clearly, there were many other materials that did not work, causing at the least perhaps pain and discomfort to patients and at the worse

TABLE 11.3 Polymers Used in Medical Devices

Polymer	Medical Device Applications
Polyethylene	Hip, tendon/ligament implants and facial implants
Polyethylene terphthalate	Aortic, tendon/ligament and facial implants
Polymethylmethacrylate	Intraocular lens, contact lenses, and bone cement
Polydimethylsiloxane	Breast, facial, and tendon implants
Polyurethane	Breast, vascular, and skin implants

Source: Adapted from Ref. 1 and references therein.

TABLE 11.4 Metals Used in Medical Devices

Metal	Medical Device Applications
Cobalt–chromium alloys	Dental applications, fracture plates, heart valves, joint components, nails, screws
Titanium alloys	Conductive leads, joint components, pacemaker cases, nails, screws
Stainless steel	Fracture plates

Source: Adapted from Ref. 1, and references therein.

TABLE 11.5 Ceramics Used in Medical Devices

Ceramic	Medical Device Applications
Hydroxyapatite	Bone and dentin replacements
Carbons	Heart valves, ligaments, dental implants
Alumina	Total hip replacement, dental components, eye lenses
Zirconia	Orthopedics and dental materials
Glasses	Orthopedics and bone replacements

Source: Adapted from Ref. 1.

unnecessary suffering and death. Increase in litigation for problems alledgedly caused by biomaterials has caused some companies to remove products from sale in this area and may lead to a distinct shortage of available materials for device fabrication (see below). Clearly, for a variety of reasons, trial-and-error optimization is not the way forward for the production of the next generation of materials to be used in the human body. The way forward requires the systematic design of a wide range of materials for specific

TABLE 11.6 Composites Used in Medical Devices

Composite	Medical Device Applications
Glass, glass–ceramic, or HAP with PMMA	Bone cement
Tricalcium phosphate/HAP with PE	Bone substitutes
Glass–ceramic, quartz with BIS/GMA	Dental restorations
Drugs with various polymers/ceramics	Drug delivery
Carbon or glass fiber with PMMA and other matrices	All of the applications above but with increased strength and/or stiffness

Source: Ref. 2.

purposes and functions. The goal is to produce biomaterials that smoothly integrate into living systems rather than fighting against biology. There are many factors that have to be considered and understood to determine how this might be accomplished. These include the structure of the original material to be replaced or augmented, and its physiology, anatomy, biochemistry, and biomechanics, including pathophysiological changes that have necessitated use of a substitute biomaterial. In addition, as devices are often present in the body for considerable periods of time, it is necessary to understand the natural degenerative processes of normal tissues, particularly in relation to the biomaterial substitute. The latter area is at present very poorly understood. All of the above clearly affect the design and development of materials for clinical usage.

11.1.3 Biomaterials Crisis[3]

Several major suppliers of materials used in medical devices, including Dow Chemical, Dow Corning, and DuPont, have recently decided not to sell certain materials (including those listed below) to device manufacturers. Biomaterials that may become unavailable for medical device manufacture include:

- *Polyurethane* for heart valves, pacemaker and defibrillator leads, tubing, and catheters
- *Polyethylene* for artificial hips, shoulders, and knees
- *Silicone* for implants, load-bearing or drug-loaded implants, and gynecological, obstetrical, cosmetic, and contraceptive applications
- *Silicone adhesive*
- *Polyester* for grafts, catheters, leads, heart valves, sutures, and vascular stents
- *Nickel–titanium alloy*
- *Tantalum* x-ray marker beads
- *Surgical stainless steel* and *titanium* for implants
- *All* of DuPont's *engineering polymers* for use in permanent implants or in permanent contact with internal body fluids or organs, including Teflon (PTFE), Dacron (polyester), Delrin (polyacetal), Zytel (nylon), Rynite (polyester), and Hytrel (thermoplastic elastomer), with assorted uses from batteries to bone cement

Short-term effects are probably going to be minimal because of manufacturers stockpiles, but unless alternatives are found, certain devices will not be available. The withdrawal of these materials for sale to device manufacturers has arisen because of the number of lawsuits filed against materials suppliers by persons claiming to have been injured by medical devices. An example are those who have had silicone breast implants and claim that the silicone-gel filling causes or has caused damage to patients' immune system and other disorders. Another group of lawsuits has been filed against a company (now bankrupt) that manufactured temporomandibular joints. These implants contained perhaps 5 cents worth of DuPont's Teflon (PTFE) in its coating, which fragmented and caused damage to surrounding tissue. Lawyers for patients affected by this problem sued the parent company and then went on to sue DuPont as codefendent, charging that the company should have known that Teflon was unsuited for the implant and should have refused to sell it for that purpose. As these companies are faced with very expensive lawsuits for little profit, the decision to stop providing materials for use in the manufacture of medical devices was a sensible one (from the company's point of view) and hence shortages of suitable materials will become apparent with time. This is clearly a very important indicator of the fact that materials need to be developed with a *clear* understanding of the nature and extent of interactions between the device (whatever it is) and the surrounding tissue. The importance of biocompatibility in the development of the next generation of materials for applications in a biological environment cannot be emphasized too strongly.

11.2 KEY TISSUES IN THE HUMAN BODY

In this section the basic biochemical components used to construct tissues in the human body are described, together with their mechanical properties, which result from the specific combination and spatial disposition of a range of molecules within the tissue.

11.2.1 Crystalline Polymers[4]

The key to the mechanical properties of fibrous materials (collagen, silks, chitin, and cellulose are all examples) lies in their crystalline nature. These and the synthetic polymers polyethylene, nylon, and tetrafluoroethylene (Teflon) aggregate into regions of crystalline order. In general, the Young's modulus (the stress–strain curve) of a bulk crystallized polymer is two to three orders of magnitude greater than that of an amorphous polymer above its glass transition temperature (e.g., rubber), and the modulus of oriented crystalline fibers is another order of magnitude greater still. The crystalline regions in linear polymers are usually interpreted to be areas where extended or perhaps helically coiled polymer chains are loosely packed in parallel arrays. It is possible for a single polymer chain to be incorporated into several crystalline regions and into intervening amorphous regions as well. For crystalline polymers to form, they must have great regularity of chain structure and be linear in form. Stereoisomerism can create irregularities of structure. This can create problems in synthetic polymers such as polyethylene but is not a problem in natural polymers such as the protein collogen, or carbohydrates such as cellulose or chitin, as only one stereoisomer, L- for amino acids and D-for saccharides is utilized in building these structures). Each polymer has a different crystalline structure, which can depend on how and when it has crystallized (e.g., under applied stress or not). Crystallinity in polymeric materials has a very important effect on the mechanical properties. When the degree of crystallinity is less than 20%, crystalline regions act as cross-links in an amophous polymer network, and these materials are mechanically similar to cross-linked rubbers. Above 40% crystallinity, the materials become quite rigid and the mechanical properties of these polymers are then quite time independent. For these materials, when stress is applied, the rate at which it decays is very low. This is due to the structural changes in the crystalline regions, where polymer chains slip past one another (weak bonds are broken and new ones formed), and in the amorphous regions, where whole sections slip past one another or rotate relative to one another. The behavior of drawn fibers can be somewhat different from that of the bulk material, as the regions of amorphous character are greatly extended and aligned perpendicular to the stress direction. A prime example of an important crystalline polymer in mammals is collagen, which is the basis of both rigid (e.g., bone) and pliant materials (e.g., tendon, skin, and cartilage).

Collagen Collagen is the basic structural fiber of the animal kingdom. Detailed information on collagen chemistry is covered in reviews and books by Gould[5] and, Bailey.[6] There are 14 genetic variants on the basic structure of the collagen molecule, together with a wide range of post-translational modifications, but all the variants have similar conformations and physical properties. The composition of various collagens and the locations in which they are found is given in Table 11.7. Each collagen chain is about 1000 amino acids in length and is dominated by a 338-contiguous repeating triplet sequence, in which every third amino acid is glycine.[7] Proline and hydroxyproline together account for about 20% of the total amino acids. The structure of collagen is based on the helical arrangement of three noncoaxial, helical polypeptides, stabilized by interchain and intrachain hydrogen bonds. A view of this is shown in Fig. 11.1. All the glycines face the center of the

TABLE 11.7 Compositions and Locations of Different Varieties of Collagen[a]

Type	Composition	Location
I	$[\alpha 1(\text{I})]_2\alpha 2$	Skin, tendon, bone, cornea
II	$[\alpha 1(\text{II})]_3$	Cartilage, intervetebral disk, vitreous body
III	$[\alpha 1(\text{III})]_3$	Fetal skin, cardiovascular system
IV	$[\alpha 1(\text{IV})]_3$	Basement membrane

[a]Each chain in a collagen molecule is in an α-helical conformation, but the identity of the chains differs from one collagen type to another, denoted I through IV.

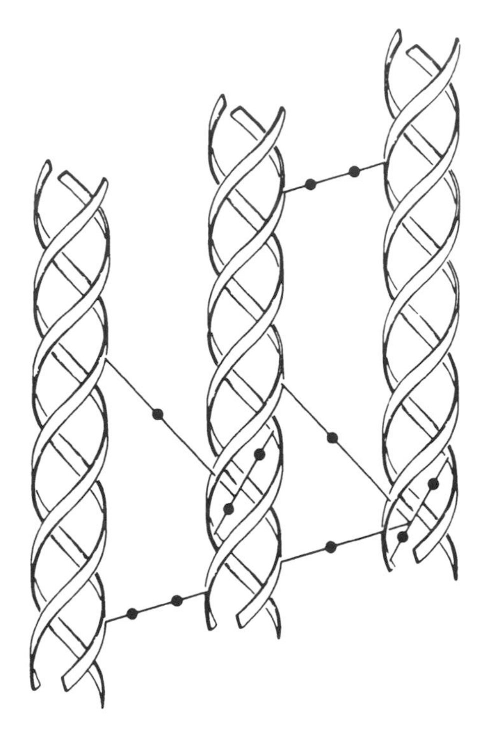

Figure 11.1 Triple-helical structure of collagen showing the intramolecular and intermolecular cross-links. (From Ref. 8.)

triple helix. Collagen molecules are about 280 nm in length and 1.5 nm in width, but the resultant coiled structure shows a characteristic 64-nm banding structure in the electron microscope, which arises from the transposition of adjacent molecules by one-fourth of their length in the axial direction (Fig. 11.2).

11.2.2 Pliant Materials

Mechanically, these materials, which are present in skin, aorta, uterine, cervix, and so on, can be deformed to large strains without breaking and often show long-range, reversible elastic properties similar to those of rubber. There are three groups of pliant biomaterials. The first are proteins with mechanical properties very similar to those of rubber. They are usually single-phase amorphous polymers such as the elastin found in skin. The second group is diverse and contains materials normally referred to as soft connective tissues. They are strictly pliant composites, as they contain some amorphous polymer component in association with a relatively inextensible, high-modulus fiber. Collagen provides the fiber component, and hydrated protein–polysaccharide complexes or a rubbery protein provides the amorphous component. The third group is borderline between pliant and rigid materials and includes materials such as cartilage, the behavior of which depends on the water-binding properties of a hydrated amorphous polymer.

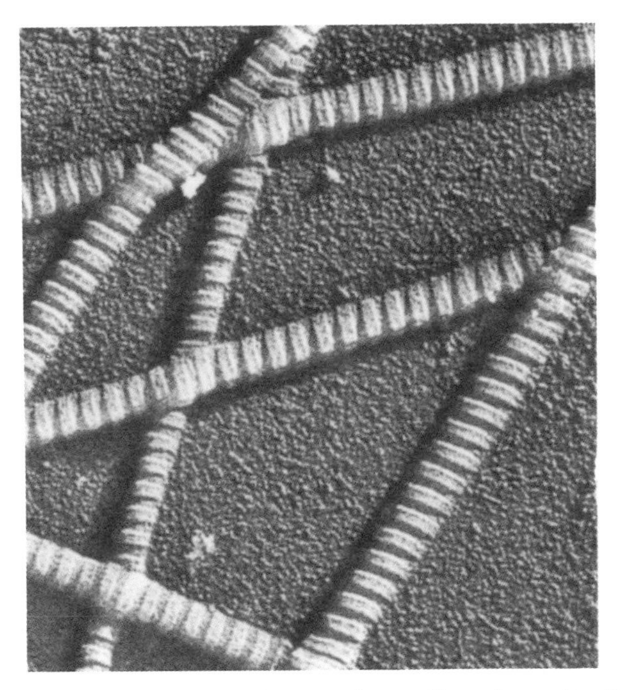

Figure 11.2 Electron micrograph of intact collagen fibers from skin. The banding along the fiber axis is 64 nm.

Elastin[4] Elastin is only found as fine fibers and is usually found in association with collagen and glycosaminoglycans and collagen. Elastin is a hydrophobic protein (95% of the amino acids have nonpolar side chains) and can be thought of as a network of globular protein molecules that are compressed or restrained by the tendency of the hydrophobic amino acids to exclude water. When elastin is stretched, the hydrophobic region must unfold, with the consequence that (1) the water exposed to nonpolar side groups becomes ordered, and (2) the opening of the hydrophobic region removes some of the restraints on the polymer chain. The result of this is that as elastin is stretched, the water becomes more ordered, and the polymer becomes less ordered.

Mucopolysaccharides These include glycosaminoglycans (polysaccharides containing amino acids), proteoglycans (covalently bonded complexes of protein and glycosaminoglycan in which the polysaccharide is the dominant feature), and protein–polysaccharide complexes, where a complex is held together through noncovalent linkages. The family of glycosaminoglycans include chondroitin 4-sulfate, chondroitin 6-sulfate, dermatan sulfate, keratin sulfate, and hyaluronic acid. Several of these structures are depicted in Fig. 11.3. These polymers are nearly always found in association with protein, but knowledge of the associated protein molecules is limited. The properties

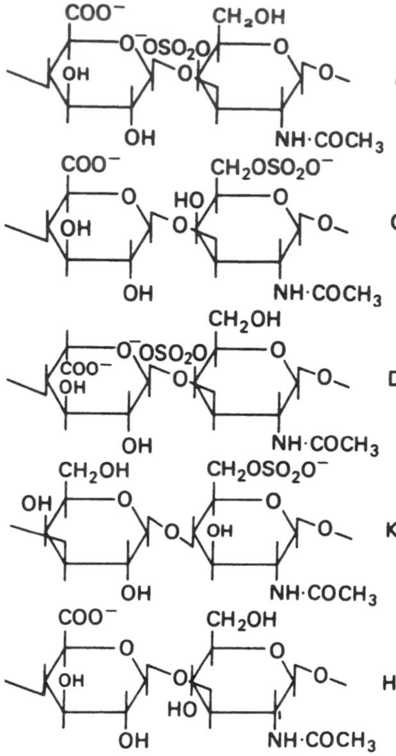

Figure 11.3 Chemical structures of the repeating units in glycosaminoglycans.

of these molecules and their protein–polysaccharide complexes are largely determined by the polyanionic character of the glycosaminoglycans. Both carboxyl and sulfated groups are ionized at physiological pH, giving highly charged polymeric molecules where the molecule takes on a highly expanded form in solution, with the extent of expansion being dictated by both pH and the ions present. The mechanical significance of this is that the molecules interact with one another at low concentrations to form entanglement networks and produce solutions with viscoelastic properties. In addition, if some permanent cross-bridges can form, gel-like structures with rubber–elastic properties can result. Situations in the human body where this occurs are in synovial fluid (lubricant for joints), in mucus from the lungs and saliva.

11.2.3 Pliant Composites

In the pliant composites, viscoelastic, gel-forming proteoglycans and protein–polysaccharide complexes together with protein rubbers are combined with high-modulus fibers to produce materials that can be stretched and bent as the animal moves and changes shape. Pliant composites are frequently found as thin skins, cuticles, or membranes that enclose and support the soft parts of organisms. Most of these materials contain collagen as the fibrillar component and although there are differences in the thickness of these fibers (15 to 150 nm) they all have a similar Young's modulus, can be extended reversibly by 3 to 4%, and all break at extensions of 8 to 10%. Any variations that there are relate to differences in cross-linking between the collagen molecules and the nature and organization of the fibrillar component and the matrix in which it is found. Examples of tendon, skin, and cartilage serve to illustrate the effect that fibrillar orientation and different compositions have on the mechanical properties of the material.

Tendon Tendon is the structure that provides the rigid attachment of muscle to bone, and as such, it must transmit the muscle force with a minimum of loss. This is achieved through the parallel arrangement of collagen fibers to form ropelike structures with a high modulus of elasticity and high tensile strength. Tendon comprises 70 to 80% (dry weight) of collagen, fibroblasts, and other cells, noncollagenous proteins, polysaccharide, and inorganic salts.[9] In its wet or functional form a more detailed breakdown of composition is given as water (60 to 65%), type I collagen (25 to 31%), type III collagen (2%), type V collagen ($<$ 1%), elastin (1 to 4%), cells (3 to 6%), and proteoglycans (1%).[10] It is suggested[4] that in tendons under long-term stress, such as the Achilles tendon, the collagen fibers are more highly cross-linked, to reduce the rate of stress relaxation to an insignificant level.

Skin Skin is a complex tissue layer made up of a thick collagenous layer (the dermis), a basement membrane, and an overlying keratinised epidermal layer. The mechanical properties of skin arise principally from the dermis, which is a three-dimensional feltwork of continuous collagen fibers embedded in a protein–polysaccharide matrix. Elastin fibers are also present either distributed throughout the tissue or concentrated in the lower layers of the dermis. The arrangement of elastin fibers within a collagen framework results in a material showing rubber–elastic properties at small extensions but is limited at longer extensions by the dimensions of the collagen framework. The major components of the matrix phase are dermatan sulfate and hyaluronic acid. The former is suggested to aid in orientation of the collagen molecules during synthesis of skin and the latter acts as a permeability barrier to large molecules and may act as a lubricant to reduce frictional wear between the collagen fibers. The tensile strength of skin appears to depend primarily on the size and cross-linking of the collagen framework. Although the fiber system in skin is an apparently random, three-dimensional feltwork, skin is not uniform over the surface of an animal and varies with the mechanical function of the skin in a particular part of the body.

Cartilage Cartilage acts as a material that maintains the shape of ears, the nose, and the invertebral disk. There are three different types of cartilage:[11] hyaline, fibrocartilage, and elastic cartilage. All contain collagen fibers (type

II), a proteoglycan matrix phase, and sometimes elastin fibers, and yet the material must be able to resist compression and bending forces. Cartilage can be thought of as a hydrostatic system in which the fluid element is provided by the hydration water of the proteoglycan gel and the container provided by the collagen fiber meshwork, which immobilizes the molecules of this gel. Thus the rigidity of the system arises from the osmotic swelling of the proteoglycan gel against the constraints imposed by the collagen fiber system. Cartilage is very rich in the proteoglycans chondroitin 4- and 6-sulfate. The factors influencing the mechanical properties of cartilage are collagen cross-link density, collagen fiber diameter, orientation of collagen fibers, number of zones or layers, proteoglycan content, rate of deformation, and source of tissue.[10]

11.2.4 Mineralized Tissues: An Introduction

Vertebrates construct their skeletal and dental hard parts from few mineralized materials, being largely apatitic in form. More unusual are the calcium carbonates used for balance organs and egg shells and magnetite for magnetofield reception, but these will not be discussed further here. The apatitic mineral structures are described, as an understanding of their structure is of paramount importance in the design of biomaterials with improved biocompatibility.

Bone and dentin have very similar structures and are largely composed of apatite crystals and collagen fibers. Although comprising the same basic chemical components, mineralized cartilage is quite different in terms of ultrastructural organization. Tooth enamel is distinct from the above in that its organic phase does not include collagen and its apatite crystals are much larger than for the three other mineralized tissues. An intermediate form bridging enamel and bone or dentin is enameloid, a tissue with enamel-like apatite crystals associated with collagen fibrils. The structure and mechanical characteristics of such materials is discussed in detail after the general mechanisms involved in the formation of minerals have been presented.

11.3 BIOMINERALIZATION: THE CONTROLLED FORMATION OF BIOLOGICAL COMPOSITES

Biomineralization refers to the processes by which organisms form minerals. These include single-celled organisms all the way through to humans. Biogenic minerals are formed on a huge scale in the biosphere, have a major impact on ocean chemistry, and are important components of marine sediments and ultimately of many sedimentary rocks as well. A major function of biogenic minerals is to provide mechanical strength to skeletal hard parts and teeth, and when organisms evolved the ability to form mineralized hard parts, they provided themselves with a major adaptational advantage. The vertebrate skeleton fulfills a variety of functions, but there are associated health problems, such as dental caries, bone fractures, mineral loss from bone, and kidney stones, to name but a few. The known biogenic minerals include carbonates, phosphates, halides, sulfates, citrates, and oxalates of principally group IIA of the periodic table, and in particular calcium; oxides of iron, silicon, and manganese; sulfides of iron, zinc, and lead and a whole range of organic crystals, including urates, tartrates, and waxes. Table 11.8 lists the most important biominerals, their chemical composition, and their function, together with examples. More than 60% of the known minerals contain hydroxyl groups and/or bound water and can be dissolved readily to release ions. The crystal lattice of the group of minerals that includes metal phosphates is particularly amenable to the incorporation of various additional ions, such as fluoride, carbonate, hydroxylate, and magnesium. This allows for modification of the crystalline structure of the material and hence the mechanical properties of the material and has lead to the family of calcium phosphates being used for a wide range of applications.

Most mineralization processes that are controlled by the organism tend to have associated macromolecules. These macromolecules fulfill important functions during the formation of the tissue and in modifying the biomechanical properties of the final product. In general, although there may be tens or hundreds of different associated macromolecules, they are generally classified into two types: framework macromolecules and acidic macromolecules. The major framework macromolecules include collagen, α- and β-chitin, and chitin–protein complexes. More detail on the structure of collagen is provided in Section 11.4. The major acidic macromolecules are poorly defined in some organisms but include glycoproteins, proteoglycans, Gla-containing proteins, and acidic polysaccharides. Very little is known about the secondary conformations of the acidic macromolecules except that for acidic glycoproteins rich in glutamic and aspartic acid all partially adopt the β-sheet conformation *in vitro* in the presence of calcium.[12] Although the composition of these macromolecules varies little between species, this is not the case for the framework macromolecules. These vary from tissue to tissue, and a number of mineralized hard parts do not *appear* to have any framework macromolecules at all. It is suggested that these molecules are required as a substrate for the acidic macromolecules. As they are absent from some tissues, it is suggested that their role may lie in modification of the mechanical properties of the final product rather than having any role to play in the regulation of biomineralization.

The wisdom of many millions of years is encoded in the composite mineral structures that result. The mineral phases formed by organisms can also be produced in the laboratory or geochemically. The conditions under which they form are, however, very different, with significantly milder conditions prevailing in the biological environment. Of more significance is the fact that the biogenic minerals generally differ from their inorganic counterparts in two very specific ways: morphology and the way they are ordered within the

TABLE 11.8 Most Important Biominerals

Chemical Composition	Mineral Phase	Function and Examples
Calcium carbonate $CaCO_3$	Calcite Aragonite Vaterite Amorphous	Exoskeletons (e.g., egg shells, corals, mollusks, sponge spicules)
Calcium phosphates		
$Ca_{10}(OH)_2(PO_4)_6$	Hydroxyapatite (HAP)	Endoskeletons (bones and teeth)
$Ca_{10-x}(HPO_4)_x(PO_4)_{6-x}(OH)_{2-x}$	Defect apatites	
$Ca_{10}F_2(PO_4)_6$	Fluoroapatite	
$Ca_2(HPO_4)_2 \cdot 2H_2O$	Dicalcium phosphate dihydrate (DCPD)	
$Ca_2(HPO_4)_2$	Dicalcium phosphate (DCPA)	
$Ca_8(HPO_4)_2(PO_4)_4 \cdot H_2O$	Octacalcium phosphate (OCP)	
$Ca_3(PO_4)_2$	β-Tricalcium phosphate (TCP)	
Calcium oxalate		
$Ca_2C_2O_4 \cdot (1 \text{ or } 2)H_2O$	Whewellite Wheddelite	Calcium storage and passive deposits in plants calculi of excretory tracts
Metal sulfates		
$CaSO_4 \cdot 2H_2O$	Gypsum	Gravity sensors
$SrSO_4$	Celestite	Exoskeletons (acantharia)
$BaSO_4$	Baryte	Gravity sensors
Amorphous silica		
$SiO_n(OH)_{4-2n}$	Amorphous (opal)	Defense in plants, diatom valves, sponge spicules, and radiolarian tests
Iron oxides		
Fe_3O_4	Magnetite	Chiton teeth, magnetic sensors
α,γ-Fe(O)OH	Goethite, lepidocrocite	Chiton teeth
$5Fe_2O_3 \cdot 9H_2O$	Ferrihydrite	Chiton teeth, iron storage

biological system (the terms used are *texture* or *ultrastructure*). It is likely that common mechanisms exist for the formation of these minerals, and if one could understand *the common principles* underlying the formation of such materials, it may be possible to generate new materials or modify existing materials for a wide range of applications in materials science, including the biomaterials field.

11.3.1 Mineralization Processes

The formation of an inorganic solid from aqueous solution is achieved by a combination of three fundamental physicochemical steps: supersaturation, nucleation, and crystal growth or maturation. The spectrum of biomineralization processes can in principle be divided into those for which there appears to be some control over mineralization and those for which there is not. Three different terms are used to describe the processes of biomineralization;

1. Biologically induced mineralization,[13] where mineralization is the simple removal of ionic/waste compounds from the cell.
2. Biologically controlled mineralization,[14] where a specific "machinery" is set up for the purposes of mineralization and distinguishes the mineralization process from the more specific term.
3. Organic-matrix-mediated mineralization,[13] which refers specifically to mineralization within a preformed organic framework.

Biologically Induced Mineralization Mineralization occurs in the open environment and not in a space particularly delineated for the purpose. No specialized cellular or macromolecular machinery is set up to induce mineralization, which may occur simply as a result of a small physical or chemical perturbation of the system. The crystals that form usually result in random aggregates, with a range of particle sizes and morphologies similar to those formed by their inorganic counterparts. In addition, the type of mineral formed is a function just as much of the environmental conditions in which the organism lives as it is of the biological processes involved in its formation, as the same organism can form different minerals in different environments. This process is particularly common in single-celled organisms and fungi, but examples are also found in higher species, including animals. For example, no fewer than seven different sulfide minerals have been shown to form in laboratory culture when the organism *Desulfovibrio* is present.[15]

Biologically Controlled Mineralization Figure 11.4 shows examples of the complexity of mineral morphology that can occur when mineralization is controlled. An essential element for controlled biological mineralization is spatial localization, which may occur either through the use of membrane-bounded compartments or in specific cell wall locations. This compartmentalization allows local regulation and control over physicochemical factors through selectivity in biochemical processes such as ion and molecular transport. Spatial controls in the production of organized amorphous materials may have special significance.

Space delineation is used to seal off the mineralization compartment to control the composition of the mother liquor from which the mineral forms.

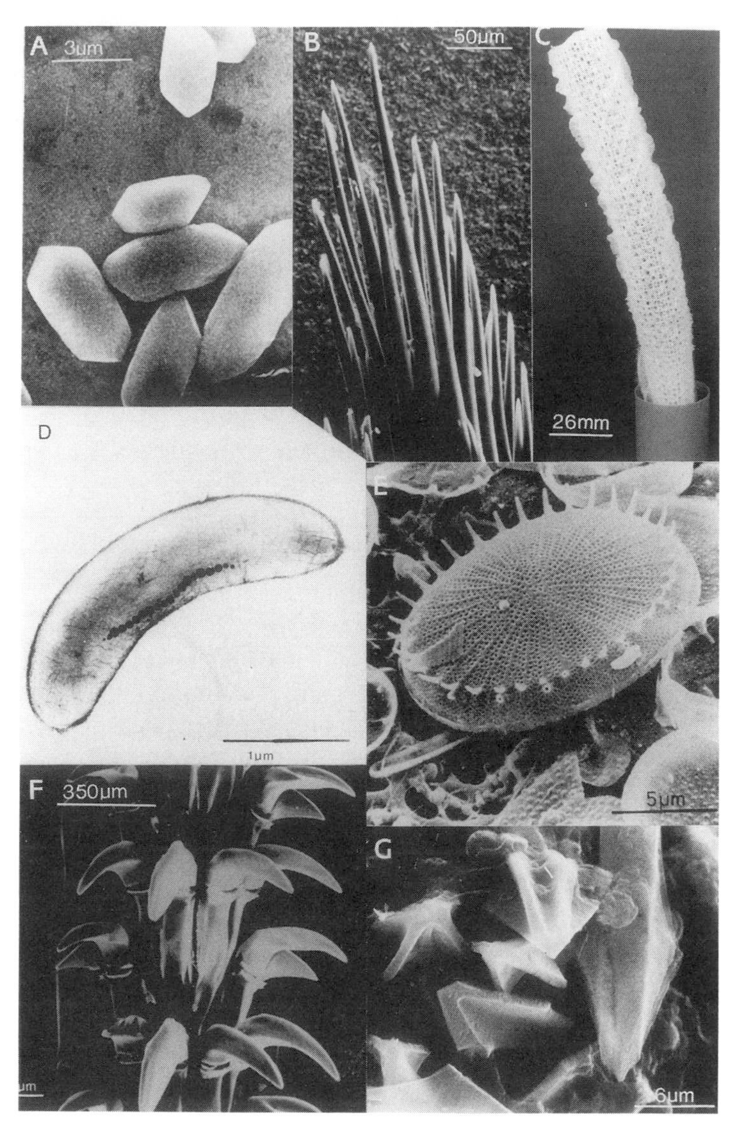

Figure 11.4 Electron micrographs of biominerals produced under biological control: (A) calcium carbonate crystals from otoconia in the inner ear; (B) siliceous plant hairs containing at least three different silica aggregate structures; (C) siliceous sponge (even though the organism exhibits a recognizable engineered structure, the components are all totally amorphous); (D) single crystals of magnetite in a magnetotactic bacterium; (E) siliceous diatom containing gel-like silica; (F) limpet teeth containing both amorphous silica and iron oxide (goethite) crystallites; (G) single crystals of strontium sulfate from an acantharian. Scale bars as shown in the individual figures.

The compartment may be defined by a lipid bilayer, which may have acidic functionality and/or proteins present within the membrane layer. These constituents may have a role to play in mineralization. Alternatively, and less frequently, the space may be defined by an impervious layer of water-insoluble macromolecules such as protein and/or polysaccharide. Further subdivision of the reaction volume may also occur where cells involved in space delineation synthesize an array of macromolecules which are then secreted into the extracellular space, where they self-assemble into a three-dimensional framework. An example from a study of rat incisor formation by Weinstock and Leblond[16] showed that collagen I (the framework protein) was secreted and self-assembled about 24 h before the acidic proteins were secreted into the incisor. For minerals to form, the solubility product of the mineral in question must be exceeded. To do this, the compartments in which the mineral is formed allow passive diffusion of ions and/or the active accumulation of ions against concentration gradients. Ion-specific pumps and channels are therefore necessary components of the machinery required for biomineralization. It should be noted that the vast majority of biogenic minerals formed under controlled conditions precipitate from solutions whose compositions are very closely monitored by the responsible cells, with the result that the trace element and stable isotope composition of many mineralized hard parts are *not* in equilibrium with the medium from which they were formed.

Nucleation in controlled biomineralization will involve low supersaturation in conjunction with active interfaces. The former may be regulated by transport and reaction-mediated processes (inhibitors, etc.) and the latter generated by organic substrates in the mineralization zone. In the indirect sense, molecules present in solution (additives) can specifically inhibit the formation of nuclei of one material phase and by so doing allow another phase to develop.[17] Direct control over nucleation usually results from the provision of specific surfaces on which nucleation occurs. Charged surfaces are considered to be important, although it is known that provision of a highly charged surface may not be the best way to promote nucleation from specific crystal faces, as binding between ions approaching the surface and the surface will be too strong and not allow for the rearrangements of ions that are often necessary. Although nucleation surfaces are envisaged as being solid, they will of necessity possess a degree of fluidity. Nucleation may occur on specific surfaces or within cavities or pores (the smallest imaginable being the meeting of two ions of opposite charge in the presence of, for example, a protein.

At the most basic level, crystal growth depends on the supply of material to the newly formed interface. Low supersaturation conditions will again be favored, to reduce the number of nuclei and to limit secondary nucleation, thereby reducing the disorder within the crystalline phase, with any concomitant effects on the mechanical properties of the mineral phase. Figure 11.5 shows a high-resolution electron microscope image of part of a single crystal

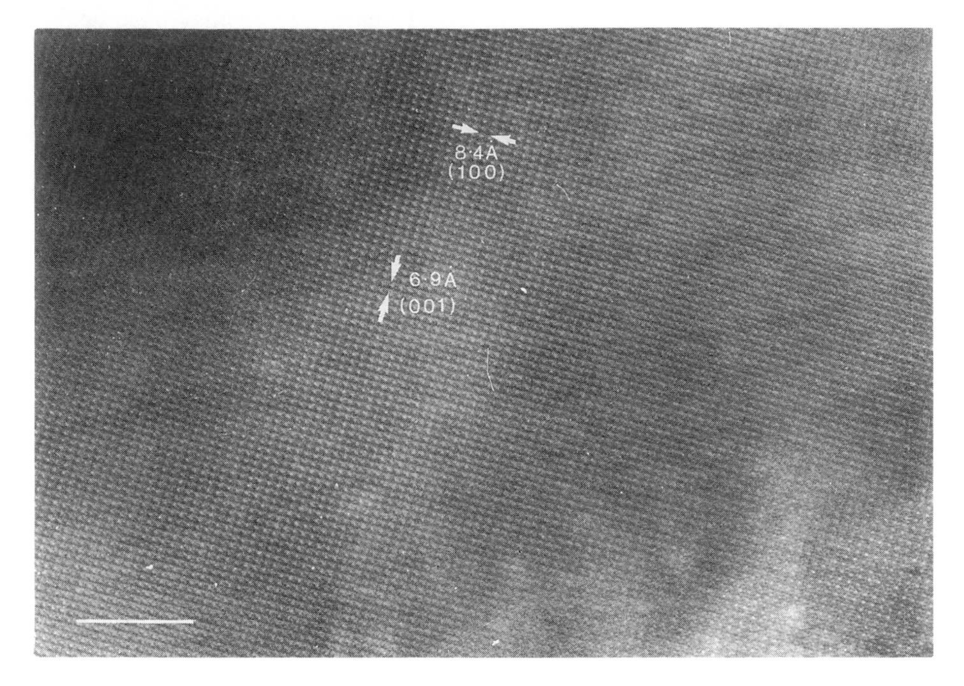

Figure 11.5 High-resolution transmission electron micrograph of part of a strontium sulfate spicule shaft from acantharia. The plane of the spicule shaft, containing the major axis of the rounded rhombic spicule cross section and the principal spicule axis, lie perpendicular to the [010] zone axis. Lattice spacings for (001) and (100) crystal planes are depicted on the figure. Notice the virtual absence of defects in the image, a feat rarely achieved in laboratory-grown crystals. Scale bar represents 10 nm.

of strontium sulfate where this control is shown and results in the development of large defect-free areas within the mineral phase. Under low supersaturation conditions the growth rate will be determined by the rate of attachment of ions to the surface. Under these conditions extraneous ions and small and large biomolecules may be incorporated onto the surface, thereby modifying crystal growth and causing modifications in crystal morphology to occur.

The ultimate stage in the formation of a biomineral is the cessation of mineral growth. This may be accomplished simply by the lack of a supply of ions to the mineralization site, by the crystal coming into contact with another crystal (the development of spherulites), or by the mineral coming into contact with the preformed organic phase. An outcome of the latter method of control is that crystals formed in the presence of vesicles (membrane-bounded phospholipid compartments) tend to have curved surfaces such as that depicted for a coccolith in Fig. 11.6, whereas those formed in the

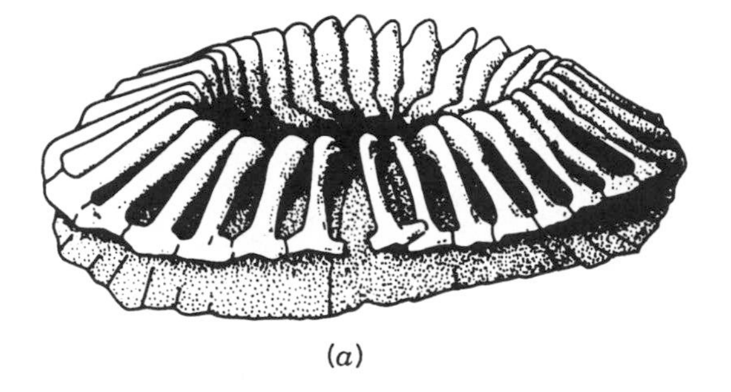

(*a*)

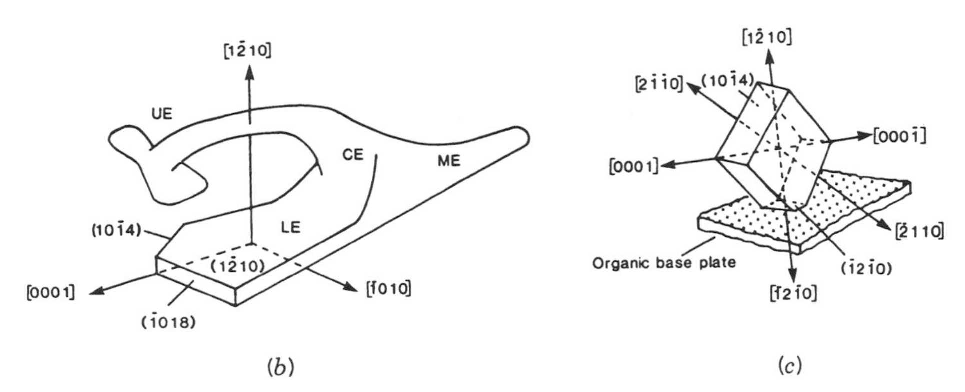

(*b*) (*c*)

Figure 11.6 Schematic drawing of a (*a*) coccolith together with (*b*) the elements that make up the coccolith. The arrows indicate the crystallographic directions and faces of the calcite single crystal in the mature coccolith. (*c*) Corresponding crystallographic orientation of a rhombohedral calcite nucleus formed at the surface of an organic baseplate.

presence of compartments delineated by matrix macromolecules tend to form more regular crystals (e.g., bone and enamel/enameloid, Fig. 11.7). It should be noted that very little is known about the structures of the lipid bilayers or macromolecular sheets that are responsible for crystal growth inhibition.

Amorphous phases are not common in biology unless the activation energy barrier to crystallization cannot be overcome. One such example is hydrated biogenic silica. Although silica exhibits no long-range crystallographic order, microscopic morphological order is often observed.[18] Such order may arise during either the nucleation or the growth processes, since the silica network has similar surface properties to a crystalline system. Energy considerations support the formation of a silica aggregate with a densely packed, covalently bonded core and a highly hydrated surface. The surface is likely to interact with organic substrates in a biological environment in an manner analogous to crystal interactions, thereby lowering the free energy of aggregate forma-

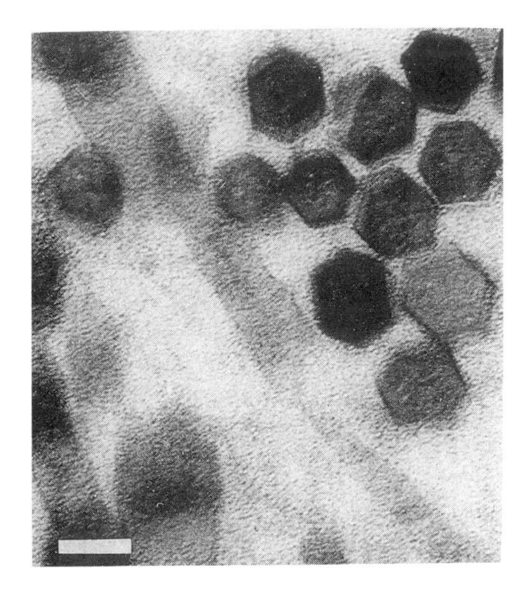

Figure 11.7 Transverse view of enameloid crystals. Scale bar represents 33 nm. (From Ref. 15.)

tion and controlling aggregate morphology on the microscopic scale. During the growth stage the chemical nature of the mineralization zone or the physical forces under which the mineral is deposited may control the mineral morphology.[18] Control over these processes leads to the formation of intricate composite mineral structures where the physicomechanical properties of the resultant phase are distinct from the properties of the components. A specific example is that of the sea urchin spine, which shows that incorporation of organic molecules at crystal domain boundaries[19] leads to a material with fracture behavior more akin to glass than to a simple crystal.

One of the most complex biomineralized structures is bone, which is described here in some detail, together with enamel, as these are materials for which prosthetic devices are required which are biocompatible and which assist in the natural regeneration of tissue when in place in the human body. What is often not appreciated is that the formation of a mineral may require the various components to pass through temporal related changes in structure which may involve the presence of trace levels of other ions and molecules for the mineralization process to occur. One specific example is the growth of bone, where localized concentrations of silicon-containing materials are found where there is active bone growth.[20] The reason for this is not known, and our ongoing research is aimed at gaining an understanding of the processes with which silicon-containing species may be involved, as this may explain the increased activity of certain hydroxyapatite phases and gel glasses in promoting bone cell proliferation and bone formation.

11.4 BONE AND OTHER MINERALIZED TISSUES

Bone has unusual physical and mechanical properties, as it is able to support its own weight, withstand acute forces, bend without shattering, and can be flexible without breaking within predefined limits. Bone also acts as an ion reservoir for both cations and anions. In material terms, bone is a three-phase material: The organic fibers (collagen) can be compared to steel cables in reinformed cement, the inorganic crystalline phase (carbonated hydroxyapatite) to a heat-treated ceramic, and the bone matrix to a ground substance that performs various cellular functions. The unique physical and mechanical properties of bone are a direct result of the atomic and molecular interactions intrinsic to this unusual composite material.

The stiffness, strength, and toughness of bone are related to the mineral content of the bone, as shown in Fig. 11.8. Although strength and stiffness increase linearly with increase in mineral content, toughness does not show the same relationship and there is an optimal mineral composition, which leads to maximal toughness of bone. Clearly, this pattern of behavior is why bone contains a restricted amount of mineral in relation to the organic

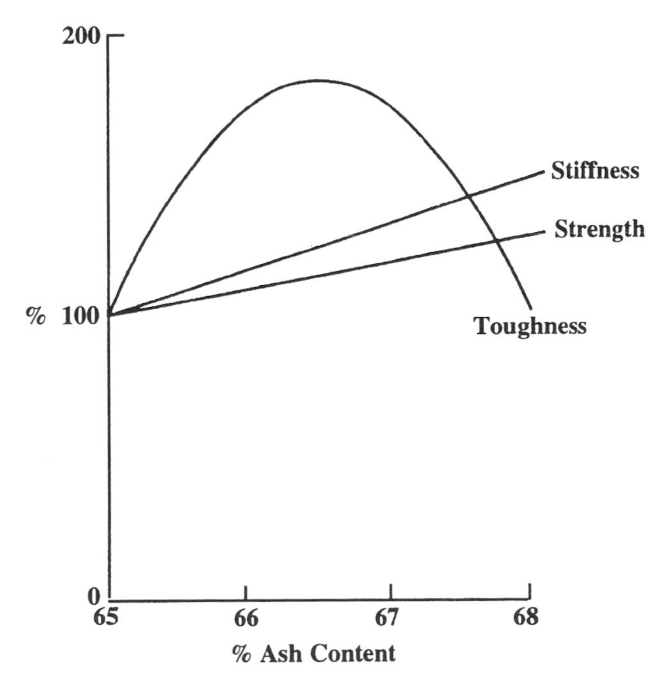

Figure 11.8 Graph showing the relationship between ash content of bone and its stiffness, strength, and toughness. Ordinate: each value of a quality is compared with its value at 65% ash. (Redrawn from Ref. 4.)

matrix.[4] The structure of the bone can be understood in terms of the different levels of organization that exist within the material.[21] The lowest level of organization describes the crystals, the organic framework (mostly collagen fibrils), and the relationship between the framework and the crystals. The next level of organizations (tens of microns) describes the longer-range order of collagen and associated crystals. Three distinct forms can be recognized. In *woven bone* the mineralized collagen fibrils show little preferred orientation and are almost randomly distributed over distances greater than a few microns. *Lamellar bone* is much more ordered and comprises sheets about 3 μm thick built up from collagen fibrils aligned over long distances. Orientation between fibrils in adjacent lamellae can be offset by a few degrees up to 90°. *Parallel-fibered bone* is intermediate between woven and lamellar and is not widely distributed. There is a further level of organization where recognizable structural units extend over distances of several millimeters. Two of the types correspond to woven or lamellar bone textures extending over much longer distances. In the third type, known as the *Haversian system*, concentric cylinders of bone are laid down when resorption of bone by active blood vessel components leads to a secondary deposition stage. The result of this process is that "mature" bone may contain recently mineralized collagen fibrils. In the fourth structure type, known as *fibrolamellar bone*, parallel fibered bone is laid down first, followed by lamellar bone. The highest level of organization describes structures extending over several centimeters and differentiates between cortical or compact (solid) and cancellous (trabecular or spongy) bone. Figure 11.9 shows some of these structural types. Because of its lower density, cancellous bone has a lower modulus of elasticity and higher strain to failure than cortical bone and will therefore differ in its response to implanted materials.

11.4.1 Molecular Organization of Bone

Bone is composed of type I collagen fibrils intimately associated in an orderly fashion with calcium phosphate crystals. Minor components include small molecules associated with the mineral phase and macromolecules associated with the collagen phase. Bone crystals are extremely small with an average length of 50 nm (range 20 to 150 nm), width 25 nm (range 10 to 80 nm), and thickness of 2 to 5 nm. Clearly, the consequence of this is that a significant proportion of each crystal is surface and hence able to interact with its surroundings. The hydroxyapatite phase contains 4 to 6 wt % carbonate and is most correctly described as the mineral dahllite. The mineral composition changes with age[22] and is always calcium deficient, with both phosphate and carbonate ions in the crystal lattice. The following formula is needed to generally represent bone mineral: $Ca_{8.3}(PO_4)_{4.3}(CO)_{3x}(HPO_4)_y(OH)_{0.3}$; y decreases and x increases with increasing age, whereas $x + y$ remains constant and equal to 1.7!

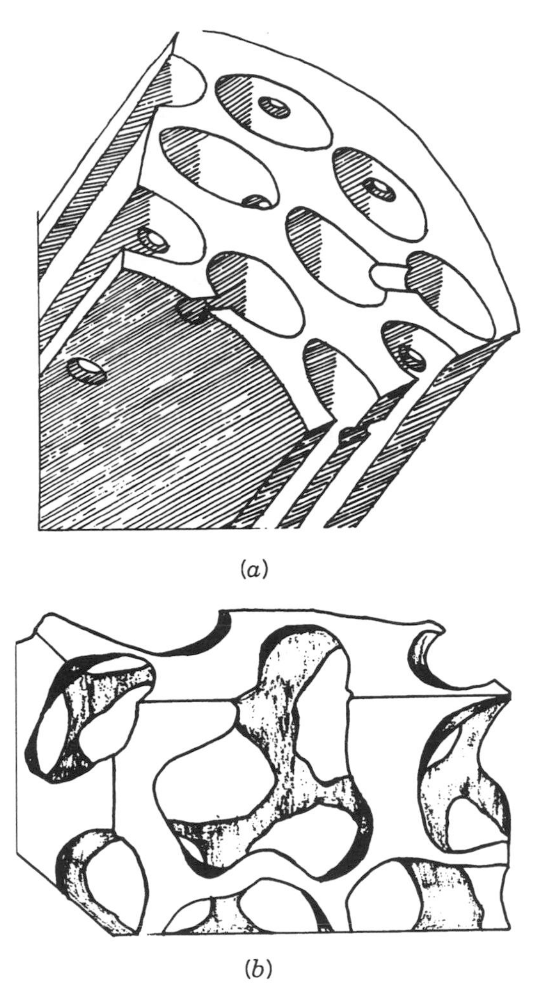

Figure 11.9 Schematic drawings of different bone types: (*a*) human cortical bone; (*b*) human cancellous bone.

Organic Matrix The major components of bone, type I collagen, usually comprises between 85 and 90% of the total protein. The noncollagenous fraction has more than 200 components, including serum proteins, and is discussed further below.

Collagen Crystal Relations An intimate association exists between the crystals and collagen in bone. A key question is whether or not all of the crystals in bone are associated with collagen. There is evidence that at least a major portion of the crystals in bone are associated at the gap level within the collagen fibrils. The suggestion is that in some tissues (e.g., turkey

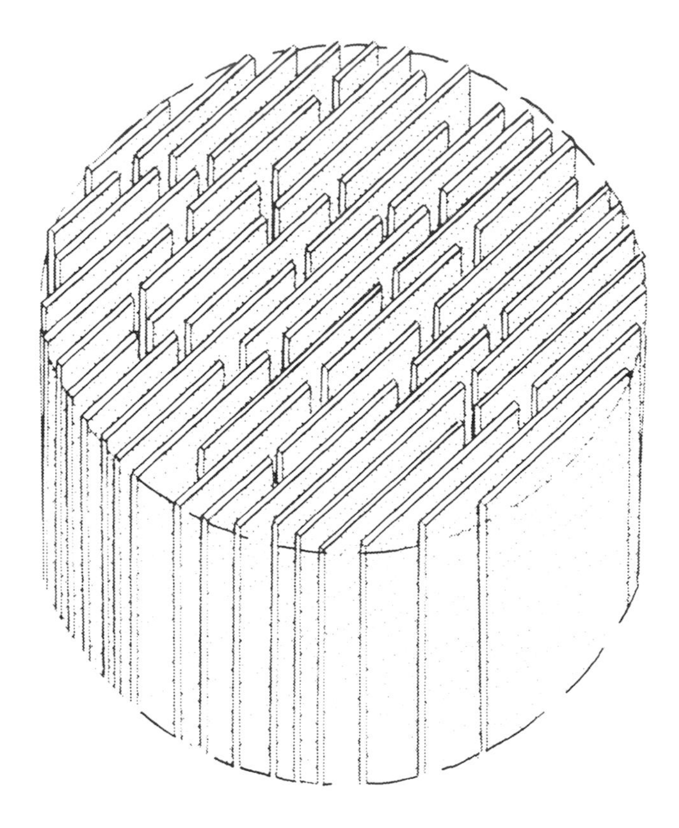

Figure 11.10 Schematic illustration of the arrangement of crystals in a section through a collagen fibril, following the observations of Weiner and Traub.[23]

tendon) the crystals are packed more or less like a pack of cards across a collagen fibril diameter.[23] Figure 11.10 shows the suggested arrangement of crystals through a collagen fibril. Individual molecules align themselves in two and three dimensions to form holes and grooves where the gaps between the molecules are on the order of 36 nm. The idea that a single crystal can fit into an individual hole[24] cannot be correct, as the sizes do not match (the crystals are generally bigger). There must, therefore, be some special structural arrangement of the triple helical molecules in type I bone collagen that can accommodate the crystals.

Noncollagenous Proteins There are three main types of noncollagenous proteins (NCPs) in addition to the low levels of "factors" which are known to induce various changes in cell behavior and the serum proteins, which are ubiquitous. The three major classes of NCPs are acidic glycoproteins, proteoglycans, and Gla-(γ-carboxyglutamic acid) proteins. The acidic glycoproteins contain considerable amounts of the amino acids phosphoserine, phosphothreonine, and γ-carboxyglutamic acid. The phosphoproteins are intimately associated with the initiation and regulation of crystal growth and may serve as a source of inorganic phosphate on enzymatic release by phosphatases. The proteoglycans have one or more (negatively) charged glycosaminoglycan chains attached to the main protein chain. They may serve to inhibit crystal growth by their negative charge, and as they structure many times their mass of water, they may serve to reserve the extracellular space for future calcium phosphate crystal growth. Both these classes of proteins, together with alkaline phosphatase, are found in a range of mineralized tissues, and their wide distribution suggests that they have a basic role to play in controlling mineralization systems. Bone Gla-containing proteins are unique to bone and dentin and as such are expected to have a specific functional role to fulfill in these tissues.

11.4.2 Stages of Bone Mineralization

Stages in the formation of bone are as follows:

1. Synthesis and extracellular assembly of the organic matrix framework
2. Mineralization of the framework
3. Secondary mineralization as the bone constantly forms and re-forms

The initial site of mineralization is not associated with the collagen; rather, precipitation occurs within membrane-bounded vesicles.[25] Those farthest from the mineralization front are not mineralized, those closer contain amorphous deposits, and those close to the mineralization front contain crystalline deposits with some suggestion that they have a needlelike structure. It has been suggested that these crystals are translocated from the vesicles onto and/or into the collagen fibril.[26] However, there is no evidence for any biological process in bone that transports the crystals and slots them into the collagen fibril (even assuming that they were of the correct size!!). It seems less likely that a physicochemical process could be responsible. A more likely function of these matrix vesicles[15] is as a temporary storage site for calcium and phosphate ions needed for collagen mineralization. The first mineral phase formed is probably hydroxyapatite (HAP), although there is a large body of evidence to support the initial formation of octacalcium phosphate (OCP).[27] Where these crystals form is also a matter of conjecture, although there is evidence that mineralization starts at a precise position toward the N-terminus side gap in collagen.[28,29] The question about where the crystals form (within and without the collagen fibrils) is also largely unanswered. It is possible that mineralization can be considered to occur in two stages, with crystal growth in the gap/groove volumes constituting the rapid-growth phase and crystal maturation and growth between fibrils the slow secondary mineralization stage, where aggregates are found to form. It is suggested that the breakdown of NCPs during late-stage mineralization may be related to aggregate formation.[30]

11.4.3 Other Mineralized Tissues

Cartilage Cartilage provides a mold around which the major bones of the body, except the skull, form. It may be unmineralized or mineralized during the process that leads to bone formation, but it is later replaced by marrow and vascular elements.[31]

Unmineralized Cartilage Unmineralized cartilage is composed primarily of type II collagen fibrils, with low levels of types IX and X also being present, proteoglycans, and about 60 to 80 wt % water. The change from type I to type II collagen results in the formation of thinner fibrils than are found in bone. These are also not arranged in preferred orientational patterns except in the human intervertebral disk.

Mineralized Cartilage Mineralized cartilage has basically the same assemblage of macromolecular constituents as nonmineralized cartilage together with hydroxyapatite crystals. Phosphorylated amino acids are found in regulatory glycoproteins similar to those proposed for bone. The crystals do not form an association with the collagen fibrils, and there is no regular organization of the crystallites with respect to the collagen matrix. It can be inferred that the mineralization process is poorly controlled. In cartilage before crystal formation occurs, the matrix is loaded with calcium and then just prior to mineralization with phosphorous.[32] It is suggested that the fine tuning of the timing of mineralization may be the controlled increase in phosphate concentration. Mineralization occurs both within matrix vesicles and within the matrix itself. Within the matrix there is association of the forming crystals with proteoglycans and a C-propeptide of collagen II, and it is suggested[15] that in cartilage, the sulfated side chains of the proteoglycans concentrate calcium and nucleation sites are provided either by the surfaces of crystals released from the matrix vesicles or from another ordered acidic macromolecule, such as the C-propeptide.

Enamel and Enameloid In bony fish, sharks, and tadpole-stage frogs, the tooth covering is called enameloid. In adult frogs, reptiles, and mammals the tooth covering is termed enamel. Enameloid contains relatively small and aprismatic calcium hydroxyapatite (prototypic hydroxyapatite), whereas enamel contains large, highly ordered, and prismatic enamel crystals. These crystals may be of the order of 100 μm in length but are only 0.05 μm wide, with a hexagonal cross section (Fig. 11.7). The crystals are the products of the combined activities of cells and extracellular matrix components, and it is

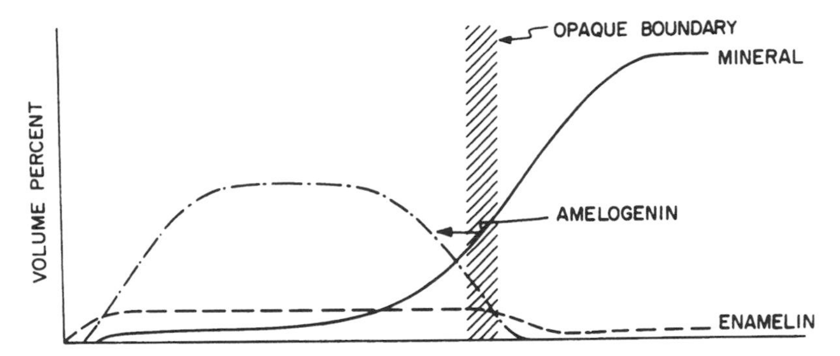

Figure 11.11 Schematic illustration of the changes in composition of the organic and inorganic components of enamel. The opaque boundary is the zone in which rapid changes in the proportions of these constituents take place. (Redrawn from Ref. 15.)

suggested that although the crystals formed are different in morphology, similar crystal formation processes may be involved in both tissues. Both tissues are formed in the presence of the acidic enamelins, but the framework proteins for each tissue are different. Enameloid contains collagen and enamel contains bydrophobic amelogenins. In enameloid, crystals form between the collagen fibrils, and in enamel they form parallel arrays in the absence of collagen. It is suggested that the crystals grow into preformed spaces (subdivided from the collagen fibril structure in the case of enameloid) created by an organic sheath which is probably, in part, built up from the enamelins.[33] Another difference between the two tissues is that in enameloid the entire extracellular matrix is preformed prior to the start of mineralization, but in enamel mineralization starts almost immediately after matrix elaboration.[34] Enamel formation is also extremely dynamic, as Fig. 11.11 suggests, in that the volume percent of the major macromolecular phases and the mineral phase changes with time as the matrix is broken down and the mineral matures. It is suggested that the major role of the matrix is to control the cross-sectional dimensions and the continual elongation of the crystals. During the forming stages crystals are well separated from each other by the organic material and have very well defined and uniform shapes.[35] When the organic matrix is removed from the tissue during maturation, the crystals undergo a growth spurt and eventually form a massive structure with a fused appearance which is likely to be a result of the organic tubules (formed from enamelins) losing their coherent structure,[36] although this cannot be proved.

The sequential stages involved in enamel production are:

1. Noncollagenous and collagenous proteins are deposited in the enamel extracellular matrix.
2. These extracellular matrix molecules (initially associated with matrix vesicles and subsequent nonvesicle sites) serve as templates to nucleate

HAP formation and also provide subsequent enzymic and dehydration functions.

3. Amorphous calcium phosphate is converted to large oriented and ordered HAP crystals.
4. Enamel HAP crystals growth along the *c*-axis is regulated by amelogenins and coincides with sequential removal of water and protein, resulting in the production of a composite ceramic enamel.

11.4.4 General Repair Mechanisms

Bone repair is a regenerative process and scar tissue is not normally formed (contrast skin). Regeneration occurs either through formation of membranous bone or through mineralization of cartilage. In facial bones, clavicle, mandible, and subperiosteal bones, membranous bone growth involves calcification of osteoid tissue (endochondral bone formation). In long bones the stages of repair include induction, inflammation, soft callus formation, callus calcification, and remodeling.

Cartilage can repair itself,[37] although the degree to which this can occur depends on the age of the patient, protection against abrasion, proximity to blood vessels, and wound depth and location. Repair tissue consists of fibrocartilage containing type I collagen (compared to type II collagen found in normal cartilage), and repair proceeds in a pattern which follows that observed for skin.

Skin repair usually results in the formation of scar tissue. Repair of the dermis precedes that of the endodermis. Dermal repair involves inflammation, immunity, blood clotting, platelet aggregation, fibrinolysis, and activation of complement and kinin systems. In the absence of a chronic inflammatory response, dermal wounds are repaired through deposition and remodeling of collagen to form scar tissue.

An example of repair mechanisms in ocular tissues is corneal wound healing, which occurs through the deposition of scar tissue that contains large-diameter collagen fibrils, and trimers of type I collagen are synthesized at the expense of the usual type V collagen.

It is important to note that there is no biological process to repair decayed or destroyed enamel, and hence materials compatible with enamel are required for repair of dental caries.

11.4.5 Hard Tissues and Biomaterial Compatibility

It is clear from the discussion above that the "hard" parts produced by the body are extremely complex materials where both mineral and macromolecular components have their own distinctive role to play in the generation of a material with specific physicomechanical characteristics. It is clear that the materials formed are in dynamic equilibrium with their environment and change with age. Significantly, there is much less information on how these materials change in the diseased state and how they are affected by the presence of foreign materials in close proximity. All of these factors are of considerable importance in the design of new materials and in the development of existing materials for biomaterial applications. It is to be hoped that an understanding of the structures of the materials that are to be augmented or replaced will aid in the development of suitable materials that will be compatible with the materials they are seeking to replace.

For soft tissues the response to an implanted synthetic material is slightly different, in that it is an inflammatory reaction to a foreign body, and factors that minimize this inflammation will maximize biocompatibility. The materials chosen for the implant should be nontoxic, nonantigenic, and if porous, allow immune and phagocytic cells and native tissue growth.[38]

11.5 MATERIALS OF CONSTRUCTION

All medical implants are composed of either polymers, metals, ceramics, or mixtures and composites of these materials. Tissue replacement with synthetic materials requires selection of a material or materials with physical properties most similar to those of the natural tissue. This information is given in Table 11.2 and shown pictorially in Fig. 11.12.

11.5.1 Polymers

Polymers are the most widely used materials in health care and are used in almost all phases of medical and/or dental treatment. Typical polymers in use are listed in Table 11.3. Polymers may be formed by condensation reactions between complementary functional groups to make poly(esters), poly(amides), and poly(urethanes). They may alternatively be formed by free-radical polymerization of unsaturated compounds to give addition polymers. Examples of this class include poly(ethylene), poly(vinyl chloride) and poly(methyl methacrylate). Any application that involves a polymeric material will require characterization of the polymer phase by a variety of methods[40] in order to understand its probable behavior on processing to form a device. However, to produce an end product, polymers are usually compounded with several other ingredients as indicated in Table 11.9.

Plasticizers, antioxidants, and fillers are added to lower the cost and improve the mechanical properties of the polymer device. There are, however, problems associated with their use because of their low molecular weight (in relation to the polymer phase), as they may be leached from the device, causing deleterious effects on the device itself and on the human body. The mechanical properties of polymers are a consequence of the degree of crystallinity (see Section 11.2.1) and the transition temperature at which they change from a viscoelastic material to a rigid glass. They cover a

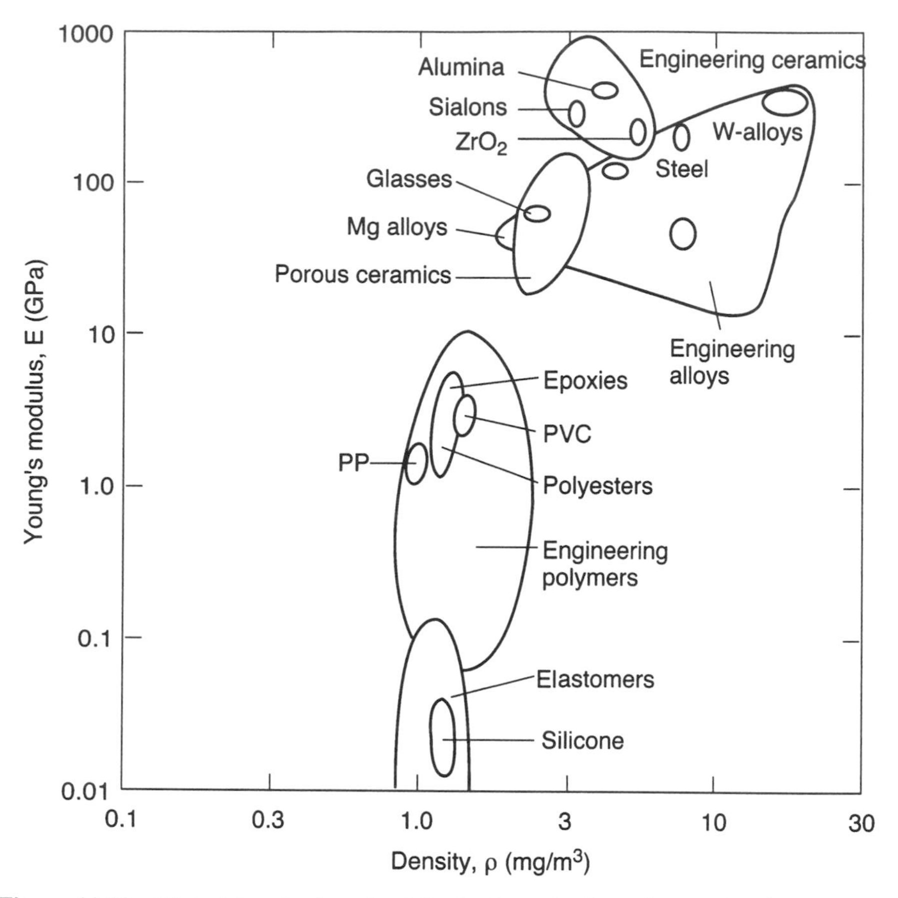

Figure 11.12 Materials selection chart for both synthetic and natural polymer-based materials. (Redrawn and modified from Ref. 39.)

wide range of strengths (as measured by values for Young's modulus) and can therefore be used for a wide range of applications. The material selection charts in Fig. 11.12 compare the strength-to-weight characteristics for a range of polymers and the materials they are designed to replace and/or augment.

Applications of Polymers As examples, the use of polymers in ophthalmology, skin wound treatments, and breast treatments are described.

Applications in Ophthamology[42] Ophthamology employs the use of a large number of materials for correction of vision, intraocular lens replacement, correction of the curvature of the cornea, indentation of a detached retina, insertion of intraocular lenses, and for cataract removal. Each of these applications requires knowledge of the properties of ocular tissues as well as

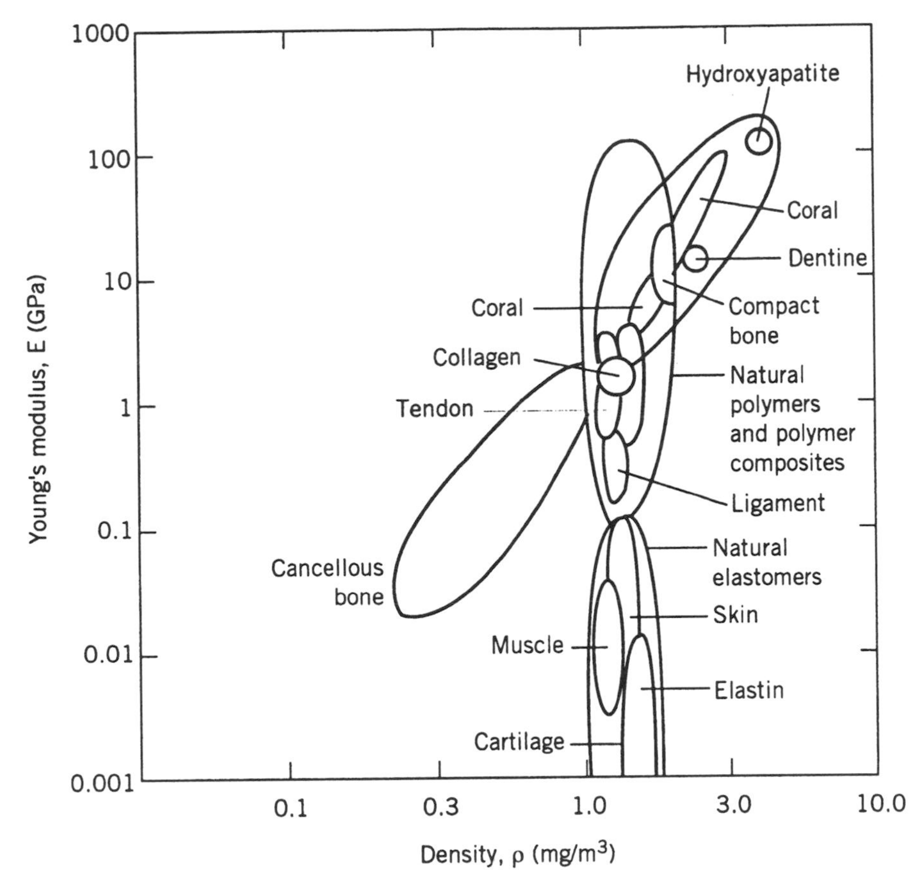

Figure 11.12 *Continued)*

TABLE 11.9 Additives Used to Process Polymers into Engineering Materials

Additive	Function
Accelerators	Increase the kinetics of cross-linking
Antioxidants	Minimize cracking of device when exposed to oxidants
Cross-linking agent	Prevents viscous flow of final product
Plasticizers	Facilitate flow of polymer into shape desired
Reinforcing agents	Improve mechanical properties of polymers

Source: Ref. 41.

selection of materials that are biocompatible and produce the desired changes in the eye.

Ophthamology is a field that has advanced rapidly as a result of the development of new techniques and materials. Correction of vision using contact lenses has been achieved since the 1960s and will eventually be replaced using a procedure in which a biomaterial and laser surgery are used to change the curvature of the cornea. Biomaterials are an important component of the procedures that are used to improve and maintain vision.

As with all components of the human body, the eye is extremely complex, with 10 components, each of which has a specific function necessary for correct functioning of the unit and enabling the human body to see objects. What is particularly significant are the large number of structures in the eye which are composed of collagenous tissue together with other macromolecules. Hence it is important that materials used to replace and augment damaged tissues have some of the same physicomechanical properties as those of the collagen-containing materials naturally found in eye tissues.

Intraocular lenses (IOLs) are used to replace natural lenses that have become affected by cataracts. They have been in use since 1949 and most are made from PMMA. IOLs consist of an optic portion through which light passes, along with loop or attachment regions. The composition can be complex and the optical portion can be built up from PMMA, polyhydroxyethylacrylate (PHEMA), or silicones, the loop from metals, nylon, glass, poly(imide), and polyethylene. Included in the polymer portion are antioxidants, ultraviolet light absorbers, and other agents to retard the decomposition of the lens. Tissue response to IOLs has been monitored over 30 years and results show that even PMMA lenses induce a foreign body type of encapsulation with fibrovascular tissue, and the membrane surrounding the implant can be coated with cells and cell aggregates, including foreign-body giant cells. In addition, low amounts of free monomer can be released from implanted lenses. PHEMA appears to show improved compatibility over PMMA.[43]

Contact lens materials used include cellulose acetate butyrate, PMMA, siloxane methacrylates, and silicones. PMMA was introduced because of its superior optical properties; however, because of its chemistry and rigid chain structure, the oxygen permeability of this material is very low. Oxygen transport to the cornea is achieved via tear fluid circulation and lenses have been developed that allow for oxygen diffusion through the lens. These materials, which often have negatively charged surfaces and small pores are vulnerable to deposition of lipomucoprotein from the teat film and to the growth of fungi and bacteria. The latter problem has led to the development of rigid gas-permeable lenses, where modification of silicone polymers by copolymerizing methyl methacrylate and alkylsiloxanyl methacrylate into a silicone acrylate has been achieved.

Viscoelastic solutions protect cells from mechanical damage, maintain and create tissue spaces, separate and lubricate tissues, allow manipulation of tissues while limiting mechanical damage, and prevent and control cell movement and activity. Since the 1970s these materials have been used in anterior or segment surgery. Materials used include hyaluronic acid (HA), chondroitin sulfate (CS), polyacrylamide, hydroxypropylmethyl cellulose (HPMC), collagen, and mixtures of these materials. All except collagen and polyacrylamide are polysaccharide derivatives and exhibit conformational freedom such that they can form random coils in solution, a property necessary for the required physical behavior of these materials.

Applications in Skin and Wound Healing Skin is a multicomponent composite of cells and macromolecules.[44] The two major components, the dermis and the epidermis, both contain large amounts of collagen and elastin. In addition, the epidermis (outer layer) contains large amounts of keratin. Cell types in these layers are involved in protein production, sensory transmission, protection against radiation damage, immune system activation, and inflammatory responses. When dealing with skin repairs, the functioning of the cellular fraction, in addition to the rebuilding of the extracellular matrix, should be borne in mind when deciding on appropriate materials for construction.

Wound dressings and artificial skin are used to repair and replace areas of skin damaged by localized pressure (bed sores), burns, cancer excision sites, and complications of conditions including diabetes (diabetic skin ulcers) and insufficient cardiac output (venous stasis skin ulcers). Materials used in the treatment of burns include biological (porcine skin, cadaver skin, amniotic membranes), biosynthetic (BioBrane), and other materials. Wound dressings include biologically derived collagen sponges, skin grafts, amnion, placenta, pericardium, cell-seeded collagen, Bell's skin equivalent, Yannas' collagen/GAG membrane (glycosaminoglycans) and synthetic hydrocolloids, foams, and membranes.

Some of the best materials for the treatment of burn wounds appear to be reconstituted collagenous materials, as they do not stimulate rejection and can be seeded with cells to form a permanent skin replacement. Allograft skin is used for coverage of full-thickness burn wounds after removal of necrotic tissues. Biodegradable dermal substitutes can be prepared from cell-free extracts of collagen to make cell-free xenografts. In the body such materials are biodegraded and replaced by host tissue. The rate of biodegradation is controlled by varying the cross-link density. Glycosaminoglycans (GAGs) are often added, as these modify the mechanical and cross-link density and produce a material with an open pore structure.[45]

The term *living skin equivalent* (artificial skin) has been coined to describe skin cells grown on collagen substrates to yield a tissuelike product.[46] A number of tissues have been engineered by growth of one or more cell types on collagen matrices. These cells are derived from skin, liver, thyroid, cartilage, bone, pancreas, cornea, cardiovascular, adipose, mammary, and nervous tissues. Autologous sheets of epithelial cells have been produced to facilitate the treatment of full-thickness skin defects. Their advantages lie in their inherent simplicity and their biocompatibility; however, there is a lag time for preparation prior to implantation. Skin grafts produced on collagen sponges overcome the problem of implant fragility while inducing rapid dermal ingrowth. Cell-inoculated collagen sponges initiate rapid wound closure with no lag time prior to grafting. The limiting factor for preparation of these grafts is the availability of autologous epithelial cells. In addition, skin equivalents consisting of fibroblast-seeded collagen gels overlaid with epithelial cells have been investigated.

Applications in Breast Implants Current materials use for breast implants involves the use of silicone polymers for fabrication of the shell or shells, silicone gel or saline for filling the inner sac, and polyurethane for covering the outer shell,[47] although, as in all areas of science, the materials used are ever changing in response to research by the materials scientists and problems caused by existing materials. Saline-filled implants are currently the only alternative for cosmetic augmentation of breast tissue. The silicone material used in the sac has a silicone network with a small amount of silicone oil added to control the viscosity of the gel. Amorphous silica can be added to the shell coating to improve mechanical properties of the sac. Contaminants found in the implant silicone materials due to synthesis include siloxane monomers, platinum catalysts, and peroxide derivatives. The polyurethanes used may contain a wide range of functional groups, such as aromatic/aliphatic hydrocarbons, epoxies, poly(eters), poly(esters), poly(olefins), silicones, and ureas and may be susceptible to hydrolysis during implantation, thereby weakening the outer shells used to protect the silicone or saline filling from the biological environment.[48] In some cases the use of this material has been banned and the texture required on the outer surface of the implant is being provided by "rough"-machined silicone-based materials.

Millions of breast implant procedures have been carried out both for medical and cosmetic reasons. It is only now that the effects of prolonged implant–host tissue interactions are being evaluated. Infection, capsular contraction, and implant leakage are among the most common problems associated with these devices. Clearly, there is room for improvement with the materials used in the fabrication of these devices.

Polymer Cell Interactions The ability of polymer surfaces to support the attachment and growth of anchorage-dependent cells (which is required for favorable biologic response in most cases) varies widely. Polystyrene, polyethylene and terphthalate, polyfluoroethylene, and perfluorinated ethylene propylene copolymer, for instance, are poor supports, while other polymeric materials suitable for the attachment and proliferation of cells are now commercially available (see the section "New Generation of Synthetic Polymeric Materials"). The attachment of cells is mediated by adhesive glycoproteins such as fibronectin and vitronectin, which compete with other proteins (in extracellular space there may be on the order of 100 proteins all competing for the same site, and therefore there has to be a very significant interaction for specific binding to occur). In addition, it is probable that an organized distribution of protein on a surface is required so that good/efficient recognition can occur for adsorption on polymer surfaces. Much attention has been directed toward identifying the role of particular chemical groups on polymer surfaces in the attachment of proteins and cells, and surface modification by gas plasma (glow discharge) methods has become popular as a versatile means of fabricating various surface chemistries for evalulation in biological tests. Gas plasma methods produce surfaces with functionalities rich in oxygen groups including O, OH, surface sulfonate, and carboxyl, which are, in principle, more compatible with biological tissues than the general functionalities present in the parent polymers. Polymer surfaces are also being treated with albumin and heparin to reduce problems with thrombogenicity.[49] There are, however, circumstances when polymer–protein binding should be minimized, and materials are now being generated with permanent hydrophilic surfaces but with low protein adsorption characteristics.[50] The market for these materials is in the production of blood-contacting devices, chromatographic supports, coatings to minimize biofouling, separation membranes, contact lenses, immunoassays, protein drug-contacting materials, and so on. PEO has been used to modify artificial surfaces by chemical grafting, physical coating, and the establishment of PEO-surface and PEO-interpenetrating networks. An alternative is to use the melt blend technique, which is simple and cost-efficient. In this approach a small amount of protein compatible additives, such as PEO, poly(vinyl alcohol) (PVA), poly(ethyl oxazoline) (PEOX), and poly(vinyl pyrrolidine) (PNVP) have been used to modify the substrate polymer. The base polymer can be a single polymer or a mixture of ethylene–vinyl acetate (EVA), polypropylene (PP), glycol-modified poly(ethylene terephthalate) (PETG), poly(methyl methacrylate) (PMMA), styrene–butadiene copolymer, and polyamide-based copolymers. Materials based on a wide variety of compositions show reduced protein adsorption and are currently being developed for drug release.

Surface Modification of Polymers Most conventional materials currently used as biomaterials do not meet the demands required with respect to both their surface and bulk properties. Materials, based on polymer systems are often used after purification to remove toxic substances from the surface but without further surface modification. Now that there is a need for biomaterials which are used in contact with water or moisture, materials with biocompatible surfaces which may also exhibit physiological activity in an aqueous environment are required. It is surprising that surface properties of polymers have not previously been modified, as the technology for doing this has been available for more than 50 years in the industrial sector. Surfaces of polymers can be modified by grafting techniques. A grafted surface can be produced primarily either by graft polymerization (often a free-radical process) of monomers or the covalent coupling of existing polymer molecules onto the substrate polymer surface. A terminal group that is reactive to the functional groups present on the substrate polymer surface is required for the polymer chains to be used in the coupling reaction, while the graft polymerization method needs active species on the substrate polymer to initiate radical polymerization. An apparently grafted surface can also be produced by physical adsorption of existing polymers. The adsorbed chains are readily desorbed upon immersion in solvents or washing with surfactants unless the adsorptive force is remarkably strong.

A novel surface is produced only when the chemical chains on the surface are in contact with their solvent or swelling agents. It is hoped that biocompatibility and the functionality of existing biomaterials can be improved using this approach.

New Generation of Synthetic Polymeric Materials Many materials are currently being synthesized and tested for biologic compatibility. In the following section we detail some of these materials, most of which are in the early stages of development for potential use in the medical environment.

Biodegradable Materials Materials being developed for use in sutures, small bone fixation devices, and drug delivery systems include lactide/glycolide polymers,[51] polydioxanone,[52] polyorthoesters,[53] polyanhydrides,[54] derivatives of poly(L-glutamate),[55] polyphosphazenes,[56] poly(ϵ-caprolactone),[57] pseudo polyamino acids,[58] and tyrosine–polycarbonates.[59] Future applications of degradable polymers may involve their use as scaffolds for the regeneration of specific tissues, such as liver, bone, cartilage, or vascular walls. In addition to processibility to produce devices of varying shape, the materials must be adhesive substrates for cells, promote cell growth, and allow retention of differentiated cell function. High porosity is a natural requirement, as it allows for cell seeding, growth, and ECM production together with the development of an organized network of cell constituents. [Some of these materials, including the polylactic materials, have been fabricated into three-dimensional porous biodegradable foams to mimic the natural extracellular matrix. The synthetic scaffold acts as a physical support and as an adhesive substrate for isolated cells during *in vitro* culture and subsequent implantation. Concurrently, the scaffold degrades continuously and is eliminated as the need for an artificial support diminishes.[60]]

Copolymers of PEO and poly(γ-benzyl-*L*-glutamate (PBLG) encapped with lactose have been produced.[61] This material has the same α-helical structural characteristics as the PBLG homopolymer and shows good adhesion to hepatocytes. The structure of such a polymer is shown in Fig. 11.13. It is suggested that such a molecule will be useful for investigating biological recognition phenomena and may have application in liver-targetable drug

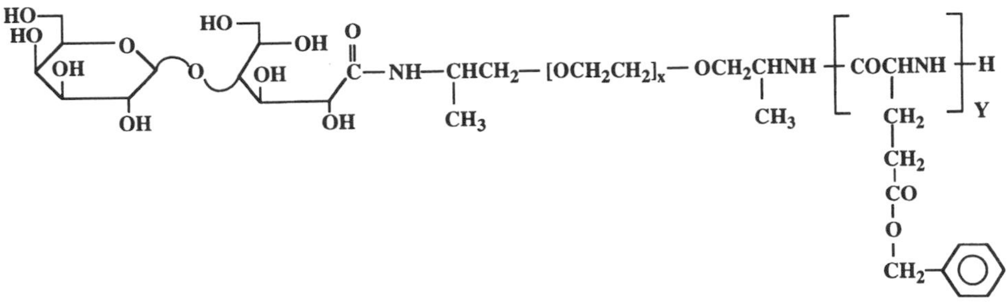

$$\text{Lactose}-C(=O)-NH-CH(CH_3)CH_2-[OCH_2CH_2]_x-OCH_2CH(CH_3)NH-[COCH(CH_2CH_2COOCH_2C_6H_5)NH]_Y-H$$

Figure 11.13 Polymer based on the copolymerization of PEO and poly(γ-benzyl-L-glutamate) (PBLG) encapped with lactose.

carriers. It is clear that the new families of synthetic polymers are being produced from an increased understanding of the needs for biocompatibility.

Poly(ethylene oxide) Star Molecules[62] Poly(ethylene oxide) (PEO) in a range of forms is known for its lack of interaction with biological material. It is soluble in both aqueous and nonaqueous media, including chlorinated hydrocarbons and benzene, it is able to solvate monovalent metal ions in nonpolar solvents, and in water its relative viscocity increases as its concentration decreases. It provides an inert surface in biological applications, including blood contact, although issues including knowledge of the packing density required to prevent ingress of other biopolymers is not known. In addition, if it is to be used as a molecular leash for bioactive molecules, including peptides, oligosaccharides, and antibodies, the concentration and availability of terminal hydrogel groups is important. If star molecules are considered, the properties of isolated star molecules diverge increasingly from those of linear counterparts as the number of arms on the star increases beyond three or four. This has an effect of increasing the number of arms within a given hydrodynamic volume (increased density) and increasing the number of terminal groups available for functionalization. Figure 11.14 shows a comparison of a linear PEO molecule as a random coil as compared with a PEO star. If external functionality is required, the core is synthesized first and the functional groups added afterward. Potential applications for these materials include their use as immobilized layers, which may serve only to prevent nonspecific adsorption of biological entities onto the underlying support, or in addition to serve as a leash to connect a bioactive ligand to a support. Applications for nonimmobilized PEO star molecules might be in the carriage of small bioactive molecules to be injected into the bloodstream. These could be size selected by ultrafiltration so that they are not rapidly eliminated by the kidney and hence remain within the body for a longer period. The free PEO star molecules could also be used as enhancers of antibody–antigen reactions.

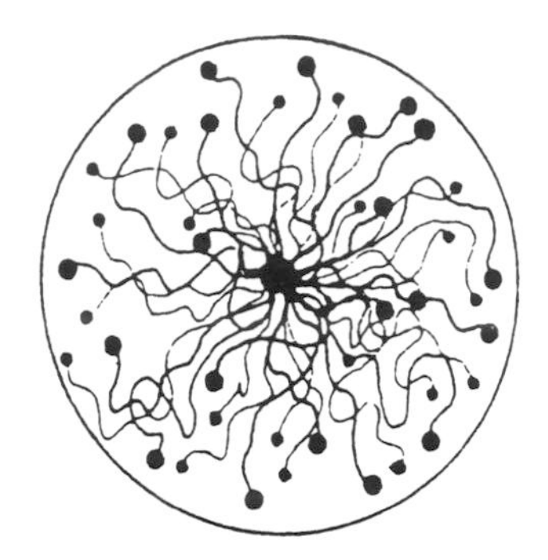

Figure 11.14 Comparison of linear PEO molecule as a random coil with a PEO star. (Redrawn from Ref. 62.)

Polymers from Natural Sources A complementary route to the design of new materials for biological applications is to take naturally occurring polymers and modify the structure of these materials. For this chapter, only materials based on collagen and hyaluronic acid are discussed. The major source of collagen for these materials is from calf skin. The major sources of hyaluronic acid are from calves and from shark fins.

Materials Based on Collagen Collagen is used for a variety of end uses in wound healing, ophthalmology, and dentistry and is used in film, membrane, solution, gel, and sponge forms. It is also used in conjunction with glucosaminoglycans, tricalcium phosphate, and allogenic bone, and with drugs such as tetracycline. When used in the treatment of skin ulcers, the collagen sponge matrix is found to enhance repair by organizing the spatial deposition of newly synthesized collagen and also accelerates the process of remodeling.[63] In the dental area, collagen membranes have the capacity to support regeneration of periodontal tissues, with the membranes either being incorporated during the healing tissues or being degraded during the healing process.[64]

Most research on collagen-based implants have required the implant to be a permanent feature in the human body. The materials used were partially degraded, enzyme-extracted collagen, or stabilized the collagen by cross-linking it with cytotoxic glutaraldehyde or chromium salts, or else assembled the collagen into nonnatural polymeric structures such as films and sponges. An alternative approach is aimed at maintaining as much of the biological and organizational structure of collagen as possible.[65] A novel method has been developed to make strong, continuous collagen threads at rates that permit the formation of collagen fibers. Collagen braids of up to 576 plies can be produced with successive braiding operations and tricot collagen fabrics produced. Both single-, double-, and triple-bar materials have been formed, each with different bulk and extensibility properties. The materials have been tested *in vivo* using a canine ligament model and a rat abdominal repair model. The biologic response after implantation is typical of that found as part of a normal wound-healing response, with the implanted devices continuing to perform their required function during resorption and tissue remodeling.

Materials Based on Hyaluronic Acid Hyaluronic acid is a nonsulfonated glycosaminoglycan consisting of alternate residues of *N*-acetyl-D-glucosamine and D-glucuronic acid, (Fig. 11.3). It occurs naturally and in highest concentration in soft connective tissue. The materials can be esterified (either 100% or fractional amounts), and the acid esters so produced have physicochemical properties which are significantly different from those of hyaluronic acid. Ethyl hyaluronic acid ester (HYAFF-07) and benzyl hyaluronic acid ester (HYAFF-11) seem to be promising biomaterials for applications in replacing short nerve segments and in wound healing. The materials can be used as threads, films, fabrics, and sponges, and additional applications are expected in plastic surgery and orthopedics.[66]

11.5.2 Metals

Metals have a large range of applications, including devices for fracture fixation, partial and total joint replacement, surgical instruments, external splints, braces, and traction apparatus as well as dental amalgams. The high modulus and yield point, coupled with the ductility of metals, makes them suitable for bearing large loads without leading to large deformations and permanent dimensional changes. They can be processed into final parts using a number of conventional techniques and have good resistance to degradation in a biological environment. Metals are crystalline materials and it is possible to incorporate a wide range of different atoms either within the crystal lattice, in interstitial sites, or at grain boundaries, and thus a multitude of metal-containing materials can be made. The deformation characteristics of the material are determined by the grain size of the material as crystal imperfections are concentrated at the grain boundaries. Mixing of metal atoms of different sizes as in the production of an alloy can serve to modify the properties of the metallic phase. Although the properties of metals can be modified by the chemical addition of different atoms to the alloy mixture, physical treatments can produce additional effects, as listed in Table 11.10.

The metals and alloys that were originally used in the medical environment were selected on the basis of their strength and ductility. Alloys may originally have been developed for earthmoving, aero- and space applications but have proved excellent in the biomaterials field. Metals are prone to corrosion and the compositions of the materials devised required resistance to corrosion under saline conditions. Alloying can be used to render materials passive to corrosion. Another method is to promote the formation of an

TABLE 11.10 Metal Treatment Processes

Treatment	Process and Result
Annealing	Thermal treatment below melting temperature; increases modulus and toughness
Grinding, polishing, and sand blasting	Removal of surface inpurities to improve surface aesthetics and properties
Precipitation hardening (carburization)	Formation of an oxide layer on heating a metal to elevated temperatures; increases tensile strength
Tempering	Thermal treatment by annealing followed by rapid cooling; increases toughness of brittle alloys
Work hardening	Mechanical conditioning below melting temperature; increases modulus

Source: Adapted from Ref. 41.

adherent oxide layer, often by acid treatment of the metallic phase. Both methods are currently used.

Steels were the first modern metallic alloys to be used in orthopedics.[41] In the initial stages of use there was a problem with corrosion, but this was overcome by modification of the composition of the steel by the addition of carbon, chromium, and molybdenum. Carbon is added at low concentrations (about 0.03 to 0.8%) to initiate carbide formation, while chromium (17 to 19%) facilitates formation of a stable surface oxide layer and molybdenum (2.0 to 3.0%) controls corrosion. Standard stainless steels are designated by ASTM for sheet and strip (F56, F139), bar and wire for surgical implants (F55, F138), where F55, F56, F138 and F139 have the same compositions. In addition to the elements above, the steels contain slightly differing levels of some or all of the following elements: manganese, phosphorus, sulfur, silicon, nickel, nitrogen, copper, and iron. The steels have high ductility and can undergo extensive processing after casting.

There are at least four compositions of cobalt-base alloys in use, which are designated by ASTM as F75, F90, F562, and F563. Again, these differ in the relative composition of the following elements: manganese, silicon, chromium, nickel, molybdenum, carbon, iron, phosphorus, sulfur, tungsten, titanium, and cobalt. These alloys are used because of their superior strengths, but they are difficult to machine and much more expensive to produce. Titanium-base alloys are found in many commercial medical devices and they are also used as coatings. Two alloy compositions are covered by ASTM specifications and are denoted F67 and F136. The latter phase is often referred to as Ti6A14V because it contains an average 6% aluminum and 4% vanadium. There is, however, some concern about the biocompatibility of this phase.

Applications in Orthopedics An area where these materials have been much used is in total-hip and total-knee replacements. Hip replacement consists of a femoral component that is ball mounted on a shaft and an acetabular component that has a socket into which the ball sits. Both cobalt–chrome and titanium–aluminum–vanadium (Ti6A14V) are used for the femoral component as well as with high-molecular-weight polyethylene to cover the socket. Knee replacements require a more complex design because of the complex loading pattern of the knee. The tibial component of the total knee replacement is fixed in the cancellous (spongy) bone of the tibia. The requirements for a successful knee replacement are for the femoral component to consist of a fairly thin, rigid shell with an attached fixation system to bone.[67] In some cases a central mechanical stabilizing element is present. The geometry of the femoral shell requires a stiff, high-strength, low-wear-rate material, which is typically a metal. The fixation system may be either a PMMA cemented type or biologic ingrowth type. The tibial portion consists of a broad plateau covering the tibia, after the bone of the subchondral plate is removed. In most designs, a composite, consisting of a stiff metal tray supporting a polymeric or fiber-reinforced polymer, is used. Fixation is achieved using PMMA cement, and therefore loss of fixation leads to mechanical failure. Problems associated with this sort of treatment include corrosion of the alloys and fatigue separation of coating elements from their base material. Other problems include component loosening at the bone, infection, mechanical instability, and restricted range of motion. Ceramics are being used for these applications to address some of the problems associated with the metallic phase. Problems with loosening of cemented components has led to the development of devices that are fixed to bone by porous ingrowth.

Applications in Dentistry In the mouth, a variety of metal alloys are used for repair of cavities and in the preparation of inlays, crowns, and bridges. Metal amalgams containing mercury are used extensively. The metals used include copper, silver, zinc, and tin, and a variety of compositions and mixtures are used. Failure of these amalgams is a result of excessive expansion, leading to interfacial separation between the enamel and the alloy. Noble metal alloys are also used in dental appliance construction and contain gold, silver, copper, palladium, platinum, and zinc atoms.

11.5.3 Bioceramics

During the last 40 years a revolution has occurred in the use of ceramics. The revolution is the development of specially designed and fabricated ceramics for the repair and reconstruction of diseased, damaged, and "worn-out" parts of the human body. Most clinical applications of bioceramics relate to the repair of the skeletal system, composed of bones, joints, and teeth, and to augment both hard and soft tissues. Ceramics are also used to replace parts of the cardiovasular system, especially heart valves, and also in the treatment of tumors.[68] A schematic of the clinical uses of bioceramics is shown in Fig. 11.15.

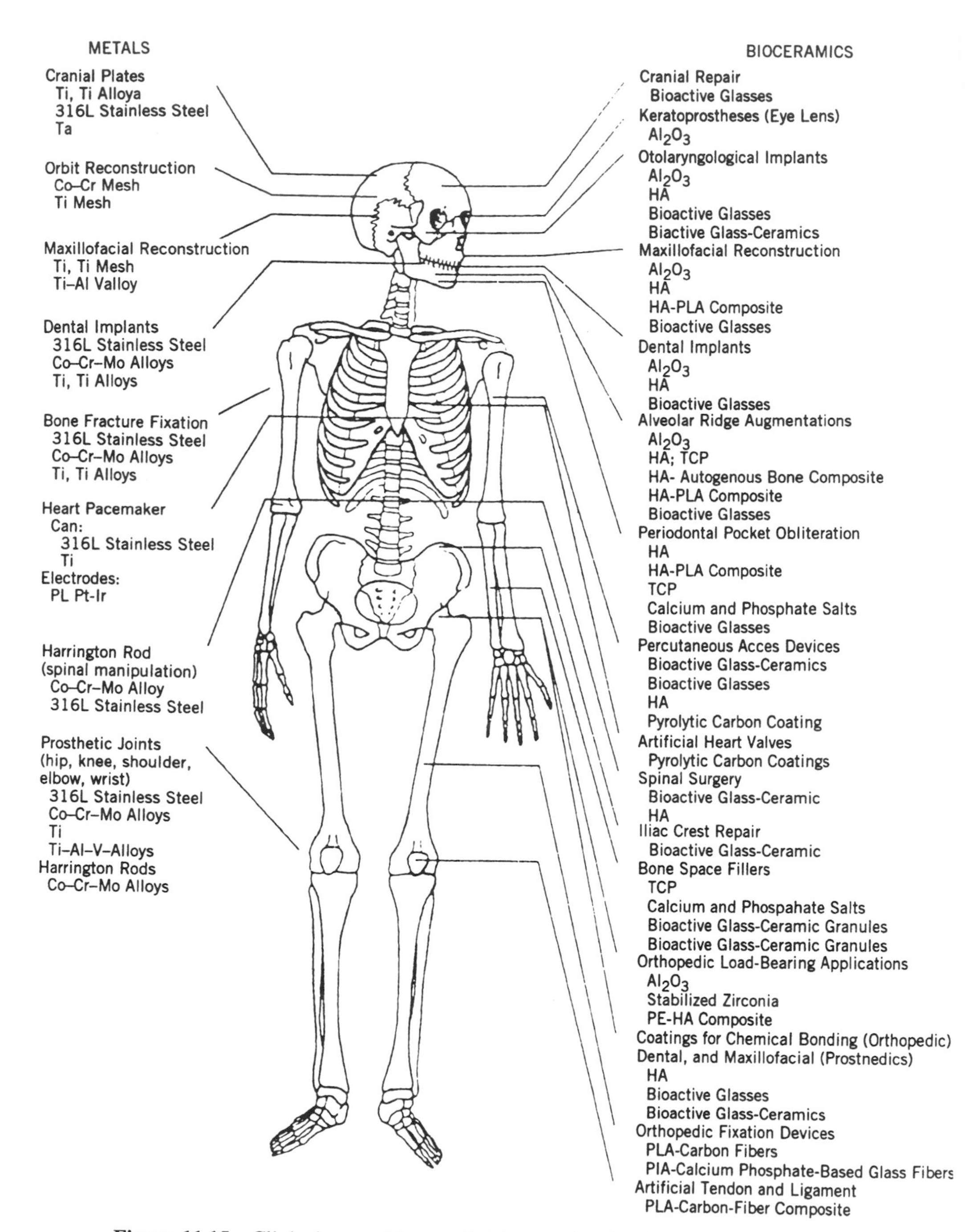

Figure 11.15 Clinical uses of inorganic bioceramics. (Redrawn from Ref. 69.)

Ceramics include inorganic materials such as silica or other metal oxides in crystalline and glassy phases similar to polymers. Salts such as hydroxyapatite are considered as ceramics but are dealt with principally in Section 11.5.4. Carbons are another class of ceramics. Ceramics are stiff, brittle materials that are insoluble in water. Their primary advantages include high strength and hardness, but their downside is that they deform minimally under loading and therefore fracture easily. These materials can only be used where medical applications require limited deformability. When these materials are used in the human body, they are termed bioceramics and are made in a variety of forms (powder, coating, and bulk) and phases. They can be single crystals (sapphire), polycrystalline (alumina or hydroxyapatite), glass (Bioglass), glass–ceramics (A/W glass–ceramic), or composites (polyethylene–hydroxyapatite). The material used depends on the application required. For example, single-crystal sapphire is used as a dental implant because of its high strength. A/W glass–ceramic is used to replace vertebrae because it has high strength and bonds to bone. Bioactive glasses have low strength but bond rapidly to bone so are used to augment the repair of boney defects. It should be noted that in all applications, implants must remain within the human body for many years and the long-term stability of the materials used (in the biological environment in which they are found) must always be considered.

Bioinert Versus Biocompatible Materials Early studies on bioceramics were based on the idea that as ceramics were in an oxidized state (e.g., oxide phases) they would not degrade in the human body, nor would they be included in biological reactions. Hence the search was for bioinert ceramic materials. As ceramic materials are also characterized by a high hardness, they were used for the mechanical parts of joint prostheses in addition to the prostheses themselves. The material of choice was alumina, which was first used in total hip replacement in 1970.[70,71] Its necessary strength and wear resistance can be achieved only by using highly pure and highly dense materials with 3 to 4 μm grains. In addition, manufacture and implantation must obey strict criteria for optimal success. Alumina–alumina combinations provide very good resistance to wear compared with metal–polyethylene and alumina–polyethylene combinations, but not all patients are suitable to receive such implants.

However, no material implanted in living tissue is inert; all materials elicit a response from the host tissue, and therefore materials used should be *biocompatible* with the host tissue. Hench[72] proposed the hypothesis that the biocompatibility of an implant material would be optimal if it promoted the formation of normal tissue at its surface, and in addition, if it would establish a contiguous interface capable of supporting the loads that normally occur at the site of implantation. Hence the current goal is to produce materials that are recognized and assimulated by the body. A high degree of bioactivity is desirable and is measured by the "ability of a material to form surface hydroxyapatite in a controlled *in vitro* environment."

Tissue Response to Bioceramics Tissue response to diverse bioceramic compositions leads to four primary categories of behavior.

1. For toxic materials, the surrounding tissue dies.
2. For nontoxic materials and those that dissolve, the surrounding tissue replaces the implant.
3. For biologically nontoxic and inactive materials, a fibrous tissue capsule of variable thickness forms.
4. For nontoxic and biologically active materials, an interfacial bond forms.

Clearly, category 4 behavior would be favored for almost all applications where an implant is in contact with living tissue.

Conventional Ceramics These include oxides and salts of metallic elements and silicon together with carbons. The primary use of inorganic biomaterials is the repair, replacement, or augmentation of diseased or damaged parts of the musculoskeletal system, such as bones, joints, and teeth. Materials are often used as implants or prostheses and the form of the biomaterial depends on its intended function in the body. Load-bearing implants are usually made from bulk, nonporous materials, but coatings or composite structures may also be used to achieve improved mechanical and interfacial chemical properties. Implants that serve only to fill space or augment existing bone tissue are used in the form of powders, particulates, or porous materials. Other clinical uses include pyrolytic carbon coatings on prosthetic heart valves and as radioactive glass beads in the treatment of tumors. A major problem, as with all implants, is the reduced lifetime of the implants, which can, for the large part, be traced to the instability of the interface between the implant and its host tissue. This is due to a chemical and mechanical mismatch between the implant and living tissues. For metallic implants corrosion leads to toxicity and migration of metallic ions leads to additional toxic responses, not necessarily at the site of the implant.

Ceramics, especially oxides, eliminate the problem of metallic corrosion, and high-purity alumina (Al_2O_3) was one of the first oxide materials to be developed for use in load-bearing orthopedic prostheses. This material does not, however, offer any advantage over metallic implants with respect to interfacial stability, as a thin fibrous capsule forms at the interface with bone.

Porous implants improve interfacial stability, as bone will grow in pores greater than 100 μm in diameter and a blood supply can be maintained. A downside of such materials is that large pores severely diminish the strength and toughness of the implant. Porous coatings applied to metals in some ways provide the best of both worlds, as ingrowth of bone occurs at the porous interface and the mechanical load is carried by the bulk–metal substrate.

The other ceramic widely used is hydroxyapatite (see also Section 11.5.4), which is conventionally prepared by high-temperature routes. The calcium phosphate phase found in bone is a microcrystalline, carbonate-containing (4 to 6%) calcium-deficient, structurally disordered hydroxyapatite. The imperfect crystalline structure results in the bone mineral being soluble and reactive with respect to body fluids. In contrast, the calcium phosphates used as implant materials, as a result of their being prepared at high temperatures, are carbonate free and are made up of crystals that are much larger and more perfect than those found in biological apatites. A consequence of this is that materials are much less reactive than those found in living tissue and problems with biocompatibility can arise.

There are other ceramic materials that are being developed for medical applications, including materials based on zirconia (partially stabilized zirconia), carbon-composite ceramics, boroaluminosilicate ceramics (Macor and derivatives, and silicon/aluminum nitride/aluminum oxide (Sialon) materials.[73]

Partially Stabilized Zirconia Pure zirconia is impossible to sinter because of unfavorable phase transitions which occur on cooling the oxide from the molten state to room temperature. Below 1000°C a phase change from tetragonal to monoclinic results in an increase in unit cell dimensions and concommitant decrease in density, making it impossible to densify the material fully. Stabilization of the oxide by the addition of magnesia or yttria (3 to 9%) results in a cubic phase that can be sintered to full density. The ceramics produced have lower Young's modulus, higher strength and fracture toughness, and smaller grain size[73] than for the single-element oxide alone.

Carbons and Carbon-Composite Ceramics Glassy or vitreous carbons (showing different degrees of short-range order) and pyrolytic carbons are used in the production of artificial heart valves. For orthopedic applications, three different kinds of carbon and carbon-containing composites have been developed. High-strength isotropic carbon is a glassy carbon that retains some porosity and has relatively good wear resistance. It is manufactured by pyrolysis of a semicoke. The silicon carbide/carbon composites are manufactured by impregnating a porous carbon with liquid silicon. Carbon-reinforced carbons and other reinforced carbons are made by impregnating the filaments, after orientation if desired, with a pitch, compressing, heat-treating, and carbonizing or graphitizing. Composites such as CFSiC admixed with SiC lead to materials with Young's moduli close to that of bone. They exhibit biological and mechanical stability, and they are being investigated as candidates for hip joint replacement.

Mica Glass Ceramics[74] The first machinable glass–ceramic containing mica as the main crystal phase was developed by Beall et al.,[75] and is termed Macor. The composition of the base glass is 47.2 SiO_2, 8.5 B_2O_3, 16.7 Al_2O_3,

14.5 MgO, 9.5 K_2O, and 6.3 F^-. Another glass–ceramic phase that has been developed is Dicor, which contains significant amounts of the alkaline-earth oxides. The material shows excellent machinability, translucency, and good bending strength. It is specifically used in the production of crowns for dental work. Bioverit II is a phlogopite-type glass–ceramic with two major crystal phases, mica and cordierite present in the glassy ceramic. The translucency, mechanical properties, and thermal expansion coefficient can all be regulated in accordance with the required medical use. The material is biocompatible but not bioactive and is used in middle-ear implants and in dental work. Bioverit I is a mica–apatite glass–ceramic and offers the possibility of a machinable glass together with bioactivity (see the section "Bioactive Ceramics" below). These glassy ceramics contain the following oxides in variable amounts: SiO_2—Al_2O_3—MgO—Na_2O—K_2O—F^-—CaO—P_2O_5. Both Bioverit materials have been used successfully in head and neck surgery and in orthopedic surgery.

Silicon Aluminum Nitride / Aluminum Oxide (Sialon) Sialon ceramics are based on the β-Si_3N_4 phase, with relatively large amounts of aluminum oxide in solid solution. The addition of the aluminum oxide results in an increase in unit cell dimensions for the nitride phase, and the material has higher fracture toughness than for the nitride phase alone and is currently under investigation in both fiber and powder form.

Bioactive Ceramics Bioactive ceramics are defined as those which are nontoxic and biologically active and which favor the development of an interfacial bond between implant and tissue. Four major categories of materials have been developed: (1) dense hydroxy(l)apatite (HAP) ceramics (see Section 11.5.4), (2) bioactive glasses, (3) bioactive glass–ceramics, and (4) bioactive composites. Each of these materials develops a bond to living bone, with bond thickness varying from 0.01 to 200 μm. The physical properties and biomechanics at the implant–tissue interface vary so much that all four types of material have found use in a variety of clinical applications.

Bioactive Glass: Bone Bonding The general theory of bone bonding of bioactive glasses is due to the establishment of a competitive advantage for osteogenic precursor cells over fibroblasts. It is suggested that this is due to the presence of biologically active silanol, calcium, and phosphate sites on the surface. When a bioglass is exposed to water or body fluids, several key reactions occur. Cation exchange of sodium and calcium ions in the glass for protons from the surrounding solution results in hydrolysis of surface groups and leads to interfacial dissolution. As the solution becomes more alkaline, repolymerization of the silanol groups occurs, producing a silica-rich surface layer. Another direct consequence of the high pH at the glass solution interface is that calcium and phosphorous oxides which have been released into solution during network dissolution crystallize into a mixed hydroxycarbonate apatite on the surface. It is proposed that crystallites of HCA phases bond to interfacial metabolites such as mucopolysaccharides and collagen. It is hypothesized that this incorporation of organic biological constituents within the growing HCA- and silica-rich layers appears to be the initial step in establishing bioactivity and bonding to tissues.

Conventional Bioactive Glasses Bioactive glasses are conventionally prepared by the traditional methods of mixing particles of the oxides or carbonates and then melting and homogenizing at temperatures of 1250 to 1400°C. The molten glass is cast into steel or graphite molds to make bulk implants. A final grind and polish are often required. If powders are required, these are produced by grinding and then sieving to achieve the desired particle-size characteristics. A common characteristic of the materials that have been used is the chemical components, which include CaO, P_2O_5, Na_2O, and SiO_2. The bonding to bone has been associated with the formation of hydroxyapatite (HAP) on the surface of the implant. Although a range of compositions can be used (up to 60% silica), an even narrower range of compositions bond to soft tissues. A characteristic of the soft-tissue bonding compositions is the very rapid rate of HAP formation. This has previously been attributed to the presence of Na_2O or other alkali cations in the glass composition, which increases the solution pH at the implant–tissue interface and thereby enhances the precipitation and crystallization of HAP. The rate of HAP formation has also been shown to be strongly dependent on the ratio of SiO_2, the glass network former to Na_2O, the network modifier in the glass. When the glass contains over 60% SiO_2 or more, bonding to tissues is no longer observed. The solubility and chemistry (including diffusion of Na^+ ions, for example by the addition of La_2O_3) of the glass phase can be modified by the incorporation of other phases.

Problems that have been associated with the conventional high-temperature method of production arise from:

1. Highly charged impurities such as Al^{3+}, Zr^{4+}, Sb^{3+}, Ti^{4+}, and Ta^{5+}, which can be picked up at any stage of the preparation process. This phenomenon is also related to the low silica and high alkali content of the traditional bioactive glass compositions. The incorporation of impurity ions leads to dramatic reductions in bioactivity.
2. Processing steps such as grinding and polishing expose the bioactive powder to potential contaminants.
3. There is a compositional limitation on materials prepared by the conventional high-temperature methods, due to the extremely high equilibrium liquidus temperature of SiO_2, 1713°C, and the extremely high viscosity of silicate melts with a high silica content.
4. High-temperature processing leads to increased processing costs.

Sol-Gel Routes to Bioactive Glasses Low-temperature sol-gel processing (see Chapter 9) offers an alternative to conventional high-temperature mixing and melting methods. The methods used to make sol-gel materials include (1) gelation of collodial powders, (2) hypercritical drying, and (3) controlled hydrolysis and condensation of metal alkoxide percursors followed by drying at ambient pressure. Advantages include ease of powder production, a potentially broader range of bioactivity due to changes in composition and/or microstructure through manipulation of the processing parameters. The materials so produced can be sintered at relatively low (600 to 700°C) temperatures, thus giving much better control over the purity of the resultant materials.[72, 76] It has been possible to reduce the number of chemical components required to produce bioactivity from four to three by this route. $CaO—P_2O_5—SiO_2$ gels have been produced with silica content from 50 to 90% and calcium oxide from 46 to 1%, all with 4% phosphorous pentoxide. *In vitro* studies of such materials using the rate at which hydroxyapatite is formed on the surface of the particles have shown that high-calcium-content gels produce higher levels of bioactivity over an extended period of time. This will be of particular importance for the bonding of biogels to soft-tissue samples.

Materials produced by the sol-gel route show much higher levels of bioactivity for correspondingly lower levels of calcium oxide within the glass. The high bioactivity is associated with materials of high surface area (> 100 m^2/g for binding to bone) and pore volume (0.3 to 0.6 cm^3/g). It is thought that these ultrastructural features may give rise to an increased density of potential nucleation sites for formation of the hydroxyapatite layer. It may also be the case that the porous materials provide conveniently sized loci for bone cells to grow and spread in.

Problems of Longevity of Implant Use: Toward Resorbable Implants Biomaterials used to repair the body need to last as long as the patient does. At present this is not the case, and some people may face several hip replacement operations, for example, each time there being less bone material (or less healthy bone material) for incorporation of devices. The current life expectancy of such replacements is on the order of 10 years. This needs to be doubled or tripled in the future. None of the materials described above is able to address the problem of tissue alteration with age and disease. The skeletal system has the capacity to repair itself, this ability diminishing with age and disease state of the material. The ideal solution to this problem is to use biomaterials to augment the body's own reparative process. Certain of the resorbable implants, such as tricalcium phosphate, and some bioactive glasses are based on this concept. Problems that exist with the development of resorbable materials are that (1) the products of resorption must be compatible with cellular metabolic processes and (2) the rate of resorption must also be matched by the capacities of the body to process and transport the products of this process. In addition, as the material is resorbed and new material formed, the properties of both phases will alter and compatibility must be maintained at all times. This is difficult to achieve.

Questions Relating to the Use of Bioceramics in the Human Body

Literally hundreds of questions remain unanswered with respect to the use of bioceramics in the human body. The following areas were defined by workshops held at a bioceramics meeting in 1986 and published as a volume by the Annals of the New York Academy of Sciences.[77]

1. *Bioactive glasses, aluminum oxide and titanium:* (a) ion transport phenomena and surface analysis, and (b) biochemistry of the interface
2. *Significance of the porosity and physical chemistry of calcium phosphate ceramics:* (a) biodegradation–bioresorption, (b) dental and other head and neck uses, and (c) orthopedic uses
3. *Biomechanical stability and design:* (a) stiffness and remodeling, (b) strength, and (c) wear.

Questions Relating to the Biochemistry of the Tissue–Implant Interface

Detailed questions that were proposed for future research in the area of the biochemistry at the interface between tissue and implant for oxide phases are given below.[78]

Questions relating to the *formation of the initial interface* include:

- How does the nature of the surface influence the process of hydration and the organization of water?
- How does the gel layer that forms at the surface modulate ion flux between the implant and the cells?
- What are the adsorption–desorption characteristics of organic constituents at the surface?
- How does the initial gel layer alter the subsequent composition of the organic and inorganic components of the gel layer?
- Do the chemical and physical characteristics of the gel layer influence the migration, attachment, and differentiation of cells?
- By what mechanisms do the cells influence the maturation of the interfacial material?

Biomaterial–tissue interfaces are dynamic throughout the life of the implant because they are continually modulated by cell metabolism and tissue remodeling. After the effects of implantation and associated inflammation (which always occur) have dispersed, the interface is more-or-less stabilized, but many questions pertain to this stabilized interface. Questions include:

- How firmly does the interfacial material attach to the surfaces of biomaterials?
- Is there a quantitative or qualitative difference between the proteoglycans and glycosaminoglycans on the surfaces of biomaterials and those between natural tissues?
- Is there epithelial attachment to bioactive surfaces other than calcium phosphate ceramics?
- What is the nature of the interface *in vivo* at the ultrastructural level and biochemically?
- What is the relative contribution of epithelium and connective tissue to successful function of the permucosal and percutaneous implants?

With respect to bone:

- Does the presence of an implant material alter the kinetics of bone remodeling?
- Bone GLA proteins appear to accumulate on bioactive but not on inert surfaces; is this coincident with the observed change in collagen from type III to type I or with mineralization of the interface?
- How do bone-inductive proteins interact with active surfaces? How do they interact with inert surfaces? How might they affect the dynamics of bone modeling?
- Can these interactions be used to facilitate treatment of osteoporotic and other diseased bone?
- Is the mineral phase of bone continuous with the implant surface?
- What is the role of noncollagenous matrix proteins in the process of mineralization? What is the role, if any, of silicon in the mineralization of the interface? How is this ion transported to the interface, and how long does it take to accumulate at the mineralization front? Can this accumulation be aided by silicon-containing biomaterials?

Finally, questions relating to implant effect on blood:

- How do the surfaces of biomaterials moderate the deposition and turnover of plasma proteins?
- Can alterations in the surfaces of biomaterials (e.g., chemical and electrical changes) control this?
- Does the presence of a biomaterial affect the range and quantity of extracellular materials produced by the cells at the interface?

- Does the presence of a ceramic biomaterial affect the normal process of revascularization in the adjacent tissue?

Clearly, we are a decade from the date of this meeting and the workshops that formulated these questions. Some of the questions posed are applicable to *all* types of biomaterial, whether of inorganic or organic origin, and there is no doubt that there are at least partial answers to many of the questions proposed above, but much remains unknown.

11.5.4 Composites

Composite materials are used clinically to take advantage of the desirable properties of each of the constituent materials, while limiting the undesirable or deleterious properties of the individual phases. Composites cover a wide range of compositions, and representative materials are listed in Table 11.6. Most of what is discussed in this section will relate to bioceramic composites. Bioceramic composites are either bioinert, bioactive, or biodegradable. Examples of each of these classes of composite are given in Table 11.11.

The ceramic phase can be the reinforcing material, the matrix material, or both. The incorporation of high-strength fibers increases the mechanical strength of the composites while maintaining the bioactivity of the material.[80] In the case of glass doped materials, the fracture toughness of the material increases dramatically and renders materials suitable for dental implantation and hip replacement therapy. The chemical composition of the fibers can be important in establishing continuity between the metallic component and the coating, with titanium being an especially good candidate to achieve such an effect.

TABLE 11.11 Bioceramic Composites

Category	Examples
Inert	Carbon fiber–reinforced carbon
	Carbon fiber polymeric matrix materials [polysulfone, poly (aryl)ether ketone]
	Carbon fiber–reinforced bone cement
Bioactive	A/W glass–ceramic
	Stainless steel fiber–reinforced Bioglass
	Titanium fiber–reinforced bioactive glass
	Zirconia-reinforced A/W glass–ceramic
	Calcium phosphate particle–reinforced polyethylene
	Calcium phosphate fiber–and particle-reinforced bone cement
Resorbable	Calcium phosphate fiber–reinforced polylactic acid

Source: Ref. 79.

Composites Based on HAP There are many applications for calcium phosphate bioceramics, such as:

1. *Craniofacial applications*
   a. Augmentation
      - Ridge
      - Mandibular
      - Zygomatic
      - Chin
   b. Reconstruction
      - Peridonatal reconstruction
      - Mandibular reconstruction
      - Orthognathy
      - Bone grafting
      - Cranioplasty
      - Orbital floor reconstruction
      - Anterior nasal spine reconstruction
2. *Prosthetic implants*
   a. Subperiosteal implants
   b. Endosteal implants
      - Endosseous implants
      - Endontic pins
      - Orthodontic pins
   c. Transosseous implants
      - Transmandibular implants
3. *Otological applications*
   a. Ossicular reconstruction
   b. Canal wall prosthesis

The form of calcium phosphate used in orthopedic clinical applications is usually based on hydroxyapatite and β-tricalcium phosphate. The materials are widely used in composite formulations together with:

1. *Ceramics:* mixed calcium phosphates, calcium sulfates, zinc calcium phosphates, aluminum calcium phosphates, metacalcium phosphates, sodium metacalcium phosphate, calcium carbonate, magnesium calcium carbonate, and magnesium carbonate
2. *Biological derivatives:* bone derivatives (autografts, allografts, and xenografts), collagen, dura, fibrin, amino acids, polyfunctional acids, inductive factors (bone morphogenic protein), growth factors (bone, epidermal tissue, cartilage, platelet, insulin)
3. *Therapeutic agents:* hormones, antibiotics, chemotherapeutic drugs

4. *Synthetic polymers:* polylactic acid (PLA), polyglycolic acid (PGA), polycaprolactone (PCA), polyamino acids, polyethylene (PE) and high-molecular-weight derivatives, polysulfone, polyhydroxybutyrate
5. *Metals:* titanium-, cobalt-, and iron-based alloys

These materials can be developed in the form of particulates with a range of porosities, moldable forms, block forms, scaffolds, fibers, and coatings. The clinical applications range from bone graft substitutes for augmentation and replacement, to fracture fixation materials and coatings for fixation to materials for drug delivery implants. Figure 11.16 illustrates the use of HAP together with metal plates and pins in the treatment of broken bones. Although most methods of use require the HAP to be prepared prior to implantation, a process has recently been developed for the in situ formation of the mineral phase of bone.[81] In this newly developed approach, inorganic calcium and phosphate sources are combined to form a paste that is surgically implanted by injection. Under physiological conditions, the material hardens in minutes and dahllite formation is almost complete within 12 h. The material is remodeled *in vivo* and is being tested in human trials for various applications, including the repair of acute fractures.

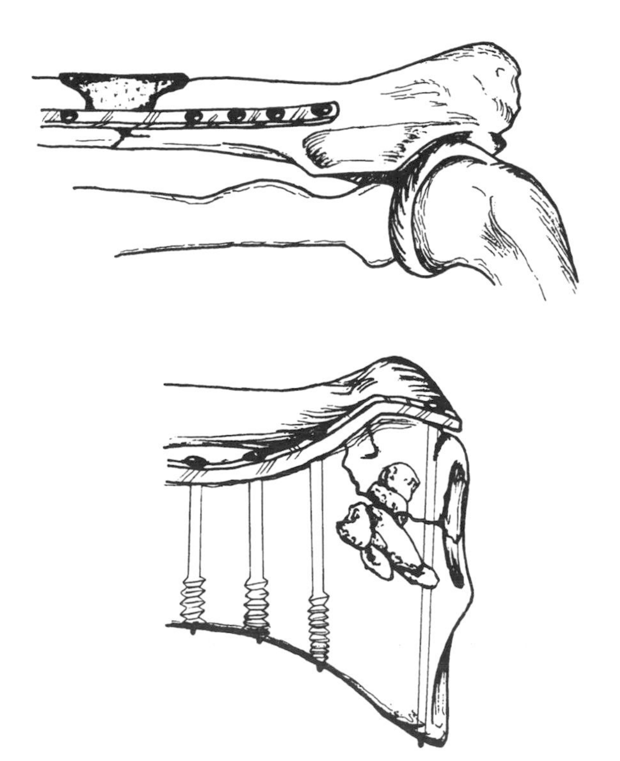

Figure 11.16 Fracture repair with metal plate and pins together with HAP filler.

Bone Graft Materials Bone graft substitutes are available based on alumina chemistry (see Section 11.5.3), silica (glass-ionomer and bioactive glasses; Sections 11.5.3), synthetic and natural calcium salts (phosphate, carbonate, sulfate, and hydroxide), and these materials combined with natural polymers (collagen, Sections 11.2.1 and 11.5.1), and synthetic polymers (PMMA, PHEMA, UHMWPE; see Section 11.5.1). The subject is well reviewed by Constantino and Friedman.[82] Both sintered and nonsintered materials based on calcium phosphate are available, with the nonsintered versions showing greater biocompatibility (simply due to better resorption characteristics). Materials can be produced with a high degree of porosity, thus mimicking natural bone and allowing cells to permeate the implanted material over time. These can be produced from natural corals, where the biomineralized skeleton of calcium carbonate is replaced by calcium phosphate. Examples include Interpre 200 and 500, manufactured by Interpore International, Irvine, California, and the materials are nearly nonresorbable. The same coral-based materials (Biocoral, manufactured by Inoteb CY, Saint-Gonnery, Noyal Pontivy, France) can be used in their calcium carbonate form without modification and the same materials can be resorbed and replaced by bone (fibro-osseus tissue). Calcium sulfate has been used in clinical medicine for over 100 years, usually as a casting material for fractures. It could also be used internally in filling bone cysts and for cranial defect reconstruction, although there is current FDA approval for use in the latter area of research. There is FDA approval for use in dental repairs together with porous ceramic hydroxyapatite granules (Hapset). The calcium sulfate is resorbed and can be replaced with the osseous tissue growing around the HAP granules and holding them in place. Another biomaterial that makes use of calcium hydroxide together with PMMA coated with PHEMA is the hard-tissue replacement polymer HTR. The implant material consists of PMMA beads sintered together to give a porous mass, which is then coated with PHEMA and calcium hydroxide. The PHEMA coating absorbs a lot of water, and a gel is formed at the surface containing calcium ions. This material is very biocompatible.

Other alternatives for implants are based on natural bone rather than the synthetic derivatives. For example, ashed bone in combination with ultrahigh-molecular weight polyethylene (UHMWPE) is being evaluated for human implant applications, including total hip surgery.[83] This material is designed to be used as a coating on a porous implant for the purpose of biological fixation in total hip replacement. Biological fixation is conceived as a potentially more promising technique than the acrylic cement fixation method, particularly if the problem of micromotion which leads to the formation of fibrous tissue in place of bone ingrowth in the early stage of healing is fully overcome. Shock waves caused by heel strike are sufficient to cause osteoarthritic changes and are a serious factor contributing to hip joint loosening. The problems associated with micromotion and shock wave transients may be overcome by the employment of (1) the enhancement of bone ingrowth at the early stage of healing postimplantation and (2) the application of a high-damping material to attenuate all vibrations and transmitted shock waves at the bone–implant interface. A composite of ashed bone–UHMWPE satisfies both criteria. Ashed bone provides the needed bioactive properties for this application, and UHMWPE affords the required high-damping properties. The presence of the HAP, a bioactive material with resorption properties, enables a strong interfacial bond to develop which stabilizes the implant for an extended life. Biologically produced HAP appears better than synthetic HAP, probably due to the impurities present in the material—clearly an area for investigation. UHMWPE is the polymer of choice for the matrix material because of its abrasion resistance, impact resistance, corrosion resistance, and zero moisture absorption.

Research Directions in Biocomposites Important subjects for research in this area include the need to understand the details of the process of bone formation and to identify the key components in order to attempt to mimic this complex phenomenon and to produce a composite organic ceramic with characteristics similar to bone but in the absence of living cells. Several approaches can be identified, including (1) investigations into methods of alignment of isolated osteoblasts (the bone-forming cells) into sheets of cells that secrete sheets of mineralizable matrix, (2) natural matrices such as those found in turkey tendon or ligament used to concentrate inorganic molecules on or in these natural matrices to form bonelike substances,[84] and (3) the use of artificial matrices, including collagen gels, to understand the principles by which calcium phosphate can form within such materials. The goal is to fabricate matrices that both support and control mineral deposition to create an artificial composite ceramic with the physical characteristics of bone without the interface of living cell.[85]

In this area it is suggested that long-term fundamental materials research be devoted to studies of the mechanisms of the exchanges between the biomaterials and the tissues. To this end it will be necessary to utilize synthetic models (as in biomineralization studies) which progressively approach the natural systems. In the clinical environment extensive research is needed on the various substances and combinations of substances to establish safety and efficacy for the proposed dental and orthopedic applications.

11.6 OTHER APPLICATIONS

Although many conventional materials, such as polymers, metals, ceramics, and glasses, are used in the production of biomaterials, biology is able to provide new routes to advanced and "smart" materials for a wide range of applications in the chemical, pharmaceutical, and electronics industries. Two examples are given for illustrative purposes.

11.6.1 Drug Delivery Systems

Drug delivery technology requires the pharmaceutical, chemical engineering, biomaterials, and medical communities to work together. The ultimate objectives of the technology are to enhance the safety and effectiveness of the pharmaceutical agents. Three main approaches to drug delivery have been developed.

1. The chemical conversion of a drug to an inactive or pro-drug form to enable storage, introduction, and delivery to the target area. Conversion to the active drug is controlled by intrinsic physiological processes, which therefore reduces inappropriate drug activity.
2. The drug is immobilized on simple soluble macromolecules that exhibit intrinsic targeting properties. Delivery of the active drug bound to the macromolecular carrier is controlled by processes intrinsic to the carrier and endogenous ligands for the carrier molecules.
3. The drug is immobilized or entrained by more complex particulate multicomponent structures which act largely as protective devices and shield the drug from degradative processes during transit.

In the latter category are drugs immobilized by organic moieties, drugs immobilized within wholly inorganic matrices, and drugs immobilized on inorganic–organic composite materials. The volume edited by Tirrell et al.[86] offers detailed information on the use of liposomes, polysaccharides, poly(amino acids), and other polymeric materials in the targeting of drugs to specific sites within the body. Collagen in a variety of forms is used in drug delivery. Collagen-based drug delivery systems include injectable microspheres based on gelatin (degraded form of collagen), implantable collagen–synthetic polymer hydrogels, interpenetrating networks of collagen, synthetic polymer collagen membranes for opthalmic delivery, and collagen–liposomal composites.[87] More recent research has looked at the possibility of using activated charcoal,[88] calcium hydroxapatite,[89] and zeolites[90] in drug delivery. The greatest limiting factor in the development of such systems is the inherent bioincompatibility of the inorganic phase and in the usually destructive interactions that occur between the drug carrier and the drug. Current thought tends to suggest that it may be possible to overcome this problem by coating the surface of inorganic particulates with an organic-based "natural" coating that will be recognized by both the mineral phase and the drug phase. These materials may work well because of their ability to provide a surface of structured water that will enable carriage of the "drug" in its natural state. This situation will maintain the molecules in their correct conformation and maximize the potential for drug–receptor interaction at the required site.

11.6.2 Engineered Protein Thin Films

An alternative definition of the term *biomaterials* as used within the materials community arises from the use of molecular or solid-state biological structures in the development of functional or "smart" materials for a range of applications, including electronic devices. One example from the area of protein thin film assembly will serve to illustrate the potential.

Biological macromolecules such as proteins show extremely high selectivity in recognition processes and as such make excellent building blocks on the nanometer scale in the construction of molecular devices for sensing applications. Although biology provides some systems, such as the photosynthetic reaction center, which can be used more or less as they are in molecular devices, most proteins do not have the required physical and chemical properties for use as molecular building blocks. Recombinant DNA technology can be used to provide the essential synthetic control over protein structure–function relationships and enables the manipulation and optimization of key features in biomolecular assemblies. The electrical and optical properties of the components (protein molecules) and the assembly patterns of the biomolecules can all be regulated. This is done by site-directed mutagenesis of particular amino acid residues and/or changes to the prosthetic groups. These may then be combined on surfaces with the hydrophilic and hydrophobic characteristics of the amino acid side chains determining the pattern of interaction of the molecular components.[91]

11.7 BIOCOMPATIBILITY TESTING

The development of materials that are intended for use in medical devices is a complicated process, as most devices are produced using design criteria not based on sound scientific principles. Most of the materials used currently have been adapted from aviation, marine, and automotive engineering. A limit has been reached in adapting these old materials to new applications, and the next generation of materials used in medical devices will be designed on analysis of properties of normal tissues, hence the emphasis on the structural components and processes involved in the formation of tissues within the body. The length of time required to introduce any new material into general clinical practice can vary from several to over 10 years, depending on the application.

11.7.1 Safety and Efficacy Testing

The testing conducted on a biomaterial intended for use in a medical device must address safety and effectiveness criteria as outlined in several recently published texts.[41, 92] The specific tests required vary with the type of device

TABLE 11.12 Biologic Tests Used to Evaluate Biocompatibility

Cell culture cytotoxicity	Skin irritation
Intramuscular and subcutaneous implantation	Blood compatibility
Hemolysis	Carcinogenesis
Long-term implantation	Mucous membrane irritation
Systemic injection acute toxicity	Intracutaneous injection
Sensitization	Mutagenicity
Pyrogenicity	

and application, but Table 11.12 lists potentially applicable biocompatibility tests. Information on the theory of these tests and the biological systems that are involved can be found in Chapter 4 of *Biocompatibility*.[93] All of the tests listed in Table 11.12 have ASTM standards that must be achieved if a biomaterial is to be used in a medical device. Additionally, medical devices and hence materials to be used in particular medical devices come in three categories, and testing is different for all three (Table 11.13).

Table 11.14 shows an example of the acceptable outcome for the preclinical tests on a wound dressing submitted for a premarket approval (PMA) application. All devices not equivalent to a device on the market before 1976 are automatically considered as class III if they are implantable. As can clearly be seen, the tests required for biocompatibility testing will vary with the material in question and its intended end use. A consequence of present litigation in respect to materials implanted in the body is that the tests required for the acceptance of new biomaterials are more stringent, and hence new materials are more costly to develop for the marketplace. An understanding of the chemistry and biology of both the materials and their intended host will expedite the process and is likely to produce fewer long-term problems with respect to implant–biomaterial rejection.

TABLE 11.13 Classification of Medical Devices

Class	Types of Device	FDA Filing[a]
I	Crutches, bedpans, depressors, adhesive bandages, hospital beds	PMN/510(k)
II	Hearing aids, blood pumps, catheters, contact lenses, electrodes, catheters	510(k)
III	Cardiac pacemakers, intrauterine devices, intraocular lenses, heart valves, orthopedic devices, skin implants(?)	PMA

Source: Ref. 94.

[a]FDA, U.S. Food and Drug Administration; PMN, premarket notification; 510(k), substantial equivalence to pre-1976 device; PMA, premarket approval.

TABLE 11.14 Outcome of Preclinical Tests on a Wound Dressing

Test	Acceptable Outcome
Acute system toxicity	No signs of respiratory, motor, convulsive, ocular, salivary, piloerectal, gastrointestinal, or skin reactions
Agar overlay cytotoxicity	No detectable zone of cell reactivity
Hemolysis	Less than 5% hemolysis
MEM cytotoxicity	No cytotoxicity of test material
Primary skin irritation	No primary skin irritation
USP rabbit pyrogen	No rabbit shows individual temperature rise of 0.6°C, or for three rabbits the temperature rise is below 1.4°C
Inhibition of cell growth	No inhibition of cell growth at all extract concentrations
Limulus amebocyte lysate (LAL) testing	No LAL activity present

Source: Ref. 41.

11.8 FUNDING FOR RESEARCH: THE WAY FORWARD

This chapter has not been able to address all areas of activity in the biomaterials field, but the author hopes that interest may have been stimulated in those unfamiliar with aspects of biomaterials as a distinct branch of materials chemistry. As in all areas of scientific discovery, areas of active research require funding from external bodies. Due to the funding strategies, these bodies in part dictate the avenues for exploration but if wise will always be on the lookout for novel approaches and areas for research that could potentially yield high dividends for the community at large. In the United States there are many federal programs in biotechnology and materials. Agencies involved include:

- The Food and Drug Administration
- The National Institute of Standards and Technology
- The National Institute of Health
- The National Institute of Dental Research
- The National Science Foundation

The NSF has an ongoing foundation-wide effort to support research in biomolecular materials in the following areas: (1) genetic or other modification of natural synthetic pathways to produce materials with novel structures and optical, mechanical, or electrical properties; (2) biomolecular self-organization and phase behavior to develop new materials not found in nature; (3) novel catalyst, sensor, or transducer materials based on biochemical and biophysical processes; (4) materials aspects of *in vivo* processing of biopolymers and other naturally occurring materials; and (5) complex macromolecular structures that mimic naturally occurring composites such as bone, muscle, and photoreceptor arrays as well as materials that are biodegradable or recyclable. Research is also sponsored in advanced materials and processing. Researchers are urged to address the critical need for biomaterials that provide longer-term clinical benefit and fewer complications.

To address the need for understanding and developing the production of structural and functional biomaterials using biological organisms, such as bioceramics, biopolymers, and fibers (such as silk), bioadhesives, biosensors, and functional molecular arrays (such as light-transducing materials), the following are recommended: (1) the development of new expression systems (including plant systems) to improve both the quality and quantity of "designer biomolecular materials" and improved existing yeast, bacilovirus, and vaccinia systems; (2) the establishment of new interdisciplinary research groups at universities, incorporating provisions for coordination and liasion with industry partners; (3) the development of new biosensor and transduction materials using immobilized enzymes (such as bacteriorhodopsin); establishing their feasibility in applications such as information storage technology and in situ monitoring of cell function; and (4) focusing efforts on molecular details of bioceramic nucleation at surfaces and interfaces in nature; within 5 years, determination of key structural details present in naturally occurring nucleation molecules for application to analogous bioceramic materials; and initiation of studies of long-term stability and efficacy of bioceramic coatings for orthopedic implants. (It is also recognized that there is a need to understand interactions occurring at the solution–solid interface and of transport of reactive elements through the reaction volume, whatever that may be.)

Many of these aims seem to point to the need to have real interdisciplinary teams, but these would need to be large to really have a chance of addressing the problems that exist. There is a realization that all scientists and technologists need to learn the same language and to publish their data in a variety of journals so that their findings will reach the appropriate audiences rather than the expected audience. It is only then that substantial breakthroughs will be made and ideas generated that will lead to materials being produced for biomaterials use in the twenty-first century.

REFERENCES

1. F. H. Silver, *Biomaterials, Medical Devices and Tissue Engineering*, Chapman & Hall, London, 1994.
2. K. de Groot, A. Tencer, P. Waite, J. Nichols, J. Kay, in *Bioceramics: Material Characteristics Versus in Vivo Behavior*, P. Ducheyne, J. E. Lemons, Eds., *Ann. N.Y. Acad. Sci.* **523**, 272 (1988).
3. J. H. Felder, *IEEE Eng. Med. Biol.* **14**, 439 (1995).
4. S. A. Wainwright, W. D. Biggs, J. D. Currey, J. M. Gosline, *The Mechanical Design of Organisms*, Edward Arnold, London, 1976.

5. B. S. Gould, Ed., *Treatise on Collagen*, 3 vols. Academic Press, San Diego, CA, 1968.
6. A. J. Bailey, in *Comprehensive Biochemistry*, Vol. 26-B, M. Florkin, E. H. Stotz, Eds., Elsevier, 1968, pp. 297–423.
7. H. Hofmann, K. Kuhn, in *Structural Analysis of Recognition and Assembly in Biological Macromolecules*, M. Balaban, J. L. Sussman, W. Traub, A. Yonath, Eds., Balaban ISS, Rehovot, Israel, 1981, pp. 403–426.
8. L. Stryer, *Biochemistry*, 2nd ed., W. H. Freeman, San Francisco, 1981.
9. D. H. Elliott, *Biol. Rev.* **40**, 392 (1965).
10. F. H. Silver, G. Pins, *J. Long Term Effects Med. Implants* **2**, 67 (1992).
11. A. J. Wasserman, M. G. Dunn, in *Applications of Biominerals in Facial Plastic Surgery*, A. I. Glasgold, F. H. Silver, Eds., CRC Press, Boca Raton, FL, 1991.
12. L. Addadi, S. Weiner, *Proc. Natl. Acad. Sci. USA* **82**, 4110 (1985).
13. H. A. Lowenstam, *Science* **211**, 1126 (1981).
14. S. Mann, *Struct. Bonding* **54**, 125 (1983).
15. H. A. Lowenstam, S. Weiner, *On Biomineralization*, Oxford University Press, Oxford, 1989, and references therein.
16. M. Weinstock, C. P. Leblond, *J. Cell. Biol.* **56**, 838 (1973).
17. A. S. Posner, F. Betts, N. C. Blumenthal, *Metab. Bone Sid. Rel. Res.* **1**, 179 (1978).
18. C. C. Perry, in *Biomineralization: Chemical and Biochemical Perspectives*, S. Mann, J. Webb, R. J. P. Williams, Eds., VCH, New York, 1989, pp. 223–254.
19. A. Berman, L. Addadi, S. Weiner, *Nature (London)* **331**, 546 (1988).
20. E. M. Carlisle, *Science* **178**, 619 (1972).
21. J. Currey, *The Mechanical Adaptations of Bone*, Princeton University Press, Princeton, NJ, 1984.
22. R. Legeros, N. Balmain, G. Bonel, *Calcif. Tissue Int.* **41**, 137 (1987).
23. S. Weiner, W. Traub, *FEBS Lett.* **206**, 262 (1986).
24. M. J. Glimcher, *Philos. Trans. R. Soc. London Ser. B* **304**, 479 (1984).
25. G. W. Bernard, D. C. Pease, *Am. J. Anat.* **125**, 271 (1969).
26. W. T. Butler, *The Chemistry and Biology of Mineralized Tissues*, EBSCO Media, Birmingham, AL, 1985.
27. D. G. A. Nelson, G. J. Wood, J. C. Barry, *Ultramicroscopy* **19**, 253 (1986).
28. C. Berthet-Colominas, A. Miller, S. W. White, *J. Mol. Biol.* **134**, 431 (1979).
29. A. Miller, *Philos. Trans. R. Soc. London Ser. B* **304**, 455 (1984).
30. L. W. Fisher, J. D. Termine, *Clin. Orthop.* **280**, 362 (1985).
31. A. I. Caplan, D. G. Pechak, in *Bone and Mineral Research*, W. A. Peck, Ed., Elsevier, Amsterdam, 1987, pp. 117–183.
32. I. M. Shapiro, A. Boyde, *Metab. Bone Dis.* **5**, 317 (1984).
33. N. E. Kemp, *J. Morphol.* **184**, 215 (1985).
34. E. J. Reith, E. O. Butcher, in *Structural and Chemical Organization of Teeth*, A. E. W. Miles, Ed., Academic Press, New York, 1967, pp. 371–397.
35. M. U. Nylen, E. D. Evans, K. A. Omnel, *J. Cell. Biol.* **18**, 109 (1963).
36. S. Weiner, *CRC Crit. Rev. Biochem.* **20**, 365 (1986).

37. F. H. Silver, J. R. Parsons, in *Applications of Biomaterials in Facial Plastic Surgery*, A. I. Glasgold, F. H. Silver, Eds., CRC Press, Boca Raton, FL, 1991.
38. J. M. Morehead, G. R. Holt, *Otolaryngol. Clin. North Am.* **27**, 195 (1994).
39. L. J. Gibson, M. F. Ashby, *Mater. Res. Soc. Symp. Proc.* **255**, 343 (1992).
40. F. H. Silver, *Biological Materials: Structure, Mechanical Properties, and Modelling of Soft Tissues*, New York University Press, New York, 1987.
41. J. Black, *Orthopedic Biomaterials in Research and Practice*, Churchill Livingston, Edinburgh, 1988.
42. F. H. Silver, *Biomaterials, Medical Devices and Tissue Engineering*, Chapman & Hall, London, 1994, and references therein.
43. B. Tighe, P. Corkhill, *J. Macromol. Sci. Appl. Chem.* **A31**, 707 (1994).
44. J. C. Geesin, R. A. Berg, in *Applications of Biomaterials in Facial Plastic Surgery*, A. I. Glasgold, F. H. Silver, Eds., CRC Press, Boca Raton, FL, 1991.
45. I. V. Yannas, J. F. Burke, *J. Biomed. Mater. Res.* **14**, 65 (1980).
46. E. Bell, H. P. Ehrlich, S. Sher, C. Merrill, R. Sarber, B. Hull, T. Nakatsuji, D. Church, D. J. Buttle, *Plast. Reconstr. Surg.* **76**, 386 (1981).
47. C. Batich, D. DePalma, *Long Term Eff. Med. Implants* **1**, 253 (1992).
48. C. Batich, J. Williams, T. King, *J. Biomed. Mater. Res. Appl. Biomater.* **23**, 311 (1989).
49. M. Amiji, K. Park, *J. Biomater. Sci. Polym. Educ.* **3**, 217 (1993).
50. B. E. Rabinow, Y. S. Ding, C. Qin, M. L. McHalsky, J. H. Schneider, K. A. Ashline, T. L. Shelbourn, R. M. Albrecht, *J. Biomater. Sci. Polym. Educ.* **6**, 91 (1994).
51. D. K. Gilding, A. M. Reed, *Polymer* **20**, 1459 (1979).
52. J. A. Ray, N. Doddi, J. A. Regula, J. A. Williams, A. Melveger, *Surg. Gynecol. Obstet.* **153**, 497 (1981).
53. J. Heller, R. V. Sparer, G. M. Zentner, in *Biodegradable Polymers as Drug Delivery Systems*, M. Chasin, R. Langer, Eds., Marcel Dekker, New York, 1990, pp. 121–162.
54. M. Chasin, A. Domb, E. Ron, E. Mathiowitz, R. Langer, K. Leong, C. Laurencin, in *Biodegradable Polymers as Drug Delivery Systems*, M. Chaisin, R. Marcel, Eds., Marcel Dekker, New York, 1990, pp. 43–70.
55. J. M. Anderson, A. Hiltner, K. Schodt, R. Woods, *J. Biomed. Mater. Res. Symp.* **3**, 25 (1972).
56. H. R. Allcock, in *Biodegradable Polymers as Drug Delivery Systems*, M. Chaisen, R. Marcel, Eds., Marcel Dekker, New York, 1990, pp. 163–193.
57. C. G. Pitt, in *Biodegradable Polymers as Drug Delivery Systems*, M. Chaisen, R. Marcel, Eds., Marcel Dekker, New York, 1990, pp. 71–120.
58. J. Kohn, *Trends Polym. Sci.* **1**, 206 (1993).
59. S. I. Ertel, J. Kohn, *J. Biomed. Mater. Res.* **28**, 919 (1994).
60. A. G. Mikos, G. Sarakinos, S. M. Leite, J. P. Vacanti, R. Langer, *Biomaterials* **14**, 323 (1993).
61. C.-S. Cho, S.-J. Chung, M. Goto, A. Kobayashi, T. Akaike, *Chem. Lett. Jpn.*, 1817 (1994).

62. E. W. Merrill, *J. Biomater. Sci. Polym. Educ.* **5**, 1 (1993).
63. C. J. Doillon, M. G. Dunn, R. A. Berg, F. H. Silver, *Scanning Microsc.* **II**, 897 (1985).
64. S. Pitaru, H. Tal, M. Soldinger, A. Grosskopf, M. Noff, *J. Clin. Periodontol.* **59**, 380 (1988).
65. J. F. Cavallaro, P. D. Kemp, K. H. Kruas, *Biotechnol. Bioeng.* **43**, 781 (1994).
66. R. Barbucci, A. Magnani, A. Baszkin, M. L. Da Costa, H. Bauser, G. Hellwig, E. Martuscelli, S. Cimmino, *J. Biomater. Sci. Polym. Educ.* **4**, 245 (1993).
67. J. Black, 1989.
68. L. L. Hench, J. Wilson, *Introduction to Bioceramics*, World Scientific, River Edge, NJ, 1993.
69. J. Black, *Orthopedic Clinics of North America* **20**, 1–13 (1989).
70. P. Boutin, *Presse Med.* **79**, 639 (1970).
71. P. Boutin, *Rev. Chir. Orthop.* **58**, 229 (1972).
72. L. L. Hench, Örjan Andersson, in *An Introduction to Bioceramics*, L. L. Hench, J. Wilson, Eds., Advanced Series in Ceramics, Vol. 1, World Scientific, River Edge, NJ, 1993, Chapter 3, pp. 41–62.
73. P. Christel, in *Biomechanics: Current Interdisciplinary Research*, S. M. Perren, E. Schneider, Eds., Martinus Nijhoff, Dordrecht, The Netherlands, 1985, pp. 61–72.
74. W. Höland, W. Gland, W. Götz, G. Carl, W. Vogel, *Cells Mater.* **2**, 105 (1992), and references therein.
75. G. H. Beall, M. R. Montierth, P. Smith, *Glas-Email-Keramo-Tech.* **22**, 409 (1971).
76. L. L. Hench, E. C. Ethridge, *Biomaterials: An Interfacial Approach*, Academic Press, San Diego, CA, 1982.
77. P. Ducheyne, J. E. Lemons, *Bioceramics: Material Characteristics Versus in Vivo Behavior, Ann. N.Y. Acad. Sci.* **523** (1988).
78. B. Boyan, E. Schepers, T. Yamamuro, J. Wilson, U. Gross, R. Reck, P. Vast, D. Steflik, J. R. Tsai, A. Yamagami, T. Kitsugi, in *Bioceramics: Material Characteristics Versus in Vivo Behavior*, P. Ducheyne, J. E. Lemons, Eds., *Ann. N.Y. Acad. Sci.* **523**, 262 (1988).
79. P. Ducheyne, M. Marcolongo, E. Schepers, in *An Introduction to Bioceramics*, L. L. Hench, J. Wilson, Eds., World Scientific, River Edge, NJ, 1993, Chapter 15, pp. 281–297.
80. U. Soltész, in *Bioceramics: Material Characteristics Versus in Vivo Behavior*, P. Ducheyne, J. E. Lemons, Eds., *Ann. N.Y. Acad. Sci.* **523**, 137 (1988), and references therein.
81. B. R. Constantz, I. C. Ison, M. T. Fulmer, R. D. Poser, S. T. Poser, M. Vanwagoner, J. Ross, S. A. Goldstein, J. B. Jupiter, D. I. Rosenthal, *Science* **267**, 1796 (1995).
82. P. D. Constantino, C. D. Friedman, *Otolaryngol. Clin. North Am.* **27**, 1037 (1994).
83. A. S. Nash, L. Dahl, M. Shepler, *Mater. Res. Soc. Symp. Proc.* **218**, 257 (1991).
84. M. J. Glimcher, D. Brickley-Parsons, D. Kossiva, *Calif. Tissue Int.* **27**, 281 (1979).
85. A. I. Caplan, *Mater. Res. Symp. Proc.* **174**, 9 (1990).

86. D. A. Tirrell, L. G. Donaruma, A. B. Twek, Eds., *Macromolecules as Drugs and as Carriers for Biologically Active Materials*, *Ann. N.Y. Acad. Sci.* **446** (1985).
87. K. P. Rao, *J. Biomater. Sci. Polym. Educ.* **7**, 623 (1995).
88. K. Ito, K. Kiriyama, T. Watanabe, M. Yamauki, K. Akiyama, K. Kondin, H. Takagi, *ASAIO Trans.* **36**, 199 (1990).
89. Y. Shinto, A. Uchida, N. Araki, K. Ono, *Gan To Kagaku Richo* **18**, 221 (1991).
90. C. S. Uglea, I. Albu, A. Vatajanu, M. Croitoru, S. Antoniu, L. Panaitescu, R. M. Ottenbrite, *J. Biomater. Sci. Polym. Educ.* **6**, 633 (1994).
91. P. S. Stayton, J. M. Ollinger, P. W. Bohn, S. G. Sligar, *J. Am. Chem. Soc.* **114**, 9298 (1992).
92. A. A. Ciarowski, in *Handbook of Biomaterials Applications*, A. F. von Recum, Ed., Macmillan, New York, 1986, Chapter 42.
93. F. Silver, C. Doillon, *Biocompatibility: Interactions of Biological and Implantable Materials*, Vol. 1, *Polymers*, VCH, New York, 1989, Chapter 4.
94. J. R. Phelps, R. A. Dormer, in *Handbook of Biomaterials Applications*, A. F. von Recum, Ed., Macmillan, New York, 1986.

INDEX